SINGLE LINEAR-BIVARIATE CUBIC SYSTEMS

Advisory Board

Series on Complexity, Nonlinearity and Chaos

ISSN 2010-0019

Aims and Scope

The books in this series will focus on the recent developments, findings and progress on fundamental theories and principles, analytical and symbolic approaches, computational techniques in nonlinear physical science and nonlinear mathematics.

Topics of interest in Complexity, Nonlinearity and Chaos include but not limited to:

- New findings and discoveries in nonlinear physics and mathematics,
- Complexity and mathematical structures in nonlinear physics,
- Nonlinear phenomena and observations in nature,
- Computational methods and simulations in complex systems,
- New theories, and principles and mathematical methods,
- Stability, bifurcation, chaos and fractals in nonlinear physical science.

Vol. 6 *Single Linear-Bivariate Cubic Systems*
Albert C. J. Luo

Vol. 5 *Fractional Calculus: Models and Numerical Methods (Second Edition)*
D. Baleanu, K. Diethelm, E. Scalas & J. J. Trujillo

Vol. 4 *Asymptotic Integration and Stability: For Ordinary, Functional and Discrete Differential Equations of Fractional Order*
D. Baleanu & O. G. Mustafa

Vol. 3 *Fractional Calculus: Models and Numerical Methods*
D. Baleanu, K. Diethelm & J. J. Trujillo

Vol. 2 *Applications of Lie Group Analysis in Geophysical Fluid Dynamics*
N. H. Ibragimov & R. N. Ibragimov

Vol. 1 *Ray and Wave Chaos in Ocean Acoustics: Chaos in Waveguides*
D. Makarov, S. Prants, A. Virovlyansky & G. Zaslavsky

CNC Series on Complexity, Nonlinearity and Chaos – Vol. 6

SINGLE LINEAR-BIVARIATE CUBIC SYSTEMS

Albert C J Luo
Southern Illinois University at Edwardsville, USA

World Scientific

NEW JERSEY • LONDON • SINGAPORE • BEIJING • SHANGHAI • HONG KONG • TAIPEI • CHENNAI • TOKYO

Published by

World Scientific Publishing Co. Pte. Ltd.
5 Toh Tuck Link, Singapore 596224
USA office: 27 Warren Street, Suite 401-402, Hackensack, NJ 07601
UK office: 57 Shelton Street, Covent Garden, London WC2H 9HE

British Library Cataloguing-in-Publication Data
A catalogue record for this book is available from the British Library.

Series on Complexity, Nonlinearity and Chaos — Vol. 6
SINGLE LINEAR-BIVARIATE CUBIC SYSTEMS

ISBN 978-981-98-1839-6 (hardcover)
ISBN 978-981-98-1840-2 (ebook for institutions)
ISBN 978-981-98-1841-9 (ebook for individuals)

For any available supplementary material, please visit
https://www.worldscientific.com/worldscibooks/10.1142/14455#t=suppl

Typeset by Stallion Press
Email: enquiries@stallionpress.com

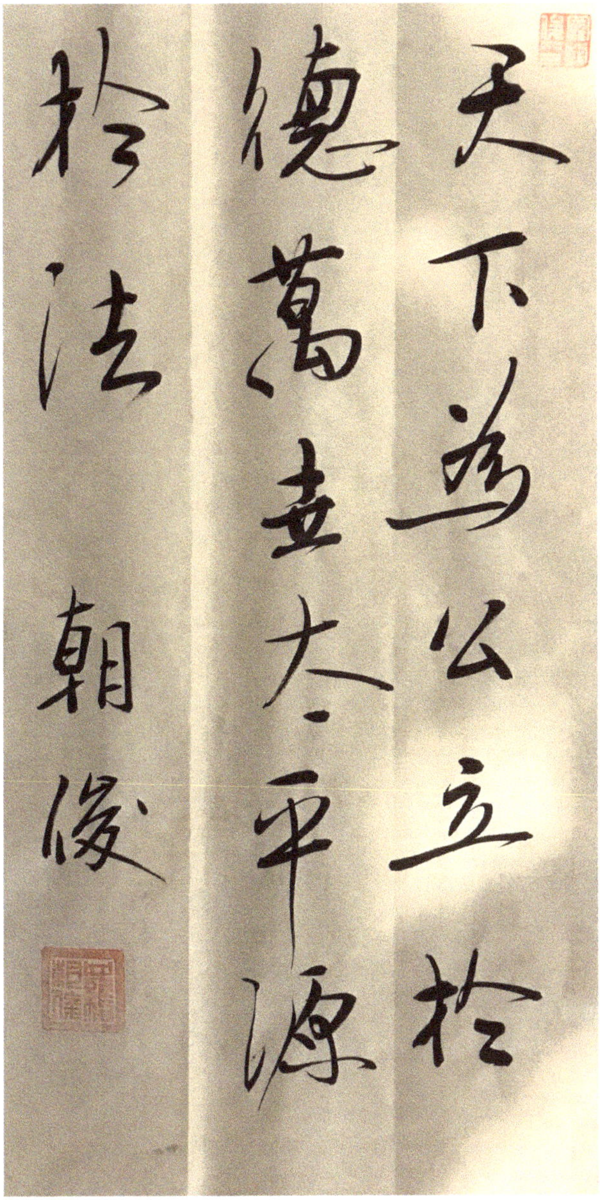

天下为公立于德，万世太平源于法

——朝

A just world is established through virtue; eternal peace is rooted in law.

Preface

This book is about nonlinear dynamics of single-linear-bivariate cubic dynamical systems. For such cubic dynamical systems, the inflection flows and third-order parabola flows exist for appearing bifurcations. The inflection flows are for appearing bifurcations of two parabola flows on the same direction. The third-order parabola flows are for the appearing bifurcation of inflection and parabola flows, and for the appearing bifurcations of up, down, and up-parabola flows or down, up, and down-parabola flows. Third-order parabola flows are for the appearing bifurcation among the up and down-parabola flows. There are four types of infinite-equilibriums: (i) The inflection-source and sink infinite-equilibriums are for the switching bifurcations of two parabola flows on the two directions. (ii) The parabola-source and sink infinite-equilibriums are for the switching bifurcations of parabola and inflection flows on the two directions. (iii) The inflection-saddle infinite-equilibriums are for the switching bifurcation of two inflection flows in two directions. (iv) The parabola-saddle infinite-equilibriums are for the switching bifurcations of parabola and inflection flows on the two directions are the switching bifurcations for parabola and inflection flows. Such switching bifurcations for one-dimensional flow are based on the infinite-equilibriums, which will help one understand global dynamics in nonlinear cubic systems.

In Chapter 1, two-dimensional nonlinear systems possessing constant and single-linear-bivariate cubic vector fields are discussed, and the appearing and switching bifurcations for up-parabola and down-parabola flows are discussed herein. The upper-saddle and lower-saddle flows on the static specific lines are the appearing bifurcations for sink and source flows. The increasing-inflection and deceasing-inflection flows are the appearing bifurcations for up-parabola and down-parabola flows.

In Chapter 2, two-dimensional nonlinear systems possessing single-linear-bivariate linear and cubic vector fields are discussed, and the appearing and switching bifurcations for the one-dimensional flows (i.e., up-parabola and down-parabola flows) are discussed. The third-order parabola flows are the appearing bifurcations for up-parabola, down-parabola, and up-parabola flows or for down-parabola, up-parabola, and down-parabola flows. The increasing-inflection and deceasing-inflection flows are the appearing bifurcations for up-parabola and down-parabola flows. The inflection-source and sink infinite-equilibriums are for switching bifurcations of vertical and horizontal parabola flows. The parabola-source and sink infinite-parabola flows are for the switching bifurcations of parabola flows and inflection flows and for the appearing bifurcations of the vertical and horizontal parabola flows.

In Chapter 3, two-dimensional nonlinear systems possessing single-linear-bivariate quadratic and cubic vector fields are discussed, and the appearing and switching bifurcations for the one-dimensional flows (i.e., up-parabola and down-parabola flows, and inflection flows) are discussed herein. The increasing-inflection and deceasing-inflection flows are the appearing bifurcations for up-parabola and down-parabola flows. The third-order parabola flows are the appearing bifurcations among up-parabola and down-parabola flows, and for parabola and inflection flows. The inflection-sink and source infinite-equilibriums are the switching bifurcations of two parabola flows in the two directions and for two inflections in the two directions. The parabola-sink and source infinite-equilibriums are the switching bifurcations of parabola and inflection flows in the two directions and for inflection and third-order parabola flows in the two directions. The second-order inflection-source/sink infinite-equilibriums are the switching bifurcations of parabola and third-order parabola flows in the two directions.

In Chapter 4, two-dimensional nonlinear systems possessing two single-linear-bivariate cubic vector fields are discussed, and the appearing and switching bifurcations for the one-dimensional flows (i.e., the first-order and third-order up-parabola and down-parabola flows, and inflection flows) are discussed herein. The increasing-inflection and deceasing-inflection flows are the appearing bifurcations for up-parabola and down-parabola flows. The third-order parabola flows are the appearing bifurcations among up-parabola and down-parabola flows, and for parabola and inflection flows. The inflection-sink and source infinite-equilibriums are the switching bifurcations of two parabola flows in the two directions and for two inflections in the two directions. The parabola-sink and source infinite-equilibriums are the switching bifurcations of parabola and inflection flows in the two directions and for inflection and third-order parabola flows in the two directions. The second-order inflection-source/sink infinite-equilibriums are the switching bifurcations of parabola and third-order parabola flows in the two directions. The parabola-saddle infinite-equilibriums are the switching bifurcations of the parabola and inflection flows. The inflection-third source/sink infinite-equilibriums are the switching bifurcations for the third-order up-parabola flows and the third-order down-parabola flows and for the third-order down-parabola flows and the third-order up-parabola flows.

Finally, the author hopes the materials presented herein can provide a better understanding of appearing and switching bifurcations in single-linear-bivariate cubic systems.

Albert C.J. Luo
Edwardsville, Illinois

Contents

Preface vii

Chapter 1. Constant and Single-Linear-Bivariate Cubic Systems **1**

1.1 Constant and Single-Linear-Bivariate Cubic Vector Fields 1
1.2 Proof of Theorem 1.1 13
1.3 Inflection and Parabola Flows 29
1.4 Parabola Flows Series 37

Chapter 2. Single-Linear-Bivariate Linear and Cubic Systems **45**

2.1 Single-Linear-Bivariate Linear and Cubic Vector Fields 45
2.2 Theorem Proof 75
2.3 Third-Order Parabola Flows with Bifurcations 111
2.3.1 Parabola flows with third-order singularity 112
2.3.2 Switching bifurcations 125
2.4 Second-Order Singular Single-Linear-Bivariate Cubic Systems 131
2.4.1 Inflection flows and parabola-sources (sinks) . . . 131
2.4.2 Switching bifurcations 144
2.5 Parabola Flows' Series and Switching Bifurcations 154
2.5.1 Series of parabola flows 154
2.5.2 Inflection-sink (source) switching bifurcations . . . 163

Chapter 3. Single-Linear-Bivariate Quadratic and Cubic Systems **177**

3.1 Single-Linear-Bivariate Quadratic and Cubic Vector Fields 177
3.2 Proof of Theorem 3.1 209
3.3 Second-Order Singular Quadratic and Cubic Systems . . 257
3.3.1 Inflection flows and appearing bifurcations 257
3.3.2 Inflection-saddle and parabola sources and sinks 270
3.4 Singular Parabola Flows and Infinite-Equilibriums 279
3.4.1 Third-order parabola and inflection flows 279
3.4.2 Parabola-saddle infinite-equilibriums 282

Chapter 4. Single-Linear-Bivariate Double-Cubic Systems **289**

4.1 Two Single-Linear-Bivariate Cubic Vector Fields 289
4.2 Proof of Theorem 4.1 352
4.3 Singular Single-Linear-Bivariate Cubic Systems 449
4.3.1 Singular parabola and inflection flows 449
4.3.2 Inflection-saddle and parabola-sources and sinks 462
4.4 Singular Parabola Flows and Infinite-Equilibriums 471
4.4.1 Third-order parabola flows 471
4.4.2 Inflection-third-order sources and sinks 475

Chapter 1

Constant and Single-Linear-Bivariate Cubic Systems

In this chapter, two-dimensional nonlinear systems possessing constant and single-linear-bivariate cubic vector fields are discussed, and the appearing and switching bifurcations for up-parabola and down-parabola flows are discussed herein. The upper-saddle and lower-saddle flows on the static-specific lines are the appearing bifurcations for sink and source flows. The increasing-inflection and deceasing-inflection flows are the appearing bifurcations for up-parabola and down-parabola flows.

1.1 Constant and Single-Linear-Bivariate Cubic Vector Fields

In this section, consider a quadratic dynamical system possessing constant and linear-bivariate cubic vector fields. The corresponding dynamical behaviors and bifurcations are presented through the following theorem.

Theorem 1.1. *Consider a two-dimensional, nonlinear dynamical system as*

$$\begin{aligned}\dot{x}_{j_1} &= f_{j_1}(\mathbf{x},\mathbf{p}) = A_{j_1j_13} + A_{j_1j_12}y_{j_1} + A_{j_1j_11}y_{j_1}^2 + A_{j_1j_10}y_{j_1}^3,\\ \dot{x}_{j_2} &= f_{j_2}(\mathbf{x},\mathbf{p}) = A_{j_2j_23} + A_{j_2j_22}y_{j_1} + A_{j_2j_21}y_{j_1}^2 + A_{j_2j_20}y_{j_1}^3\end{aligned} \tag{1.1}$$

with

$$y_{j_1} = \alpha_{j_1j_1}x_{j_1} + \alpha_{j_1j_2}x_{j_2} \quad \text{for } j_1, j_2 \in \{1,2\}, \quad j_1 \neq j_2. \tag{1.2}$$

(*i*) *For* $A_{j_2j_22} = A_{j_2j_21} = A_{j_2j_20} = 0(j_1, j_2 \in \{1,2\}, j_1 \neq j_2)$,

$$\begin{aligned}\dot{x}_{j_2} &= A_{j_2j_23} = 0,\\ \dot{x}_{j_1} &= A_{j_1j_13} + A_{j_1j_12}y_{j_1} + A_{j_1j_11}y_{j_1}^2 + A_{j_1j_10}y_{j_1}^3\\ &= a_{j_1j_10}(y_{j_1} - b_{j_1j_11})[y_{j_1}^2 + B_{j_1j_1}y_{j_1} + C_{j_1j_1}].\end{aligned} \tag{1.3}$$

If $A_{j_3j_33} = 0$, the two-dimensional dynamical system becomes a one-dimensional cubic nonlinear system in the x_{j_1}-direction.

(i_1) For $\Delta_{j_1j_1} < 0$, the one-dimensional system becomes

$$\begin{aligned}\dot{x}_{j_2} &= A_{j_2j_23} = 0,\\ \dot{x}_{j_1} &= a_{j_1j_10}(y_{j_1} - a_{j_1j_11})[(y_{j_1} - a_{j_1j_1})^2 + b_{j_1j_1}]\end{aligned} \tag{1.4}$$

with

$$\begin{aligned}a_{j_1j_1} &= -\frac{1}{2}B_{j_1j_1}, b_{j_1j_1} = -\frac{1}{4}C_{j_1j_1}, a_{j_1j_11} = b_{j_1j_11},\\ \Delta_{j_1j_1} &= (B_{j_1j_1})^2 - 4C_{j_1j_1}, \quad a_{j_1j_10} = A_{j_1j_10}.\end{aligned} \tag{1.5}$$

With an initial condition $y_{j_1} = y_{j_10}$ at $t = t_0$, the solution is

$$\begin{aligned}t - t_0 = &\frac{1}{\alpha_{j_1j_1}a_{j_1j_10}[(a_{j_1j_11} - a_{j_1j_1})^2 + b_{j_1j_1}]}\left[\ln\frac{|y_{j_1} - a_{j_1j_11}|}{|y_{j_10} - a_{j_1j_11}|}\right.\\ &-\frac{1}{2}\ln\frac{|(y_{j_1} - a_{j_1j_1})^2 + b_{j_1j_1}|}{|(y_{j_10} - a_{j_1j_1})^2 + b_{j_1j_1}|}\\ &\left.+\frac{(a_{j_1j_11} - a_{j_1j_1})}{\sqrt{b_{j_1j_1}}}\left(\arctan\frac{y_{j_1} - a_{j_1j_1}}{\sqrt{b_{j_1j_1}}} - \arctan\frac{y_{j_10} - a_{j_1j_1}}{\sqrt{b_{j_1j_1}}}\right)\right].\end{aligned} \tag{1.6}$$

The infinite-equilibrium of $y_{j_1} = a_{j_1j_11}$ has the following properties:

- *For $a_{j_1j_10}\alpha_{j_1j_1} > 0$, the line of $y_{j_1} = a_{j_1j_11}$ in the x_{j_1}-direction is a source infinite-equilibrium.*
- *For $a_{j_1j_10}\alpha_{j_1j_1} < 0$, the line of $y_{j_1} = a_{j_1j_11}$ in the x_{j_1}-direction is a sink infinite-equilibrium.*

(i_2) For $\Delta_{j_1j_1} = 0$ and $a_{j_1j_1s_1} \neq a_{j_1j_1s_2}$, the one-dimensional system becomes

$$\begin{aligned}\dot{x}_{j_2} &= A_{j_2j_23} = 0,\\ \dot{x}_{j_1} &= a_{j_1j_10}(y_{j_1} - a_{j_1j_1s_1})(y_{j_1} - a_{j_1j_1s_2})^2\end{aligned} \tag{1.7}$$

with

$$a_{j_1j_1s_2} = -\frac{1}{2}B_{j_1j_1}, \quad a_{j_1j_1s_1} = b_{j_1j_11}, \quad a_{j_1j_10} = A_{j_1j_10}. \tag{1.8}$$

With an initial condition $y_{j_1} = y_{j_10}$ *at* $t = t_0$, *the solution is*

$$
\begin{aligned}
t - t_0 = {} & \frac{1}{\alpha_{j_1j_1}a_{j_1j_10}(a_{j_1j_1s_1} - a_{j_1j_1s_2})^2}\Bigg[\ln\frac{|y_{j_1} - a_{j_1j_1s_1}|}{|y_{j_10} - a_{j_1j_1s_1}|} \\
& - \ln\frac{|y_{j_1} - a_{j_1j_1s_2}|}{|y_{j_10} - a_{j_1j_1s_2}|} + \frac{1}{(a_{j_1j_1s_2} - a_{j_1j_1s_1})} \\
& \times\left(\frac{1}{y_{j_1} - a_{j_1j_1s_2}} - \frac{1}{y_{j_10} - a_{j_1j_1s_2}}\right)\Bigg].
\end{aligned}
\tag{1.9}
$$

(i_{2a}) *The equilibrium line of* $y_{j_1} = a_{j_1j_1s_2}$ *has the following properties*:

- *For* $a_{j_1j_10}\alpha_{j_1j_1}(a_{j_1j_1s_1} - a_{j_1j_1s_2})^2 > 0$, *the line of* $y_{j_1} = a_{j_1j_1s_1}$ *in the* x_{j_1}*-direction is a source infinite-equilibrium.*
- *For* $a_{j_1j_10}\alpha_{j_1j_1}(a_{j_1j_1s_1} - a_{j_1j_1s_2})^2 < 0$, *the line of* $y_{j_1} = a_{j_1j_1s_1}$ *in the* x_{j_1}*-direction is a sink infinite-equilibrium.*

(i_{2b}) *The equilibrium line of* $y_{j_1} = a_{j_1j_1s_2}$ *has the following properties*:

- *For* $a_{j_1j_10}(\alpha_{j_1j_1})^2(a_{j_1j_1s_2} - a_{j_1j_1s_1}) > 0$, *the line of* $y_{j_1} = a_{j_1j_1s_2}$ *in the* x_{j_1}*-direction is an upper-saddle infinite-equilibrium.*
- *For* $a_{j_1j_10}(\alpha_{j_1j_1})^2(a_{j_1j_1s_2} - a_{j_1j_1s_1}) < 0$, *the line of* $y_{j_1} = a_{j_1j_1s_2}$ *in the* x_{j_1}*-direction is a lower-saddle infinite-equilibrium.*

(i_3) *For* $\Delta_{j_1j_1} = 0$ *and* $a_{j_1j_1s_1} = a_{j_1j_1s_2} = a_{j_1j_11}$, *the one-dimensional system becomes*

$$
\begin{aligned}
\dot{x}_{j_2} &= A_{j_2j_23} = 0, \\
\dot{x}_{j_1} &= a_{j_1j_10}(y_{j_1} - a_{j_1j_11})^3
\end{aligned}
\tag{1.10}
$$

with

$$
a_{j_1j_1s_2} = -\frac{1}{2}B_{j_1j_1}, \quad a_{j_1j_1s_1} = b_{j_1j_11}, \quad a_{j_1j_10} = A_{j_1j_10}. \tag{1.11}
$$

With an initial condition $y_{j_1} = y_{j_10}$ *at* $t = t_0$, *the solution is*

$$
t - t_0 = -\frac{1}{2\alpha_{j_1j_1}a_{j_1j_10}}\left[\frac{1}{(y_{j_1} - a_{j_1j_11})^2} - \frac{1}{(y_{j_10} - a_{j_1j_11})^2}\right]. \tag{1.12}
$$

- *For* $a_{j_1j_10}(\alpha_{j_1j_1})^3 > 0$, *the line of* $y_{j_1} = a_{j_1j_11}$ *in the* x_{j_1}*-direction is a third-order source infinite-equilibrium.*
- *For* $a_{j_1j_10}(\alpha_{j_1j_1})^3 < 0$, *the line of* $y_{j_1} = a_{j_1j_11}$ *in the* x_{j_1}*-direction is a third-order sink infinite-equilibrium.*

(i_4) *For* $\Delta_{j_1j_1} > 0$, *the one-dimensional system becomes*

$$\begin{aligned}\dot{x}_{j_2} &= A_{j_2j_23} = 0,\\ \dot{x}_{j_1} &= a_{j_1j_10}(y_{j_1} - a_{j_1j_11})(y_{j_1} - a_{j_1j_12})(y_{j_1} - a_{j_1j_13})\end{aligned} \tag{1.13}$$

with

$$\begin{aligned}&\{a_{j_1j_11}, a_{j_1j_12}, a_{j_1j_13}\}\\ &\quad = \text{sort}\{b_{j_1j_11}, b_{j_1j_12}, b_{j_1j_13}\}, \quad a_{j_1j_11} < a_{j_1j_12} < a_{j_1j_13},\\ &b_{j_1j_12}, b_{j_1j_13} = -\frac{1}{2}(B_{j_1j_1} \pm \sqrt{\Delta_{j_1j_1}}), \quad a_{j_1j_10} = A_{j_1j_10}.\end{aligned} \tag{1.14}$$

With an initial condition $y_{j_1} = y_{j_10}$ *at* $t = t_0$, *the solution is*

$$t - t_0 = \frac{1}{a_{j_1j_10}\alpha_{j_1j_1}} \sum_{s_1=1}^{3} \frac{1}{\prod_{s_2=1, s_2\neq s_1}^{3} (a_{j_1j_1s_1} - a_{j_1j_1s_2})} \ln \frac{|y_{j_1} - a_{j_1j_1s_1}|}{|y_{j_10} - a_{j_1j_1s_1}|}. \tag{1.15}$$

- *For* $a_{j_1j_10}\alpha_{j_1j_1}\prod_{s_2=1,s_2\neq s_1}^{3}(a_{j_1j_1s_1} - a_{j_1j_1s_2}) > 0$, *the lines of* $y_{j_1} = a_{j_1j_1s_1}$ *in the* x_{j_1}*-direction are a source infinite-equilibrium.*
- *For* $a_{j_1j_10}\alpha_{j_1j_1}\prod_{s_2=1,s_2\neq s_1}^{3}(a_{j_1j_1s_1} - a_{j_1j_1s_2}) < 0$, *the lines of* $y_{j_1} = a_{j_1j_1s_1}$ *in the* x_{j_1}*-direction are a sink infinite-equilibrium.*

(ii) *If*

$$A_{j_2j_23} \neq 0 \quad \text{and} \quad \Delta_{j_1j_1} < 0, \tag{1.16}$$

the standard form of the dynamical system is

$$\begin{aligned}\dot{x}_{j_2} &= A_{j_2j_23},\\ \dot{x}_{j_1} &= a_{j_1j_10}(y_{j_1} - a_{j_1j_11})[(y_{j_1} - a_{j_1j_1})^2 + b_{j_1j_1}].\end{aligned} \tag{1.17}$$

The dynamical system in the y_{j_1}*-direction is*

$$\dot{y}_{j_1} = \alpha_{j_1j_2}A_{j_2j_23} + \alpha_{j_1j_1}a_{j_1j_10}(y_{j_1} - a_{j_1j_11})[(y_{j_1} - a_{j_1j_1})^2 + b_{j_1j_1}], \tag{1.18}$$

with a dynamical system in the $y_{j_1}^{\perp}$*-direction as*

$$\dot{y}_{j_1}^{\perp} = \beta_{j_1j_2}A_{j_2j_23} + \beta_{j_1j_1}a_{j_1j_10}(y_{j_1} - a_{j_1j_11})[(y_{j_1} - a_{j_1j_1})^2 + b_{j_1j_1}], \tag{1.19}$$

where

$$\begin{aligned}y_{j_1}^{\perp} &= \beta_{j_1j_1}x_{j_1} + \beta_{j_1j_2}x_{j_2}, \mathbf{n}_{j_1}^{\perp} = (\beta_{j_1j_1}, \beta_{j_1j_2})^{\mathrm{T}},\\ \mathbf{n}_{j_1} \cdot \mathbf{n}_{j_1}^{\perp} &= \alpha_{j_1j_1}\beta_{j_1j_1} + \alpha_{j_1j_2}\beta_{j_1j_2} = 0.\end{aligned} \tag{1.20}$$

The two-dimensional dynamical system has the following properties:

- *For* $a_{j_1j_10}\alpha_{j_1j_2} > 0$ *and* $A_{j_2j_23} > 0$,

$$(a_{j_1j_11}, \dot{x}^*_{j_2}) = \underbrace{(\text{UP,pF})}_{\text{up-parabola flow }(+)}. \tag{1.21}$$

- *For* $a_{j_1j_10}\alpha_{j_1j_2} < 0$ *and* $A_{j_2j_23} > 0$,

$$(a_{j_1j_11}, \dot{x}^*_{j_2}) = \underbrace{(\text{DP,pF})}_{\text{down-parabola flow }(+)}. \tag{1.22}$$

- *For* $a_{j_1j_10}\alpha_{j_1j_2} > 0$ *and* $A_{j_2j_23} < 0$,

$$(a_{j_1j_11}, \dot{x}^*_{j_2}) = \underbrace{(\text{DP,nF})}_{\text{down-parabola flow }(-)}. \tag{1.23}$$

- *For* $a_{j_1j_10}\alpha_{j_1j_2} < 0$ *and* $A_{j_2j_23} < 0$,

$$(a_{j_1j_11}, \dot{x}^*_{j_2}) = \underbrace{(\text{UP,nF})}_{\text{up-parabola flow }(-)}. \tag{1.24}$$

(iii) If

$$A_{j_2j_23} \neq 0 \quad \text{and} \quad \Delta_{j_1j_1} = 0, \quad a_{j_1j_1s_1} \neq a_{j_1j_1s_2}, \tag{1.25}$$

the corresponding dynamical system becomes

$$\begin{aligned} \dot{x}_{j_2} &= A_{j_2j_23}, \\ \dot{x}_{j_1} &= a_{j_1j_10}(y_{j_1} - a_{j_1j_1s_1})(y_{j_1} - a_{j_1j_1s_2})^2. \end{aligned} \tag{1.26}$$

There is a dynamical system in the y_{j_1}*-direction as*

$$\dot{y}_{j_1} = \alpha_{j_1j_2}A_{j_2j_23} + \alpha_{j_1j_1}a_{j_1j_10}(y_{j_1} - a_{j_1j_1s_1})(y_{j_1} - a_{j_1j_1s_2})^2, \tag{1.27}$$

with a dynamical system in the $y^{\perp}_{j_1}$*-direction as*

$$\dot{y}^{\perp}_{j_1} = \beta_{j_1j_2}\alpha_{j_1j_2}A_{j_2j_23} + \beta_{j_1j_1}a_{j_1j_10}(y_{j_1} - a_{j_1j_1s_1})(y_{j_1} - a_{j_1j_1s_2})^2, \tag{1.28}$$

where

$$\begin{aligned} y^{\perp}_{j_1} &= \beta_{j_1j_1}x_{j_1} + \beta_{j_1j_2}x_{j_2}, \quad \mathbf{n}^{\perp}_{j_1} = (\beta_{j_1j_1}, \beta_{j_1j_2})^{\mathrm{T}}, \\ \mathbf{n}_{j_1} \cdot \mathbf{n}^{\perp}_{j_1} &= \alpha_{j_1j_1}\beta_{j_1j_1} + \alpha_{j_1j_2}\beta_{j_1j_2} = 0. \end{aligned} \tag{1.29}$$

(iii_1) The equilibrium on the line of $y^*_{j_1} = a_{j_1 j_1 s_1}$ for the x_{j_1}-direction with $\dot{x}^*_{j_2} = A_{j_2 j_2 3}$ possesses the following properties:

- *For* $a_{j_1 j_1 0}\alpha_{j_1 j_2}(a_{j_1 j_1 s_1} - a_{j_1 j_1 s_2})^2 > 0$ *and* $A_{j_2 j_2 3} > 0$,

$$(a_{j_1 j_1 s_1}, \dot{x}^*_{j_2}) = \underbrace{(\text{UP,pF})}_{\text{up-parabola flow }(+)}. \tag{1.30}$$

- *For* $a_{j_1 j_1 0}\alpha_{j_1 j_2}(a_{j_1 j_1 s_1} - a_{j_1 j_1 s_2})^2 < 0$ *and* $A_{j_2 j_2 3} > 0$,

$$(a_{j_1 j_1 s_1}, \dot{x}^*_{j_2}) = \underbrace{(\text{DP,pF})}_{\text{down-parabola flow }(+)}. \tag{1.31}$$

- *For* $a_{j_1 j_1 0}\alpha_{j_1 j_2}(a_{j_1 j_1 s_1} - a_{j_1 j_1 s_2})^2 > 0$ *and* $A_{j_2 j_2 3} < 0$,

$$(a_{j_1 j_1 s_1}, \dot{x}^*_{j_2}) = \underbrace{(\text{DP,nF})}_{\text{down-parabola flow }(-)}. \tag{1.32}$$

- *For* $a_{j_1 j_1 0}\alpha_{j_1 j_2}(a_{j_1 j_1 s_1} - a_{j_1 j_1 s_2})^2 < 0$ *and* $A_{j_2 j_2 3} < 0$,

$$(a_{j_1 j_1 s_1}, \dot{x}^*_{j_2}) = \underbrace{(\text{UP,nF})}_{\text{up-parabola flow }(-)}. \tag{1.33}$$

(iii_2) The equilibrium on the line of $y^*_{j_1} = a_{j_1 j_1 s_2}$ for the x_{j_1}-direction with $\dot{x}^*_{j_2} = A_{j_2 j_2 3}$ possesses the following properties:

- *For* $a_{j_1 j_1 0}(\alpha_{j_1 j_2})^2(a_{j_1 j_1 s_2} - a_{j_1 j_1 s_1}) > 0$ *and* $A_{j_2 j_2 3} > 0$,

$$(a_{j_1 j_1 s_2}, \dot{x}^*_{j_2}) = \underbrace{(\text{II,pF})}_{\text{increasing-inflection flow }(+)}. \tag{1.34}$$

- *For* $a_{j_1 j_1 0}(\alpha_{j_1 j_2})^2(a_{j_1 j_1 s_2} - a_{j_1 j_1 s_1}) < 0$ *and* $A_{j_2 j_2 3} > 0$,

$$(a_{j_1 j_1 s_2}, \dot{x}^*_{j_2}) = \underbrace{(\text{DI,pF})}_{\text{decreasing-inflection flow }(+)}. \tag{1.35}$$

- *For* $a_{j_1 j_1 0}(\alpha_{j_1 j_2})^2(a_{j_1 j_1 s_2} - a_{j_1 j_1 s_1}) > 0$ *and* $A_{j_2 j_2 3} < 0$,

$$(a_{j_1 j_1 s_2}, \dot{x}^*_{j_2}) = \underbrace{(\text{DI,nF})}_{\text{decreasing-inflection flow }(-)}. \tag{1.36}$$

- *For* $a_{j_1 j_1 0}(\alpha_{j_1 j_2})^2(a_{j_1 j_1 s_2} - a_{j_1 j_1 s_1}) < 0$ *and* $A_{j_2 j_2 3} < 0$,

$$(a_{j_1 j_1 1}, \dot{x}^*_{j_2}) = \underbrace{(\text{II,nF})}_{\text{increasing-inflection flow }(-)}. \tag{1.37}$$

The corresponding increasing-inflection and decreasing-inflection flows are the appearing bifurcations for up-parabola flow and down-parabola flow.

(iv) If

$$A_{j_2 j_2 3} \neq 0 \quad \text{and} \quad \Delta_{j_1 j_1} = 0, \quad a_{j_1 j_1 s_1} = a_{j_1 j_1 s_2} = a_{j_1 j_1 1}, \tag{1.38}$$

the corresponding dynamical systems becomes

$$\begin{aligned} \dot{x}_{j_2} &= A_{j_2 j_2 3}, \\ \dot{x}_{j_1} &= a_{j_1 j_1 0}(y_{j_1} - a_{j_1 j_1 1})^3. \end{aligned} \tag{1.39}$$

There is a dynamical system in the y_{j_1}-direction as

$$\dot{y}_{j_1} = \alpha_{j_1 j_2} A_{j_2 j_2 3} + \alpha_{j_1 j_1} a_{j_1 j_1 0}(y_{j_1} - a_{j_1 j_1 s_2})^3, \tag{1.40}$$

with a dynamical system in the $y_{j_1}^{\perp}$-direction as

$$\dot{y}_{j_1}^{\perp} = \beta_{j_1 j_2} \alpha_{j_1 j_2} A_{j_2 j_2 3} + \beta_{j_1 j_1} a_{j_1 j_1 0}(y_{j_1} - a_{j_1 j_1 1})^3, \tag{1.41}$$

where

$$\begin{aligned} y_{j_1}^{\perp} &= \beta_{j_1 j_1} x_{j_1} + \beta_{j_1 j_2} x_{j_2}, \quad \mathbf{n}_{j_1}^{\perp} = (\beta_{j_1 j_1}, \beta_{j_1 j_2})^{\mathrm{T}}, \\ \mathbf{n}_{j_1} \cdot \mathbf{n}_{j_1}^{\perp} &= \alpha_{j_1 j_1} \beta_{j_1 j_1} + \alpha_{j_1 j_2} \beta_{j_1 j_2} = 0. \end{aligned} \tag{1.42}$$

The equilibrium on the line of $y_{j_1}^* = a_{j_1 j_1 s_1}$ for the x_{j_1}-direction with $\dot{x}_{j_2}^* = A_{j_2 j_2 3}$ possesses the following properties:

- *For $a_{j_1 j_1 0}(\alpha_{j_1 j_2})^3 > 0$ and $A_{j_2 j_2 3} > 0$,*

$$(a_{j_1 j_1 s_1}, \dot{x}_{j_2}^*) = \underbrace{(3^{\mathrm{rd}}\mathrm{UP,pF})}_{\text{third-order up-parabola flow }(+)}. \tag{1.43}$$

- *For $a_{j_1 j_1 0}(\alpha_{j_1 j_2})^3 < 0$ and $A_{j_2 j_2 3} > 0$,*

$$(a_{j_1 j_1 s_1}, \dot{x}_{j_2}^*) = \underbrace{(3^{\mathrm{rd}}\mathrm{DP,pF})}_{\text{third-order down-parabola flow }(+)}. \tag{1.44}$$

- *For $a_{j_1 j_1 0}(\alpha_{j_1 j_2})^3 > 0$ and $A_{j_2 j_2 3} < 0$,*

$$(a_{j_1 j_1 s_1}, \dot{x}_{j_2}^*) = \underbrace{(3^{\mathrm{rd}}\mathrm{DP,nF})}_{\text{third-order down-parabola flow }(-)}. \tag{1.45}$$

- *For* $a_{j_1j_10}(\alpha_{j_1j_2})^3 < 0$ *and* $A_{j_2j_23} < 0$,

$$(a_{j_1j_1s_1}, \dot{x}^*_{j_2}) = \underbrace{(3^{\text{rd}}\text{UP,nF})}_{\text{third-order up-parabola flow }(-)}. \tag{1.46}$$

– *The corresponding third-order up-parabola flows are the appearing bifurcations for up-parabola, down-parabola, and up-parabola flows.*
– *The corresponding third-order down-parabola flows are the appearing bifurcations for down-parabola, up-parabola, and down-parabola flows.*

(*v*) *If*

$$A_{j_2j_23} \neq 0 \quad \text{and} \quad \Delta_{j_1j_1} > 0, \tag{1.47}$$

the corresponding dynamical system becomes

$$\begin{aligned} \dot{x}_{j_2} &= A_{j_2j_23}, \\ \dot{x}_{j_1} &= a_{j_1j_10}(y_{j_1} - a_{j_1j_11})(y_{j_1} - a_{j_1j_12})(y_{j_1} - a_{j_1j_13}), \end{aligned} \tag{1.48}$$

where

$$\begin{gathered} \{a_{j_1j_11}, a_{j_1j_12}, a_{j_1j_13}\} = \text{sort}\{b_{j_1j_11}, b_{j_1j_12}, b_{j_1j_13}\}, \\ a_{j_1j_11} < a_{j_1j_12} < a_{j_1j_13}, \\ b_{j_1j_12}, b_{j_1j_13} = -\frac{1}{2}(B_{j_1j_1} \pm \sqrt{\Delta_{j_1j_1}}), \quad a_{j_1j_10} = A_{j_1j_10}. \end{gathered} \tag{1.49}$$

There is a dynamical system in the y_1*-direction as*

$$\dot{y}_{j_1} = \alpha_{j_1j_2}A_{j_2j_23} + \alpha_{j_1j_1}a_{j_1j_10}(y_{j_1} - a_{j_1j_11})(y_{j_1} - a_{j_1j_12})(y_{j_1} - a_{j_1j_13}), \tag{1.50}$$

with a dynamical system in the $y^{\perp}_{j_1}$*-direction as*

$$\dot{y}^{\perp}_{j_1} = \beta_{j_1j_1}A_{j_2j_23} + \beta_{j_1j_2}a_{j_1j_10}(y_{j_1} - a_{j_1j_11})(y_{j_1} - a_{j_1j_12})(y_{j_1} - a_{j_1j_13}). \tag{1.51}$$

The equilibrium of $y^*_{j_1} = a_{j_1j_1s_1}$ $(s_1, s_2 \in \{1, 2, 3\},\ s_1 \neq s_2)$ *for the* x_{j_1}*-direction with* $\dot{x}^*_{j_2} = A_{j_2j_23}$ *possesses the following properties*:

- *For* $a_{j_1j_10}\alpha_{j_1j_2}\prod_{s_2=1,s_2\neq s_1}^{3}(a_{j_1j_1s_1} - a_{j_1j_1s_2}) > 0$ *and* $A_{j_2j_23} > 0$,

$$(a_{j_1j_1s_1}, \dot{x}^*_{j_2}) = \underbrace{(\text{UP,pF})}_{\text{up-parabola flow }(+)}. \tag{1.52}$$

- *For* $a_{j_1j_10}\alpha_{j_1j_2}\prod_{s_2=1,s_2\neq s_1}^{3}(a_{j_1j_1s_1} - a_{j_1j_1s_2}) < 0$ *and* $A_{j_2j_23} > 0$,

$$(a_{j_1j_1s_1}, \dot{x}^*_{j_2}) = \underbrace{(\text{DP,pF})}_{\text{down-parabola flow }(+)}. \tag{1.53}$$

- *For* $a_{j_1j_10}\alpha_{j_1j_2}\prod_{s_2=1,s_2\neq s_1}^{3}(a_{j_1j_1s_1}-a_{j_1j_1s_2})>0$ *and* $A_{j_2j_23}<0$,

$$(a_{j_1j_1s_1},\dot{x}^*_{j_2})=\underbrace{\text{(DP,nF)}}_{\text{down-parabola flow }(-)}. \tag{1.54}$$

- *For* $a_{j_1j_10}\alpha_{j_1j_2}\prod_{s_2=1,s_2\neq s_1}^{3}(a_{j_1j_1s_1}-a_{j_1j_1s_2})<0$ *and* $A_{j_2j_23}<0$,

$$(a_{j_1j_1s_1},\dot{x}^*_{j_2})=\underbrace{\text{(UP,nF)}}_{\text{up-parabola flow }(-)}. \tag{1.55}$$

(*vi*) *Under the following condition of*

$$A_{j_2j_23}\neq 0 \quad \text{and} \quad \Delta_{j_1j_1}=0, \tag{1.56}$$

the following appearing bifurcations of parabola flows exist:

- *For* $a_{j_1j_10}(\alpha_{j_1j_2})^2(a_{j_1j_1s_2}-a_{j_1j_1s_1})>0$ *and* $A_{j_2j_23}>0$,

$$\begin{gathered}\underbrace{(\dot{x}_{j_1},\dot{x}_{j_2})}_{\text{(pF,pF)-flow}}\rightleftharpoons\underbrace{(a_{j_1j_1s_2},\dot{x}_{j_2})}_{\substack{\text{(II,pF)-flow}\\ \text{appearing bifurcation}}}\rightleftharpoons\left\{\begin{matrix}\underbrace{(a_{j_1j_1(s_2)_1},\dot{x}_{j_2})}_{\text{(DP,pF)-flow}}\\ \underbrace{(a_{j_1j_1(s_2)_2},\dot{x}_{j_2})}_{\text{(UP,pF)-flow}}\end{matrix}\right\} \text{ for } \alpha_{j_1j_2}>0,\\ \underbrace{(\dot{x}_{j_1},\dot{x}_{j_2})}_{\text{(pF,pF)-flow}}\rightleftharpoons\underbrace{(a_{j_1j_1s_2},\dot{x}_{j_2})}_{\substack{\text{(II,pF)-flow}\\ \text{appearing bifurcation}}}\rightleftharpoons\left\{\begin{matrix}\underbrace{(a_{j_1j_1(s_2)_1},\dot{x}_{j_2})}_{\text{(UP,pF)-flow}}\\ \underbrace{(a_{j_1j_1(s_2)_2},\dot{x}_{j_2})}_{\text{(DP,pF)-flow}}\end{matrix}\right\} \text{ for } \alpha_{j_1j_2}<0.\end{gathered} \tag{1.57}$$

- *For* $a_{j_1j_10}(\alpha_{j_1j_2})^2(a_{j_1j_1s_2}-a_{j_1j_1s_1})<0$ *and* $A_{j_2j_23}>0$,

$$\begin{gathered}\underbrace{(\dot{x}_{j_1},\dot{x}_{j_2})}_{\text{(nF,pF)-flow}}\rightleftharpoons\underbrace{(a_{j_1j_1s_2},\dot{x}_{j_2})}_{\substack{\text{(DI,pF)-flow}\\ \text{appearing bifurcation}}}\rightleftharpoons\left\{\begin{matrix}\underbrace{(a_{j_1j_1(s_2)_1},\dot{x}_{j_2})}_{\text{(UP,pF)-flow}}\\ \underbrace{(a_{j_1j_1(s_2)_2},\dot{x}_{j_2})}_{\text{(DP,pF)-flow}}\end{matrix}\right\} \text{ for } \alpha_{j_1j_2}>0,\\ \underbrace{(\dot{x}_{j_1},\dot{x}_{j_2})}_{\text{(nF,pF)-flow}}\rightleftharpoons\underbrace{(a_{j_1j_1s_2},\dot{x}_{j_2})}_{\substack{\text{(DI,pF)-flow}\\ \text{appearing bifurcation}}}\rightleftharpoons\left\{\begin{matrix}\underbrace{(a_{j_1j_1(s_2)_1},\dot{x}_{j_2})}_{\text{(DP,pF)-flow}}\\ \underbrace{(a_{j_1j_1(s_2)_2},\dot{x}_{j_2})}_{\text{(UP,pF)-flow}}\end{matrix}\right\} \text{ for } \alpha_{j_1j_2}<0.\end{gathered} \tag{1.58}$$

- *For* $a_{j_1j_10}(\alpha_{j_1j_2})^2(a_{j_1j_1s_2}-a_{j_1j_1s_1})>0$ *and* $A_{j_2j_23}<0$,

$$\begin{aligned}&\underbrace{(\dot{x}_{j_1},\dot{x}_{j_2})}_{\text{(pF,nF)-flow}}\rightleftharpoons\underbrace{(a_{j_1j_1s_2},\dot{x}_{j_2})}_{\substack{\text{(DI,nF)-flow}\\\text{appearing bifurcation}}}\rightleftharpoons\left\{\begin{matrix}\underbrace{(a_{j_1j_1(s_2)_1},\dot{x}_{j_2})}_{\text{(UP,nF)-flow}},\\\underbrace{(a_{j_1j_1(s_2)_2},\dot{x}_{j_2})}_{\text{(DP,nF)-flow}}\end{matrix}\right\}\text{ for }\alpha_{j_1j_2}>0,\\&\underbrace{(\dot{x}_{j_1},\dot{x}_{j_2})}_{\text{(pF,nF)-flow}}\rightleftharpoons\underbrace{(a_{j_1j_1s_2},\dot{x}_{j_2})}_{\substack{\text{(DI,nF)-flow}\\\text{appearing bifurcation}}}\rightleftharpoons\left\{\begin{matrix}\underbrace{(a_{j_1j_1(s_2)_1},\dot{x}_{j_2})}_{\text{(DP,nF)-flow}},\\\underbrace{(a_{j_1j_1(s_2)_2},\dot{x}_{j_2})}_{\text{(UP,nF)-flow}}\end{matrix}\right\}\text{ for }\alpha_{j_1j_2}<0.\end{aligned}\tag{1.59}$$

- *For* $a_{j_1j_10}(\alpha_{j_1j_2})^2(a_{j_1j_1s_2}-a_{j_1j_1s_1})<0$ *and* $A_{j_2j_23}<0$,

$$\begin{aligned}&\underbrace{(\dot{x}_{j_1},\dot{x}_{j_2})}_{\text{(pF,pF)-flow}}\rightleftharpoons\underbrace{(a_{j_1j_1s_2},\dot{x}_{j_2})}_{\substack{\text{(II,nF)-flow}\\\text{appearing bifurcation}}}\rightleftharpoons\left\{\begin{matrix}\underbrace{(a_{j_1j_1(s_2)_1},\dot{x}_{j_2})}_{\text{(DP,nF)-flow}}\\\underbrace{(a_{j_1j_1(s_2)_2},\dot{x}_{j_2})}_{\text{(UP,nF)-flow}}\end{matrix}\right\}\text{ for }\alpha_{j_1j_2}>0,\\&\underbrace{(\dot{x}_{j_1},\dot{x}_{j_2})}_{\text{(pF,pF)-flow}}\rightleftharpoons\underbrace{(a_{j_1j_1s_2},\dot{x}_{j_2})}_{\substack{\text{(II,nF)-flow}\\\text{appearing bifurcation}}}\rightleftharpoons\left\{\begin{matrix}\underbrace{(a_{j_1j_1(s_2)_1},\dot{x}_{j_2})}_{\text{(UP,nF)-flow}}\\\underbrace{(a_{j_1j_1(s_2)_2},\dot{x}_{j_2})}_{\text{(DP,nF)-flow}}\end{matrix}\right\}\text{ for }\alpha_{j_1j_2}<0.\end{aligned}\tag{1.60}$$

(*vii*) *Consider*

$$\begin{aligned}\dot{x}_{j_2}&=A_{j_2j_23},\\\dot{x}_{j_1}&=a_{j_1j_10}(x_{j_1}-a_{j_1j_1s_1})(x_{j_1}-a_{j_1j_1s_2})(x_{j_1}-a_{j_1j_1s_3}),\end{aligned}\tag{1.61}$$

with

$$a^-_{j_1j_1s_1}\rightleftharpoons a^+_{j_1j_1s_2}\quad\text{and}\quad a^-_{j_1j_1s_2}\rightleftharpoons a^+_{j_1j_1s_1}.\tag{1.62}$$

Under

$$\begin{aligned}A_{j_2j_23}\neq 0,\quad\text{and}\quad\Delta_{j_1j_1}&=(a_{j_1j_1s_1}-a_{j_1j_1s_2})^2=0,\\a_{j_1j_1s_1}&=a_{j_1j_1s_2}\equiv a_{j_1j_11},\end{aligned}\tag{1.63}$$

the following switching bifurcations of parabola flows exist:

- *For* $a_{j_1j_10}(\alpha_{j_1j_2})^2(a_{j_1j_11}-a_{j_1j_1s_3})>0$ *and* $A_{j_2j_23}>0$,

$$\left.\begin{array}{l}\left.\begin{array}{c}\underbrace{(a^{-}_{j_1j_1s_1},\dot{x}_{j_2})}_{\text{(DP,pF)-flow}}\\ \underbrace{(a^{-}_{j_1j_1s_2},\dot{x}_{j_2})}_{\text{(UP,pF)-flow}}\end{array}\right\}\rightleftharpoons\underbrace{(a_{j_1j_11},\dot{x}_{j_2})}_{\substack{\text{(II,pF)-flow}\\ \text{switching bifurcation}}}\rightleftharpoons\left\{\begin{array}{c}\underbrace{(a^{+}_{j_1j_1s_1},\dot{x}_{j_2})}_{\text{(DP,pF)-flow}}\\ \underbrace{(a^{+}_{j_1j_1s_2},\dot{x}_{j_2})}_{\text{(UP,pF)-flow}}\end{array}\right. \text{ for } \alpha_{j_1j_2}>0,\\ \left.\begin{array}{c}\underbrace{(a^{-}_{j_1j_1s_1},\dot{x}_{j_2})}_{\text{(UP,pF)-flow}}\\ \underbrace{(a^{-}_{j_1j_1s_2},\dot{x}_{j_2})}_{\text{(DP,pF)-flow}}\end{array}\right\}\rightleftharpoons\underbrace{(a_{j_1j_11},\dot{x}_{j_2})}_{\substack{\text{(II,pF)-flow}\\ \text{switching bifurcation}}}\rightleftharpoons\left\{\begin{array}{c}\underbrace{(a^{+}_{j_1j_1s_1},\dot{x}_{j_2})}_{\text{(UP,pF)-flow}}\\ \underbrace{(a^{+}_{j_1j_1s_2},\dot{x}_{j_2})}_{\text{(DP,pF)-flow}}\end{array}\right. \text{ for } \alpha_{j_1j_2}<0.\end{array}\right. \tag{1.64}$$

- *For* $a_{j_1j_10}(\alpha_{j_1j_2})^2(a_{j_1j_11}-a_{j_1j_1s_3})<0$ *and* $A_{j_2j_23}>0$,

$$\left.\begin{array}{l}\left.\begin{array}{c}\underbrace{(a^{-}_{j_1j_1s_1},\dot{x}_{j_2}),}_{\text{(UP,pF)-flow}}\\ \underbrace{(a^{-}_{j_1j_1s_2},\dot{x}_{j_2})}_{\text{(DP,pF)-flow}}\end{array}\right\}\rightleftharpoons\underbrace{(a_{j_1j_11},\dot{x}_{j_2})}_{\substack{\text{(DI,pF)-flow}\\ \text{switching bifurcation}}}\rightleftharpoons\left\{\begin{array}{c}\underbrace{(a^{+}_{j_1j_1s_1},\dot{x}_{j_2}),}_{\text{(UP,pF)-flow}}\\ \underbrace{(a^{+}_{j_1j_1s_2},\dot{x}_{j_2})}_{\text{(DF,pF)-flow}}\end{array}\right. \text{ for } \alpha_{j_1j_2}>0,\\ \left.\begin{array}{c}\underbrace{(a^{-}_{j_1j_1s_1},\dot{x}_{j_2}),}_{\text{(DP,pF)-flow}}\\ \underbrace{(a^{-}_{j_1j_1s_2},\dot{x}_{j_2})}_{\text{(UP,pF)-flow}}\end{array}\right\}\rightleftharpoons\underbrace{(a_{j_1j_11},\dot{x}_{j_2})}_{\substack{\text{(DI,pF)-flow}\\ \text{switching bifurcation}}}\rightleftharpoons\left\{\begin{array}{c}\underbrace{(a^{+}_{j_1j_1s_1},\dot{x}_{j_2}),}_{\text{(DP,pF)-flow}}\\ \underbrace{(a^{+}_{j_1j_1s_2},\dot{x}_{j_2})}_{\text{(UP,pF)-flow}}\end{array}\right. \text{ for } \alpha_{j_1j_2}<0.\end{array}\right. \tag{1.65}$$

- *For* $a_{j_1j_10}(\alpha_{j_1j_2})^2(a_{j_1j_11}-a_{j_1j_1s_3})>0$ *and* $A_{j_2j_23}<0$,

$$\left.\begin{array}{l}\left.\begin{array}{c}\underbrace{(a^{-}_{j_1j_1s_1},\dot{x}_{j_2}),}_{\text{(UP,nF)-flow}}\\ \underbrace{(a^{-}_{j_1j_1s_2},\dot{x}_{j_2})}_{\text{(DP,nF)-flow}}\end{array}\right\}\rightleftharpoons\underbrace{(a_{j_1j_11},\dot{x}_{j_2})}_{\substack{\text{(DI,nF)-flow}\\ \text{switching bifurcation}}}\rightleftharpoons\left\{\begin{array}{c}\underbrace{(a^{+}_{j_1j_1s_1},\dot{x}_{j_2}),}_{\text{(UP,nF)-flow}}\\ \underbrace{(a^{+}_{j_1j_1s_2},\dot{x}_{j_2})}_{\text{(DP,nF)-flow}}\end{array}\right. \text{ for } \alpha_{j_1j_2}>0,\\ \left.\begin{array}{c}\underbrace{(a^{-}_{j_1j_1s_1},\dot{x}_{j_2}),}_{\text{(DP,nF)-flow}}\\ \underbrace{(a^{-}_{j_1j_1s_2},\dot{x}_{j_2})}_{\text{(UP,nF)-flow}}\end{array}\right\}\rightleftharpoons\underbrace{(a_{j_1j_11},\dot{x}_{j_2})}_{\substack{\text{(DI,nF)-flow}\\ \text{switching bifurcation}}}\rightleftharpoons\left\{\begin{array}{c}\underbrace{(a^{+}_{j_1j_1s_1},\dot{x}_{j_2}),}_{\text{(DP,nF)-flow}}\\ \underbrace{(a^{+}_{j_1j_1s_2},\dot{x}_{j_2})}_{\text{(UP,nF)-flow}}\end{array}\right. \text{ for } \alpha_{j_1j_2}<0.\end{array}\right. \tag{1.66}$$

- *For* $a_{j_1j_10}(\alpha_{j_1j_2})^2(a_{j_1j_11}-a_{j_1j_1s_3})<0$ *and* $A_{j_2j_23}<0$,

$$\left\{\begin{matrix}\underbrace{(a^-_{j_1j_1s_1},\dot{x}_{j_2})}_{\text{(DP,nF)-flow}}\\ \underbrace{(a^-_{j_1j_1s_2},\dot{x}_{j_2})}_{\text{(UP,nF)-flow}}\end{matrix}\right\}\rightleftharpoons\underbrace{(a_{j_1j_11},\dot{x}_{j_2})}_{\substack{\text{(II,nF)-flow}\\ \text{switching bifurcation}}}\rightleftharpoons\left\{\begin{matrix}\underbrace{(a^+_{j_1j_1s_1},\dot{x}_{j_2})}_{\text{(DP,nF)-flow}}\\ \underbrace{(a^+_{j_1j_1s_2},\dot{x}_{j_2})}_{\text{(UP,nF)-flow}}\end{matrix}\right\}\quad\text{for } \alpha_{j_1j_2}>0,$$

$$\left\{\begin{matrix}\underbrace{(a^-_{j_1j_1s_1},\dot{x}_{j_2})}_{\text{(UP,nF)-flow}}\\ \underbrace{(a^-_{j_1j_1s_2},\dot{x}_{j_2})}_{\text{(DP,nF)-flow}}\end{matrix}\right\}\rightleftharpoons\underbrace{(a_{j_1j_11},\dot{x}_{j_2})}_{\substack{\text{(II,nF)-flow}\\ \text{switching bifurcation}}}\rightleftharpoons\left\{\begin{matrix}\underbrace{(a^+_{j_1j_1s_1},\dot{x}_{j_2})}_{\text{(UP,nF)-flow}}\\ \underbrace{(a^+_{j_1j_1s_2},\dot{x}_{j_2})}_{\text{(DP,nF)-flow}}\end{matrix}\right\}\quad\text{for } \alpha_{j_1j_2}<0. \tag{1.67}$$

(viii) Under the following condition of

$$A_{j_2j_22}\neq 0\quad\text{and}\quad\Delta_{j_1j_1}=0,\quad a_{j_1j_1s_1}=a_{j_1j_1s_2}=a_{j_1j_11}, \tag{1.68}$$

the following appearing bifurcations of parabola flows exist:

- *For* $a_{j_1j_10}(\alpha_{j_1j_2})^3>0$ *and* $A_{j_2j_23}>0$,

$$\underbrace{(a_{j_1j_11},\dot{x}_{j_2})}_{\text{(UP,pF)-flow}}\rightleftharpoons\underbrace{(a_{j_1j_11},\dot{x}_{j_2})}_{\substack{(3^{\text{rd}}\text{UP,pF})\text{-flow}\\ \text{appearing bifurcation}}}\rightleftharpoons\left\{\begin{matrix}\underbrace{(a_{j_1j_11},\dot{x}_{j_2})}_{\text{(UP,pF)-flow}}\\ \underbrace{(a_{j_1j_12},\dot{x}_{j_2})}_{\text{(DP,pF)-flow}}\\ \underbrace{(a_{j_1j_13},\dot{x}_{j_2})}_{\text{(UP,pF)-flow}}\end{matrix}\right\}. \tag{1.69}$$

- *For* $a_{j_1j_10}(\alpha_{j_1j_2})^3<0$ *and* $A_{j_2j_23}>0$,

$$\underbrace{(a_{j_1j_11},\dot{x}_{j_2})}_{\text{(DP,pF)-flow}}\rightleftharpoons\underbrace{(a_{j_1j_11},\dot{x}_{j_2})}_{\substack{(3^{\text{rd}}\text{DP,pF})\text{-flow}\\ \text{appearing bifurcation}}}\rightleftharpoons\left\{\begin{matrix}\underbrace{(a_{j_1j_11},\dot{x}_{j_2})}_{\text{(DP,pF)-flow}}\\ \underbrace{(a_{j_1j_12},\dot{x}_{j_2})}_{\text{(UP,pF)-flow}}\\ \underbrace{(a_{j_1j_13},\dot{x}_{j_2})}_{\text{(DP,pF)-flow}}\end{matrix}\right\}. \tag{1.70}$$

- *For* $a_{j_1j_10}(\alpha_{j_1j_2})^3 > 0$ *and* $A_{j_2j_23} < 0$,

$$\underbrace{(a_{j_1j_11}, \dot{x}_{j_2})}_{\text{(DP,nF)-flow}} \rightleftharpoons \underbrace{(a_{j_1j_11}, \dot{x}_{j_2})}_{\substack{\text{(3}^{\text{rd}}\text{DP,nF)-flow} \\ \text{appearing bifurcation}}} \rightleftharpoons \left\{ \begin{array}{c} \underbrace{(a_{j_1j_11}, \dot{x}_{j_2})}_{\text{(DP,nF)-flow}} \\ \underbrace{(a_{j_1j_12}, \dot{x}_{j_2})}_{\text{(UP,nF)-flow}} \\ \underbrace{(a_{j_1j_13}, \dot{x}_{j_2})}_{\text{(DP,nF)-flow}} \end{array} \right\}. \tag{1.71}$$

- *For* $a_{j_1j_10}(\alpha_{j_1j_2})^3 < 0$ *and* $A_{j_2j_23} < 0$,

$$\underbrace{(a_{j_1j_11}, \dot{x}_{j_2})}_{\text{(UP,nF)-flow}} \rightleftharpoons \underbrace{(a_{j_1j_11}, \dot{x}_{j_2})}_{\substack{\text{(3}^{\text{rd}}\text{UP,nF)-flow} \\ \text{appearing bifurcation}}} \rightleftharpoons \left\{ \begin{array}{c} \underbrace{(a_{j_1j_11}, \dot{x}_{j_2})}_{\text{(UP,nF)-flow}} \\ \underbrace{(a_{j_1j_12}, \dot{x}_{j_2})}_{\text{(DP,nF)-flow}} \\ \underbrace{(a_{j_1j_13}, \dot{x}_{j_2})}_{\text{(UP,nF)-flow}} \end{array} \right\}. \tag{1.72}$$

1.2 Proof of Theorem 1.1

Consider a two-dimensional, nonlinear dynamical system as

$$\begin{aligned} \dot{x}_{j_1} &= f_{j_1}(\mathbf{x}, \mathbf{p}) = A_{j_1j_13} + A_{j_1j_12}y_{j_1} + A_{j_1j_11}y_{j_1}^2 + A_{j_1j_10}y_{j_1}^3, \\ \dot{x}_{j_2} &= f_{j_2}(\mathbf{x}, \mathbf{p}) = A_{j_2j_23} + A_{j_2j_22}y_{j_1} + A_{j_2j_21}y_{j_1}^2 + A_{j_2j_20}y_{j_1}^3 \end{aligned}$$

with

$$y_{j_1} = \alpha_{j_1j_1}x_{j_1} + \alpha_{j_1j_2}x_{j_2} \quad \text{for } j_1, j_2 \in \{1, 2\}, \quad j_1 \neq j_2.$$

(i) For $A_{j_2j_22} = A_{j_2j_21} = A_{j_2j_20} = 0$ $(j_1, j_2 \in \{1, 2\}, j_1 \neq j_2)$,

$$\begin{aligned} \dot{x}_{j_2} &= A_{j_2j_23} = 0, \\ \dot{x}_{j_1} &= A_{j_1j_13} + A_{j_1j_12}y_{j_1} + A_{j_1j_11}y_{j_1}^2 + A_{j_1j_10}y_{j_1}^3 \\ &= a_{j_1j_10}(y_{j_1} - b_{j_1j_11})[y_{j_1}^2 + B_{j_1j_1}y_{j_1} + C_{j_1j_1}]. \end{aligned}$$

For $A_{j_3j_33} = 0$, the two-dimensional dynamical system becomes a one-dimensional cubic nonlinear system in the x_{j_1}-direction.

(i_1) For $\Delta_{j_1j_1} < 0$, the one-dimensional system becomes

$$\begin{aligned}\dot{x}_{j_2} &= A_{j_2j_23} = 0,\\ \dot{x}_{j_1} &= a_{j_1j_10}(y_{j_1} - a_{j_1j_11})[(y_{j_1} - a_{j_1j_1})^2 + b_{j_1j_1}],\end{aligned}$$

with

$$\begin{gathered}a_{j_1j_1} = -\frac{1}{2}B_{j_1j_1}, \quad b_{j_1j_1} = -\frac{1}{4}C_{j_1j_1}, \quad a_{j_1j_11} = b_{j_1j_11},\\ \Delta_{j_1j_1} = (B_{j_1j_1})^2 - 4C_{j_1j_1}, \quad a_{j_1j_10} = A_{j_1j_10}.\end{gathered}$$

From the above equations, the following equation exists:

$$\dot{y}_{j_1} = \alpha_{j_1j_1}a_{j_1j_10}(y_{j_1} - a_{j_1j_11})[(y_{j_1} - a_{j_1j_1})^2 + b_{j_1j_1}].$$

The above equation deformation yields

$$\begin{aligned}\alpha_{j_1j_1}a_{j_1j_10}dt &= \frac{dy_{j_1}}{(a_{j_1j_11} - a_{j_1j_1})^2 + b_{j_1j_1}}\\ &\quad\times\left[\frac{1}{y_{j_1} - a_{j_1j_11}} - \frac{y_{j_1} - a_{j_1j_1} - (a_{j_1j_1} - a_{j_1j_11})}{(y_{j_1} - a_{j_1j_1})^2 + b_{j_1j_1}}\right].\end{aligned}$$

With an initial condition $y_{j_1} = y_{j_10}$ at $t = t_0$, integration of the foregoing equation gives a solution through the following equation:

$$\begin{aligned}t - t_0 = &\frac{1}{\alpha_{j_1j_1}a_{j_1j_10}[(a_{j_1j_11} - a_{j_1j_1})^2 + b_{j_1j_1}]}\Bigg[\ln\frac{|y_{j_1} - a_{j_1j_11}|}{|y_{j_10} - a_{j_1j_11}|}\\ &-\frac{1}{2}\ln\frac{|(y_{j_1} - a_{j_1j_1})^2 + b_{j_1j_1}|}{|(y_{j_10} - a_{j_1j_1})^2 + b_{j_1j_1}|}\\ &+\frac{(a_{j_1j_11} - a_{j_1j_1})}{\sqrt{b_{j_1j_1}}}\left(\arctan\frac{y_{j_1} - a_{j_1j_1}}{\sqrt{b_{j_1j_1}}} - \arctan\frac{y_{j_10} - a_{j_1j_1}}{\sqrt{b_{j_1j_1}}}\right)\Bigg].\end{aligned}$$

For $y^*_{j_1} = a_{j_1j_11}$, the variational equation is

$$\Delta\dot{y}_{j_1} = \alpha_{j_1j_1}a_{j_1j_10}[(a_{j_1j_11} - a_{j_1j_1})^2 + b_{j_1j_1}]\Delta y_{j_1}$$

and

$$\Delta\dot{x}_{j_1} = \alpha_{j_1j_1}a_{j_1j_10}[(a_{j_1j_11} - a_{j_1j_1})^2 + b_{j_1j_1}]\Delta x_{j_1}.$$

Due to

$$(a_{j_1j_11} - a_{j_1j_1})^2 + b_{j_1j_1} > 0,$$

the equilibrium line of $y^*_{j_1} = a_{j_1j_1 1}$ is a source for $\alpha_{j_1j_1} a_{j_1j_1 0} > 0$ and a sink for $\alpha_{j_1j_1} a_{j_1j_1 0} < 0$, respectively. Due to $x_{j_2} = C$ (const), therefore, the equilibrium line of $y^*_{j_1} = a_{j_1j_1 1}$ has the following properties:

- For $a_{j_1j_1 0}\alpha_{j_1j_1} > 0$, the line of $y^*_{j_1} = a_{j_1j_1 1}$ in the x_{j_1}-direction is a source infinite-equilibrium.
- For $a_{j_1j_1 0}\alpha_{j_1j_1} < 0$, the line of $y^*_{j_1} = a_{j_1j_1 1}$ in the x_{j_1}-direction is a sink infinite-equilibrium.

(i$_2$) For $\Delta_{j_1j_1} = 0$ and $a_{j_1j_1 s_1} \neq a_{j_1j_1 s_2}$, the one-dimensional system becomes

$$\begin{aligned}
\dot{x}_{j_2} &= A_{j_2j_2 3} = 0,\\
\dot{x}_{j_1} &= a_{j_1j_1 0}(y_{j_1} - a_{j_1j_1 s_1})(y_{j_1} - a_{j_1j_1 s_2})^2,
\end{aligned}$$

with

$$a_{j_1j_1 s_2} = -\frac{1}{2}B_{j_1j_1}, \quad a_{j_1j_1 s_1} = b_{j_1j_1 1}, \quad a_{j_1j_1 0} = A_{j_1j_1 0}.$$

From the above equations, the following equation exists:

$$\dot{y}_{j_1} = \alpha_{j_1j_1} a_{j_1j_1 0}(y_{j_1} - a_{j_1j_1 s_1})(y_{j_1} - a_{j_1j_1 s_2})^2.$$

From the above equations, the following equation exists:

$$\dot{y}_{j_1} = \alpha_{j_1j_1} a_{j_1j_1 0}(y_{j_1} - a_{j_1j_1 s_1})(y_{j_1} - a_{j_1j_1 s_2})^2.$$

The above equation deformation yields

$$\begin{aligned}
\alpha_{j_1j_1} a_{j_1j_1 0} dt = {} & \frac{dy_{j_1}}{(a_{j_1j_1 s_1} - a_{j_1j_1 s_2})^2}\left[\frac{1}{y_{j_1} - a_{j_1j_1 s_1}} - \frac{1}{y_{j_1} - a_{j_1j_1 s_2}}\right.\\
& \left. + \frac{(a_{j_1j_1 s_2} - a_{j_1j_1 s_1})}{(y_{j_1} - a_{j_1j_1 s_2})^2}\right].
\end{aligned}$$

With an initial condition $y_{j_1} = y_{j_1 0}$ at $t = t_0$, integration of the foregoing equation gives a solution through the following equation:

$$\begin{aligned}
t - t_0 = {} & \frac{1}{\alpha_{j_1j_1} a_{j_1j_1 0}(a_{j_1j_1 s_1} - a_{j_1j_1 s_2})^2}\left[\ln\frac{|y_{j_1} - a_{j_1j_1 s_1}|}{|y_{j_1 0} - a_{j_1j_1 s_1}|}\right.\\
& - \ln\frac{|y_{j_1} - a_{j_1j_1 s_2}|}{|y_{j_1 0} - a_{j_1j_1 s_2}|} - (a_{j_1j_1 s_2} - a_{j_1j_1 s_1})\\
& \left. \times\left(\frac{1}{y_{j_1} - a_{j_1j_1 s_2}} - \frac{1}{y_{j_1 0} - a_{j_1j_1 s_2}}\right)\right].
\end{aligned}$$

(i_{2a}) For $y_{j_1}^* = a_{j_1 j_1 s_1}$, the variational equation is

$$\Delta \dot{y}_{j_1} = \alpha_{j_1 j_1} a_{j_1 j_1 0} (a_{j_1 j_1 s_1} - a_{j_1 j_1 s_2})^2 \Delta y_{j_1}.$$

The equilibrium of $y_{j_1}^* = a_{j_1 j_1 s_1}$ is a source for $\alpha_{j_1 j_1} a_{j_1 j_1 0}(a_{j_1 j_1 s_1} - a_{j_1 j_1 s_2})^2 > 0$ and a sink for $\alpha_{j_1 j_1} a_{j_1 j_1 0}(a_{j_1 j_1 s_1} - a_{j_1 j_1 s_2})^2 < 0$, respectively. Due to $x_{j_2} = C$ (const), therefore, the equilibrium line of $y_{j_1}^* = a_{j_1 j_1 s_1}$ has the following properties:

- For $a_{j_1 j_1 0} \alpha_{j_1 j_1} (a_{j_1 j_1 s_1} - a_{j_1 j_1 s_2})^2 > 0$, the line of $y_{j_1}^* = a_{j_1 j_1 s_1}$ in the in the x_{j_1}-direction is a source infinite-equilibrium.
- For $a_{j_1 j_1 0} \alpha_{j_1 j_1} (a_{j_1 j_1 s_1} - a_{j_1 j_1 s_2})^2 < 0$, the line of $y_{j_1}^* = a_{j_1 j_1 s_1}$ in the in the x_{j_1}-direction is a sink infinite-equilibrium.

(i_{2b}) For $y_{j_1}^* = a_{j_1 j_1 s_2}$, the variational equation is

$$\Delta \dot{y}_{j_1} = \alpha_{j_1 j_1} a_{j_1 j_1 0} (a_{j_1 j_1 s_2} - a_{j_1 j_1 s_1}) (\Delta y_{j_1})^2$$

and

$$\Delta \dot{x}_{j_1} = (\alpha_{j_1 j_1})^2 a_{j_1 j_1 0} (a_{j_1 j_1 s_2} - a_{j_1 j_1 s_1}) (\Delta x_{j_1})^2.$$

The equilibrium of $y_{j_1}^* = a_{j_1 j_1 s_2}$ is an upper-saddle for $\alpha_{j_1 j_1} a_{j_1 j_1 0}(a_{j_1 j_1 s_2} - a_{j_1 j_1 s_1}) > 0$ and a lower-saddle for $\alpha_{j_1 j_1} a_{j_1 j_1 0}(a_{j_1 j_1 s_2} - a_{j_1 j_1 s_1}) < 0$, respectively. Due to $x_{j_2} = C$ (const), therefore, the equilibrium line of $y_{j_1}^* = a_{j_1 j_1 s_2}$ has the following properties:

- For $a_{j_1 j_1 0} (\alpha_{j_1 j_1})^2 (a_{j_1 j_1 s_2} - a_{j_1 j_1 s_1}) > 0$, the line of $y_{j_1}^* = a_{j_1 j_1 s_2}$ in the x_{j_1}-direction is an upper-saddle infinite-equilibrium.
- For $a_{j_1 j_1 0} (\alpha_{j_1 j_1})^2 (a_{j_1 j_1 s_2} - a_{j_1 j_1 s_1}) < 0$, the line of $y_{j_1}^* = a_{j_1 j_1 s_2}$ in the x_{j_1}-direction is a lower-saddle infinite-equilibrium.

(i_3) For $\Delta_{j_1 j_1} = 0$ and $a_{j_1 j_1 s_1} = a_{j_1 j_1 s_2} = a_{j_1 j_1 1}$, the one-dimensional system becomes

$$\begin{aligned} \dot{x}_{j_2} &= A_{j_2 j_2 3} = 0, \\ \dot{x}_{j_1} &= a_{j_1 j_1 0} (y_{j_1} - a_{j_1 j_1 1})^3, \end{aligned}$$

with

$$a_{j_1 j_1 s_2} = -\frac{1}{2} B_{j_1 j_1}, \quad a_{j_1 j_1 s_1} = b_{j_1 j_1 1}, \quad a_{j_1 j_1 0} = A_{j_1 j_1 0}.$$

From the above equations, the following equation exists:

$$\dot{y}_{j_1} = \alpha_{j_1 j_1} a_{j_1 j_1 0} (y_{j_1} - a_{j_1 j_1 1})^3.$$

The above equation deformation yields

$$\alpha_{j_1j_1}a_{j_1j_10}dt = \frac{dy_{j_1}}{(y_{j_1} - a_{j_1j_11})^3}.$$

With an initial condition $y_{j_1} = y_{j_10}$ at $t = t_0$, integration of the foregoing equation gives a solution through the following equation:

$$t - t_0 = -\frac{1}{2\alpha_{j_1j_1}a_{j_1j_10}}\left[\frac{1}{(y_{j_1} - a_{j_1j_11})^2} - \frac{1}{(y_{j_10} - a_{j_1j_11})^2}\right].$$

For $y^*_{j_1} = a_{j_1j_1s_2}$, the variational equation is

$$\Delta\dot{y}_{j_1} = \alpha_{j_1j_1}a_{j_1j_10}(\Delta y_{j_1})^3$$

and

$$\Delta\dot{x}_{j_1} = (\alpha_{j_1j_1})^3 a_{j_1j_10}(\Delta x_{j_1})^3.$$

The equilibrium of $y^*_{j_1} = a_{j_1j_1s_2}$ is an upper-saddle for $\alpha_{j_1j_1}a_{j_1j_10} > 0$ and a lower-saddle for $\alpha_{j_1j_1}a_{j_1j_10} < 0$, respectively. Due to $x_{j_2} = C$ (const), therefore, the equilibrium line of $y^*_{j_1} = a_{j_1j_1s_2}$ has the following properties:

- For $a_{j_1j_10}(\alpha_{j_1j_1})^3 > 0$, the line of $y_{j_1} = a_{j_1j_11}$ in the x_{j_1}-direction is a third-order source infinite-equilibrium.
- For $a_{j_1j_10}(\alpha_{j_1j_1})^3 < 0$, the line of $y_{j_1} = a_{j_1j_11}$ in the x_{j_1}-direction is a third-order sink infinite-equilibrium.

(i$_4$) For $\Delta_{j_1j_1} > 0$, the one-dimensional system becomes

$$\dot{x}_{j_2} = A_{j_2j_23} = 0,$$
$$\dot{x}_{j_1} = a_{j_1j_10}(y_{j_1} - a_{j_1j_11})(y_{j_1} - a_{j_1j_12})(y_{j_1} - a_{j_1j_13}),$$

with

$$\{a_{j_1j_11}, a_{j_1j_12}, a_{j_1j_13}\} = \text{sort}\{b_{j_1j_11}, b_{j_1j_12}, b_{j_1j_13}\}, \quad a_{j_1j_11} < a_{j_1j_12} < a_{j_1j_13},$$
$$b_{j_1j_12}, b_{j_1j_13} = -\frac{1}{2}(B_{j_1j_1} \pm \sqrt{\Delta_{j_1j_1}}), \quad a_{j_1j_10} = A_{j_1j_10}.$$

With an initial condition $y_{j_1} = y_{j_10}$ at $t = t_0$, the solution is

$$t - t_0 = \frac{1}{a_{j_1j_10}\alpha_{j_1j_1}}\sum_{s_1=1}^{3}\frac{1}{\prod_{s_2=1,s_2\neq s_1}^{3}(a_{j_1j_1s_1} - a_{j_1j_1s_2})}\ln\frac{|y_{j_1} - a_{j_1j_1s_1}|}{|y_{j_10} - a_{j_1j_1s_1}|}.$$

For $y^*_{j_1} = a_{j_1j_1s_2}$, the variational equation is

$$\Delta\dot{y}_{j_1} = \alpha_{j_1j_1}a_{j_1j_10}\prod_{s_2=1,s_2\neq s_1}^{3}(a_{j_1j_1s_1} - a_{j_1j_1s_2})\Delta y_{j_1}$$

and

$$\Delta \dot{x}_{j_1} = \alpha_{j_1 j_1} a_{j_1 j_1 0} \prod_{s_2=1, s_2 \neq s_1}^{3} (a_{j_1 j_1 s_1} - a_{j_1 j_1 s_2}) \Delta x_{j_1}.$$

The equilibrium of $y_{j_1}^* = a_{j_1 j_1 s_2}$ is a source for $\alpha_{j_1 j_1} a_{j_1 j_1 0} \prod_{s_2=1, s_2 \neq s_1}^{3} (a_{j_1 j_1 s_1} - a_{j_1 j_1 s_2}) > 0$ and a sink for $\alpha_{j_1 j_1} a_{j_1 j_1 0} \prod_{s_2=1, s_2 \neq s_1}^{3} (a_{j_1 j_1 s_1} - a_{j_1 j_1 s_2}) < 0$, respectively. Due to $x_{j_2} = C$ (const), therefore, the equilibrium line of $y_{j_1}^* = a_{j_1 j_1 s_2}$ has the following properties:

- For $a_{j_1 j_1 0} \alpha_{j_1 j_1} \prod_{s_2=1, s_2 \neq s_1}^{3} (a_{j_1 j_1 s_1} - a_{j_1 j_1 s_2}) > 0$, the lines of $y_{j_1} = a_{j_1 j_1 s_1}$ in the x_{j_1}-direction are a source infinite-equilibrium.
- For $a_{j_1 j_1 0} \alpha_{j_1 j_1} \prod_{s_2=1, s_2 \neq s_1}^{3} (a_{j_1 j_1 s_1} - a_{j_1 j_1 s_2}) < 0$, the on the lines of $y_{j_1} = a_{j_1 j_1 s_1}$ in the x_{j_1}-direction are a sink infinite-equilibrium.

(ii) If

$$A_{j_2 j_2 3} \neq 0 \quad \text{and} \quad \Delta_{j_1 j_1} < 0,$$

the standard form of the dynamical system is

$$\begin{aligned} \dot{x}_{j_2} &= A_{j_2 j_2 3}, \\ \dot{x}_{j_1} &= a_{j_1 j_1 0}(y_{j_1} - a_{j_1 j_1 1})[(y_{j_1} - a_{j_1 j_1})^2 + b_{j_1 j_1}]. \end{aligned}$$

The dynamical system in the y_{j_1}-direction is

$$\dot{y}_{j_1} = \alpha_{j_1 j_2} A_{j_2 j_2 3} + \alpha_{j_1 j_1} a_{j_1 j_1 0}(y_{j_1} - a_{j_1 j_1 1})[(y_{j_1} - a_{j_1 j_1})^2 + b_{j_1 j_1}],$$

with a dynamical system in the $y_{j_1}^{\perp}$-direction as

$$\dot{y}_{j_1}^{\perp} = \beta_{j_1 j_2} A_{j_2 j_2 3} + \beta_{j_1 j_1} a_{j_1 j_1 0}(y_{j_1} - a_{j_1 j_1 1})[(y_{j_1} - a_{j_1 j_1})^2 + b_{j_1 j_1}],$$

where

$$\begin{aligned} y_{j_1}^{\perp} &= \beta_{j_1 j_1} x_{j_1} + \beta_{j_1 j_2} x_{j_2}, \quad \mathbf{n}_{j_1}^{\perp} = (\beta_{j_1 j_1}, \beta_{j_1 j_2})^{\mathrm{T}}, \\ \mathbf{n}_{j_1} \cdot \mathbf{n}_{j_1}^{\perp} &= \alpha_{j_1 j_1} \beta_{j_1 j_1} + \alpha_{j_1 j_2} \beta_{j_1 j_2} = 0. \end{aligned}$$

In phase space, at $y_{j_1}^* = a_{j_1 j_1 1}$,

$$\left. \frac{dx_{j_1}}{dx_{j_2}} \right|_{y_{j_1}^* = a_{j_1 j_1 1}} = \left. \frac{a_{j_1 j_1 0}(y_{j_1} - a_{j_1 j_1 1})[(y_{j_1} - a_{j_1 j_1})^2 + b_{j_1 j_1}]}{A_{j_2 j_2 3}} \right|_{y_{j_1}^* = a_{j_1 j_1 1}} = 0.$$

If

$$\left.\frac{d^2x_{j_1}}{dx_{j_2}^2}\right|_{y_{j_1}^*=a_{j_1j_11}} = \frac{\alpha_{j_1j_2}a_{j_1j_10}[(a_{j_1j_11}-a_{j_1j_1})^2+b_{j_1j_1}]}{A_{j_2j_23}} > 0,$$

the flow at $y_{j_1}^* = a_{j_1j_11}$ is an up-parabola flow in the x_{j_1}-direction. If

$$\left.\frac{d^2x_{j_1}}{dx_{j_2}^2}\right|_{y_{j_1}^*=a_{j_1j_11}} = \frac{\alpha_{j_1j_2}a_{j_1j_10}[(a_{j_1j_11}-a_{j_1j_1})^2+b_{j_1j_1}]}{A_{j_2j_23}} < 0,$$

the flow at $y_{j_1}^* = a_{j_1j_11}$ is a down-parabola flow in the x_{j_1}-direction.

Due to $\dot{x}_{j_2} = A_{j_2j_23}$, the parabola flows in the x_{j_1}-direction are positive and negative for $A_{j_2j_23} > 0$ and $A_{j_2j_23} < 0$, respectively.

Therefore, the two-dimensional dynamical system has the following properties as in Eqs. (1.21)–(1.24):

- For $a_{j_1j_10}\alpha_{j_1j_2} > 0$ and $A_{j_2j_23} > 0$,

$$(a_{j_1j_11}, \dot{x}_{j_2}^*) = \underbrace{(\text{UP,pF})}_{\text{up-parabola flow }(+)}.$$

- For $a_{j_1j_10}\alpha_{j_1j_2} < 0$ and $A_{j_2j_23} > 0$,

$$(a_{j_1j_11}, \dot{x}_{j_2}^*) = \underbrace{(\text{DP,pF})}_{\text{down-parabola flow }(+)}.$$

- For $a_{j_1j_10}\alpha_{j_1j_2} > 0$ and $A_{j_2j_23} < 0$,

$$(a_{j_1j_11}, \dot{x}_{j_2}^*) = \underbrace{(\text{DP,nF})}_{\text{down-parabola flow }(-)}.$$

- For $a_{j_1j_10}\alpha_{j_1j_2} < 0$ and $A_{j_2j_23} < 0$,

$$(a_{j_1j_11}, \dot{x}_{j_2}^*) = \underbrace{(\text{UP,nF})}_{\text{up-parabola flow }(-)}.$$

(iii) If

$$A_{j_2j_23} \neq 0 \quad \text{and} \quad \Delta_{j_1j_1} = 0, a_{j_1j_1s_1} \neq a_{j_1j_1s_2},$$

the corresponding dynamical system becomes

$$\begin{aligned}\dot{x}_{j_2} &= A_{j_2j_23},\\ \dot{x}_{j_1} &= a_{j_1j_10}(y_{j_1}-a_{j_1j_1s_1})(y_{j_1}-a_{j_1j_1s_2})^2.\end{aligned}$$

There is a dynamical system in the y_{j_1}-direction as

$$\dot{y}_{j_1} = \alpha_{j_1j_2}A_{j_2j_23} + \alpha_{j_1j_1}a_{j_1j_10}(y_{j_1} - a_{j_1j_1s_1})(y_{j_1} - a_{j_1j_1s_2})^2,$$

with a dynamical system in the $y_{j_1}^{\perp}$-direction as

$$\dot{y}_{j_1}^{\perp} = \beta_{j_1j_2}\alpha_{j_1j_2}A_{j_2j_23} + \beta_{j_1j_1}a_{j_1j_10}(y_{j_1} - a_{j_1j_1s_1})(y_{j_1} - a_{j_1j_1s_2})^2,$$

where

$$y_{j_1}^{\perp} = \beta_{j_1j_1}x_{j_1} + \beta_{j_1j_2}x_{j_2}, \quad \mathbf{n}_{j_1}^{\perp} = (\beta_{j_1j_1}, \beta_{j_1j_2})^{\mathrm{T}},$$
$$\mathbf{n}_{j_1} \cdot \mathbf{n}_{j_1}^{\perp} = \alpha_{j_1j_1}\beta_{j_1j_1} + \alpha_{j_1j_2}\beta_{j_1j_2} = 0.$$

(iii$_1$) In phase space, at $y_{j_1}^* = a_{j_1j_1s_1} \neq a_{j_1j_1s_2}$,

$$\left.\frac{dx_{j_1}}{dx_{j_2}}\right|_{y_{j_1}^*=a_{j_1j_1s_1}} = \left.\frac{a_{j_1j_10}(y_{j_1} - a_{j_1j_1s_1})(y_{j_1} - a_{j_1j_1s_2})^2}{A_{j_2j_23}}\right|_{y_{j_1}^*=a_{j_1j_1s_1}} = 0.$$

If

$$\left.\frac{d^2x_{j_1}}{dx_{j_2}^2}\right|_{y_{j_1}^*=a_{j_1j_1s_1}} = \frac{\alpha_{j_1j_2}a_{j_1j_10}(a_{j_1j_1s_1} - a_{j_1j_1s_2})^2}{A_{j_2j_23}} > 0,$$

the flow at $y_{j_1}^* = a_{j_1j_1s_1}$ is an up-parabola flow in the x_{j_1}-direction. If

$$\left.\frac{d^2x_{j_1}}{dx_{j_2}^2}\right|_{y_{j_1}^*=a_{j_1j_1s_1}} = \frac{\alpha_{j_1j_2}a_{j_1j_10}(a_{j_1j_1s_1} - a_{j_1j_1s_2})^2}{A_{j_2j_23}} < 0,$$

the flow at $y_{j_1}^* = a_{j_1j_1s_1}$ is a down-parabola flow in the x_{j_1}-direction.

Due to $\dot{x}_{j_2} = A_{j_2j_23}$, the parabola flows in the x_{j_1}-direction are positive and negative for $A_{j_2j_23} > 0$ and $A_{j_2j_23} < 0$, respectively.

Therefore, the equilibrium on the line of $y_{j_1}^* = a_{j_1j_1s_1}$ for the x_{j_1}-direction with $\dot{x}_{j_2}^* = A_{j_2j_23}$ possesses the following properties as in Eqs. (1.30)–(1.33):

- For $a_{j_1j_10}\alpha_{j_1j_2}(a_{j_1j_1s_1} - a_{j_1j_1s_2})^2 > 0$ and $A_{j_2j_23} > 0$,

$$(a_{j_1j_1s_1}, \dot{x}_{j_2}^*) = \underbrace{(\text{UP,pF})}_{\text{up-parabola flow }(+)}.$$

- For $a_{j_1j_10}\alpha_{j_1j_2}(a_{j_1j_1s_1} - a_{j_1j_1s_2})^2 < 0$ and $A_{j_2j_23} > 0$,

$$(a_{j_1j_1s_1}, \dot{x}_{j_2}^*) = \underbrace{(\text{DP,pF})}_{\text{down-parabola flow }(+)}.$$

- For $a_{j_1j_10}\alpha_{j_1j_2}(a_{j_1j_1s_1}-a_{j_1j_1s_2})^2>0$ and $A_{j_2j_23}<0$,

$$(a_{j_1j_1s_1},\dot{x}^*_{j_2})=\underbrace{(\text{DP,nF})}_{\text{down-parabola flow }(-)}.$$

- For $a_{j_1j_10}\alpha_{j_1j_2}(a_{j_1j_1s_1}-a_{j_1j_1s_2})^2<0$ and $A_{j_2j_23}<0$,

$$(a_{j_1j_1s_1},\dot{x}^*_{j_2})=\underbrace{(\text{UP,nF})}_{\text{up-parabola flow }(-)}.$$

(iii$_2$) In phase space, at $y^*_{j_1}=a_{j_1j_1s_2}\neq a_{j_1j_1s_1}$,

$$\left.\frac{dx_{j_1}}{dx_{j_2}}\right|_{y^*_{j_1}=a_{j_1j_1s_2}}=\left.\frac{a_{j_1j_10}(y_{j_1}-a_{j_1j_1s_1})(y_{j_1}-a_{j_1j_1s_2})^2}{A_{j_2j_23}}\right|_{y^*_{j_1}=a_{j_1j_1s_2}}=0,$$

$$\left.\frac{d^2x_{j_1}}{dx^2_{j_2}}\right|_{y^*_{j_1}=a_{j_1j_1s_2}}=0.$$

If

$$\left.\frac{d^3x_{j_1}}{dx^3_{j_2}}\right|_{y^*_{j_1}=a_{j_1j_1s_2}}=\frac{(\alpha_{j_1j_2})^2a_{j_1j_10}(a_{j_1j_1s_2}-a_{j_1j_1s_1})}{A_{j_2j_23}}>0,$$

the flow at $y^*_{j_1}=a_{j_1j_1s_1}$ is an increasing-inflection flow in the x_{j_1}-direction. If

$$\left.\frac{d^3x_{j_1}}{dx^3_{j_2}}\right|_{y^*_{j_1}=a_{j_1j_1s_2}}=\frac{(\alpha_{j_1j_2})^2a_{j_1j_10}(a_{j_1j_1s_2}-a_{j_1j_1s_1})}{A_{j_2j_23}}<0,$$

the flow at $y^*_{j_1}=a_{j_1j_1s_1}$ is a decreasing-inflection flow in the x_{j_1}-direction.

Due to $\dot{x}_{j_2}=A_{j_2j_23}$, the inflection flows in the x_{j_1}-direction are positive and negative for $A_{j_2j_23}>0$ and $A_{j_2j_23}<0$, respectively.

Therefore, the equilibrium on the line of $y^*_{j_1}=a_{j_1j_1s_2}$ for the x_{j_1}-direction with $\dot{x}^*_{j_2}=A_{j_2j_23}$ possesses the following properties as in Eqs. (1.34)–(1.37):

- For $a_{j_1j_10}(\alpha_{j_1j_2})^2(a_{j_1j_1s_2}-a_{j_1j_1s_1})>0$ and $A_{j_2j_23}>0$,

$$(a_{j_1j_1s_2},\dot{x}^*_{j_2})=\underbrace{(\text{II,pF})}_{\text{increasing-inflection flow }(+)}.$$

- For $a_{j_1j_10}(\alpha_{j_1j_2})^2(a_{j_1j_1s_2}-a_{j_1j_1s_1})<0$ and $A_{j_2j_23}>0$,

$$(a_{j_1j_1s_2},\dot{x}^*_{j_2})=\underbrace{(\text{DI,pF})}_{\text{decreasing-inflection flow }(+)}.$$

- For $a_{j_1j_10}(\alpha_{j_1j_2})^2(a_{j_1j_1s_2} - a_{j_1j_1s_1}) > 0$ and $A_{j_2j_23} < 0$,

$$(a_{j_1j_1s_2}, \dot{x}^*_{j_2}) = \underbrace{\text{(DI,nF)}}_{\text{decreasing-inflection flow }(-)}.$$

- For $a_{j_1j_10}(\alpha_{j_1j_2})^2(a_{j_1j_1s_2} - a_{j_1j_1s_1}) < 0$ and $A_{j_2j_23} < 0$,

$$(a_{j_1j_11}, \dot{x}^*_{j_2}) = \underbrace{\text{(II,nF)}}_{\text{increasing-inflection flow }(-)}.$$

The corresponding increasing-inflection and decreasing-inflection flows are the appearing bifurcations for up-parabola-flow and down-parabola-flow.

(iv) If

$$A_{j_2j_23} \neq 0 \quad \text{and} \quad \Delta_{j_1j_1} = 0, a_{j_1j_1s_1} = a_{j_1j_1s_2} = a_{j_1j_11},$$

the corresponding dynamical systems becomes

$$\begin{aligned}\dot{x}_{j_2} &= A_{j_2j_23},\\ \dot{x}_{j_1} &= a_{j_1j_10}(y_{j_1} - a_{j_1j_11})^3.\end{aligned}$$

There is a dynamical system in the y_{j_1}-direction as

$$\dot{y}_{j_1} = \alpha_{j_1j_2}A_{j_2j_23} + \alpha_{j_1j_1}a_{j_1j_10}(y_{j_1} - a_{j_1j_11})^3,$$

with a dynamical system in the $y^{\perp}_{j_1}$-direction as

$$\dot{y}^{\perp}_{j_1} = \beta_{j_1j_2}\alpha_{j_1j_2}A_{j_2j_23} + \beta_{j_1j_1}a_{j_1j_10}(y_{j_1} - a_{j_1j_11})^3,$$

where

$$\begin{aligned}y^{\perp}_{j_1} &= \beta_{j_1j_1}x_{j_1} + \beta_{j_1j_2}x_{j_2}, \quad \mathbf{n}^{\perp}_{j_1} = (\beta_{j_1j_1}, \beta_{j_1j_2})^{\mathrm{T}},\\ \mathbf{n}_{j_1}\cdot\mathbf{n}^{\perp}_{j_1} &= \alpha_{j_1j_1}\beta_{j_1j_1} + \alpha_{j_1j_2}\beta_{j_1j_2} = 0.\end{aligned}$$

In phase space, at $y^*_{j_1} = a_{j_1j_1s_2} \neq a_{j_1j_1s_1}$,

$$\left.\frac{dx_{j_1}}{dx_{j_2}}\right|_{y^*_{j_1}=a_{j_1j_11}} = \left.\frac{a_{j_1j_10}(y_{j_1} - a_{j_1j_11})^3}{A_{j_2j_23}}\right|_{y^*_{j_1}=a_{j_1j_11}} = 0,$$

$$\left.\frac{d^2x_{j_1}}{dx^2_{j_2}}\right|_{y^*_{j_1}=a_{j_1j_11}} = \left.\frac{3a_{j_1j_10}\alpha_{j_1j_2}(y_{j_1} - a_{j_1j_11})^2}{A_{j_2j_23}}\right|_{y^*_{j_1}=a_{j_1j_11}} = 0,$$

$$\left.\frac{d^3x_{j_1}}{dx^3_{j_2}}\right|_{y^*_{j_1}=a_{j_1j_11}} = \left.\frac{6a_{j_1j_10}(\alpha_{j_1j_2})^2(y_{j_1} - a_{j_1j_11})}{A_{j_2j_23}}\right|_{y^*_{j_1}=a_{j_1j_11}} = 0.$$

If

$$\left.\frac{d^4x_{j_1}}{dx_{j_2}^4}\right|_{y_{j_1}^*=a_{j_1j_11}} = \frac{6a_{j_1j_10}(\alpha_{j_1j_2})^3}{A_{j_2j_23}} > 0,$$

the flow at $y_{j_1}^* = a_{j_1j_1s_1}$ is a third-order up-parabola flow in the x_{j_1}-direction. If

$$\left.\frac{d^4x_{j_1}}{dx_{j_2}^4}\right|_{y_{j_1}^*=a_{j_1j_1s_2}} = \frac{6a_{j_1j_10}(\alpha_{j_1j_2})^3}{A_{j_2j_23}} < 0,$$

the flow at $y_{j_1}^* = a_{j_1j_1s_1}$ is a third-order down-parabola flow in the x_{j_1}-direction.

Due to $\dot{x}_{j_2} = A_{j_2j_23}$, the third-order parabola flows in the x_{j_1}-direction are positive and negative for$A_{j_2j_23} > 0$ and $A_{j_2j_23} < 0$, respectively.

Therefore, the equilibrium on the line of $y_{j_1}^* = a_{j_1j_1s_1}$ for the x_{j_1}-direction with $\dot{x}_{j_2}^* = A_{j_2j_23}$ possesses the following properties as in Eqs. (1.43)–(1.46):

- For $a_{j_1j_10}(\alpha_{j_1j_2})^3 > 0$ and $A_{j_2j_23} > 0$,

$$(a_{j_1j_1s_1}, \dot{x}_{j_2}^*) = \underbrace{(3^{\text{rd}}\text{UP,pF})}_{\text{third-order up-parabola flow }(+)}.$$

- For $a_{j_1j_10}(\alpha_{j_1j_2})^3 < 0$ and $A_{j_2j_23} > 0$,

$$(a_{j_1j_1s_1}, \dot{x}_{j_2}^*) = \underbrace{(3^{\text{rd}}\text{DP,pF})}_{\text{third-order down-parabola flow }(+)}.$$

- For $a_{j_1j_10}(\alpha_{j_1j_2})^3 > 0$ and $A_{j_2j_23} < 0$,

$$(a_{j_1j_1s_1}, \dot{x}_{j_2}^*) = \underbrace{(3^{\text{rd}}\text{DP,nF})}_{\text{third-order down-parabola flow }(-)}.$$

- For $a_{j_1j_10}(\alpha_{j_1j_2})^3 < 0$ and $A_{j_2j_23} < 0$,

$$(a_{j_1j_1s_1}, \dot{x}_{j_2}^*) = \underbrace{(3^{\text{rd}}\text{UP,nF})}_{\text{third-order up-parabola flow }(-)}.$$

– The corresponding third-order up-parabola flows are the appearing bifurcations for up-parabola, down-parabola, and up-parabola flows.
– The corresponding third-order down-parabola flows are the appearing bifurcations for down-parabola, up-parabola, and down-parabola flows.

(v) If

$$A_{j_2j_23} \neq 0 \quad \text{and} \quad \Delta_{j_1j_1} > 0,$$

the corresponding dynamical system becomes

$$\dot{x}_{j_2} = A_{j_2j_23},$$
$$\dot{x}_{j_1} = a_{j_1j_10}(y_{j_1} - a_{j_1j_11})(y_{j_1} - a_{j_1j_12})(y_{j_1} - a_{j_1j_13}).$$

where

$$\{a_{j_1j_11}, a_{j_1j_12}, a_{j_1j_13}\} = \text{sort}\{b_{j_1j_11}, b_{j_1j_12}, b_{j_1j_13}\}, \quad a_{j_1j_11} < a_{j_1j_12} < a_{j_1j_13},$$
$$b_{j_1j_12}, b_{j_1j_13} = -\frac{1}{2}(B_{j_1j_1} \pm \sqrt{\Delta_{j_1j_1}}), \quad a_{j_1j_10} = A_{j_1j_10}.$$

There is a dynamical system in the y_1-direction as

$$\dot{y}_{j_1} = \alpha_{j_1j_2}A_{j_2j_23} + \alpha_{j_1j_1}a_{j_1j_10}(y_{j_1} - a_{j_1j_11})(y_{j_1} - a_{j_1j_12})(y_{j_1} - a_{j_1j_13}),$$

with a dynamical system in the $y_{j_1}^{\perp}$-direction as

$$\dot{y}_{j_1}^{\perp} = \beta_{j_1j_1}A_{j_2j_23} + \beta_{j_1j_2}a_{j_1j_10}(y_{j_1} - a_{j_1j_11})(y_{j_1} - a_{j_1j_12})(y_{j_1} - a_{j_1j_13}).$$

In phase space, at $y_{j_1}^* = a_{j_1j_1s_1} \neq a_{j_1j_1s_2}$,

$$\left.\frac{dx_{j_1}}{dx_{j_2}}\right|_{y_{j_1}^* = a_{j_1j_1s_1}} = \left.\frac{a_{j_1j_10}(y_{j_1} - a_{j_1j_1s_1})\prod_{s_2=1,s_2\neq s_1}^{3}(y_{j_1} - a_{j_1j_1s_2})}{A_{j_2j_23}}\right|_{y_{j_1}^* = a_{j_1j_1s_1}} = 0.$$

If

$$\left.\frac{d^2x_{j_1}}{dx_{j_2}^2}\right|_{y_{j_1}^* = a_{j_1j_1s_1}} = \frac{\alpha_{j_1j_2}a_{j_1j_10}\prod_{s_2=1,s_2\neq s_1}^{3}(a_{j_1j_1s_1} - a_{j_1j_1s_2})}{A_{j_2j_23}} > 0,$$

the flow at $y_{j_1}^* = a_{j_1j_1s_1}$ is an up-parabola flow in the x_{j_1}-direction. If

$$\left.\frac{d^2x_{j_1}}{dx_{j_2}^2}\right|_{y_{j_1}^* = a_{j_1j_1s_1}} = \frac{\alpha_{j_1j_2}a_{j_1j_10}\prod_{s_2=1,s_2\neq s_1}^{3}(a_{j_1j_1s_1} - a_{j_1j_1s_2})}{A_{j_2j_23}} < 0,$$

the flow at $y_{j_1}^* = a_{j_1j_1s_1}$ is a down-parabola flow in the x_{j_1}-direction.

Due to $\dot{x}_{j_2} = A_{j_2j_23}$, the parabola flows in the x_{j_1}-direction are positive and negative for $A_{j_2j_23} > 0$ and $A_{j_2j_23} < 0$, respectively.

Therefore, the equilibrium of $y_{j_1}^* = a_{j_1j_1s_1}$ ($s_1, s_2 \in \{1, 2, 3\}$, $s_1 \neq s_2$) for the x_{j_1}-direction with $\dot{x}_{j_2}^* = A_{j_2j_23}$ possesses the following properties as in Eqs. (1.52)–(1.55):

- For $a_{j_1j_10}\alpha_{j_1j_2}\prod_{s_2=1,s_2\neq s_1}^{3}(a_{j_1j_1s_1}-a_{j_1j_1s_2})>0$ and $A_{j_2j_23}>0$,

$$(a_{j_1j_1s_1},\dot{x}_{j_2}^*)=\underbrace{(\text{UP,pF})}_{\text{up-parabola flow }(+)}.$$

- For $a_{j_1j_10}\alpha_{j_1j_2}\prod_{s_2=1,s_2\neq s_1}^{3}(a_{j_1j_1s_1}-a_{j_1j_1s_2})<0$ and $A_{j_2j_23}>0$,

$$(a_{j_1j_1s_1},\dot{x}_{j_2}^*)=\underbrace{(\text{DP,pF})}_{\text{down-parabola flow }(+)}.$$

- For $a_{j_1j_10}\alpha_{j_1j_2}\prod_{s_2=1,s_2\neq s_1}^{3}(a_{j_1j_1s_1}-a_{j_1j_1s_2})>0$ and $A_{j_2j_23}<0$,

$$(a_{j_1j_1s_1},\dot{x}_{j_2}^*)=\underbrace{(\text{DP,nF})}_{\text{down-parabola flow }(-)}.$$

- For $a_{j_1j_10}\alpha_{j_1j_2}\prod_{s_2=1,s_2\neq s_1}^{3}(a_{j_1j_1s_1}-a_{j_1j_1s_2})<0$ and $A_{j_2j_23}<0$,

$$(a_{j_1j_1s_1},\dot{x}_{j_2}^*)=\underbrace{(\text{UP,nF})}_{\text{up-parabola flow }(-)}.$$

(vi) Under the following condition of

$$A_{j_2j_23}\neq 0 \quad\text{and}\quad \Delta_{j_1j_1}=0,$$

from (iii$_2$) and (v), the following appearing bifurcations of parabola flows are given as in Eqs. (1.57)–(1.60):

- For $a_{j_1j_10}(\alpha_{j_1j_2})^2(a_{j_1j_1s_2}-a_{j_1j_1s_1})>0$ and $A_{j_2j_23}>0$,

$$\underbrace{(\dot{x}_{j_1},\dot{x}_{j_2})}_{\text{(pF,pF)-flow}}\rightleftharpoons\underbrace{(a_{j_1j_1s_2},\dot{x}_{j_2})}_{\substack{\text{(II,pF)-flow}\\\text{appearing bifurcation}}}\rightleftharpoons\begin{Bmatrix}\underbrace{(a_{j_1j_1(s_2)_1},\dot{x}_{j_2})}_{\text{(DP,pF)-flow}}\\\underbrace{(a_{j_1j_1(s_2)_2},\dot{x}_{j_2})}_{\text{(UP,pF)-flow}}\end{Bmatrix}\quad\text{for }\alpha_{j_1j_2}>0,$$

$$\underbrace{(\dot{x}_{j_1},\dot{x}_{j_2})}_{\text{(pF,pF)-flow}}\rightleftharpoons\underbrace{(a_{j_1j_1s_2},\dot{x}_{j_2})}_{\substack{\text{(II,pF)-flow}\\\text{appearing bifurcation}}}\rightleftharpoons\begin{Bmatrix}\underbrace{(a_{j_1j_1(s_2)_1},\dot{x}_{j_2})}_{\text{(UP,pF)-flow}}\\\underbrace{(a_{j_1j_1(s_2)_2},\dot{x}_{j_2})}_{\text{(DP,pF)-flow}}\end{Bmatrix}\quad\text{for }\alpha_{j_1j_2}<0.$$

- For $a_{j_1j_10}(\alpha_{j_1j_2})^2(a_{j_1j_1s_2} - a_{j_1j_1s_1}) < 0$ and $A_{j_2j_23} > 0$,

$$\underbrace{(\dot{x}_{j_1}, \dot{x}_{j_2})}_{\text{(nF,pF)-flow}} \rightleftharpoons \underbrace{(a_{j_1j_1s_2}, \dot{x}_{j_2})}_{\substack{\text{(DI,pF)-flow}\\ \text{appearing bifurcation}}} \rightleftharpoons \left\{\begin{matrix} \underbrace{(a_{j_1j_1(s_2)_1}, \dot{x}_{j_2})}_{\text{(UP,pF)-flow}} \\ \underbrace{(a_{j_1j_1(s_2)_2}, \dot{x}_{j_2})}_{\text{(DP,pF)-flow}} \end{matrix}\right\} \text{ for } \alpha_{j_1j_2} > 0,$$

$$\underbrace{(\dot{x}_{j_1}, \dot{x}_{j_2})}_{\text{(nF,pF)-flow}} \rightleftharpoons \underbrace{(a_{j_1j_1s_2}, \dot{x}_{j_2})}_{\substack{\text{(DI,pF)-flow}\\ \text{appearing bifurcation}}} \rightleftharpoons \left\{\begin{matrix} \underbrace{(a_{j_1j_1(s_2)_1}, \dot{x}_{j_2})}_{\text{(DP,pF)-flow}} \\ \underbrace{(a_{j_1j_1(s_2)_2}, \dot{x}_{j_2})}_{\text{(UP,pF)-flow}} \end{matrix}\right\} \text{ for } \alpha_{j_1j_2} < 0.$$

- For $a_{j_1j_10}(\alpha_{j_1j_2})^2(a_{j_1j_1s_2} - a_{j_1j_1s_1}) > 0$ and $A_{j_2j_23} < 0$,

$$\underbrace{(\dot{x}_{j_1}, \dot{x}_{j_2})}_{\text{(pF,nF)-flow}} \rightleftharpoons \underbrace{(a_{j_1j_1s_2}, \dot{x}_{j_2})}_{\substack{\text{(DI,nF)-flow}\\ \text{appearing bifurcation}}} \rightleftharpoons \left\{\begin{matrix} \underbrace{(a_{j_1j_1(s_2)_1}, \dot{x}_{j_2}),}_{\text{(UP,nF)-flow}} \\ \underbrace{(a_{j_1j_1(s_2)_2}, \dot{x}_{j_2})}_{\text{(DP,nF)-flow}} \end{matrix}\right\} \text{ for } \alpha_{j_1j_2} > 0,$$

$$\underbrace{(\dot{x}_{j_1}, \dot{x}_{j_2})}_{\text{(pF,nF)-flow}} \rightleftharpoons \underbrace{(a_{j_1j_1s_2}, \dot{x}_{j_2})}_{\substack{\text{(DI,nF)-flow}\\ \text{appearing bifurcation}}} \rightleftharpoons \left\{\begin{matrix} \underbrace{(a_{j_1j_1(s_2)_1}, \dot{x}_{j_2}),}_{\text{(DP,nF)-flow}} \\ \underbrace{(a_{j_1j_1(s_2)_2}, \dot{x}_{j_2})}_{\text{(UP,nF)-flow}} \end{matrix}\right\} \text{ for } \alpha_{j_1j_2} < 0.$$

- For $a_{j_1j_10}(\alpha_{j_1j_2})^2(a_{j_1j_1s_2} - a_{j_1j_1s_1}) < 0$ and $A_{j_2j_22} < 0$,

$$\underbrace{(\dot{x}_{j_1}, \dot{x}_{j_2})}_{\text{(pF,pF)-flow}} \rightleftharpoons \underbrace{(a_{j_1j_1s_2}, \dot{x}_{j_2})}_{\substack{\text{(II,nF)-flow}\\ \text{appearing bifurcation}}} \rightleftharpoons \left\{\begin{matrix} \underbrace{(a_{j_1j_1(s_2)_1}, \dot{x}_{j_2})}_{\text{(DP,nF)-flow}} \\ \underbrace{(a_{j_1j_1(s_2)_2}, \dot{x}_{j_2})}_{\text{(UP,nF)-flow}} \end{matrix}\right\} \text{ for } \alpha_{j_1j_2} > 0,$$

$$\underbrace{(\dot{x}_{j_1}, \dot{x}_{j_2})}_{\text{(pF,pF)-flow}} \rightleftharpoons \underbrace{(a_{j_1j_1s_2}, \dot{x}_{j_2})}_{\substack{\text{(II,nF)-flow}\\ \text{appearing bifurcation}}} \rightleftharpoons \left\{\begin{matrix} \underbrace{(a_{j_1j_1(s_2)_1}, \dot{x}_{j_2})}_{\text{(UP,nF)-flow}} \\ \underbrace{(a_{j_1j_1(s_2)_2}, \dot{x}_{j_2})}_{\text{(DP,nF)-flow}} \end{matrix}\right\} \text{ for } \alpha_{j_1j_2} < 0.$$

(vii) Consider

$$\begin{aligned}\dot{x}_{j_2} &= A_{j_2 j_2 3},\\ \dot{x}_{j_1} &= a_{j_1 j_1 0}(x_{j_1} - a_{j_1 j_1 s_1})(x_{j_1} - a_{j_1 j_1 s_2})(x_{j_1} - a_{j_1 j_1 s_3}).\end{aligned}$$

with

$$a^-_{j_1 j_1 s_1} \rightleftharpoons a^+_{j_1 j_1 s_2} \quad \text{and} \quad a^-_{j_1 j_1 s_2} \rightleftharpoons a^+_{j_1 j_1 s_1}.$$

Under

$$A_{j_2 j_2 3} \neq 0, \quad \text{and} \quad \Delta_{j_1 j_1} = (a_{j_1 j_1 s_1} - a_{j_1 j_1 s_2})^2 = 0,$$

$$a_{j_1 j_1 s_1} = a_{j_1 j_1 s_2} \equiv a_{j_1 j_1 1},$$

from (iii$_2$) and (v), the following switching bifurcations of parabola flows are given as in Eqs. (1.64)–(1.67):

- For $a_{j_1 j_1 0}(\alpha_{j_1 j_2})^2(a_{j_1 j_1 1} - a_{j_1 j_1 s_3}) > 0$ and $A_{j_2 j_2 3} > 0$,

$$\left\{\begin{array}{c}\underbrace{(a^-_{j_1 j_1 s_1}, \dot{x}_{j_2})}_{\text{(DP,pF)-flow}}\\ \underbrace{(a^-_{j_1 j_1 s_2}, \dot{x}_{j_2})}_{\text{(UP,pF)-flow}}\end{array}\right\} \rightleftharpoons \underbrace{(a_{j_1 j_1 1}, \dot{x}_{j_2})}_{\substack{\text{(II,pF)-flow}\\ \text{switching bifurcation}}} \rightleftharpoons \left\{\begin{array}{c}\underbrace{(a^+_{j_1 j_1 s_1}, \dot{x}_{j_2})}_{\text{(DP,pF)-flow}}\\ \underbrace{(a^+_{j_1 j_1 s_2}, \dot{x}_{j_2})}_{\text{(UP,pF)-flow}}\end{array}\right\} \quad \text{for } \alpha_{j_1 j_2} > 0,$$

$$\left\{\begin{array}{c}\underbrace{(a^-_{j_1 j_1 s_1}, \dot{x}_{j_2})}_{\text{(UP,pF)-flow}}\\ \underbrace{(a^-_{j_1 j_1 s_2}, \dot{x}_{j_2})}_{\text{(DP,pF)-flow}}\end{array}\right\} \rightleftharpoons \underbrace{(a_{j_1 j_1 1}, \dot{x}_{j_2})}_{\substack{\text{(II,pF)-flow}\\ \text{switching bifurcation}}} \rightleftharpoons \left\{\begin{array}{c}\underbrace{(a^+_{j_1 j_1 s_1}, \dot{x}_{j_2})}_{\text{(UP,pF)-flow}}\\ \underbrace{(a^+_{j_1 j_1 s_2}, \dot{x}_{j_2})}_{\text{(DP,pF)-flow}}\end{array}\right\} \quad \text{for } \alpha_{j_1 j_2} < 0.$$

- For $a_{j_1 j_1 0}(\alpha_{j_1 j_2})^2(a_{j_1 j_1 1} - a_{j_1 j_1 s_3}) < 0$ and $A_{j_2 j_2 3} > 0$,

$$\left\{\begin{array}{c}\underbrace{(a^-_{j_1 j_1 s_1}, \dot{x}_{j_2}),}_{\text{(UP,pF)-flow}}\\ \underbrace{(a^-_{j_1 j_1 s_2}, \dot{x}_{j_2})}_{\text{(DP,pF)-flow}}\end{array}\right\} \rightleftharpoons \underbrace{(a_{j_1 j_1 1}, \dot{x}_{j_2})}_{\substack{\text{(DI,pF)-flow}\\ \text{switching bifurcation}}} \rightleftharpoons \left\{\begin{array}{c}\underbrace{(a^+_{j_1 j_1 s_1}, \dot{x}_{j_2}),}_{\text{(UP,pF)-flow}}\\ \underbrace{(a^+_{j_1 j_1 s_2}, \dot{x}_{j_2})}_{\text{(DF,pF)-flow}}\end{array}\right\} \quad \text{for } \alpha_{j_1 j_2} > 0,$$

$$\left\{\begin{array}{c}\underbrace{(a^-_{j_1 j_1 s_1}, \dot{x}_{j_2}),}_{\text{(DP,pF)-flow}}\\ \underbrace{(a^-_{j_1 j_1 s_2}, \dot{x}_{j_2})}_{\text{(UP,pF)-flow}}\end{array}\right\} \rightleftharpoons \underbrace{(a_{j_1 j_1 1}, \dot{x}_{j_2})}_{\substack{\text{(DI,pF)-flow}\\ \text{switching bifurcation}}} \rightleftharpoons \left\{\begin{array}{c}\underbrace{(a^+_{j_1 j_1 s_1}, \dot{x}_{j_2}),}_{\text{(DP,pF)-flow}}\\ \underbrace{(a^+_{j_1 j_1 s_2}, \dot{x}_{j_2})}_{\text{(UP,pF)-flow}}\end{array}\right\} \quad \text{for } \alpha_{j_1 j_2} < 0.$$

- For $a_{j_1j_10}(\alpha_{j_1j_2})^2(a_{j_1j_11}-a_{j_1j_1s_3})>0$ and $A_{j_2j_23}<0$,

$$\left\{\begin{matrix}\underbrace{(a^-_{j_1j_1s_1},\dot{x}_{j_2}),}_{\text{(UP,nF)-flow}}\\ \underbrace{(a^-_{j_1j_1s_2},\dot{x}_{j_2})}_{\text{(DP,nF)-flow}}\end{matrix}\right\} \rightleftharpoons \underbrace{(a_{j_1j_11},\dot{x}_{j_2})}_{\substack{\text{(DI,nF)-flow}\\ \text{switching bifurcation}}} \rightleftharpoons \left\{\begin{matrix}\underbrace{(a^+_{j_1j_1s_1},\dot{x}_{j_2}),}_{\text{(UP,nF)-flow}}\\ \underbrace{(a^+_{j_1j_1s_2},\dot{x}_{j_2})}_{\text{(DP,nF)-flow}}\end{matrix}\right\} \quad \text{for } \alpha_{j_1j_2}>0,$$

$$\left\{\begin{matrix}\underbrace{(a^-_{j_1j_1s_1},\dot{x}_{j_2}),}_{\text{(DP,nF)-flow}}\\ \underbrace{(a^-_{j_1j_1s_2},\dot{x}_{j_2})}_{\text{(UP,nF)-flow}}\end{matrix}\right\} \rightleftharpoons \underbrace{(a_{j_1j_11},\dot{x}_{j_2})}_{\substack{\text{(DI,nF)-flow}\\ \text{switching bifurcation}}} \rightleftharpoons \left\{\begin{matrix}\underbrace{(a^+_{j_1j_1s_1},\dot{x}_{j_2}),}_{\text{(DP,nF)-flow}}\\ \underbrace{(a^+_{j_1j_1s_2},\dot{x}_{j_2})}_{\text{(UP,nF)-flow}}\end{matrix}\right\} \quad \text{for } \alpha_{j_1j_2}<0.$$

- For $a_{j_1j_10}(\alpha_{j_1j_2})^2(a_{j_1j_11}-a_{j_1j_1s_3})<0$ and $A_{j_2j_23}<0$,

$$\left\{\begin{matrix}\underbrace{(a^-_{j_1j_1s_1},\dot{x}_{j_2})}_{\text{(DP,nF)-flow}}\\ \underbrace{(a^-_{j_1j_1s_2},\dot{x}_{j_2})}_{\text{(UP,nF)-flow}}\end{matrix}\right\} \rightleftharpoons \underbrace{(a_{j_1j_11},\dot{x}_{j_2})}_{\substack{\text{(II,nF)-flow}\\ \text{switching bifurcation}}} \rightleftharpoons \left\{\begin{matrix}\underbrace{(a^+_{j_1j_1s_1},\dot{x}_{j_2})}_{\text{(DP,nF)-flow}}\\ \underbrace{(a^+_{j_1j_1s_2},\dot{x}_{j_2})}_{\text{(UP,nF)-flow}}\end{matrix}\right\} \quad \text{for } \alpha_{j_1j_2}>0,$$

$$\left\{\begin{matrix}\underbrace{(a^-_{j_1j_1s_1},\dot{x}_{j_2})}_{\text{(UP,nF)-flow}}\\ \underbrace{(a^-_{j_1j_1s_2},\dot{x}_{j_2})}_{\text{(DP,nF)-flow}}\end{matrix}\right\} \rightleftharpoons \underbrace{(a_{j_1j_11},\dot{x}_{j_2})}_{\substack{\text{(II,nF)-flow}\\ \text{switching bifurcation}}} \rightleftharpoons \left\{\begin{matrix}\underbrace{(a^+_{j_1j_1s_1},\dot{x}_{j_2})}_{\text{(UP,nF)-flow}}\\ \underbrace{(a^+_{j_1j_1s_2},\dot{x}_{j_2})}_{\text{(DP,nF)-flow}}\end{matrix}\right\} \quad \text{for } \alpha_{j_1j_2}<0.$$

(viii) Under the following condition of

$$A_{j_2j_23}\neq 0 \quad \text{and} \quad \Delta_{j_1j_1}=0, \quad a_{j_1j_1s_1}=a_{j_1j_1s_2}=a_{j_1j_11},$$

from (ii), (iv) and (v), the following appearing bifurcations of parabola flows are given as in Eqs. (1.69)–(1.72):

- For $a_{j_1j_10}(\alpha_{j_1j_2})^3>0$ and $A_{j_2j_23}>0$,

$$\underbrace{(a_{j_1j_11},\dot{x}_{j_2})}_{\text{(UP,pF)-flow}} \rightleftharpoons \underbrace{(a_{j_1j_11},\dot{x}_{j_2})}_{\substack{(3^{\text{rd}}\text{UP,pF})\text{-flow}\\ \text{appearing bifurcation}}} \rightleftharpoons \left\{\begin{matrix}\underbrace{(a_{j_1j_11},\dot{x}_{j_2})}_{\text{(UP,pF)-flow}}\\ \underbrace{(a_{j_1j_12},\dot{x}_{j_2})}_{\text{(DP,pF)-flow}}\\ \underbrace{(a_{j_1j_13},\dot{x}_{j_2})}_{\text{(UP,pF)-flow}}\end{matrix}\right\}.$$

- For $a_{j_1j_10}(\alpha_{j_1j_2})^3 < 0$ and $A_{j_2j_23} > 0$,

$$\underbrace{(a_{j_1j_11}, \dot{x}_{j_2})}_{\text{(DP,pF)-flow}} \rightleftharpoons \underbrace{(a_{j_1j_11}, \dot{x}_{j_2})}_{\substack{(3^{\text{rd}}\text{DP,pF)-flow} \\ \text{appearing bifurcation}}} \rightleftharpoons \left\{ \begin{array}{c} \underbrace{(a_{j_1j_11}, \dot{x}_{j_2})}_{\text{(DP,pF)-flow}} \\ \underbrace{(a_{j_1j_12}, \dot{x}_{j_2})}_{\text{(UP,pF)-flow}} \\ \underbrace{(a_{j_1j_13}, \dot{x}_{j_2})}_{\text{(DP,pF)-flow}} \end{array} \right\}.$$

- For $a_{j_1j_10}(\alpha_{j_1j_2})^3 > 0$ and $A_{j_2j_23} < 0$,

$$\underbrace{(a_{j_1j_11}, \dot{x}_{j_2})}_{\text{(DP,nF)-flow}} \rightleftharpoons \underbrace{(a_{j_1j_11}, \dot{x}_{j_2})}_{\substack{(3^{\text{rd}}\text{DP,nF)-flow} \\ \text{appearing bifurcation}}} \rightleftharpoons \left\{ \begin{array}{c} \underbrace{(a_{j_1j_11}, \dot{x}_{j_2})}_{\text{(DP,nF)-flow}} \\ \underbrace{(a_{j_1j_12}, \dot{x}_{j_2})}_{\text{(UP,nF)-flow}} \\ \underbrace{(a_{j_1j_13}, \dot{x}_{j_2})}_{\text{(DP,nF)-flow}} \end{array} \right\}.$$

- For $a_{j_1j_10}(\alpha_{j_1j_2})^3 < 0$ and $A_{j_2j_23} < 0$,

$$\underbrace{(a_{j_1j_11}, \dot{x}_{j_2})}_{\text{(UP,nF)-flow}} \rightleftharpoons \underbrace{(a_{j_1j_11}, \dot{x}_{j_2})}_{\substack{(3^{\text{rd}}\text{UP,nF)-flow} \\ \text{appearing bifurcation}}} \rightleftharpoons \left\{ \begin{array}{c} \underbrace{(a_{j_1j_11}, \dot{x}_{j_2})}_{\text{(UP,nF)-flow}} \\ \underbrace{(a_{j_1j_12}, \dot{x}_{j_2})}_{\text{(DP,nF)-flow}} \\ \underbrace{(a_{j_1j_13}, \dot{x}_{j_2})}_{\text{(UP,nF)-flow}} \end{array} \right\}.$$

In the end, the theorem is proved from the above local analysis. ■

1.3 Inflection and Parabola Flows

Consider a dynamical system as

$$\begin{aligned} \dot{x}_2 &= A_{223}, \\ \dot{x}_1 &= a_{110}(y_1 - a_{111})(y_1 - a_{112})^2 \end{aligned} \tag{1.73}$$

with

$$y_1 = \alpha_{11}x_1 + \alpha_{12}x_2. \tag{1.74}$$

The corresponding dynamical system on the y_1-direction is

$$\dot{y}_1 = \alpha_{11}a_{110}(y_1 - a_{111})(y_1 - a_{112})^2 + \alpha_{12}A_{223} \tag{1.75}$$

and

$$\dot{y}_1^{\perp} = \beta_{11}a_{110}(y_1 - a_{111})(y_1 - a_{112})^2 + \beta_{12}A_{223}, \tag{1.76}$$

where

$$\begin{aligned} y_1^{\perp} &= \beta_{11}x_1 + \beta_{12}x_2, \quad \mathbf{n}_1^{\perp} = (\beta_{11}, \beta_{12})^{\mathrm{T}}, \\ (\mathbf{n}_1)^{\mathrm{T}} \cdot \mathbf{n}_1^{\perp} &= \alpha_{11}\beta_{11} + \alpha_{12}\beta_{12} = 0. \end{aligned} \tag{1.77}$$

For $\alpha_{11} > 0$ and $\alpha_{12} < 0$, the phase portraits on the lines of $y_1 = a_{111}, a_{112}$ for the four cases of increasing- and decreasing-inflection flows for the x_1-direction are presented in Figs. 1.1(a)–1.1(d).

- For $a_{110} > 0$ and $A_{223} > 0$,

$$\left\{\begin{matrix} (a_{111}, \dot{x}_2) \\ (a_{112}, \dot{x}_2) \end{matrix}\right\} = \left\{\begin{matrix} \underbrace{(\text{DP,pF})}_{\text{down-parabola flow }(+)} \\ \underbrace{(\text{II,pF})}_{\text{increasing-inflection flow }(+)} \end{matrix}\right\} \text{ for } \alpha_{j_1j_2} < 0. \tag{1.78}$$

- For $a_{110} < 0$ and $A_{223} > 0$,

$$\left\{\begin{matrix} (a_{111}, \dot{x}_2) \\ (a_{112}, \dot{x}_2) \end{matrix}\right\} = \left\{\begin{matrix} \underbrace{(\text{UP,pF})}_{\text{up-parabola flow }(+)} \\ \underbrace{(\text{DI,pF})}_{\text{decreasing-inflection flow }(+)} \end{matrix}\right\} \text{ for } \alpha_{j_1j_2} < 0. \tag{1.79}$$

- For $a_{110} > 0$ and $A_{223} < 0$,

$$\left\{\begin{matrix} (a_{111}, \dot{x}_2) \\ (a_{112}, \dot{x}_2) \end{matrix}\right\} = \left\{\begin{matrix} \underbrace{(\text{UP,nF})}_{\text{up-parabola flow }(-)} \\ \underbrace{(\text{DI,nF})}_{\text{decreasing-inflection flow }(-)} \end{matrix}\right\} \text{ for } \alpha_{j_1j_2} < 0. \tag{1.80}$$

- For $a_{110} > 0$ and $a_{223} > 0$,

$$\left\{\begin{matrix} (a_{111}, \dot{x}_2) \\ (a_{112}, \dot{x}_2) \end{matrix}\right\} = \left\{\begin{matrix} \underbrace{(\text{DP,nF})}_{\text{down-parabola flow }(-)} \\ \underbrace{(\text{II,nF})}_{\text{Increasing-inflection flow }(-)} \end{matrix}\right\} \text{ for } \alpha_{j_1j_2} < 0. \tag{1.81}$$

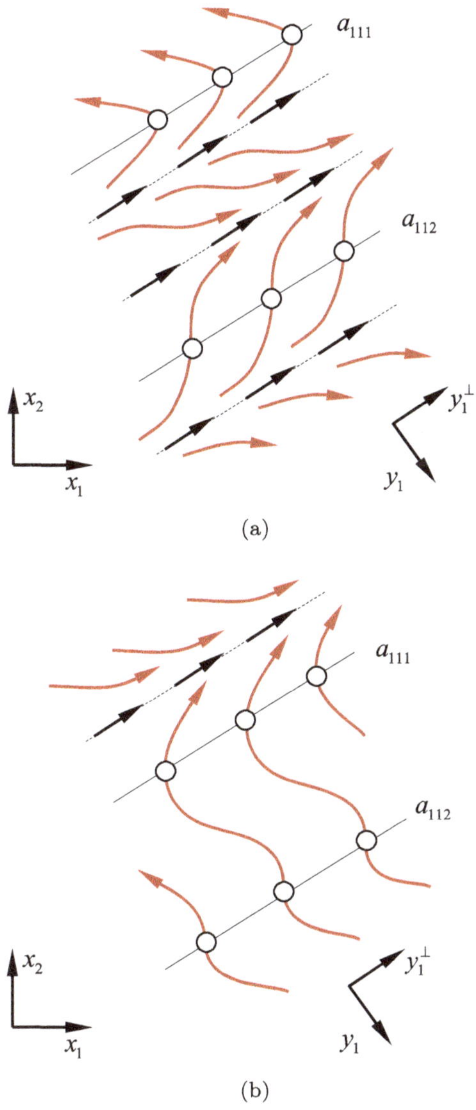

Fig. 1.1. Phase portraits for two-dimensional systems on the x_1-direction with $y_1^* = a_{111}, a_{112}$ and on the x_2-direction with $\dot{x}_2 = A_{223} \neq 0$ $(\alpha_{11} > 0, \alpha_{12} < 0)$: (a) (DP,pF)-positive down-parabola flow and (II,pF)-positive increasing inflection flow $(a_{110} > 0, A_{223} > 0)$, (b) (UP,pF)-positive up-parabola flow and (DI,pF)-positive deceasing-inflection flows $(a_{110} < 0, A_{223} > 0)$, (c) (UP,nF)-negative up-parabola flow and (DI,nF)-negative increasing-inflection low $(a_{110} > 0, A_{223} < 0)$, (d) (DP,nF)-negative down-parabola flow and (II,nF)-negative increasing-inflection flow $(a_{110} < 0,$ $A_{223} < 0)$.

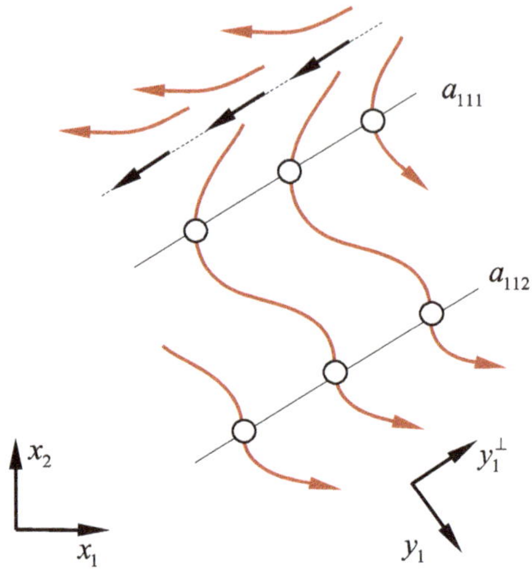

(c)

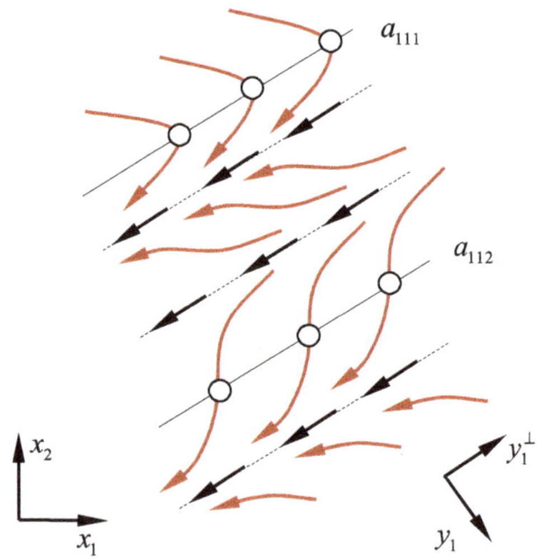

(d)

Fig. 1.1. *(Continued)*

In Fig. 1.1(a), the (DP,pF)-positive down-parabola flow and (II,pF)-positive increasing-inflection flow on the line of $y_1^* = a_{111}, a_{112}$ $(a_{110} > 0, A_{223} > 0)$ are presented. In the x_1-direction, the equilibrium on the line of $y_1^* = a_{111}$ is a source, and in the x_2-direction, the constant speed on the line of $y_1^* = a_{111}$ is positive. In the x_1-direction, the equilibrium on the line of $y_1^* = a_{112}$ is an upper-saddle, and in the x_2-direction, the constant speed on the line of $y_1^* = a_{112}$ is positive. From $\dot{y}_1 = 0$, the corresponding flow at $y_1 = y_1^*$ is determined and the positive and negative directions of the flow are determined by $\dot{y}_1^{\perp} > 0$ and $\dot{y}_1^{\perp} < 0$, respectively. There are three infinite-equilibrium flows in the y_1-direction from Eq. (1.75) with $\dot{y}_1 = 0$.

In Fig. 1.1(b), the (UP,pF)-*positive* up-parabola flow and (DI,pF)-*positive* decreasing-inflection flow on the line of $y_1^* = a_{111}, a_{112}$ ($a_{110} > 0, A_{223} > 0$) are presented. In the x_1-direction, the equilibrium on the line of $y_1^* = a_{111}$ is a sink, and in the x_2-direction, the constant speed on the line of $y_1^* = a_{111}$ is positive. In the x_1-direction, the equilibrium on the line of $y_1^* = a_{112}$ is a lower-saddle, and in the x_2-direction, the constant speed on the line of $y_1^* = a_{112}$ is positive. At least, there is also an infinite-equilibrium flow in the y_1-direction.

In Fig. 1.1(c), the (UP,nF)-*negative* up-parabola flow and (DI,nF)-*negative* decreasing-inflection flow on the line of $y_1^* = a_{111}, a_{112}$ ($a_{110} > 0, A_{223} > 0$) are presented. In the x_1-direction, the equilibrium on the line of $y_1^* = a_{111}$ is a source, and in the x_2-direction, the constant speed on the line of $y_1^* = a_{111}$ is negative. In the x_1-direction, the equilibrium on the line of $y_1^* = a_{112}$ is a lower-saddle, and in the x_2-direction, the constant speed on the line of $y_1^* = a_{112}$ is negative. At least, there is also an infinite-equilibrium flow in the y_1-direction.

In Fig. 1.1(d), the (DP,nF)-*negative* down-parabola flow and (II,nF)-*negative* increasing-inflection flow on the line of $y_1^* = a_{111}, a_{112}$ ($a_{110} > 0, A_{223} > 0$) are presented. In the x_1-direction, the equilibrium on the line of $y_1^* = a_{111}$ is a sink, and in the x_2-direction, the constant speed on the line of $y_1^* = a_{111}$ is negative. In the x_1-direction, the equilibrium on the line of $y_1^* = a_{112}$ is an upper-saddle, and in the x_2-direction, the constant speed on the line of $y_1^* = a_{112}$ is negative. There are three infinite-equilibrium flows in the y_1-direction.

For $\alpha_{11} > 0$ and $\alpha_{12} > 0$, the phase portraits on the line of $y_1 = a_{111}, a_{112}$ for the four cases of increasing- and decreasing-inflection flows for the x_1-direction are presented in Figs. 1.2(a)–1.2(d).

- For $a_{110} > 0$ and $A_{223} > 0$,

$$\begin{Bmatrix} (a_{111}, \dot{x}_2) \\ (a_{112}, \dot{x}_2) \end{Bmatrix} = \begin{Bmatrix} \underbrace{\text{(UP,pF)}}_{\text{up-parabola flow }(+)} \\ \underbrace{\text{(II,pF)}}_{\text{increasing-inflection flow }(+)} \end{Bmatrix} \quad \text{for } \alpha_{j_1 j_2} > 0. \tag{1.82}$$

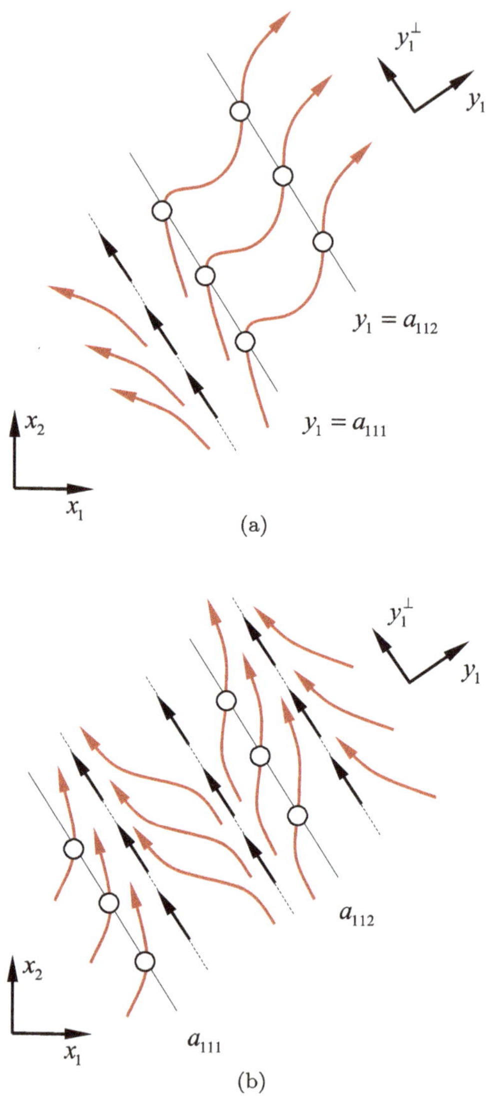

Fig. 1.2. Phase portraits for two-dimensional systems on the x_1-direction with $y_1^* = a_{111}, a_{112}$ and on the x_2-direction with $\dot{x}_2 = A_{223} \neq 0$ $(\alpha_{11} > 0, \alpha_{12} > 0)$: (a) (UP,pF)-positive up-parabola flow and (II,pF)-positive increasing-inflection flow $(a_{110} > 0, A_{223} > 0)$, (b) (DP,pF)-positive down-parabola flow and (DI,pF)-positive deceasing-inflection flows $(a_{110} < 0, A_{223} > 0)$, (c) (DP,nF)-negative down-parabola flow and (DI,nF)-negative increasing-inflection low $(a_{110} > 0, A_{223} < 0)$, (d) (UP,nF)-negative up-parabola flow and (II,nF)-negative increasing-inflection flow $(a_{110} < 0, A_{223} < 0)$.

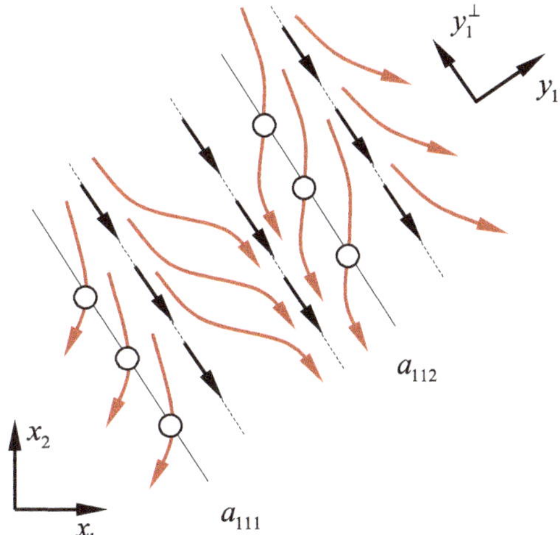

(c)

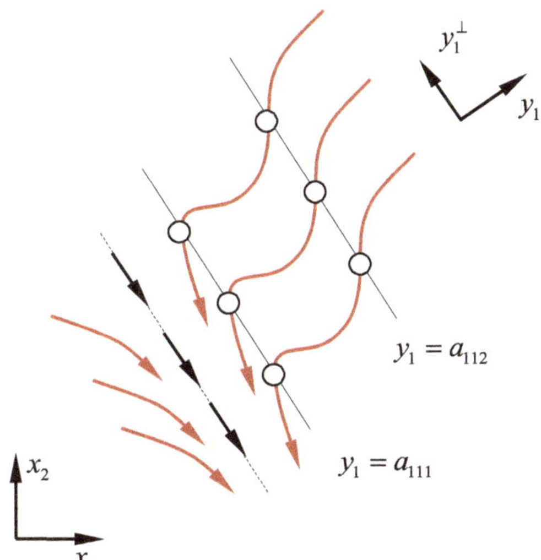

(d)

Fig. 1.2. *(Continued)*

- For $a_{110} < 0$ and $A_{223} > 0$,

$$\begin{Bmatrix} (a_{111}, \dot{x}_2) \\ (a_{112}, \dot{x}_2) \end{Bmatrix} = \left\{ \begin{matrix} \underbrace{(\text{DP,pF})}_{\text{down-parabola flow}\,(+)} \\ \underbrace{(\text{DI,pF})}_{\text{decreasing-inflection flow}\,(+)} \end{matrix} \right\} \text{ for } \alpha_{j_1 j_2} > 0. \quad (1.83)$$

- For $a_{110} > 0$ and $A_{223} < 0$,

$$\begin{Bmatrix} (a_{111}, \dot{x}_2) \\ (a_{112}, \dot{x}_2) \end{Bmatrix} = \left\{ \begin{array}{c} \underbrace{(\text{DP,nF})}_{\text{down-parabola flow } (-)} \\ \underbrace{(\text{DI,nF})}_{\text{decreasing-inflection flow } (-)} \end{array} \right\} \text{ for } \alpha_{j_1 j_2} > 0. \quad (1.84)$$

- For $a_{110} > 0$ and $A_{223} > 0$,

$$\begin{Bmatrix} (a_{111}, \dot{x}_2) \\ (a_{112}, \dot{x}_2) \end{Bmatrix} = \left\{ \begin{array}{c} \underbrace{(\text{UP,nF})}_{\text{up-parabola flow } (-)} \\ \underbrace{(\text{II,nF})}_{\text{Increasing-inflection flow } (-)} \end{array} \right\} \text{ for } \alpha_{j_1 j_2} > 0. \quad (1.85)$$

In Fig. 1.2(a), the (UP,pF)-positive up-parabola flow and (II,pF)-positive increasing-inflection flow on the line of $y_1^* = a_{111}, a_{112}$ ($a_{110} > 0, A_{223} > 0$) are presented. From $\dot{y}_1 = 0$, the corresponding flow at $y_1 = y_1^*$ is determined and the positive and negative directions of the flow are determined by $\dot{y}_1^{\perp} > 0$ and $\dot{y}_1^{\perp} < 0$, respectively. At least, there is an infinite-equilibrium flow in the y_1-direction from Eq. (1.75) with $\dot{y}_1 = 0$. In Fig. 1.2(b), the (UP,pF)-*positive* up-parabola flow and (DI,pF)-*positive* decreasing-inflection flow on the line of $y_1^* = a_{111}, a_{112}$ ($a_{110} > 0, A_{223} > 0$) are presented. There are three infinite-equilibrium flows in the y_1-direction. In Fig. 1.2(c), the (UP,nF)-*negative* up-parabola flow and (DI,nF)-*negative* decreasing-inflection flow on the line of $y_1^* = a_{111}, a_{112}$ ($a_{110} > 0, A_{223} > 0$) are presented. There are also three infinite-equilibrium flows in the y_1-direction. In Fig. 1.2(d), the (DP,nF)-*negative* down-parabola flow and (II,nF)-*negative* increasing-inflection flow on the line of $y_1^* = a_{111}, a_{112}$ ($a_{110} > 0, A_{223} > 0$) are presented. At least, there is also an infinite-equilibrium flow in the y_1-direction.

Consider a dynamical system as

$$\begin{aligned} \dot{x}_2 &= A_{223}, \\ \dot{x}_1 &= a_{110}(y_1 - a_{111})^2(y_1 - a_{112}) \end{aligned} \quad (1.86)$$

with

$$y_1 = \alpha_{11}x_1 + \alpha_{12}x_2. \quad (1.87)$$

The corresponding dynamical system on the y_1-direction is

$$\dot{y}_1 = \alpha_{11}a_{110}(y_1 - a_{111})^2(y_1 - a_{112}) + \alpha_{12}A_{223} \quad (1.88)$$

and

$$\dot{y}_1^{\perp} = \beta_{11}a_{110}(y_1 - a_{111})^2(y_1 - a_{112}) + \beta_{12}A_{223}, \tag{1.89}$$

where

$$\begin{aligned} y_1^{\perp} &= \beta_{11}x_1 + \beta_{12}x_2, \quad \mathbf{n}_1^{\perp} = (\beta_{11}, \beta_{12})^{\mathrm{T}}, \\ (\mathbf{n}_1)^{\mathrm{T}} &\cdot \mathbf{n}_1^{\perp} = \alpha_{11}\beta_{11} + \alpha_{12}\beta_{12} = 0. \end{aligned} \tag{1.90}$$

Readers can complete the inflection and parabola flows for two lines of $y_1 = a_{111}, a_{112}$.

1.4 Parabola Flows Series

Consider the dynamical system as

$$\begin{aligned} \dot{x}_2 &= A_{223}, \\ \dot{x}_1 &= a_{110}(y_1 - a_{111})(y_1 - a_{112})(y_1 - a_{113}) \end{aligned} \tag{1.91}$$

and

$$\dot{y}_1 = \alpha_{12}A_{223} + \alpha_{11}a_{110}(y_1 - a_{111})(y_1 - a_{112})(y_1 - a_{113}), \tag{1.92}$$

with a dynamical system in the $y_1^{\perp}$-direction as

$$\dot{y}_1^{\perp} = \beta_{12}A_{223} + \beta_{11}a_{110}(y_1 - a_{111})(y_1 - a_{112})(y_1 - a_{113}). \tag{1.93}$$

For $\alpha_{11} > 0$ and $\alpha_{12} < 0$, the corresponding parabola flows in the phase portraits are presented in Figs. 1.3(a)–1.3(d).

- For $a_{110} > 0$ and $A_{223} > 0$,

$$\begin{Bmatrix} (a_{111}, \dot{x}_2) \\ (a_{112}, \dot{x}_2) \\ (a_{111}, \dot{x}_2) \end{Bmatrix} = \begin{Bmatrix} \underbrace{(\text{DP,pF})}_{\text{down-parabola flow }(+)} \\ \underbrace{(\text{UP,pF})}_{\text{up-parabola flow }(+)} \\ \underbrace{(\text{DP,pF})}_{\text{down-parabola flow }(+)} \end{Bmatrix} \text{ for } \alpha_{12} < 0. \tag{1.94}$$

- For $a_{110} < 0$ and $A_{223} > 0$,

$$\begin{Bmatrix} (a_{111}, \dot{x}_2) \\ (a_{112}, \dot{x}_2) \\ (a_{111}, \dot{x}_2) \end{Bmatrix} = \begin{Bmatrix} \underbrace{(\text{UP,pF})}_{\text{up-parabola flow }(+)} \\ \underbrace{(\text{DP,pF})}_{\text{down-parabola flow }(+)} \\ \underbrace{(\text{UP,pF})}_{\text{up-parabola flow }(+)} \end{Bmatrix} \text{ for } \alpha_{12} < 0. \tag{1.95}$$

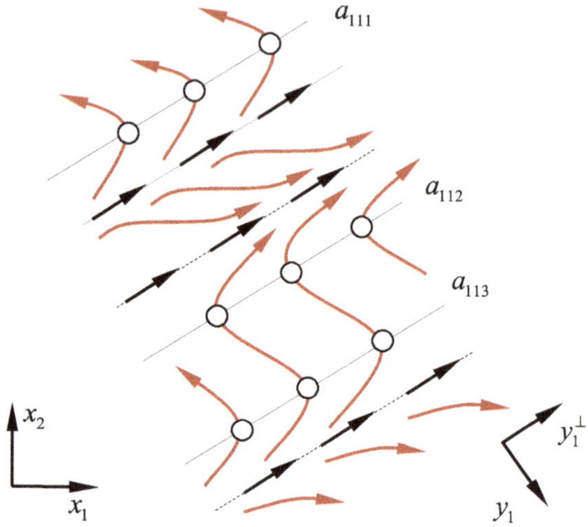

(a)

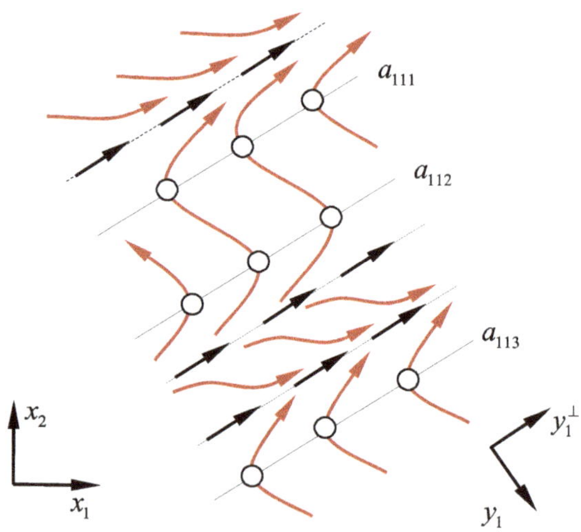

(b)

Fig. 1.3. Phase portraits for two-dimensional systems on the x_1-direction with $y_1^* = a_{111}, a_{112}, a_{113}$ and on the x_2-direction with $\dot{x}_2 = A_{223} \neq 0(\alpha_{11} > 0, \alpha_{12} < 0)$: (a) (DP,pF), (UP,pF), and (DP,pF)-flows ($a_{110} > 0, A_{223} > 0$), (b) (UP,pF), (DP,pF), and (UP,pF)-flows ($a_{110} < 0, A_{223} > 0$), (c) (UP,nF), (DP,nF), and (UP,nF)-flows ($a_{110} > 0, A_{223} < 0$), (d) (DP,nF), (UP,nF), and (DP,nF)-flows ($a_{110} < 0, A_{223} < 0$).

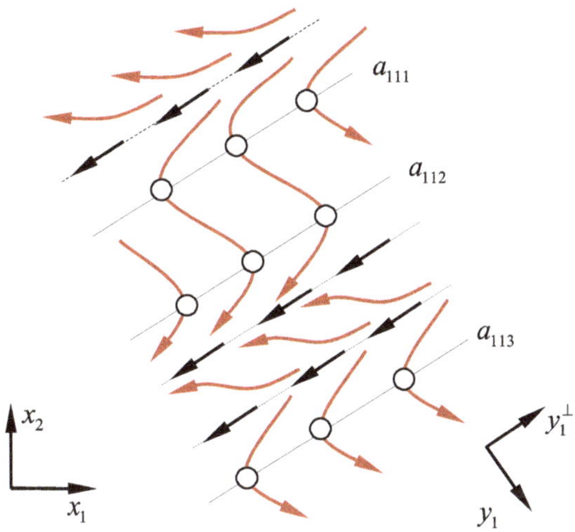

(c)

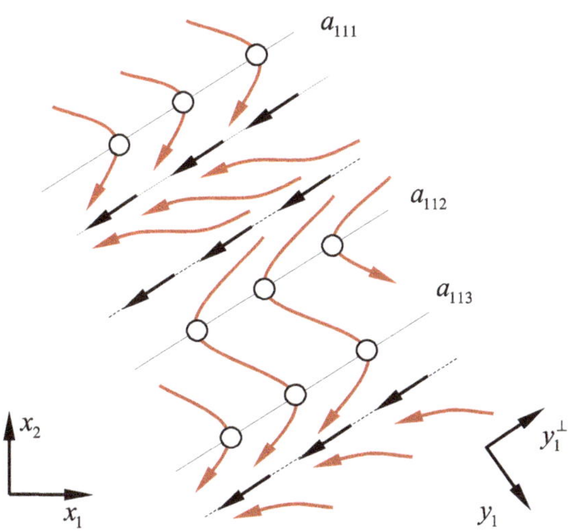

(d)

Fig. 1.3. (*Continued*)

- For $a_{110} > 0$ and $A_{223} < 0$,

$$\left\{\begin{array}{c}(a_{111}, \dot{x}_2)\\(a_{112}, \dot{x}_2)\\(a_{111}, \dot{x}_2)\end{array}\right\} = \left\{\begin{array}{c}\underbrace{\text{(UP,pF)}}_{\text{up-parabola flow }(-)}\\\underbrace{\text{(DP,pF)}}_{\text{down-parabola flow }(-)}\\\underbrace{\text{(UP,pF)}}_{\text{up-parabola flow }(-)}\end{array}\right\} \text{ for } \alpha_{12} < 0. \tag{1.96}$$

- For $a_{110} > 0$ and $A_{223} > 0$,

$$\left\{\begin{array}{c}(a_{111}, \dot{x}_2)\\(a_{112}, \dot{x}_2)\\(a_{111}, \dot{x}_2)\end{array}\right\} = \left\{\begin{array}{c}\underbrace{\text{(DP,nF)}}_{\text{down-parabola flow }(-)}\\\underbrace{\text{(UP,nF)}}_{\text{up-parabola flow}(-)}\\\underbrace{\text{(DP,nF)}}_{\text{down-parabola flow }(-)}\end{array}\right\} \text{ for } \alpha_{12} < 0. \tag{1.97}$$

The (DP,pF)-positive down-parabola flow, (UP,pF)-positive up-parabola flow, and (DP,pF)-positive down-parabola flow on the lines of $y_1^* = a_{111}, a_{112}, a_{113}$ $(a_{110} > 0, A_{223} > 0)$ are presented in Fig. 1.3(a). There are three infinite-equilibrium flows in the y_1-direction, which are source, sink, and source flows. The (UP,pF)-positive up-parabola flow, (DP,pF)-positive down-parabola flow, and (UP,pF)-positive up-parabola flow on the lines of $y_1^* = a_{111}, a_{112}, a_{113}$ $(a_{110} < 0, A_{223} > 0)$ are presented in Fig. 1.3(b). There are also three infinite-equilibrium flows in the y_1-direction, which are sink, source, and sink flows. In Fig. 1.3(c), the (UP,nF)-negative up-parabola flow, (DP,nF)-negative down-parabola flow, (UP,nF)-negative up-parabola flow on the lines of $y_1^* = a_{111}, a_{112}, a_{113}$ $(a_{110} > 0, A_{223} < 0)$ are presented. Three infinite-equilibrium flows in the y_1-direction exist, which are sink, source, and sink flows. In Fig. 1.3(d), presented are the (DP,nF)-negative down-parabola flow, (UP,nF)-negative up-parabola flow, and (DP,nF)-negative down-parabola flow on the lines of $y_1^* = a_{111}, a_{112,}, a_{113}$ $(a_{110} < 0, A_{223} < 0)$. Three infinite-equilibrium flows in the y_1-direction exist, which are source, sink, and source flows.

For $\alpha_{11} > 0$ and $\alpha_{12} > 0$, the corresponding sink and source flows in the phase portraits are presented in Figs. 1.4(a)–1.4(d).

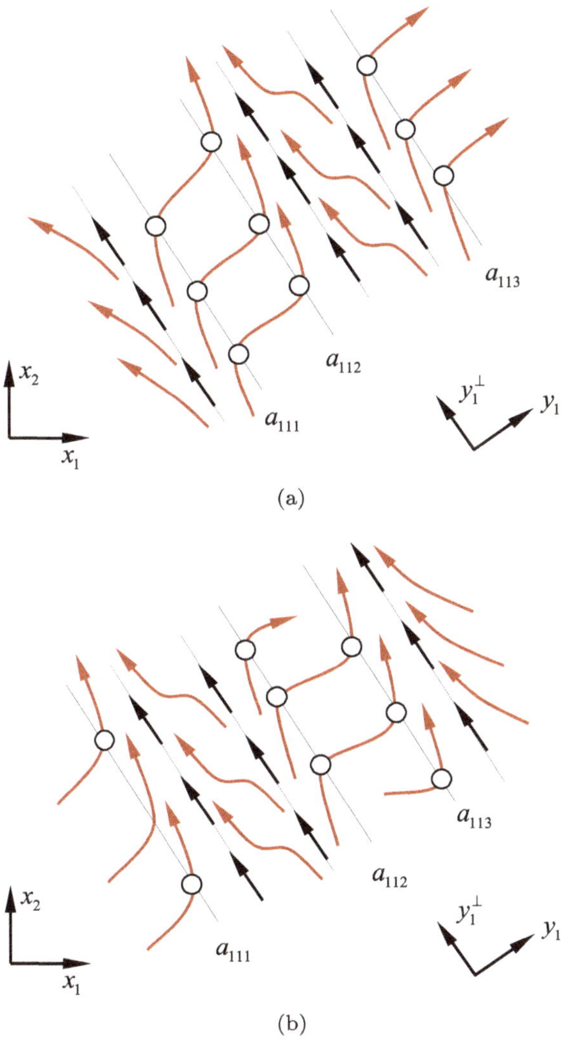

Fig. 1.4. Phase portraits for two-dimensional systems on the x_1-direction with $y_1^* = a_{111}, a_{112}, a_{113}$ and on the x_2-direction with $\dot{x}_2 = A_{223} \neq 0$ ($\alpha_{11} > 0, \alpha_{12} > 0$): (a) (UP,pF), (DP,pF), and (UP,pF)-flows ($a_{110} > 0, A_{223} > 0$), (b) (DP,pF), (UP,pF), and (DP,pF)-flows ($a_{110} < 0, A_{223} > 0$), (c) (DP,nF), (UP,nF), and (DP,nF)-flows ($a_{110} > 0, A_{223} < 0$), (d) (UP,nF), (DP,nF), and (UP,nF)-flows ($a_{110} < 0, A_{223} < 0$).

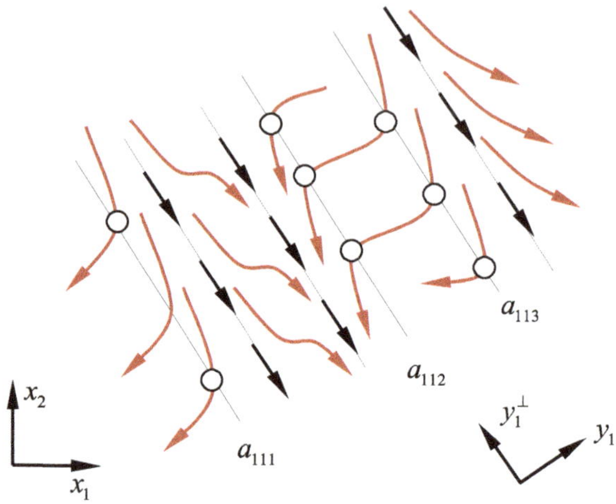

(c)

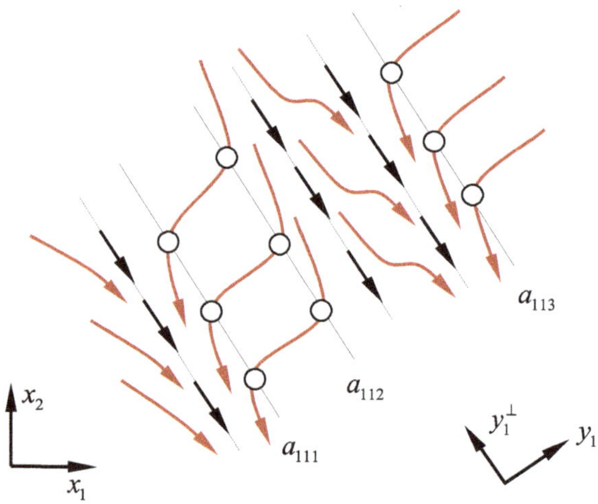

(d)

Fig. 1.4. (*Continued*)

- For $a_{110} > 0$ and $A_{223} > 0$,

$$\begin{Bmatrix} (a_{111}, \dot{x}_2) \\ (a_{112}, \dot{x}_2) \\ (a_{111}, \dot{x}_2) \end{Bmatrix} = \begin{Bmatrix} \underbrace{\text{(UP,pF)}}_{\text{up-parabola flow }(+)} \\ \underbrace{\text{(DP,pF)}}_{\text{down-parabola flow }(+)} \\ \underbrace{\text{(UP,pF)}}_{\text{up-parabola flow }(+)} \end{Bmatrix} \text{ for } \alpha_{12} > 0. \tag{1.98}$$

- For $a_{110} < 0$ and $A_{223} > 0$,

$$\begin{Bmatrix} (a_{111}, \dot{x}_2) \\ (a_{112}, \dot{x}_2) \\ (a_{111}, \dot{x}_2) \end{Bmatrix} = \begin{Bmatrix} \underbrace{\text{(DP,pF)}}_{\text{down-parabola flow }(+)} \\ \underbrace{\text{(UP,pF)}}_{\text{up-parabola flow }(+)} \\ \underbrace{\text{(DP,pF)}}_{\text{down-parabola flow }(+)} \end{Bmatrix} \text{ for } \alpha_{12} > 0. \tag{1.99}$$

- For $a_{110} > 0$ and $A_{223} < 0$,

$$\begin{Bmatrix} (a_{111}, \dot{x}_2) \\ (a_{112}, \dot{x}_2) \\ (a_{111}, \dot{x}_2) \end{Bmatrix} = \begin{Bmatrix} \underbrace{\text{(DP,nF)}}_{\text{down-parabola flow }(-)} \\ \underbrace{\text{(UP,nF)}}_{\text{up-parabola flow}(-)} \\ \underbrace{\text{(DP,nF)}}_{\text{down-parabola flow }(-)} \end{Bmatrix} \text{ for } \alpha_{12} > 0. \tag{1.100}$$

- For $a_{110} > 0$ and $A_{223} > 0$,

$$\begin{Bmatrix} (a_{111}, \dot{x}_2) \\ (a_{112}, \dot{x}_2) \\ (a_{111}, \dot{x}_2) \end{Bmatrix} = \begin{Bmatrix} \underbrace{\text{(UP,pF)}}_{\text{up-parabola flow }(-)} \\ \underbrace{\text{(DP,pF)}}_{\text{down-parabola flow }(-)} \\ \underbrace{\text{(UP,pF)}}_{\text{up-parabola flow }(-)} \end{Bmatrix} \text{ for } \alpha_{12} > 0. \tag{1.101}$$

The (UP,pF)-positive up-parabola flow, (DP,pF)-positive down-parabola flow, and (UP,pF)-positive up-parabola flow on the lines of $y_1^* = a_{111}, a_{112}, a_{113}$ $(a_{110} < 0, A_{223} > 0)$ are presented in Fig. 1.4(a). There are also three infinite-equilibrium flows in the y_1-direction, which are sink, source, and sink flows. The (DP,pF)-positive down-parabola flow, (UP,pF)-positive up-parabola flow, and (DP,pF)-positive down-parabola flow on the lines of $y_1^* = a_{111}, a_{112}, a_{113}$ $(a_{110} > 0, A_{223} > 0)$ are presented in Fig. 1.4b(a). There are three infinite-equilibrium flows in the y_1-direction, which are source, sink, and source flows. In Fig. 1.4(c), presented are the (DP,nF)-negative down-parabola flow, (UP,nF)-negative up-parabola flow, and (DP,nF)-negative down-parabola flow on the lines of $y_1^* = a_{111}, a_{112,}, a_{113}$ $(a_{110} < 0, A_{223} < 0)$. Three infinite-equilibrium flows in the y_1-direction exist, which are source, sink, and source flows. In Fig. 1.4(d), the (UP,nF)-negative up-parabola flow, (DP,nF)-negative down-parabola flow, and (UP,nF)-negative up-parabola flow on the lines of $y_1^* = a_{111}, a_{112}, a_{113}$ $(a_{110} > 0, A_{223} < 0)$ are presented. Three infinite-equilibrium flows in the y_1-direction exist, which are sink, source, and sink flows.

Chapter 2

Single-Linear-Bivariate Linear and Cubic Systems

In this chapter, two-dimensional nonlinear systems possessing single-linear-bivariate linear and cubic vector fields, and the appearing and switching bifurcations for the one-dimensional flows (i.e., up-parabola and down-parabola flows) are discussed. The third-order parabola flows are the appearing bifurcations for up-parabola, down-parabola, and up-parabola flows or for down-parabola, up-parabola, and down-parabola flows. The increasing-inflection and deceasing-inflection flows are the appearing bifurcations for up-parabola and down-parabola flows. The inflection-source and sink infinite-equilibriums are for switching bifurcations of vertical and horizontal parabola flows. The parabola-source and sink infinite-equilibrium flows are for the switching bifurcations of parabola flows and inflection flows and also for the appearing bifurcations of the vertical and horizontal parabola flows.

2.1 Single-Linear-Bivariate Linear and Cubic Vector Fields

In this section, the dynamics of single-bivariate linear and quadratic systems are presented. The singular flows are presented and the corresponding switching and appearing bifurcations for the regular flows (i.e., sink, source, and parabola flows) are discussed.

Theorem 2.1. *Consider a two-dimensional, dynamical system on the x_i-direction ($i \in \{1,2\}$) as*

$$\begin{aligned} \dot{x}_{j_1} &= f_{j_1}(\mathbf{x},\mathbf{p}) = A_{j_1j_13} + A_{j_1j_12}y_{j_1} + A_{j_1j_11}y_{j_1}^2 + A_{j_1j_10}y_{j_1}^3, \\ \dot{x}_{j_2} &= f_{j_2}(\mathbf{x},\mathbf{p}) = A_{j_2j_23} + A_{j_2j_22}y_{j_1} + A_{j_2j_21}y_{j_1}^2 + A_{j_2j_20}y_{j_1}^3 \end{aligned} \tag{2.1}$$

with

$$y_{j_1} = \alpha_{j_1j_1}x_{j_1} + \alpha_{j_1j_2}x_{j_2} \quad \text{for } j_1, j_2 \in \{1,2\}, j_1 \neq j_2. \tag{2.2}$$

For $A_{j_2j_20} = A_{j_2j_21} = 0$ *with* $(A_{j_2j_22} \neq 0, A_{j_2j_23} \neq 0)$, *the two-dimensional dynamical system is*

$$\begin{aligned} \dot{x}_{j_2} &= A_{j_2j_23} + A_{j_2j_22} y_{j_1}, \\ \dot{x}_{j_1} &= a_{j_1j_10}(y_{j_1} - b_{j_1j_11})(y_{j_1}^2 + B_{j_1j_1} y_{j_1} + C_{j_1j_1}). \end{aligned} \tag{2.3}$$

(i) If

$$A_{j_2j_22} \neq 0, \quad A_{j_2j_23} \neq 0 \quad \text{and} \quad \Delta_{j_1j_1} < 0, \tag{2.4}$$

the corresponding dynamical system is

$$\begin{aligned} \dot{x}_{j_2} &= a_{j_2j_20}(y_{j_1} - a_{j_2j_21}), \\ \dot{x}_{j_1} &= a_{j_1j_10}(y_{j_1} - a_{j_1j_11})[(y_{j_1} - a_{j_1j_1})^2 + b_{j_1j_1}], \end{aligned} \tag{2.5}$$

where

$$\begin{aligned} a_{j_2j_21} &= -\frac{1}{A_{j_2j_22}} A_{j_2j_23}, \quad a_{j_2j_20} = A_{j_2j_22}, \\ a_{j_1j_1} &= -\frac{1}{2} B_{j_1j_1}, \quad b_{j_1j_1} = -\frac{1}{4} C_{j_1j_1}, \quad a_{j_1j_11} = b_{j_1j_11}, \\ \Delta_{j_1j_1} &= (B_{j_1j_1})^2 - 4C_{j_1j_1}, \quad a_{j_1j_10} = A_{j_1j_10}. \end{aligned} \tag{2.6}$$

There is a dynamical system in the y_{j_1}*-direction as*

$$\begin{aligned} \dot{y}_{j_1} &= \alpha_{j_1j_1} a_{j_1j_10}(y_{j_1} - a_{j_1j_11})[(y_{j_1} - a_{j_1j_1})^2 + b_{j_1j_1}] \\ &\quad + \alpha_{j_1j_2} a_{j_2j_20}(y_{j_1} - a_{j_2j_21}) \end{aligned} \tag{2.7}$$

with a dynamical system in the $y_{j_1}^{\perp}$*-direction as*

$$\begin{aligned} \dot{y}_{j_1}^{\perp} &= \beta_{j_1j_1} a_{j_1j_10}(y_{j_1} - a_{j_1j_11})[(y_{j_1} - a_{j_1j_1})^2 + b_{j_1j_1}] \\ &\quad + \beta_{j_1j_2} a_{j_2j_20}(y_{j_1} - a_{j_2j_21}), \end{aligned} \tag{2.8}$$

where

$$\begin{aligned} y_{j_1}^{\perp} &= \beta_{j_1j_1} x_{j_1} + \beta_{j_1j_2} x_{j_2}, \quad \mathbf{n}_{j_1}^{\perp} = (\beta_{j_1j_1}, \beta_{j_1j_2})^{\mathrm{T}}, \\ \mathbf{n}_{j_1} \cdot \mathbf{n}_{j_1}^{\perp} &= \alpha_{j_1j_1}\beta_{j_1j_1} + \alpha_{j_1j_2}\beta_{j_1j_2} = 0. \end{aligned} \tag{2.9}$$

(i_{1a}) *The equilibrium of* $y_{j_1}^* = a_{j_1j_11}$ *in the* x_{j_1}*-direction has the following properties:*

- *For* $a_{j_1j_10}\alpha_{j_1j_2} > 0$ *and* $a_{j_2j_20}(a_{j_1j_11} - a_{j_2j_21}) > 0$,

$$(a_{j_1j_11}, \dot{x}_{j_2}^*) = \underbrace{(\mathrm{UP}, \mathrm{pF})}_{\text{up-parabola flow } (+)}. \tag{2.10}$$

- *For* $a_{j_1j_10}\alpha_{j_1j_2} < 0$ *and* $a_{j_2j_20}(a_{j_1j_11} - a_{j_2j_21}) > 0$,

$$(a_{j_1j_11}, \dot{x}^*_{j_2}) = \underbrace{(\text{DP}, \text{pF})}_{\text{down-parabola flow } (+)}. \tag{2.11}$$

- *For* $a_{j_1j_10}\alpha_{j_1j_2} > 0$ *and* $a_{j_2j_20}(a_{j_1j_11} - a_{j_2j_21}) < 0$,

$$(a_{j_1j_11}, \dot{x}^*_{j_2}) = \underbrace{(\text{DP}, \text{nF})}_{\text{down-parabola flow } (-)}. \tag{2.12}$$

- *For* $a_{j_1j_10}\alpha_{j_1j_2} < 0$ *and* $a_{j_2j_20}(a_{j_1j_11} - a_{j_2j_21}) < 0$,

$$(a_{j_1j_11}, \dot{x}^*_{j_2}) = \underbrace{(\text{UP}, \text{nF})}_{\text{up-parabola flow } (-)}. \tag{2.13}$$

(i_{1b}) *The equilibrium of* $y^*_{j_1} = a_{j_2j_21}$ *in the* x_{j_2}*-direction has the following properties*:

- *For* $a_{j_1j_10}(a_{j_2j_21} - a_{j_1j_11}) > 0$ *and* $\alpha_{j_1j_1}a_{j_2j_20} > 0$,

$$(\dot{x}^*_{j_1}, a_{j_2j_21}) = \underbrace{(\text{pF}, \text{UP})}_{\text{up-parabola flow } (+)}. \tag{2.14}$$

- *For* $a_{j_1j_10}(a_{j_2j_21} - a_{j_1j_11}) < 0$ *and* $\alpha_{j_1j_1}a_{j_2j_20} > 0$,

$$(\dot{x}^*_{j_1}, a_{j_2j_21}) = \underbrace{(\text{nF}, \text{DP})}_{\text{down-parabola flow } (-)}. \tag{2.15}$$

- *For* $a_{j_1j_10}(a_{j_2j_21} - a_{j_1j_11}) > 0$ *and* $\alpha_{j_1j_1}a_{j_2j_20} < 0$,

$$(\dot{x}^*_{j_1}, a_{j_2j_21}) = \underbrace{(\text{pF}, \text{DP})}_{\text{down-parabola flow } (+)}. \tag{2.16}$$

- *For* $a_{j_1j_10}(a_{j_2j_21} - a_{j_1j_11}) < 0$ *and* $\alpha_{j_1j_1}a_{j_2j_20} < 0$,

$$(\dot{x}^*_{j_1}, a_{j_2j_21}) = \underbrace{(\text{nF}, \text{UP})}_{\text{up-parabola flow } (-)}. \tag{2.17}$$

(i_2) For $a_{j_1j_1 1} = a_{j_2j_2 1}$,

$$\begin{aligned}\dot{x}_{j_2} &= a_{j_2j_2 0}(y_{j_1} - a_{j_2j_2 1}),\\ \dot{x}_{j_1} &= a_{j_1j_1 0}(y_{j_1} - a_{j_1j_1 1})[(y_{j_1} - a_{j_1j_1})^2 + b_{j_1j_1}].\end{aligned} \tag{2.18}$$

There is a dynamical system in the y_{j_1}-direction as

$$\dot{y}_{j_1} = \alpha_{j_1j_1} a_{j_1j_1 0}(y_{j_1} - a_{j_1j_1 1})[(y_{j_1} - a_{j_1j_1})^2 + b_{j_1j_1} + c_{j_1j_1}], \tag{2.19}$$

with a dynamical system in the $y_{j_1}^{\perp}$-direction as

$$\dot{y}_{j_1}^{\perp} = \beta_{j_1j_1} a_{j_1j_1 0}(y_{j_1} - a_{j_1j_1 1})[(y_{j_1} - a_{j_1j_1})^2 + b_{j_1j_1} + d_{j_1j_1}], \tag{2.20}$$

where

$$c_{j_1j_1} = \frac{\alpha_{j_1j_2} a_{j_2j_2 0}}{\alpha_{j_1j_1} a_{j_1j_1 0}}, \quad d_{j_1j_1} = \frac{\beta_{j_1j_2} a_{j_2j_2 0}}{\beta_{j_1j_1} a_{j_1j_1 0}}. \tag{2.21}$$

Let

$$\begin{aligned}G_{j_1j_1 0} &= \frac{\partial}{\partial y_{j_1}}[\mathbf{n}_{j_1}^{\mathrm{T}} \cdot \mathbf{F}(y_{j_1}, \mathbf{p})]\bigg|_{y_{j_1} = a_{jj_1j_1}}\\ &= \alpha_{j_1j_1} a_{j_1j_1 0}[(a_{j_1j_1 1} - a_{j_1j_1})^2 + b_{j_1j_1}] + \alpha_{j_1j_2} a_{j_2j_2 0}.\end{aligned} \tag{2.22}$$

(i_{2a}) The infinite-equilibrium of $y_{j_1}^* = a_{j_1j_1 1} = a_{j_2j_2 1}$ possesses the following properties:

- *If*

$$\frac{a_{j_1j_1 0}}{a_{j_2j_2 0}} > 0 \quad \text{and} \quad G_{j_1j_1 0} > 0, \tag{2.23}$$

then

$$(a_{j_1j_1 1}, a_{j_2j_2 1}) = \underbrace{(\mathrm{II}, \mathrm{SO})}_{\text{increasing-inflection source}}. \tag{2.24}$$

- *If*

$$\frac{a_{j_1j_1 0}}{a_{j_2j_2 0}} < 0 \quad \text{and} \quad G_{j_1j_1 0} > 0, \tag{2.25}$$

then

$$(a_{j_1j_1 1}, a_{j_2j_2 1}) = \underbrace{(\mathrm{DI}, \mathrm{SO})}_{\text{decreasing-inflection source}}. \tag{2.26}$$

• *If*

$$\frac{a_{j_1 j_1 0}}{a_{j_2 j_2 0}} < 0 \quad \text{and} \quad G_{j_1 j_1 0} < 0, \tag{2.27}$$

then

$$(a_{j_1 j_1 1}, a_{j_2 j_2 1}) = \underbrace{(\mathrm{DI}, \mathrm{SI})}_{\text{decreasing-inflection sink}}. \tag{2.28}$$

• *If*

$$\frac{a_{j_1 j_1 0}}{a_{j_2 j_2 0}} > 0 \quad \text{and} \quad G_{j_1 j_1 0} < 0, \tag{2.29}$$

then

$$(a_{j_1 j_1 1}, a_{j_2 j_2 1}) = \underbrace{(\mathrm{II}, \mathrm{SI})}_{\text{increasing-inflection sink}}. \tag{2.30}$$

(i_{2b}) *The inflection-source and sink infinite-equilibriums at* $y^*_{j_1} = a_{j_1 j_1 1} = a_{j_2 j_2 1}$ *are the switching bifurcations for the parabola flows at* $y^*_{j_1} = a_{j_1 j_1 1}$ *and* $y^*_{j_1} = a_{j_2 j_2 1}$.

• *For* $a_{j_1 j_1 0} > 0$ *and* $a_{j_2 j_2 0} > 0$,

$$\begin{aligned}
&\left\{\begin{matrix} \underbrace{(\mathrm{nF}, \mathrm{DP})}_{\text{down-parabola }(-)} \\ \underbrace{(\mathrm{UP}, \mathrm{pF})}_{\text{up-parabola }(+)} \end{matrix}\right\} \rightleftharpoons \underbrace{(\mathrm{II}, \mathrm{SO})}_{\text{switching bifurcation}} \rightleftharpoons \left\{\begin{matrix} \underbrace{(\mathrm{DP}, \mathrm{nF})}_{\text{down-parabola }(-)} \\ \underbrace{(\mathrm{pF}, \mathrm{UP})}_{\text{up-parabola }(+)} \end{matrix}\right\} \\
&\text{for } (\alpha_{j_1 j_1}, \alpha_{j_1 j_2}) = (+,+) \quad \text{and} \quad G_{j_1 j_1 0} > 0, \\
&\left\{\begin{matrix} \underbrace{(\mathrm{nF}, \mathrm{DP})}_{\text{down-parabola }(-)} \\ \underbrace{(\mathrm{DP}, \mathrm{pF})}_{\text{down-parabola }(+)} \end{matrix}\right\} \rightleftharpoons \underbrace{(\mathrm{II}, \mathrm{SO})}_{\text{switching bifurcation}} \rightleftharpoons \left\{\begin{matrix} \underbrace{(\mathrm{UP}, \mathrm{nF})}_{\text{up-parabola }(-)} \\ \underbrace{(\mathrm{pF}, \mathrm{UP})}_{\text{up-parabola }(+)} \end{matrix}\right\} \\
&\text{for } (\alpha_{j_1 j_1}, \alpha_{j_1 j_2}) = (+,-) \quad \text{and} \quad G_{j_1 j_1 0} > 0, \\
&\left\{\begin{matrix} \underbrace{(\mathrm{UP}, \mathrm{nF})}_{\text{up-parabola }(-)} \\ \underbrace{(\mathrm{pF}, \mathrm{UP})}_{\text{up-parabola }(+)} \end{matrix}\right\} \rightleftharpoons \underbrace{(\mathrm{II}, \mathrm{SI})}_{\text{switching bifurcation}} \rightleftharpoons \left\{\begin{matrix} \underbrace{(\mathrm{nF}, \mathrm{DP})}_{\text{down-parabola }(-)} \\ \underbrace{(\mathrm{DP}, \mathrm{pF})}_{\text{down-parabola }(+)} \end{matrix}\right\} \\
&\text{for } (\alpha_{j_1 j_1}, \alpha_{j_1 j_2}) = (+,-) \quad \text{and} \quad G_{j_1 j_1 0} < 0.
\end{aligned} \tag{2.31}$$

- *For* $a_{j_1 j_1 0} < 0$ *and* $a_{j_2 j_2 0} > 0$,

$$\left\{\begin{array}{c}\underbrace{(\mathrm{pF},\mathrm{UP})}_{\text{up-parabola }(+)}\\ \underbrace{(\mathrm{DP},\mathrm{pF})}_{\text{down-parabola }(+)}\end{array}\right\} \rightleftharpoons \underbrace{(\mathrm{DI},\mathrm{SI})}_{\text{switching bifurcation}} \rightleftharpoons \left\{\begin{array}{c}\underbrace{(\mathrm{UP},\mathrm{nF})}_{\text{up-parabola }(-)}\\ \underbrace{(\mathrm{nF},\mathrm{DP})}_{\text{down-parabola }(-)}\end{array}\right\}$$

$$\text{for } (\alpha_{j_1 j_1}, \alpha_{j_1 j_2}) = (+,+) \quad \text{and} \quad G_{j_1 j_1 0} < 0;$$

$$\left\{\begin{array}{c}\underbrace{(\mathrm{UP},\mathrm{nF})}_{\text{up-parabola }(-)}\\ \underbrace{(\mathrm{nF},\mathrm{DP})}_{\text{down-parabola }(-)}\end{array}\right\} \rightleftharpoons \underbrace{(\mathrm{DI},\mathrm{SO})}_{\text{switching bifurcation}} \rightleftharpoons \left\{\begin{array}{c}\underbrace{(\mathrm{pF},\mathrm{UP})}_{\text{up-parabola }(+)}\\ \underbrace{(\mathrm{DP},\mathrm{pF})}_{\text{down-parabola }(+)}\end{array}\right\}$$

$$\text{for } (\alpha_{j_1 j_1}, \alpha_{j_1 j_2}) = (+,+) \quad \text{and} \quad G_{j_1 j_1 0} > 0;$$

$$\left\{\begin{array}{c}\underbrace{(\mathrm{pF},\mathrm{UP})}_{\text{up-parabola }(+)}\\ \underbrace{(\mathrm{UP},\mathrm{pF})}_{\text{up-parabola }(+)}\end{array}\right\} \rightleftharpoons \underbrace{(\mathrm{DI},\mathrm{SI})}_{\text{switching bifurcation}} \rightleftharpoons \left\{\begin{array}{c}\underbrace{(\mathrm{DP},\mathrm{nF})}_{\text{down-parabola }(-)}\\ \underbrace{(\mathrm{nF},\mathrm{DP})}_{\text{down-parabola }(-)}\end{array}\right\}$$

$$\text{for } (\alpha_{j_1 j_1}, \alpha_{j_1 j_2}) = (+,-) \quad \text{and} \quad G_{j_1 j_1 0} < 0. \tag{2.32}$$

- *For* $a_{j_1 j_1 0} > 0$ *and* $a_{j_2 j_2 0} < 0$,

$$\left\{\begin{array}{c}\underbrace{(\mathrm{nF},\mathrm{UP})}_{\text{up-parabola }(-)}\\ \underbrace{(\mathrm{DP},\mathrm{nF})}_{\text{down-parabola }(-)}\end{array}\right\} \rightleftharpoons \underbrace{(\mathrm{DI},\mathrm{SO})}_{\text{switching bifurction}} \rightleftharpoons \left\{\begin{array}{c}\underbrace{(\mathrm{UP},\mathrm{pF})}_{\text{up-parabola }(+)}\\ \underbrace{(\mathrm{pF},\mathrm{DP})}_{\text{down-parabola }(+)}\end{array}\right\}$$

$$\text{for } (\alpha_{j_1 j_1}, \alpha_{j_1 j_2}) = (+,+) \quad \text{and} \quad G_{j_1 j_1 0} > 0;$$

$$\left\{\begin{array}{c}\underbrace{(\mathrm{UP},\mathrm{pF})}_{\text{up-parabola }(+)}\\ \underbrace{(\mathrm{pF},\mathrm{DP})}_{\text{down-parabola }(+)}\end{array}\right\} \rightleftharpoons \underbrace{(\mathrm{DI},\mathrm{SI})}_{\text{switching bifurction}} \rightleftharpoons \left\{\begin{array}{c}\underbrace{(\mathrm{nF},\mathrm{UP})}_{\text{up-parabola }(-)}\\ \underbrace{(\mathrm{DP},\mathrm{nF})}_{\text{down-parabola }(-)}\end{array}\right\}$$

$$\text{for } (\alpha_{j_1 j_1}, \alpha_{j_1 j_2}) = (+,+) \quad \text{and} \quad G_{j_1 j_1 0} < 0$$

$$\left\{\begin{array}{c}\underbrace{(\mathrm{nF},\mathrm{UP})}_{\text{up-parabola }(-)}\\ \underbrace{(\mathrm{UP},\mathrm{nF})}_{\text{up-parabola }(-)}\end{array}\right\} \rightleftharpoons \underbrace{(\mathrm{DI},\mathrm{SO})}_{\text{switching bifurction}} \rightleftharpoons \left\{\begin{array}{c}\underbrace{(\mathrm{DP},\mathrm{pF})}_{\text{down-parabola }(+)}\\ \underbrace{(\mathrm{pF},\mathrm{DP})}_{\text{down-parabola }(+)}\end{array}\right\}$$

$$\text{for } (\alpha_{j_1 j_1}, \alpha_{j_1 j_2}) = (+,-) \quad \text{and} \quad G_{j_1 j_1 0} > 0. \tag{2.33}$$

- *For* $a_{j_1j_10} < 0$ *and* $a_{j_2j_20} < 0$,

$$\left\{\begin{array}{c}\underbrace{(\mathrm{pF},\mathrm{DP})}_{\text{down-parabola }(+)}\\ \underbrace{(\mathrm{UP},\mathrm{nF})}_{\text{up-parabola }(-)}\end{array}\right\} \rightleftharpoons \underbrace{(\mathrm{II},\mathrm{SI})}_{\text{switching bifurcation}} \rightleftharpoons \left\{\begin{array}{c}\underbrace{(\mathrm{DP},\mathrm{pF})}_{\text{down-parabola }(+)}\\ \underbrace{(\mathrm{nF},\mathrm{UP})}_{\text{up-p-parabola }(-)}\end{array}\right\}$$

$$\text{for } (\alpha_{j_1j_1},\alpha_{j_1j_2}) = (+,+) \quad \text{and} \quad G_{j_1j_10} < 0;$$

$$\left\{\begin{array}{c}\underbrace{(\mathrm{pF},\mathrm{DP})}_{\text{down-parabola }(+)}\\ \underbrace{(\mathrm{DP},\mathrm{nF})}_{\text{down-parabola }(-)}\end{array}\right\} \rightleftharpoons \underbrace{(\mathrm{II},\mathrm{SI})}_{\text{switching bifurcation}} \rightleftharpoons \left\{\begin{array}{c}\underbrace{(\mathrm{UP},\mathrm{pF})}_{\text{up-parabola }(+)}\\ \underbrace{(\mathrm{nF},\mathrm{UP})}_{\text{up-parabola }(-)}\end{array}\right\}$$

$$\text{for } (\alpha_{j_1j_1},\alpha_{j_1j_2}) = (+,-) \quad \text{and} \quad G_{j_1j_10} < 0;$$

$$\left\{\begin{array}{c}\underbrace{(\mathrm{UP},\mathrm{pF})}_{\text{up-parabola }(+)}\\ \underbrace{(\mathrm{nF},\mathrm{UP})}_{\text{up-p-parabola }(-)}\end{array}\right\} \rightleftharpoons \underbrace{(\mathrm{II},\mathrm{SO})}_{\text{switching bifurcation}} \rightleftharpoons \left\{\begin{array}{c}\underbrace{(\mathrm{pF},\mathrm{DP})}_{\text{down-parabola }(+)}\\ \underbrace{(\mathrm{DP},\mathrm{nF})}_{\text{down-parabola }(-)}\end{array}\right\}$$

$$\text{for } (\alpha_{j_1j_1},\alpha_{j_1j_2}) = (+,-) \quad \text{and} \quad G_{j_1j_10} > 0. \tag{2.34}$$

(*ii*) *If*

$$A_{j_2j_21} \neq 0, \quad A_{j_2j_22} \neq 0 \quad \text{and} \quad \Delta_{j_1j_1} = 0, \tag{2.35}$$

the two-dimensional single-bivariate dynamical system becomes

$$\begin{aligned}\dot{x}_{j_2} &= a_{j_2j_20}(y_{j_1} - a_{j_2j_21}),\\ \dot{x}_{j_1} &= a_{j_1j_10}(y_{j_1} - a_{j_1j_1s_1})(y_{j_1} - a_{j_1j_1s_2})^2.\end{aligned} \tag{2.36}$$

(ii_1) *For* $a_{j_2j_21} \neq a_{j_1j_1s_1}, a_{j_1j_1s_2}$, *there is a dynamical system in the* y_{j_1}-*direction as*

$$\begin{aligned}\dot{y}_{j_1} &= \alpha_{j_1j_1}a_{j_1j_10}(y_{j_1} - a_{j_1j_1s_1})(y_{j_1} - a_{j_1j_1s_2})^2\\ &\quad + \alpha_{j_1j_2}a_{j_2j_20}(y_{j_1} - a_{j_2j_21}),\end{aligned} \tag{2.37}$$

with a dynamical system in the $y_{j_1}^{\perp}$-*direction as*

$$\begin{aligned}\dot{y}_{j_1}^{\perp} &= \beta_{j_1j_1}a_{j_1j_10}(y_{j_1} - a_{j_1j_1s_1})(y_{j_1} - a_{j_1j_1s_2})^2\\ &\quad + \beta_{j_1j_2}a_{j_2j_20}(y_{j_1} - a_{j_2j_21}).\end{aligned} \tag{2.38}$$

(ii_{1a}) *The equilibrium on the line of* $y_{j_1}^* = a_{j_1j_1s_1}$ *for the* x_{j_1}-*direction with* $\dot{x}_{j_2}^* = a_{j_2j_20}(a_{j_1j_1s_1} - a_{j_2j_21})$ *possesses the following properties:*

- *For* $a_{j_1j_10}\alpha_{j_1j_2}(a_{j_1j_1s_1}-a_{j_1j_1s_2})^2>0$ *and* $a_{j_2j_20}(a_{j_1j_1s_1}-a_{j_2j_21})>0$,

$$(a_{j_1j_1s_1},\dot{x}^*_{j_2})=\underbrace{(\mathrm{UP},\mathrm{pF})}_{\text{up-parabola flow }(+)}. \tag{2.39}$$

- *For* $a_{j_1j_10}\alpha_{j_1j_2}(a_{j_1j_1s_1}-a_{j_1j_1s_2})^2<0$ *and* $a_{j_2j_20}(a_{j_1j_1s_1}-a_{j_2j_21})>0$,

$$(a_{j_1j_1s_1},\dot{x}^*_{j_2})=\underbrace{(\mathrm{DP},\mathrm{pF})}_{\text{down-parabola flow }(+)}. \tag{2.40}$$

- *For* $a_{j_1j_10}\alpha_{j_1j_2}(a_{j_1j_1s_1}-a_{j_1j_1s_2})^2>0$ *and* $a_{j_2j_20}(a_{j_1j_1s_1}-a_{j_2j_21})<0$,

$$(a_{j_1j_1s_1},\dot{x}^*_{j_2})=\underbrace{(\mathrm{DP},\mathrm{nF})}_{\text{down-parabola flow }(-)}. \tag{2.41}$$

- *For* $a_{j_1j_10}\alpha_{j_1j_2}(a_{j_1j_1s_1}-a_{j_1j_1s_2})^2<0$ *and* $a_{j_2j_20}(a_{j_1j_1s_1}-a_{j_2j_21})<0$,

$$(a_{j_1j_1s_1},\dot{x}^*_{j_2})=\underbrace{(\mathrm{UP},\mathrm{nF})}_{\text{up-parabola flow }(-)}. \tag{2.42}$$

(ii_{1b}) The equilibrium on the line of $y^*_{j_1}=a_{j_1j_1s_2}$ *for the* x_{j_1}*-direction with* $\dot{x}^*_{j_2}=a_{j_2j_20}(a_{j_1j_1s_2}-a_{j_2j_21})$ *possesses the following properties:*

- *For* $a_{j_1j_10}\alpha^2_{j_1j_2}(a_{j_1j_1s_2}-a_{j_1j_1s_1})>0$ *and* $a_{j_2j_20}(a_{j_1j_1s_2}-a_{j_2j_21})>0$,

$$(a_{j_1j_1s_2},\dot{x}^*_{j_2})=\underbrace{(\mathrm{II},\mathrm{pF})}_{\text{increasing-inflection flow }(+)}. \tag{2.43}$$

- *For* $a_{j_1j_10}\alpha^2_{j_1j_2}(a_{j_1j_1s_2}-a_{j_1j_1s_1})<0$ *and* $a_{j_2j_20}(a_{j_1j_1s_2}-a_{j_2j_21})>0$,

$$(a_{j_1j_1s_2},\dot{x}^*_{j_2})=\underbrace{(\mathrm{DI},\mathrm{pF})}_{\text{decreasing-inflection flow }(+)}. \tag{2.44}$$

- *For* $a_{j_1j_10}\alpha^2_{j_1j_2}(a_{j_1j_1s_2}-a_{j_1j_1s_1})>0$ *and* $a_{j_2j_20}(a_{j_1j_1s_2}-a_{j_2j_21})<0$,

$$(a_{j_1j_1s_2},\dot{x}^*_{j_2})=\underbrace{(\mathrm{DI},\mathrm{nF})}_{\text{decreasing-inflection flow }(-)}. \tag{2.45}$$

- *For* $a_{j_1j_10}\alpha^2_{j_1j_2}(a_{j_1j_1s_2}-a_{j_1j_1s_1})<0$ *and* $a_{j_2j_20}(a_{j_1j_1s_2}-a_{j_2j_21})<0$,

$$(a_{j_1j_1s_2},\dot{x}^*_{j_2})=\underbrace{(\mathrm{II},\mathrm{nF})}_{\text{increasing-inflection flow }(-)}. \tag{2.46}$$

– *The corresponding increasing-inflection and decreasing-inflection flows are the appearing bifurcations for up-parabola flow and down-parabola flow.*

(ii_{1c}) *With*

$$\dot{x}_{j_1}^{*} = a_{j_1 j_1 0}(a_{j_2 j_2 1} - a_{j_1 j_1 s_1})(a_{j_2 j_2 1} - a_{j_1 j_1 s_2})^2, \tag{2.47}$$

the simple equilibrium of $y_{j_1}^{*} = a_{j_2 j_2 1}$ *in the* x_{j_2}*-direction for such a system has the following properties:*

- *For* $a_{j_1 j_1 0}(a_{j_2 j_2 1} - a_{j_1 j_1 s_1})(a_{j_2 j_2 1} - a_{j_1 j_1 s_2})^2 > 0$ *and* $\alpha_{j_1 j_1} a_{j_2 j_2 0} > 0$,
$$(\dot{x}_{j_1}^{*}, a_{j_2 j_2 1}) = \underbrace{(\mathrm{pF}, \mathrm{UP})}_{\text{up-parabola flow } (+)}. \tag{2.48}$$
- *For* $a_{j_1 j_1 0}(a_{j_2 j_2 1} - a_{j_1 j_1 s_1})(a_{j_2 j_2 1} - a_{j_1 j_1 s_2})^2 < 0$ *and* $\alpha_{j_1 j_1} a_{j_2 j_2 0} > 0$,
$$(\dot{x}_{j_1}^{*}, a_{j_2 j_2 1}) = \underbrace{(\mathrm{nF}, \mathrm{DP})}_{\text{down-parabola flow } (-)}. \tag{2.49}$$
- *For* $a_{j_1 j_1 0}(a_{j_2 j_2 1} - a_{j_1 j_1 s_1})(a_{j_2 j_2 1} - a_{j_1 j_1 s_2})^2 > 0$ *and* $\alpha_{j_1 j_1} a_{j_2 j_2 0} < 0$,
$$(\dot{x}_{j_1}^{*}, a_{j_2 j_2 1}) = \underbrace{(\mathrm{pF}, \mathrm{DP})}_{\text{down-parabola flow } (+)}. \tag{2.50}$$
- *For* $a_{j_1 j_1 0}(a_{j_2 j_2 1} - a_{j_1 j_1 s_1})(a_{j_2 j_2 1} - a_{j_1 j_1 s_2})^2 < 0$ *and* $\alpha_{j_1 j_1} a_{j_2 j_2 0} < 0$,
$$(\dot{x}_{j_1}^{*}, a_{j_2 j_2 1}) = \underbrace{(\mathrm{nF}, \mathrm{UP})}_{\text{up-parabola flow } (-)}. \tag{2.51}$$

(ii_2) *For* $a_{j_2 j_2 1} = a_{j_1 j_1 s_1}$ *and* $a_{j_2 j_2 1} \neq a_{j_1 j_1 s_2}$, *the two-dimensional single-bivariate dynamical system becomes*

$$\begin{aligned} \dot{x}_{j_2} &= a_{j_2 j_2 0}(y_{j_1} - a_{j_2 j_2 1}), \\ \dot{x}_{j_1} &= a_{j_1 j_1 0}(y_{j_1} - a_{j_1 j_1 s_1})(y_{j_1} - a_{j_1 j_1 s_2})^2. \end{aligned} \tag{2.52}$$

There is a dynamical system in the y_{j_1}*-direction as*

$$\dot{y}_{j_1} = \alpha_{j_1 j_1} a_{j_1 j_1 0}(y_{j_1} - a_{j_1 j_1 s_1})[(y_{j_1} - a_{j_1 j_1 s_2})^2 + c_{j_1 j_1}] \tag{2.53}$$

and a dynamical system in the $y_{j_1}^{\perp}$*-direction as*

$$\dot{y}_{j_1}^{\perp} = \beta_{j_1 j_1} a_{j_1 j_1 0}(y_{j_1} - a_{j_1 j_1 s_1})[(y_{j_1} - a_{j_1 j_1 s_2})^2 + d_{j_1 j_1}]. \tag{2.54}$$

Let

$$\begin{aligned} G_{j_1 j_1 0} &= \frac{\partial}{\partial y_{j_1}}[\mathbf{n}_{j_1}^{\mathrm{T}} \cdot \mathbf{F}(y_{j_1}, \mathbf{p})]\bigg|_{y_{j_1}^{*} = a_{j_1 j_1 s_1}} \\ &= \alpha_{j_1 j_2} a_{j_1 j_2 0}(a_{j_1 j_1 s_1} - a_{j_1 j_1 s_2})^2 + \alpha_{j_1 j_2} a_{j_1 j_2 0}. \end{aligned} \tag{2.55}$$

(ii_{2a}) The infinite-equilibrium of $y_{j_1}^* = a_{j_1 j_1 s_1} = a_{j_2 j_2 1}$ possesses the following properties:

- *If*

$$\frac{a_{j_1 j_1 0}(a_{j_1 j_1 s_1} - a_{j_1 j_1 s_2})^2}{a_{j_2 j_2 0}} > 0 \quad \text{and} \quad G_{j_1 j_1 0} > 0, \tag{2.56}$$

then

$$(a_{j_1 j_1 s_1}, a_{j_2 j_2 1}) = \underbrace{(\text{II}, \text{SO})}_{\text{increasing-inflection source}}. \tag{2.57}$$

- *If*

$$\frac{a_{j_1 j_1 0}(a_{j_1 j_1 s_1} - a_{j_1 j_1 s_2})^2}{a_{j_2 j_2 0}} < 0 \quad \text{and} \quad G_{j_1 j_1 0} > 0, \tag{2.58}$$

then

$$(a_{j_1 j_1 s_1}, a_{j_2 j_2 1}) = \underbrace{(\text{DI}, \text{SO})}_{\text{decreasing-inflection source}}. \tag{2.59}$$

- *If*

$$\frac{a_{j_1 j_1 0}(a_{j_1 j_1 s_1} - a_{j_1 j_1 s_2})^2}{a_{j_2 j_2 0}} < 0 \quad \text{and} \quad G_{j_1 j_1 0} < 0, \tag{2.60}$$

then

$$(a_{j_1 j_1 s_1}, a_{j_2 j_2 1}) = \underbrace{(\text{DI}, \text{SI})}_{\text{decreasing-inflection sink}}. \tag{2.61}$$

- *If*

$$\frac{a_{j_1 j_1 0}(a_{j_1 j_1 s_1} - a_{j_1 j_1 s_2})^2}{a_{j_2 j_2 0}} > 0 \quad \text{and} \quad G_{j_1 j_1 0} < 0, \tag{2.62}$$

then

$$(a_{j_1 j_1 s_1}, a_{j_2 j_2 1}) = \underbrace{(\text{II}, \text{SI})}_{\text{increasing-inflection sink}}. \tag{2.63}$$

(ii_{2b}) The inflection-source and sink infinite equilibriums at $y_{j_1}^* = a_{j_1 j_1 s_1} = a_{j_2 j_2 1}$ are the switching bifurcations for the parabola flows at $y_{j_1}^* = a_{j_1 j_1 s_1}$ and $y_{j_1}^* = a_{j_2 j_2 1}$:

- *For* $a_{j_1 j_1 0}(a_{j_1 j_1 s_1} - a_{j_1 j_1 s_2})^2 > 0$ *and* $a_{j_2 j_2 0} > 0$,

$$\begin{aligned}
&\left\{\begin{matrix}\underbrace{\text{(nF,DP)}}_{\text{down-parabola }(-)}\\ \underbrace{\text{(UP,pF)}}_{\text{up-parabola }(+)}\end{matrix}\right\} \rightleftharpoons \underbrace{\text{(II,SO)}}_{\text{switching bifurcation}} \rightleftharpoons \left\{\begin{matrix}\underbrace{\text{(DP,nF)}}_{\text{down-parabola }(-)}\\ \underbrace{\text{(pF,UP)}}_{\text{up-parabola }(+)}\end{matrix}\right\}\\
&\text{for}\quad (\alpha_{j_1 j_1}, \alpha_{j_1 j_2}) = (+,+) \quad \text{and} \quad G_{j_1 j_1 0} > 0;\\
&\left\{\begin{matrix}\underbrace{\text{(nF,DP)}}_{\text{down-parabola }(-)}\\ \underbrace{\text{(DP,pF)}}_{\text{down-parabola }(+)}\end{matrix}\right\} \rightleftharpoons \underbrace{\text{(II,SO)}}_{\text{switching bifurcation}} \rightleftharpoons \left\{\begin{matrix}\underbrace{\text{(UP,nF)}}_{\text{up-parabola }(-)}\\ \underbrace{\text{(pF,UP)}}_{\text{up-parabola }(+)}\end{matrix}\right\}\\
&\text{for}\quad (\alpha_{j_1 j_1}, \alpha_{j_1 j_2}) = (+,-) \quad \text{and} \quad G_{j_1 j_1 0} > 0;\\
&\left\{\begin{matrix}\underbrace{\text{(UP,nF)}}_{\text{up-parabola }(-)}\\ \underbrace{\text{(pF,UP)}}_{\text{up-parabola }(+)}\end{matrix}\right\} \rightleftharpoons \underbrace{\text{(II,SI)}}_{\text{switching bifurcation}} \rightleftharpoons \left\{\begin{matrix}\underbrace{\text{(nF,DP)}}_{\text{down-parabola }(-)}\\ \underbrace{\text{(DP,pF)}}_{\text{down-parabola }(+)}\end{matrix}\right\}\\
&\text{for } (\alpha_{j_1 j_1}, \alpha_{j_1 j_2}) = (+,-) \quad \text{and} \quad G_{j_1 j_1 0} < 0.
\end{aligned} \tag{2.64}$$

- *For* $a_{j_1 j_1 0}(a_{j_1 j_1 s_1} - a_{j_1 j_1 s_2})^2 < 0$ *and* $a_{j_2 j_2 0} > 0$,

$$\begin{aligned}
&\left\{\begin{matrix}\underbrace{\text{(pF, UP)}}_{\text{up-parabola }(+)}\\ \underbrace{\text{(DP, pF)}}_{\text{down-parabola }(+)}\end{matrix}\right\} \rightleftharpoons \underbrace{\text{(DI, SI)}}_{\text{switching bifurcation}} \rightleftharpoons \left\{\begin{matrix}\underbrace{\text{(UP, nF)}}_{\text{up-parabola }(-)}\\ \underbrace{\text{(nF, DP)}}_{\text{down-parabola }(-)}\end{matrix}\right\}\\
&\text{for}\quad (\alpha_{j_1 j_1}, \alpha_{j_1 j_2}) = (+,+) \quad \text{and} \quad G_{j_1 j_1 0} < 0;\\
&\left\{\begin{matrix}\underbrace{\text{(UP, nF)}}_{\text{up-parabola }(-)}\\ \underbrace{\text{(nF, DP)}}_{\text{down-parabola }(-)}\end{matrix}\right\} \rightleftharpoons \underbrace{\text{(DI, SO)}}_{\text{switching bifurcation}} \rightleftharpoons \left\{\begin{matrix}\underbrace{\text{(pF, UP)}}_{\text{up-parabola }(+)}\\ \underbrace{\text{(DP, pF)}}_{\text{down-parabola }(+)}\end{matrix}\right\}\\
&\text{for } (\alpha_{j_1 j_1}, \alpha_{j_1 j_2}) = (+,+) \quad \text{and} \quad G_{j_1 j_1 0} > 0;
\end{aligned}$$

$$\left\{\begin{array}{c}\underbrace{(\mathrm{pF},\mathrm{UP})}_{\text{up-parabola }(+)}\\ \underbrace{(\mathrm{UP},\mathrm{pF})}_{\text{up-parabola }(+)}\end{array}\right\} \rightleftharpoons \underbrace{(\mathrm{DI},\mathrm{SI})}_{\text{switching bifurcation}} \rightleftharpoons \left\{\begin{array}{c}\underbrace{(\mathrm{DP},\mathrm{nF})}_{\text{down-parabola }(-)}\\ \underbrace{(\mathrm{nF},\mathrm{DP})}_{\text{down-parabola }(-)}\end{array}\right\}$$

$$\text{for } (\alpha_{j_1j_1},\alpha_{j_1j_2}) = (+,-) \quad \text{and} \quad G_{j_1j_10} < 0. \tag{2.65}$$

- *For* $a_{j_1j_10}(a_{j_1j_1s_1} - a_{j_1j_1s_2})^2 > 0$ *and* $a_{j_2j_20} < 0$,

$$\begin{array}{l}\left\{\begin{array}{c}\underbrace{(\mathrm{nF},\mathrm{UP})}_{\text{up-parabola }(-)}\\ \underbrace{(\mathrm{DP},\mathrm{nF})}_{\text{down-parabola }(-)}\end{array}\right\} \rightleftharpoons \underbrace{(\mathrm{DI},\mathrm{SO})}_{\text{switching bifurction}} \rightleftharpoons \left\{\begin{array}{c}\underbrace{(\mathrm{UP},\mathrm{pF})}_{\text{up-parabola }(+)}\\ \underbrace{(\mathrm{pF},\mathrm{DP})}_{\text{down-parabola }(+)}\end{array}\right\}\\ \text{for } (\alpha_{j_1j_1},\alpha_{j_1j_2}) = (+,+) \quad \text{and} \quad G_{j_1j_10} > 0;\\ \left\{\begin{array}{c}\underbrace{(\mathrm{UP},\mathrm{pF})}_{\text{up-parabola }(+)}\\ \underbrace{(\mathrm{pF},\mathrm{DP})}_{\text{down-parabola }(+)}\end{array}\right\} \rightleftharpoons \underbrace{(\mathrm{DI},\mathrm{SI})}_{\text{switching bifurction}} \rightleftharpoons \left\{\begin{array}{c}\underbrace{(\mathrm{nF},\mathrm{UP})}_{\text{up-parabola }(-)}\\ \underbrace{(\mathrm{DP},\mathrm{nF})}_{\text{down-parabola }(-)}\end{array}\right\}\\ \text{for } (\alpha_{j_1j_1},\alpha_{j_1j_2}) = (+,+) \quad \text{and} \quad G_{j_1j_10} < 0;\\ \left\{\begin{array}{c}\underbrace{(\mathrm{nF},\mathrm{UP})}_{\text{up-parabola }(-)}\\ \underbrace{(\mathrm{UP},\mathrm{nF})}_{\text{up-parabola }(-)}\end{array}\right\} \rightleftharpoons \underbrace{(\mathrm{DI},\mathrm{SO})}_{\text{switching bifurction}} \rightleftharpoons \left\{\begin{array}{c}\underbrace{(\mathrm{DP},\mathrm{pF})}_{\text{down-parabola }(+)}\\ \underbrace{(\mathrm{pF},\mathrm{DP})}_{\text{down-parabola }(+)}\end{array}\right\}\\ \text{for } (\alpha_{j_1j_1},\alpha_{j_1j_2}) = (+,-) \quad \text{and} \quad G_{j_1j_10} > 0.\end{array} \tag{2.66}$$

- *For* $a_{j_1j_10}(a_{j_1j_1s_1} - a_{j_1j_1s_2})^2 < 0$ *and* $a_{j_2j_20} < 0$,

$$\left\{\begin{array}{c}\underbrace{(\mathrm{pF},\mathrm{DP})}_{\text{down-parabola }(+)}\\ \underbrace{(\mathrm{UP},\mathrm{nF})}_{\text{up-parabola }(-)}\end{array}\right\} \rightleftharpoons \underbrace{(\mathrm{II},\mathrm{SI})}_{\text{switching bifurcation}} \rightleftharpoons \left\{\begin{array}{c}\underbrace{(\mathrm{DP},\mathrm{pF})}_{\text{down-parabola }(+)}\\ \underbrace{(\mathrm{nF},\mathrm{UP})}_{\text{up-parabola }(-)}\end{array}\right\}$$

$$\text{for } (\alpha_{j_1j_1},\alpha_{j_1j_2}) = (+,+) \quad \text{and} \quad G_{j_1j_10} < 0;$$

$$\left\{\begin{array}{c}\underbrace{(\mathrm{pF},\mathrm{DP})}_{\text{down-parabola }(+)}\\ \underbrace{(\mathrm{DP},\mathrm{nF})}_{\text{down-parabola }(-)}\end{array}\right\} \rightleftharpoons \underbrace{(\mathrm{II},\mathrm{SI})}_{\text{switching bifurcation}} \rightleftharpoons \left\{\begin{array}{c}\underbrace{(\mathrm{UP},\mathrm{pF})}_{\text{up-parabola }(+)}\\ \underbrace{(\mathrm{nF},\mathrm{UP})}_{\text{up-parabola }(-)}\end{array}\right\}$$

$$\text{for } (\alpha_{j_1j_1},\alpha_{j_1j_2}) = (+,-) \quad \text{and} \quad G_{j_1j_10} < 0;$$

$$\left\{\begin{array}{c}\underbrace{(\mathrm{UP},\mathrm{pF})}_{\text{up-parabola }(+)}\\ \underbrace{(\mathrm{nF},\mathrm{UP})}_{\text{up-parabola }(-)}\end{array}\right\} \rightleftharpoons \underbrace{(\mathrm{II},\mathrm{SO})}_{\text{switching bifurcation}} \rightleftharpoons \left\{\begin{array}{c}\underbrace{(\mathrm{pF},\mathrm{DP})}_{\text{down-parabola }(+)}\\ \underbrace{(\mathrm{DP},\mathrm{nF})}_{\text{down-parabola }(-)}\end{array}\right\}$$

$$\text{for } (\alpha_{j_1j_1},\alpha_{j_1j_2}) = (+,-) \quad \text{and} \quad G_{j_1j_10} > 0. \tag{2.67}$$

(ii_3) For $a_{j_2j_21} = a_{j_1j_1s_2}$ and $a_{j_2j_21} \neq a_{j_1j_1s_1}$, the two-dimensional single-bivariate dynamical system becomes

$$\begin{aligned}\dot{x}_{j_2} &= a_{j_2j_20}(y_{j_1} - a_{j_1j_11}),\\ \dot{x}_{j_1} &= a_{j_1j_10}(y_{j_1} - a_{j_1j_1s_1})(y_{j_1} - a_{j_1j_1s_2})^2.\end{aligned} \tag{2.68}$$

There is a dynamical system in the y_{j_1}-direction as

$$\dot{y}_{j_1} = \alpha_{j_1j_1}a_{j_1j_10}(y_{j_1} - a_{j_1j_1s_2})[(y_{j_1} - a_{j_1j_1s_1})(y_{j_1} - a_{j_1j_1s_2}) + c_{j_1j_1}), \tag{2.69}$$

and a dynamical system in the $y_{j_1}^{\perp}$-direction as

$$\dot{y}_{j_1}^{\perp} = \beta_{j_1j_1}a_{j_1j_10}(y_{j_1} - a_{j_1j_1s_2})[(y_{j_1} - a_{j_1j_1s_1})(y_{j_1} - a_{j_1j_1s_2}) + d_{j_1j_1}). \tag{2.70}$$

Let

$$G_{j_1j_10} = \frac{\partial}{\partial y_{j_1}}[\mathbf{n}_{j_1}^{\mathrm{T}} \cdot \mathbf{F}(y_{j_1},\mathbf{p})]\bigg|_{y_{j_1}^* = a_{j_1j_1s_2}} = \alpha_{j_1j_2}a_{j_2j_20}. \tag{2.71}$$

(ii_{3a}) The infinite-equilibrium of $y_{j_1}^* = a_{j_1j_1s_2}$ for the x_{j_1}-direction and $y_{j_1}^* = a_{j_2j_21}$ for the x_{j_2}-direction possess the following properties:

- *For $a_{j_1j_10}\alpha_{j_1j_2}^2(a_{j_1j_1s_2} - a_{j_1j_1s_1}) > 0$ and $a_{j_2j_20}\alpha_{j_1j_2} > 0$,*

$$(a_{j_1j_1s_2}, a_{j_2j_21}) = \underbrace{(\mathrm{UP},\mathrm{SO})}_{\text{up-parabola source}}. \tag{2.72}$$

- *For $a_{j_1j_10}\alpha_{j_1j_2}^2(a_{j_1j_1s_2} - a_{j_1j_1s_1}) < 0$ and $a_{j_2j_20}\alpha_{j_1j_2} > 0$,*

$$(a_{j_1j_1s_2}, a_{j_2j_21}) = \underbrace{(\mathrm{DP},\mathrm{SO})}_{\text{down-parabola source}}. \tag{2.73}$$

- *For* $a_{j_1j_10}\alpha^2_{j_1j_2}(a_{j_1j_1s_2} - a_{j_1j_1s_1}) > 0$ *and* $a_{j_2j_20}\alpha_{j_1j_2} < 0$,

$$(a_{j_1j_1s_2}, a_{j_2j_21}) = \underbrace{(\mathrm{DP}, \mathrm{SI})}_{\text{down-parabola sink}}. \tag{2.74}$$

- *For* $a_{j_1j_10}\alpha^2_{j_1j_2}(a_{j_1j_1s_2} - a_{j_1j_1s_1}) < 0$ *and* $a_{j_2j_20}\alpha_{j_1j_2} < 0$,

$$(a_{j_1j_1s_2}, a_{j_2j_21}) = \underbrace{(\mathrm{UP}, \mathrm{SI})}_{\text{up-parabola sink}}. \tag{2.75}$$

(ii_{3b}) *For* $a_{j_2j_21} = a_{j_1j_11}$, *the infinite-equilibrium of* $y^*_{j_1} = a_{j_1j_11}$ *is a switching bifurcation*:

- *For* $a_{j_1j_10}\alpha^2_{j_1j_2}(a_{j_1j_1s_2} - a_{j_1j_1s_1}) > 0$ *and* $a_{j_2j_20}\alpha_{j_1j_2} > 0$,

$$\left\{\begin{array}{c}\underbrace{(\mathrm{pF}, \mathrm{UP})}_{\text{up-parabola flow}}\\ \underbrace{(\mathrm{II}, \mathrm{pF})}_{\text{increasing-inflection flow}}\end{array}\right\} \rightleftharpoons \underbrace{(\mathrm{UP}, \mathrm{SO})}_{\text{switching bifurcation}} \rightleftharpoons \left\{\begin{array}{c}\underbrace{(\mathrm{DI}, \mathrm{nF})}_{\text{decreasing-inflection flow}}\\ \underbrace{(\mathrm{pF}, \mathrm{UP})}_{\text{up-parabola flow}}\end{array}\right\}. \tag{2.76}$$

- *For* $a_{j_1j_10}\alpha^2_{j_1j_2}(a_{j_1j_1s_2} - a_{j_1j_1s_1}) < 0$ *and* $a_{j_2j_20}\alpha_{j_1j_2} > 0$,

$$\left\{\begin{array}{c}\underbrace{(\mathrm{nF}, \mathrm{DP})}_{\text{down-parabola flow}}\\ \underbrace{(\mathrm{DI}, \mathrm{pF})}_{\text{decreasing-inflection flow}}\end{array}\right\} \underbrace{(\mathrm{DP}, \mathrm{SO})}_{\text{switching bifurcation}} \rightleftharpoons \left\{\begin{array}{c}\underbrace{(\mathrm{II}, \mathrm{nF})}_{\text{increasing-inflection flow}}\\ \underbrace{(\mathrm{nF}, \mathrm{DP})}_{\text{down-parabola flow}}\end{array}\right\}. \tag{2.77}$$

- *For* $a_{j_1j_10}\alpha^2_{j_1j_2}(a_{j_1j_1s_2} - a_{j_1j_1s_1}) > 0$ *and* $a_{j_2j_20}\alpha_{j_1j_2} < 0$,

$$\left\{\begin{array}{c}\underbrace{(\mathrm{pF}, \mathrm{DP})}_{\text{down-parabola flow}}\\ \underbrace{(\mathrm{DI}, \mathrm{nF})}_{\text{decreasing-inflection flow}}\end{array}\right\} \underbrace{(\mathrm{DP}, \mathrm{SI})}_{\text{switching bifurcation}} \rightleftharpoons \left\{\begin{array}{c}\underbrace{(\mathrm{II}, \mathrm{pF})}_{\text{increasing-inflection flow}}\\ \underbrace{(\mathrm{pF}, \mathrm{DP})}_{\text{down-parabola flow}}\end{array}\right\}. \tag{2.78}$$

- *For* $a_{j_1j_10}\alpha^2_{j_1j_2}(a_{j_1j_1s_2} - a_{j_1j_1s_1}) < 0$ *and* $a_{j_2j_20}\alpha_{j_1j_2} < 0$,

$$\left\{\begin{array}{c}\underbrace{(\mathrm{nF}, \mathrm{UP})}_{\text{up-parabola flow}}\\ \underbrace{(\mathrm{II}, \mathrm{nF})}_{\text{increasing-inflection flow}}\end{array}\right\} \rightleftharpoons \underbrace{(\mathrm{UP}, \mathrm{SI})}_{\text{switching bifurcation}} \rightleftharpoons \left\{\begin{array}{c}\underbrace{(\mathrm{DI}, \mathrm{pF})}_{\text{decreasing-inflection flow}}\\ \underbrace{(\mathrm{nF}, \mathrm{UP})}_{\text{up-parabola flow}}\end{array}\right\}. \tag{2.79}$$

(iii) For $a_{j_1j_1s_1} = a_{j_1j_1s_2} = a_{j_1j_11}$, the two-dimensional single-bivariate dynamical system becomes

$$\begin{aligned}\dot{x}_{j_2} &= a_{j_2j_20}(y_{j_1} - a_{j_2j_21}),\\ \dot{x}_{j_1} &= a_{j_1j_10}(y_{j_1} - a_{j_1j_11})^3.\end{aligned} \tag{2.80}$$

(iii$_1$) For $a_{j_2j_21} \neq a_{j_1j_11}$, there is a dynamical system in the y_{j_1}-direction as

$$\dot{y}_{j_1}^{\perp} = \alpha_{j_1j_1}a_{j_1j_10}(y_{j_1} - a_{j_1j_11})^3 + \alpha_{j_1j_2}a_{j_2j_20}(y_{j_1} - a_{j_2j_21}) \tag{2.81}$$

with a dynamical system in the $y_{j_1}^{\perp}$-direction as

$$\dot{y}_{j_1}^{\perp} = \beta_{j_1j_1}a_{j_1j_10}(y_{j_1} - a_{j_1j_11})^3 + \beta_{j_1j_2}a_{j_2j_20}(y_{j_1} - a_{j_2j_21}). \tag{2.82}$$

(iii$_{1a}$) The equilibrium on the line of $y_{j_1}^* = a_{j_1j_1s_1}$ for the x_{j_1}-direction with $\dot{x}_{j_2}^* = a_{j_2j_20}(a_{j_1j_11} - a_{j_2j_21})$ possesses the following properties:

- *For $a_{j_1j_10}\alpha_{j_1j_2}^3 > 0$ and $a_{j_2j_20}(a_{j_1j_1s_1} - a_{j_2j_21}) > 0$,*

$$(a_{j_1j_1s_1}, \dot{x}_{j_2}^*) = \underbrace{(3^{\text{rd}}\text{UP}, \text{pF})}_{\text{third-order up-parabola flow } (+)}. \tag{2.83}$$

- *For $a_{j_1j_10}\alpha_{j_1j_2}^3 < 0$ and $a_{j_2j_20}(a_{j_1j_1s_1} - a_{j_2j_21}) > 0$,*

$$(a_{j_1j_1s_1}, \dot{x}_{j_2}^*) = \underbrace{(3^{\text{rd}}\text{DP}, \text{pF})}_{\text{third-order down-parabola flow } (+)}. \tag{2.84}$$

- *For $a_{j_1j_10}\alpha_{j_1j_2}^3 > 0$ and $a_{j_2j_20}(a_{j_1j_1s_1} - a_{j_2j_21}) < 0$,*

$$(a_{j_1j_1s_1}, \dot{x}_{j_2}^*) = \underbrace{(3^{\text{rd}}\text{DP}, \text{nF})}_{\text{third-order down-parabola flow } (-)}. \tag{2.85}$$

- *For $a_{j_1j_10}\alpha_{j_1j_2}^3 < 0$ and $a_{j_2j_20}(a_{j_1j_1s_1} - a_{j_2j_21}) < 0$,*

$$(a_{j_1j_1s_1}, \dot{x}_{j_2}^*) = \underbrace{(3^{\text{rd}}\text{UP}, \text{nF})}_{\text{third-order up-parabola flow } (-)}. \tag{2.86}$$

The third-order up-parabola flows are the appearing bifurcations from an up-parabola flow to up-parabola, down-parabola, and up-parabola flows.

The third-order down-parabola flows are the appearing bifurcations from a down-parabola flow to down-parabola, up-parabola, and down-parabola flows.

(iii_{1b}) The simple equilibrium of $y^*_{j_1} = a_{j_2j_21}$ in the x_{j_2}-direction for such a system with $\dot{x}^*_{j_1} = a_{j_1j_10}(a_{j_2j_21} - a_{j_1j_11})^3$ has the following properties:

- *For $\alpha_{j_1j_1}a_{j_1j_10}(a_{j_2j_21} - a_{j_1j_11})^3 > 0$ and $\alpha_{j_1j_1}a_{j_2j_20} > 0$,*

$$(\dot{x}^*_{j_1}, a_{j_2j_21}) = \underbrace{(\text{pF}, \text{UP})}_{\text{up-parabola flow } (+)}. \tag{2.87}$$

- *For $\alpha_{j_1j_1}a_{j_1j_10}(a_{j_2j_21} - a_{j_1j_11})^3 < 0$ and $\alpha_{j_1j_1}a_{j_2j_20} > 0$,*

$$(\dot{x}^*_{j_1}, a_{j_2j_21}) = \underbrace{(\text{nF}, \text{DP})}_{\text{down-parabola flow } (-)}. \tag{2.88}$$

- *For $\alpha_{j_1j_1}a_{j_1j_10}(a_{j_2j_21} - a_{j_1j_11})^3 > 0$ and $\alpha_{j_1j_1}a_{j_2j_20} < 0$,*

$$(\dot{x}^*_{j_1}, a_{j_2j_21}) = \underbrace{(\text{pF}, \text{DP})}_{\text{down-parabola flow } (+)}. \tag{2.89}$$

- *For $\alpha_{j_1j_1}a_{j_1j_10}(a_{j_2j_21} - a_{j_1j_11})^3 < 0$ and $\alpha_{j_1j_1}a_{j_2j_20} < 0$,*

$$(\dot{x}^*_{j_1}, a_{j_2j_21}) = \underbrace{(\text{nF}, \text{UP})}_{\text{up-parabola flow } (-)}. \tag{2.90}$$

(iii_2) For $a_{j_2j_21} = a_{j_1j_11}$, there is a dynamical system in the y_{j_1}-direction as

$$\dot{y}_{j_1} = \alpha_{j_1j_1}a_{j_1j_10}(y_{j_1} - a_{j_1j_11})[(y_{j_1} - a_{j_1j_11})^2 + c_{j_1j_1}] \tag{2.91}$$

with a dynamical system in the $y^{\perp}_{j_1}$-direction as

$$\dot{y}^{\perp}_{j_1} = \beta_{j_1j_1}a_{j_1j_10}(y_{j_1} - a_{j_1j_11})[(y_{j_1} - a_{j_1j_11})^2 + d_{j_1j_1}]. \tag{2.92}$$

Let

$$G_{j_1j_10} = \frac{\partial}{\partial y_{j_1}}[\mathbf{n}^{\mathrm{T}}_{j_1} \cdot \mathbf{F}(y_{j_1}, \mathbf{p})]\Big|_{y_{j_1} = a_{j_1j_11}} = \alpha_{j_1j_2}a_{j_2j_20}. \tag{2.93}$$

(iii_{2a}) The infinite-equilibrium of $y^*_{j_1} = a_{j_1j_11}$ for the x_{j_1}-direction and $y^*_{j_1} = a_{j_2j_21}$ for the x_{j_2}-direction possess the following properties:

- *For $\alpha^3_{j_1j_2}a_{j_1j_10} > 0$ and $\alpha_{j_1j_2}a_{j_2j_20} > 0$,*

$$(a_{j_1j_11}, a_{j_2j_21}) = \underbrace{(2^{\text{nd}}\text{II}, \text{SO})}_{\text{second-order increasing-inflection source}}. \tag{2.94}$$

- *For $\alpha^3_{j_1j_2}a_{j_1j_10} < 0$ and $\alpha_{j_1j_2}a_{j_2j_20} > 0$,*

$$(a_{j_1j_11}, a_{j_2j_21}) = \underbrace{(2^{\text{nd}}\text{DI}, \text{SO})}_{\text{second-order decreasing-inflection source}}. \tag{2.95}$$

- *For* $\alpha^3_{j_1j_2}a_{j_1j_10} > 0$ *and* $\alpha_{j_1j_2}a_{j_2j_20} < 0$,

$$(a_{j_1j_11}, a_{j_2j_21}) = \underbrace{(2^{\text{nd}}\text{DI}, \text{SI})}_{\text{second-order decreasing-inflection sink}}. \tag{2.96}$$

- *For* $\alpha^3_{j_1j_2}a_{j_1j_10} < 0$ *and* $\alpha_{j_1j_2}a_{j_2j_20} < 0$,

$$(a_{j_1j_11}, a_{j_2j_21}) = \underbrace{(2^{\text{nd}}\text{II}, \text{SI})}_{\text{second-order increasing-inflection sink}}. \tag{2.97}$$

(iii_{2b}) *For* $a_{j_2j_21} = a_{j_1j_11}$, *the infinite-equilibrium of* $y^*_{j_1} = a_{j_1j_11}$ *is a switching bifurcation*:

- *For* $\alpha^3_{j_1j_2}a_{j_1j_10} > 0$ *and* $\alpha_{j_1j_2}a_{j_2j_20} > 0$,

$$\left\{\begin{matrix} \underbrace{(\text{nF}, \text{DP})}_{\text{down-parabola flow }(-)} \\ \underbrace{(3^{\text{rd}}\text{UP}, \text{pF})}_{\text{third-order up-parabola flow }(+)} \end{matrix}\right\} \rightleftharpoons \underbrace{(2^{\text{nd}}\text{II}, \text{SO})}_{\text{switching bifurcation}}$$

$$\rightleftharpoons \left\{\begin{matrix} \underbrace{(3^{\text{rd}}\text{DP}, \text{nF})}_{\text{third-order down-parabola flow }(-)} \\ \underbrace{(\text{pF}, \text{UP})}_{\text{up-parabola flow }(+)} \end{matrix}\right\}$$

for $\alpha_{j_1j_2} > 0$; (2.98)

$$\left\{\begin{matrix} \underbrace{(\text{nF}, \text{DP})}_{\text{down-parabola flow }(-)} \\ \underbrace{(3^{\text{rd}}\text{DP}, \text{pF})}_{\text{third-order down-parabola flow }(+)} \end{matrix}\right\} \rightleftharpoons \underbrace{(2^{\text{nd}}\text{II}, \text{SO})}_{\text{switching bifurcation}}$$

$$\rightleftharpoons \left\{\begin{matrix} \underbrace{(3^{\text{rd}}\text{UP}, \text{nF})}_{\text{third-order up-parabola flow }(-)} \\ \underbrace{(\text{pF}, \text{UP})}_{\text{up-parabola flow }(+)} \end{matrix}\right\}$$

for $\alpha_{j_1j_2} < 0$. (2.99)

- *For* $\alpha_{j_1j_2}^3 a_{j_1j_10} < 0$ *and* $\alpha_{j_1j_2} a_{j_2j_20} > 0$,

$$\left\{\begin{array}{c}\underbrace{(\mathrm{pF},\mathrm{UP})}_{\text{up-parabola flow }(+)}\\ \underbrace{(3^{\mathrm{rd}}\mathrm{DP},\mathrm{pF})}_{\text{third-order down-parabola flow }(+)}\end{array}\right\} \rightleftharpoons \underbrace{(2^{\mathrm{nd}}\mathrm{DI},\mathrm{SO})}_{\text{switching bifurcation}}$$

$$\rightleftharpoons \left\{\begin{array}{c}\underbrace{(3^{\mathrm{rd}}\mathrm{UP},\mathrm{nF})}_{\text{third-order up-parabola flow }(-)}\\ \underbrace{(\mathrm{nF},\mathrm{DP})}_{\text{down-parabola flow }(-)}\end{array}\right\}$$

for $\alpha_{j_1j_2} > 0$; (2.100)

$$\left\{\begin{array}{c}\underbrace{(\mathrm{pF},\mathrm{UP})}_{\text{up-parabola flow }(+)}\\ \underbrace{(3^{\mathrm{rd}}\mathrm{UP},\mathrm{pF})}_{\text{third-order up-parabola flow }(+)}\end{array}\right\} \rightleftharpoons \underbrace{(2^{\mathrm{nd}}\mathrm{DI},\mathrm{SO})}_{\text{switching bifurcation}}$$

$$\rightleftharpoons \left\{\begin{array}{c}\underbrace{(3^{\mathrm{rd}}\mathrm{DP},\mathrm{nF})}_{\text{third-order down-parabola flow }(-)}\\ \underbrace{(\mathrm{nF},\mathrm{DP})}_{\text{down-parabola flow }(-)}\end{array}\right\} \quad (2.101)$$

- *For* $\alpha_{j_1j_2}^3 a_{j_1j_10} > 0$ *and* $\alpha_{j_1j_2} a_{j_2j_20} < 0$,

$$\left\{\begin{array}{c}\underbrace{(\mathrm{nF},\mathrm{UP})}_{\text{up-parabola flow }(-)}\\ \underbrace{(3^{\mathrm{rd}}\mathrm{DP},\mathrm{nF})}_{\text{third-order down-parabola flow }(-)}\end{array}\right\} \rightleftharpoons \underbrace{(2^{\mathrm{nd}}\mathrm{DI},\mathrm{SI})}_{\text{switching bifurcation}}$$

$$\rightleftharpoons \left\{\begin{array}{c}\underbrace{(3^{\mathrm{rd}}\mathrm{UP},\mathrm{pF})}_{\text{third-order up-parabola flow }(+)}\\ \underbrace{(\mathrm{pF},\mathrm{DP})}_{\text{down-parabola flow }(+)}\end{array}\right\}$$

for $\alpha_{j_1j_2} > 0$; (2.102)

$$\left\{\begin{array}{c}\underbrace{(\mathrm{nF},\mathrm{UP})}_{\text{up-parabola flow }(-)}\\ \underbrace{(3^{\mathrm{rd}}\mathrm{UP},\mathrm{nF})}_{\text{third-order up-parabola flow }(-)}\end{array}\right\} \rightleftharpoons \underbrace{(2^{\mathrm{nd}}\mathrm{DI},\mathrm{SI})}_{\text{switching bifurcation}}$$

$$\rightleftharpoons \left\{\begin{array}{c}\underbrace{(3^{\mathrm{rd}}\mathrm{DP},\mathrm{pF})}_{\text{third-order down-parabola flow }(+)}\\ \underbrace{(\mathrm{pF},\mathrm{DP})}_{\text{down-parabola flow }(+)}\end{array}\right\}$$

for $\alpha_{j_1j_2} < 0$. (2.103)

- *For* $\alpha^3_{j_1j_2}a_{j_1j_10} < 0$ *and* $\alpha_{j_1j_2}a_{j_2j_20} < 0$,

$$\left\{\begin{array}{c}\underbrace{(\mathrm{pF},\mathrm{DP})}_{\text{down-parabola flow }(+)}\\ \underbrace{(3^{\mathrm{rd}}\mathrm{UP},\mathrm{nF})}_{\text{third-order up-parabola flow }(-)}\end{array}\right\} \rightleftharpoons \underbrace{(2^{\mathrm{nd}}\mathrm{II},\mathrm{SI})}_{\text{switching bifurcation}}$$

$$\rightleftharpoons \left\{\begin{array}{c}\underbrace{(3^{\mathrm{rd}}\mathrm{DP},\mathrm{pF})}_{\text{third-order down-parabola flow }(+)}\\ \underbrace{(\mathrm{nF},\mathrm{UP})}_{\text{up-parabola flow }(-)}\end{array}\right\}$$

for $\alpha_{j_1j_2} > 0$; (2.104)

$$\left\{\begin{array}{c}\underbrace{(\mathrm{pF},\mathrm{DP})}_{\text{down-parabola flow }(+)}\\ \underbrace{(3^{\mathrm{rd}}\mathrm{DP},\mathrm{nF})}_{\text{third-order down-parabola flow }(-)}\end{array}\right\} \rightleftharpoons \underbrace{(2^{\mathrm{nd}}\mathrm{II},\mathrm{SI})}_{\text{switching bifurcation}}$$

$$\rightleftharpoons \left\{\begin{array}{c}\underbrace{(3^{\mathrm{rd}}\mathrm{UP},\mathrm{pF})}_{\text{third-order up-parabola flow }(+)}\\ \underbrace{(\mathrm{nF},\mathrm{UP})}_{\text{up-parabola flow }(-)}\end{array}\right\}$$

for $\alpha_{j_1j_2} < 0$. (2.105)

(iii_{2c}) For $a_{j_2j_21} = a_{j_1j_11}$, the infinite-equilibrium of $y^*_{j_1} = a_{j_1j_11}$ is an appearing bifurcation as follows:

- *For $\alpha^3_{j_1j_2}a_{j_1j_10} > 0$ and $\alpha_{j_1j_2}a_{j_2j_20} > 0$,*

$$\left\{\begin{array}{c}\underbrace{(\mathrm{nF},\mathrm{DP})}_{\text{down-parabola flow }(-)}\\ \underbrace{(\mathrm{UP},\mathrm{pF})}_{\text{up-parabola flow }(+)}\end{array}\right\} \rightleftharpoons \underbrace{(2^{\mathrm{nd}}\mathrm{II},\mathrm{SO})}_{\text{switching bifurcation}} \rightleftharpoons \left\{\begin{array}{c}\underbrace{(\mathrm{DP},\mathrm{nF})}_{\text{down-parabola flow }(-)}\\ \underbrace{(\mathrm{UP},\mathrm{nF})}_{\text{up-parabola flow }(-)}\\ \underbrace{(\mathrm{nF},\mathrm{DP})}_{\text{down-parabola flow }(-)}\\ \underbrace{(\mathrm{UP},\mathrm{pF})}_{\text{up-parabola flow }(+)}\end{array}\right\};$$

$$\left\{\begin{array}{c}\underbrace{(\mathrm{DP},\mathrm{nF})}_{\text{down-parabola flow }(-)}\\ \underbrace{(\mathrm{pF},\mathrm{UP})}_{\text{up-parabola flow }(+)}\end{array}\right\} \rightleftharpoons \underbrace{(2^{\mathrm{nd}}\mathrm{II},\mathrm{SO})}_{\text{switching bifurcation}} \rightleftharpoons \left\{\begin{array}{c}\underbrace{(\mathrm{DP},\mathrm{nF})}_{\text{down-parabola flow }(-)}\\ \underbrace{(\mathrm{pF},\mathrm{UP})}_{\text{up-parabola flow }(+)}\\ \underbrace{(\mathrm{DP},\mathrm{pF})}_{\text{down-parabola flow }(+)}\\ \underbrace{(\mathrm{UP},\mathrm{pF})}_{\text{up-parabola flow }(+)}\end{array}\right\}$$

for $\alpha_{j_1j_2} > 0$. (2.106)

$$\left\{\begin{array}{c}\underbrace{(\mathrm{nF},\mathrm{DP})}_{\text{down-parabola flow }(-)}\\ \underbrace{(\mathrm{DP},\mathrm{pF})}_{\text{down-parabola flow }(+)}\end{array}\right\} \rightleftharpoons \underbrace{(2^{\mathrm{nd}}\mathrm{II},\mathrm{SO})}_{\text{switching bifurcation}} \rightleftharpoons \left\{\begin{array}{c}\underbrace{(\mathrm{DP},\mathrm{nF})}_{\text{down-parabola flow }(-)}\\ \underbrace{(\mathrm{UP},\mathrm{nF})}_{\text{up-parabola flow }(-)}\\ \underbrace{(\mathrm{nF},\mathrm{DP})}_{\text{down-parabola flow }(-)}\\ \underbrace{(\mathrm{DP},\mathrm{pF})}_{\text{down-parabola flow }(+)}\end{array}\right\};$$

$$
\left\{\begin{array}{c}\underbrace{(\mathrm{UP},\mathrm{nF})}_{\text{up-parabola flow }(-)}\\ \underbrace{(\mathrm{pF},\mathrm{UP})}_{\text{up-parabola flow }(+)}\end{array}\right\} \rightleftharpoons \underbrace{(2^{\mathrm{nd}}\mathrm{II},\mathrm{SO})}_{\text{switching bifurcation}} \rightleftharpoons \left\{\begin{array}{c}\underbrace{(\mathrm{UP},\mathrm{nF})}_{\text{up-parabola flow }(-)}\\ \underbrace{(\mathrm{pF},\mathrm{UP})}_{\text{up-parabola flow }(+)}\\ \underbrace{(\mathrm{UP},\mathrm{pF})}_{\text{up-parabola flow }(+)}\\ \underbrace{(\mathrm{DP},\mathrm{pF})}_{\text{down-parabola flow }(+)}\end{array}\right\}
$$

for $\alpha_{j_1 j_2} < 0.$ (2.107)

- *For* $\alpha^3_{j_1 j_2} a_{j_1 j_1 0} < 0$ *and* $\alpha_{j_1 j_2} a_{j_2 j_2 0} > 0$,

$$
\left\{\begin{array}{c}\underbrace{(\mathrm{pF},\mathrm{UP})}_{\text{up-parabola flow }(+)}\\ \underbrace{(\mathrm{DP},\mathrm{pF})}_{\text{down-parabola flow }(+)}\end{array}\right\} \rightleftharpoons \underbrace{(2^{\mathrm{nd}}\mathrm{DI},\mathrm{SO})}_{\text{switching bifurcation}} \rightleftharpoons \left\{\begin{array}{c}\underbrace{(\mathrm{UP},\mathrm{nF})}_{\text{up-parabola flow }(-)}\\ \underbrace{(\mathrm{DP},\mathrm{nF})}_{\text{down-parabola flow }(-)}\\ \underbrace{(\mathrm{pF},\mathrm{UP})}_{\text{up-parabola flow }(+)}\\ \underbrace{(\mathrm{DP},\mathrm{pF})}_{\text{down-parabola flow }(+)}\end{array}\right\};
$$

$$
\left\{\begin{array}{c}\underbrace{(\mathrm{UP},\mathrm{nF})}_{\text{up-parabola flow }(-)}\\ \underbrace{(\mathrm{nF},\mathrm{DP})}_{\text{down-parabola flow }(-)}\end{array}\right\} \rightleftharpoons \underbrace{(2^{\mathrm{nd}}\mathrm{DI},\mathrm{SO})}_{\text{switching bifurcation}} \rightleftharpoons \left\{\begin{array}{c}\underbrace{(\mathrm{UP},\mathrm{nF})}_{\text{up-parabola flow }(-)}\\ \underbrace{(\mathrm{nF},\mathrm{DP})}_{\text{down-parabola flow }(-)}\\ \underbrace{(\mathrm{UP},\mathrm{pF})}_{\text{up-parabola flow }(+)}\\ \underbrace{(\mathrm{DP},\mathrm{pF})}_{\text{down-parabola flow }(+)}\end{array}\right\}
$$

for $\alpha_{j_1 j_2} > 0.$ (2.108)

$$\left\{\begin{array}{c}\underbrace{(\mathrm{pF},\mathrm{UP})}_{\text{up-parabola flow }(+)}\\\underbrace{(\mathrm{UP},\mathrm{pF})}_{\text{up-parabola flow }(+)}\end{array}\right\}\rightleftharpoons\underbrace{(2^{\mathrm{nd}}\mathrm{DI},\mathrm{SO})}_{\text{switching bifurcation}}\rightleftharpoons\left\{\begin{array}{c}\underbrace{(\mathrm{DP},\mathrm{nF})}_{\text{down-parabola flow }(-)}\\\underbrace{(\mathrm{UP},\mathrm{nF})}_{\text{up-parabola flow }(-)}\\\underbrace{(\mathrm{pF},\mathrm{UP})}_{\text{up-parabola flow }(+)}\\\underbrace{(\mathrm{UP},\mathrm{pF})}_{\text{up-parabola flow }(+)}\end{array}\right\};$$

$$\left\{\begin{array}{c}\underbrace{(\mathrm{DP},\mathrm{nF})}_{\text{down-parabola flow }(-)}\\\underbrace{(\mathrm{nF},\mathrm{DP})}_{\text{down-parabola flow }(-)}\end{array}\right\}\rightleftharpoons\underbrace{(2^{\mathrm{nd}}\mathrm{DI},\mathrm{SO})}_{\text{switching bifurcation}}\rightleftharpoons\left\{\begin{array}{c}\underbrace{(\mathrm{DP},\mathrm{nF})}_{\text{down-parabola flow }(-)}\\\underbrace{(\mathrm{nF},\mathrm{DP})}_{\text{down-parabola flow }(-)}\\\underbrace{(\mathrm{DP},\mathrm{pF})}_{\text{up-parabola flow }(+)}\\\underbrace{(\mathrm{UP},\mathrm{pF})}_{\text{up-parabola flow }(+)}\end{array}\right\}$$

for $\alpha_{j_1j_2} < 0$. (2.109)

- *For* $\alpha^3_{j_1j_2}a_{j_1j_10} > 0$ *and* $\alpha_{j_1j_2}a_{j_2j_20} < 0$,

$$\left\{\begin{array}{c}\underbrace{(\mathrm{nF},\mathrm{UP})}_{\text{up-parabola flow }(-)}\\\underbrace{(3^{\mathrm{rd}}\mathrm{DP},\mathrm{nF})}_{\text{down-parabola flow }(-)}\end{array}\right\}\rightleftharpoons\underbrace{(2^{\mathrm{nd}}\mathrm{DI},\mathrm{SI})}_{\text{switching bifurcation}}\rightleftharpoons\left\{\begin{array}{c}\underbrace{(\mathrm{UP},\mathrm{pF})}_{\text{up-parabola flow }(+)}\\\underbrace{(\mathrm{DP},\mathrm{pF})}_{\text{down-parabola flow }(+)}\\\underbrace{(\mathrm{nF},\mathrm{UP})}_{\text{up-parabola flow }(-)}\\\underbrace{(\mathrm{DP},\mathrm{nF})}_{\text{down-parabola flow }(-)}\end{array}\right\};$$

$$\left\{\begin{array}{c}\underbrace{(\mathrm{UP},\mathrm{pF})}_{\text{up-parabola flow }(+)}\\\underbrace{(\mathrm{pF},\mathrm{DP})}_{\text{down-parabola flow }(+)}\end{array}\right\}\rightleftharpoons\underbrace{(2^{\mathrm{nd}}\mathrm{DI},\mathrm{SI})}_{\text{switching bifurcation}}\rightleftharpoons\left\{\begin{array}{c}\underbrace{(\mathrm{UP},\mathrm{pF})}_{\text{up-parabola flow }(+)}\\\underbrace{(\mathrm{pF},\mathrm{DP})}_{\text{down-parabola flow }(+)}\\\underbrace{(\mathrm{UP},\mathrm{nF})}_{\text{up-parabola flow }(-)}\\\underbrace{(\mathrm{DP},\mathrm{nF})}_{\text{down-parabola flow }(-)}\end{array}\right\}$$

for $\alpha_{j_1j_2}>0$. (2.110)

$$\left\{\begin{array}{c}\underbrace{(\mathrm{nF},\mathrm{UP})}_{\text{up-parabola flow }(-)}\\\underbrace{(\mathrm{UP},\mathrm{nF})}_{\text{up-parabola flow }(-)}\end{array}\right\}\rightleftharpoons\underbrace{(2^{\mathrm{nd}}\mathrm{DI},\mathrm{SI})}_{\text{switching bifurcation}}\rightleftharpoons\left\{\begin{array}{c}\underbrace{(\mathrm{DP},\mathrm{pF})}_{\text{down-parabola flow }(+)}\\\underbrace{(\mathrm{UP},\mathrm{pF})}_{\text{up-parabola flow }(+)}\\\underbrace{(\mathrm{nF},\mathrm{UP})}_{\text{up-parabola flow }(-)}\\\underbrace{(\mathrm{UP},\mathrm{nF})}_{\text{up-parabola flow }(-)}\end{array}\right\};$$

$$\left\{\begin{array}{c}\underbrace{(\mathrm{DP},\mathrm{pF})}_{\text{down-parabola flow }(+)}\\\underbrace{(\mathrm{pF},\mathrm{DP})}_{\text{down-parabola flow }(+)}\end{array}\right\}\rightleftharpoons\underbrace{(2^{\mathrm{nd}}\mathrm{DI},\mathrm{SI})}_{\text{switching bifurcation}}\rightleftharpoons\left\{\begin{array}{c}\underbrace{(\mathrm{DP},\mathrm{pF})}_{\text{down-parabola flow }(+)}\\\underbrace{(\mathrm{pF},\mathrm{DP})}_{\text{down-parabola flow }(+)}\\\underbrace{(\mathrm{DP},\mathrm{nF})}_{\text{up-parabola flow }(-)}\\\underbrace{(\mathrm{UP},\mathrm{nF})}_{\text{up-parabola flow }(-)}\end{array}\right\}$$

for $\alpha_{j_1j_2}<0$. (2.111)

- *For* $\alpha^3_{j_1j_2}a_{j_1j_10} < 0$ *and* $\alpha_{j_1j_2}a_{j_2j_20} < 0$,

$$\left\{\begin{matrix}\underbrace{(\text{pF},\text{DP})}_{\text{down-parabola flow }(+)}\\ \underbrace{(\text{UP},\text{nF})}_{\text{up-parabola flow }(-)}\end{matrix}\right\} \rightleftharpoons \underbrace{(2^{\text{nd}}\text{II},\text{SI})}_{\text{switching bifurcation}} \rightleftharpoons \left\{\begin{matrix}\underbrace{(\text{DP},\text{pF})}_{\text{down-parabola flow }(+)}\\ \underbrace{(\text{UP},\text{pF})}_{\text{up-parabola flow }(+)}\\ \underbrace{(\text{pF},\text{DP})}_{\text{down-parabola flow }(+)}\\ \underbrace{(\text{UP},\text{nF})}_{\text{up-parabola flow }(-)}\end{matrix}\right\}$$

$$\left\{\begin{matrix}\underbrace{(\text{DP},\text{pF})}_{\text{down-parabola flow }(+)}\\ \underbrace{(\text{nF},\text{UP})}_{\text{up-parabola flow }(-)}\end{matrix}\right\} \rightleftharpoons \underbrace{(2^{\text{nd}}\text{II},\text{SI})}_{\text{switching bifurcation}} \rightleftharpoons \left\{\begin{matrix}\underbrace{(\text{DP},\text{pF})}_{\text{down-parabola flow }(+)}\\ \underbrace{(\text{nF},\text{UP})}_{\text{up-parabola flow }(-)}\\ \underbrace{(\text{DP},\text{nF})}_{\text{down-parabola flow }(-)}\\ \underbrace{(\text{UP},\text{nF})}_{\text{up-parabola flow }(-)}\end{matrix}\right\}$$

for $\alpha_{j_1j_2} > 0$. (2.112)

$$\left\{\begin{matrix}\underbrace{(\text{pF},\text{DP})}_{\text{down-parabola flow }(+)}\\ \underbrace{(\text{DP},\text{nF})}_{\text{down-parabola flow }(-)}\end{matrix}\right\} \rightleftharpoons \underbrace{(2^{\text{nd}}\text{II},\text{SI})}_{\text{switching bifurcation}} \rightleftharpoons \left\{\begin{matrix}\underbrace{(\text{DP},\text{pF})}_{\text{down-parabola flow }(+)}\\ \underbrace{(\text{UP},\text{pF})}_{\text{up-parabola flow }(+)}\\ \underbrace{(\text{pF},\text{DP})}_{\text{down-parabola flow }(+)}\\ \underbrace{(\text{DP},\text{nF})}_{\text{down-parabola flow }(-)}\end{matrix}\right\};$$

$$\left\{\begin{array}{c}\underbrace{(\mathrm{UP},\mathrm{pF})}_{\text{up-parabola flow }(+)}\\ \underbrace{(\mathrm{nF},\mathrm{UP})}_{\text{up-parabola flow }(-)}\end{array}\right\} \rightleftharpoons \underbrace{(2^{\mathrm{nd}}\mathrm{II},\mathrm{SI})}_{\text{switching bifurcation}} \rightleftharpoons \left\{\begin{array}{c}\underbrace{(\mathrm{UP},\mathrm{pF})}_{\text{up-parabola flow }(+)}\\ \underbrace{(\mathrm{nF},\mathrm{UP})}_{\text{up-parabola flow }(-)}\\ \underbrace{(\mathrm{UP},\mathrm{nF})}_{\text{up-parabola flow }(-)}\\ \underbrace{(\mathrm{DP},\mathrm{nF})}_{\text{down-parabola flow }(-)}\end{array}\right\}$$

for $\alpha_{j_1j_2} < 0$. (2.113)

(iv) If

$$A_{j_2j_21} \neq 0, A_{j_2j_22} \neq 0 \quad \text{and} \quad \Delta_{j_1j_1} > 0, \tag{2.114}$$

the two-dimensional dynamical system has two simple equilibriums for the y_{j_1}-direction on the x_i-direction, i.e.,

$$\begin{aligned} y_{j_1}^* &= a_{j_1j_11}, a_{j_1j_12}, a_{j_1j_13} \text{ for the } x_{j_1}\text{-direction},\\ y_{j_1}^* &= a_{j_2j_21} \text{ for the } x_{j_2}\text{-direction}, \end{aligned} \tag{2.115}$$

where

$$\begin{aligned} &\{a_{j_1j_11}, a_{j_1j_12}\} = \text{sort}\{b_{j_1j_11}, b_{j_1j_12}, b_{j_1j_13}\}, a_{j_1j_11} < a_{j_1j_12},\\ &b_{j_1j_12}, b_{j_1j_13} = -\frac{1}{2}(B_{j_1j_1} \pm \sqrt{\Delta_{j_1j_1}}), \end{aligned} \tag{2.116}$$

and the corresponding dynamical system becomes

$$\begin{aligned} \dot{x}_{j_2} &= a_{j_2j_20}(y_{j_1} - a_{j_2j_21}),\\ \dot{x}_{j_1} &= a_{j_1j_10}(y_{j_1} - a_{j_1j_11})(y_{j_1} - a_{j_1j_12})(y_{j_1} - a_{j_1j_13}). \end{aligned} \tag{2.117}$$

There is a dynamical system in the y_{j_1}-direction as

$$\begin{aligned} \dot{y}_{j_1} = \alpha_{j_1j_1}a_{j_1j_10}[&(y_{j_1} - a_{j_1j_11})(y_{j_1} - a_{j_1j_12})(y_{j_1} - a_{j_1j_13})\\ &+ c_{j_1j_1}(y_{j_1} - a_{j_2j_21})] \end{aligned} \tag{2.118}$$

with a dynamical system in the $y_{j_1}^{\perp}$-direction as

$$\begin{aligned} \dot{y}_{j_1}^{\perp} = \beta_{j_1j_1}a_{j_1j_10}[&(y_{j_1} - a_{j_1j_11})(y_{j_1} - a_{j_1j_12})(y_{j_1} - a_{j_1j_13})\\ &+ d_{j_1j_1}(y_{j_1} - a_{j_2j_21})]. \end{aligned} \tag{2.119}$$

(iv_{1a}) *The equilibriums of* $y^*_{j_1} = a_{j_1j_1s_1}(s_1, s_2 \in \{1,2,3\},\ s_1 \neq s_2)$ *for the* x_{j_1}*-direction with* $\dot{x}^*_{j_2} = a_{j_2j_20}(a_{j_1j_1s_1} - a_{j_2j_21})$ *possess the following properties*:

- *For* $\alpha_{j_1j_2}a_{j_1j_10}\prod_{s_2=1,s_2\neq s_1}^{3}(a_{j_1j_1s_1} - a_{j_1j_1s_2}) > 0$ *and* $a_{j_2j_20}(a_{j_1j_1s_1} - a_{j_2j_21}) > 0$,

$$(a_{j_1j_1s}, \dot{x}^*_{j_2}) = \underbrace{(\mathrm{UP}, \mathrm{pF})}_{\text{up-parabola flow}}. \tag{2.120}$$

- *For* $\alpha_{j_1j_2}a_{j_1j_10}\prod_{s_2=1,s_2\neq s_1}^{3}(a_{j_1j_1s_1} - a_{j_1j_1s_2}) < 0$ *and* $a_{j_2j_20}(a_{j_1j_1s_1} - a_{j_2j_21}) > 0$,

$$(a_{j_1j_1s}, \dot{x}^*_{j_2}) = \underbrace{(\mathrm{DP}, \mathrm{pF})}_{\text{down-parabola flow}}. \tag{2.121}$$

- *For* $\alpha_{j_1j_2}a_{j_1j_10}\prod_{s_2=1,s_2\neq s_1}^{3}(a_{j_1j_1s_1} - a_{j_1j_1s_2}) > 0$ *and* $a_{j_2j_20}(a_{j_1j_1s_1} - a_{j_2j_21}) < 0$,

$$a_{j_1j_1s}, \dot{x}^*_{j_2} = \underbrace{(\mathrm{DP}, \mathrm{nF})}_{\text{down-parabola flow}}. \tag{2.122}$$

- *For* $\alpha_{j_1j_2}a_{j_1j_10}\prod_{s_2=1,s_2\neq s_1}^{3}(a_{j_1j_1s_1} - a_{j_1j_1s_2}) < 0$ *and* $a_{j_2j_20}(a_{j_1j_1s_1} - a_{j_2j_21}) < 0$,

$$(a_{j_1j_1s}, \dot{x}^*_{j_2}) = \underbrace{(\mathrm{UP}, \mathrm{nF})}_{\text{up-parabola flow}}. \tag{2.123}$$

(iv_{1b}) *The equilibriums of* $y^*_{j_1} = a_{j_2j_21}$ *for the* x_{j_2}*-direction possess the following properties*:

- *For* $a_{j_1j_10}\prod_{s=1}^{3}(a_{j_2j_21} - a_{j_1j_1s}) > 0$ *and* $\alpha_{j_1j_1}a_{j_2j_20} > 0$,

$$(\dot{x}^*_{j_1}, a_{j_2j_21}) = \underbrace{(\mathrm{pF}, \mathrm{UP})}_{\text{up-parabola flow}}. \tag{2.124}$$

- *For* $a_{j_1j_10}\prod_{s=1}^{3}(a_{j_2j_21} - a_{j_1j_1s}) < 0$ *and* $\alpha_{j_1j_1}a_{j_2j_20} > 0$,

$$(\dot{x}^*_{j_1}, a_{j_2j_21}) = \underbrace{(\mathrm{nF}, \mathrm{DP})}_{\text{down-parabola flow}}. \tag{2.125}$$

- *For* $a_{j_1j_10}\prod_{s=1}^{3}(a_{j_2j_21} - a_{j_1j_1s}) > 0$ *and* $\alpha_{j_1j_1}a_{j_2j_20} < 0$,

$$(\dot{x}^*_{j_1}, a_{j_2j_21}) = \underbrace{(\mathrm{pF}, \mathrm{DP})}_{\text{down-parabola flow}}. \tag{2.126}$$

- *For* $a_{j_1j_10}\prod_{s=1}^{3}(a_{j_2j_21}-a_{j_1j_1s})<0$ *and* $\alpha_{j_1j_1}a_{j_2j_20}<0$,

$$(\dot{x}^*_{j_1},a_{j_2j_21})=\underbrace{(\text{nF,UP})}_{\text{up-parabola flow}}. \quad (2.127)$$

(iv_2) *For* $a_{j_2j_21}=a_{j_1j_1s_1}(s_1,s_2\in\{1,2,3\},s_1\neq s_2)$, *the corresponding dynamical system becomes*

$$\begin{aligned}\dot{x}_{j_2}&=a_{j_2j_20}(y_{j_1}-a_{j_2j_21}),\\ \dot{x}_{j_1}&=a_{j_1j_10}(y_{j_1}-a_{j_1j_1s_1})\prod_{s_2=1,s_2\neq s_1}^{3}(y_{j_1}-a_{j_1j_1s_2}).\end{aligned} \quad (2.128)$$

There is a dynamical system in the y_{j_1}*-direction as*

$$\dot{y}_{j_1}=\alpha_{j_1j_1}a_{j_1j_10}(y_{j_1}-a_{j_1j_1s_1})\Big[\prod_{s_2=1,s_2\neq s_1}^{3}(y_{j_1}-a_{j_1j_1s_2})+c_{j_1j_1}\Big] \quad (2.129)$$

with a dynamical system in the $y^{\perp}_{j_1}$*-direction as*

$$\dot{y}^{\perp}_{j_1}=\beta_{j_1j_1}(y_{j_1}-a_{j_1j_1s_1})\Big[\prod_{s_2=1,s_2\neq s_1}^{3}(y_{j_1}-a_{j_1j_1s_2})+d_{j_1j_1}\Big]. \quad (2.130)$$

Let

$$\begin{aligned}G_{j_1j_10}&=\frac{\partial}{\partial y_{j_1}}[\mathbf{n}^{\mathrm{T}}_{j_1}\cdot\mathbf{F}(y_{j_1},\mathbf{p})]\Big|_{y_{j_1}=a_{j_1j_1s_1}}\\ &=\alpha_{j_1j_1}a_{j_1j_10}\prod_{s_2=1,s_2\neq s_1}^{3}(a_{j_1j_1s_1}-a_{j_1j_1s_2})+\alpha_{j_1j_2}a_{j_2j_20}.\end{aligned} \quad (2.131)$$

(iv_{2a}) *The infinite-equilibrium of* $y^*_{j_1}=a_{j_1j_1s_1}=a_{j_2j_21}(s_1,s_2\in\{1,2,3\},$ $s_1\neq s_2)$ *possesses the following properties*:

- *If*

$$\frac{a_{j_1j_10}\prod_{s_2=1,s_2\neq s_1}^{3}(a_{j_1j_1s_1}-a_{j_1j_1s_2})}{a_{j_2j_20}}>0 \quad \text{and} \quad G_{j_1j_10}>0, \quad (2.132)$$

  then

$$(a_{11s_1},a_{j_2j_21})=\underbrace{(\text{II, SO})}_{\text{infinite-equilibrium}}. \quad (2.133)$$

- *If*

$$\frac{a_{j_1j_10}\prod_{s_2=1,s_2\neq s_1}^{3}(a_{j_1j_1s_1}-a_{j_1j_1s_2})}{a_{j_2j_20}}<0 \quad \text{and} \quad G_{j_1j_10}>0, \tag{2.134}$$

then

$$(a_{11s_1},a_{j_2j_21})=\underbrace{(\text{DI},\text{SO})}_{\text{infinite-equilibrium}}. \tag{2.135}$$

- *If*

$$\frac{a_{j_1j_10}\prod_{s_2=1,s_2\neq s_1}^{3}(a_{j_1j_1s_1}-a_{j_1j_1s_2})}{a_{j_2j_20}}<0 \quad \text{and} \quad G_{j_1j_10}<0, \tag{2.136}$$

then

$$(a_{11s_1},a_{j_2j_21})=\underbrace{(\text{DI},\text{SI})}_{\text{infinite-equilibrium}}. \tag{2.137}$$

- *If*

$$\frac{a_{j_1j_10}\prod_{s_2=1,s_2\neq s_1}^{3}(a_{j_1j_1s_1}-a_{j_1j_1s_2})}{a_{j_2j_20}}>0 \quad \text{and} \quad G_{j_1j_10}<0, \tag{2.138}$$

then

$$(a_{11s_1},a_{j_2j_21})=\underbrace{(\text{II},\text{SI})}_{\text{infinite-equilibrium}}. \tag{2.139}$$

(iv_{2b}) *The inflection-source and sink infinite-equilibriums at* $y_{j_1}^*=a_{j_1j_1s_1}=a_{j_2j_21}$ *are the switching bifurcations for the parabola flows at* $y_{j_1}^*=a_{j_1j_1s_1}$ *and* $y_{j_1}^*=a_{j_2j_21}$:

- *For* $a_{j_11j_10}\prod_{s_2=1,s_2\neq s_1}^{3}(a_{j_1j_1s_1}-a_{j_1j_1s_2})>0$ *and* $a_{j_2j_20}>0$,

$$\left\{\begin{array}{c}\underbrace{(\text{nF},\text{DP})}_{\text{down-parabola }(-)}\\ \underbrace{(\text{UP},\text{pF})}_{\text{up-parabola }(+)}\end{array}\right\}\rightleftharpoons\underbrace{(\text{II},\text{SO})}_{\text{switching bifurcation}}\rightleftharpoons\left\{\begin{array}{c}\underbrace{(\text{DP},\text{nF})}_{\text{down-parabola }(-)}\\ \underbrace{(\text{pF},\text{UP})}_{\text{up-parabola }(+)}\end{array}\right\}$$

for $(\alpha_{j_1j_1},\alpha_{j_1j_2})=(+,+)$ and $G_{j_1j_10}>0$;

$$\left\{\begin{array}{c}\underbrace{(\mathrm{nF},\mathrm{DP})}_{\text{down-parabola }(-)}\\ \underbrace{(\mathrm{DP},\mathrm{pF})}_{\text{down-parabola }(+)}\end{array}\right\} \rightleftharpoons \underbrace{(\mathrm{II},\mathrm{SO})}_{\text{switching bifurcation}} \rightleftharpoons \left\{\begin{array}{c}\underbrace{(\mathrm{UP},\mathrm{nF})}_{\text{up-parabola }(-)}\\ \underbrace{(\mathrm{pF},\mathrm{UP})}_{\text{up-parabola }(+)}\end{array}\right\}$$

$$\text{for } (\alpha_{j_1j_1},\alpha_{j_1j_2})=(+,-) \quad \text{and} \quad G_{j_1j_10}>0;$$

$$\left\{\begin{array}{c}\underbrace{(\mathrm{UP},\mathrm{nF})}_{\text{up-parabola }(-)}\\ \underbrace{(\mathrm{pF},\mathrm{UP})}_{\text{up-parabola }(+)}\end{array}\right\} \rightleftharpoons \underbrace{(\mathrm{II},\mathrm{SI})}_{\text{switching bifurcation}} \rightleftharpoons \left\{\begin{array}{c}\underbrace{(\mathrm{nF},\mathrm{DP})}_{\text{down-parabola }(-)}\\ \underbrace{(\mathrm{DP},\mathrm{pF})}_{\text{down-parabola }(+)}\end{array}\right\}$$

$$\text{for } (\alpha_{j_1j_1},\alpha_{j_1j_2})=(+,-) \quad \text{and} \quad G_{j_1j_10}<0. \tag{2.140}$$

- *For* $a_{j_11j_10}\prod_{s_2=1,s_2\neq s_1}^{3}(a_{j_1j_1s_1}-a_{j_1j_1s_2})<0$ *and* $a_{j_2j_20}>0$,

$$\left\{\begin{array}{c}\underbrace{(\mathrm{pF},\mathrm{UP})}_{\text{up-parabola }(+)}\\ \underbrace{(\mathrm{DP},\mathrm{pF})}_{\text{down-parabola }(+)}\end{array}\right\} \rightleftharpoons \underbrace{(\mathrm{DI},\mathrm{SI})}_{\text{switching bifurcation}} \rightleftharpoons \left\{\begin{array}{c}\underbrace{(\mathrm{UP},\mathrm{nF})}_{\text{up-parabola }(-)}\\ \underbrace{(\mathrm{nF},\mathrm{DP})}_{\text{down-parabola }(-)}\end{array}\right\}$$

$$\text{for } (\alpha_{j_1j_1},\alpha_{j_1j_2})=(+,+) \quad \text{and} \quad G_{j_1j_10}<0;$$

$$\left\{\begin{array}{c}\underbrace{(\mathrm{UP},\mathrm{nF})}_{\text{up-parabola }(-)}\\ \underbrace{(\mathrm{nF},\mathrm{DP})}_{\text{down-parabola }(-)}\end{array}\right\} \rightleftharpoons \underbrace{(\mathrm{DI},\mathrm{SO})}_{\text{switching bifurcation}} \rightleftharpoons \left\{\begin{array}{c}\underbrace{(\mathrm{pF},\mathrm{UP})}_{\text{up-parabola }(+)}\\ \underbrace{(\mathrm{DP},\mathrm{pF})}_{\text{down-parabola }(+)}\end{array}\right\}$$

$$\text{for } (\alpha_{j_1j_1},\alpha_{j_1j_2})=(+,+) \quad \text{and} \quad G_{j_1j_10}>0;$$

$$\left\{\begin{array}{c}\underbrace{(\mathrm{pF},\mathrm{UP})}_{\text{up-parabola }(+)}\\ \underbrace{(\mathrm{UP},\mathrm{pF})}_{\text{up-parabola }(+)}\end{array}\right\} \rightleftharpoons \underbrace{(\mathrm{DI},\mathrm{SI})}_{\text{switching bifurcation}} \rightleftharpoons \left\{\begin{array}{c}\underbrace{(\mathrm{DP},\mathrm{nF})}_{\text{down-parabola }(-)}\\ \underbrace{(\mathrm{nF},\mathrm{DP})}_{\text{down-parabola }(-)}\end{array}\right\}$$

$$\text{for } (\alpha_{j_1j_1},\alpha_{j_1j_2})=(+,-) \quad \text{and} \quad G_{j_1j_10}<0. \tag{2.141}$$

- *For* $a_{j_{11}j_10}\prod_{s_2=1,s_2\neq s_1}^{3}(a_{j_1j_1s_1}-a_{j_1j_1s_2})>0$ *and* $a_{j_2j_20}<0$,

$$\left\{\begin{array}{c}\underbrace{(\mathrm{nF},\mathrm{UP})}_{\text{up-parabola }(-)}\\ \underbrace{(\mathrm{DP},\mathrm{nF})}_{\text{down-parabola }(-)}\end{array}\right\}\rightleftharpoons\underbrace{(\mathrm{DI},\mathrm{SO})}_{\text{switching bifurction}}\rightleftharpoons\left\{\begin{array}{c}\underbrace{(\mathrm{UP},\mathrm{pF})}_{\text{up-parabola }(+)}\\ \underbrace{(\mathrm{pF},\mathrm{DP})}_{\text{down-parabola }(+)}\end{array}\right\}$$

$$\text{for } (\alpha_{j_1j_1},\alpha_{j_1j_2})=(+,+)\quad\text{and}\quad G_{j_1j_10}>0;$$

$$\left\{\begin{array}{c}\underbrace{(\mathrm{UP},\mathrm{pF})}_{\text{up-parabola }(+)}\\ \underbrace{(\mathrm{pF},\mathrm{DP})}_{\text{down-parabola }(+)}\end{array}\right\}\rightleftharpoons\underbrace{(\mathrm{DI},\mathrm{SI})}_{\text{switching bifurction}}\rightleftharpoons\left\{\begin{array}{c}\underbrace{(\mathrm{nF},\mathrm{UP})}_{\text{up-parabola }(-)}\\ \underbrace{(\mathrm{DP},\mathrm{nF})}_{\text{down-parabola }(-)}\end{array}\right\}$$

$$\text{for } (\alpha_{j_1j_1},\alpha_{j_1j_2})=(+,+)\quad\text{and}\quad G_{j_1j_10}<0;$$

$$\left\{\begin{array}{c}\underbrace{(\mathrm{nF},\mathrm{UP})}_{\text{up-parabola }(-)}\\ \underbrace{(\mathrm{UP},\mathrm{nF})}_{\text{up-parabola }(-)}\end{array}\right\}\rightleftharpoons\underbrace{(\mathrm{DI},\mathrm{SO})}_{\text{switching bifurction}}\rightleftharpoons\left\{\begin{array}{c}\underbrace{(\mathrm{DP},\mathrm{pF})}_{\text{down-parabola }(+)}\\ \underbrace{(\mathrm{pF},\mathrm{DP})}_{\text{down-parabola }(+)}\end{array}\right\}$$

$$\text{for } (\alpha_{j_1j_1},\alpha_{j_1j_2})=(+,-)\quad\text{and}\quad G_{j_1j_10}>0. \tag{2.142}$$

- *For* $a_{j_{11}j_10}\prod_{s_2=1,s_2\neq s_1}^{3}(a_{j_1j_1s_1}-a_{j_1j_1s_2})<0$ *and* $a_{j_2j_20}<0$,

$$\left\{\begin{array}{c}\underbrace{(\mathrm{pF},\mathrm{DP})}_{\text{down-parabola }(+)}\\ \underbrace{(\mathrm{UP},\mathrm{nF})}_{\text{up-parabola }(-)}\end{array}\right\}\rightleftharpoons\underbrace{(\mathrm{II},\mathrm{SI})}_{\text{switching bifurcation}}\rightleftharpoons\left\{\begin{array}{c}\underbrace{(\mathrm{DP},\mathrm{pF})}_{\text{down-parabola }(+)}\\ \underbrace{(\mathrm{nF},\mathrm{UP})}_{\text{up-parabola }(-)}\end{array}\right\}$$

$$\text{for } (\alpha_{j_1j_1},\alpha_{j_1j_2})=(+,+)\quad\text{and}\quad G_{j_1j_10}<0;$$

$$\left\{\begin{array}{c}\underbrace{(\mathrm{pF},\mathrm{DP})}_{\text{down-parabola }(+)}\\ \underbrace{(\mathrm{DP},\mathrm{nF})}_{\text{down-parabola }(-)}\end{array}\right\}\rightleftharpoons\underbrace{(\mathrm{II},\mathrm{SO})}_{\text{switching bifurcation}}\rightleftharpoons\left\{\begin{array}{c}\underbrace{(\mathrm{UP},\mathrm{pF})}_{\text{up-parabola }(+)}\\ \underbrace{(\mathrm{nF},\mathrm{UP})}_{\text{up-parabola }(-)}\end{array}\right\}$$

$$\text{for } (\alpha_{j_1j_1},\alpha_{j_1j_2})=(+,-)\quad\text{and}\quad G_{j_1j_10}>0;$$

$$\left\{\begin{array}{c}\underbrace{(\mathrm{UP},\mathrm{pF})}_{\text{up-parabola }(+)}\\ \underbrace{(\mathrm{nF},\mathrm{UP})}_{\text{up-parabola }(-)}\end{array}\right\} \rightleftharpoons \underbrace{(\mathrm{II},\mathrm{SI})}_{\text{switching bifurcation}} \rightleftharpoons \left\{\begin{array}{c}\underbrace{(\mathrm{pF},\mathrm{DP})}_{\text{down-parabola }(+)}\\ \underbrace{(\mathrm{DP},\mathrm{nF})}_{\text{down-parabola }(-)}\end{array}\right\}$$
$$\text{for } (\alpha_{j_1j_1},\alpha_{j_1j_2}) = (+,-) \quad \text{and} \quad G_{j_1j_10} < 0. \tag{2.143}$$

2.2 Theorem Proof

Consider a cubic system in the x_i-direction ($i \in \{1,2\}$) as

$$\begin{aligned}\dot{x}_{j_1} &= f_{j_1}(\mathbf{x},\mathbf{p}) = A_{j_1j_13} + A_{j_1j_12}y_{j_1} + A_{j_1j_11}y_{j_1}^2 + A_{j_1j_10}y_{j_1}^3,\\ \dot{x}_{j_2} &= f_{j_2}(\mathbf{x},\mathbf{p}) = A_{j_2j_23} + A_{j_2j_22}y_{j_1} + A_{j_2j_21}y_{j_1}^2 + A_{j_2j_20}y_{j_1}^3,\end{aligned}$$

with

$$y_{j_1} = \alpha_{j_1j_1}x_{j_1} + \alpha_{j_1j_2}x_{j_2} \quad \text{for } j_1,j_2 \in \{1,2\},\ j_1 \neq j_2.$$

For $A_{j_1j_10} = A_{j_2j_21} = 0$, $A_{j_2j_22} \neq 0, A_{j_2j_23} \neq 0$, the two-diemnsional dynamical system is

$$\begin{aligned}\dot{x}_{j_2} &= A_{j_2j_23} + A_{j_2j_22}y_{j_1},\\ \dot{x}_{j_1} &= a_{110}(y_{j_1} - b_{j_1j_11})(y_{j_1}^2 + B_{j_1j_1}y_{j_1} + C_{j_1j_1}).\end{aligned}$$

(i) If

$$A_{j_2j_22} \neq 0, A_{j_2j_23} \neq 0 \quad \text{and} \quad \Delta_{j_1j_1} < 0,$$

the corresponding dynamical system is

$$\begin{aligned}\dot{x}_{j_2} &= a_{j_2j_20}(y_{j_1} - a_{j_2j_21}),\\ \dot{x}_{j_1} &= a_{j_1j_10}(y_{j_1} - a_{j_1j_11})[(y_{j_1} - a_{j_1j_1})^2 + b_{j_1j_1}],\end{aligned}$$

where

$$\begin{aligned}&a_{j_2j_21} = -\frac{1}{A_{j_2j_22}}A_{j_2j_23}, \quad a_{j_2j_20} = A_{j_2j_22};\\ &a_{j_1j_1} = -\frac{1}{2}B_{j_1j_1}, \quad b_{j_1j_1} = -\frac{1}{4}C_{j_1j_1}, \quad a_{j_1j_11} = b_{j_1j_11};\\ &\Delta_{j_1j_1} = (B_{j_1j_1})^2 - 4C_{j_1j_1}, \quad a_{j_1j_10} = A_{j_1j_10}.\end{aligned}$$

There is a dynamical system in the y_{j_1}-direction as

$$\begin{aligned}\dot{y}_{j_1} &= \alpha_{j_1j_1}a_{j_1j_10}(y_{j_1} - a_{j_1j_11})[(y_{j_1} - a_{j_1j_1})^2 + b_{j_1j_1}] \\ &\quad + \alpha_{j_1j_2}a_{j_2j_20}(y_{j_1} - a_{j_2j_21})\end{aligned}$$

with a dynamical system in the $y_{j_1}^{\perp}$-direction as

$$\begin{aligned}\dot{y}_{j_1}^{\perp} &= \beta_{j_1j_1}a_{j_1j_10}(y_{j_1} - a_{j_1j_11})[(y_{j_1} - a_{j_1j_1})^2 + b_{j_1j_1}] \\ &\quad + \beta_{j_1j_2}a_{j_2j_20}(y_{j_1} - a_{j_2j_21}),\end{aligned}$$

where

$$\begin{aligned}y_{j_1}^{\perp} &= \beta_{j_1j_1}x_{j_1} + \beta_{j_1j_2}x_{j_2}, \quad \mathbf{n}_{j_1}^{\perp} = (\beta_{j_1j_1}, \beta_{j_1j_2})^{\mathrm{T}}, \\ \mathbf{n}_{j_1} \cdot \mathbf{n}_{j_1}^{\perp} &= \alpha_{j_1j_1}\beta_{j_1j_1} + \alpha_{j_1j_2}\beta_{j_1j_2} = 0.\end{aligned}$$

($\mathrm{i}_{1\mathrm{a}}$) In phase space, for $a_{j_1j_11} \neq a_{j_2j_21}$, at $y_{j_1}^* = a_{j_1j_11}$,

$$\left.\frac{dx_{j_1}}{dx_{j_2}}\right|_{y_{j_1}^* = a_{j_1j_11}} = \frac{a_{j_1j_10}}{a_{j_2j_20}} \left.\frac{(y_{j_1} - a_{j_1j_11})[(y_{j_1} - a_{j_1j_1})^2 + b_{j_1j_1}]}{(y_{j_1} - a_{j_2j_21})}\right|_{y_{j_1}^* = a_{j_1j_11}} = 0.$$

If

$$\left.\frac{d^2x_{j_1}}{dx_{j_2}^2}\right|_{y_{j_1}^* = a_{j_1j_11}} = \frac{a_{j_1j_10}}{a_{j_2j_20}} \frac{a_{j_1j_2}[(a_{j_1j_11} - a_{j_1j_1})^2 + b_{j_1j_1}]}{(a_{j_1j_11} - a_{j_2j_21})} > 0,$$

there is an up-parabola flow at $y_{j_1}^* = a_{j_1j_11}$ in the x_{j_1}-direction. If

$$\left.\frac{d^2x_{j_1}}{dx_{j_2}^2}\right|_{y_{j_1}^* = a_{jj_1j_1}} = \frac{a_{j_1j_10}}{a_{j_2j_20}} \frac{a_{j_1j_2}[(a_{j_1j_11} - a_{j_1j_1})^2 + b_{j_1j_1}]}{(a_{j_1j_11} - a_{j_2j_21})} < 0,$$

there is a down-parabola flow at $y_{j_1}^* = a_{j_1j_11}$ in the x_{j_1}-direction. In addition,

$$(a_{j_1j_11} - a_{j_1j_1})^2 + b_{j_1j_1} > 0.$$

Therefore, the simple equilibrium of $y_{j_1}^* = a_{j_1j_11}$ in the x_{j_1}-direction has the following properties as in Eqs. (2.10)–(2.13):

- For $a_{j_1j_10}\alpha_{j_1j_2} > 0$ and $a_{j_2j_20}(a_{j_1j_11} - a_{j_2j_21}) > 0$,

$$(a_{j_1j_11}, \dot{x}_{j_2}^*) = \underbrace{(\mathrm{UP}, \mathrm{pF})}_{\text{up-parabola flow } (+)}.$$

- For $a_{j_1j_10}\alpha_{j_1j_2} < 0$ and $a_{j_2j_20}(a_{j_1j_11} - a_{j_2j_21}) > 0$,

$$(a_{j_1j_11}, \dot{x}^*_{j_2}) = \underbrace{(\mathrm{DP}, \mathrm{pF})}_{\text{down-parabola flow }(+)}.$$

- For $a_{j_1j_10}\alpha_{j_1j_2} > 0$ and $a_{j_2j_20}(a_{j_1j_11} - a_{j_2j_21}) < 0$,

$$(a_{j_1j_11}, \dot{x}^*_{j_2}) = \underbrace{(\mathrm{DP}, \mathrm{nF})}_{\text{down-parabola flow }(-)}.$$

- For $a_{j_1j_10}\alpha_{j_1j_2} < 0$ and $a_{j_2j_20}(a_{j_1j_11} - a_{j_2j_21}) < 0$,

$$(a_{j_1j_11}, \dot{x}^*_{j_2}) = \underbrace{(\mathrm{UP}, \mathrm{nF})}_{\text{up-parabola flow }(-)}.$$

(i_{1b}) In phase space, for $a_{j_2j_21} \neq a_{j_1j_11}$, at $y^*_{j_1} = a_{j_2j_21}$,

$$\left.\frac{dx_{j_2}}{dx_{j_1}}\right|_{y^*_{j_1}=a_{j_2j_21}} = \frac{a_{j_2j_20}}{a_{j_1j_10}} \left.\frac{(y_{j_1} - a_{j_2j_21})}{(y_{j_1} - a_{j_1j_1}1)[(y_{j_1} - a_{j_1j_1})^2 + b_{j_1j_1}]}\right|_{y^*_{j_1}=a_{j_2j_21}} = 0.$$

If

$$\left.\frac{d^2x_{j_2}}{dx^2_{j_1}}\right|_{y^*_{j_2}=a_{j_2j_21}} = \frac{a_{j_2j_20}}{a_{j_1j_10}} \frac{\alpha_{j_1j_1}}{(a_{j_2j_21} - a_{j_1j_11})[(a_{j_2j_21} - a_{j_1j_1})^2 + b_{j_1j_1}]} > 0,$$

there is an up-parabola flow at $y^*_{j_1} = a_{j_2j_21}$ in the x_{j_1}-direction. If

$$\left.\frac{d^2x_{j_2}}{dx^2_{j_1}}\right|_{y^*_{j_2}=a_{jj_2j_1}} = \frac{a_{j_2j_20}}{a_{j_1j_10}} \frac{\alpha_{j_1j_1}}{(a_{j_2j_21} - a_{j_1j_11})[(a_{j_2j_21} - a_{j_1j_1})^2 + b_{j_1j_1}]} < 0,$$

there is a down-parabola flow at $y^*_{j_1} = a_{j_2j_21}$ in the x_{j_1}-direction. In addition,

$$(a_{j_2j_21} - a_{j_1j_1})^2 + b_{j_1j_1} > 0.$$

The flows at $y^*_{j_1} = a_{j_2j_21}$ are positive and negative for $a_{j_1j_10}(a_{j_2j_21} - a_{j_1j_11}) > 0$ and $a_{j_1j_10}(a_{j_2j_21} - a_{j_1j_11}) < 0$, respectively.

Therefore, the equilibrium of $y^*_{j_1} = a_{j_2j_21}$ in the x_{j_2}-direction has the following properties as in Eqs. (2.14)–(2.17):

- For $a_{j_1j_10}(a_{j_2j_21} - a_{j_1j_11}) > 0$ and $\alpha_{j_1j_1}a_{j_2j_20} > 0$,

$$(\dot{x}^*_{j_1}, a_{j_2j_21}) = \underbrace{(\mathrm{pF}, \mathrm{UP})}_{\text{up-parabola flow }(+)}.$$

- For $a_{j_1j_10}(a_{j_2j_21} - a_{j_1j_11}) < 0$ and $\alpha_{j_1j_1}a_{j_2j_20} > 0$,

$$(\dot{x}^*_{j_1}, a_{j_2j_21}) = \underbrace{(\text{nF}, \text{DP})}_{\text{down-parabola flow } (-)}.$$

- For $a_{j_1j_10}(a_{j_2j_21} - a_{j_1j_11}) > 0$ and $\alpha_{j_1j_1}a_{j_2j_20} < 0$,

$$(\dot{x}^*_{j_1}, a_{j_2j_21}) = \underbrace{(\text{pF}, \text{DP})}_{\text{down-parabola flow } (+)}.$$

- For $a_{j_1j_10}(a_{j_2j_21} - a_{j_1j_11}) < 0$ and $\alpha_{j_1j_1}a_{j_2j_20} < 0$,

$$(\dot{x}^*_{j_1}, a_{j_2j_21}) = \underbrace{(\text{nF}, \text{UP})}_{\text{up-parabola flow } (-)}.$$

(i_2) For $a_{j_1j_11} = a_{j_2j_21}$,

$$\begin{aligned}
\dot{x}_{j_2} &= a_{j_2j_20}(y_{j_1} - a_{j_2j_21}),\\
\dot{x}_{j_1} &= a_{j_1j_10}(y_{j_1} - a_{j_1j_11})[(y_{j_1} - a_{j_1j_1})^2 + b_{j_1j_1}].
\end{aligned}$$

There is a dynamical system in the y_{j_1}-direction as

$$\dot{y}_{j_1} = \alpha_{j_1j_1}a_{j_1j_10}(y_{j_1} - a_{j_1j_11})[(y_{j_1} - a_{j_1j_1})^2 + b_{j_1j_1} + c_{j_1j_1}],$$

with a dynamical system in the $y^{\perp}_{j_1}$-direction as

$$\dot{y}^{\perp}_{j_1} = \beta_{j_1j_1}a_{j_1j_10}(y_{j_1} - a_{j_1j_11})[(y_{j_1} - a_{j_1j_1})^2 + b_{j_1j_1} + d_{j_1j_1}],$$

where

$$c_{j_1j_1} = \frac{\alpha_{j_1j_2}a_{j_2j_20}}{\alpha_{j_1j_1}a_{j_1j_10}}, \quad d_{j_1j_1} = \frac{\beta_{j_1j_2}a_{j_2j_20}}{\beta_{j_1j_1}a_{j_1j_10}}.$$

($\text{i}_{2\text{a}}$) In phase space, at $y^*_{j_1} = a_{j_1j_11} = a_{j_2j_21}$, if

$$\left.\frac{dx_{j_1}}{dx_{j_2}}\right|_{y^*_{j_1} = a_{j_1j_11}} = \frac{a_{j_1j_10}[(a_{j_1j_11} - a_{j_1j_1})^2 + b_{j_1j_1}]}{a_{j_2j_20}} > 0,$$

there is an increasing-inflection flow. If

$$\left.\frac{dx_{j_1}}{dx_{j_2}}\right|_{y^*_{j_1} = a_{j_1,j_11}} = \frac{a_{j_1j_10}[(a_{j_1j_11} - a_{j_1j_1})^2 + b_{j_1j_1}]}{a_{j_2j_20}} < 0,$$

there is a decreasing-inflection flow. In addition, the variational equation at $y^*_{j_1} = a_{j_1j_11} = a_{j_2j_21}$ is

$$\begin{aligned}
\Delta\dot{y}_{j_1} &= \left\{\alpha_{j_1j_1}a_{j_1j_10}[(a_{j_1j_11} - a_{j_1j_1})^2 + b_{j_1j_1}] + \alpha_{j_1j_2}a_{j_2j_20}\right\}\Delta y_{j_1}\\
&= G_{j_1j_10}\Delta y_{j_1},
\end{aligned}$$

with

$$G_{j_1 j_1 0} = \frac{\partial}{\partial y_{j_1}}[\mathbf{n}_{j_1}^T \cdot \mathbf{F}(y_{j_1}, \mathbf{p})]\Big|_{y_{j_1}^* = a_{jj_1 1}}$$

$$= \alpha_{j_1 j_1} a_{j_1 j_1 0}[(a_{j_1 j_1 1} - a_{j_1 j_1})^2 + b_{j_1 j_1}] + \alpha_{j_1 j_2} a_{j_2 j_2 0}.$$

The flows at $y_{j_1}^* = a_{j_1 j_1 1} = a_{j_2 j_2 1}$ are source and sink in the x_{j_1}-direction for $G_{j_1 j_1 0} > 0$ and $G_{j_1 j_1 0} < 0$, respectively.

Therefore, the infinite-equilibrium of $y_{j_1}^* = a_{j_1 j_1 1} = a_{j_2 j_2 1}$ possesses the following properties as in Eqs. (2.23)–(2.30):

- If

$$\frac{a_{j_1 j_1 0}}{a_{j_2 j_2 0}} > 0 \quad \text{and} \quad G_{j_1 j_1 0} > 0,$$

then

$$(a_{j_1 j_1 1}, a_{j_2 j_2 1}) = \underbrace{(\text{II}, \text{SO})}_{\text{increasing-inflection source}}.$$

- If

$$\frac{a_{j_1 j_1 0}}{a_{j_2 j_2 0}} < 0 \quad \text{and} \quad G_{j_1 j_1 0} > 0,$$

then

$$(a_{j_1 j_1 1}, a_{j_2 j_2 1}) = \underbrace{(\text{DI}, \text{SO})}_{\text{decreasing-inflection source}}.$$

- If

$$\frac{a_{j_1 j_1 0}}{a_{j_2 j_2 0}} < 0 \quad \text{and} \quad G_{j_1 j_1 0} < 0,$$

then

$$(a_{j_1 j_1 1}, a_{j_2 j_2 1}) = \underbrace{(\text{DI}, \text{SI})}_{\text{decreasing-inflection sink}}.$$

- If

$$\frac{a_{j_1 j_1 0}}{a_{j_2 j_2 0}} > 0 \quad \text{and} \quad G_{j_1 j_1 0} < 0,$$

then

$$(a_{j_1 j_1 1}, a_{j_2 j_2 1}) = \underbrace{(\text{II}, \text{SI})}_{\text{increasing-inflection sink}}.$$

($\mathrm{i_{2b}}$) From ($\mathrm{i_{1a,b}}$) and ($\mathrm{i_{2a}}$), it proved is that the inflection-source and sink infinite-equilibriums at $y_{j_1}^* = a_{j_1j_1 1} = a_{j_2j_2 1}$ are the switching bifurcations for the parabola flows at $y_{j_1}^* = a_{j_1j_1 1}$ and $y_{j_1}^* = a_{j_2j_2 1}$ as in Eqs. (2.31)–(2.34):

- *For* $a_{j_1j_1 0} > 0$ and $a_{j_2j_2 0} > 0$,

$$\left\{\begin{array}{c}\underbrace{(\mathrm{nF},\mathrm{DP})}_{\text{down-parabola }(-)}\\ \underbrace{(\mathrm{UP},\mathrm{pF})}_{\text{up-parabola }(+)}\end{array}\right\} \rightleftharpoons \underbrace{(\mathrm{II},\mathrm{SO})}_{\text{switching bifurcation}} \rightleftharpoons \left\{\begin{array}{c}\underbrace{(\mathrm{DP},\mathrm{nF})}_{\text{down-parabola }(-)}\\ \underbrace{(\mathrm{pF},\mathrm{UP})}_{\text{up-parabola }(+)}\end{array}\right\}$$

for $(\alpha_{j_1j_1}, \alpha_{j_1j_2}) = (+,+)$ and $G_{j_1j_1 0} > 0$,

$$\left\{\begin{array}{c}\underbrace{(\mathrm{nF},\mathrm{DP})}_{\text{down-parabola }(-)}\\ \underbrace{(\mathrm{DP},\mathrm{pF})}_{\text{down-parabola }(+)}\end{array}\right\} \rightleftharpoons \underbrace{(\mathrm{II},\mathrm{SO})}_{\text{switching bifurcation}} \rightleftharpoons \left\{\begin{array}{c}\underbrace{(\mathrm{UP},\mathrm{nF})}_{\text{up-parabola }(-)}\\ \underbrace{(\mathrm{pF},\mathrm{UP})}_{\text{up-parabola }(+)}\end{array}\right\}$$

for $(\alpha_{j_1j_1}, \alpha_{j_1j_2}) = (+,-)$ and $G_{j_1j_1 0} > 0$,

$$\left\{\begin{array}{c}\underbrace{(\mathrm{UP},\mathrm{nF})}_{\text{up-parabola }(-)}\\ \underbrace{(\mathrm{pF},\mathrm{UP})}_{\text{up-parabola }(+)}\end{array}\right\} \rightleftharpoons \underbrace{(\mathrm{II},\mathrm{SI})}_{\text{switching bifurcation}} \rightleftharpoons \left\{\begin{array}{c}\underbrace{(\mathrm{nF},\mathrm{DP})}_{\text{down-parabola }(-)}\\ \underbrace{(\mathrm{DP},\mathrm{pF})}_{\text{down-parabola }(+)}\end{array}\right\}$$

for $(\alpha_{j_1j_1}, \alpha_{j_1j_2}) = (+,-)$ and $G_{j_1j_1 0} < 0$.

- *For* $a_{j_1j_1 0} < 0$ and $a_{j_2j_2 0} > 0$,

$$\left\{\begin{array}{c}\underbrace{(\mathrm{pF},\mathrm{UP})}_{\text{up-parabola }(+)}\\ \underbrace{(\mathrm{DP},\mathrm{pF})}_{\text{down-parabola }(+)}\end{array}\right\} \rightleftharpoons \underbrace{(\mathrm{DI},\mathrm{SI})}_{\text{switching bifurcation}} \rightleftharpoons \left\{\begin{array}{c}\underbrace{(\mathrm{UP},\mathrm{nF})}_{\text{up-parabola }(-)}\\ \underbrace{(\mathrm{nF},\mathrm{DP})}_{\text{down-parabola }(-)}\end{array}\right\}$$

for $(\alpha_{j_1j_1}, \alpha_{j_1j_2}) = (+,+)$ and $G_{j_1j_1 0} < 0$;

$$\left\{\begin{array}{c}\underbrace{(\mathrm{UP},\mathrm{nF})}_{\text{up-parabola }(-)}\\ \underbrace{(\mathrm{nF},\mathrm{DP})}_{\text{down-parabola }(-)}\end{array}\right\} \rightleftharpoons \underbrace{(\mathrm{DI},\mathrm{SO})}_{\text{switching bifurcation}} \rightleftharpoons \left\{\begin{array}{c}\underbrace{(\mathrm{pF},\mathrm{UP})}_{\text{up-parabola }(+)}\\ \underbrace{(\mathrm{DP},\mathrm{pF})}_{\text{down-parabola }(+)}\end{array}\right\}$$

for $(\alpha_{j_1j_1}, \alpha_{j_1j_2}) = (+,+)$ and $G_{j_1j_1 0} > 0$;

$$\left\{\begin{array}{c}\underbrace{(\mathrm{pF},\mathrm{UP})}_{\text{up-parabola }(+)}\\\underbrace{(\mathrm{UP},\mathrm{pF})}_{\text{up-parabola }(+)}\end{array}\right\}\rightleftharpoons\underbrace{(\mathrm{DI},\mathrm{SI})}_{\text{switching bifurcation}}\rightleftharpoons\left\{\begin{array}{c}\underbrace{(\mathrm{DP},\mathrm{nF})}_{\text{down-parabola }(-)}\\\underbrace{(\mathrm{nF},\mathrm{DP})}_{\text{down-parabola }(-)}\end{array}\right\}$$

$$\text{for } (\alpha_{j_1j_1},\alpha_{j_1j_2})=(+,-)\quad\text{and}\quad G_{j_1j_10}<0.$$

- *For* $a_{j_1j_10}>0$ and $a_{j_2j_20}<0$,

$$\left\{\begin{array}{c}\underbrace{(\mathrm{nF},\mathrm{UP})}_{\text{up-parabola }(-)}\\\underbrace{(\mathrm{DP},\mathrm{nF})}_{\text{down-parabola }(-)}\end{array}\right\}\rightleftharpoons\underbrace{(\mathrm{DI},\mathrm{SO})}_{\text{switching bifurction}}\rightleftharpoons\left\{\begin{array}{c}\underbrace{(\mathrm{UP},\mathrm{pF})}_{\text{up-parabola }(+)}\\\underbrace{(\mathrm{pF},\mathrm{DP})}_{\text{down-parabola }(+)}\end{array}\right\}$$

$$\text{for } (\alpha_{j_1j_1},\alpha_{j_1j_2})=(+,+)\quad\text{and}\quad G_{j_1j_10}>0;$$

$$\left\{\begin{array}{c}\underbrace{(\mathrm{UP},\mathrm{pF})}_{\text{up-parabola }(+)}\\\underbrace{(\mathrm{pF},\mathrm{DP})}_{\text{down-parabola }(+)}\end{array}\right\}\rightleftharpoons\underbrace{(\mathrm{DI},\mathrm{SI})}_{\text{switching bifurction}}\rightleftharpoons\left\{\begin{array}{c}\underbrace{(\mathrm{nF},\mathrm{UP})}_{\text{up-parabola }(-)}\\\underbrace{(\mathrm{DP},\mathrm{nF})}_{\text{down-parabola }(-)}\end{array}\right\}$$

$$\text{for } (\alpha_{j_1j_1},\alpha_{j_1j_2})=(+,+)\quad\text{and}\quad G_{j_1j_10}<0;$$

$$\left\{\begin{array}{c}\underbrace{(\mathrm{nF},\mathrm{UP})}_{\text{up-parabola }(-)}\\\underbrace{(\mathrm{UP},\mathrm{nF})}_{\text{up-parabola }(-)}\end{array}\right\}\rightleftharpoons\underbrace{(\mathrm{DI},\mathrm{SO})}_{\text{switching bifurction}}\rightleftharpoons\left\{\begin{array}{c}\underbrace{(\mathrm{DP},\mathrm{pF})}_{\text{down-parabola }(+)}\\\underbrace{(\mathrm{pF},\mathrm{DP})}_{\text{down-parabola }(+)}\end{array}\right\}$$

$$\text{for } (\alpha_{j_1j_1},\alpha_{j_1j_2})=(+,-)\quad\text{and}\quad G_{j_1j_10}>0.$$

- *For* $a_{j_1j_10}<0$ and $a_{j_2j_20}<0$,

$$\left\{\begin{array}{c}\underbrace{(\mathrm{pF},\mathrm{DP})}_{\text{down-parabola }(+)}\\\underbrace{(\mathrm{UP},\mathrm{nF})}_{\text{up-parabola }(-)}\end{array}\right\}\rightleftharpoons\underbrace{(\mathrm{II},\mathrm{SI})}_{\text{switching bifurcation}}\rightleftharpoons\left\{\begin{array}{c}\underbrace{(\mathrm{DP},\mathrm{pF})}_{\text{down-parabola }(+)}\\\underbrace{(\mathrm{nF},\mathrm{UP})}_{\text{up-parabola }(-)}\end{array}\right\}$$

$$\text{for } (\alpha_{j_1j_1},\alpha_{j_1j_2})=(+,+)\quad\text{and}\quad G_{j_1j_10}<0;$$

$$\left\{\begin{array}{c}\underbrace{(\mathrm{pF},\mathrm{DP})}_{\text{down-parabola }(+)}\\ \underbrace{(\mathrm{DP},\mathrm{nF})}_{\text{down-parabola }(-)}\end{array}\right\} \rightleftharpoons \underbrace{(\mathrm{II},\mathrm{SI})}_{\text{switching bifurcation}} \rightleftharpoons \left\{\begin{array}{c}\underbrace{(\mathrm{UP},\mathrm{pF})}_{\text{up-parabola }(+)}\\ \underbrace{(\mathrm{nF},\mathrm{UP})}_{\text{up-parabola }(-)}\end{array}\right\}$$

$$\text{for } (\alpha_{j_1j_1},\alpha_{j_1j_2}) = (+,-) \quad \text{and} \quad G_{j_1j_10} < 0;$$

$$\left\{\begin{array}{c}\underbrace{(\mathrm{UP},\mathrm{pF})}_{\text{up-parabola }(+)}\\ \underbrace{(\mathrm{nF},\mathrm{UP})}_{\text{up-parabola }(-)}\end{array}\right\} \rightleftharpoons \underbrace{(\mathrm{II},\mathrm{SO})}_{\text{switching bifurcation}} \rightleftharpoons \left\{\begin{array}{c}\underbrace{(\mathrm{pF},\mathrm{DP})}_{\text{down-parabola }(+)}\\ \underbrace{(\mathrm{DP},\mathrm{nF})}_{\text{down-parabola }(-)}\end{array}\right\}$$

$$\text{for } (\alpha_{j_1j_1},\alpha_{j_1j_2}) = (+,-) \quad \text{and} \quad G_{j_1j_10} > 0.$$

(ii) If

$$A_{j_2j_21} \neq 0, A_{j_2j_22} \neq 0 \quad \text{and} \quad \Delta_{j_1j_1} = 0,$$

the two-dimensional single-bivariate dynamical system becomes

$$\begin{aligned}\dot{x}_{j_2} &= a_{j_2j_20}(y_{j_1} - a_{j_2j_21}),\\ \dot{x}_{j_1} &= a_{j_1j_10}(y_{j_1} - a_{j_1j_1s_1})(y_{j_1} - a_{j_1j_1s_2})^2.\end{aligned}$$

(ii$_1$) For $a_{j_2j_21} \neq a_{j_1j_1s_1}, a_{j_1j_1s_2}$, there is a dynamical system in the y_{j_1}-direction as

$$\begin{aligned}\dot{y}_{j_1} &= \alpha_{j_1j_1}a_{j_1j_10}(y_{j_1} - a_{j_1j_1s_1})(y_{j_1} - a_{j_1j_1s_2})^2\\ &\quad + \alpha_{j_1j_2}a_{j_2j_20}(y_{j_1} - a_{j_2j_21})\end{aligned}$$

with a dynamical system in the $y_{j_1}^{\perp}$-direction as

$$\begin{aligned}\dot{y}_{j_1}^{\perp} &= \beta_{j_1j_1}a_{j_1j_10}(y_{j_1} - a_{j_1j_1s_1})(y_{j_1} - a_{j_1j_1s_2})^2\\ &\quad + \beta_{j_1j_2}a_{j_2j_20}(y_{j_1} - a_{j_2j_21}).\end{aligned}$$

(ii$_{1a}$) In phase space, for $a_{j_1j_1s_1} \neq a_{j_1j_1s_2}, a_{j_2j_21}$, at $y_{j_1}^* = a_{j_1j_1s_1}$,

$$\left.\frac{dx_{j_1}}{dx_{j_2}}\right|_{y_{j_1}^*=a_{j_1j_1s_1}} = \frac{a_{j_1j_10}}{a_{j_2j_20}}\left.\frac{(y_{j_1} - a_{j_1j_1s_1})(y_{j_1} - a_{j_1j_1s_2})^2}{(y_{j_1} - a_{j_2j_21})}\right|_{y_{j_1}^*=a_{j_1j_1s_1}} = 0.$$

If

$$\left.\frac{d^2x_{j_1}}{dx_{j_2}^2}\right|_{y_{j_1}^*=a_{j_1j_1s_1}} = \frac{a_{j_1j_10}}{a_{j_2j_20}}\frac{(a_{j_1j_1s_1} - a_{j_1j_1s_2})^2}{(a_{j_1j_1s_1} - a_{j_2j_21})} > 0,$$

there is an up-parabola flow at $y_{j_1}^* = a_{j_1j_1s_1}$ in the x_{j_1}-direction. If

$$\left.\frac{d^2x_{j_1}}{dx_{j_2}^2}\right|_{y_{j_1}^*=a_{j_1j_1s_1}} = \frac{a_{j_1j_10}}{a_{j_2j_20}}\frac{(a_{j_1j_1s_1} - a_{j_1j_1s_2})^2}{(a_{j_1j_1s_1} - a_{j_2j_21})} < 0,$$

there is a down-parabola flow at $y^*_{j_1} = a_{j_1j_1s_1} \neq a_{j_1j_1s_2}, a_{j_2j_21}$ in the x_{j_1}-direction. Let

$$\dot{x}^*_{j_2} = a_{j_2j_20}(a_{j_1j_1s_1} - a_{j_2j_21}).$$

For $\dot{x}^*_{j_2} > 0$ and $\dot{x}^*_{j_2} < 0$, the parabola flows are positive and negative, respectively.

Therefore, the equilibrium on the line of $y^*_{j_1} = a_{j_1j_1s_1}$ for the x_{j_1}-direction with $\dot{x}^*_{j_2} = a_{j_2j_20}(a_{j_1j_1s_1} - a_{j_2j_21})$ possesses the following properties as in Eqs. (2.39)–(2.42):

- For $a_{j_1j_10}\alpha_{j_1j_2}(a_{j_1j_1s_1} - a_{j_1j_1s_2})^2 > 0$ and $a_{j_2j_20}(a_{j_1j_1s_1} - a_{j_2j_21}) > 0$,

$$(a_{j_1,j_1s_1}, \dot{x}^*_{j_2}) = \underbrace{(\mathrm{UP}, \mathrm{pF})}_{\text{up-parabola flow }(+)}.$$

- For $a_{j_1j_10}\alpha_{j_1j_2}(a_{j_1j_1s_1} - a_{j_1j_1s_2})^2 < 0$ and $a_{j_2j_20}(a_{j_1j_1s_1} - a_{j_2j_21}) > 0$,

$$(a_{j_1j_1s_1}, \dot{x}^*_{j_2}) = \underbrace{(\mathrm{DP}, \mathrm{pF})}_{\text{down-parabola flow }(+)}.$$

- For $a_{j_1j_10}\alpha_{j_1j_2}(a_{j_1j_1s_1} - a_{j_1j_1s_2})^2 > 0$ and $a_{j_2j_20}(a_{j_1j_1s_1} - a_{j_2j_21}) < 0$,

$$(a_{j_1j_1s_1}, \dot{x}^*_{j_2}) = \underbrace{(\mathrm{DP}, \mathrm{nF})}_{\text{down-parabola flow }(-)}.$$

- For $a_{j_1j_10}\alpha_{j_1j_2}(a_{j_1j_1s_1} - a_{j_1j_1s_2})^2 < 0$ and $a_{j_2j_20}(a_{j_1j_1s_1} - a_{j_2j_21}) < 0$,

$$(a_{j_1j_1s_1}, \dot{x}^*_{j_2}) = \underbrace{(\mathrm{UP}, \mathrm{nF})}_{\text{up-parabola flow }(-)}.$$

(ii_{1b}) In phase space, for $a_{j_1j_1s_2} \neq a_{j_1j_1s_1}, a_{j_2j_21}$, at $y^*_{j_1} = a_{j_1j_1s_2}$,

$$\left.\frac{dx_{j_1}}{dx_{j_2}}\right|_{y^*_{j_1}=a_{j_1j_1j_2}} = \frac{a_{j_1j_10}}{a_{j_2j_20}} \left.\frac{(y_{j_1} - a_{j_1j_1s_1})(y_{j_1} - a_{j_1j_1s_2})^2}{(y_{j_1} - a_{j_2j_21})}\right|_{y^*_{j_1}=a_{j_1j_1s_2}} = 0,$$

$$\left.\frac{d^2x_{j_1}}{dx^2_{j_2}}\right|_{y^*_{j_1}=a_{j_1j_1s_2}} = 0.$$

If

$$\left.\frac{d^3x_{j_1}}{dx^3_{j_2}}\right|_{y^*_{j_1}=a_{j_1j_1s_2}} = \frac{a_{j_1j_10}}{a_{j_2j_20}} \frac{2\alpha^2_{j_1j_2}(a_{j_1j_1s_2} - a_{j_1j_1s_1})}{(a_{j_1j_1s_1} - a_{j_2j_21})} > 0,$$

there is an increasing-inflection flow at $y^*_{j_1} = a_{j_1j_1s_2}$ in the x_{j_1}-direction. If

$$\left.\frac{d^3x_{j_1}}{dx^3_{j_2}}\right|_{y^*_{j_1}=a_{j_1j_1j_2}} = \frac{a_{j_1j_10}}{a_{j_2j_20}}\frac{2\alpha^2_{j_1j_2}(a_{j_1j_1s_2}-a_{j_1j_1s_1})}{(a_{j_1j_1s_1}-a_{j_2j_21})} < 0,$$

there is a decreasing-inflection flow at $y^*_{j_1} = a_{j_1j_1s_2}$ in the x_{j_1}-direction. *Let*

$$\dot{x}^*_{j_2} = a_{j_2j_20}(a_{j_1j_1s_2} - a_{j_2j_21}).$$

For $\dot{x}^*_{j_2} > 0$ and $\dot{x}^*_{j_2} < 0$, the inflection flows are positive and negative, respectively.

Therefore, the equilibrium on the line of $y^*_{j_1} = a_{j_1j_1s_2}$ for the x_{j_1}-direction with $\dot{x}^*_{j_2} = a_{j_2j_20}(a_{j_1j_1s_2} - a_{j_2j_21})$ possesses the following properties as in Eqs. (2.43)–(2.46):

- For $a_{j_1j_10}\alpha^2_{j_1j_2}(a_{j_1j_1s_2} - a_{j_1j_1s_1}) > 0$ and $a_{j_2j_20}(a_{j_1j_1s_2} - a_{j_2j_21}) > 0$,

$$(a_{j_1j_1s_2}, \dot{x}^*_{j_2}) = \underbrace{(\text{II}, \text{pF})}_{\text{increasing-inflection flow }(+)}.$$

- For $a_{j_1j_10}\alpha^2_{j_1j_2}(a_{j_1j_1s_2} - a_{j_1j_1s_1}) < 0$ and $a_{j_2j_20}(a_{j_1j_1s_2} - a_{j_2j_21}) > 0$,

$$(a_{j_1j_1s_2}, \dot{x}^*_{j_2}) = \underbrace{(\text{DI}, \text{pF})}_{\text{decreasing-inflection flow }(+)}.$$

- For $a_{j_1j_10}\alpha^2_{j_1j_2}(a_{j_1j_1s_2} - a_{j_1j_1s_1}) > 0$ and $a_{j_2j_20}(a_{j_1j_1s_2} - a_{j_2j_21}) < 0$,

$$(a_{j_1j_1s_2}, \dot{x}^*_{j_2}) = \underbrace{(\text{DI,nF})}_{\text{decreasing-inflection flow }(-)}.$$

- For $a_{j_1j_10}\alpha^2_{j_1j_2}(a_{j_1j_1s_2} - a_{j_1j_1s_1}) < 0$ and $a_{j_2j_20}(a_{j_1j_1s_2} - a_{j_2j_21}) < 0$,

$$(a_{j_1j_1s_2}, \dot{x}^*_{j_2}) = \underbrace{(\text{II}, \text{nF})}_{\text{increasing-inflection flow }(-)}.$$

– The corresponding increasing-inflection and decreasing-inflection flows are the appearing bifurcations for up-parabola flow and down-parabola flow.

(ii_{1c}) In phase space, for $a_{j_2j_21} \neq a_{j_1j_1s_1}, a_{j_1j_1s_2}$, at $y_{j_1}^* = a_{j_2j_21}$,

$$\left.\frac{dx_{j_2}}{dx_{j_1}}\right|_{y_{j_1}^*=a_{j_2j_21}} = \frac{a_{j_2j_20}}{a_{j_1j_10}} \left.\frac{(y_{j_1} - a_{j_2j_21})}{(y_{j_1} - a_{j_1j_1s_1})(y_{j_1} - a_{j_1j_1s_2})^2}\right|_{y_{j_1}^*=a_{j_2j_2j_1}} = 0.$$

If

$$\left.\frac{d^2x_{j_2}}{dx_{j_1}^2}\right|_{y_{j_1}^*=a_{j_2j_21}} = \frac{a_{j_2j_20}}{a_{j_1j_10}} \frac{\alpha_{j_1j_1}}{(a_{j_2j_21} - a_{j_1j_1s_1})(a_{j_2j_21} - a_{j_1j_1s_2})^2} > 0,$$

there is an up-parabola flow at $y_{j_2}^* = a_{j_2j_21}$ in the x_{j_1}-direction. If

$$\left.\frac{d^2x_{j_2}}{dx_{j_1}^2}\right|_{y_{j_1}^*=a_{j_2j_21}} = \frac{a_{j_2j_20}}{a_{j_1j_10}} \frac{\alpha_{j_1j_1}}{(a_{j_2j_21} - a_{j_1j_1s_1})(a_{j_2j_21} - a_{j_1j_1s_2})^2} < 0,$$

there is a down-parabola flow at $y_{j_1}^* = a_{j_1j_11}$ in the x_{j_1}-direction. With

$$\dot{x}_{j_1}^* = a_{j_1j_10}(a_{j_2j_21} - a_{j_1j_1s_1})(a_{j_2j_21} - a_{j_1j_1s_2})^2,$$

therefore, the simple equilibrium of $y_{j_1}^* = a_{j_2j_21}$ in the x_{j_2}-direction for such a system has the following properties as in Eqs. (2.48)–(2.51):

- For $a_{j_1j_10}(a_{j_2j_21} - a_{j_1j_1s_1})(a_{j_2j_21} - a_{j_1j_1s_2})^2 > 0$ and $\alpha_{j_1j_1}a_{j_2j_20} > 0$,

$$(\dot{x}_{j_1}^*, a_{j_2j_21}) = \underbrace{(\text{pF}, \text{UP})}_{\text{up-parabola flow } (+)}.$$

- For $a_{j_1j_10}(a_{j_2j_21} - a_{j_1j_1s_1})(a_{j_2j_21} - a_{j_1j_1s_2})^2 < 0$ and $\alpha_{j_1j_1}a_{j_2j_20} > 0$,

$$(\dot{x}_{j_1}^*, a_{j_2j_21}) = \underbrace{(\text{nF}, \text{DP})}_{\text{down-parabola flow } (-)}.$$

- For $a_{j_1j_10}(a_{j_2j_21} - a_{j_1j_1s_1})(a_{j_2j_21} - a_{j_1j_1s_2})^2 > 0$ and $\alpha_{j_1j_1}a_{j_2j_20} < 0$,

$$(\dot{x}_{j_1}^*, a_{j_2j_21}) = \underbrace{(\text{pF}, \text{DP})}_{\text{down-parabola flow } (+)}.$$

- For $a_{j_1j_10}(a_{j_2j_21} - a_{j_1j_1s_1})(a_{j_2j_21} - a_{j_1j_1s_2})^2 < 0$ and $\alpha_{j_1j_1}a_{j_2j_20} < 0$,

$$(\dot{x}_{j_1}^*, a_{j_2j_21}) = \underbrace{(\text{nF}, \text{UP})}_{\text{up-parabola flow } (-)}.$$

(ii$_2$) For $a_{j_2j_21} = a_{j_1j_1s_1}$ and $a_{j_2j_21} \neq a_{j_1j_1s_2}$, the two-dimensional single-bivariate dynamical system becomes

$$\dot{x}_{j_2} = a_{j_2j_20}(y_{j_1} - a_{j_2j_21}),$$
$$\dot{x}_{j_1} = a_{j_1j_10}(y_{j_1} - a_{j_1j_1s_1})(y_{j_1} - a_{j_1j_1s_2})^2.$$

There is a dynamical system in the y_{j_1}-direction as

$$\dot{y}_{j_1} = \alpha_{j_1j_1}a_{j_1j_10}(y_{j_1} - a_{j_1j_1s_1})[(y_{j_1} - a_{j_1j_1s_1})(y_{j_1} - a_{j_1j_1s_2}) + c_{j_1j_1}],$$

and a dynamical system in the $y_{j_1}^{\perp}$-direction as

$$\dot{y}_{j_1}^{\perp} = \beta_{j_1j_1}a_{j_1j_10}(y_{j_1} - a_{j_1j_1s_1})[(y_{j_1} - a_{j_1j_1s_1})(y_{j_1} - a_{j_1j_1s_2}) + d_{j_1j_1}].$$

(ii$_{2a}$) In phase space, at $y_{j_1}^* = a_{j_1j_1s_1} = a_{j_2j_21}$, if

$$\left.\frac{dx_{j_1}}{dx_{j_2}}\right|_{y_{j_1}^* = a_{j_1j_1s_2}} = \frac{a_{j_1j_10}(a_{j_1j_1s_1} - a_{j_1j_1s_2})^2}{a_{j_2j_20}} > 0,$$

there is an increasing-inflection flow. If

$$\left.\frac{dx_{j_1}}{dx_{j_2}}\right|_{y_{j_1}^* = a_{j_1j_1s_2}} = \frac{a_{j_1j_10}(a_{j_1j_1s_1} - a_{j_1j_1s_2})^2}{a_{j_2j_20}} < 0,$$

there is a decreasing-inflection flow. In addition, the variational equation at $y_{j_1}^* = a_{j_1j_1s_1} = a_{j_2j_21}$ is

$$\Delta\dot{y}_{j_1} = [a_{j_1j_10}\alpha_{j_1j_1}(a_{j_1j_1s_1} - a_{j_1j_1s_2})^2 + a_{j_2j_20}\alpha_{j_1j_2}]\Delta y_{j_1} = G_{j_1j_10}\Delta y_{j_1},$$

with

$$G_{j_1j_10} = \left.\frac{\partial}{\partial y_{j_1}}[\mathbf{n}_{j_1}^{\mathrm{T}} \cdot \mathbf{F}(y_{j_1}, \mathbf{p})]\right|_{y_{j_1} = a_{j_1j_1s_1}}$$
$$= a_{j_1j_10}\alpha_{j_1j_1}(a_{j_1j_1s_1} - a_{j_1j_1s_2})^2 + a_{j_2j_20}\alpha_{j_1j_2}.$$

The flows at $y_{j_1}^* = a_{j_1j_1s_1} = a_{j_2j_21}$ are source and sink in the y_{j_1}-direction for $G_{j_1j_10} > 0$ and $G_{j_1j_10} < 0$, respectively.

Therefore, the infinite-equilibrium of $y_{j_1}^* = a_{j_1j_1s_1} = a_{j_2j_21}$ possesses the following properties as in Eqs. (2.56)–(2.63):

- If

$$\frac{a_{j_1j_10}(a_{j_1j_1s_1} - a_{j_1j_1s_2})^2}{a_{j_2j_20}} > 0 \quad \text{and} \quad G_{j_1j_10} > 0,$$

then

$$(a_{j_1j_1s_1}, a_{j_2j_21}) = \underbrace{(\mathrm{II}, \mathrm{SO})}_{\text{increasing-inflection source}}.$$

- If

$$\frac{a_{j_1j_10}(a_{j_1j_1s_1}-a_{j_1j_1s_2})^2}{a_{j_2j_20}}<0 \quad \text{and} \quad G_{j_1j_10}>0,$$

then

$$(a_{j_1j_1s_1},a_{j_2j_21})=\underbrace{(\text{DI},\text{SO})}_{\text{decreasing-inflection source}}.$$

- If

$$\frac{a_{j_1j_10}(a_{j_1j_1s_1}-a_{j_1j_1s_2})^2}{a_{j_2j_20}}<0 \quad \text{and} \quad G_{j_1j_10}<0,$$

then

$$(a_{j_1j_1s_1},a_{j_2j_21})=\underbrace{(\text{DI},\text{SI})}_{\text{decreasing-inflection sink}}.$$

- If

$$\frac{a_{j_1j_10}(a_{j_1j_1s_1}-a_{j_1j_1s_2})^2}{a_{j_2j_20}}>0 \quad \text{and} \quad G_{j_1j_10}<0,$$

then

$$(a_{j_1j_1s_1},a_{j_2j_21})=\underbrace{(\text{II},\text{SI})}_{\text{increasing-inflection sink}}.$$

($\text{ii}_{2\text{b}}$) From ($\text{ii}_{1\text{a,c}}$) and ($\text{ii}_{2\text{a}}$), the inflection-source and sink infinite-equilibriums at $y_{j_1}^*=a_{j_1j_1s_1}=a_{j_2j_21}$ are the switching bifurcations for the parabola flows at $y_{j_1}^*=a_{j_1j_1s_1}$ and $y_{j_1}^*=a_{j_2j_21}$ as in Eqs. (2.64)–(2.67):

- *For* $a_{j_1j_10}(a_{j_1j_1s_1}-a_{j_1j_1s_2})^2>0$ and $a_{j_2j_20}>0$,

$$\left\{\begin{matrix}\underbrace{(\text{nF},\text{DP})}_{\text{down-parabola }(-)}\\ \underbrace{(\text{UP},\text{pF})}_{\text{up-parabola }(+)}\end{matrix}\right\}\rightleftharpoons\underbrace{(\text{II},\text{SO})}_{\text{switching bifurcation}}\rightleftharpoons\left\{\begin{matrix}\underbrace{(\text{DP},\text{nF})}_{\text{down-parabola }(-)}\\ \underbrace{(\text{pF},\text{UP})}_{\text{up-parabola }(+)}\end{matrix}\right\}$$

for $(\alpha_{j_1j_1},\alpha_{j_1j_2})=(+,+)$ and $G_{j_1j_10}>0$;

$$\left\{\begin{array}{c}\underbrace{(\mathrm{nF},\mathrm{DP})}_{\text{down-parabola }(-)}\\ \underbrace{(\mathrm{DP},\mathrm{pF})}_{\text{down-parabola }(+)}\end{array}\right\} \rightleftharpoons \underbrace{(\mathrm{II},\mathrm{SO})}_{\text{switching bifurcation}} \rightleftharpoons \left\{\begin{array}{c}\underbrace{(\mathrm{UP},\mathrm{nF})}_{\text{up-parabola }(-)}\\ \underbrace{(\mathrm{pF},\mathrm{UP})}_{\text{up-parabola }(+)}\end{array}\right\}$$

for $(\alpha_{j_1j_1},\alpha_{j_1j_2})=(+,-)$ and $G_{j_1j_10}>0$;

$$\left\{\begin{array}{c}\underbrace{(\mathrm{UP},\mathrm{nF})}_{\text{up-parabola }(-)}\\ \underbrace{(\mathrm{pF},\mathrm{UP})}_{\text{up-parabola }(+)}\end{array}\right\} \rightleftharpoons \underbrace{(\mathrm{II},\mathrm{SI})}_{\text{switching bifurcation}} \rightleftharpoons \left\{\begin{array}{c}\underbrace{(\mathrm{nF},\mathrm{DP})}_{\text{down-parabola }(-)}\\ \underbrace{(\mathrm{DP},\mathrm{pF})}_{\text{down-parabola }(+)}\end{array}\right\}$$

for $(\alpha_{j_1j_1},\alpha_{j_1j_2})=(+,-)$ and $G_{j_1j_10}<0$.

- *For* $a_{j_1j_10}(a_{j_1j_1s_1}-a_{j_1j_1s_2})^2<0$ and $a_{j_2j_20}>0$,

$$\left\{\begin{array}{c}\underbrace{(\mathrm{pF},\mathrm{UP})}_{\text{up-p-parabola }(+)}\\ \underbrace{(\mathrm{DP},\mathrm{pF})}_{\text{down-parabola }(+)}\end{array}\right\} \rightleftharpoons \underbrace{(\mathrm{DI},\mathrm{SI})}_{\text{switching bifurcation}} \rightleftharpoons \left\{\begin{array}{c}\underbrace{(\mathrm{UP},\mathrm{nF})}_{\text{up-parabola }(-)}\\ \underbrace{(\mathrm{nF},\mathrm{DP})}_{\text{down-parabola }(-)}\end{array}\right\}$$

for $(\alpha_{j_1j_1},\alpha_{j_1j_2})=(+,+)$ and $G_{j_1j_10}<0$;

$$\left\{\begin{array}{c}\underbrace{(\mathrm{UP},\mathrm{nF})}_{\text{up-parabola }(-)}\\ \underbrace{(\mathrm{nF},\mathrm{DP})}_{\text{down-parabola }(-)}\end{array}\right\} \rightleftharpoons \underbrace{(\mathrm{DI},\mathrm{SO})}_{\text{switching bifurcation}} \rightleftharpoons \left\{\begin{array}{c}\underbrace{(\mathrm{pF},\mathrm{UP})}_{\text{up-parabola }(+)}\\ \underbrace{(\mathrm{DP},\mathrm{pF})}_{\text{down-parabola }(+)}\end{array}\right\}$$

for $(\alpha_{j_1j_1},\alpha_{j_1j_2})=(+,+)$ and $G_{j_1j_10}>0$;

$$\left\{\begin{array}{c}\underbrace{(\mathrm{pF},\mathrm{UP})}_{\text{up-parabola }(+)}\\ \underbrace{(\mathrm{UP},\mathrm{pF})}_{\text{up-parabola }(+)}\end{array}\right\} \rightleftharpoons \underbrace{(\mathrm{DI},\mathrm{SI})}_{\text{switching bifurcation}} \rightleftharpoons \left\{\begin{array}{c}\underbrace{(\mathrm{DP},\mathrm{nF})}_{\text{down-prarabola }(-)}\\ \underbrace{(\mathrm{nF},\mathrm{DP})}_{\text{down-parabola }(-)}\end{array}\right\}$$

for $(\alpha_{j_1j_1},\alpha_{j_1j_2})=(+,-)$ and $G_{j_1j_10}<0$.

- *For* $a_{j_1j_10}(a_{j_1j_1s_1}-a_{j_1j_1s_2})^2>0$ and $a_{j_2j_20}<0$,

$$\left\{\begin{matrix}\underbrace{(\mathrm{nF},\mathrm{UP})}_{\text{up-parabola }(-)}\\ \underbrace{(\mathrm{DP},\mathrm{pF})}_{\text{down-parabola }(+)}\end{matrix}\right\} \rightleftharpoons \underbrace{(\mathrm{DI},\mathrm{SO})}_{\text{switching bifurction}} \rightleftharpoons \left\{\begin{matrix}\underbrace{(\mathrm{UP},\mathrm{pF})}_{\text{up-parabola }(+)}\\ \underbrace{(\mathrm{pF},\mathrm{DP})}_{\text{down-parabola }(+)}\end{matrix}\right\}$$

for $(\alpha_{j_1j_1},\alpha_{j_1j_2})=(+,+)$ and $G_{j_1j_10}>0$;

$$\left\{\begin{matrix}\underbrace{(\mathrm{UP},\mathrm{pF})}_{\text{up-parabola }(+)}\\ \underbrace{(\mathrm{pF},\mathrm{DP})}_{\text{down-parabola }(+)}\end{matrix}\right\} \rightleftharpoons \underbrace{(\mathrm{DI},\mathrm{SI})}_{\text{switching bifurction}} \rightleftharpoons \left\{\begin{matrix}\underbrace{(\mathrm{nF},\mathrm{UP})}_{\text{up-parabola }(-)}\\ \underbrace{(\mathrm{DP},\mathrm{pF})}_{\text{down-parabola }(+)}\end{matrix}\right\}$$

for $(\alpha_{j_1j_1},\alpha_{j_1j_2})=(+,+)$ and $G_{j_1j_10}<0$;

$$\left\{\begin{matrix}\underbrace{(\mathrm{nF},\mathrm{UP})}_{\text{up-parabola }(-)}\\ \underbrace{(\mathrm{UP},\mathrm{nF})}_{\text{up-parabola }(-)}\end{matrix}\right\} \rightleftharpoons \underbrace{(\mathrm{DI},\mathrm{SO})}_{\text{switching bifurction}} \rightleftharpoons \left\{\begin{matrix}\underbrace{(\mathrm{DP},\mathrm{pF})}_{\text{down-parabola }(+)}\\ \underbrace{(\mathrm{pF},\mathrm{DP})}_{\text{down-parabola }(+)}\end{matrix}\right\}$$

for $(\alpha_{j_1j_1},\alpha_{j_1j_2})=(+,-)$ and $G_{j_1j_10}>0$.

- *For* $a_{j_1j_10}(a_{j_1j_1s_1}-a_{j_1j_1s_2})^2<0$ and $a_{j_2j_20}<0$,

$$\left\{\begin{matrix}\underbrace{(\mathrm{pF},\mathrm{DP})}_{\text{down-parabola }(+)}\\ \underbrace{(\mathrm{UP},\mathrm{nF})}_{\text{up-parabola }(-)}\end{matrix}\right\} \rightleftharpoons \underbrace{(\mathrm{II},\mathrm{SI})}_{\text{switching bifurcation}} \rightleftharpoons \left\{\begin{matrix}\underbrace{(\mathrm{DP},\mathrm{pF})}_{\text{down-parabola }(+)}\\ \underbrace{(\mathrm{nF},\mathrm{UP})}_{\text{up-parabola }(-)}\end{matrix}\right\}$$

for $(\alpha_{j_1j_1},\alpha_{j_1j_2})=(+,+)$ and $G_{j_1j_10}<0$;

$$\left\{\begin{matrix}\underbrace{(\mathrm{pF},\mathrm{DP})}_{\text{down-parabola }(+)}\\ \underbrace{(\mathrm{DP},\mathrm{nF})}_{\text{down-parabola }(-)}\end{matrix}\right\} \rightleftharpoons \underbrace{(\mathrm{II},\mathrm{SI})}_{\text{switching bifurcation}} \rightleftharpoons \left\{\begin{matrix}\underbrace{(\mathrm{UP},\mathrm{pF})}_{\text{up-parabola }(+)}\\ \underbrace{(\mathrm{nF},\mathrm{UP})}_{\text{up-parabola }(-)}\end{matrix}\right\}$$

for $(\alpha_{j_1j_1},\alpha_{j_1j_2})=(+,-)$ and $G_{j_1j_10}<0$;

$$\left\{\begin{array}{c}\underbrace{(\mathrm{UP},\mathrm{pF})}_{\text{up-parabola }(+)}\\ \underbrace{(\mathrm{nF},\mathrm{UP})}_{\text{up-parabola }(-)}\end{array}\right\} \rightleftharpoons \underbrace{(\mathrm{II},\mathrm{SO})}_{\text{switching bifurcation}} \rightleftharpoons \left\{\begin{array}{c}\underbrace{(\mathrm{pF},\mathrm{DP})}_{\text{down-parabola }(+)}\\ \underbrace{(\mathrm{DP},\mathrm{nF})}_{\text{down-parabola }(-)}\end{array}\right\}$$

$$\text{for } (\alpha_{j_1j_1},\alpha_{j_1j_2}) = (+,-) \quad \text{and} \quad G_{j_1j_10} > 0.$$

(ii$_3$) For $a_{j_2j_21} = a_{j_1j_1s_2}$ and $a_{j_2j_21} \neq a_{j_1j_1s_1}$, the two-dimensional single-bivariate dynamical system becomes

$$\begin{aligned}\dot{x}_{j_2} &= a_{j_2j_20}(y_{j_1} - a_{j_1j_11}),\\ \dot{x}_{j_1} &= a_{j_1j_10}(y_{j_1} - a_{j_1j_1s_1})(y_{j_1} - a_{j_1j_1s_2})^2.\end{aligned}$$

There is a dynamical system in the y_{j_1}-direction as

$$\dot{y}_{j_1} = (y_{j_1} - a_{j_1j_1s_2})[\alpha_{j_1j_1}a_{j_1j_10}(y_{j_1} - a_{j_1j_1s_1})(y_{j_1} - a_{j_1j_1s_2}) + \alpha_{j_1j_2}a_{j_2j_20})],$$

and a dynamical system in the $y_{j_1}^{\perp}$-direction as

$$\dot{y}_{j_1}^{\perp} = (y_{j_1} - a_{j_1j_1s_2})[\beta_{j_1j_1}a_{j_1j_10}(y_{j_1} - a_{j_1j_1s_1})(y_{j_1} - a_{j_1j_1s_2}) + \beta_{j_1j_2}a_{j_2j_20})].$$

(ii$_{3\mathrm{a}}$) In phase space, at $y_{j_1}^* = a_{j_1j_1s_2} = a_{j_2j_21}$,

$$\left.\frac{dx_{j_1}}{dx_{j_2}}\right|_{y_{j_1}^*=a_{j_1j_1s_2}} = \frac{a_{j_1j_10}(y_{j_1} - a_{j_1j_1s_1})(y_{j_1} - a_{j_1j_1s_2})}{a_{j_2j_20}} = 0.$$

If

$$\begin{aligned}\left.\frac{d^2x_{j_1}}{dx_{j_2}^2}\right|_{y_{j_1}^*=a_{j_1j_1s_2}} &= \frac{a_{j_1j_10}\alpha_{j_1j_2}(a_{j_1j_1s_2} - a_{j_1j_1s_1})}{a_{j_2j_20}}\\ &= \frac{a_{j_1j_10}\alpha_{j_1j_2}^2(a_{j_1j_1s_2} - a_{j_1j_1s_1})}{a_{j_2j_20}\alpha_{j_1j_2}} > 0,\end{aligned}$$

there is an up-parabola flow. If

$$\left.\frac{d^2x_{j_1}}{dx_{j_2}^2}\right|_{y_{j_1}^*=a_{j_1j_1s_2}} = \frac{a_{j_1j_10}\alpha_{j_1j_2}^2(a_{j_1j_1s_2} - a_{j_1j_1s_1})}{a_{j_2j_20}\alpha_{j_1j_2}} < 0,$$

there is a down-parabola flow. In addition, the variational equation at $y_{j_1}^* = a_{j_1j_1s_2} = a_{j_2j_21}$ is

$$\begin{aligned}\Delta\dot{y}_{j_1} &= [2\alpha_{j_1j_1}a_{j_1j_10}(a_{j_1j_1s_2} - a_{j_1j_1s_1})(a_{j_1j_1s_2} - a_{j_1j_1s_2}) + \alpha_{j_1j_2}a_{j_2j_20}]\Delta y_{j_1}\\ &= \alpha_{j_1j_2}a_{j_2j_20}\Delta y_{j_1} = G_{j_1j_10}\Delta y_{j_1},\end{aligned}$$

where

$$G_{j_1j_10} = \frac{\partial}{\partial y_{j_1}}[\mathbf{n}_{j_1}^{\mathrm{T}} \cdot \mathbf{F}(y_{j_1}, \mathbf{p})]\bigg|_{y_{j_1}=a_{j_1j_1s_2}} = a_{j_2j_20}\alpha_{j_1j_2}.$$

The flows at $y_{j_1}^* = a_{j_1j_1s_1} = a_{j_2j_21}$ are source and sink in the x_{j_2}-direction for $G_{j_1j_10} > 0$ and $G_{j_1j_10} < 0$, respectively.

Therefore, the infinite-equilibrium of $y_{j_1}^* = a_{j_1j_1s_2}$ for the x_{j_1}-direction and $y_{j_1}^* = a_{j_2j_21}$ for the x_{j_2}-direction possess the following properties as in Eqs. (2.72)–(2.75):

- For $a_{j_1j_10}\alpha_{j_1j_2}^2(a_{j_1j_1s_2} - a_{j_1j_1s_1}) > 0$ and $a_{j_2j_20}\alpha_{j_1j_2} > 0$,

$$(a_{j_1j_1s_2}, a_{j_2j_21}) = \underbrace{(\mathrm{UP}, \mathrm{SO})}_{\text{up-parabola source}}.$$

- For $a_{j_1j_10}\alpha_{j_1j_2}^2(a_{j_1j_1s_2} - a_{j_1j_1s_1}) < 0$ and $a_{j_2j_20}\alpha_{j_1j_2} > 0$,

$$(a_{j_1j_1s_2}, a_{j_2j_21}) = \underbrace{(\mathrm{DP}, \mathrm{SO})}_{\text{down-parabola source}}.$$

- For $a_{j_1j_10}\alpha_{j_1j_2}^2(a_{j_1j_1s_2} - a_{j_1j_1s_1}) > 0$ and $a_{j_2j_20}\alpha_{j_1j_2} < 0$,

$$(a_{j_1j_1s_2}, a_{j_2j_21}) = \underbrace{(\mathrm{DP}, \mathrm{SI})}_{\text{down-parabola sink}}.$$

- For $a_{j_1j_10}\alpha_{j_1j_2}^2(a_{j_1j_1s_2} - a_{j_1j_1s_1}) < 0$ and $a_{j_2j_20}\alpha_{j_1j_2} < 0$,

$$(a_{j_1j_1s_2}, a_{j_2j_21}) = \underbrace{(\mathrm{UP}, \mathrm{SI})}_{\text{up-parabola sink}}.$$

(ii$_{3\mathrm{b}}$) From (ii$_{1\mathrm{b,c}}$) and (ii$_{3\mathrm{a}}$), for $a_{j_2j_21} = a_{j_1j_11}$, the infinite-equilibrium of $y_{j_1}^* = a_{j_1j_11}$ is a switching bifurcation as in Eqs. (2.76)–(2.79):

- For $a_{j_1j_10}\alpha_{j_1j_2}^2(a_{j_1j_1s_2} - a_{j_1j_1s_1}) > 0$ and $a_{j_2j_20}\alpha_{j_1j_2} > 0$,

$$\left\{\begin{matrix} \underbrace{(\mathrm{pF}, \mathrm{UP})}_{\text{up-parabola flow}} \\ \underbrace{(\mathrm{II}, \mathrm{pF})}_{\text{increasing-inflection flow}} \end{matrix}\right\} \rightleftharpoons \underbrace{(\mathrm{UP}, \mathrm{SO})}_{\text{switching bifurcation}} \rightleftharpoons \left\{\begin{matrix} \underbrace{(\mathrm{DI}, \mathrm{nF})}_{\text{decreasing-inflection flow}} \\ \underbrace{(\mathrm{pF}, \mathrm{UP})}_{\text{up-parabola flow}} \end{matrix}\right\}.$$

- For $a_{j_1j_10}\alpha_{j_1j_2}^2(a_{j_1j_1s_2} - a_{j_1j_1s_1}) < 0$ and $a_{j_2j_20}\alpha_{j_1j_2} > 0$,

$$\left\{\begin{array}{c}\underbrace{(\mathrm{nF},\mathrm{DP})}_{\text{down-parabola flow}}\\ \underbrace{(\mathrm{DI},\mathrm{pF})}_{\text{decreasing-inflection flow}}\end{array}\right\} \rightleftharpoons \underbrace{(\mathrm{DP},\mathrm{SO})}_{\text{iswitching bifurcation}} \rightleftharpoons \left\{\begin{array}{c}\underbrace{(\mathrm{II},\mathrm{nF})}_{\text{increasing-inflection flow}}\\ \underbrace{(\mathrm{nF},\mathrm{DP})}_{\text{down-parabola flow}}\end{array}\right\}.$$

- For $a_{j_1j_10}\alpha_{j_1j_2}^2(a_{j_1j_1s_2} - a_{j_1j_1s_1}) > 0$ and $a_{j_2j_20}\alpha_{j_1j_2} < 0$,

$$\left\{\begin{array}{c}\underbrace{(\mathrm{pF},\mathrm{DP})}_{\text{down-parabola flow}}\\ \underbrace{(\mathrm{DI},\mathrm{nF})}_{\text{decreasing-inflection flow}}\end{array}\right\} \rightleftharpoons \underbrace{(\mathrm{DP},\mathrm{SI})}_{\text{iswitching bifurcation}} \rightleftharpoons \left\{\begin{array}{c}\underbrace{(\mathrm{II},\mathrm{pF})}_{\text{increasing-inflection flow}}\\ \underbrace{(\mathrm{pF},\mathrm{DP})}_{\text{down-parabola flow}}\end{array}\right\}.$$

- For $a_{j_1j_10}\alpha_{j_1j_2}^2(a_{j_1j_1s_2} - a_{j_1j_1s_1}) < 0$ and $a_{j_2j_20}\alpha_{j_1j_2} < 0$,

$$\left\{\begin{array}{c}\underbrace{(\mathrm{nF},\mathrm{UP})}_{\text{up-parabola flow}}\\ \underbrace{(\mathrm{II},\mathrm{nF})}_{\text{increasing-inflection flow}}\end{array}\right\} \rightleftharpoons \underbrace{(\mathrm{UP},\mathrm{SI})}_{\text{switching bifurcation}} \rightleftharpoons \left\{\begin{array}{c}\underbrace{(\mathrm{DI},\mathrm{pF})}_{\text{decreasing-inflection flow}}\\ \underbrace{(\mathrm{nF},\mathrm{UP})}_{\text{up-parabola flow}}\end{array}\right\}.$$

(iii) For $a_{j_1j_1s_1} = a_{j_1j_1s_2} = a_{j_1j_1 1}$, the two-dimensional single-bivariate dynamical system becomes

$$\begin{aligned}\dot{x}_{j_2} &= a_{j_2j_20}(y_{j_1} - a_{j_2j_21}),\\ \dot{x}_{j_1} &= a_{j_1j_10}(y_{j_1} - a_{j_1j_11})^3.\end{aligned}$$

(iii$_1$) For $a_{j_2j_21} \neq a_{j_1j_11}$, there is a dynamical system in the y_{j_1}-direction as

$$\dot{y}_{j_1}^{\perp} = \alpha_{j_1j_1}a_{j_1j_10}(y_{j_1} - a_{j_1j_11})^3 + \alpha_{j_1j_2}a_{j_2j_20}(y_{j_1} - a_{j_2j_21})$$

with a dynamical system in the $y_{j_1}^{\perp}$-direction as

$$\dot{y}_{j_1}^{\perp} = \beta_{j_1j_1}a_{j_1j_10}(y_{j_1} - a_{j_1j_11})^3 + \beta_{j_1j_2}a_{j_2j_20}(y_{j_1} - a_{j_2j_21}).$$

(iii_{1a}) *In phase space*, for $a_{j_1j_1} \neq a_{j_2j_21}$, at $y^*_{j_1} = a_{j_1j_11}$,

$$\left.\frac{dx_{j_1}}{dx_{j_2}}\right|_{y^*_{j_1}=a_{j_1j_11}} = \left.\frac{a_{j_1j_10}}{a_{j_2j_20}}\frac{(y_{j_1}-a_{j_1j_11})^3}{(y_{j_1}-a_{j_2j_21})}\right|_{y^*_{j_1}=a_{j_1j_11}} = 0,$$

$$\left.\frac{d^2x_{j_1}}{dx^2_{j_2}}\right|_{y^*_{j_1}=a_{j_1j_11}} = \left.\frac{d^3x_{j_1}}{dx^3_{j_2}}\right|_{y^*_{j_1}=a_{j_1j_11}} = 0.$$

If

$$\left.\frac{d^4x_{j_1}}{dx^4_{j_2}}\right|_{y^*_{j_1}=a_{j_1j_11}} = \frac{a_{j_1j_10}}{a_{j_2j_20}}\frac{6\alpha^3_{j_1j_2}}{(a_{j_1j_11}-a_{j_2j_21})} > 0,$$

there is a third-order up-parabola flow at $y^*_{j_1} = a_{j_1j_11}$ in the x_{j_1}-direction. If

$$\left.\frac{d^4x_{j_1}}{dx^4_{j_2}}\right|_{y^*_{j_1}=a_{j_1j_11}} = \frac{a_{j_1j_10}}{a_{j_2j_20}}\frac{6\alpha^3_{j_1j_2}}{(a_{j_1j_11}-a_{j_2j_21})} < 0,$$

there is a third-order down-parabola flow at $y^*_{j_1} = a_{j_1j_11}$ in the x_{j_1}-direction. *Let*

$$\dot{x}^*_{j_2} = a_{j_2j_20}(a_{j_1j_11} - a_{j_2j_21}).$$

For $\dot{x}^*_{j_2} > 0$ and $\dot{x}^*_{j_2} < 0$, the third-order parabola flows are positive and negative, respectively.

Therefore, the equilibrium on the line of $y^*_{j_1} = a_{j_1j_11}$ for the x_{j_1}-direction with $\dot{x}^*_{j_2} = a_{j_2j_20}(a_{j_1j_11} - a_{j_2j_21})$ possesses the following properties as in Eqs. (2.83)–(2.86):

- For $a_{j_1j_10}a^3_{j_1j_2} > 0$ and $a_{j_2j_20}(a_{j_1j_11} - a_{j_2j_21}) > 0$,

$$(a_{j_1j_1s_1}, \dot{x}^*_{j_2}) = \underbrace{(3^{\text{rd}}\text{UP}, \text{pF})}_{\text{third-order up-parabola flow }(+)}.$$

- For $a_{j_1j_10}a^3_{j_1j_2} < 0$ and $a_{j_2j_20}(a_{j_1j_11} - a_{j_2j_21}) > 0$,

$$(a_{j_1j_1s_1}, \dot{x}^*_{j_2}) = \underbrace{(3^{\text{rd}}\,\text{DP}, \text{pF})}_{\text{third-order down-parabola flow }(+)}.$$

- For $a_{j_1j_10}a^3_{j_1j_2} > 0$ and $a_{j_2j_20}(a_{j_1j_11} - a_{j_2j_21}) < 0$,

$$(a_{j_1j_1s_1}, \dot{x}^*_{j_2}) = \underbrace{(3^{\text{rd}}\,\text{DP}, \text{nF})}_{\text{third-order down-parabola flow }(-)}.$$

- For $a_{j_1j_10}a^3_{j_1j_2} < 0$ and $a_{j_2j_20}(a_{j_1j_11} - a_{j_2j_21}) < 0$,

$$(a_{j_1j_1s_1}, \dot{x}^*_{j_2}) = \underbrace{(3^{\text{rd}}\ \text{UP}, \text{nF})}_{\text{third-order up-parabola flow }(-)}.$$

– The third-order up-parabola flows are the appearing bifurcations from an up-parabola flow to up-parabola, down-parabola, and up-parabola flows.
– The third-order down-parabola flows are the appearing bifurcations from a down-parabola flow to down-parabola, up-parabola, and down-parabola flows.

(iii$_{1b}$) In phase space, for $y^*_{j_1} = a_{j_2j_21} \neq a_{j_1j_11}$,

$$\left.\frac{dx_{j_2}}{dx_{j_1}}\right|_{y^*_{j_1}=a_{j_2j_21}} = \left.\frac{a_{j_2j_20}}{a_{j_1j_10}} \frac{(y_{j_1} - a_{j_2j_21})}{(y_{j_1} - a_{j_1j_11})^3}\right|_{y^*_{j_1}=a_{j_2j_21}} = 0.$$

If

$$\left.\frac{d^2x_{j_2}}{dx^2_{j_1}}\right|_{y^*_{j_1}=a_{j_2j_21}} = \frac{a_{j_2j_20}}{a_{j_1j_10}} \frac{\alpha_{j_1j_1}}{(a_{j_2j_21} - a_{j_1j_11})^3} > 0,$$

there is an up-parabola flow at $y^*_{j_2} = a_{j_2j_21}$ in the x_{j_1}-direction. If

$$\left.\frac{d^2x_{j_2}}{dx^2_{j_1}}\right|_{y^*_{j_1}=a_{j_1j_2s_1}} = \frac{a_{j_2j_20}}{a_{j_1j_10}} \frac{a_{j_1j_1}}{(a_{j_2j_21} - a_{j_1j_11})^3} < 0,$$

there is a down-parabola flow at $y^*_{j_1} = a_{j_1j_11}$ in the x_{j_1}-direction. With

$$\dot{x}^*_{j_1} = a_{j_1j_10}(a_{j_2j_21} - a_{j_1j_11})^3,$$

therefore, the simple equilibrium of $y^*_{j_1} = a_{j_2j_21}$ in the x_{j_2}-direction for such a system has the following properties as in Eqs. (2.87)–(2.90):

- For $a_{j_1j_10}(a_{j_2j_21} - a_{j_1j_11})^3 > 0$ and $\alpha_{j_1j_1}a_{j_2j_20} > 0$,

$$(\dot{x}^*_{j_1}, a_{j_2j_21}) = \underbrace{(\text{pF}, \text{UP})}_{\text{up-parabola flow }(+)}.$$

- For $a_{j_1j_10}(a_{j_2j_21} - a_{j_1j_1s_1})(a_{j_2j_21} - a_{j_1j_1s_2})^2 < 0$ and $\alpha_{j_1j_1}a_{j_2j_20} > 0$,

$$(\dot{x}^*_{j_1}, a_{j_2j_21}) = \underbrace{(\text{nF}, \text{DP})}_{\text{down-parabola flow }(-)}.$$

- For $a_{j_1j_10}(a_{j_2j_21}-a_{j_1j_1s_1})(a_{j_2j_21}-a_{j_1j_1s_2})^2>0$ and $\alpha_{j_1j_1}a_{j_2j_20}<0$,

$$(\dot{x}^*_{j_1}, a_{j_2j_21}) = \underbrace{(\text{pF}, \text{DP})}_{\text{down-parabola flow }(+)}.$$

- For $a_{j_1j_10}(a_{j_2j_21}-a_{j_1j_1s_1})(a_{j_2j_21}-a_{j_1j_1s_2})^2<0$ and $\alpha_{j_1j_1}a_{j_2j_20}<0$,

$$(\dot{x}^*_{j_1}, a_{j_2j_21}) = \underbrace{(\text{nF}, \text{UP})}_{\text{up-parabola flow }(-)}.$$

(iii$_2$) For $a_{j_2j_21}=a_{j_1j_11}$, there is a dynamical system in the y_{j_1}-direction as

$$\dot{y}_{j_1} = \alpha_{j_1j_1}a_{j_1j_10}(y_{j_1}-a_{j_1j_1})[(y_{j_1}-a_{j_1j_1})^2+c_{j_1j_1}]$$

with a dynamical system in the $y^{\perp}_{j_1}$-direction as

$$\dot{y}^{\perp}_{j_1} = \beta_{j_1j_1}a_{j_1j_10}(y_{j_1}-a_{j_1j_11})[(y_{j_1}-a_{j_1j_11})^2+d_{j_1j_1}].$$

(iii$_{2a}$) In phase space, for $y^*_{j_1}=a_{j_1j_11}=a_{j_2j_21}$,

$$\left.\frac{dx_{j_1}}{dx_{j_2}}\right|_{y^*_{j_1}=a_{j_1j_11}} = \left.\frac{a_{j_1j_10}(y_{j_1}-a_{j_1j_11})^2}{a_{j_2j_20}}\right|_{y^*_{j_1}=a_{j_1j_11}} = 0,$$

$$\left.\frac{d^2x_{j_1}}{dx^2_{j_2}}\right|_{y^*_{j_1}=a_{j_1j_11}} = 0.$$

If

$$\left.\frac{d^3x_{j_1}}{dx^3_{j_2}}\right|_{y^*_{j_1}=a_{j_1j_11}} = \frac{2a_{j_1j_10}\alpha^2_{j_1j_2}}{a_{j_2j_20}} > 0,$$

there is an increasing-inflection flow at $y^*_{j_1}=a_{j_1j_11}=a_{j_2j_21}$ in the x_{j_1}-direction. If

$$\left.\frac{d^3x_{j_1}}{dx^3_{j_2}}\right|_{y^*_{j_1}=a_{j_1j_11}} = \frac{2\alpha^2_{j_1j_2}a_{j_1j_10}}{a_{j_2j_20}} < 0,$$

there is a decreasing-inflection flow at $y^*_{j_1}=a_{j_1j_11}=a_{j_2j_21}$ in the x_{j_1}-direction. The variant equation is

$$\begin{aligned}\Delta\dot{y}_{j_1} &= [3\alpha_{j_1j_1}a_{j_1j_10}(a_{j_1j_11}-a_{j_1j_11})^2+\alpha_{j_1j_2}a_{j_2j_20}]\Delta y_{j_1}\\ &= \alpha_{j_1j_2}a_{j_2j_20}\Delta y_{j_1} = G_{j_1j_10}\Delta y_{j_1},\end{aligned}$$

where

$$G_{j_1 j_1 0} = \frac{\partial}{\partial y_{j_1}}[\mathbf{n}_{j_1}^{\mathrm{T}} \cdot \mathbf{F}(y_{j_1}, \mathbf{p})]\bigg|_{y_{j_1}=a_{j_1 j_1 1}} = a_{j_2 j_2 0}\alpha_{j_1 j_2}.$$

The flows at $y_{j_1}^* = a_{j_1 j_1 1} = a_{j_2 j_2 1}$ are source and sink in the y_{j_1}-direction for $G_{j_1 j_1 0} > 0$ and $G_{j_1 j_1 0} < 0$, respectively.

(iii_{2a}) The infinite-equilibriums of $y_{j_1}^* = a_{j_1 j_1 1}$ for the x_{j_1}-direction and $y_{j_1}^* = a_{j_2 j_2 1}$ for the x_{j_2}-direction possess the following properties as in Eqs. (2.94)–(2.97):

- For $\alpha_{j_1 j_2}^3 a_{j_1 j_1 0} > 0$ and $\alpha_{j_1 j_2} a_{j_2 j_2 0} > 0$,

$$(a_{j_1 j_1 1}, a_{j_2 j_2 1}) = \underbrace{(2^{\text{nd}}\text{II}, \text{SO})}_{\text{second-order increasing-inflection source}}.$$

- For $\alpha_{j_1 j_2}^3 a_{j_1 j_1 0} < 0$ and $\alpha_{j_1 j_2} a_{j_2 j_2 0} > 0$,

$$(a_{j_1 j_1 1}, a_{j_2 j_2 1}) = \underbrace{(2^{\text{nd}}\text{DI}, \text{SO})}_{\text{second-order decreasing-inflection source}}.$$

- For $\alpha_{j_1 j_2}^3 a_{j_1 j_1 0} > 0$ and $\alpha_{j_1 j_2} a_{j_2 j_2 0} < 0$,

$$(a_{j_1 j_1 1}, a_{j_2 j_2 1}) = \underbrace{(2^{\text{nd}}\text{DI}, \text{SI})}_{\text{second-order decreasing-inflection sink}}.$$

- For $\alpha_{j_1 j_2}^3 a_{j_1 j_1 0} < 0$ and $\alpha_{j_1 j_2} a_{j_2 j_2 0} < 0$,

$$(a_{j_1 j_1 1}, a_{j_2 j_2 1}) = \underbrace{(2^{\text{nd}}\text{II}, \text{SI})}_{\text{second-order increasing-inflection sink}}.$$

(iii_{2b}) For $a_{j_2 j_2 1} = a_{j_1 j_1 1}$, the infinite-equilibrium of $y_{j_1}^* = a_{j_1 j_1 1}$ is a switching bifurcation as in Eqs. (2.98)–(2.105):

- For $\alpha^3_{j_1j_2}a_{j_1j_10} > 0$ and $\alpha_{j_1j_2}a_{j_2j_20} > 0$,

$$\left\{\begin{array}{c}\underbrace{(\mathrm{nF},\mathrm{DP})}_{\text{down-parabola flow }(-)}\\ \underbrace{(3^{\mathrm{rd}}\mathrm{UP},\mathrm{pF})}_{\text{third-order up-parabola flow }(+)}\end{array}\right\} \rightleftharpoons \underbrace{(2^{\mathrm{nd}}\mathrm{II},\mathrm{SO})}_{\text{switching bifurcation}} \rightleftharpoons \left\{\begin{array}{c}\underbrace{(3^{\mathrm{rd}}\mathrm{DP},\mathrm{nF})}_{\text{third-order down-parabola flow }(-)}\\ \underbrace{(\mathrm{pF},\mathrm{UP})}_{\text{up-parabola flow }(+)}\end{array}\right\}$$

for $\alpha_{j_1j_2} > 0$;

$$\left\{\begin{array}{c}\underbrace{(\mathrm{nF},\mathrm{DP})}_{\text{down-parabola flow }(-)}\\ \underbrace{(3^{\mathrm{rd}}\mathrm{DP},\mathrm{pF})}_{\text{third-order down-parabola flow }(+)}\end{array}\right\} \rightleftharpoons \underbrace{(2^{\mathrm{nd}}\mathrm{II},\mathrm{SO})}_{\text{switching bifurcation}} \rightleftharpoons \left\{\begin{array}{c}\underbrace{(3^{\mathrm{rd}}\mathrm{UP},\mathrm{nF})}_{\text{third-order up-parabola flow }(-)}\\ \underbrace{(\mathrm{pF},\mathrm{UP})}_{\text{up-parabola flow }(+)}\end{array}\right\}$$

for $\alpha_{j_1j_2} < 0$.

- For $\alpha^3_{j_1j_2}a_{j_1j_10} < 0$ and $\alpha_{j_1j_2}a_{j_2j_20} > 0$,

$$\left\{\begin{array}{c}\underbrace{(\mathrm{pF},\mathrm{UP})}_{\text{up-parabola flow }(+)}\\ \underbrace{(2^{\mathrm{nd}}\mathrm{DP},\mathrm{pF})}_{\text{third-order down-parabola flow }(+)}\end{array}\right\} \rightleftharpoons \underbrace{(2^{\mathrm{nd}}\mathrm{DI},\mathrm{SO})}_{\text{switching bifurcation}} \rightleftharpoons \left\{\begin{array}{c}\underbrace{(3^{\mathrm{rd}}\mathrm{UP},\mathrm{nF})}_{\text{third-order up-parabola flow }(-)}\\ \underbrace{(\mathrm{nF},\mathrm{DP})}_{\text{down-parabola flow }(-)}\end{array}\right\}$$

for $\alpha_{j_1j_2} > 0$;

$$\left\{\begin{array}{c}\underbrace{(\mathrm{pF},\mathrm{UP})}_{\text{up-parabola flow }(+)}\\ \underbrace{(3^{\mathrm{rd}}\mathrm{UP},\mathrm{pF})}_{\text{third-order up-parabola flow }(+)}\end{array}\right\}\rightleftharpoons\underbrace{(2^{\mathrm{nd}}\mathrm{DI},\mathrm{SO})}_{\text{switching bifurcation}}$$

$$\rightleftharpoons\left\{\begin{array}{c}\underbrace{(3^{\mathrm{rd}}\mathrm{DP},\mathrm{nF})}_{\text{third-order down-parabola flow }(-)}\\ \underbrace{(\mathrm{nF},\mathrm{DP})}_{\text{down-parabola flow }(-)}\end{array}\right\}$$

- For $\alpha^{3}_{j_1j_2}a_{j_1j_10}>0$ and $\alpha_{j_1j_2}a_{j_2j_20}<0$,

$$\left\{\begin{array}{c}\underbrace{(\mathrm{nF},\mathrm{UP})}_{\text{up-parabola flow }(-)}\\ \underbrace{(3^{\mathrm{rd}}\mathrm{DP},\mathrm{nF})}_{\text{third-order down-parabola flow }(-)}\end{array}\right\}\rightleftharpoons\underbrace{(2^{\mathrm{nd}}\mathrm{DI},\mathrm{SI})}_{\text{switching bifurcation}}$$

$$\rightleftharpoons\left\{\begin{array}{c}\underbrace{(3^{\mathrm{rd}}\mathrm{UP},\mathrm{pF})}_{\text{third-order up-parabola flow }(+)}\\ \underbrace{(\mathrm{pF},\mathrm{DP})}_{\text{down-parabola flow }(+)}\end{array}\right\}$$

for $\alpha_{j_1j_2}>0$;

$$\left\{\begin{array}{c}\underbrace{(\mathrm{nF},\mathrm{UP})}_{\text{up-parabola flow }(-)}\\ \underbrace{(3^{\mathrm{rd}}\mathrm{UP},\mathrm{nF})}_{\text{third-order up-parabola flow }(-)}\end{array}\right\}\rightleftharpoons\underbrace{(2^{\mathrm{nd}}\mathrm{DI},\mathrm{SI})}_{\text{switching bifurcation}}$$

$$\rightleftharpoons\left\{\begin{array}{c}\underbrace{(3^{\mathrm{rd}}\mathrm{DP},\mathrm{pF})}_{\text{third-order down-parabola flow }(+)}\\ \underbrace{(\mathrm{pF},\mathrm{DP})}_{\text{down-parabola flow }(+)}\end{array}\right\}$$

for $\alpha_{j_1j_2}<0$.

- For $\alpha_{j_1j_2}^3 a_{j_1j_10} < 0$ and $\alpha_{j_1j_2} a_{j_2j_20} < 0$,

$$\left\{\begin{array}{c}\underbrace{(\text{pF},\text{DP})}_{\text{down-parabola flow }(+)}\\ \underbrace{(3^{\text{rd}}\text{UP},\text{nF})}_{\text{third-order up-parabola flow }(-)}\end{array}\right\} \rightleftharpoons \underbrace{(2^{\text{nd}}\text{II},\text{SI})}_{\text{switching bifurcation}}$$

$$\rightleftharpoons \left\{\begin{array}{c}\underbrace{(3^{\text{rd}}\text{DP},\text{pF})}_{\text{third-order down-parabola flow }(+)}\\ \underbrace{(\text{nF},\text{UP})}_{\text{up-parabola flow }(-)}\end{array}\right\}$$

for $\alpha_{j_1j_2} > 0$;

$$\left\{\begin{array}{c}\underbrace{(\text{pF},\text{DP})}_{\text{down-parabola flow }(+)}\\ \underbrace{(3^{\text{rd}}\text{DP},\text{nF})}_{\text{third-order down-parabola flow }(-)}\end{array}\right\} \rightleftharpoons \underbrace{(2^{\text{nd}}\text{II},\text{SI})}_{\text{switching bifurcation}}$$

$$\rightleftharpoons \left\{\begin{array}{c}\underbrace{(3^{\text{rd}}\text{UP},\text{pF})}_{\text{third-order up-parabola flow }(+)}\\ \underbrace{(\text{nF},\text{UP})}_{\text{up-parabola flow }(-)}\end{array}\right\}$$

for $\alpha_{j_1j_2} < 0$.

(iii_{2c}) For $a_{j_2j_21} = a_{j_1j_11}$, the infinite-equilibrium of $y_{j_1}^* = a_{j_1j_11}$ is an appearing bifurcation as in Eqs. (2.106)–(2.113):

- For $\alpha_{j_1j_2}^3 a_{j_1j_10} > 0$ and $\alpha_{j_1j_2} a_{j_2j_20} > 0$,

$$\left\{\begin{array}{c}\underbrace{(\text{nF},\text{DP})}_{\text{down-parabola flow }(-)}\\ \underbrace{(\text{UP},\text{pF})}_{\text{up-parabola flow }(+)}\end{array}\right\} \rightleftharpoons \underbrace{(2^{\text{nd}}\text{II},\text{SO})}_{\text{switching bifurcation}} \rightleftharpoons \left\{\begin{array}{c}\underbrace{(\text{DP},\text{nF})}_{\text{down-parabola flow }(-)}\\ \underbrace{(\text{UP},\text{nF})}_{\text{up-parabola flow }(-)}\\ \underbrace{(\text{nF},\text{DP})}_{\text{down-parabola flow }(-)}\\ \underbrace{(\text{UP},\text{pF})}_{\text{up-parabola flow }(+)}\end{array}\right\};$$

$$\left\{\begin{array}{c}\underbrace{(\mathrm{DP},\mathrm{nF})}_{\text{down-parabola flow }(-)}\\ \underbrace{(\mathrm{pF},\mathrm{UP})}_{\text{up-parabola flow }(+)}\end{array}\right\}\rightleftharpoons\underbrace{(2^{\mathrm{nd}}\mathrm{II},\mathrm{SO})}_{\text{switching bifurcation}}\rightleftharpoons\left\{\begin{array}{c}\underbrace{(\mathrm{DP},\mathrm{nF})}_{\text{down-parabola flow }(-)}\\ \underbrace{(\mathrm{pF},\mathrm{UP})}_{\text{up-parabola flow }(+)}\\ \underbrace{(\mathrm{DP},\mathrm{pF})}_{\text{down-parabola flow }(+)}\\ \underbrace{(\mathrm{UP},\mathrm{pF})}_{\text{up-parabola flow }(+)}\end{array}\right\}$$

for $\alpha_{j_1j_2} > 0$.

$$\left\{\begin{array}{c}\underbrace{(\mathrm{nF},\mathrm{DP})}_{\text{down-parabola flow }(-)}\\ \underbrace{(\mathrm{DP},\mathrm{pF})}_{\text{down-parabola flow }(+)}\end{array}\right\}\rightleftharpoons\underbrace{(2^{\mathrm{nd}}\mathrm{II},\mathrm{SO})}_{\text{switching bifurcation}}\rightleftharpoons\left\{\begin{array}{c}\underbrace{(\mathrm{DP},\mathrm{nF})}_{\text{down-parabola flow }(-)}\\ \underbrace{(\mathrm{UP},\mathrm{nF})}_{\text{up-parabola flow }(-)}\\ \underbrace{(\mathrm{nF},\mathrm{DP})}_{\text{down-parabola flow }(-)}\\ \underbrace{(\mathrm{DP},\mathrm{pF})}_{\text{down-parabola flow }(+)}\end{array}\right\};$$

$$\left\{\begin{array}{c}\underbrace{(\mathrm{UP},\mathrm{nF})}_{\text{up-parabola flow }(-)}\\ \underbrace{(\mathrm{pF},\mathrm{UP})}_{\text{up-parabola flow }(+)}\end{array}\right\}\rightleftharpoons\underbrace{(2^{\mathrm{nd}}\mathrm{II},\mathrm{SO})}_{\text{switching bifurcation}}\rightleftharpoons\left\{\begin{array}{c}\underbrace{(\mathrm{UP},\mathrm{nF})}_{\text{up-parabola flow }(-)}\\ \underbrace{(\mathrm{pF},\mathrm{UP})}_{\text{up-parabola flow }(+)}\\ \underbrace{(\mathrm{UP},\mathrm{pF})}_{\text{up-parabola flow }(+)}\\ \underbrace{(\mathrm{DP},\mathrm{pF})}_{\text{down-parabola flow }(+)}\end{array}\right\}$$

for $\alpha_{j_1j_2} < 0$.

- *For* $\alpha^3_{j_1j_2}a_{j_1j_10} < 0$ and $\alpha_{j_1j_2}a_{j_2j_20} > 0$,

$$\left\{\begin{matrix}\underbrace{(\mathrm{pF},\mathrm{UP})}_{\text{up-parabola flow }(+)}\\ \underbrace{(\mathrm{DP},\mathrm{pF})}_{\text{down-parabola flow }(+)}\end{matrix}\right\} \rightleftharpoons \underbrace{(2^{\mathrm{nd}}\mathrm{DI},\mathrm{SO})}_{\text{switching bifurcation}} \rightleftharpoons \left\{\begin{matrix}\underbrace{(\mathrm{UP},\mathrm{nF})}_{\text{up-parabola flow }(-)}\\ \underbrace{(\mathrm{DP},\mathrm{nF})}_{\text{down-parabola flow }(-)}\\ \underbrace{(\mathrm{pF},\mathrm{UP})}_{\text{up-parabola flow }(+)}\\ \underbrace{(\mathrm{DP},\mathrm{pF})}_{\text{down-parabola flow }(+)}\end{matrix}\right\};$$

$$\left\{\begin{matrix}\underbrace{(\mathrm{UP},\mathrm{nF})}_{\text{up-parabola flow }(-)}\\ \underbrace{(\mathrm{nF},\mathrm{DP})}_{\text{down-parabola flow }(-)}\end{matrix}\right\} \rightleftharpoons \underbrace{(2^{\mathrm{nd}}\mathrm{DI},\mathrm{SO})}_{\text{switching bifurcation}} \rightleftharpoons \left\{\begin{matrix}\underbrace{(\mathrm{UP},\mathrm{nF})}_{\text{up-parabola flow }(-)}\\ \underbrace{(\mathrm{nF},\mathrm{DP})}_{\text{down-parabola flow }(-)}\\ \underbrace{(\mathrm{UP},\mathrm{pF})}_{\text{up-parabola flow }(+)}\\ \underbrace{(\mathrm{DP},\mathrm{pF})}_{\text{down-parabola flow }(+)}\end{matrix}\right\}$$

for $\alpha_{j_1j_2} > 0$.

$$\left\{\begin{matrix}\underbrace{(\mathrm{pF},\mathrm{UP})}_{\text{up-parabola flow }(+)}\\ \underbrace{(\mathrm{UP},\mathrm{pF})}_{\text{up-parabola flow }(+)}\end{matrix}\right\} \rightleftharpoons \underbrace{(2^{\mathrm{nd}}\mathrm{DI},\mathrm{SO})}_{\text{switching bifurcation}} \rightleftharpoons \left\{\begin{matrix}\underbrace{(\mathrm{DP},\mathrm{nF})}_{\text{down-parabola flow }(-)}\\ \underbrace{(\mathrm{UP},\mathrm{nF})}_{\text{up-parabola flow }(-)}\\ \underbrace{(\mathrm{pF},\mathrm{UP})}_{\text{up-parabola flow }(+)}\\ \underbrace{(\mathrm{UP},\mathrm{pF})}_{\text{up-parabola flow }(+)}\end{matrix}\right\};$$

$$\left\{\begin{array}{c}\underbrace{(\mathrm{DP},\mathrm{nF})}_{\text{down-parabola flow }(-)}\\\underbrace{(\mathrm{nF},\mathrm{DP})}_{\text{down-parabola flow }(-)}\end{array}\right\}\rightleftharpoons\underbrace{(2^{\mathrm{nd}}\mathrm{DI},\mathrm{SO})}_{\text{switching bifurcation}}\rightleftharpoons\left\{\begin{array}{c}\underbrace{(\mathrm{DP},\mathrm{nF})}_{\text{down-parabola flow }(-)}\\\underbrace{(\mathrm{nF},\mathrm{DP})}_{\text{down-parabola flow }(-)}\\\underbrace{(\mathrm{DP},\mathrm{pF})}_{\text{up-parabola flow }(+)}\\\underbrace{(\mathrm{UP},\mathrm{pF})}_{\text{up-parabola flow }(+)}\end{array}\right\}$$

for $\alpha_{j_1j_2} < 0$;

- For $\alpha_{j_1j_2}^3 a_{j_1j_10} > 0$ and $\alpha_{j_1j_2} a_{j_2j_20} < 0$,

$$\left\{\begin{array}{c}\underbrace{(\mathrm{nF},\mathrm{UP})}_{\text{up-parabola flow }(-)}\\\underbrace{(3^{\mathrm{rd}}\mathrm{DP},\mathrm{nF})}_{\text{down-parabola flow }(-)}\end{array}\right\}\rightleftharpoons\underbrace{(2^{\mathrm{nd}}\mathrm{DI},\mathrm{SI})}_{\text{switching bifurcation}}\rightleftharpoons\left\{\begin{array}{c}\underbrace{(\mathrm{UP},\mathrm{pF})}_{\text{up-parabola flow }(+)}\\\underbrace{(\mathrm{DP},\mathrm{pF})}_{\text{down-parabola flow }(+)}\\\underbrace{(\mathrm{nF},\mathrm{UP})}_{\text{up-parabola flow }(-)}\\\underbrace{(\mathrm{DP},\mathrm{nF})}_{\text{down-parabola flow }(-)}\end{array}\right\};$$

$$\left\{\begin{array}{c}\underbrace{(\mathrm{UP},\mathrm{pF})}_{\text{up-parabola flow }(+)}\\\underbrace{(\mathrm{pF},\mathrm{DP})}_{\text{down-parabola flow }(+)}\end{array}\right\}\rightleftharpoons\underbrace{(2^{\mathrm{nd}}\mathrm{DI},\mathrm{SI})}_{\text{switching bifurcation}}\rightleftharpoons\left\{\begin{array}{c}\underbrace{(\mathrm{UP},\mathrm{pF})}_{\text{up-parabola flow }(+)}\\\underbrace{(\mathrm{pF},\mathrm{DP})}_{\text{down-parabola flow }(+)}\\\underbrace{(\mathrm{UP},\mathrm{nF})}_{\text{up-parabola flow }(-)}\\\underbrace{(\mathrm{DP},\mathrm{nF})}_{\text{down-parabola flow }(-)}\end{array}\right\}$$

for $\alpha_{j_1j_2} > 0$.

$$\left\{\begin{array}{c}\underbrace{(\mathrm{nF},\mathrm{UP})}_{\text{up-parabola flow }(-)}\\ \underbrace{(\mathrm{UP},\mathrm{nF})}_{\text{up-parabola flow }(-)}\end{array}\right\} \rightleftharpoons \underbrace{(2^{\mathrm{nd}}\mathrm{DI},\mathrm{SI})}_{\text{switching bifurcation}} \rightleftharpoons \left\{\begin{array}{c}\underbrace{(\mathrm{DP},\mathrm{pF})}_{\text{down-parabola flow }(+)}\\ \underbrace{(\mathrm{UP},\mathrm{pF})}_{\text{up-parabola flow }(+)}\\ \underbrace{(\mathrm{nF},\mathrm{UP})}_{\text{up-parabola flow }(-)}\\ \underbrace{(\mathrm{UP},\mathrm{nF})}_{\text{up-parabola flow }(-)}\end{array}\right\};$$

$$\left\{\begin{array}{c}\underbrace{(\mathrm{DP},\mathrm{pF})}_{\text{down-parabola flow }(+)}\\ \underbrace{(\mathrm{pF},\mathrm{DP})}_{\text{down-parabola flow }(+)}\end{array}\right\} \rightleftharpoons \underbrace{(2^{\mathrm{nd}}\mathrm{DI},\mathrm{SI})}_{\text{switching bifurcation}} \rightleftharpoons \left\{\begin{array}{c}\underbrace{(\mathrm{DP},\mathrm{pF})}_{\text{down-parabola flow }(+)}\\ \underbrace{(\mathrm{pF},\mathrm{DP})}_{\text{down-parabola flow }(+)}\\ \underbrace{(\mathrm{DP},\mathrm{nF})}_{\text{up-parabola flow }(-)}\\ \underbrace{(\mathrm{UP},\mathrm{nF})}_{\text{up-parabola flow }(-)}\end{array}\right\}$$

for $\alpha_{j_1 j_2} < 0$.

- For $\alpha_{j_1 j_2}^3 a_{j_1 j_1 0} < 0$ and $\alpha_{j_1 j_2} a_{j_2 j_2 0} < 0$,

$$\left\{\begin{array}{c}\underbrace{(\mathrm{pF},\mathrm{DP})}_{\text{down-parabola flow }(+)}\\ \underbrace{(\mathrm{UP},\mathrm{nF})}_{\text{up-parabola flow }(-)}\end{array}\right\} \rightleftharpoons \underbrace{(2^{\mathrm{nd}}\mathrm{II},\mathrm{SI})}_{\text{switching bifurcation}} \rightleftharpoons \left\{\begin{array}{c}\underbrace{(\mathrm{DP},\mathrm{pF})}_{\text{down-parabola flow }(+)}\\ \underbrace{(\mathrm{UP},\mathrm{pF})}_{\text{up-parabola flow }(+)}\\ \underbrace{(\mathrm{pF},\mathrm{DP})}_{\text{down-parabola flow }(+)}\\ \underbrace{(\mathrm{UP},\mathrm{nF})}_{\text{up-parabola flow }(-)}\end{array}\right\};$$

$$\left\{\begin{array}{c}\underbrace{(\mathrm{DP},\mathrm{pF})}_{\text{down-parabola flow }(+)}\\ \underbrace{(\mathrm{nF},\mathrm{UP})}_{\text{up-parabola flow }(-)}\end{array}\right\}\rightleftharpoons\underbrace{(2^{\mathrm{nd}}\mathrm{II},\mathrm{SI})}_{\text{switching bifurcation}}\rightleftharpoons\left\{\begin{array}{c}\underbrace{(\mathrm{DP},\mathrm{pF})}_{\text{down-parabola flow }(+)}\\ \underbrace{(\mathrm{nF},\mathrm{UP})}_{\text{up-parabola flow }(-)}\\ \underbrace{(\mathrm{DP},\mathrm{nF})}_{\text{down-parabola flow }(-)}\\ \underbrace{(\mathrm{UP},\mathrm{nF})}_{\text{up-parabola flow }(-)}\end{array}\right\}$$

for $\alpha_{j_1 j_2} > 0$.

$$\left\{\begin{array}{c}\underbrace{(\mathrm{pF},\mathrm{DP})}_{\text{down-parabola flow }(+)}\\ \underbrace{(\mathrm{DP},\mathrm{nF})}_{\text{down-parabola flow }(-)}\end{array}\right\}\rightleftharpoons\underbrace{(2^{\mathrm{nd}}\mathrm{II},\mathrm{SI})}_{\text{switching bifurcation}}\rightleftharpoons\left\{\begin{array}{c}\underbrace{(\mathrm{DP},\mathrm{pF})}_{\text{down-parabola flow }(+)}\\ \underbrace{(\mathrm{UP},\mathrm{pF})}_{\text{up-parabola flow }(+)}\\ \underbrace{(\mathrm{pF},\mathrm{DP})}_{\text{down-parabola flow }(+)}\\ \underbrace{(\mathrm{DP},\mathrm{nF})}_{\text{down-parabola flow }(-)}\end{array}\right\};$$

$$\left\{\begin{array}{c}\underbrace{(\mathrm{UP},\mathrm{pF})}_{\text{up-parabola flow }(+)}\\ \underbrace{(\mathrm{nF},\mathrm{UP})}_{\text{up-parabola flow }(-)}\end{array}\right\}\rightleftharpoons\underbrace{(2^{\mathrm{nd}}\mathrm{II},\mathrm{SI})}_{\text{switching bifurcation}}\rightleftharpoons\left\{\begin{array}{c}\underbrace{(\mathrm{UP},\mathrm{pF})}_{\text{up-parabola flow }(+)}\\ \underbrace{(\mathrm{nF},\mathrm{UP})}_{\text{up-parabola flow }(-)}\\ \underbrace{(\mathrm{UP},\mathrm{nF})}_{\text{up-parabola flow }(-)}\\ \underbrace{(\mathrm{DP},\mathrm{nF})}_{\text{down-parabola flow }(-)}\end{array}\right\}$$

for $\alpha_{j_1 j_2} < 0$.

(iv) If

$$A_{j_2 j_2 1} \neq 0, \quad A_{j_2 j_2 2} \neq 0 \quad \text{and} \quad \Delta_{j_1 j_1} > 0,$$

the two-dimensional dynamical system has two simple equilibriums for the y_{j_1}-direction on the x_i-direction, i.e.,

$$y_{j_1}^* = a_{j_1 j_1 1}, \quad a_{j_1 j_1 2}, \quad a_{j_1 j_1 3} \text{ for the } x_{j_1}\text{-direction},$$

$$y_{j_1}^* = a_{j_2 j_2 1} \text{ for the } x_{j_2}\text{-direction},$$

where

$$\{a_{j_1j_1 1}, a_{j_1j_1 2}\} = \text{sort}\{b_{j_1j_1 1}, b_{j_1j_1 2}, b_{j_1j_1 3}\}, \quad a_{j_1j_1 1} < a_{j_1j_1 2},$$

$$b_{j_1j_1 2}, b_{j_1j_1 3} = -\frac{1}{2}(B_{j_1j_1} \pm \sqrt{\Delta_{j_1j_1}}),$$

and the corresponding dynamical system becomes

$$\dot{x}_{j_2} = a_{j_2j_2 0}(y_{j_1} - a_{j_2j_2 1}),$$
$$\dot{x}_{j_1} = a_{j_1j_1 0}(y_{j_1} - a_{j_1j_1 1})(y_{j_1} - a_{j_1j_1 2})(y_{j_1} - a_{j_1j_1 3}).$$

The corresponding dynamical system in the y_{j_1}-direction is constructed by

$$\dot{y}_{j_1} = \alpha_{j_1j_1} a_{j_1j_1 0} \prod_{s_1=1}^{3} (y_{j_1} - a_{j_1j_1 s_1}) + \alpha_{j_1j_2} a_{j_2j_2 0}(y_{j_1} - a_{j_2j_2 1}),$$

and a dynamical system in the $y_{j_1}^{\perp}$-direction is constructed by

$$\dot{y}_{j_1}^{\perp} = \beta_{j_1j_1} a_{j_1j_1 0} \prod_{s_1=1}^{3} (y_{j_1} - a_{j_1j_1 s_1}) + \beta_{j_1j_2} a_{j_2j_2 0}(y_{j_1} - a_{j_2j_2 1}).$$

(iv$_{1a}$) For $y_{j_1}^* = a_{j_1j_1 s}$ with $y_{j_1} \neq a_{j_2j_2 1}$,

$$\left.\frac{dx_{j_1}}{dx_{j_2}}\right|_{y_{j_1}^* = a_{j_1j_1 s_1}} = \left.\frac{a_{j_1j_1 0}(y_{j_1} - a_{j_1j_1 s_1}) \prod_{s_2=1, s_2\neq s_1}^{3} (y_{j_1} - a_{j_1j_1 s_2})}{a_{j_2j_2 0}(y_{j_1} - a_{j_2j_2 1})}\right|_{y_{j_1}^* = a_{j_1j_1 s_1}} = 0.$$

If

$$\left.\frac{d^2x_{j_1}}{dx_{j_2}^2}\right|_{y_{j_1}^* = a_{j_1j_1 s_1}} = \frac{\alpha_{j_1j_2} a_{j_1j_1 0} \prod_{s_2=1, s_2\neq s_1}^{3} (a_{j_1j_1 s_1} - a_{j_1j_1 s_2})}{a_{j_2j_2 0}(a_{j_1j_1 s_1} - a_{j_2j_2 1})} > 0,$$

there is an up-parabola flow near $y_{j_1}^* = a_{j_1j_1 s_1}$. If

$$\left.\frac{d^2x_{j_1}}{dx_{j_2}^2}\right|_{y_{j_1}^* = a_{j_1j_1 s_1}} = \frac{\alpha_{j_1j_2} a_{j_1j_1 0} \prod_{s_2=1, s_2\neq s_1}^{3} (a_{j_1j_1 s_1} - a_{j_1j_1 s_2})}{a_{j_2j_2 0}(a_{j_1j_1 s_1} - a_{j_2j_2 1})} < 0,$$

there is a down-parabola flow near $y_{j_1}^* = a_{j_1j_1 s_1}$. Let

$$\dot{x}_{j_2}^* = a_{j_2j_2 0}(a_{j_1j_1 s_1} - a_{j_2j_2 1}).$$

For $\dot{x}_{j_2}^* > 0$ and $\dot{x}_{j_2}^* < 0$, the flow at $y_{j_1}^* = a_{j_1j_1 s_1}$ in the x_{j_1}-direction is positive and negative, respectively.

Therefore, the equilibriums of $y_{j_1}^* = a_{j_1j_1s_1}$ $(s_1, s_2 \in \{1,2,3\}, s_1 \neq s_2)$ for the x_{j_1}-direction possess the following properties as in Eqs. (2.120)–(2.123):

- For $\alpha_{j_1j_2}a_{j_1j_10}\prod_{s_2=1,s_2\neq s_1}^{3}(a_{j_1j_1s_1} - a_{j_1j_1s_2}) > 0$ and $\dot{x}_{j_2}^* > 0$,
$$(a_{j_1j_1s_1}, \dot{x}_{j_2}^*) = \underbrace{(\mathrm{UP}, \mathrm{pF})}_{\text{up-parabola flow}}.$$
- For $\alpha_{j_1j_2}a_{j_1j_10}\prod_{s_2=1,s_2\neq s_1}^{3}(a_{j_1j_1s_1} - a_{j_1j_1s_2}) < 0, \dot{x}_{j_2}^* > 0$,
$$(a_{j_1j_1s}, \dot{x}_{j_2}^*) = \underbrace{(\mathrm{DP}, \mathrm{pF})}_{\text{down-parabola flow}}.$$
- For $\alpha_{j_1j_2}a_{j_1j_10}\prod_{s_2=1,s_2\neq s_1}^{3}(a_{j_1j_1s_1} - a_{j_1j_1s_2}) > 0, \dot{x}_{j_2}^* < 0$,
$$a_{j_1j_1s}, \dot{x}_{j_2}^* = \underbrace{(\mathrm{DP}, \mathrm{nF})}_{\text{down-parabola flow}}.$$
- For $\alpha_{j_1j_2}a_{j_1j_10}\prod_{s_2=1,s_2\neq s_1}^{3}(a_{j_1j_1s_1} - a_{j_1j_1s_2}) < 0, \dot{x}_{j_2}^* < 0$,
$$(a_{j_1j_1s}, \dot{x}_{j_2}^*) = \underbrace{(\mathrm{UP}, \mathrm{nF})}_{\text{up-parabola flow}}.$$

($\mathrm{iv}_{1\mathrm{b}}$) For $y_{j_1}^* = a_{j_2j_21}$ with $y_{j_1} \neq a_{j_1j_11}, a_{j_1j_12}, a_{j_1j_13}$

$$\left.\frac{dx_{j_2}}{dx_{j_1}}\right|_{y_{j_1}^*=a_{j_2j_21}} = \left.\frac{a_{j_2j_20}(y_{j_1} - a_{j_2j_21})}{a_{j_1j_10}\prod_{s_1=1}^{3}(y_{j_1} - a_{j_1j_1s_1})}\right|_{y_{j_1}^*=a_{j_2j_21}} = 0.$$

If

$$\left.\frac{d^2x_{j_2}}{dx_{j_1}^2}\right|_{y_{j_1}^*=a_{j_2j_21}} = \frac{a_{j_2j_20}\alpha_{j_1j_1}}{a_{j_1j_10}\prod_{s_1=1}^{3}(a_{j_2j_21} - a_{j_1j_1s_1})} > 0,$$

there is an up-parabola flow at $y_{j_1}^* = a_{j_2j_21}$ in the y_{j_2}-direction. If

$$\left.\frac{d^2x_{j_2}}{dx_{j_1}^2}\right|_{y_{j_1}^*=a_{j_2j_21}} = \frac{a_{j_2j_20}\alpha_{j_1j_1}}{a_{j_1j_10}\prod_{s_1=1}^{3}(a_{j_2j_21} - a_{j_1j_1s_1})} < 0,$$

there is a down-parabola flow at $y_{j_1}^* = a_{j_2j_21}$ in the y_{j_2}-direction. Let

$$\dot{x}_{j_1}^* = a_{j_1j_10}\prod_{s=1}^{3}(a_{j_2j_21} - a_{j_1j_1s}).$$

For $\dot{x}_{j_1}^* > 0$ and $\dot{x}_{j_1}^* < 0$, the flow at $y_{j_1}^* = a_{j_2j_21}$ in the x_{j_2}-direction is positive and negative, respectively.

Therefore, the equilibriums of $y_{j_1}^* = a_{j_2j_21}$ for the x_{j_2}-direction possess the following properties as in Eqs. (2.124)–(2.127).

- For $\dot{x}_{j_1}^{*} > 0$ and $a_{j_2j_20}\alpha_{j_1j_1} > 0$,

$$(\dot{x}_{j_1}^{*}, a_{j_2j_21}) = \underbrace{(\text{pF,UP})}_{\text{up-parabola flow}}.$$

- For $\dot{x}_{j_1}^{*} < 0$ and $a_{j_2j_20}\alpha_{j_1j_1} > 0$,

$$(\dot{x}_{j_1}^{*}, a_{j_2j_21}) = \underbrace{(\text{nF, DP})}_{\text{down-parabola flow}}.$$

- For $\dot{x}_{j_1}^{*} > 0$ and $a_{j_2j_20}\alpha_{j_1j_1} < 0$,

$$(\dot{x}_{j_1}^{*}, a_{j_2j_21}) = \underbrace{(\text{pF, DP})}_{\text{down-parabola flow}}.$$

- For $\dot{x}_{j_1}^{*} < 0$ and $a_{j_2j_20}\alpha_{j_1j_1} < 0$,

$$(\dot{x}_{j_1}^{*}, a_{j_2j_21}) = \underbrace{(\text{nF,UP})}_{\text{up-parabola flow}}.$$

(iv$_2$) For $a_{j_2j_21} = a_{j_1j_1s_1}(s_1, s_2 \in \{1,2,3\}, s_1 \neq s_2)$, consider a dynamical system as

$$\dot{x}_{j_2} = a_{j_2j_20}(y_{j_1} - a_{j_2j_21}),$$

$$\dot{x}_{j_1} = a_{j_1j_10}(y_{j_1} - a_{j_1j_1s_1}) \prod_{s_2=1, s_2\neq s_1}^{3} (y_{j_1} - a_{j_1j_1s_2}).$$

The corresponding dynamical system in the y_{j_1}-direction is

$$\dot{y}_{j_1} = (y_{j_1} - a_{j_1j_1s_1})\Big[\alpha_{j_1j_1}a_{j_1j_10} \prod_{s_2=1, s_2\neq s_1}^{3} (y_{j_1} - a_{j_1j_1s_1}) + \alpha_{j_1j_2}a_{j_2j_20}\Big],$$

and a dynamical system in the $y_{j_1}^{\perp}$-direction is

$$\dot{y}_{j_1}^{\perp} = (y_{j_1} - a_{j_1j_1s_1})\Big[\beta_{j_1j_1}a_{j_1j_10} \prod_{s_2=1, s_2\neq s_1}^{3} (y_{j_1} - a_{j_1j_1s_1}) + \beta_{j_1j_2}a_{j_2j_20}\Big].$$

Thus, the equilibrium of $y_{j_1}^{*} = a_{j_1j_1s_1} = a_{j_2j_21}$ is an infinite-equilibrium.

(iv$_2$) For $y_{j_1}^{*} = a_{j_1j_1s_1} = a_{j_2j_21}(s_1, s_2 \in \{1,2,3\}, s_1 \neq s_2)$, if

$$\left.\frac{dx_{j_1}}{dx_{j_2}}\right|_{y_{j_1}^{*} = a_{j_1j_1s_1}} = \frac{\prod_{s_2=1, s_2\neq s_1}^{3} (a_{j_1j_1s_1} - a_{j_1j_1s_2})}{a_{j_2j_20}} > 0,$$

there is an increasing-inflection flow for $y_{j_1}^* = a_{j_1j_1s_1} = a_{j_2j_21}$, and if

$$\frac{dx_{j_1}}{dx_{j_2}}\bigg|_{y_{j_1}^*=a_{j_1j_1s_1}} = \frac{\prod_{s_2=1,s_2\neq s_1}^{3}(a_{j_1j_1s_1}-a_{j_1j_1s_2})}{a_{j_2j_20}} < 0,$$

there is a decreasing-inflection flow for $y_{j_1}^* = a_{j_1j_1s_1} = a_{j_2j_21}$.

In the vicinity of $y_{j_1}^* = a_{j_1j_1s_1} = a_{j_2j_21}$, we have

$$\begin{aligned}\Delta\dot{y}_{j_1} &= \left[\alpha_{j_1j_1}a_{j_1j_10}\prod_{s_2=1,s_2\neq s_1}^{3}(a_{j_1j_1s_1}-a_{j_1j_1s_1})+\alpha_{j_1j_2}a_{j_2j_20}\right]\Delta y_{j_1}\\ &= G_{j_1j_10}\Delta y_{j_1},\end{aligned}$$

with

$$\begin{aligned}G_{j_1j_10} &= \frac{\partial}{\partial y_{j_1}}[\mathbf{n}_{j_1}^{\mathrm{T}}\cdot\mathbf{F}(y_{j_1},\mathbf{p})]\bigg|_{y_{j_1}=a_{j_1j_1s_1}}\\ &= \alpha_{j_1j_1}a_{j_1j_10}\prod_{s_2=1,s_2\neq s_1}^{3}(a_{j_1j_1s_1}-a_{j_1j_1s_2})+\alpha_{j_1j_2}a_{j_2j_20}.\end{aligned}$$

The flows at $y_{j_1}^* = a_{j_1j_1s_1} = a_{j_2j_21}$ are the source and sink for $G_{j_1j_10} > 0$ and $G_{j_1j_10} < 0$, respectively.

Therefore, the infinite-equilibrium of $y_{j_1}^* = a_{j_1j_1s_1} = a_{j_2j_21}$ $(s_1, s_2 \in \{1,2,3\})$ possesses the following properties as in Eqs. (2.132)–(2.139):

- If

$$\frac{a_{j_1j_10}\prod_{s_2=1,s_2\neq s_1}^{3}(a_{j_1j_1s_1}-a_{j_1j_1s_2})}{\alpha_{j_1j_2}a_{j_2j_20}} > 0 \quad\text{and}\quad G_{j_1j_10} > 0,$$

  then

$$(a_{11s_1}, a_{j_2j_21}) = \underbrace{(\mathrm{II},\mathrm{SO})}_{\text{infinite-equilibrium}}.$$

- If

$$\frac{a_{j_1j_10}\prod_{s_2=1,s_2\neq s_1}^{3}(a_{j_1j_1s_1}-a_{j_1j_1s_2})}{a_{j_2j_20}} < 0 \quad\text{and}\quad G_{j_1j_10} > 0,$$

  then

$$(a_{11s_1}, a_{j_2j_21}) = \underbrace{(\mathrm{DI},\mathrm{SO})}_{\text{infinite-equilibrium}}.$$

- If

$$\frac{a_{j_1j_10}\prod_{s_2=1,s_2\neq s_1}^{3}(a_{j_1j_1s_1}-a_{j_1j_1s_2})}{a_{j_2j_20}}<0 \quad \text{and} \quad G_{j_1j_10}<0,$$

then

$$(a_{11s_1},a_{j_2j_21})=\underbrace{(\mathrm{DI},\mathrm{SI})}_{\text{infinite-equilibrium}}.$$

- If

$$\frac{a_{j_1j_10}\prod_{s_2=1,s_2\neq s_1}^{3}(a_{j_1j_1s_1}-a_{j_1j_1s_2})}{a_{j_2j_20}}>0 \quad \text{and} \quad G_{j_1j_10}<0,$$

then

$$(a_{11s_1},a_{j_2j_21})=\underbrace{(\mathrm{II},\mathrm{SI})}_{\text{infinite-equilibrium}}.$$

($\mathrm{iv}_{2\mathrm{b}}$) From ($\mathrm{iv}_{1a,b}$) and ($\mathrm{iv}_{2a}$), the inflection-source and sink infinite-equilibriums at $y_{j_1}^*=a_{j_1j_1s_1}=a_{j_2j_21}$ are the switching bifurcations for the parabola flows at $y_{j_1}^*=a_{j_1j_1s_1}$ and $y_{j_1}^*=a_{j_2j_21}$ as in Eqs. (2.140)–(2.143):

- *For* $a_{j_11j_10}\prod_{s_2=1,s_2\neq s_1}^{3}(a_{j_1j_1s_1}-a_{j_1j_1s_2})>0$ and $a_{j_2j_20}>0$,

$$\left\{\begin{array}{c}\underbrace{(\mathrm{nF},\mathrm{DP})}_{\text{down-parabola }(-)}\\ \underbrace{(\mathrm{UP},\mathrm{pF})}_{\text{up-parabola }(+)}\end{array}\right\}\rightleftharpoons\underbrace{(\mathrm{II},\mathrm{SO})}_{\text{switching bifurcation}}\rightleftharpoons\left\{\begin{array}{c}\underbrace{(\mathrm{DP},\mathrm{nF})}_{\text{down-parabola }(-)}\\ \underbrace{(\mathrm{pF},\mathrm{UP})}_{\text{up-parabola }(+)}\end{array}\right\}$$

for $(\alpha_{j_1j_1},\alpha_{j_1j_2})=(+,+)$ and $G_{j_1j_10}>0$;

$$\left\{\begin{array}{c}\underbrace{(\mathrm{nF},\mathrm{DP})}_{\text{down-parabola }(-)}\\ \underbrace{(\mathrm{DP},\mathrm{pF})}_{\text{down-parabola }(+)}\end{array}\right\}\rightleftharpoons\underbrace{(\mathrm{II},\mathrm{SO})}_{\text{switching bifurcation}}\rightleftharpoons\left\{\begin{array}{c}\underbrace{(\mathrm{UP},\mathrm{nF})}_{\text{up-parabola }(-)}\\ \underbrace{(\mathrm{pF},\mathrm{UP})}_{\text{up-parabola }(+)}\end{array}\right\}$$

for $(\alpha_{j_1j_1},\alpha_{j_1j_2})=(+,-)$ and $G_{j_1j_10}>0$;

$$\left\{\begin{array}{c}\underbrace{(\mathrm{UP},\mathrm{nF})}_{\text{up-parabola }(-)}\\ \underbrace{(\mathrm{pF},\mathrm{UP})}_{\text{up-parabola }(+)}\end{array}\right\}\rightleftharpoons\underbrace{(\mathrm{II},\mathrm{SI})}_{\text{switching bifurcation}}\rightleftharpoons\left\{\begin{array}{c}\underbrace{(\mathrm{nF},\mathrm{DP})}_{\text{down-parabola }(-)}\\ \underbrace{(\mathrm{DP},\mathrm{pF})}_{\text{down-parabola }(+)}\end{array}\right\}$$

for $(\alpha_{j_1j_1},\alpha_{j_1j_2})=(+,-)$ and $G_{j_1j_10}<0$.

- *For* $a_{j_1 1 j_1 0}\prod_{s_2=1,s_2\neq s_1}^{3}(a_{j_1 j_1 s_1}-a_{j_1 j_1 s_2})<0$ and $a_{j_2 j_2 0}>0$,

$$\left\{\begin{array}{c}\underbrace{(\mathrm{pF},\mathrm{UP})}_{\text{up-parabola }(+)}\\ \underbrace{(\mathrm{DP},\mathrm{pF})}_{\text{down-parabola }(+)}\end{array}\right\} \rightleftharpoons \underbrace{(\mathrm{DI},\mathrm{SI})}_{\text{switching bifurcation}} \rightleftharpoons \left\{\begin{array}{c}\underbrace{(\mathrm{UP},\mathrm{nF})}_{\text{up-parabola }(-)}\\ \underbrace{(\mathrm{nF},\mathrm{DP})}_{\text{down-parabola }(-)}\end{array}\right\}$$

for $(\alpha_{j_1 j_1},\alpha_{j_1 j_2})=(+,+)$ and $G_{j_1 j_1 0}<0$;

$$\left\{\begin{array}{c}\underbrace{(\mathrm{UP},\mathrm{nF})}_{\text{up-parabola }(-)}\\ \underbrace{(\mathrm{nF},\mathrm{DP})}_{\text{down-parabola }(-)}\end{array}\right\} \rightleftharpoons \underbrace{(\mathrm{DI},\mathrm{SO})}_{\text{switching bifurcation}} \rightleftharpoons \left\{\begin{array}{c}\underbrace{(\mathrm{pF},\mathrm{UP})}_{\text{up-parabola }(+)}\\ \underbrace{(\mathrm{DP},\mathrm{pF})}_{\text{down-parabola }(+)}\end{array}\right\}$$

for $(\alpha_{j_1 j_1},\alpha_{j_1 j_2})=(+,+)$ and $G_{j_1 j_1 0}>0$;

$$\left\{\begin{array}{c}\underbrace{(\mathrm{pF},\mathrm{UP})}_{\text{up-parabola }(+)}\\ \underbrace{(\mathrm{UP},\mathrm{pF})}_{\text{up-parabola }(+)}\end{array}\right\} \rightleftharpoons \underbrace{(\mathrm{DI},\mathrm{SI})}_{\text{switching bifurcation}} \rightleftharpoons \left\{\begin{array}{c}\underbrace{(\mathrm{DP},\mathrm{nF})}_{\text{down-parabola }(-)}\\ \underbrace{(\mathrm{nF},\mathrm{DP})}_{\text{down-parabola }(-)}\end{array}\right\}$$

for $(\alpha_{j_1 j_1},\alpha_{j_1 j_2})=(+,-)$ and $G_{j_1 j_1 0}<0$.

- *For* $a_{j_1 1 j_1 0}\prod_{s_2=1,s_2\neq s_1}^{3}(a_{j_1 j_1 s_1}-a_{j_1 j_1 s_2})>0$ and $a_{j_2 j_2 0}<0$,

$$\left\{\begin{array}{c}\underbrace{(\mathrm{nF},\mathrm{UP})}_{\text{up-parabola }(-)}\\ \underbrace{(\mathrm{DP},\mathrm{nF})}_{\text{down-parabola }(-)}\end{array}\right\} \rightleftharpoons \underbrace{(\mathrm{DI},\mathrm{SO})}_{\text{switching bifurction}} \rightleftharpoons \left\{\begin{array}{c}\underbrace{(\mathrm{UP},\mathrm{pF})}_{\text{up-parabola }(+)}\\ \underbrace{(\mathrm{pF},\mathrm{DP})}_{\text{down-parabola }(+)}\end{array}\right\}$$

for $(\alpha_{j_1 j_1},\alpha_{j_1 j_2})=(+,+)$ and $G_{j_1 j_1 0}>0$;

$$\left\{\begin{array}{c}\underbrace{(\mathrm{UP},\mathrm{pF})}_{\text{up-parabola }(+)}\\ \underbrace{(\mathrm{pF},\mathrm{DP})}_{\text{down-parabola }(+)}\end{array}\right\} \rightleftharpoons \underbrace{(\mathrm{DI},\mathrm{SI})}_{\text{switching bifurction}} \rightleftharpoons \left\{\begin{array}{c}\underbrace{(\mathrm{nF},\mathrm{UP})}_{\text{up-parabola }(-)}\\ \underbrace{(\mathrm{DP},\mathrm{nF})}_{\text{down-parabola}(-)}\end{array}\right\}$$

for $(\alpha_{j_1 j_1},\alpha_{j_1 j_2})=(+,+)$ and $G_{j_1 j_1 0}<0$;

$$\left\{\begin{array}{c}\underbrace{(\mathrm{nF},\mathrm{UP})}_{\text{up-parabola }(-)}\\ \underbrace{(\mathrm{UP},\mathrm{nF})}_{\text{up-parabola }(-)}\end{array}\right\} \rightleftharpoons \underbrace{(\mathrm{DI},\mathrm{SO})}_{\text{switching bifurction}} \rightleftharpoons \left\{\begin{array}{c}\underbrace{(\mathrm{DP},\mathrm{pF})}_{\text{down-parabola }(+)}\\ \underbrace{(\mathrm{pF},\mathrm{DP})}_{\text{down-parabola }(+)}\end{array}\right\}$$
$$\text{for } (\alpha_{j_1j_1},\alpha_{j_1j_2}) = (+,-) \quad \text{and} \quad G_{j_1j_10} > 0.$$

- *For* $a_{j_11j_10}\prod_{s_2=1,s_2\neq s_1}^{3}(a_{j_1j_1s_1} - a_{j_1j_1s_2}) < 0$ and $a_{j_2j_20} < 0$,

$$\left\{\begin{array}{c}\underbrace{(\mathrm{pF},\mathrm{DP})}_{\text{down-parabola }(+)}\\ \underbrace{(\mathrm{UP},\mathrm{nF})}_{\text{up-parabola }(-)}\end{array}\right\} \rightleftharpoons \underbrace{(\mathrm{II},\mathrm{SI})}_{\text{switching bifurcation}} \rightleftharpoons \left\{\begin{array}{c}\underbrace{(\mathrm{DP},\mathrm{pF})}_{\text{down-parabola }(+)}\\ \underbrace{(\mathrm{nF},\mathrm{UP})}_{\text{up-parabola }(-)}\end{array}\right\}$$
$$\text{for } (\alpha_{j_1j_1},\alpha_{j_1j_2}) = (+,+) \quad \text{and} \quad G_{j_1j_10} < 0;$$

$$\left\{\begin{array}{c}\underbrace{(\mathrm{pF},\mathrm{DP})}_{\text{down-parabola }(+)}\\ \underbrace{(\mathrm{DP},\mathrm{nF})}_{\text{down-parabola }(-)}\end{array}\right\} \rightleftharpoons \underbrace{(\mathrm{II},\mathrm{SO})}_{\text{switching bifurcation}} \rightleftharpoons \left\{\begin{array}{c}\underbrace{(\mathrm{UP},\mathrm{pF})}_{\text{up-parabola }(+)}\\ \underbrace{(\mathrm{nF},\mathrm{UP})}_{\text{up-parabola }(-)}\end{array}\right\}$$
$$\text{for } (\alpha_{j_1j_1},\alpha_{j_1j_2}) = (+,-) \quad \text{and} \quad G_{j_1j_10} > 0;$$

$$\left\{\begin{array}{c}\underbrace{(\mathrm{UP},\mathrm{pF})}_{\text{up-parabola }(+)}\\ \underbrace{(\mathrm{nF},\mathrm{UP})}_{\text{up-parabola }(-)}\end{array}\right\} \rightleftharpoons \underbrace{(\mathrm{II},\mathrm{SI})}_{\text{switching bifurcation}} \rightleftharpoons \left\{\begin{array}{c}\underbrace{(\mathrm{pF},\mathrm{DP})}_{\text{down-parabola }(+)}\\ \underbrace{(\mathrm{DP},\mathrm{nF})}_{\text{down-parabola }(-)}\end{array}\right\}$$
$$\text{for } (\alpha_{j_1j_1},\alpha_{j_1j_2}) = (+,-) \quad \text{and} \quad G_{j_1j_10} < 0.$$

In the end, this theorem is proved. □

2.3 Third-Order Parabola Flows with Bifurcations

In this section, a third-order vertical parabola flow with a horizontal parabola flow is presented, and the third-order parabola flows are the appearing bifurcations for three parabola flows and for a parabola and an inflection flow. The switching bifurcations between the third-order vertical parabola and appearing bifurcations are discussed.

2.3.1 Parabola flows with third-order singularity

Consider a two-dimensional system with third-order singularity as

$$\begin{aligned}\dot{x}_1 &= a_{110}(y_1 - a_{111})^3,\\ \dot{x}_2 &= a_{220}(y_1 - a_{221})\end{aligned} \tag{2.144}$$

with

$$y_1 = \alpha_{11}x_1 + \alpha_{12}x_2. \tag{2.145}$$

There is a dynamical system in the y_1-direction as

$$\begin{aligned}\dot{y}_1 &= \alpha_{11}a_{110}(y_1 - a_{111})^3 + \alpha_{12}a_{220}(y_1 - a_{221})\\ &= \alpha_{11}a_{110}(y_1 - a_{111})^3 + \alpha_{12}a_{220}(y_1 - a_{111}) + \alpha_{12}a_{220}(a_{111} - a_{221}),\end{aligned} \tag{2.146}$$

with a dynamical system in the $y_1^{\perp}$-direction as

$$\dot{y}_1^{\perp} = \beta_{11}a_{110}(y_1 - a_{111})^3 + \beta_{12}a_{220}(y_1 - a_{221}), \tag{2.147}$$

where

$$\begin{aligned}&y_1^{\perp} = \beta_{11}x_1 + \beta_{12}x_2, \mathbf{n}_1^{\perp} = (\beta_{11}, \beta_{12})^{\mathrm{T}},\\ &(\mathbf{n}_1)^{\mathrm{T}} \cdot \mathbf{n}_1^{\perp} = \alpha_{11}\beta_{11} + \alpha_{12}\beta_{12} = 0.\end{aligned} \tag{2.148}$$

For $\alpha_{12}^3 a_{110}/a_{220}(a_{111} - a_{221}) > 0$ and $\alpha_{12}^3 a_{110}/a_{220}(a_{111} - a_{221}) < 0$, there exist the third-order up-parabola and third-order down-parabola flows at $y_1^* = a_{111}$, respectively. The positive and negative third-order parabola flows are determined through $a_{220}(a_{111} - a_{221}) > 0$ and $a_{220}(a_{111} - a_{221}) < 0$, respectively. However, for $a_{220}\alpha_{11}/a_{110}(a_{221} - a_{111})^3 > 0$ and $a_{220}\alpha_{11}/a_{110}(a_{221} - a_{111})^3 < 0$, there exist the up-parabola and down-parabola flows at $y_1^* = a_{111}$, respectively. The positive and negative parabola flows are determined through $a_{110}(a_{221} - a_{111})^3 > 0$ and $a_{220}(a_{221} - a_{111})^3 < 0$, respectively.

(i) For $a_{111} > a_{221}$ with $\alpha_{11} > 0$ and $\alpha_{12} > 0$, presented in Fig. 2.1 are the one-dimensional flows of $(y_1^*, \dot{x}_2) = (a_{111}, \dot{x}_2)$ and $(\dot{x}_1, y_1^*) = (\dot{x}_1, a_{221})$. The corresponding properties for $\alpha_{11} > 0$ and $\alpha_{12} > 0$ are presented as follows:

$$\begin{Bmatrix} (a_{111}, \dot{x}_2) \\ (\dot{x}_1, a_{221}) \end{Bmatrix} = \begin{Bmatrix} \underbrace{(3^{\text{rd}}\text{UP}, \text{pF})}_{\text{third-order up-parabola flow } (+)} \\ \underbrace{(\text{nF}, \text{DP})}_{\text{down-parabola flow } (-)} \end{Bmatrix}$$

$$\text{for } a_{110} > 0 \text{ and } a_{220} > 0, \tag{2.149}$$

$$\begin{Bmatrix} (a_{111}, \dot{x}_2) \\ (\dot{x}_1, a_{221}) \end{Bmatrix} = \begin{Bmatrix} \underbrace{(3^{\text{rd}}\text{DP}, \text{pF})}_{\text{third-order down-parabola flow } (+)} \\ \underbrace{(\text{pF}, \text{UP})}_{\text{up-parabola flow } (+)} \end{Bmatrix}$$

$$\text{for } a_{110} < 0 \text{ and } a_{220} > 0, \tag{2.150}$$

$$\begin{Bmatrix} (a_{111}, \dot{x}_2) \\ (\dot{x}_1, a_{221}) \end{Bmatrix} = \begin{Bmatrix} \underbrace{(3^{\text{rd}}\text{DP}, \text{nF})}_{\text{third-order down-parabola flow } (-)} \\ \underbrace{(\text{nF}, \text{UP})}_{\text{up-parabola flow } (-)} \end{Bmatrix}$$

$$\text{for } a_{110} > 0 \text{ and } a_{220} < 0, \tag{2.151}$$

$$\begin{Bmatrix} (a_{111}, \dot{x}_2) \\ (\dot{x}_1, a_{221}) \end{Bmatrix} = \begin{Bmatrix} \underbrace{(3^{\text{rd}}\text{UP}, \text{nF})}_{\text{third-order up-parabola flow } (-)} \\ \underbrace{(\text{pF}, \text{DP})}_{\text{down-parabola flow } (+)} \end{Bmatrix}$$

$$\text{for } a_{110} < 0 \text{ and } a_{220} < 0. \tag{2.152}$$

In Fig. 2.1(a), the (3^{rd}UP,pF)-flow of $y_1^* = a_{111}$ and (nF,DP)-flow of $y_1^* = a_{221}$ $(a_{110} > 0, a_{220} > 0)$ with $(\alpha_{11} > 0, \alpha_{12} > 0)$ are presented:

$$\begin{aligned} &\left.\begin{aligned} &\alpha_{11}a_{110}(y_1 - a_{111})^3 > 0, \\ &\alpha_{12}a_{220}(y_1 - a_{221}) > 0 \end{aligned}\right\} \quad \text{for } y_1 > a_{111}, \\ &\left.\begin{aligned} &\alpha_{11}a_{110}(y_1 - a_{111})^3 < 0, \\ &\alpha_{12}a_{220}(y_1 - a_{221}) > 0, \end{aligned}\right\} \quad \text{for } y_1 \in (a_{221}, a_{111}), \\ &\left.\begin{aligned} &\alpha_{11}a_{110}(y_1 - a_{111})^3 < 0, \\ &\alpha_{12}a_{220}(y_1 - a_{221}) < 0 \end{aligned}\right\} \quad \text{for } y_1 < a_{221}. \end{aligned} \tag{2.153}$$

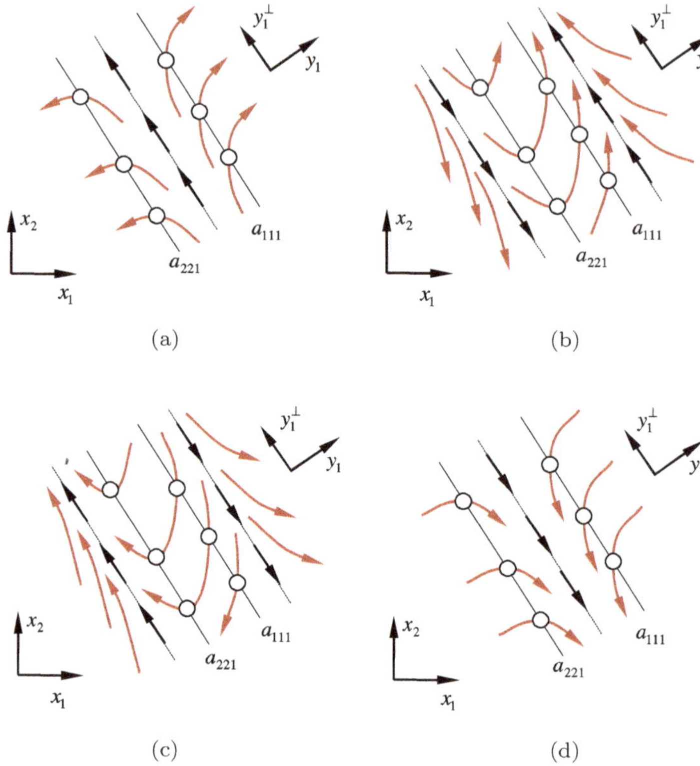

Fig. 2.1. Phase portraits ($a_{111} > a_{221}$ and $\alpha_{11} > 0, \alpha_{12} > 0$) for Two-dimensional systems on the x_1-direction with $y_1^* = a_{111}$ and on the x_2-direction with $y_1^* = a_{221}$: (a) (3^{rd}UP,pF)-flow of $y_1^* = a_{111}$ and (nF,DP)-flow of $y_1^* = a_{221}$ ($a_{110} > 0, a_{220} > 0$), (b) (3^{rd}DP,pF)-flow of $y_1^* = a_{111}$ and (pF,UP)-flow of $y_1^* = a_{221}$ ($a_{110} < 0, a_{220} > 0$), (c) (3^{rd}DP,nF)-flow of $y_1^* = a_{111}$ and (nF,UP)-flow of $y_1^* = a_{221}$ ($a_{110} > 0, a_{220} < 0$), (d) (3^{rd}UP,nF)-flow of $y_1^* = a_{111}$ and (pF,DP)-flow of $y_1^* = a_{221}$ ($a_{110} < 0, a_{220} < 0$).

Thus, there is at least an equilibrium flow at $y_1 = \Delta_1 \in (a_{221}, a_{111})$ with $\dot{y}_1 = 0$, i.e.,

$$\dot{y}_1 = \alpha_{11}a_{110}(\Delta_1 - a_{111})^3 + \alpha_{12}a_{220}(\Delta_1 - a_{221}) = 0, \tag{2.154}$$

and the direction of the equilibrium flow is determined through

$$\begin{aligned} \dot{x}_1 &= a_{110}(\Delta_1 - a_{111})^3 < 0, \\ \dot{x}_2 &= a_{220}(\Delta_1 - a_{221}) > 0. \end{aligned} \tag{2.155}$$

For $y_1 > a_{111}$ and $y_1 < a_{221}$, we have $\dot{y}_1 > 0$ and $\dot{y}_1 < 0$ always.

In Fig. 2.1(b), the ($3^{\rm rd}$DP,pF)-flow of $y_1^* = a_{111}$ and (pF,UP)-flow of $y_1^* = a_{221}$ $(a_{110} < 0, a_{220} > 0)$ with $(\alpha_{11} > 0, \alpha_{12} > 0)$ are presented:

$$\left.\begin{array}{l}\alpha_{11}a_{110}(y_1 - a_{111})^3 < 0,\\ \alpha_{12}a_{220}(y_1 - a_{221}) > 0\end{array}\right\} \text{ for } y_1 > a_{111},$$
$$\left.\begin{array}{l}\alpha_{11}a_{110}(y_1 - a_{111})^3 > 0,\\ \alpha_{12}a_{220}(y_1 - a_{221}) > 0,\end{array}\right\} \text{ for } y_1 \in (a_{221}, a_{111}), \tag{2.156}$$
$$\left.\begin{array}{l}\alpha_{11}a_{110}(y_1 - a_{111})^3 > 0,\\ \alpha_{12}a_{220}(y_1 - a_{221}) < 0\end{array}\right\} \text{ for } y_1 < a_{221}.$$

Thus, there are two equilibrium flows at $y_1 = \Delta_1 \notin (a_{221}, a_{111})$ with $\dot{y}_1 = 0$, i.e.,

$$\dot{y}_1 = \alpha_{11}a_{110}(\Delta_1 - a_{111})^3 + \alpha_{12}a_{220}(\Delta_1 - a_{221}) = 0, \tag{2.157}$$

and the direction of the equilibrium flow is determined by

$$\left.\begin{array}{l}\dot{x}_1 = a_{110}(\Delta_1 - a_{111})^3 < 0,\\ \dot{x}_2 = a_{220}(\Delta_1 - a_{221}) > 0\end{array}\right\} \text{ for } \Delta_1 > a_{111};$$
$$\left.\begin{array}{l}\dot{x}_1 = a_{110}(\Delta_1 - a_{111})^3 > 0,\\ \dot{x}_2 = a_{220}(\Delta_1 - a_{221}) < 0]\end{array}\right\} \text{ for } \Delta_1 < a_{221}. \tag{2.158}$$

For $y_1 \in (a_{221}, a_{111})$, $\dot{y}_1 > 0$ always.

In Fig. 2.1(c), the ($3^{\rm rd}$DP,nF)-flow of $y_1^* = a_{111}$ and (nF,UP)-flow of $y_1^* = a_{221}$ $(a_{110} > 0, a_{220} < 0)$ with $(\alpha_{11} > 0, \alpha_{12} > 0)$ are presented:

$$\left.\begin{array}{l}\alpha_{11}a_{110}(y_1 - a_{111})^3 > 0,\\ \alpha_{12}a_{220}(y_1 - a_{221}) < 0\end{array}\right\} \text{ for } y_1 > a_{111},$$
$$\left.\begin{array}{l}\alpha_{11}a_{110}(y_1 - a_{111})^3 < 0,\\ \alpha_{12}a_{220}(y_1 - a_{221}) < 0,\end{array}\right\} \text{ for } y_1 \in (a_{221}, a_{111}), \tag{2.159}$$
$$\left.\begin{array}{l}\alpha_{11}a_{110}(y_1 - a_{111})^3 < 0,\\ \alpha_{12}a_{220}(y_1 - a_{221}) > 0\end{array}\right\} \text{ for } y_1 < a_{221}.$$

Thus, there are two equilibrium flows at $y_1 = \Delta_1 \notin (a_{221}, a_{111})$ with $\dot{y}_1 = 0$, i.e.,

$$\dot{y}_1 = \alpha_{11}a_{110}(\Delta_1 - a_{111})^3 + \alpha_{12}a_{220}(\Delta_1 - a_{221}) = 0, \tag{2.160}$$

and the direction of the equilibrium flow is determined through

$$\left.\begin{array}{l}\dot{x}_1 = a_{110}(\Delta_1 - a_{111})^3 > 0,\\ \dot{x}_2 = a_{220}(\Delta_1 - a_{221}) < 0\end{array}\right\} \text{ for } \Delta_1 > a_{111},$$
$$\left.\begin{array}{l}\dot{x}_1 = a_{110}(\Delta_1 - a_{111})^3 < 0,\\ \dot{x}_2 = a_{220}(\Delta_1 - a_{221}) > 0\end{array}\right\} \text{ for } \Delta_1 < a_{221}. \tag{2.161}$$

For $y_1 \in (a_{221}, a_{111})$, $\dot{y}_1 < 0$ always.

In Fig. 2.1(d), the (3^{rd}UP,nF)-flow of $y_1^* = a_{111}$ and (pF,DP)-flow of $y_1^* = a_{221}$ $(a_{110} < 0, a_{220} < 0)$ with $(\alpha_{11} > 0, \alpha_{12} > 0)$ are presented:

$$\left.\begin{array}{l}\alpha_{11}a_{110}(y_1 - a_{111})^3 < 0,\\ \alpha_{12}a_{220}(y_1 - a_{221}) < 0\end{array}\right\} \text{ for } y_1 > a_{111},$$
$$\left.\begin{array}{l}\alpha_{11}a_{110}(y_1 - a_{111})^3 > 0,\\ \alpha_{12}a_{220}(y_1 - a_{221}) < 0,\end{array}\right\} \text{ for } y_1 \in (a_{221}, a_{111}), \tag{2.162}$$
$$\left.\begin{array}{l}\alpha_{11}a_{110}(y_1 - a_{111})^3 > 0,\\ \alpha_{12}a_{220}(y_1 - a_{221}) > 0\end{array}\right\} \text{ for } y_1 < a_{221}.$$

Thus, there is at least an equilibrium flow at $y_1 = \Delta_1 \in (a_{221}, a_{111})$ with $\dot{y}_1 = 0$, i.e.,

$$\dot{y}_1 = \alpha_{11}a_{110}(\Delta_1 - a_{111})^3 + \alpha_{12}a_{220}(\Delta_1 - a_{221}) = 0, \tag{2.163}$$

and the direction of the equilibrium flow is determined through

$$\begin{aligned}\dot{x}_1 &= a_{110}(\Delta_1 - a_{111})^3 > 0,\\ \dot{x}_2 &= a_{220}(\Delta_1 - a_{221}) < 0.\end{aligned} \tag{2.164}$$

For $y_1 > a_{111}$ and $y_1 < a_{221}$, we have $\dot{y}_1 < 0$ and $\dot{y}_1 > 0$ always.

(ii) For $a_{111} < a_{221}$ with $\alpha_{11} > 0$ and $\alpha_{12} > 0$, presented in Fig. 2.2 are the one-dimensional flows of $(y_1^*, \dot{x}_2) = (a_{111}, \dot{x}_2)$ and $(\dot{x}_1, y_1^*) = (\dot{x}_1, a_{221})$. The corresponding properties for $\alpha_{11} > 0$ and $\alpha_{12} > 0$ are presented as follows:

$$\left\{\begin{array}{l}(a_{111}, \dot{x}_2)\\ (\dot{x}_1, a_{221})\end{array}\right\} = \left\{\begin{array}{c}\underbrace{(3^{\text{rd}}\text{DP}, \text{pF})}_{\text{third-order down-parabola flow } (+)}\\ \underbrace{(\text{nF}, \text{DP})}_{\text{down-parabola flow } (-)}\end{array}\right\}$$
$$\text{for } a_{110} > 0 \text{ and } a_{220} > 0, \tag{2.165}$$

$$\left\{\begin{array}{l}(a_{111}, \dot{x}_2)\\ (\dot{x}_1, a_{221})\end{array}\right\} = \left\{\begin{array}{c}\underbrace{(3^{\text{rd}}\text{UP}, \text{pF})}_{\text{third-order up-parabola flow } (+)}\\ \underbrace{(\text{pF}, \text{UP})}_{\text{up-parabola flow } (+)}\end{array}\right\}$$
$$\text{for } a_{110} < 0 \text{ and } a_{220} > 0, \tag{2.166}$$

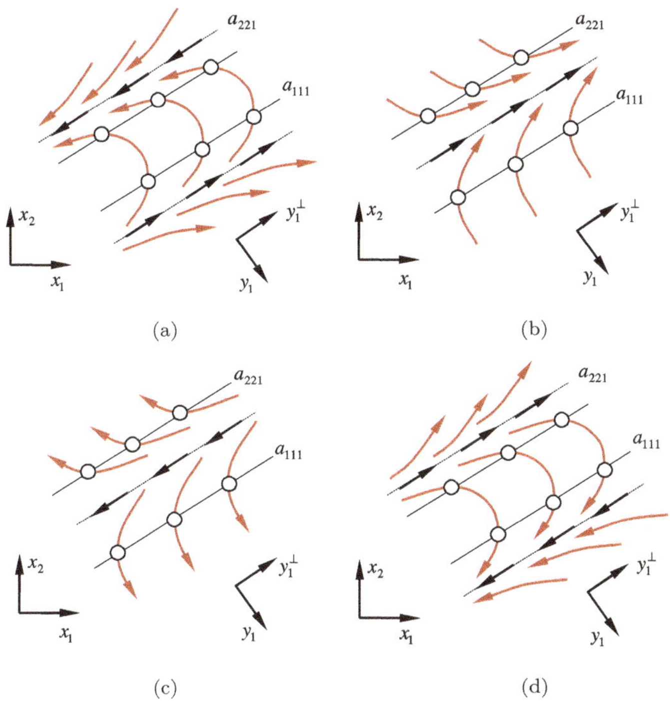

Fig. 2.2. Phase portraits ($a_{111} > a_{221}$ and $\alpha_{11} > 0, \alpha_{12} < 0$) for two-dimensional systems on the x_1-direction with $y_1^* = a_{111}$ and on the x_2-direction with $y_1^* = a_{221}$: (a) (3^{rd}DP,pF)-flow of $y_1^* = a_{111}$ and (nF,DP)-flow of $y_1^* = a_{221}$ ($a_{110} > 0, a_{220} > 0$), (b) (3^{rd}UP,pF)-flow of $y_1^* = a_{111}$ and (pF,UP)-flow of $y_1^* = a_{221}$ ($a_{110} < 0, a_{220} > 0$), (c) (3^{rd}UP,nF)-flow of $y_1^* = a_{111}$ and (nF,UP)-flow of $y_1^* = a_{221}$ ($a_{110} > 0, a_{220} < 0$), (d) (3^{rd}DP,nF)-flow of $y_1^* = a_{111}$ and (pF,DP)-flow of $y_1^* = a_{221}$ ($a_{110} < 0, a_{220} < 0$).

$$\begin{Bmatrix} (a_{111}, \dot{x}_2) \\ (\dot{x}_1, a_{221}) \end{Bmatrix} = \begin{Bmatrix} \underbrace{(3^{\text{rd}}\text{UP}, \text{nF})}_{\text{third-order up-parabola flow } (-)} \\ \underbrace{(\text{nF}, \text{UP})}_{\text{up-parabola flow } (-)} \end{Bmatrix}$$

$$\text{for } a_{110} > 0 \text{ and } a_{220} < 0, \tag{2.167}$$

$$\begin{Bmatrix} (a_{111}, \dot{x}_2) \\ (\dot{x}_1, a_{221}) \end{Bmatrix} = \begin{Bmatrix} \underbrace{(3^{\text{rd}}\text{DP}, \text{nF})}_{\text{third-order down-parabola flow } (-)} \\ \underbrace{(\text{pF}, \text{DP})}_{\text{down-parabola flow } (+)} \end{Bmatrix}$$

$$\text{for } a_{110} < 0 \text{ and } a_{220} < 0. \tag{2.168}$$

In Fig. 2.2(a), the ($3^{\rm rd}$DP,pF)-flow of $y_1^* = a_{111}$ and (nF,DP)-flow of $y_1^* = a_{221}$ $(a_{110} > 0, a_{220} > 0)$ with $(\alpha_{11} > 0, \alpha_{12} < 0)$ are presented:

$$
\left.\begin{array}{l}\alpha_{11}a_{110}(y_1 - a_{111})^3 > 0,\\ \alpha_{12}a_{220}(y_1 - a_{221}) < 0\end{array}\right\} \quad \text{for } y_1 > a_{111},
$$
$$
\left.\begin{array}{l}\alpha_{11}a_{110}(y_1 - a_{111})^3 < 0,\\ \alpha_{12}a_{220}(y_1 - a_{221}) < 0,\end{array}\right\} \quad \text{for } y_1 \in (a_{221}, a_{111}), \tag{2.169}
$$
$$
\left.\begin{array}{l}\alpha_{11}a_{110}(y_1 - a_{111})^3 < 0,\\ \alpha_{12}a_{220}(y_1 - a_{221}) > 0\end{array}\right\} \quad \text{for } y_1 < a_{221}.
$$

Thus, there are two equilibrium flows at $y_1 = \Delta_1 \notin (a_{221}, a_{111})$ with $\dot{y}_1 = 0$, i.e.,

$$
\dot{y}_1 = \alpha_{11}a_{110}(\Delta_1 - a_{111})^3 + \alpha_{12}a_{220}(\Delta_1 - a_{221}) = 0, \tag{2.170}
$$

and the directions of the two equilibrium flows are determined by

$$
\begin{array}{l}
\left.\begin{array}{l}\dot{x}_1 = a_{110}(\Delta_1 - a_{111})^3 > 0,\\ \dot{x}_2 = a_{220}(\Delta_1 - a_{221}) > 0\end{array}\right\} \quad \text{for } \Delta_1 > a_{111};\\
\left.\begin{array}{l}\dot{x}_1 = a_{110}(\Delta_1 - a_{111})^3 < 0,\\ \dot{x}_2 = a_{220}(\Delta_1 - a_{221}) < 0\end{array}\right\} \quad \text{for } \Delta_1 < a_{221}.
\end{array} \tag{2.171}
$$

For $y_1 \in (a_{221}, a_{111})$, we have $\dot{y}_1 < 0$ always.

In Fig. 2.2(b), the ($3^{\rm rd}$UP,pF)-flow of $y_1^* = a_{111}$ and (pF,UP)-flow of $y_1^* = a_{221}$ $(a_{110} < 0, a_{220} > 0)$ with $(\alpha_{11} > 0, \alpha_{12} < 0)$ are presented:

$$
\left.\begin{array}{l}\alpha_{11}a_{110}(y_1 - a_{111})^3 < 0,\\ \alpha_{12}a_{220}(y_1 - a_{221}) < 0\end{array}\right\} \quad \text{for } y_1 > a_{111},
$$
$$
\left.\begin{array}{l}\alpha_{11}a_{110}(y_1 - a_{111})^3 > 0,\\ \alpha_{12}a_{220}(y_1 - a_{221}) < 0,\end{array}\right\} \quad \text{for } y_1 \in (a_{221}, a_{111}), \tag{2.172}
$$
$$
\left.\begin{array}{l}\alpha_{11}a_{110}(y_1 - a_{111})^3 > 0,\\ \alpha_{12}a_{220}(y_1 - a_{221}) > 0\end{array}\right\} \quad \text{for } y_1 < a_{221}.
$$

Thus, there is an equilibrium flow at $y_1 = \Delta_1 \in (a_{221}, a_{111})$ with $\dot{y}_1 = 0$, i.e.,

$$
\dot{y}_1 = \alpha_{11}a_{110}(\Delta_1 - a_{111})^3 + \alpha_{12}a_{220}(\Delta_1 - a_{221}) = 0, \tag{2.173}
$$

and the direction of the equilibrium flow is determined through

$$
\begin{array}{l}\dot{x}_1 = a_{110}(\Delta_1 - a_{111})^3 > 0,\\ \dot{x}_2 = a_{220}(\Delta_1 - a_{221}) > 0.\end{array} \tag{2.174}
$$

For $y_1 < a_{221}$ and $y_1 > a_{111}$, $\dot{y}_1 > 0$ and $\dot{y}_1 < 0$, respectively.

In Fig. 2.2(c), the ($3^{\rm rd}$UP,nF)-flow of $y_1^* = a_{111}$ and (nF,UP)-flow of $y_1^* = a_{221}$ ($a_{110} > 0, a_{220} < 0$) with ($\alpha_{11} > 0, \alpha_{12} < 0$) are presented:

$$\begin{aligned}
&\left.\begin{aligned}&\alpha_{11}a_{110}(y_1 - a_{111})^3 > 0,\\ &\alpha_{12}a_{220}(y_1 - a_{221}) > 0\end{aligned}\right\} \quad \text{for } y_1 > a_{111},\\
&\left.\begin{aligned}&\alpha_{11}a_{110}(y_1 - a_{111})^3 < 0,\\ &\alpha_{12}a_{220}(y_1 - a_{221}) > 0,\end{aligned}\right\} \quad \text{for } y_1 \in (a_{221}, a_{111}),\\
&\left.\begin{aligned}&\alpha_{11}a_{110}(y_1 - a_{111})^3 < 0,\\ &\alpha_{12}a_{220}(y_1 - a_{221}) < 0\end{aligned}\right\} \quad \text{for } y_1 < a_{221}.
\end{aligned} \tag{2.175}$$

Thus, there is an equilibrium flow at $y_1 = \Delta_1 \in (a_{221}, a_{111})$ with $\dot{y}_1 = 0$, i.e.,

$$\dot{y}_1 = \alpha_{11}a_{110}(\Delta_1 - a_{111})^3 + \alpha_{12}a_{220}(\Delta_1 - a_{221}) = 0, \tag{2.176}$$

and the direction of the equilibrium flow is determined by

$$\begin{aligned}
\dot{x}_1 &= a_{110}(\Delta_1 - a_{111})^3 < 0,\\
\dot{x}_2 &= a_{220}(\Delta_1 - a_{221}) < 0.
\end{aligned} \tag{2.177}$$

For $y_1 < a_{221}$ and $y_1 > a_{111}$, $\dot{y}_1 < 0$ and $\dot{y}_1 > 0$, respectively.

In Fig. 2.2(d), the ($3^{\rm rd}$DP,nF)-flow of $y_1^* = a_{111}$ and (pF,DP)-flow of $y_1^* = a_{221}$ ($a_{110} < 0, a_{220} < 0$) with ($\alpha_{11} > 0, \alpha_{12} < 0$) are presented:

$$\begin{aligned}
&\left.\begin{aligned}&\alpha_{11}a_{110}(y_1 - a_{111})^3 < 0,\\ &\alpha_{12}a_{220}(y_1 - a_{221}) > 0\end{aligned}\right\} \quad \text{for } y_1 > a_{111},\\
&\left.\begin{aligned}&\alpha_{11}a_{110}(y_1 - a_{111})^3 > 0,\\ &\alpha_{12}a_{220}(y_1 - a_{221}) > 0,\end{aligned}\right\} \quad \text{for } y_1 \in (a_{221}, a_{111}),\\
&\left.\begin{aligned}&\alpha_{11}a_{110}(y_1 - a_{111})^3 > 0,\\ &\alpha_{12}a_{220}(y_1 - a_{221}) < 0\end{aligned}\right\} \quad \text{for } y_1 < a_{221}.
\end{aligned} \tag{2.178}$$

Thus, there are two equilibrium flows at $y_1 = \Delta_1 \notin (a_{221}, a_{111})$ with $\dot{y}_1 = 0$, i.e.,

$$\dot{y}_1 = \alpha_{11}a_{110}(\Delta_1 - a_{111})^3 + \alpha_{12}a_{220}(\Delta_1 - a_{221}) = 0, \tag{2.179}$$

and the directions of the equilibrium flows are determined by

$$\begin{aligned}
&\left.\begin{aligned}&\dot{x}_1 = a_{110}(\Delta_1 - a_{111})^3 < 0,\\ &\dot{x}_2 = a_{220}(\Delta_1 - a_{221}) < 0\end{aligned}\right\} \quad \text{for } \Delta_1 > a_{111};\\
&\left.\begin{aligned}&\dot{x}_1 = a_{110}(\Delta_1 - a_{111})^3 > 0,\\ &\dot{x}_2 = a_{220}(\Delta_1 - a_{221}) > 0\end{aligned}\right\} \quad \text{for } \Delta_1 < a_{221}.
\end{aligned} \tag{2.180}$$

For $y_1 \in (a_{111}, a_{221})$, $\dot{y}_1 > 0$ holds.

(iii) For $a_{111} < a_{221}$ with $\alpha_{11} > 0$ and $\alpha_{12} > 0$, as for $a_{111} > a_{221}$, presented in Fig. 2.3 are the one-dimensional flows of $(y_1^*, \dot{x}_2) = (a_{111}, \dot{x}_2)$ and $(\dot{x}_1, y_1^*) = (\dot{x}_1, a_{221})$. The corresponding properties for $\alpha_{11} > 0$ and $\alpha_{12} > 0$ are presented as follows:

$$\begin{Bmatrix} (\dot{x}_1, a_{221}) \\ (a_{111}, \dot{x}_2) \end{Bmatrix} = \begin{Bmatrix} \underbrace{(\mathrm{pF}, \mathrm{UP})}_{\text{up-parabola flow } (+)} \\ \underbrace{(3^{\mathrm{rd}}\mathrm{DP}, \mathrm{nF})}_{\text{third-order down-parabola flow } (-)} \end{Bmatrix}$$

$$\text{for } a_{110} > 0 \quad \text{and} \quad a_{220} > 0, \tag{2.181}$$

$$\begin{Bmatrix} (\dot{x}_1, a_{221}) \\ (a_{111}, \dot{x}_2) \end{Bmatrix} = \begin{Bmatrix} \underbrace{(\mathrm{nF}, \mathrm{DP})}_{\text{down-parabola flow } (-)} \\ \underbrace{(3^{\mathrm{rd}}\mathrm{UP}, \mathrm{nF})}_{\text{third-order up-parabola flow } (-)} \end{Bmatrix}$$

$$\text{for } a_{110} < 0 \quad \text{and } a_{220} > 0, \tag{2.182}$$

$$\begin{Bmatrix} (\dot{x}_1, a_{221}) \\ (a_{111}, \dot{x}_2) \end{Bmatrix} = \begin{Bmatrix} \underbrace{(\mathrm{pF}, \mathrm{DP})}_{\text{down-parabola flow } (+)} \\ \underbrace{(3^{\mathrm{rd}}\mathrm{UP}, \mathrm{pF})}_{\text{third-order up-parabola flow } (+)} \end{Bmatrix}$$

$$\text{for } a_{110} > 0 \quad \text{and} \quad a_{220} < 0, \tag{2.183}$$

$$\begin{Bmatrix} (\dot{x}_1, a_{221}) \\ (a_{111}, \dot{x}_2) \end{Bmatrix} = \begin{Bmatrix} \underbrace{(\mathrm{nF}, \mathrm{UP})}_{\text{up-parabola flow } (-)} \\ \underbrace{(3^{\mathrm{rd}}\mathrm{DP}, \mathrm{pF})}_{\text{third-order down-parabola flow } (+)} \end{Bmatrix}$$

$$\text{for } a_{110} < 0 \quad \text{and} \quad a_{220} < 0. \tag{2.184}$$

In Fig. 2.3(a), the $(3^{\mathrm{rd}}\mathrm{DP},\mathrm{nF})$-flow of $y_1^* = a_{111}$ and (pF,DP)-flow of $y_1^* = a_{221}$ $(a_{110} > 0, a_{220} > 0)$ with $(\alpha_{11} > 0, \alpha_{12} > 0)$ are presented. There is at least an equilibrium flow at $y_1 = \Delta_1 \in (a_{111}, a_{221})$ with $\dot{y}_1 = 0$, and the direction of the equilibrium flow is determined from $\dot{x}_1 = a_{110}(\Delta_1 - a_{111})^3 > 0$ and $\dot{x}_2 = a_{220}(\Delta_1 - a_{221}) < 0$. For $y_1 > a_{221}$ and $y_1 < a_{111}$, $\dot{y}_1 > 0$ and $\dot{y}_1 < 0$, respectively.

In Fig. 2.3(b), the $(3^{\mathrm{rd}}\mathrm{UP},\mathrm{nF})$-flow of $y_1^* = a_{111}$ and (nF,DP)-flow of $y_1^* = a_{221}$ $(a_{110} < 0, a_{220} > 0)$ with $(\alpha_{11} > 0, \alpha_{12} > 0)$ are presented. Thus,

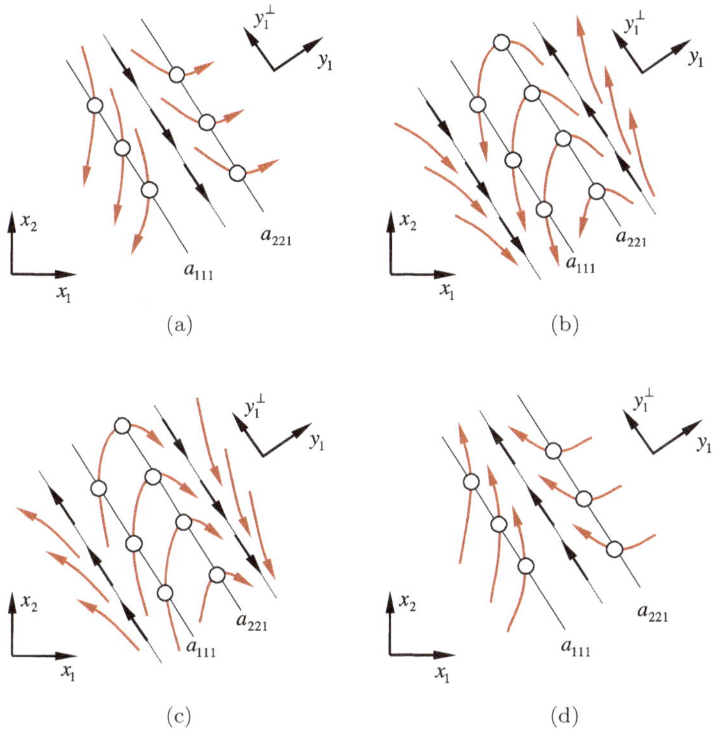

Fig. 2.3. Phase portraits ($a_{111} < a_{221}$ and $\alpha_{11} > 0, \alpha_{12} > 0$) for two-dimensional systems on the x_1-direction with $y_1^* = a_{111}$ and on the x_2-direction with $y_1^* = a_{221}$: (a) (3^{rd}DP,nF)-flow of $y_1^* = a_{111}$ and (pF,UP)-flow of $y_1^* = a_{221}$ ($a_{110} > 0, a_{220} > 0$), (b) (3^{rd}UP,nF)-flow of $y_1^* = a_{111}$ and (nF,DP)-flow of $y_1^* = a_{221}$ ($a_{110} < 0, a_{220} > 0$), (c) (3^{rd}UP,pF)-flow of $y_1^* = a_{111}$ and (pF,DP)-flow of $y_1^* = a_{221}$ ($a_{110} > 0, a_{220} < 0$), (d) (3^{rd}UP,pF)-flow of $y_1^* = a_{111}$ and (nF,UP)-flow of $y_1^* = a_{221}$ ($a_{110} < 0, a_{220} < 0$).

there are at least two equilibrium flows at $y_1 = \Delta_1 \notin (a_{111}, a_{221})$ with $\dot{y}_1 = 0$, and the directions of the equilibrium flows are determined from ($\dot{x}_1 < 0$ and $\dot{x}_2 > 0$) for $\Delta_1 > a_{211}$ and ($\dot{x}_1 > 0$ and $\dot{x}_2 < 0$) for $\Delta_1 < a_{111}$. For $y_1 \in (a_{111}, a_{221})$, $\dot{y}_1 < 0$ always.

In Fig. 2.3(c), the (3^{rd}UP,pF)-flow of $y_1^* = a_{111}$ and (pF,DP)-flow of $y_1^* = a_{221}$ ($a_{110} > 0, a_{220} < 0$) with ($\alpha_{11} > 0, \alpha_{12} > 0$) are presented. There are at least two equilibrium flows at $y_1 = \Delta_1 \notin (a_{111}, a_{221})$ with $\dot{y}_1 = 0$, and the direction of the equilibrium flow is determined from ($\dot{x}_1 > 0$ and $\dot{x}_2 < 0$) for $\Delta_1 > a_{211}$ and ($\dot{x}_1 < 0$ and $\dot{x}_2 > 0$) for $\Delta_1 < a_{111}$. For $y_1 \in (a_{111}, a_{221})$, $\dot{y}_1 > 0$ always.

In Fig. 2.3(d), the ($3^{\rm rd}$DP,pF)-flow of $y_1^* = a_{111}$ and (nF,UP)-flow of $y_1^* = a_{221}$ ($a_{110} < 0, a_{220} < 0$) with ($\alpha_{11} > 0, \alpha_{12} > 0$) are presented. There is at least an equilibrium flow at $y_1 = \Delta_1 \in (a_{111}, a_{221})$ with $\dot{y}_1 = 0$, and the direction of the two equilibrium flows is determined from $\dot{x}_1 < 0$ and $\dot{x}_2 > 0$. For $y_1 > a_{221}$ and $y_1 < a_{111}$, $\dot{y}_1 < 0$ and $\dot{y}_1 > 0$, respectively.

(iv) For $a_{111} < a_{221}$ with $\alpha_{11} > 0$ and $\alpha_{12} < 0$, as for $a_{111} < a_{221}$, presented in Fig. 2.4 are the one-dimensional flows of $(y_1^*, \dot{x}_2) = (a_{111}, \dot{x}_2)$ and $(\dot{x}_1, y_1^*) = (\dot{x}_1, a_{221})$. The corresponding properties for $\alpha_{11} > 0$ and $\alpha_{12} > 0$ are presented as follows:

$$\begin{Bmatrix} (\dot{x}_1, a_{221}) \\ (a_{111}, \dot{x}_2) \end{Bmatrix} = \begin{Bmatrix} \underbrace{(\text{pF}, \text{UP})}_{\text{up-parabola flow } (+)} \\ \underbrace{(3^{\text{rd}}\text{UP}, \text{nF})}_{\text{third-order up-parabola } (-)} \end{Bmatrix}$$

$$\text{for } a_{110} > 0 \quad \text{and} \quad a_{220} > 0, \tag{2.185}$$

$$\begin{Bmatrix} (\dot{x}_1, a_{221}) \\ (a_{111}, \dot{x}_2) \end{Bmatrix} = \begin{Bmatrix} \underbrace{(\text{nF}, \text{DP})}_{\text{down-parabola flow } (-)} \\ \underbrace{(3^{\text{rd}}\text{DP}, \text{nF})}_{\text{third-order down-parabola } (-)} \end{Bmatrix}$$

$$\text{for } a_{110} < 0 \quad \text{and} \quad a_{220} > 0, \tag{2.186}$$

$$\begin{Bmatrix} (\dot{x}_1, a_{221}) \\ (a_{111}, \dot{x}_2) \end{Bmatrix} = \begin{Bmatrix} \underbrace{(\text{pF}, \text{DP})}_{\text{down-parabola flow } (+)} \\ \underbrace{(3^{\text{rd}}\text{DP}, \text{pF})}_{\text{third-order down-parabola } (+)} \end{Bmatrix}$$

$$\text{for } a_{110} > 0 \quad \text{and} \quad a_{220} < 0, \tag{2.187}$$

$$\begin{Bmatrix} (\dot{x}_1, a_{221}) \\ (a_{111}, \dot{x}_2) \end{Bmatrix} = \begin{Bmatrix} \underbrace{(\text{nF}, \text{UP})}_{\text{up-parabola flow } (-)} \\ \underbrace{(3^{\text{rd}}\text{UP}, \text{pF})}_{\text{third-order up-parabola } (+)} \end{Bmatrix}$$

$$\text{for } a_{110} < 0 \quad \text{and} \quad a_{220} < 0. \tag{2.188}$$

In Fig. 2.4(a), the ($3^{\rm rd}$UP,nF)-flow of $y_1^* = a_{111}$ and (pF,UP)-flow of $y_1^* = a_{221}$ ($a_{110} > 0, a_{220} > 0$) with ($\alpha_{11} > 0, \alpha_{12} < 0$) are presented. There are at least two equilibrium flows at $y_1 = \Delta_1 \notin (a_{111}, a_{221})$ with $\dot{y}_1 = 0$, and

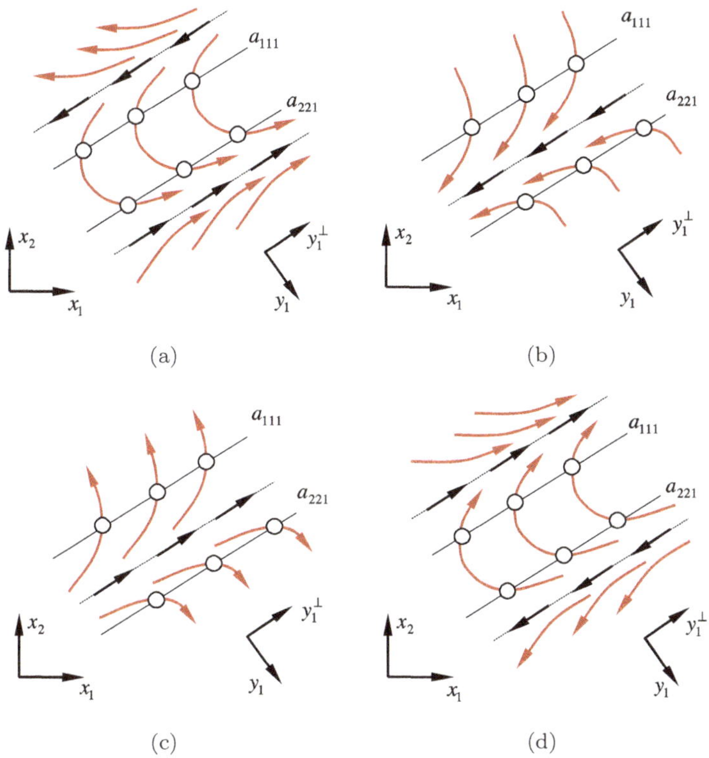

Fig. 2.4. Phase portraits ($a_{111} < a_{221}$ and $\alpha_{11} > 0, \alpha_{12} < 0$) for two-dimensional systems on the x_1-direction with $y_1^* = a_{111}$ and on the x_2-direction with $y_1^* = a_{221}$: (a) (3^{rd}UP,nF)-flow of $y_1^* = a_{111}$ and (pF,UP)-flow of $y_1^* = a_{221}$ ($a_{110} > 0, a_{220} > 0$), (b) (3^{rd}DP,nF)-flow of $y_1^* = a_{111}$ and (nF,DP)-flow of $y_1^* = a_{221}$ ($a_{110} < 0, a_{220} > 0$), (c) (3^{rd}DP,pF)-flow of $y_1^* = a_{111}$ and (pF,DP)-flow of $y_1^* = a_{221}$ ($a_{110} > 0, a_{220} < 0$), (d) (3^{rd}UP,pF)-flow of $y_1^* = a_{111}$ and (nF,UP)-flow of $y_1^* = a_{221}$ ($a_{110} < 0, a_{220} < 0$).

the directions of the two equilibrium flows are determined from ($\dot{x}_1 > 0$ and $\dot{x}_2 > 0$) for $y_1 = \Delta_1 > a_{221}$ and ($\dot{x}_1 < 0$ and $\dot{x}_2 < 0$) for $y_1 = \Delta_1 < a_{111}$. For $y_1 \in (a_{111}, a_{221})$, $\dot{y}_1 > 0$ holds.

In Fig. 2.4(b), the (3^{rd}DP,nF)-flow of $y_1^* = a_{111}$ and (nF,DP)-flow of $y_1^* = a_{221}$ ($a_{110} < 0, a_{220} > 0$) with ($\alpha_{11} > 0, \alpha_{12} < 0$) are presented. There is at least an equilibrium flow at $y_1 = \Delta_1 \in (a_{111}, a_{221})$ with $\dot{y}_1 = 0$, and the direction of the equilibrium flows is determined from ($\dot{x}_1 < 0$ and $\dot{x}_2 < 0$). For $y_1 > a_{221}$ and $y_1 > a_{111}$, $\dot{y}_1 < 0$ and $\dot{y}_1 > 0$ exist, respectively.

In Fig. 2.3(c), the (3^{rd}DP,pF)-flow of $y_1^* = a_{111}$ and (pF,DP)-flow of $y_1^* = a_{221}$ ($a_{110} > 0, a_{220} < 0$) with ($\alpha_{11} > 0, \alpha_{12} < 0$) are presented. There is at least an equilibrium flow at $y_1 = \Delta_1 \in (a_{111}, a_{221})$ with $\dot{y}_1 = 0$, and the direction of the equilibrium flows is determined from ($\dot{x}_1 > 0$ and $\dot{x}_2 > 0$). For $y_1 > a_{221}$ and $y_1 > a_{111}$, $\dot{y}_1 > 0$ and $\dot{y}_1 < 0$ exist, respectively.

In Fig. 2.4(d), the (3^{rd}UP,pF)-flow of $y_1^* = a_{111}$ and (nF,UP)-flow of $y_1^* = a_{221}$ $(a_{110} < 0, a_{220} < 0)$ with $(\alpha_{11} > 0, \alpha_{12} < 0)$ are presented. There are at least two equilibrium flows at $y_1 = \Delta_1 \notin (a_{111}, a_{221})$ with $\dot{y}_1 = 0$, and the directions of the two equilibrium flows are determined from ($\dot{x}_1 < 0$ and $\dot{x}_2 < 0$) for $y_1 = \Delta_1 > a_{221}$ and ($\dot{x}_1 > 0$ and $\dot{x}_2 > 0$) for $y_1 = \Delta_1 < a_{111}$. For $y_1 \in (a_{111}, a_{221})$, $\dot{y}_1 < 0$ holds.

The corresponding appearing bifurcations are given as follows:

$$\underbrace{(3^{\text{rd}}\text{UP}, \text{pF})}_{\text{appearing bifurcation}} \rightleftharpoons \left\{ \begin{array}{c} \underbrace{(\text{UP}, \text{pF})}_{\text{up-parabola flow}} \\ \underbrace{(\text{DP}, \text{pF})}_{\text{down-parabola flow}} \\ \underbrace{(\text{UP}, \text{pF})}_{\text{up-parabola flow}} \end{array} \right\}, \tag{2.189}$$

$$\underbrace{(3^{\text{rd}}\text{UP}, \text{nF})}_{\text{appearing bifurcation}} \rightleftharpoons \left\{ \begin{array}{c} \underbrace{(\text{UP}, \text{nF})}_{\text{up-parabola flow}} \\ \underbrace{(\text{DP}, \text{nF})}_{\text{down-parabola flow}} \\ \underbrace{(\text{UP}, \text{nF})}_{\text{up-parabola flow}} \end{array} \right\}, \tag{2.190}$$

$$\underbrace{(3^{\text{rd}}\text{DP}, \text{pF})}_{\text{appearing bifurcation}} \rightleftharpoons \left\{ \begin{array}{c} \underbrace{(\text{DP}, \text{pF})}_{\text{down-parabola flow}} \\ \underbrace{(\text{UP}, \text{pF})}_{\text{up-parabola flow}} \\ \underbrace{(\text{DP}, \text{pF})}_{\text{down-parabola flow}} \end{array} \right\}, \tag{2.191}$$

$$\underbrace{(3^{\text{rd}}\text{DP}, \text{nF})}_{\text{appearing bifurcation}} \rightleftharpoons \left\{ \begin{array}{c} \underbrace{(\text{DP}, \text{nF})}_{\text{down-parabola flow}} \\ \underbrace{(\text{UP}, \text{nF})}_{\text{up-parabola flow}} \\ \underbrace{(\text{DP}, \text{nF})}_{\text{down-parabola flow}} \end{array} \right\}. \tag{2.192}$$

2.3.2 *Switching bifurcations*

Consider a two-dimensional system for $a_{221} = a_{111}$ as

$$\begin{aligned}\dot{x}_1 &= a_{110}(y_1 - a_{111})^3,\\ \dot{x}_2 &= a_{220}(y_1 - a_{221}).\end{aligned} \tag{2.193}$$

There is a dynamical system in the y_1-direction as

$$\dot{y}_1 = (y_1 - a_{111})[\alpha_{11}a_{110}(y_1 - a_{111})^2 + \alpha_{12}a_{220}], \tag{2.194}$$

with a dynamical system in the $y_1^{\perp}$-direction as

$$\dot{y}_1^{\perp} = (y_1 - a_{111})[\beta_{11}a_{110}(y_1 - a_{111})^2 + \beta_{12}a_{220}], \tag{2.195}$$

with

$$G_{110} = \alpha_{12}a_{220}. \tag{2.196}$$

The increasing-inflection and decreasing-inflection infinite-equilibriums of $y_1^* = a_{111} = a_{221}$ are determined by $\alpha_{12}^3 a_{110}/\alpha_{12}a_{220} > 0$ and $\alpha_{12}^3 a_{110}/\alpha_{12}a_{220} < 0$, respectively. The corresponding inflection infinite-equilibriums are source and sink through $G_{110} > 0$ and $G_{110} < 0$, accordingly. For $\alpha_{11}a_{110}/\alpha_{12}a_{220} < 0$, there are three equilibriums of $\dot{y}_1 = 0$. Except for the infinite-equilibrium, the two one-dimensional flow directions are positive and negative by $\dot{y}_1^{\perp} > 0$ and $\dot{y}_1^{\perp} < 0$, respectively. Only the infinite-equilibrium of $\dot{y}_1 = 0$ exists for $\alpha_{11}a_{110}/\alpha_{12}a_{220} > 0$.

For $\alpha_{11} > 0$ and $\alpha_{12} > 0$, the infinite-equilibriums are presented as follows:

$$(a_{111}, a_{221}) = \underbrace{(2^{\text{nd}}\text{II}, \text{SO})}_{\text{second-order increasing-inflection source}}$$

for $a_{110} > 0$ and $a_{220} > 0$,

$$(a_{111}, a_{221}) = \underbrace{(2^{\text{rd}}\text{DI}, \text{SO})}_{\text{second-order decreasing-inflection source}}$$

for $a_{110} < 0$ and $a_{220} > 0$,

$$(a_{111}, a_{221}) = \underbrace{(2^{\text{nd}}\text{DI}, \text{SI})}_{\text{second-order decreasing-inflection sink}}$$

for $a_{110} > 0$ and $a_{220} < 0$,

$$(a_{111}, a_{221}) = \underbrace{(2^{\mathrm{nd}}\mathrm{II}, \mathrm{SI})}_{\text{second-order increasing-inflection sink}}$$
$$\text{for } a_{110} > 0 \quad \text{and} \quad a_{220} > 0. \tag{2.197}$$

The infinite-equilibriums for $(\alpha_{11} > 0, \alpha_{12} > 0)$ are presented in Fig. 2.5. In Fig. 2.5(a), the (2nd II,SO)-infinite-equilibrium at $y_1^* = a_{111} = a_{221}$ is presented $(a_{110} > 0, a_{220} > 0)$. For $y_1 > a_{111}$ and $y_1 < a_{111}$, $\dot{y}_1 > 0$ and $\dot{y}_1 < 0$ hold and no equilibrium flows of $\dot{y}_1 = 0$ exist. The (2nd DI,SO)-infinite-equilibrium $(a_{110} < 0, a_{220} > 0)$ at $y_1^* = a_{111} = a_{221}$ are presented in Fig. 2.5(b). Due to $\alpha_{11}a_{110}/\alpha_{12}a_{220} < 0$, $y_1^* = a_{111} \pm \Delta_1$ with $\Delta_1 = \sqrt{|\alpha_{12}a_{220}/\alpha_{11}a_{110}|}$. The directions of the equilibrium flows of $y_1^* = a_{111} \pm \Delta_1$ are determined by $(\dot{x}_1 < 0, \dot{x}_2 > 0)$ and $(\dot{x}_1 > 0, \dot{x}_2 < 0)$. The (2nd DI,SI)-infinite-equilibrium $(a_{110} > 0, a_{220} < 0)$ at $y_1^* = a_{111} = a_{221}$ are presented in Fig. 2.5(c). Due to $\alpha_{11}a_{110}/\alpha_{12}a_{220} < 0$, $y_1^* = a_{111} \pm \Delta_1$. The corresponding directions of the equilibrium flows are determined by $(\dot{x}_1 > 0, \dot{x}_2 < 0)$ for $y_1^* = a_{111} + \Delta_1$ and $(\dot{x}_1 < 0, \dot{x}_2 > 0)$ for $y_1^* = a_{111} - \Delta_1$. In Fig. 2.5(d), the (2nd II,SI)-infinite-equilibrium at $y_1^* = a_{111} = a_{221}$ is presented $(a_{110} < 0, a_{220} < 0)$. For $y_1 > a_{111}$ and $y_1 < a_{111}$, $\dot{y}_1 < 0$ and $\dot{y}_1 > 0$ hold without any equilibrium flow.

For $\alpha_{11} > 0$ and $\alpha_{12} > 0$, the infinite-equilibriums are presented as follows:

$$(a_{111}, a_{221}) = \underbrace{(2^{\mathrm{nd}}\mathrm{II}, \mathrm{SI})}_{\text{second-order increasing-inflection sink}}$$
$$\text{for } a_{110} > 0 \quad \text{and} \quad a_{220} > 0,$$
$$(a_{111}, a_{221}) = \underbrace{(2^{\mathrm{nd}}\mathrm{DI}, \mathrm{SI})}_{\text{second-order decreasing-inflection sink}}$$
$$\text{for } a_{110} < 0 \quad \text{and} \quad a_{220} > 0,$$
$$(a_{111}, a_{221}) = \underbrace{(2^{\mathrm{rd}}\mathrm{DI}, \mathrm{SO})}_{\text{second-order decreasing-inflection source}}$$
$$\text{for } a_{110} > 0 \quad \text{and} \quad a_{220} < 0,$$

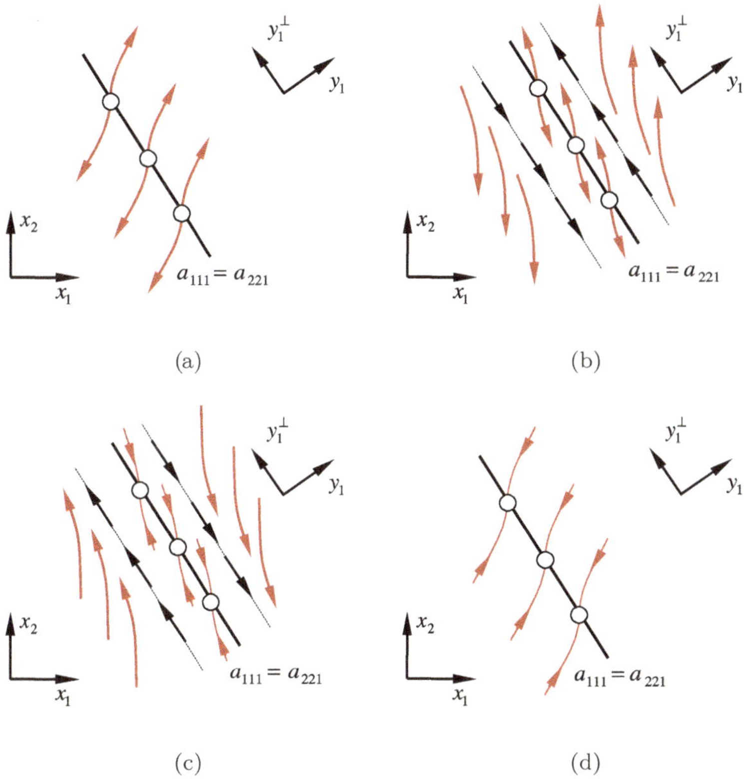

Fig. 2.5. Phase portraits $(\alpha_{11} > 0, \alpha_{12} > 0)$ for two-dimensional systems on the x_1-direction with $y_1^* = a_{111}$ and on the x_2-direction with $y_1^* = a_{111}$: (a) $(2^{\text{nd}}\text{II,SO})$-infinite equilibrium $(a_{110} > 0, a_{220} > 0)$, (b) $(2^{\text{nd}}\text{DI,SO})$-infinite equilibrium $(a_{110} < 0, a_{220} > 0)$, (c) $(2^{\text{nd}}\text{DI,SI})$-infinite equilibrium $(a_{110} < 0, a_{220} > 0)$, (d) $(2^{\text{nd}}\text{II,SI})$-infinite equilibrium $(a_{110} < 0, a_{220} < 0)$.

$$(a_{111}, a_{221}) = \underbrace{(2^{\text{rd}}\text{II}, \text{SO})}_{\text{second-order increasing-inflection source}}$$
$$\text{for } a_{110} > 0 \quad \text{and} \quad a_{220} > 0. \tag{2.198}$$

The infinite-equilibriums for $(\alpha_{11} > 0, \alpha_{12} < 0)$ are presented in Fig. 2.6. In Fig. 2.6(a), the $(2^{\text{nd}}\text{II,SI})$-infinite equilibrium at $y_1^* = a_{111} = a_{221}$ is presented $(a_{110} > 0, a_{220} > 0)$. Due to $\alpha_{11}a_{110}/\alpha_{12}a_{220} < 0$, $y_1^* = a_{111} \pm \Delta_1$ with $\Delta_1 = \sqrt{|\alpha_{12}a_{220}/\alpha_{11}a_{110}|}$. The directions of the equilibrium flows of $y_1^* = a_{111} \pm \Delta_1$ are determined by $(\dot{x}_1 < 0, \dot{x}_2 > 0)$ and $(\dot{x}_1 > 0,\ \dot{x}_2 < 0)$

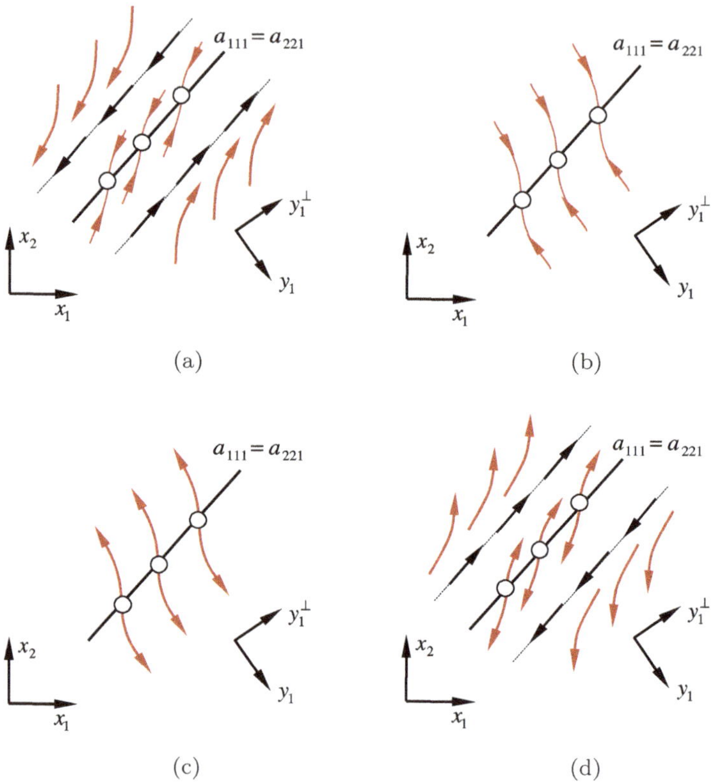

Fig. 2.6. Phase portraits ($\alpha_{11} > 0, \alpha_{12} < 0$) for two-dimensional systems on the x_1-direction with $y_1^* = a_{111}$ and on the x_2-direction with $y_1^* = a_{111}$: (a) (2^{nd}II,SI)-infinite equilibrium ($a_{110} > 0, a_{220} > 0$), (b) (2^{nd}DI,SI)-infinite equilibrium ($a_{110} < 0, a_{220} > 0$), (c) (2^{nd}DI,SO)-infinite equilibrium ($a_{110} < 0, a_{220} > 0$), (d) (2^{nd}II,SO)-infinite equilibrium ($a_{110} < 0, a_{220} < 0$).

accordingly. The (2^{nd}DI,SI)-infinite equilibrium ($a_{110} < 0, a_{220} > 0$) at $y_1^* = a_{111} = a_{221}$ is presented in Fig. 2.6(b). For $y_1 > a_{111}$ and $y_1 < a_{111}$, $\dot{y}_1 > 0$ and $\dot{y}_1 < 0$ hold and no equilibrium flows of $\dot{y}_1 = 0$ exist. The (2^{nd}DI,SO)-infinite equilibrium ($a_{110} > 0, a_{220} < 0$) at $y_1^* = a_{111} = a_{221}$ is presented in Fig. 2.6(c). For $y_1 > a_{111}$ and $y_1 < a_{111}$, $\dot{y}_1 < 0$ and $\dot{y}_1 > 0$ hold without any equilibrium flow. In Fig. 2.5(d), the (2^{nd}II,SI)-infinite equilibrium at $y_1^* = a_{111} = a_{221}$ is presented ($a_{110} < 0, a_{220} < 0$). Owing to $\alpha_{11}a_{110}/\alpha_{12}a_{220} < 0$, $y_1^* = a_{111} \pm \Delta_1$. The corresponding directions of the equilibrium flows are determined by ($\dot{x}_1 > 0, \dot{x}_2 < 0$) for $y_1^* = a_{111} + \Delta_1$ and ($\dot{x}_1 < 0, \dot{x}_2 > 0$) for $y_1^* = a_{111} - \Delta_1$.

The infinite-equilibrium of $y_1^* = a_{111} = a_{221}$ is a switching bifurcation as follows:

- For $a_{110} > 0$ and $a_{220} > 0$,

$$\left\{\begin{matrix} \underbrace{(\mathrm{nF}, \mathrm{DP})}_{\text{down-parabola flow } (-)} \\ \underbrace{(3^{\mathrm{rd}}\mathrm{UP}, \mathrm{pF})}_{\text{third-order up-parabola flow } (+)} \end{matrix}\right\} \rightleftharpoons \underbrace{(2^{\mathrm{nd}}\mathrm{II}, \mathrm{SO})}_{\text{switching bifurcation}}$$
$$\rightleftharpoons \left\{\begin{matrix} \underbrace{(3^{\mathrm{rd}}\mathrm{DP}, \mathrm{nF})}_{\text{third-order down-parabola flow } (-)} \\ \underbrace{(\mathrm{pF}, \mathrm{UP})}_{\text{up-parabola flow } (+)} \end{matrix}\right\}$$

for $\alpha_{12} > 0$,

$$\left\{\begin{matrix} \underbrace{(\mathrm{nF}, \mathrm{DP})}_{\text{down-parabola flow } (-)} \\ \underbrace{(3^{\mathrm{rd}}\mathrm{DP}, \mathrm{pF})}_{\text{third-order down-parabola flow } (+)} \end{matrix}\right\} \rightleftharpoons \underbrace{(2^{\mathrm{nd}}\mathrm{II}, \mathrm{SI})}_{\text{switching bifurcation}}$$
$$\rightleftharpoons \left\{\begin{matrix} \underbrace{(3^{\mathrm{rd}}\mathrm{UP}, \mathrm{nF})}_{\text{third-order up-parabola flow } (-)} \\ \underbrace{(\mathrm{pF}, \mathrm{UP})}_{\text{up-parabola flow } (+)} \end{matrix}\right\}$$

for $\alpha_{12} < 0$. (2.199)

- For $a_{110} < 0$ and $a_{220} > 0$,

$$\left\{\begin{matrix} \underbrace{(\mathrm{nF}, \mathrm{UP})}_{\text{up-parabola flow } (-)} \\ \underbrace{(3^{\mathrm{rd}}\mathrm{DP}, \mathrm{pF})}_{\text{third-order down-parabola flow } (+)} \end{matrix}\right\} \rightleftharpoons \underbrace{(2^{\mathrm{nd}}\mathrm{DI}, \mathrm{SO})}_{\text{switching bifurcation}}$$
$$\rightleftharpoons \left\{\begin{matrix} \underbrace{(3^{\mathrm{rd}}\mathrm{UP}, \mathrm{nF})}_{\text{third-order up-parabola flow } (-)} \\ \underbrace{(\mathrm{pF}, \mathrm{DP})}_{\text{down-parabola flow } (+)} \end{matrix}\right\}$$

for $\alpha_{12} > 0$,

$$\left\{\begin{array}{c}\underbrace{(\mathrm{pF},\mathrm{UP})}_{\text{up-parabola flow }(+)}\\ \underbrace{(3^{\mathrm{rd}}\mathrm{UP},\mathrm{pF})}_{\text{third-order up-parabola flow }(+)}\end{array}\right\}\rightleftharpoons\underbrace{(2^{\mathrm{nd}}\mathrm{DI},\mathrm{SI})}_{\text{switching bifurcation}}$$
$$\rightleftharpoons\left\{\begin{array}{c}\underbrace{(3^{\mathrm{rd}}\mathrm{DP},\mathrm{nF})}_{\text{third-order down-parabola flow }(-)}\\ \underbrace{(\mathrm{nF},\mathrm{DP})}_{\text{down-parabola flow }(-)}\end{array}\right\}$$

for $\alpha_{12} < 0$. (2.200)

- For $a_{110} > 0$ and $a_{220} < 0$,

$$\left\{\begin{array}{c}\underbrace{(\mathrm{pF},\mathrm{UP})}_{\text{up-parabola flow }(+)}\\ \underbrace{(3^{\mathrm{rd}}\mathrm{DP},\mathrm{nF})}_{\text{third-order down-parabola flow }(-)}\end{array}\right\}\rightleftharpoons\underbrace{(2^{\mathrm{nd}}\mathrm{DI},\mathrm{SI})}_{\text{switching bifurcation}}$$
$$\rightleftharpoons\left\{\begin{array}{c}\underbrace{(3^{\mathrm{rd}}\mathrm{UP},\mathrm{pF})}_{\text{third-order up-parabola flow }(+)}\\ \underbrace{(\mathrm{nF},\mathrm{DP})}_{\text{down-parabola flow }(-)}\end{array}\right\}$$

for $\alpha_{12} > 0$,

$$\left\{\begin{array}{c}\underbrace{(\mathrm{nF},\mathrm{UP})}_{\text{up-parabola flow }(-)}\\ \underbrace{(3^{\mathrm{rd}}\mathrm{UP},\mathrm{pF})}_{\text{third-order up-parabola flow }(-)}\end{array}\right\}\rightleftharpoons\underbrace{(2^{\mathrm{nd}}\mathrm{DI},\mathrm{SO})}_{\text{switching bifurcation}}$$
$$\rightleftharpoons\left\{\begin{array}{c}\underbrace{(3^{\mathrm{rd}}\mathrm{DP},\mathrm{pF})}_{\text{third-order down-parabola flow }(+)}\\ \underbrace{(\mathrm{pF},\mathrm{DP})}_{\text{down-parabola flow }(+)}\end{array}\right\}$$

for $\alpha_{12} < 0$. (2.201)

- For $a_{110} < 0$ and $a_{220} < 0$,

$$\left\{\begin{array}{c}\underbrace{(\mathrm{pF},\mathrm{DP})}_{\text{down-parabola flow }(+)}\\ \underbrace{(3^{\mathrm{rd}}\mathrm{UP},\mathrm{nF})}_{\text{third-order up-parabola flow }(-)}\end{array}\right\} \rightleftharpoons \underbrace{(2^{\mathrm{nd}}\mathrm{II},\mathrm{SI})}_{\text{switching bifurcation}}$$

$$\rightleftharpoons \left\{\begin{array}{c}\underbrace{(3^{\mathrm{rd}}\mathrm{DP},\mathrm{pF})}_{\text{third-order down-parabola flow }(+)}\\ \underbrace{(\mathrm{nF},\mathrm{UP})}_{\text{up-parabola flow }(-)}\end{array}\right\}$$

for $\alpha_{12} > 0$,

$$\left\{\begin{array}{c}\underbrace{(\mathrm{pF},\mathrm{DP})}_{\text{down-parabola flow }(+)}\\ \underbrace{(3^{\mathrm{rd}}\mathrm{DP},\mathrm{nF})}_{\text{third-order down-parabola flow }(-)}\end{array}\right\} \rightleftharpoons \underbrace{(2^{\mathrm{nd}}\mathrm{II},\mathrm{SO})}_{\text{switching bifurcation}}$$

$$\rightleftharpoons \left\{\begin{array}{c}\underbrace{(3^{\mathrm{rd}}\mathrm{UP},\mathrm{pF})}_{\text{third-order up-parabola flow }(+)}\\ \underbrace{(\mathrm{nF},\mathrm{UP})}_{\text{up-parabola flow }(-)}\end{array}\right\}$$

for $\alpha_{12} < 0$. (2.202)

2.4 Second-Order Singular Single-Linear-Bivariate Cubic Systems

The dynamics of the second-order singular single-linear-bivariate systems are discussed.

2.4.1 *Inflection flows and parabola-sources (sinks)*

Consider a two-dimensional system for $a_{221} \neq a_{111}, a_{112}$ as

$$\begin{aligned}\dot{x}_1 &= a_{110}(y_1 - a_{111})(y_1 - a_{112})^2,\\ \dot{x}_2 &= a_{220}(y_1 - a_{221})\end{aligned} \tag{2.203}$$

with

$$y_1 = \alpha_{11}x_1 + \alpha_{12}x_2. \tag{2.204}$$

There is a dynamical system in the y_1-direction as

$$\dot{y}_1 = \alpha_{11}a_{110}(y_1 - a_{111})(y_1 - a_{112})^2 + \alpha_{12}a_{220}(y_1 - a_{221}) \tag{2.205}$$

with a dynamical system in the $y_1^{\perp}$-direction as

$$\dot{y}_1^{\perp} = \beta_{11}a_{110}(y_1 - a_{111})(y_1 - a_{112})^2 + \beta_{12}a_{220}(y_1 - a_{221}), \tag{2.206}$$

where

$$\begin{gathered} y_1^{\perp} = \beta_{11}x_1 + \beta_{12}x_2, \mathbf{n}_1^{\perp} = (\beta_{11}, \beta_{12})^{\mathrm{T}}; \\ (\mathbf{n}_1)^{\mathrm{T}} \cdot \mathbf{n}_1^{\perp} = \alpha_{11}\beta_{11} + \alpha_{12}\beta_{12} = 0; \\ c_{11} = \alpha_{11}a_{110}/\alpha_{12}a_{220}. \end{gathered} \tag{2.207}$$

(i) For $a_{111} > a_{221}$ with $\alpha_{11} > 0$ and $\alpha_{12} > 0$, presented in Fig. 2.7 are the one-dimensional flows of $(y_1^*, \dot{x}_2) = (a_{111}, \dot{x}_2), (a_{112}, \dot{x}_2)$ and $(\dot{x}_1, y_1^*) = (\dot{x}_1, a_{221})$. The equilibrium flows of $\dot{y}_1 = 0$ are listed in Table 2.1, and the directions of equilibrium flows are determined by the $(\dot{x}_1, \dot{x}_2)$ signs. The corresponding properties for $\alpha_{11} > 0$ and $\alpha_{12} > 0$ are presented as follows:

$$\left\{ \begin{matrix} (a_{112}, \dot{x}_2) \\ (a_{111}, \dot{x}_2) \\ (\dot{x}_1, a_{221}) \end{matrix} \right\} = \left\{ \begin{matrix} \underbrace{(\mathrm{II}, \mathrm{pF})}_{\text{increasing-inflection flow } (+)} \\ \underbrace{(\mathrm{UP}, \mathrm{pF})}_{\text{up-parabola flow } (+)} \\ \underbrace{(\mathrm{nF}, \mathrm{DP})}_{\text{down-parabola flow } (-)} \end{matrix} \right\}$$

$$\text{for } a_{110} > 0 \text{ and } a_{220} > 0, \tag{2.208}$$

$$\left\{ \begin{matrix} (a_{112}, \dot{x}_2) \\ (a_{111}, \dot{x}_2) \\ (\dot{x}_1, a_{221}) \end{matrix} \right\} = \left\{ \begin{matrix} \underbrace{(\mathrm{DI}, \mathrm{pF})}_{\text{decreasing-inflection flow } (+)} \\ \underbrace{(\mathrm{DP}, \mathrm{pF})}_{\text{down-parabola flow } (+)} \\ \underbrace{(\mathrm{pF}, \mathrm{UP})}_{\text{up-parabola flow } (+)} \end{matrix} \right\}$$

$$\text{for } a_{110} < 0 \text{ and } a_{220} > 0, \tag{2.209}$$

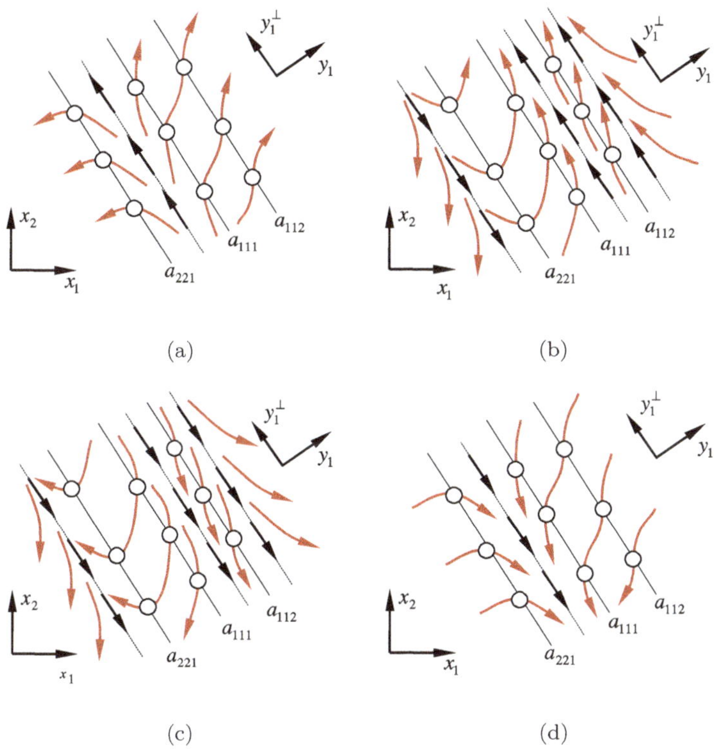

Fig. 2.7. Phase portraits ($a_{111} > a_{221}$ and $\alpha_{11} > 0, \alpha_{12} > 0$) for two-dimensional systems on the x_1-direction with $y_1^* = a_{111}, a_{112}$ and on the x_2-direction with $y_1^* = a_{221}$: (a) (UP,pF) and (II,pF)-flows of $y_1^* = a_{111}, a_{112}$ and (nF,DP)-flow of $y_1^* = a_{221}$ ($a_{110} > 0, a_{220} > 0$), (b) (DP,pF) and (DI,pF)-flows of $y_1^* = a_{111}, a_{112}$ and (pF,UP)-flow of $y_1^* = a_{221}$ ($a_{110} < 0, a_{220} > 0$), (c) (DP,nF) and (DI,nF)-flows of $y_1^* = a_{111}, a_{112}$ and (nF,UP)-flow of $y_1^* = a_{221}$ ($a_{110} > 0, a_{220} < 0$), (d) (UP,nF) and (II,nF)-flows of $y_1^* = a_{111}, a_{112}$ and (pF,DP)-flow of $y_1^* = a_{221}$ ($a_{110} < 0, a_{220} < 0$).

$$\begin{Bmatrix} (a_{112}, \dot{x}_2) \\ (a_{111}, \dot{x}_2) \\ (\dot{x}_1, a_{221}) \end{Bmatrix} = \begin{Bmatrix} \underbrace{(\text{DI}, \text{nF})}_{\text{decreasing-inflection flow } (-)} \\ \underbrace{(\text{DP}, \text{nF})}_{\text{down-parabola flow } (-)} \\ \underbrace{(\text{nF}, \text{UP})}_{\text{up-parabola flow } (-)} \end{Bmatrix}$$

$$\text{for } a_{110} > 0 \text{ and } a_{220} < 0, \tag{2.210}$$

Table 2.1. Equilibrium flows of $\dot{y}_1 = 0$ $(\alpha_{11} > 0, \alpha_{12} > 0)$.

Figure 2.7	$(\dot{x}_1, \dot{x}_2)$	y_1			
	(a_{110}, a_{220})	$(-\infty, a_{221})$	(a_{221}, a_{111})	(a_{111}, a_{112})	(a_{112}, ∞)
(a)	$(+,+)$	$\dot{y}_1 < 0$ $(-,-)$	$\dot{y}_1^* = 0$ $(-,+)$	$\dot{y}_1 > 0$ $(+,+)$	$\dot{y}_1 > 0$ $(+,+)$
(b)	$(-,+)$	$\dot{y}_1^* = 0$ $(+,-)$	$\dot{y}_1 > 0$ $(+,+)$	$\dot{y}_1^* = 0$ $(-,+)$	$\dot{y}_1^* = 0$ $(-,+)$
(c)	$(+,-)$	$\dot{y}_1^* = 0$ $(-,+)$	$\dot{y}_1 < 0$ $(-,-)$	$\dot{y}_1^* = 0$ $(+,-)$	$\dot{y}_1^* = 0$ $(+,-)$
(d)	$(-,-)$	$\dot{y}_1 > 0$ $(+,+)$	$\dot{y}_1^* = 0$ $(+,-)$	$\dot{y}_1 < 0$ $(-,-)$	$\dot{y}_1 < 0$ $(-,-)$

$$\left\{\begin{matrix} (a_{112}, \dot{x}_2) \\ (a_{111}, \dot{x}_2) \\ (\dot{x}_1, a_{221}) \end{matrix}\right\} = \left\{\begin{matrix} \underbrace{(\text{II}, \text{nF})}_{\text{increasing-inflection flow } (-)} \\ \underbrace{(\text{UP}, \text{nF})}_{\text{up-parabola flow } (-)} \\ \underbrace{(\text{pF}, \text{DP})}_{\text{down-parabola flow } (+)} \end{matrix}\right\}$$

$$\text{for } a_{110} < 0 \text{ and } a_{220} < 0. \tag{2.211}$$

(ii) For $a_{111} > a_{221}$ with $\alpha_{11} > 0$ and $\alpha_{12} < 0$, presented in Fig. 2.8 are the one-dimensional flows of $(y_1^*, \dot{x}_2) = (a_{111}, \dot{x}_2), (a_{112}, \dot{x}_2)$ and $(\dot{x}_1, y_1^*) = (\dot{x}_1, a_{221})$. The equilibrium flows of $\dot{y}_1 = 0$ are listed in Table 2.2. The corresponding properties for $\alpha_{11} > 0$ and $\alpha_{12} < 0$ are presented as follows:

$$\left\{\begin{matrix} (a_{112}, \dot{x}_2) \\ (a_{111}, \dot{x}_2) \\ (\dot{x}_1, a_{221}) \end{matrix}\right\} = \left\{\begin{matrix} \underbrace{(\text{II}, \text{pF})}_{\text{increasing-inflection flow } (+)} \\ \underbrace{(\text{DP}, \text{pF})}_{\text{down-parabola flow } (+)} \\ \underbrace{(\text{nF}, \text{DP})}_{\text{down-parabola flow } (-)} \end{matrix}\right\}$$

$$\text{for } a_{110} > 0 \text{ and } a_{220} > 0, \tag{2.212}$$

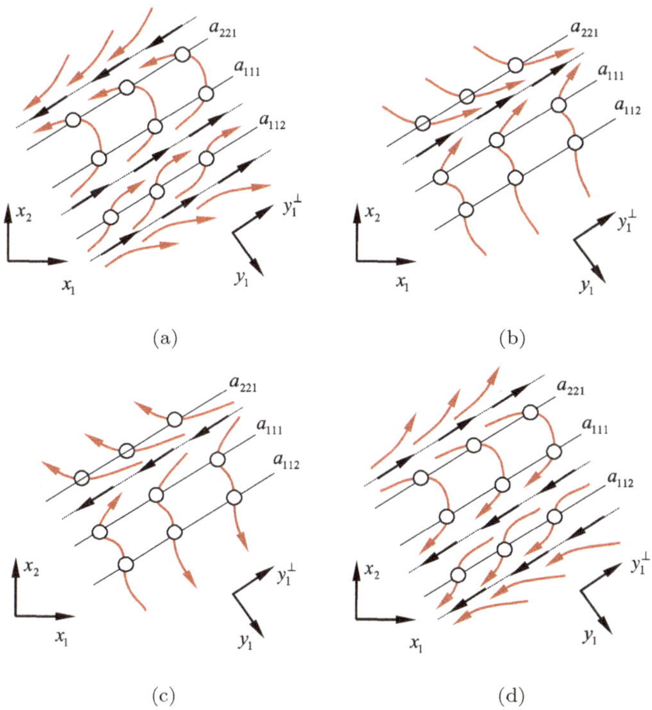

Fig. 2.8. Phase portraits ($a_{111} > a_{221}$ and $\alpha_{11} > 0, \alpha_{12} < 0$) for two-dimensional systems on the x_1-direction with $y_1^* = a_{111}, a_{112}$ and on the x_2-direction with $y_1^* = a_{221}$: (a) (DP,pF) and (II,pF)-flows of $y_1^* = a_{111}, a_{112}$ and (nF,DP)-flow of $y_1^* = a_{221}$ ($a_{110} > 0, a_{220} > 0$), (b) (DP,pF) and (DI,pF)-flows of $y_1^* = a_{111}, a_{112}$ and (nF,DP)-flow of $y_1^* = a_{221}$ ($a_{110} < 0, a_{220} > 0$), (c) (DP,nF) and (DI,nF)-flows of $y_1^* = a_{111}, a_{112}$ and (pF,DP)-flow of $y_1^* = a_{221}$ ($a_{110} > 0, a_{220} < 0$), (d) (UP,nF) and (II,nF)-flows of $y_1^* = a_{111}, a_{112}$ and (nF,UP)-flow of $y_1^* = a_{221}$ ($a_{110} < 0, a_{220} < 0$).

Table 2.2. Equilibrium flows of $\dot{y}_1 = 0$ ($\alpha_{11} > 0, \alpha_{12} < 0$).

Figure 2.8	$(\dot{x}_1, \dot{x}_2)$	y_1			
	(a_{110}, a_{220})	$(-\infty, a_{221})$	(a_{221}, a_{111})	(a_{111}, a_{112})	(a_{112}, ∞)
(a)	$(+,+)$	$\dot{y}_1^* = 0$ $(-,-)$	$\dot{y}_1 < 0$ $(-,+)$	$\dot{y}_1^* = 0$ $(+,+)$	$\dot{y}_1^* = 0$ $(+,+)$
(b)	$(-,+)$	$\dot{y}_1 < 0$ $(+,-)$	$\dot{y}_1^* = 0$ $(+,+)$	$\dot{y}_1 < 0$ $(-,+)$	$\dot{y}_1 < 0$ $(-,+)$
(c)	$(+,-)$	$\dot{y}_1 > 0$ $(-,+)$	$\dot{y}_1^* = 0$ $(-,-)$	$\dot{y}_1 > 0$ $(+,-)$	$\dot{y}_1 > 0$ $(+,-)$
(d)	$(-,-)$	$\dot{y}_1^* = 0$ $(+,+)$	$\dot{y}_1 > 0$ $(+,-)$	$\dot{y}_1^* = 0$ $(-,-)$	$\dot{y}_1^* = 0$ $(-,-)$

$$\left\{\begin{array}{l}(a_{112},\dot{x}_2)\\(a_{111},\dot{x}_2)\\(\dot{x}_1,a_{221})\end{array}\right\}=\left\{\begin{array}{c}\underbrace{(\mathrm{DI},\mathrm{pF})}_{\text{decreasing-inflection flow }(+)}\\\underbrace{(\mathrm{UP},\mathrm{pF})}_{\text{up-parabola flow }(+)}\\\underbrace{(\mathrm{pF},\mathrm{UP})}_{\text{up-parabola flow }(+)}\end{array}\right\}$$

$$\text{for } a_{110}<0 \text{ and } a_{220}>0, \tag{2.213}$$

$$\left\{\begin{array}{l}(a_{112},\dot{x}_2)\\(a_{111},\dot{x}_2)\\(\dot{x}_1,a_{221})\end{array}\right\}=\left\{\begin{array}{c}\underbrace{(\mathrm{DI},\mathrm{nF})}_{\text{decreasing-inflection flow }(-)}\\\underbrace{(\mathrm{UP},\mathrm{nF})}_{\text{up-parabola flow }(-)}\\\underbrace{(\mathrm{nF},\mathrm{UP})}_{\text{up-parabola flow }(-)}\end{array}\right\}$$

$$\text{for } a_{110}>0 \text{ and } a_{220}<0, \tag{2.214}$$

$$\left\{\begin{array}{l}(a_{112},\dot{x}_2)\\(a_{111},\dot{x}_2)\\(\dot{x}_1,a_{221})\end{array}\right\}=\left\{\begin{array}{c}\underbrace{(\mathrm{II},\mathrm{nF})}_{\text{increasing-inflection flow }(-)}\\\underbrace{(\mathrm{DP},\mathrm{nF})}_{\text{down-parabola flow }(-)}\\\underbrace{(\mathrm{pF},\mathrm{DP})}_{\text{down-parabola flow }(+)}\end{array}\right\}$$

$$\text{for } a_{110}<0 \text{ and } a_{220}<0. \tag{2.215}$$

(iii) For $a_{221} > a_{112}$ with $\alpha_{11} > 0$ and $\alpha_{12} > 0$, presented in Fig. 2.9 are the one-dimensional flows of $(y_1^*, \dot{x}_2) = (a_{111}, \dot{x}_2), (a_{112}, \dot{x}_2)$ and $(\dot{x}_1, y_1^*) = (\dot{x}_1, a_{221})$. The equilibrium flows of $\dot{y}_1 = 0$ are listed in Table 2.3. The corresponding properties for $\alpha_{11} > 0$ and $\alpha_{12} > 0$ are presented as follows:

$$\left\{\begin{array}{l}(\dot{x}_1,a_{221})\\(a_{112},\dot{x}_2)\\(a_{111},\dot{x}_2)\end{array}\right\}=\left\{\begin{array}{c}\underbrace{(\mathrm{pF},\mathrm{UP})}_{\text{up-parabola flow }(+)}\\\underbrace{(\mathrm{DI},\mathrm{nF})}_{\text{decreasing-inflection flow }(-)}\\\underbrace{(\mathrm{DP},\mathrm{nF})}_{\text{down-parabola flow }(-)}\end{array}\right\}$$

$$\text{for } a_{110}>0 \text{ and } a_{220}>0, \tag{2.216}$$

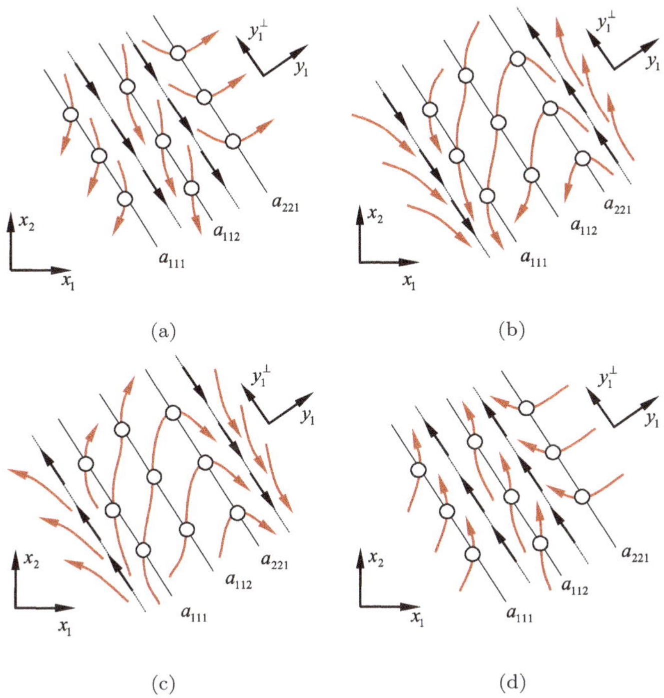

Fig. 2.9. Phase portraits ($a_{111} > a_{221}$ and $\alpha_{11} > 0, \alpha_{12} > 0$) for two-dimensional systems on the x_1-direction with $y_1^* = a_{111}, a_{112}$ and on the x_2-direction with $y_1^* = a_{221}$: (a) (DP,nF) and (DI,nF)-flows of $y_1^* = a_{111}, a_{112}$ and (pF,UP)-flow of $y_1^* = a_{221}$ ($a_{110} > 0, a_{220} > 0$), (b) (UP,nF) and (II,nF)-flows of $y_1^* = a_{111}, a_{112}$ and (nF,DP)-flow of $y_1^* = a_{221}$ ($a_{110} < 0, a_{220} > 0$), (c) (UP,pF) and (II,pF)-flows of $y_1^* = a_{111}, a_{112}$ and (pF,DP)-flow of $y_1^* = a_{221}$ ($a_{110} > 0, a_{220} < 0$), (d) (DP,pF) and (DI,pF)-flows of $y_1^* = a_{111}, a_{112}$ and (nF,UP)-flow of $y_1^* = a_{221}$ ($a_{110} < 0, a_{220} < 0$).

Table 2.3. Equilibrium flows of $\dot{y}_1 = 0(\alpha_{11} > 0, \alpha_{12} > 0)$.

Figure 2.9	$(\dot{x}_1, \dot{x}_2)$	y_1			
	(a_{110}, a_{220})	$(-\infty, a_{111})$	(a_{111}, a_{112})	(a_{112}, a_{221})	(a_{221}, ∞)
(a)	$(+,+)$	$\dot{y}_1 < 0$ $(-,-)$	$\dot{y}_1^* = 0$ $(+,-)$	$\dot{y}_1^* = 0$ $(+,-)$	$\dot{y}_1 > 0$ $(+,+)$
(b)	$(-,+)$	$\dot{y}_1^* = 0$ $(+,-)$	$\dot{y}_1 < 0$ $(-,-)$	$\dot{y}_1 < 0$ $(-,-)$	$\dot{y}_1^* = 0$ $(-,+)$
(c)	$(+,-)$	$\dot{y}_1^* = 0$ $(-,+)$	$\dot{y}_1 < 0$ $(+,+)$	$\dot{y}_1 < 0$ $(+,+)$	$\dot{y}_1^* = 0$ $(+,-)$
(d)	$(-,-)$	$\dot{y}_1 < 0$ $(+,+)$	$\dot{y}_1^* = 0$ $(-,+)$	$\dot{y}_1^* = 0$ $(-,+)$	$\dot{y}_1 > 0$ $(-,-)$

$$\begin{Bmatrix} (\dot{x}_1, a_{221}) \\ (a_{112}, \dot{x}_2) \\ (a_{111}, \dot{x}_2) \end{Bmatrix} = \begin{Bmatrix} \underbrace{(\mathrm{nF}, \mathrm{DP})}_{\text{down-parabola flow }(-)} \\ \underbrace{(\mathrm{II}, \mathrm{nF})}_{\text{increasing-inflection flow }(-)} \\ \underbrace{(\mathrm{UP}, \mathrm{nF})}_{\text{up-parabola flow }(-)} \end{Bmatrix}$$

$$\text{for } a_{110} < 0 \text{ and } a_{220} > 0, \tag{2.217}$$

$$\begin{Bmatrix} (\dot{x}_1, a_{221}) \\ (a_{112}, \dot{x}_2) \\ (a_{111}, \dot{x}_2) \end{Bmatrix} = \begin{Bmatrix} \underbrace{(\mathrm{pF}, \mathrm{DP})}_{\text{down-parabola flow }(+)} \\ \underbrace{(\mathrm{II}, \mathrm{pF})}_{\text{increasing-inflection flow }(+)} \\ \underbrace{(\mathrm{UP}, \mathrm{pF})}_{\text{up-parabola flow }(+)} \end{Bmatrix}$$

$$\text{for } a_{110} > 0 \text{ and } a_{220} < 0, \tag{2.218}$$

$$\begin{Bmatrix} (\dot{x}_1, a_{221}) \\ (a_{112}, \dot{x}_2) \\ (a_{111}, \dot{x}_2) \end{Bmatrix} = \begin{Bmatrix} \underbrace{(\mathrm{nF}, \mathrm{UP})}_{\text{up-parabola flow }(-)} \\ \underbrace{(\mathrm{DI}, \mathrm{pF})}_{\text{decreasing-inflection flow }(+)} \\ \underbrace{(\mathrm{DP}, \mathrm{pF})}_{\text{down-parabola flow }(+)} \end{Bmatrix}$$

$$\text{for } a_{110} < 0 \text{ and } a_{220} < 0. \tag{2.219}$$

(iv) For $a_{221} > a_{112}$ with $\alpha_{11} > 0$ and $\alpha_{12} < 0$, presented in Fig. 2.10 are the one-dimensional flows of $(y_1^*, \dot{x}_2) = (a_{111}, \dot{x}_2), (a_{112}, \dot{x}_2)$ and $(\dot{x}_1, y_1^*) = (\dot{x}_1, a_{221})$. The equilibrium flows of $\dot{y}_1 = 0$ are listed in Table 2.4. The corresponding properties for $\alpha_{11} > 0$ and $\alpha_{12} < 0$ are presented as follows:

$$\begin{Bmatrix} (\dot{x}_1, a_{221}) \\ (a_{112}, \dot{x}_2) \\ (a_{111}, \dot{x}_2) \end{Bmatrix} = \begin{Bmatrix} \underbrace{(\mathrm{pF}, \mathrm{UP})}_{\text{up-parabola flow }(+)} \\ \underbrace{(\mathrm{DI}, \mathrm{nF})}_{\text{decreasing-inflection flow }(-)} \\ \underbrace{(\mathrm{UP}, \mathrm{nF})}_{\text{up-parabola flow }(-)} \end{Bmatrix}$$

$$\text{for } a_{110} > 0 \text{ and } a_{220} > 0, \tag{2.220}$$

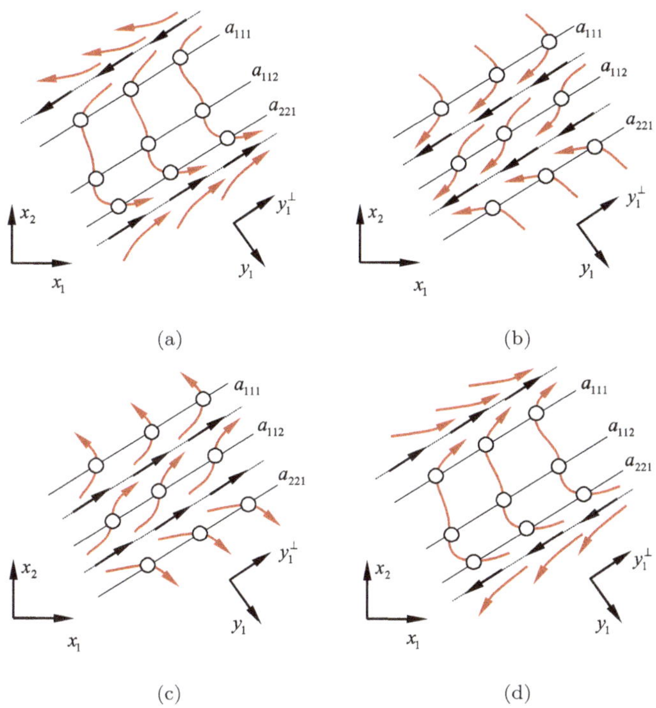

Fig. 2.10. Phase portraits ($a_{111} > a_{221}$ and $\alpha_{11} > 0, \alpha_{12} < 0$) for two-dimensional systems on the x_1-direction with $y_1^* = a_{111}, a_{112}$ and on the x_2-direction with $y_1^* = a_{221}$: (a) (UP,nF) and (DI,nF)-flows of $y_1^* = a_{111}, a_{112}$ and (pF,UP)-flow of $y_1^* = a_{221}$ ($a_{110} > 0, a_{220} > 0$), (b) (DP,nF) and (II,nF)-flows of $y_1^* = a_{111}, a_{112}$ and (nF,DP)-flow of $y_1^* = a_{221}$ ($a_{110} < 0, a_{220} > 0$), (c) (DP,pF) and (II,pF)-flows of $y_1^* = a_{111}, a_{112}$ and (pF,DP)-flow of $y_1^* = a_{221}$ ($a_{110} > 0, a_{220} < 0$), (d) (UP,pF) and (DI,pF)-flows of $y_1^* = a_{111}, a_{112}$ and (nF,UP)-flow of $y_1^* = a_{221}$ ($a_{110} < 0, a_{220} < 0$).

Table 2.4. Equilibrium flows of $\dot{y}_1 = 0$ ($\alpha_{11} > 0, \alpha_{12} < 0$).

Figure 2.10	$(\dot{x}_1, \dot{x}_2)$	y_1			
	(a_{110}, a_{220})	$(-\infty, a_{111})$	(a_{111}, a_{112})	(a_{112}, a_{221})	(a_{221}, ∞)
(a)	$(+,+)$	$\dot{y}_1^* = 0$ $(-,-)$	$\dot{y}_1 < 0$ $(+,-)$	$\dot{y}_1 > 0$ $(+,-)$	$\dot{y}_1^* = 0$ $(+,+)$
(b)	$(-,+)$	$\dot{y}_1 < 0$ $(+,-)$	$\dot{y}_1^* = 0$ $(-,-)$	$\dot{y}_1^* = 0$ $(-,-)$	$\dot{y}_1 < 0$ $(-,+)$
(c)	$(+,-)$	$\dot{y}_1 < 0$ $(-,+)$	$\dot{y}_1^* = 0$ $(+,+)$	$\dot{y}_1^* = 0$ $(+,+)$	$\dot{y}_1 > 0$ $(+,-)$
(d)	$(-,-)$	$\dot{y}_1^* = 0$ $(+,+)$	$\dot{y}_1 < 0$ $(-,+)$	$\dot{y}_1 > 0$ $(-,+)$	$\dot{y}_1^* = 0$ $(-,-)$

$$\left\{\begin{matrix}(\dot{x}_1, a_{221})\\(a_{112}, \dot{x}_2)\\(a_{111}, \dot{x}_2)\end{matrix}\right\} = \left\{\begin{matrix}\underbrace{(\mathrm{nF}, \mathrm{DP})}_{\text{down-parabola flow }(-)}\\\underbrace{(\mathrm{II}, \mathrm{nF})}_{\text{increasing-inflection flow }(-)}\\\underbrace{(\mathrm{DP}, \mathrm{nF})}_{\text{down-parabola flow }(-)}\end{matrix}\right\}$$

$$\text{for } a_{110} < 0 \text{ and } a_{220} > 0, \tag{2.221}$$

$$\left\{\begin{matrix}(\dot{x}_1, a_{221})\\(a_{112}, \dot{x}_2)\\(a_{111}, \dot{x}_2)\end{matrix}\right\} = \left\{\begin{matrix}\underbrace{(\mathrm{pF}, \mathrm{DP})}_{\text{down-parabola flow }(+)}\\\underbrace{(\mathrm{II}, \mathrm{pF})}_{\text{increasing-inflection flow }(+)}\\\underbrace{(\mathrm{DP}, \mathrm{pF})}_{\text{down-parabola flow }(+)}\end{matrix}\right\}$$

$$\text{for } a_{110} > 0 \text{ and } a_{220} < 0, \tag{2.222}$$

$$\left\{\begin{matrix}(\dot{x}_1, a_{221})\\(a_{112}, \dot{x}_2)\\(a_{111}, \dot{x}_2)\end{matrix}\right\} = \left\{\begin{matrix}\underbrace{(\mathrm{nF}, \mathrm{UP})}_{\text{up-parabola flow }(-)}\\\underbrace{(\mathrm{DI}, \mathrm{pF})}_{\text{decreasing-inflection flow }(+)}\\\underbrace{(\mathrm{UP}, \mathrm{pF})}_{\text{up-parabola flow }(+)}\end{matrix}\right\}$$

$$\text{for } a_{110} < 0 \text{ and } a_{220} < 0. \tag{2.223}$$

(v) For $a_{221} \in (a_{111}, a_{112})$ with $\alpha_{11} > 0$ and $\alpha_{12} > 0$, presented in Fig. 2.11 are the one-dimensional flows of $(y_1^*, \dot{x}_2) = (a_{111}, \dot{x}_2), (a_{112}, \dot{x}_2)$ and $(\dot{x}_1, y_1^*) = (\dot{x}_1, a_{221})$. The equilibrium flows of $\dot{y}_1 = 0$ are listed in Table 2.5. The corresponding properties for $\alpha_{11} > 0$ and $\alpha_{12} > 0$ are presented as follows:

$$\left\{\begin{matrix}(a_{112}, \dot{x}_2)\\(\dot{x}_1, a_{221})\\(a_{111}, \dot{x}_2)\end{matrix}\right\} = \left\{\begin{matrix}\underbrace{(\mathrm{II}, \mathrm{pF})}_{\text{increasing-inflection flow }(+)}\\\underbrace{(\mathrm{pF}, \mathrm{UP})}_{\text{up-parabola flow }(+)}\\\underbrace{(\mathrm{DP}, \mathrm{nF})}_{\text{down-parabola flow }(-)}\end{matrix}\right\}$$

$$\text{for } a_{110} > 0 \text{ and } a_{220} > 0, \tag{2.224}$$

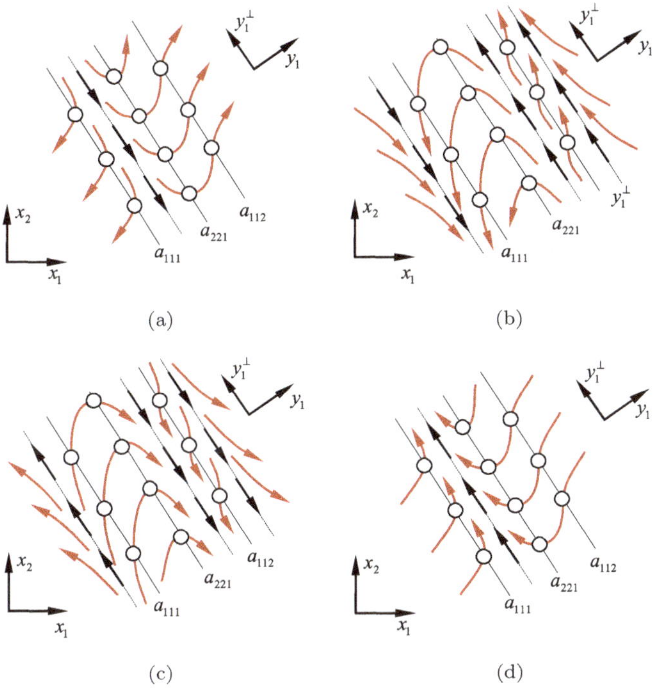

Fig. 2.11. Phase portraits ($a_{111} < a_{221} < a_{112}$ and $\alpha_{11} > 0, \alpha_{12} > 0$) for two-dimensional systems on the x_1-direction with $y_1^* = a_{111}, a_{112}$ and on the x_2-direction with $y_1^* = a_{221}$: (a) (DP, nF) and (II, pF)-flows of $y_1^* = a_{111}, a_{112}$ and (pF,UP)-flow of $y_1^* = a_{221}$ ($a_{110} > 0, a_{220} > 0$), (b) (UP,nF) and (DI, pF)-flows of $y_1^* = a_{111}, a_{112}$ and (nF, DP)-flow of $y_1^* = a_{221}$ ($a_{110} < 0,\ a_{220} < 0$), (c) (UP,pF) and (DI,nF)-flows of $y_1^* = a_{111}, a_{112}$ and (pF,DP)-flow of $y_1^* = a_{221}$ ($a_{110} > 0, a_{220} < 0$), (d) (DP, pF) and (II, nF)-flows of $y_1^* = a_{111}, a_{112}$ and (nF,UP)-flow of $y_1^* = a_{221}$ ($a_{110} < 0, a_{220} < 0$).

Table 2.5. Equilibrium flows of $\dot{y}_1 = 0$ ($\alpha_{11} > 0, \alpha_{12} > 0$).

	$(\dot{x}_1, \dot{x}_2)$	y_1			
Figure 2.11	(a_{110}, a_{220})	$(-\infty, a_{111})$	(a_{111}, a_{221})	(a_{221}, a_{112})	(a_{112}, ∞)
(a)	$(+,+)$	$\dot{y}_1 < 0$ $(-,-)$	$\dot{y}_1^* = 0$ $(+,-)$	$\dot{y}_1 > 0$ $(+,+)$	$\dot{y}_1 > 0$ $(+,+)$
(b)	$(-,+)$	$\dot{y}_1^* = 0$ $(+,-)$	$\dot{y}_1 < 0$ $(-,-)$	$\dot{y}_1^* = 0$ $(-,+)$	$\dot{y}_1^* = 0$ $(-,+)$
(c)	$(+,-)$	$\dot{y}_1^* = 0$ $(-,+)$	$\dot{y}_1 > 0$ $(+,+)$	$\dot{y}_1^* = 0$ $(+,-)$	$\dot{y}_1^* = 0$ $(+,-)$
(d)	$(-,-)$	$\dot{y}_1 > 0$ $(+,+)$	$\dot{y}_1^* = 0$ $(-,+)$	$\dot{y}_1 < 0$ $(-,-)$	$\dot{y}_1 < 0$ $(-,-)$

$$\left\{\begin{matrix}(a_{112},\dot{x}_2)\\(\dot{x}_1,a_{221})\\(a_{111},\dot{x}_2)\end{matrix}\right\}=\left\{\begin{matrix}\underbrace{(\mathrm{DI},\mathrm{pF})}_{\text{decreasing-inflection flow }(+)}\\\underbrace{(\mathrm{nF},\mathrm{DP})}_{\text{down-parabola flow }(-)}\\\underbrace{(\mathrm{UP},\mathrm{nF})}_{\text{up-parabola flow }(-)}\end{matrix}\right\}$$

$$\text{for } a_{110}<0 \text{ and } a_{220}>0, \tag{2.225}$$

$$\left\{\begin{matrix}(a_{112},\dot{x}_2)\\(\dot{x}_1,a_{221})\\(a_{111},\dot{x}_2)\end{matrix}\right\}=\left\{\begin{matrix}\underbrace{(\mathrm{DI},\mathrm{nF})}_{\text{decreasing-inflection flow }(-)}\\\underbrace{(\mathrm{pF},\mathrm{DP})}_{\text{down-parabola flow }(+)}\\\underbrace{(\mathrm{UP},\mathrm{pF})}_{\text{up-parabola flow }(+)}\end{matrix}\right\}$$

$$\text{for } a_{110}>0 \text{ and } a_{220}<0, \tag{2.226}$$

$$\left\{\begin{matrix}(a_{112},\dot{x}_2)\\(\dot{x}_1,a_{221})\\(a_{111},\dot{x}_2)\end{matrix}\right\}=\left\{\begin{matrix}\underbrace{(\mathrm{II},\mathrm{nF})}_{\text{increasing-inflection flow }(-)}\\\underbrace{(\mathrm{nF},\mathrm{UP})}_{\text{up-parabola flow }(-)}\\\underbrace{(\mathrm{DP},\mathrm{pF})}_{\text{down-parabola flow }(+)}\end{matrix}\right\}$$

$$\text{for } a_{110}<0 \text{ and } a_{220}<0. \tag{2.227}$$

(vi) For $a_{221}\in(a_{111},a_{112})$ with $\alpha_{11}>0$ and $\alpha_{12}<0$, presented in Fig. 2.12 are the one-dimensional flows of $(y_1^*,\dot{x}_2)=(a_{111},\dot{x}_2),(a_{112},\dot{x}_2)$ and $(\dot{x}_1,y_1^*)=(\dot{x}_1,a_{221})$. The equilibrium flows of $\dot{y}_1=0$ are listed in Table 2.6. The corresponding properties for $\alpha_{11}>0$ and $\alpha_{12}<0$ are presented as follows:

$$\left\{\begin{matrix}(a_{112},\dot{x}_2)\\(\dot{x}_1,a_{221})\\(a_{111},\dot{x}_2)\end{matrix}\right\}=\left\{\begin{matrix}\underbrace{(\mathrm{II},\mathrm{pF})}_{\text{increasing-inflection flow }(+)}\\\underbrace{(\mathrm{pF},\mathrm{UP})}_{\text{down-parabola flow }(+)}\\\underbrace{(\mathrm{UP},\mathrm{nF})}_{\text{up-parabola flow }(-)}\end{matrix}\right\}$$

$$\text{for } a_{110}>0 \text{ and } a_{220}>0, \tag{2.228}$$

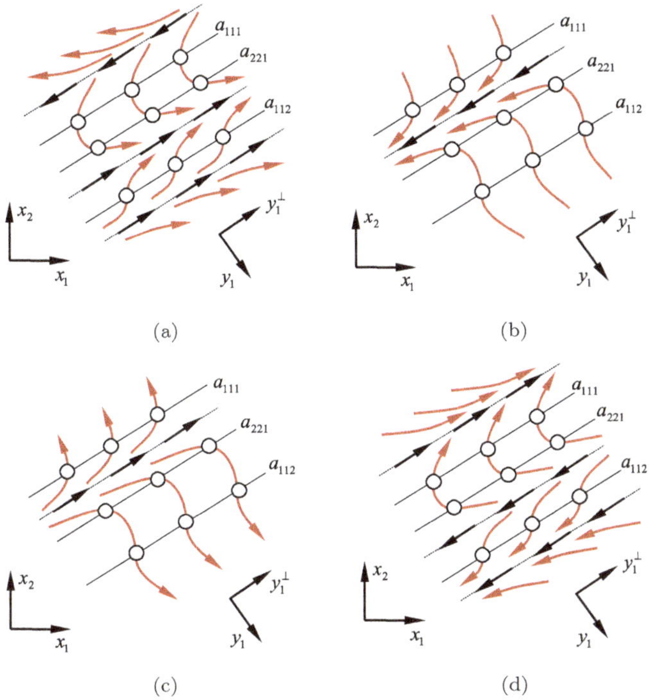

Fig. 2.12. Phase portraits ($a_{111} < a_{221} < a_{112}$ and $\alpha_{11} > 0, \alpha_{12} < 0$) for two-dimensional systems on the x_1-direction with $y_1^* = a_{111}, a_{112}$ and on the x_2-direction with $y_1^* = a_{221}$: (a) (UP,nF) and (II,pF)-flows of $y_1^* = a_{111}, a_{112}$ and (pF,UP)-flow of $y_1^* = a_{221}$ ($a_{110} > 0, a_{220} > 0$), (b) (DP,nF) and (DI,pF)-flows of $y_1^* = a_{111}, a_{112}$ and (pF,DP)-flow of $y_1^* = a_{221}$ ($a_{110} < 0, a_{220} > 0$), (c) (DP,pF) and (DI,nF)-flows of $y_1^* = a_{111}, a_{112}$ and (pF,DP)-flow of $y_1^* = a_{221}$ ($a_{110} > 0, a_{220} < 0$), (d) (UP,pF) and (II,nF)-flows of $y_1^* = a_{111}, a_{112}$ and (nF,UP)-flow of $y_1^* = a_{221}$ ($a_{110} < 0, a_{220} < 0$).

Table 2.6. Equilibrium flows of $\dot{y}_1 = 0$ ($\alpha_{11} > 0, \alpha_{12} < 0$).

Figure 2.12	$(\dot{x}_1, \dot{x}_2)$	y_1			
	(a_{110}, a_{220})	$(-\infty, a_{111})$	(a_{111}, a_{221})	(a_{221}, a_{112})	(a_{112}, ∞)
(a)	$(+,+)$	$\dot{y}_1^* = 0$ $(-,-)$	$\dot{y}_1 > 0$ $(+,-)$	$\dot{y}_1^* = 0$ $(+,+)$	$\dot{y}_1^* = 0$ $(+,+)$
(b)	$(-,+)$	$\dot{y}_1 < 0$ $(+,-)$	$\dot{y}_1^* = 0$ $(-,-)$	$\dot{y}_1 < 0$ $(-,+)$	$\dot{y}_1 < 0$ $(-,+)$
(c)	$(+,-)$	$\dot{y}_1 > 0$ $(-,+)$	$\dot{y}_1^* = 0$ $(+,+)$	$\dot{y}_1 > 0$ $(+,-)$	$\dot{y}_1 > 0$ $(+,-)$
(d)	$(-,-)$	$\dot{y}_1^* = 0$ $(+,+)$	$\dot{y}_1 < 0$ $(-,+)$	$\dot{y}_1^* = 0$ $(-,-)$	$\dot{y}_1^* = 0$ $(-,-)$

$$\left\{\begin{matrix}(a_{112},\dot{x}_2)\\(\dot{x}_1,a_{221})\\(a_{111},\dot{x}_2)\end{matrix}\right\}=\left\{\begin{matrix}\underbrace{(\mathrm{DI},\mathrm{pF})}_{\text{decreasing-inflection flow }(+)}\\\underbrace{(\mathrm{nF},\mathrm{DP})}_{\text{down-parabola flow }(-)}\\\underbrace{(\mathrm{DP},\mathrm{nF})}_{\text{down-parabola flow }(-)}\end{matrix}\right\}$$

$$\text{for } a_{110}<0 \text{ and } a_{220}>0, \tag{2.229}$$

$$\left\{\begin{matrix}(a_{112},\dot{x}_2)\\(\dot{x}_1,a_{221})\\(a_{111},\dot{x}_2)\end{matrix}\right\}=\left\{\begin{matrix}\underbrace{(\mathrm{DI},\mathrm{nF})}_{\text{decreasing-inflection flow }(-)}\\\underbrace{(\mathrm{pF},\mathrm{DP})}_{\text{down-parabola flow }(+)}\\\underbrace{(\mathrm{DP},\mathrm{pF})}_{\text{down-parabola flow }(+)}\end{matrix}\right\}$$

$$\text{for } a_{110}>0 \text{ and } a_{220}<0, \tag{2.230}$$

$$\left\{\begin{matrix}(a_{112},\dot{x}_2)\\(\dot{x}_1,a_{221})\\(a_{111},\dot{x}_2)\end{matrix}\right\}=\left\{\begin{matrix}\underbrace{(\mathrm{II},\mathrm{nF})}_{\text{increasing-inflection flow }(-)}\\\underbrace{(\mathrm{nF},\mathrm{UP})}_{\text{up-parabola flow }(-)}\\\underbrace{(\mathrm{UP},\mathrm{pF})}_{\text{up-parabola flow }(+)}\end{matrix}\right\}$$

$$\text{for } a_{110}<0 \text{ and } a_{220}<0. \tag{2.231}$$

2.4.2 *Switching bifurcations*

In this section, the switching bifurcations between two parabola flows and between parabola and inflection flows are discussed.

2.4.2.1 *Inflection-sink (source) switching bifurcations*

Consider a dynamical system as

$$\begin{aligned}\dot{x}_2&=a_{220}(y_1-a_{221}),\\\dot{x}_1&=a_{111}(y_1-a_{111})(y_1-a_{112})^2.\end{aligned} \tag{2.232}$$

For $a_{221} = a_{111}$, there is a dynamical system in the y_{j_1}-direction as

$$\dot{y}_1 = (y_1 - a_{111})[\alpha_{11}a_{110}(y_1 - a_{112})^2 + \alpha_{12}a_{220}] \tag{2.233}$$

with a dynamical system in the $y_1^{\perp}$-direction as

$$\dot{y}_1^{\perp} = (y_1 - a_{111})[\beta_{11}a_{110}(y_1 - a_{112})^2 + \beta_{12}a_{220}] \tag{2.234}$$

and

$$G_{110} = \alpha_{11}a_{110}(a_{111} - a_{112})^2 + \alpha_{12}a_{220}. \tag{2.235}$$

(i) For $\alpha_{11} > 0$ and $\alpha_{12} > 0$, the infinite-equilibrium of $y_1^* = a_{111} = a_{221}$ is presented with $\dot{x}_2^* = a_{220}(a_{112} - a_{221})$. The corresponding properties are given as follows:

- For $a_{110} > 0$ and $a_{220} > 0$,

$$\begin{Bmatrix} (a_{111}, a_{221}) \\ (a_{112}, \dot{x}_2^*) \end{Bmatrix} = \begin{Bmatrix} \underbrace{(\mathrm{II}, \mathrm{SO})}_{\text{increasing-inflection source}} \\ \underbrace{(\mathrm{II}, \mathrm{pF})}_{\text{increasing-inflection flow }(+)} \end{Bmatrix} \text{ for } G_{110} > 0. \tag{2.236}$$

- For $a_{110} < 0$ and $a_{220} > 0$,

$$\begin{aligned} \begin{Bmatrix} (a_{111}, a_{221}) \\ (a_{112}, \dot{x}_2^*) \end{Bmatrix} &= \begin{Bmatrix} \underbrace{(\mathrm{DI}, \mathrm{SI})}_{\text{decreasing-inflection sink}} \\ \underbrace{(\mathrm{DI}, \mathrm{pF})}_{\text{decreasing-inflection flow }(+)} \end{Bmatrix} \text{ for } G_{110} < 0, \\ \begin{Bmatrix} (a_{111}, a_{221}) \\ (a_{112}, \dot{x}_2^*) \end{Bmatrix} &= \begin{Bmatrix} \underbrace{(\mathrm{DI}, \mathrm{SO})}_{\text{decreasing-inflection source}} \\ \underbrace{(\mathrm{DI}, \mathrm{pF})}_{\text{decreasing-inflection flow }(+)} \end{Bmatrix} \text{ for } G_{110} > 0. \end{aligned} \tag{2.237}$$

- For $a_{110} > 0$ and $a_{220} < 0$,

$$\begin{aligned} \begin{Bmatrix} (a_{111}, a_{221}) \\ (a_{112}, \dot{x}_2^*) \end{Bmatrix} &= \begin{Bmatrix} \underbrace{(\mathrm{DI}, \mathrm{SO})}_{\text{decreasing-inflection source}} \\ \underbrace{(\mathrm{DI}, \mathrm{nF})}_{\text{decreasing-inflection flow }(-)} \end{Bmatrix} \text{ for } G_{110} > 0, \\ \begin{Bmatrix} (a_{111}, a_{221}) \\ (a_{112}, \dot{x}_2^*) \end{Bmatrix} &= \begin{Bmatrix} \underbrace{(\mathrm{DI}, \mathrm{SI})}_{\text{decreasing-inflection sink}} \\ \underbrace{(\mathrm{DI}, \mathrm{nF})}_{\text{decreasing-inflection flow }(-)} \end{Bmatrix} \text{ for } G_{110} < 0. \end{aligned} \tag{2.238}$$

- For $a_{110} < 0$ and $a_{220} < 0$,

$$\begin{Bmatrix} (a_{111}, a_{221}) \\ (a_{112}, \dot{x}_2^*) \end{Bmatrix} = \begin{Bmatrix} \underbrace{(\mathrm{II}, \mathrm{SI})}_{\text{increasing-inflection sink}} \\ \underbrace{(\mathrm{II}, \mathrm{nF})}_{\text{increasing-inflection flow }(-)} \end{Bmatrix} \quad \text{for } G_{110} < 0. \tag{2.239}$$

The inflection infinite-equilibriums and one-dimensional inflection flows are presented in Fig. 2.13. The corresponding equilibrium flows of $\dot{y}_1 = 0$ in phase space are listed in Table 2.7. For $a_{110} \times a_{220} < 0$, there are three equilibrium flows of $\dot{y}_1 = 0$. For $a_{110} \times a_{220} > 0$, there is no equilibrium flow of $\dot{y}_1 = 0$.

(ii) For $\alpha_{11} > 0$ and $\alpha_{12} < 0$, the infinite-equilibrium of $y_1^* = a_{111} = a_{221}$ is presented with $\dot{x}_2^* = a_{220}(a_{112} - a_{221})$. The corresponding properties are given as follows:

- For $a_{110} > 0$ and $a_{220} > 0$,

$$\begin{aligned} &\begin{Bmatrix} (a_{111}, a_{221}) \\ (a_{112}, \dot{x}_2^*) \end{Bmatrix} = \begin{Bmatrix} \underbrace{(\mathrm{II}, \mathrm{SO})}_{\text{decreasing-inflection source}} \\ \underbrace{(\mathrm{II}, \mathrm{pF})}_{\text{increasing-inflection flow }(+)} \end{Bmatrix} \quad \text{for } G_{110} > 0, \\ &\begin{Bmatrix} (a_{111}, a_{221}) \\ (a_{112}, \dot{x}_2^*) \end{Bmatrix} = \begin{Bmatrix} \underbrace{(\mathrm{II}, \mathrm{SI})}_{\text{decreasing-inflection sink}} \\ \underbrace{(\mathrm{II}, \mathrm{pF})}_{\text{increasing-inflection flow }(+)} \end{Bmatrix} \quad \text{for } G_{110} < 0. \end{aligned} \tag{2.240}$$

- For $a_{110} < 0$ and $a_{220} > 0$,

$$\begin{Bmatrix} (a_{111}, a_{221}) \\ (a_{112}, \dot{x}_2^*) \end{Bmatrix} = \begin{Bmatrix} \underbrace{(\mathrm{DI}, \mathrm{SI})}_{\text{decreasing-inflection sink}} \\ \underbrace{(\mathrm{DI}, \mathrm{pF})}_{\text{decreasing-inflection flow }(+)} \end{Bmatrix} \quad \text{for } G_{110} < 0. \tag{2.241}$$

- For $a_{110} > 0$ and $a_{220} < 0$,

$$\begin{Bmatrix} (a_{111}, a_{221}) \\ (a_{112}, \dot{x}_2^*) \end{Bmatrix} = \begin{Bmatrix} \underbrace{(\mathrm{DI}, \mathrm{SO})}_{\text{decreasing-inflection source}} \\ \underbrace{(\mathrm{DI}, \mathrm{nF})}_{\text{decreasing-inflection flow }(-)} \end{Bmatrix} \quad \text{for } G_{110} > 0. \tag{2.242}$$

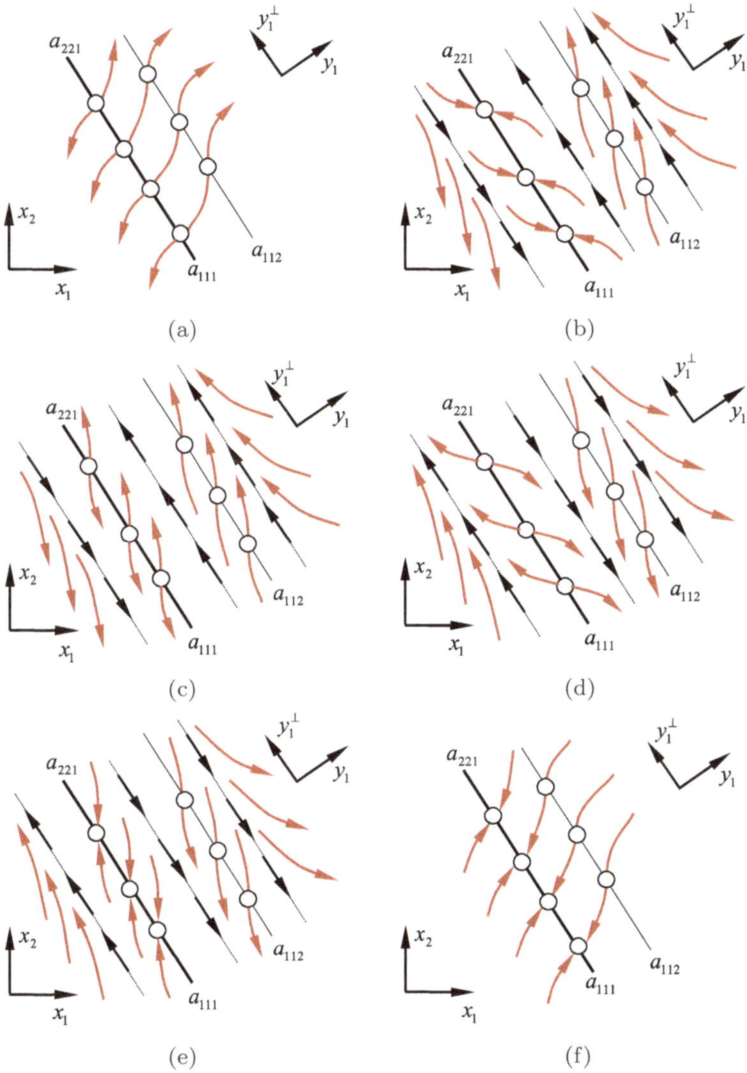

Fig. 2.13. Phase portraits ($a_{111} = a_{221}$ and $\alpha_{11} > 0$, $\alpha_{12} > 0$) for two-dimensional systems on the x_1-direction with $y_1^* = a_{111}, a_{112}$ and on the x_2-direction with $y_1^* = a_{221}$: (a) ($a_{110} > 0, a_{220} > 0$), (b, c) ($a_{110} < 0, a_{220} > 0$), (d, e) ($a_{110} > 0, a_{220} < 0$), (f) ($a_{110} < 0, a_{220} < 0$).

Table 2.7. Equilibrium flows of $\dot{y}_1 = 0$ for $a_{111} = a_{221}$ ($\alpha_{11} > 0, \alpha_{12} > 0$).

	$(\dot{x}_1, \dot{x}_2)$	y_1		
Figure 2.13	(a_{110}, a_{220})	$(-\infty, a_{111})$	(a_{111}, a_{112})	(a_{112}, ∞)
(a)	$(+,+)$	$\dot{y}_1 < 0$ $(-,-)$	$\dot{y}_1 > 0$ $(+,+)$	$\dot{y}_1 > 0$ $(+,+)$
(b,c)	$(-,+)$	$\dot{y}_1^* = 0$ $(+,-)$	$\dot{y}_1^* = 0$ $(-,+)$	$\dot{y}_1^* = 0$ $(-,+)$
(d,e)	$(+,-)$	$\dot{y}_1^* = 0$ $(-,+)$	$\dot{y}_1^* = 0$ $(+,-)$	$\dot{y}_1^* = 0$ $(+,-)$
(f)	$(-,-)$	$\dot{y}_1 > 0$ $(+,+)$	$\dot{y}_1 < 0$ $(-,-)$	$\dot{y}_1 < 0$ $(-,-)$

- For $a_{110} < 0$ and $a_{220} < 0$,

$$
\begin{aligned}
&\left\{\begin{matrix}(a_{111}, a_{221})\\ (a_{112}, \dot{x}_2^*)\end{matrix}\right\} = \left\{\begin{matrix}\underbrace{(\mathrm{II}, \mathrm{SO})}_{\text{decreasing-inflection sink}}\\ \underbrace{(\mathrm{II}, \mathrm{nF})}_{\text{increasing-inflection flow }(-)}\end{matrix}\right\} \quad \text{for } G_{110} > 0,\\
&\left\{\begin{matrix}(a_{111}, a_{221})\\ (a_{112}, \dot{x}_2^*)\end{matrix}\right\} = \left\{\begin{matrix}\underbrace{(\mathrm{II}, \mathrm{SI})}_{\text{decreasing-inflection source}}\\ \underbrace{(\mathrm{II}, \mathrm{nF})}_{\text{increasing-inflection flow }(-)}\end{matrix}\right\} \quad \text{for } G_{110} < 0.
\end{aligned} \tag{2.243}
$$

The inflection infinite-equilibriums and one-dimensional inflection flows are presented in Fig. 2.14. The corresponding equilibrium flows of $\dot{y}_1 = 0$ in phase space are listed in Table 2.8. For $a_{110} \times a_{220} > 0$, there are three equilibrium flows of $\dot{y}_1 = 0$. For $a_{110} \times a_{220} < 0$, there is not an equilibrium flow of $\dot{y}_1 = 0$.

2.4.2.2 *Parabola-sink (source) switching bifurcations*

Consider a dynamical system as

$$
\begin{aligned}
\dot{x}_2 &= a_{220}(y_1 - a_{221}),\\
\dot{x}_1 &= a_{111}(y_1 - a_{111})(y_1 - a_{112})^2.
\end{aligned} \tag{2.244}
$$

For $a_{221} = a_{112}$, there is a dynamical system in the y_{j_1}-direction as

$$
\dot{y}_1 = (y_1 - a_{112})[\alpha_{11}a_{110}(y_1 - a_{111})(y_1 - a_{112}) + \alpha_{12}a_{220}] \tag{2.245}
$$

with a dynamical system in the $y_1^{\perp}$-direction as

$$
\dot{y}_1^{\perp} = (y_1 - a_{112})[\beta_{11}a_{110}(y_1 - a_{111})(y_1 - a_{112}) + \beta_{12}a_{220}] \tag{2.246}
$$

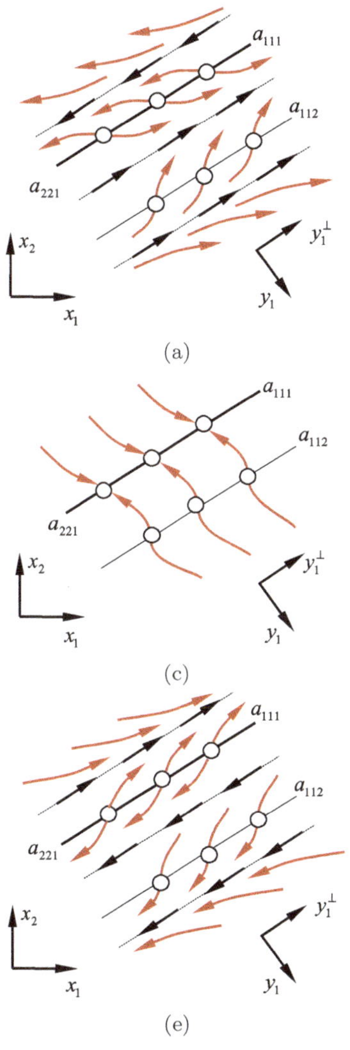

Fig. 2.14. Phase portraits ($a_{111} = a_{221}$ and $\alpha_{11} > 0, \alpha_{12} < 0$) for two-dimensional systems on the x_1-direction with $y_1^* = a_{111}, a_{112}$ and on the x_2-direction with $y_1^* = a_{221}$: (a, b) ($a_{110} > 0, a_{220} > 0$), (c) ($a_{110} < 0, a_{220} > 0$), (d) ($a_{110} > 0, a_{220} < 0$), (e, f) ($a_{110} < 0, a_{220} < 0$).

Table 2.8. Equilibrium flows of $\dot{y}_1 = 0$ for $a_{111} = a_{221}(\alpha_{11} > 0, \alpha_{12} < 0)$.

Figure 2.14	$(\dot{x}_1, \dot{x}_2)$ (a_{110}, a_{220})	y_1 $(-\infty, a_{111})$	(a_{111}, a_{112})	(a_{112}, ∞)
(a,b)	$(+,+)$	$\dot{y}_1^* = 0$ $(-,-)$	$\dot{y}_1^* = 0$ $(+,+)$	$\dot{y}_1^* = 0$ $(+,+)$
(c)	$(-,+)$	$\dot{y}_1 < 0$ $(+,-)$	$\dot{y}_1 < 0$ $(-,+)$	$\dot{y}_1 < 0$ $(-,+)$
(d)	$(+,-)$	$\dot{y}_1 > 0$ $(-,+)$	$\dot{y}_1 > 0$ $(+,-)$	$\dot{y}_1 > 0$ $(+,-)$
(e,f)	$(-,-)$	$\dot{y}_1^* = 0$ $(+,+)$	$\dot{y}_1^* = 0$ $(-,-)$	$\dot{y}_1^* = 0$ $(-,-)$

and

$$G_{110} = \alpha_{12}a_{220}. \tag{2.247}$$

(i) For $\alpha_{11} > 0$ and $\alpha_{12} > 0$, the infinite-equilibrium of $y_1^* = a_{112} = a_{221}$ is presented with $\dot{x}_2^* = a_{220}(a_{111} - a_{221})$. The corresponding properties are given as follows:

- For $a_{110} > 0$ and $a_{220} > 0$,

$$\begin{Bmatrix} (a_{111}, \dot{x}_2^*) \\ (a_{112}, a_{221}) \end{Bmatrix} = \begin{Bmatrix} \underbrace{(\mathrm{DP}, \mathrm{nF})}_{\text{down-parabola flow } (-)} \\ \underbrace{(\mathrm{UP}, \mathrm{SO})}_{\text{up-parabola source}} \end{Bmatrix} \text{ for } G_{110} = a_{220}\alpha_{12} > 0. \tag{2.248}$$

- For $a_{110} < 0$ and $a_{220} > 0$,

$$\begin{Bmatrix} (a_{111}, \dot{x}_2^*) \\ (a_{112}, a_{221}) \end{Bmatrix} = \begin{Bmatrix} \underbrace{(\mathrm{UP}, \mathrm{nF})}_{\text{up-parabola flow } (-)} \\ \underbrace{(\mathrm{DP}, \mathrm{SO})}_{\text{down-parabola source}} \end{Bmatrix} \text{ for } G_{110} = a_{220}\alpha_{12} > 0. \tag{2.249}$$

- For $a_{110} > 0$ and $a_{220} < 0$,

$$\begin{Bmatrix} (a_{111}, \dot{x}_2^*) \\ (a_{112}, a_{221}) \end{Bmatrix} = \begin{Bmatrix} \underbrace{(\mathrm{UP}, \mathrm{pF})}_{\text{up-parabola flow } (+)} \\ \underbrace{(\mathrm{DP}, \mathrm{SI})}_{\text{down-parabola sink}} \end{Bmatrix} \text{ for } G_{110} = a_{220}\alpha_{12} < 0. \tag{2.250}$$

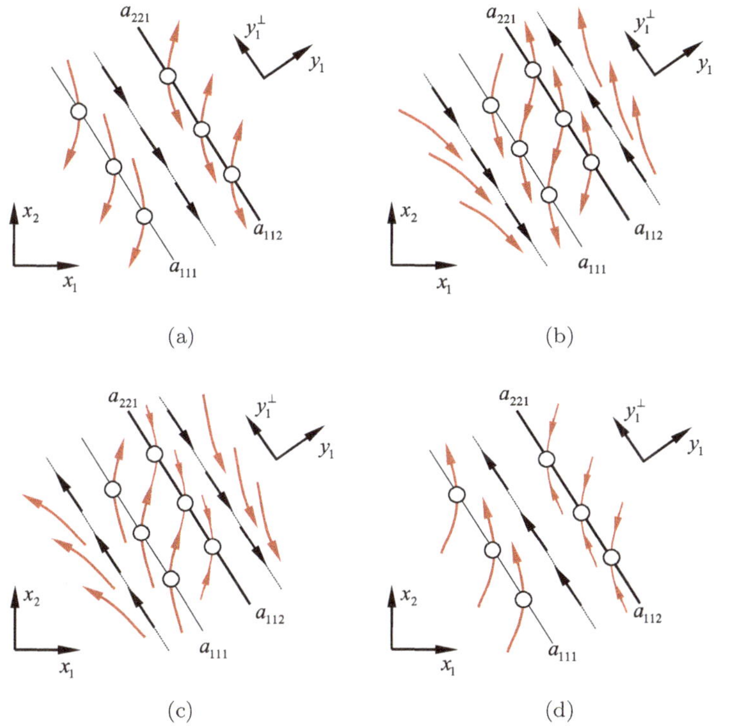

Fig. 2.15. Phase portraits ($a_{111} = a_{221}$ and $\alpha_{11} > 0, \alpha_{12} > 0$) for two-dimensional systems on the x_1-direction with $y_1^* = a_{111}, a_{112}$ and on the x_2-direction with $y_1^* = a_{221}$: (a) ($a_{110} > 0, a_{220} > 0$), (b) ($a_{110} < 0, a_{220} > 0$), (c) ($a_{110} > 0, a_{220} < 0$), (d) ($a_{110} < 0, a_{220} < 0$).

- For $a_{110} < 0$ and $a_{220} < 0$,

$$\left\{ \begin{array}{c} (a_{111}, \dot{x}_2^*) \\ (a_{112}, a_{221}) \end{array} \right\} = \left\{ \begin{array}{c} \underbrace{(\text{DP}, \text{pF})}_{\text{down flow } (+)} \\ \underbrace{(\text{UP}, \text{SI})}_{\text{up-parabola sink}} \end{array} \right\} \quad \text{for } G_{110} = a_{220}\alpha_{12} < 0. \quad (2.251)$$

The parabola infinite-equilibriums and one-dimensional parabola flows are presented in Fig. 2.15. The corresponding equilibrium flows of $\dot{y}_1 = 0$ in phase space are listed in Table 2.9. For $a_{110} \times a_{220} > 0$, there is one equilibrium flow of $\dot{y}_1 = 0$. For $a_{110} \times a_{220} < 0$, there are two equilibrium flows of $\dot{y}_1 = 0$.

Table 2.9. Equilibrium flows of $\dot{y}_1 = 0$ for $a_{111} = a_{221}(\alpha_{11} > 0, \alpha_{12} > 0)$.

Figure 2.15	$(\dot{x}_1, \dot{x}_2)$ (a_{110}, a_{220})	y_1 $(-\infty, a_{111})$	(a_{111}, a_{112})	(a_{112}, ∞)
(a)	$(+,+)$	$\dot{y}_1 < 0$ $(-,-)$	$\dot{y}_1^* = 0$ $(+,-)$	$\dot{y}_1 > 0$ $(+,+)$
(b)	$(-,+)$	$\dot{y}_1^* = 0$ $(+,-)$	$\dot{y}_1 < 0$ $(-,-)$	$\dot{y}_1^* = 0$ $(-,+)$
(c)	$(+,-)$	$\dot{y}_1^* = 0$ $(-,+)$	$\dot{y}_1 > 0$ $(+,+)$	$\dot{y}_1^* = 0$ $(+,-)$
(d)	$(-,-)$	$\dot{y}_1 > 0$ $(+,+)$	$\dot{y}_1^* = 0$ $(-,+)$	$\dot{y}_1 < 0$ $(-,-)$

(ii) For $\alpha_{11} > 0$ and $\alpha_{12} < 0$, the infinite-equilibrium of $y_1^* = a_{112} = a_{221}$ is presented with $\dot{x}_2^* = a_{220}(a_{111} - a_{221})$. The corresponding properties are given as follows:

- For $a_{110} > 0$ and $a_{220} > 0$,

$$\left\{\begin{matrix} (a_{111}, \dot{x}_2^*) \\ (a_{112}, a_{221}) \end{matrix}\right\} = \left\{\begin{matrix} \underbrace{(\text{UP}, \text{nF})}_{\text{up-parabola flow }(-)} \\ \underbrace{(\text{DP}, \text{SI})}_{\text{down-parabola sink}} \end{matrix}\right\} \text{ for } G_{110} = a_{220}\alpha_{12} < 0. \tag{2.252}$$

- For $a_{110} < 0$ and $a_{220} > 0$,

$$\left\{\begin{matrix} (a_{111}, \dot{x}_2^*) \\ (a_{112}, a_{221}) \end{matrix}\right\} = \left\{\begin{matrix} \underbrace{(\text{DP}, \text{nF})}_{\text{down-parabola flow }(-)} \\ \underbrace{(\text{UP}, \text{SI})}_{\text{up-parabola sink}} \end{matrix}\right\} \text{ for } G_{110} = a_{220}\alpha_{12} < 0. \tag{2.253}$$

- For $a_{110} > 0$ and $a_{220} < 0$,

$$\left\{\begin{matrix} (a_{111}, \dot{x}_2^*) \\ (a_{112}, a_{221}) \end{matrix}\right\} = \left\{\begin{matrix} \underbrace{(\text{DP}, \text{pF})}_{\text{down-parabola flow }(+)} \\ \underbrace{(\text{UP}, \text{SO})}_{\text{up-parabola source}} \end{matrix}\right\} \text{ for } G_{110} = a_{220}\alpha_{12} > 0. \tag{2.254}$$

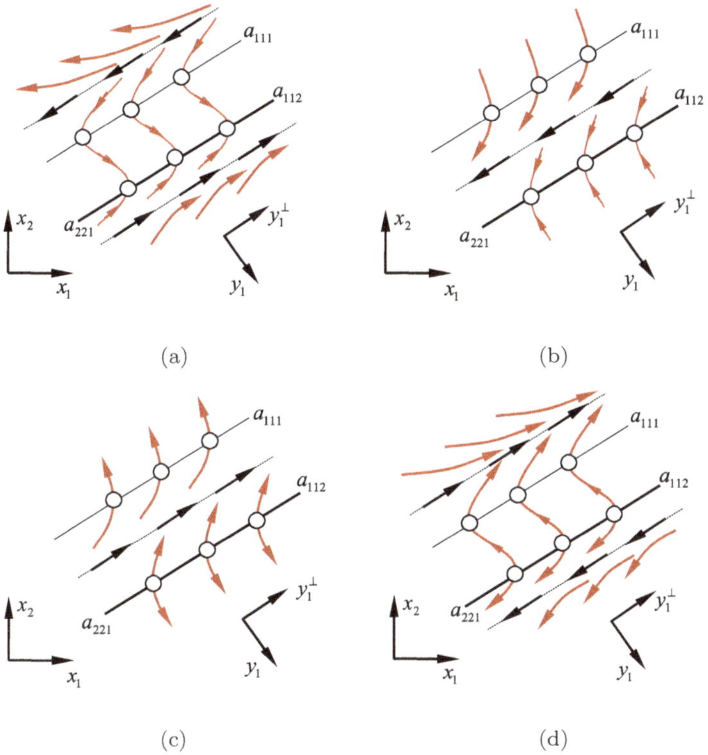

Fig. 2.16. Phase portraits ($a_{112} = a_{221}$ and $\alpha_{11} > 0, \alpha_{12} < 0$) for two-dimensional systems on the x_1-direction with $y_1^* = a_{111}, a_{112}$ and on the x_2-direction with $y_1^* = a_{221}$: (a) ($a_{110} > 0, a_{220} > 0$), (b) ($a_{110} < 0, a_{220} > 0$), (c) ($a_{110} > 0, a_{220} < 0$), (d) ($a_{110} < 0, a_{220} < 0$).

- For $a_{110} < 0$ and $a_{220} < 0$,

$$\begin{Bmatrix} (a_{111}, \dot{x}_2^*) \\ (a_{112}, a_{221}) \end{Bmatrix} = \begin{Bmatrix} \underbrace{(\text{UP}, \text{pF})}_{\text{up-parabola flow }(+)} \\ \underbrace{(\text{DP}, \text{SO})}_{\text{down-parabola source}} \end{Bmatrix} \quad \text{for } G_{110} = a_{220}\alpha_{12} > 0. \tag{2.255}$$

The parabola infinite-equilibriums and one-dimensional parabola flows are presented in Fig. 2.16. The corresponding equilibrium flows of $\dot{y}_1 = 0$ in phase space are listed in Table 2.10. For $a_{110} \times a_{220} < 0$, there is one equilibrium flow of $\dot{y}_1 = 0$. For $a_{110} \times a_{220} > 0$, there are two equilibrium flows of $\dot{y}_1 = 0$.

Table 2.10. Equilibrium flows of $\dot{y}_1 = 0$ for $a_{112} = a_{221}(\alpha_{11} > 0, \alpha_{12} < 0)$.

Figure 2.16	$(\dot{x}_1, \dot{x}_2)$	y_1		
	(a_{110}, a_{220})	$(-\infty, a_{111})$	(a_{111}, a_{112})	(a_{112}, ∞)
(a)	$(+,+)$	$\dot{y}_1^* = 0$ $(-,-)$	$\dot{y}_1 > 0$ $(+,-)$	$\dot{y}_1^* = 0$ $(+,+)$
(b)	$(-,+)$	$\dot{y}_1 > 0$ $(+,-)$	$\dot{y}_1^* = 0$ $(-,-)$	$\dot{y}_1 < 0$ $(-,+)$
(c)	$(+,-)$	$\dot{y}_1 < 0$ $(-,+)$	$\dot{y}_1^* = 0$ $(+,+)$	$\dot{y}_1 > 0$ $(+,-)$
(d)	$(-,-)$	$\dot{y}_1^* = 0$ $(+,+)$	$\dot{y}_1 < 0$ $(-,+)$	$\dot{y}_1^* = 0$ $(-,-)$

2.5 Parabola Flows Series and Switching Bifurcations

In this section, parabola flows' series and the corresponding switching bifurcation are discussed.

2.5.1 *Series of parabola flows*

Consider a dynamical system as

$$\begin{aligned} \dot{x}_2 &= a_{220}(y_1 - a_{221}), \\ \dot{x}_1 &= a_{111}(y_1 - a_{111})(y_1 - a_{112})(y_1 - a_{113}). \end{aligned} \tag{2.256}$$

There is a dynamical system in the y_{j_1}-direction as

$$\dot{y}_1 = \alpha_{11} a_{110} \prod_{s_1}^{3} (y_1 - a_{11s_1}) + \alpha_{12} a_{220}(y_1 - a_{221}) \tag{2.257}$$

with a dynamical system in the $y_1^\perp$-direction as

$$\dot{y}_1^\perp = \beta_{11} a_{110} \prod_{s_1}^{3} (y_1 - a_{11s_1}) + \beta_{12} a_{220}(y_1 - a_{221}). \tag{2.258}$$

(i) The vertical and horizontal parabola flows' series ($a_{221} < a_{111}$ with $\alpha_{11} > 0$ and $\alpha_{12} > 0$) are presented in Fig. 2.17 with $\dot{x}_2^* = a_{220}(a_{11s_1} - a_{221})$ ($s_1 \in \{1, 2, 3\}$) and $\dot{x}_1^* = a_{110} \prod_{s_1=1}^{3} (a_{221} - a_{11s_1})$. The corresponding properties are summarized in Table 2.11 and the corresponding equilibrium flows of $\dot{y}_1 = 0$ in phase space are listed in Table 2.12.

(ii) The vertical and horizontal parabola flows series ($a_{221} < a_{111}$ with $\alpha_{11} > 0$ and $\alpha_{12} < 0$) are presented in Fig. 2.18 with $\dot{x}_2^* = a_{220}(a_{11s_1} - a_{221})$ ($s_1 \in \{1, 2, 3\}$) and $\dot{x}_1^* = a_{110} \prod_{s_1=1}^{3} (a_{221} - a_{11s_1})$. The corresponding

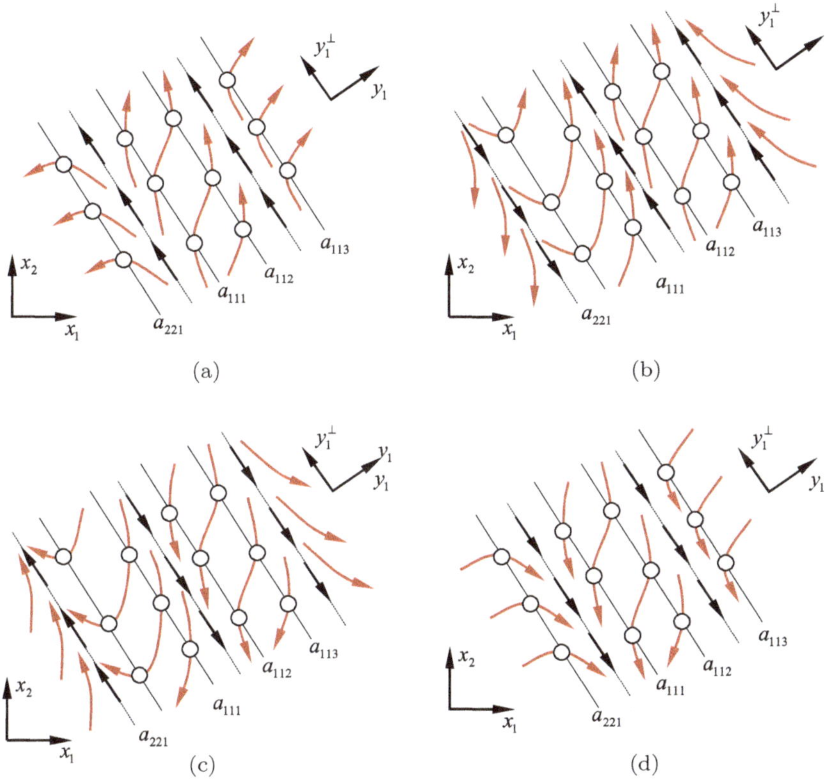

Fig. 2.17. Phase portraits ($a_{111} > a_{221}$ and $\alpha_{11} > 0, \alpha_{12} > 0$) for two-dimensional systems on the x_1-direction with $y_1^* = a_{111}, a_{112}, a_{113}$ and on the x_2-direction with $y_1^* = a_{221}$: (a) ($a_{110} > 0, a_{220} > 0$), (b) ($a_{110} < 0, a_{220} > 0$), (c) ($a_{110} > 0, a_{220} < 0$), (d) ($a_{110} < 0, a_{220} < 0$).

Table 2.11. Parabola flows for $a_{221} < a_{111}(\alpha_{11} > 0, \alpha_{12} > 0)$.

Figure 2.17	(a_{110}, a_{220})	$(\dot{x}_1^*, a_{221})$	$(a_{111}, \dot{x}_2^*)$	$(a_{112}, \dot{x}_2^*)$	$(a_{113}, \dot{x}_2^*)$
(a)	$(+,+)$	(nF,DP)	(UP,pF)	(DP,pF)	(UP,pF)
(b)	$(-,+)$	(pF,UP)	(DP,pF)	(UP,pF)	(DP,pF)
(c)	$(+,-)$	(nF,UP)	(DP,nF)	(UP,nF)	(DP,nF)
(d)	$(-,-)$	(pF,DP)	(UP,nF)	(DP,nF)	(UP,nF)

properties are summarized in Table 2.13 and the corresponding equilibrium flows of $\dot{y}_1 = 0$ in phase space are listed in Table 2.14.

(iii) The vertical and horizontal parabola flows series ($a_{221} > a_{113}$ with $\alpha_{11} > 0$ and $\alpha_{12} > 0$) are presented in Fig. 2.19 with $\dot{x}_2^* = a_{220}(a_{11s_1} - a_{221})$ ($s_1 \in \{1, 2, 3\}$) and $\dot{x}_1^* = a_{110} \prod_{s_1=1}^{3} (a_{221} - a_{11s_1})$. The corresponding

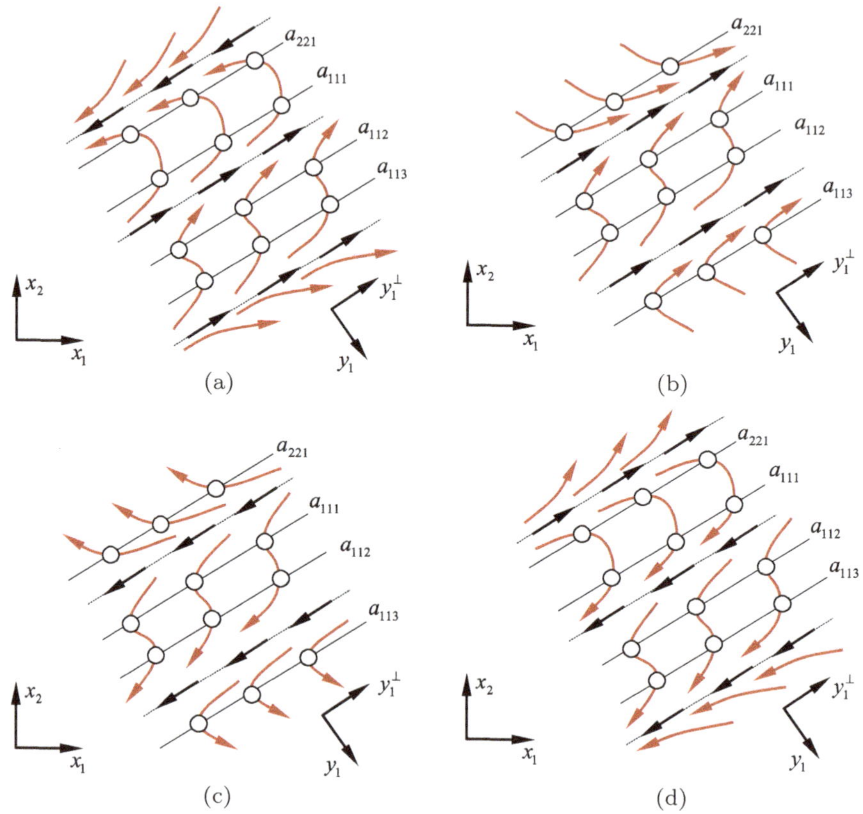

Fig. 2.18. Phase portraits ($a_{111} > a_{221}$ and $\alpha_{11} > 0, \alpha_{12} < 0$) for two-dimensional systems on the x_1-direction with $y_1^* = a_{111}, a_{112}, a_{113}$ and on the x_2-direction with $y_1^* = a_{221}$: (a) ($a_{110} > 0, a_{220} > 0$), (b) ($a_{110} < 0, a_{220} > 0$), (c) ($a_{110} > 0, a_{220} < 0$), (d) ($a_{110} < 0, a_{220} < 0$).

Table 2.12. Equilibrium flows of $\dot{y}_1 = 0$ ($\alpha_{11} > 0, \alpha_{12} > 0$).

	$(\dot{x}_1, \dot{x}_2)$	y_1				
Figure 2.17	(a_{110}, a_{220})	$(-\infty, a_{221})$	(a_{221}, a_{111})	(a_{111}, a_{112})	(a_{112}, a_{113})	(a_{113}, ∞)
(a)	$(+,+)$	$\dot{y}_1 < 0$ $(-,-)$	$\dot{y}_1^* = 0$ $(-,+)$	$\dot{y}_1 > 0$ $(+,+)$	$\dot{y}_1^* = 0$ $(-,+)$	$\dot{y}_1 > 0$ $(+,+)$
(b)	$(-,+)$	$\dot{y}_1^* = 0$ $(+,-)$	$\dot{y}_1 > 0$ $(+,+)$	$\dot{y}_1^* = 0$ $(-,+)$	$\dot{y}_1 > 0$ $(+,+)$	$\dot{y}_1^* = 0$ $(-,+)$
(c)	$(+,-)$	$\dot{y}_1^* = 0$ $(-,+)$	$\dot{y}_1 < 0$ $(-,-)$	$\dot{y}_1^* = 0$ $(+,-)$	$\dot{y}_1 < 0$ $(-,-)$	$\dot{y}_1^* = 0$ $(+,-)$
(d)	$(-,-)$	$\dot{y}_1 > 0$ $(+,+)$	$\dot{y}_1^* = 0$ $(+,-)$	$\dot{y}_1 < 0$ $(-,-)$	$\dot{y}_1^* = 0$ $(+,-)$	$\dot{y}_1 < 0$ $(-,-)$

Table 2.13. Parabola flows for $a_{221} < a_{111}$ ($\alpha_{11} > 0, \alpha_{12} < 0$).

Figure 2.18	(a_{110}, a_{220})	$(\dot{x}_1^*, a_{221})$	$(a_{111}, \dot{x}_2^*)$	$(a_{112}, \dot{x}_2^*)$	$(a_{113}, \dot{x}_2^*)$
(a)	$(+,+)$	(nF,DP)	(DP,pF)	(UP,pF)	(DP,pF)
(b)	$(-,+)$	(pF,UP)	(UP,pF)	(DP,pF)	(UP,pF)
(c)	$(+,-)$	(nF,UP)	(UP,nF)	(DP,nF)	(UP,nF)
(d)	$(-,-)$	(pF,DP)	(DP,nF)	(UP,nF)	(DP,nF)

Table 2.14. Equilibrium flows of $\dot{y}_1 = 0$ ($\alpha_{11} > 0, \alpha_{12} < 0$).

	$(\dot{x}_1, \dot{x}_2)$	y_1				
Figure 2.18	(a_{110}, a_{220})	$(-\infty, a_{221})$	(a_{221}, a_{111})	(a_{111}, a_{112})	(a_{112}, a_{113})	(a_{113}, ∞)
(a)	$(+,+)$	$\dot{y}_1^* = 0$ $(-,-)$	$\dot{y}_1 < 0$ $(-,+)$	$\dot{y}_1^* = 0$ $(+,+)$	$\dot{y}_1 < 0$ $(-,+)$	$\dot{y}_1^* = 0$ $(+,+)$
(b)	$(-,+)$	$\dot{y}_1 > 0$ $(+,-)$	$\dot{y}_1^* = 0$ $(+,+)$	$\dot{y}_1 < 0$ $(-,+)$	$\dot{y}_1^* = 0$ $(+,+)$	$\dot{y}_1 < 0$ $(-,+)$
(c)	$(+,-)$	$\dot{y}_1 < 0$ $(-,+)$	$\dot{y}_1^* = 0$ $(-,-)$	$\dot{y}_1 > 0$ $(+,-)$	$\dot{y}_1^* = 0$ $(-,-)$	$\dot{y}_1 > 0$ $(+,-)$
(d)	$(-,-)$	$\dot{y}_1^* = 0$ $(+,+)$	$\dot{y}_1 > 0$ $(+,-)$	$\dot{y}_1^* = 0$ $(-,-)$	$\dot{y}_1 > 0$ $(+,-)$	$\dot{y}_1^* = 0$ $(-,-)$

Table 2.15. Parabola flows for $a_{221} > a_{113}$ ($\alpha_{11} > 0, \alpha_{12} > 0$).

Figure 2.19	(a_{110}, a_{220})	$(a_{111}, \dot{x}_2^*)$	$(a_{112}, \dot{x}_2^*)$	$(a_{113}, \dot{x}_2^*)$	$(\dot{x}_1^*, a_{221})$
(a)	$(+,+)$	(DP,nF)	(UP,nF)	(DP,nF)	(pF,UP)
(b)	$(-,+)$	(UP,nF)	(DP,nF)	(UP,nF)	(nF,DP)
(c)	$(+,-)$	(UP,pF)	(DP,pF)	(UP,pF)	(pF,DP)
(d)	$(-,-)$	(DP,pF)	(UP,pF)	(DP,pF)	(nF,UP)

Table 2.16. Equilibrium flows of $\dot{y}_1 = 0$ ($\alpha_{11} > 0, \alpha_{12} > 0$).

	$(\dot{x}_1, \dot{x}_2)$	y_1				
Figure 2.19	(a_{110}, a_{220})	$(-\infty, a_{111})$	(a_{111}, a_{112})	(a_{112}, a_{113})	(a_{113}, a_{221})	(a_{221}, ∞)
(a)	$(+,+)$	$\dot{y}_1 < 0$ $(-,-)$	$\dot{y}_1^* = 0$ $(-,+)$	$\dot{y}_1 > 0$ $(+,+)$	$\dot{y}_1^* = 0$ $(-,+)$	$\dot{y}_1 > 0$ $(+,+)$
(b)	$(-,+)$	$\dot{y}_1^* = 0$ $(+,-)$	$\dot{y}_1 > 0$ $(+,+)$	$\dot{y}_1^* = 0$ $(-,+)$	$\dot{y}_1 > 0$ $(+,+)$	$\dot{y}_1^* = 0$ $(-,+)$
(c)	$(+,-)$	$\dot{y}_1^* = 0$ $(-,+)$	$\dot{y}_1 < 0$ $(-,-)$	$\dot{y}_1^* = 0$ $(+,-)$	$\dot{y}_1 < 0$ $(-,-)$	$\dot{y}_1^* = 0$ $(+,-)$
(d)	$(-,-)$	$\dot{y}_1 > 0$ $(+,+)$	$\dot{y}_1^* = 0$ $(+,-)$	$\dot{y}_1 < 0$ $(-,-)$	$\dot{y}_1^* = 0$ $(+,-)$	$\dot{y}_1 < 0$ $(-,-)$

properties are summarized in Table 2.15 and the corresponding equilibrium flows of $\dot{y}_1 = 0$ in phase space are listed in Table 2.16.

(iv) The vertical and horizontal parabola flows series ($a_{221} > a_{113}$ with $\alpha_{11} > 0$ and $\alpha_{12} < 0$) are presented in Fig. 2.20 with $\dot{x}_2^* = a_{220}(a_{11s_1} - a_{221})$ ($s_1 \in \{1,2,3\}$) and $\dot{x}_1^* = a_{110}\prod_{s_1=1}^{3}(a_{221} - a_{11s_1})$. The corresponding properties are summarized in Table 2.17 and the corresponding equilibrium flows of $\dot{y}_1 = 0$ in phase space are listed in Table 2.18.

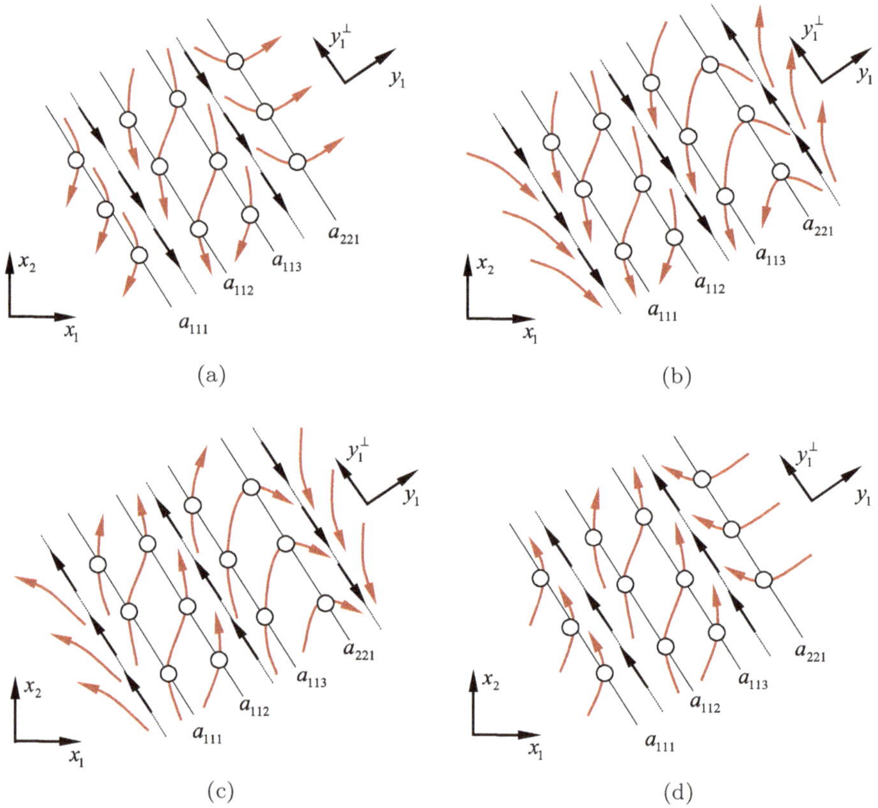

Fig. 2.19. Phase portraits ($a_{111} > a_{221}$ and $\alpha_{11} > 0, \alpha_{12} > 0$) for two-dimensional systems on the x_1-direction with $y_1^* = a_{111}, a_{112}$ and on the x_2-direction with $y_1^* = a_{221}$: (a) ($a_{110} > 0, a_{220} > 0$), (b) ($a_{110} < 0, a_{220} > 0$), (c) ($a_{110} > 0, a_{220} < 0$), (d) ($a_{110} < 0, a_{220} < 0$).

(v) The vertical and horizontal parabola flows' series ($a_{221} \in (a_{111}, a_{112})$ with $\alpha_{11} > 0$ and $\alpha_{12} > 0$) are presented in Fig. 2.21 with $\dot{x}_2^* = a_{220}(a_{11s_1} - a_{221})$ ($s_1 \in \{1, 2, 3\}$) and $\dot{x}_1^* = a_{110} \prod_{s_1=1}^{3} (a_{221} - a_{11s_1})$. The corresponding properties are summarized in Table 2.19 and the corresponding equilibrium flows of $\dot{y}_1 = 0$ in phase space are listed in Table 2.20.

(vi) The vertical and horizontal parabola flows series ($a_{221} \in (a_{111}, a_{112})$ with $\alpha_{11} > 0$ and $\alpha_{12} < 0$) are presented in Fig. 2.22 with $\dot{x}_2^* = a_{220}(a_{11s_1} - a_{221})$ ($s_1 \in \{1, 2, 3\}$) and $\dot{x}_1^* = a_{110} \prod_{s_1=1}^{3} (a_{221} - a_{11s_1})$. The corresponding

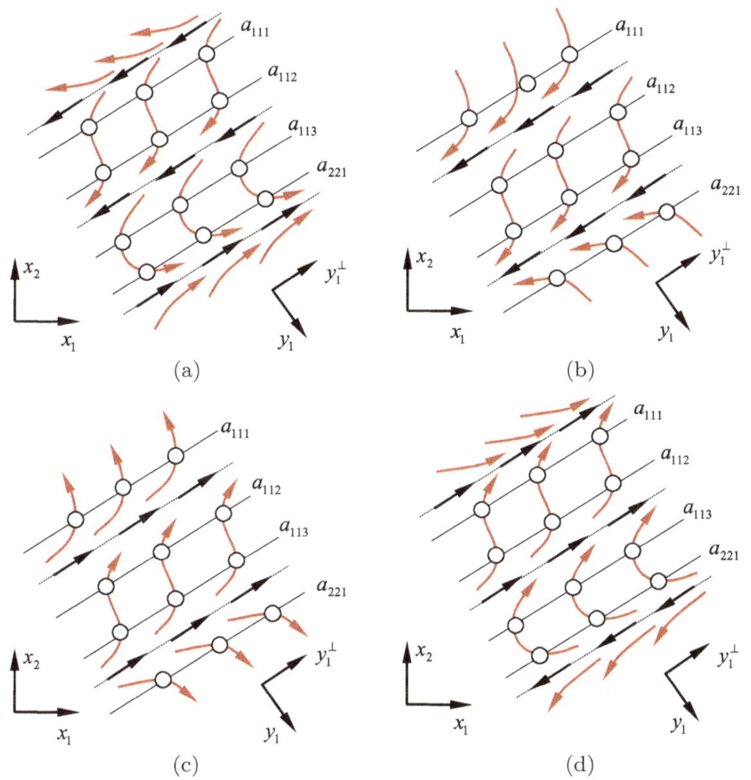

Fig. 2.20. Phase portraits ($a_{111} > a_{221}$ and $\alpha_{11} > 0, \alpha_{12} < 0$) for two-dimensional systems on the x_1-direction with $y_1^* = a_{111}, a_{112}, a_{123}$ and on the x_2-direction with $y_1^* = a_{221}$: (a) ($a_{110} > 0, a_{220} > 0$), (b) ($a_{110} < 0, a_{220} > 0$), (c) ($a_{110} > 0, a_{220} < 0$), (d) ($a_{110} < 0, a_{220} < 0$).

Table 2.17. Parabola flows for $a_{221} > a_{113}$ ($\alpha_{11} > 0, \alpha_{12} < 0$).

Figure 2.20	(a_{110}, a_{220})	$(a_{111}, \dot{x}_2^*)$	$(a_{112}, \dot{x}_2^*)$	$(a_{113}, \dot{x}_2^*)$	$(\dot{x}_1^*, a_{221})$
(a)	$(+,+)$	(UP, nF)	(DP, nF)	(UP, nF)	(pF, UP)
(b)	$(-,+)$	(DP, nF)	(UP, nF)	(DP, nF)	(nF, DP)
(c)	$(+,-)$	(DP, pF)	(UP, pF)	(DP, pF)	(pF, DP)
(d)	$(-,-)$	(UP, pF)	(DP, pF)	(UP, pF)	(nF, UP)

properties are summarized in Table 2.21 and the corresponding equilibrium flows of $\dot{y}_1 = 0$ in phase space are listed in Table 2.22.

(vii) The vertical and horizontal parabola flows series ($a_{221} \in (a_{112}, a_{113})$ with $\alpha_{11} > 0$ and $\alpha_{12} > 0$) are presented in Fig. 2.21 with $\dot{x}_2^* = a_{220}$

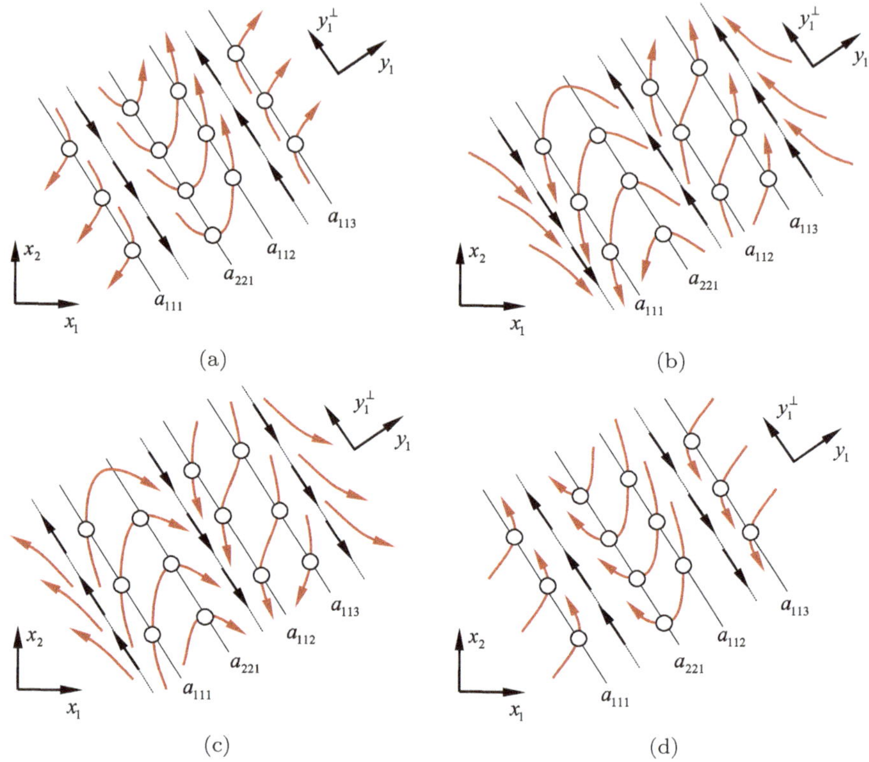

Fig. 2.21. Phase portraits ($a_{111} < a_{221} < a_{112}$ and $\alpha_{11} > 0, \alpha_{12} > 0$) for two-dimensional systems on the x_1-direction with $y_1^* = a_{111}, a_{112}, a_{113}$ and on the x_2-direction with $y_1^* = a_{221}$: (a) ($a_{110} > 0, a_{220} > 0$), (b) ($a_{110} < 0, a_{220} < 0$), (c) ($a_{110} > 0, a_{220} < 0$), (d) ($a_{110} < 0, a_{220} < 0$).

Table 2.18. Equilibrium flows of $\dot{y}_1 = 0$ ($\alpha_{11} > 0, \alpha_{12} < 0$).

Figure 2.20	$(\dot{x}_1, \dot{x}_2)$ (a_{110}, a_{220})	y_1 $(-\infty, a_{111})$	(a_{111}, a_{112})	(a_{112}, a_{113})	(a_{113}, a_{221})	(a_{221}, ∞)
(a)	$(+,+)$	$\dot{y}_1^* = 0$ $(-,-)$	$\dot{y}_1 > 0$ $(+,-)$	$\dot{y}_1^* = 0$ $(-,-)$	$\dot{y}_1 > 0$ $(+,-)$	$\dot{y}_1^* = 0$ $(+,+)$
(b)	$(-,+)$	$\dot{y}_1 > 0$ $(+,-)$	$\dot{y}_1^* = 0$ $(-,-)$	$\dot{y}_1 > 0$ $(+,-)$	$\dot{y}_1^* = 0$ $(-,-)$	$\dot{y}_1 < 0$ $(-,+)$
(c)	$(+,-)$	$\dot{y}_1 < 0$ $(-,+)$	$\dot{y}_1^* = 0$ $(+,+)$	$\dot{y}_1 < 0$ $(-,+)$	$\dot{y}_1^* = 0$ $(+,+)$	$\dot{y}_1 > 0$ $(+,-)$
(d)	$(-,-)$	$\dot{y}_1^* = 0$ $(+,+)$	$\dot{y}_1 < 0$ $(-,+)$	$\dot{y}_1^* = 0$ $(+,+)$	$\dot{y}_1 < 0$ $(-,+)$	$\dot{y}_1^* = 0$ $(-,-)$

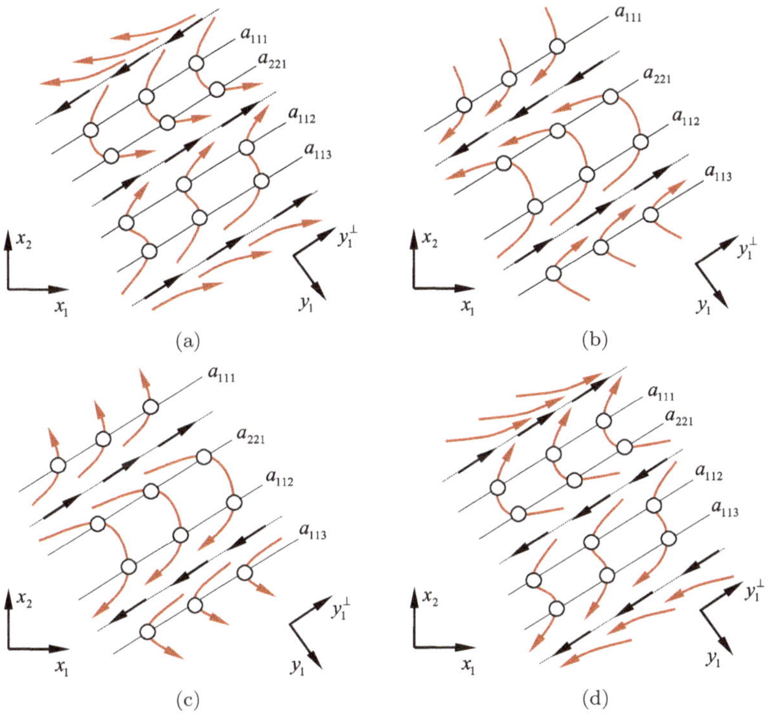

Fig. 2.22. Phase portraits ($a_{111} < a_{221} < a_{112}$ and $\alpha_{11} > 0, \alpha_{12} < 0$) for two-dimensional systems on the x_1-direction with $y_1^* = a_{111}, a_{112}, a_{113}$ and on the x_2-direction with $y_1^* = a_{221}$: (a) ($a_{110} > 0, a_{220} > 0$), (b) ($a_{110} < 0, a_{220} > 0$), (c) ($a_{110} > 0, a_{220} < 0$), (d) ($a_{110} < 0, a_{220} < 0$).

Table 2.19. Parabola flows for $a_{221} \in (a_{111}, a_{112})$ ($\alpha_{11} > 0, \alpha_{12} > 0$).

Figure 2.21	(a_{110}, a_{220})	$(a_{111}, \dot{x}_2^*)$	$(\dot{x}_1^*, a_{221})$	$(a_{112}, \dot{x}_2^*)$	$(a_{113}, \dot{x}_2^*)$
(a)	$(+,+)$	(DP,nF)	(pF,UP)	(DP,pF)	(UP,pF)
(b)	$(-,+)$	(UP,nF)	(nF,DP)	(UP,pF)	(DP,pF)
(c)	$(+,-)$	(UP,pF)	(pF,DP)	(UP,nF)	(DP,nF)
(d)	$(-,-)$	(DP,pF)	(nF,UP)	(DP,nF)	(UP,nF)

Table 2.20. Equilibrium flows of $\dot{y}_1 = 0$ ($\alpha_{11} > 0, \alpha_{12} > 0$).

	$(\dot{x}_1, \dot{x}_2)$	y_1				
Figure 2.21	(a_{110}, a_{220})	$(-\infty, a_{111})$	(a_{111}, a_{221})	(a_{221}, a_{112})	(a_{112}, a_{113})	(a_{113}, ∞)
(a)	$(+,+)$	$\dot{y}_1 < 0$ $(-,-)$	$\dot{y}_1^* = 0$ $(+,-)$	$\dot{y}_1 > 0$ $(+,+)$	$\dot{y}_1^* = 0$ $(-,+)$	$\dot{y}_1 > 0$ $(+,+)$
(b)	$(-,+)$	$\dot{y}_1^* = 0$ $(+,-)$	$\dot{y}_1 < 0$ $(-,-)$	$\dot{y}_1^* = 0$ $(-,+)$	$\dot{y}_1 > 0$ $(+,+)$	$\dot{y}_1^* = 0$ $(-,+)$
(c)	$(+,-)$	$\dot{y}_1^* = 0$ $(-,+)$	$\dot{y}_1 > 0$ $(+,+)$	$\dot{y}_1^* = 0$ $(+,-)$	$\dot{y}_1 < 0$ $(-,-)$	$\dot{y}_1^* = 0$ $(+,-)$
(d)	$(-,-)$	$\dot{y}_1 > 0$ $(+,+)$	$\dot{y}_1^* = 0$ $(-,+)$	$\dot{y}_1 < 0$ $(-,-)$	$\dot{y}_1^* = 0$ $(+,-)$	$\dot{y}_1 < 0$ $(-,-)$

Table 2.21. Parabola flows for $a_{221} \in (a_{111}, a_{112})$ $(\alpha_{11} > 0, \alpha_{12} < 0)$.

Figure 2.22	(a_{110}, a_{220})	$(a_{111}, \dot{x}_2^*)$	$(\dot{x}_1^*, a_{221})$	$(a_{112}, \dot{x}_2^*)$	$(a_{113}, \dot{x}_2^*)$
(a)	$(+,+)$	(UP,nF)	(pF,UP)	(UP,pF)	(DP,pF)
(b)	$(-,+)$	(DP,nF)	(nF,DP)	(DP,pF)	(UP,pF)
(c)	$(+,-)$	(DP,pF)	(pF,DP)	(DP,nF)	(UP,nF)
(d)	$(-,-)$	(UP,pF)	(pF,UP)	(UP,nF)	(DP,nF)

Table 2.22. Equilibrium flows of $\dot{y}_1 = 0$ $(\alpha_{11} > 0, \alpha_{12} < 0)$.

	$(\dot{x}_1, \dot{x}_2)$	y_1				
Figure 2.22	(a_{110}, a_{220})	$(-\infty, a_{111})$	(a_{111}, a_{221})	(a_{221}, a_{112})	(a_{112}, a_{113})	(a_{113}, ∞)
(a)	$(+,+)$	$\dot{y}_1^* = 0$ $(-,-)$	$\dot{y}_1 > 0$ $(+,-)$	$\dot{y}_1^* = 0$ $(+,+)$	$\dot{y}_1 < 0$ $(-,+)$	$\dot{y}_1^* = 0$ $(+,+)$
(b)	$(-,+)$	$\dot{y}_1 > 0$ $(+,-)$	$\dot{y}_1^* = 0$ $(-,-)$	$\dot{y}_1 < 0$ $(-,+)$	$\dot{y}_1^* = 0$ $(+,+)$	$\dot{y}_1 < 0$ $(-,+)$
(c)	$(+,-)$	$\dot{y}_1 < 0$ $(-,+)$	$\dot{y}_1^* = 0$ $(+,+)$	$\dot{y}_1 > 0$ $(+,-)$	$\dot{y}_1^* = 0$ $(-,-)$	$\dot{y}_1 > 0$ $(+,-)$
(d)	$(-,-)$	$\dot{y}_1^* = 0$ $(+,+)$	$\dot{y}_1 < 0$ $(-,+)$	$\dot{y}_1^* = 0$ $(-,-)$	$\dot{y}_1 > 0$ $(+,-)$	$\dot{y}_1^* = 0$ $(-,-)$

Table 2.23. Parabola flows for $a_{221} \in (a_{111}, a_{112})$ $(\alpha_{11} > 0, \alpha_{12} > 0)$.

Figure 2.23	(a_{110}, a_{220})	$(a_{111}, \dot{x}_2^*)$	$(a_{112}, \dot{x}_2^*)$	$(\dot{x}_1^*, a_{221})$	$(a_{113}, \dot{x}_2^*)$
(a)	$(+,+)$	(DP,nF)	(UP,nF)	(nF,DP)	(UP,pF)
(b)	$(-,+)$	(UP,nF)	(DP,nF)	(pF,UP)	(DP,pF)
(c)	$(+,-)$	(UP,pF)	(DP,pF)	(nF,UP)	(DP,nF)
(d)	$(-,-)$	(DP,pF)	(UP,pF)	(pF,DP)	(UP,nF)

Table 2.24. Equilibrium flows of $\dot{y}_1 = 0$ $(\alpha_{11} > 0, \alpha_{12} > 0)$.

	$(\dot{x}_1, \dot{x}_2)$	y_1				
Figure 2.23	(a_{110}, a_{220})	$(-\infty, a_{111})$	(a_{111}, a_{112})	(a_{112}, a_{221})	(a_{221}, a_{113})	(a_{113}, ∞)
(a)	$(+,+)$	$\dot{y}_1 < 0$ $(-,-)$	$\dot{y}_1^* = 0$ $(+,-)$	$\dot{y}_1 < 0$ $(-,-)$	$\dot{y}_1^* = 0$ $(-,+)$	$\dot{y}_1 > 0$ $(+,+)$
(b)	$(-,+)$	$\dot{y}_1^* = 0$ $(+,-)$	$\dot{y}_1 < 0$ $(-,-)$	$\dot{y}_1^* = 0$ $(+,-)$	$\dot{y}_1 > 0$ $(+,+)$	$\dot{y}_1^* = 0$ $(-,+)$
(c)	$(+,-)$	$\dot{y}_1^* = 0$ $(-,+)$	$\dot{y}_1 > 0$ $(+,+)$	$\dot{y}_1^* = 0$ $(-,+)$	$\dot{y}_1 < 0$ $(-,-)$	$\dot{y}_1^* = 0$ $(+,-)$
(d)	$(-,-)$	$\dot{y}_1 > 0$ $(+,+)$	$\dot{y}_1^* = 0$ $(-,+)$	$\dot{y}_1 > 0$ $(+,+)$	$\dot{y}_1^* = 0$ $(+,-)$	$\dot{y}_1 < 0$ $(-,-)$

$(a_{11s_1} - a_{221})$ $(s_1 \in \{1,2,3\})$ and $\dot{x}_1^* = a_{110} \prod_{s_1=1}^{3} (a_{221} - a_{11s_1})$. The corresponding properties are summarized in Table 2.23 and the corresponding equilibrium flows of $\dot{y}_1 = 0$ in phase space are listed in Table 2.24.

(viii) The vertical and horizontal parabola flows series ($a_{221} \in (a_{112}, a_{113})$ with $\alpha_{11} > 0$ and $\alpha_{12} < 0$) are presented in Fig. 2.24 with $\dot{x}_2^* = a_{220}(a_{11s_1} - a_{221})$ $(s_1 \in \{1,2,3\})$ and $\dot{x}_1^* = a_{110} \prod_{s_1=1}^{3} (a_{221} - a_{11s_1})$. The corresponding

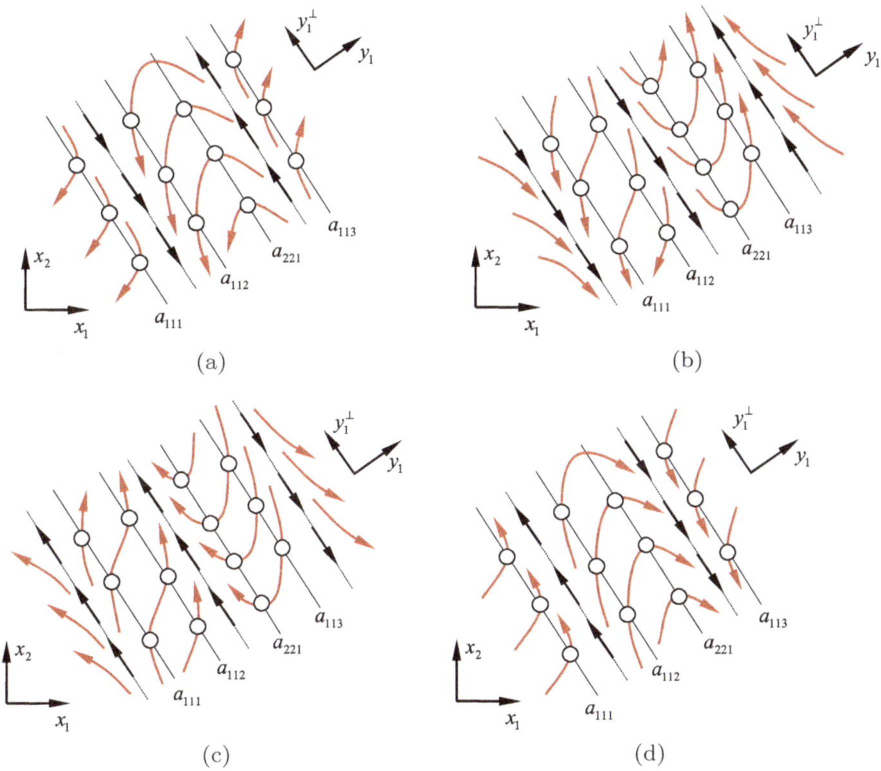

Fig. 2.23. Phase portraits ($a_{111} < a_{221} < a_{112}$ and $\alpha_{11} > 0, \alpha_{12} > 0$) for two-dimensional systems on the x_1-direction with $y_1^* = a_{111}, a_{112}, a_{113}$ and on the x_2-direction with $y_1^* = a_{221}$: (a) ($a_{110} > 0, a_{220} > 0$), (b) ($a_{110} < 0, a_{220} < 0$), (c) ($a_{110} > 0, a_{220} < 0$), (d) ($a_{110} < 0, a_{220} < 0$).

properties are summarized in Table 2.25 and the corresponding equilibrium flows of $\dot{y}_1 = 0$ in phase space are listed in Table 2.26.

2.5.2 *Inflection-sink (source) switching bifurcations*

Consider a dynamical system as

$$\begin{aligned} \dot{x}_2 &= a_{220}(y_1 - a_{221}), \\ \dot{x}_1 &= a_{111}(y_1 - a_{11s_1}) \prod_{s_2=1, s_2 \neq s_1}^{3} (y_1 - a_{11s_2}). \end{aligned} \tag{2.259}$$

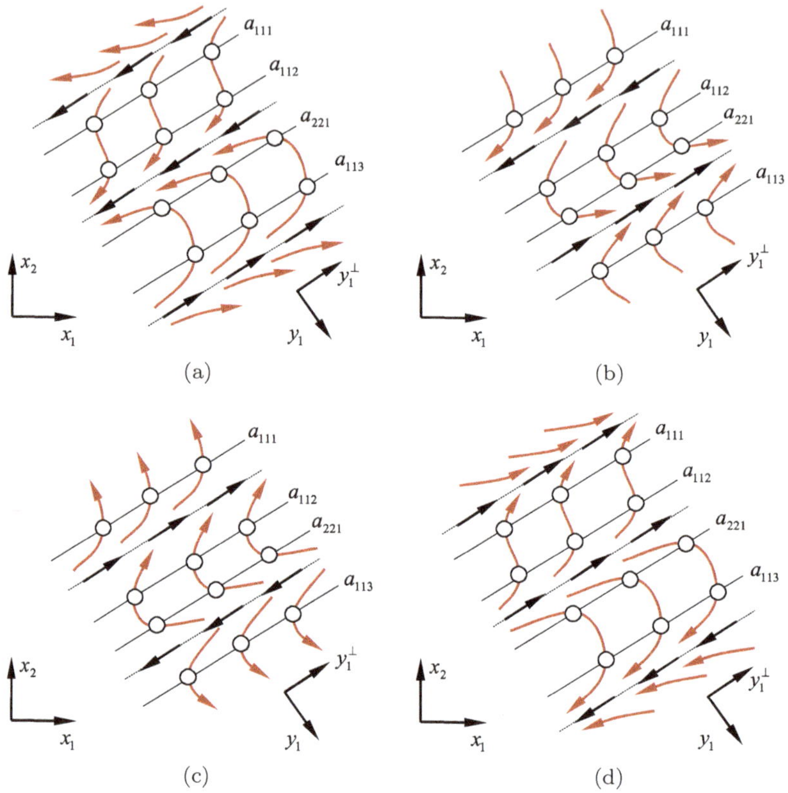

Fig. 2.24. Phase portraits ($a_{111} < a_{221} < a_{112}$ and $\alpha_{11} > 0, \alpha_{12} < 0$) for two-dimensional systems on the x_1-direction with $y_1^* = a_{111}, a_{112}, a_{113}$ and on the x_2-direction with $y_1^* = a_{221}$: (a) ($a_{110} > 0, a_{220} > 0$), (b) ($a_{110} < 0, a_{220} > 0$), (c) ($a_{110} > 0, a_{220} < 0$), (d) ($a_{110} < 0, a_{220} < 0$).

Table 2.25. Parabola flows for $a_{221} \in (a_{112}, a_{113})$ ($\alpha_{11} > 0, \alpha_{12} < 0$).

Figure 2.24	(a_{110}, a_{220})	$(a_{111}, \dot{x}_2^*)$	$(a_{112}, \dot{x}_2^*)$	$(\dot{x}_1^*, a_{221})$	$(a_{113}, \dot{x}_2^*)$
(a)	$(+,+)$	(UP, nF)	(DP, nF)	(nF, DP)	(DP, pF)
(b)	$(-,+)$	(DP, nF)	(UP, nF)	(pF, UP)	(UP, pF)
(c)	$(+,-)$	(DP, pF)	(UP, pF)	(pF, UP)	(UP, nF)
(d)	$(-,-)$	(UP, pF)	(DP, pF)	(pF, DP)	(DP, nF)

Table 2.26. Equilibrium flows of $\dot{y}_1 = 0$ ($\alpha_{11} > 0, \alpha_{12} < 0$).

	$(\dot{x}_1, \dot{x}_2)$	y_1				
Figure 2.24	(a_{110}, a_{220})	$(-\infty, a_{111})$	(a_{111}, a_{112})	(a_{112}, a_{221})	(a_{221}, a_{113})	(a_{113}, ∞)
(a)	$(+,+)$	$\dot{y}_1^* = 0$ $(-,-)$	$\dot{y}_1 > 0$ $(+,-)$	$\dot{y}_1^* = 0$ $(-,-)$	$\dot{y}_1 < 0$ $(-,+)$	$\dot{y}_1^* = 0$ $(+,+)$
(b)	$(-,+)$	$\dot{y}_1 > 0$ $(+,-)$	$\dot{y}_1'' = 0$ $(-,-)$	$\dot{y}_1 > 0$ $(+,-)$	$\dot{y}_1^* = 0$ $(+,+)$	$\dot{y}_1 < 0$ $(-,+)$
(c)	$(+,-)$	$\dot{y}_1 < 0$ $(-,+)$	$\dot{y}_1^* = 0$ $(+,+)$	$\dot{y}_1 < 0$ $(-,+)$	$\dot{y}_1^* = 0$ $(-,-)$	$\dot{y}_1 > 0$ $(+,-)$
(d)	$(-,-)$	$\dot{y}_1^* = 0$ $(+,+)$	$\dot{y}_1 < 0$ $(-,+)$	$\dot{y}_1^* = 0$ $(+,+)$	$\dot{y}_1 > 0$ $(+,-)$	$\dot{y}_1^* = 0$ $(-,-)$

For $a_{11s_1} = a_{221}$, there is a dynamical system in the y_{j_1}-direction as

$$\dot{y}_1 = (y_1 - a_{11s_1})\Big[\alpha_{11}a_{110} \prod_{s_2=1, s_2 \neq s_1}^{3} (y_1 - a_{11s_2}) + \alpha_{12}a_{220}\Big] \tag{2.260}$$

with a dynamical system in the $y_1^{\perp}$-direction as

$$\dot{y}_1^{\perp} = (y_1 - a_{11s_1})\Big[\beta_{11}a_{110} \prod_{s_2=1, s_2 \neq s_1}^{3} (y_1 - a_{11s_2}) + \beta_{12}a_{220}\Big] \tag{2.261}$$

and

$$G_{110} = \alpha_{11}a_{110} \prod_{s_2=1, s_2 \neq s_1}^{3} (a_{11s_1} - a_{11s_2}) + \alpha_{12}a_{220}. \tag{2.262}$$

The inflection infinite-equilibriums of $a_{11s_1} = a_{221}$ are source and sink for $G_{110} > 0$ and $G_{110} < 0$, respectively. The parabola flows with the inflection infinite-equilibriums are for the switching bifurcations of the vertical and horizontal parabola flow series. The inflection infinite-equilibriums ($a_{11s_1} = a_{221}$) are the switching bifurcations of $a_{11s_1} > a_{221}$ and $a_{11s_1} < a_{221}$:

(i) For $a_{111} = a_{221}$ with $\alpha_{11} > 0$ and $\alpha_{12} > 0$, the inflection infinite-equilibriums with parabola flows are presented in Fig. 2.25. With $\dot{x}_2^* = a_{220}(a_{11s_1} - a_{221})$ ($s_1 \in \{2,3\}$), the corresponding infinite-equilibriums and parabola flows are listed in Table 2.27 and the corresponding equilibrium flows of $\dot{y}_1 = 0$ in phase space are listed in Table 2.28.

(ii) For $a_{111} = a_{221}$ with $\alpha_{11} > 0$ and $\alpha_{12} < 0$, the inflection infinite-equilibriums with parabola flows in phase space are presented in Fig. 2.26. With $\dot{x}_2^* = a_{220}(a_{11s_1} - a_{221})$ ($s_1 \in \{2,3\}$), the corresponding infinite-equilibriums and parabola flows are listed in Table 2.29 and the

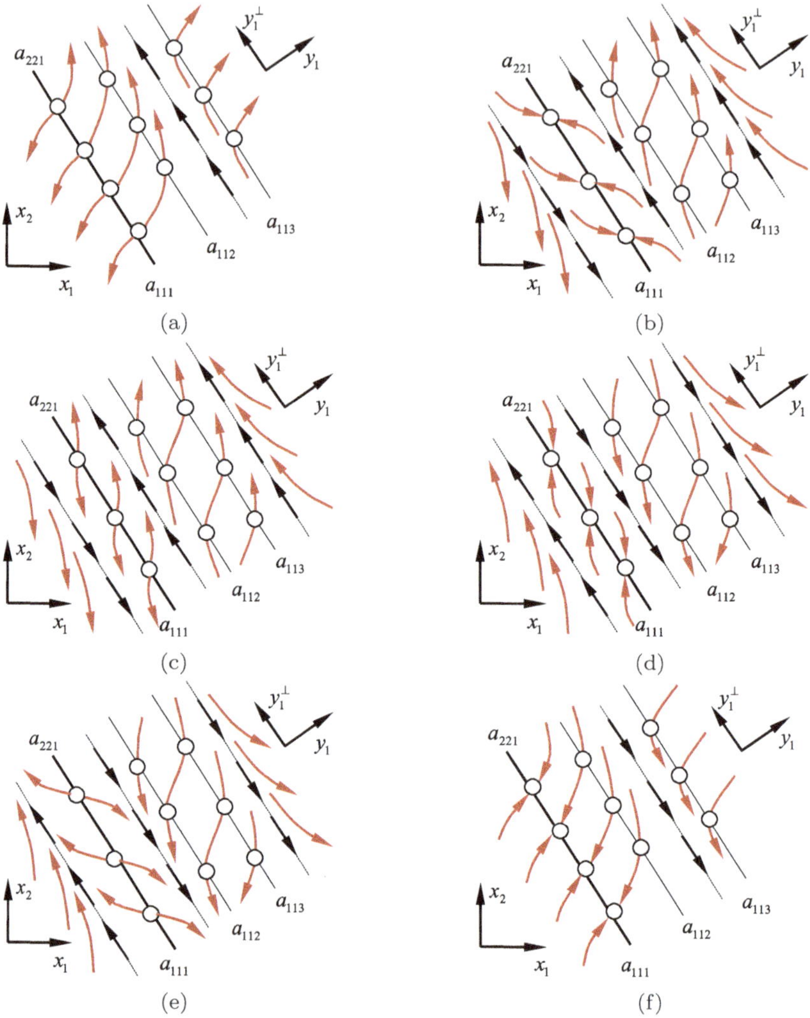

Fig. 2.25. Phase portraits ($a_{111} = a_{221}$ and $\alpha_{11} > 0, \alpha_{12} > 0$) for two-dimensional systems on the x_1-direction with $y_1^* = a_{111}, a_{112}, a_{113}$ and on the x_2-direction with $y_1^* = a_{221}$: (a) ($a_{110} > 0, a_{220} > 0$), (b,c) ($a_{110} < 0, a_{220} > 0$), (d,e) ($a_{110} > 0, a_{220} < 0$), (f) ($a_{110} < 0, a_{220} < 0$).

corresponding equilibrium flows of $\dot{y}_1 = 0$ in phase space are listed in Table 2.30.

(iii) For $a_{113} = a_{221}$ with $\alpha_{11} > 0$ and $\alpha_{12} > 0$, the inflection infinite-equilibriums with parabola flows in phase space are presented in

Table 2.27. Infinite-equilibriums and parabola flows for $a_{221} = a_{111}$ ($\alpha_{11} > 0$, $\alpha_{12} > 0$).

Figure 2.25	(a_{110}, a_{220})	(a_{111}, a_{221})		$(a_{112}, \dot{x}_2^*)$	$(a_{113}, \dot{x}_2^*)$
(a)	$(+,+)$	(II,SO)	$G_{110} > 0$	(DP,pF)	(UP,pF)
(b)	$(-,+)$	(DI,SI)	$G_{110} < 0$	(UP,pF)	(DP,pF)
(c)	$(-,+)$	(DI,SO)	$G_{110} > 0$	(UP,pF)	(DP,pF)
(d)	$(+,-)$	(DI,SI)	$G_{110} < 0$	(UP,nF)	(DP,nF)
(e)	$(+,-)$	(DI,SO)	$G_{110} > 0$	(UP,nF)	(DP,nF)
(f)	$(-,-)$	(II,SI)	$G_{110} < 0$	(DP,pF)	(DP,nF)

Table 2.28. Equilibrium flows of $\dot{y}_1 = 0$ for $a_{111} = a_{221}$ ($\alpha_{11} > 0, \alpha_{12} > 0$).

	$(\dot{x}_1, \dot{x}_2)$	y_1			
Figure 2.25	(a_{110}, a_{220})	$(-\infty, a_{111})$	(a_{111}, a_{112})	(a_{112}, a_{113})	(a_{113}, ∞)
(a)	$(+,+)$	$\dot{y}_1 < 0$ $(-,-)$	$\dot{y}_1 > 0$ $(+,+)$	$\dot{y}_1^* = 0$ $(-,+)$	$\dot{y}_1 > 0$ $(+,+)$
(b,c)	$(-,+)$	$\dot{y}_1^* = 0$ $(+,-)$	$\dot{y}_1^* = 0$ $(-,+)$	$\dot{y}_1 > 0$ $(+,+)$	$\dot{y}_1^* = 0$ $(-,+)$
(d,e)	$(+,-)$	$\dot{y}_1^* = 0$ $(-,+)$	$\dot{y}_1^* = 0$ $(+,-)$	$\dot{y}_1 < 0$ $(-,-)$	$\dot{y}_1^* = 0$ $(+,-)$
(f)	$(-,-)$	$\dot{y}_1 > 0$ $(+,+)$	$\dot{y}_1 < 0$ $(-,-)$	$\dot{y}_1^{**} = 0$ $(+,-)$	$\dot{y}_1 < 0$ $(-,-)$

Fig. 2.27. With $\dot{x}_2^* = a_{220}(a_{11s_1} - a_{221})$ ($s_1 \in \{1,2\}$), the corresponding infinite-equilibriums and parabola flows are listed in Table 2.31 and the corresponding equilibrium flows of $\dot{y}_1 = 0$ in phase space are listed in Table 2.32.

(iv) For $a_{113} = a_{221}$ with $\alpha_{11} > 0$ and $\alpha_{12} < 0$, the inflection infinite-equilibriums with parabola flows in phase space are presented in Fig. 2.28. With $\dot{x}_2^* = a_{220}(a_{11s_1} - a_{221})$ ($s_1 \in \{1,2\}$), the corresponding infinite-equilibriums and parabola flows are listed in Table 2.33 and the corresponding equilibrium flows of $\dot{y}_1 = 0$ in phase space are listed in Table 2.34.

(v) For $a_{112} = a_{221}$ with $\alpha_{11} > 0$ and $\alpha_{12} > 0$, the inflection infinite-equilibriums with parabola flows in phase space are presented in Fig. 2.29. With $\dot{x}_2^* = a_{220}(a_{11s_1} - a_{221})$ ($s_1 \in \{1,3\}$), the corresponding infinite-equilibriums and parabola flows are listed in Table 2.35 and the corresponding equilibrium flows of $\dot{y}_1 = 0$ in phase space are listed in Table 2.36.

(vi) For $a_{112} = a_{221}$ with $\alpha_{11} > 0$ and $\alpha_{12} < 0$, the inflection infinite-equilibriums with parabola flows in phase space are presented in

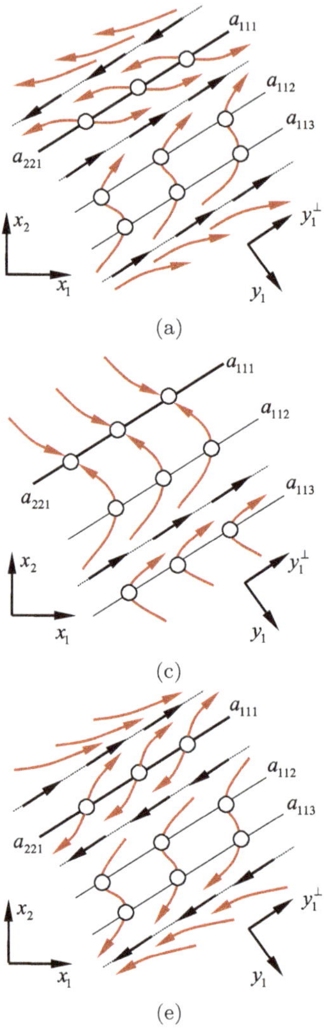

Fig. 2.26. Phase portraits ($a_{111} = a_{221}$ and $\alpha_{11} > 0, \alpha_{12} < 0$) for two-dimensional systems on the x_1-direction with $y_1^* = a_{111}, a_{112}, a_{113}$ and on the x_2-direction with $y_1^* = a_{221}$: (a,b) ($a_{110} > 0, a_{220} > 0$), (c) ($a_{110} < 0, a_{220} > 0$), (d) $y_1^* = a_{112}(a_{110} > 0, a_{220} < 0)$, (e,f) ($a_{110} < 0, a_{220} < 0$).

Fig. 2.30. With $\dot{x}_2^* = a_{220}(a_{11s_1} - a_{221})$ ($s_1 \in \{1, 3\}$), the corresponding infinite-equilibriums and parabola flows are listed in Table 2.37 and the corresponding equilibrium flows of $\dot{y}_1 = 0$ in phase space are listed in Table 2.38.

Table 2.29. Infinite-equilibrium and parabola flows for $a_{221} = a_{111}(\alpha_{11} > 0, \alpha_{12} < 0)$.

Figure 2.26	(a_{110}, a_{220})	(a_{111}, a_{221})		$(a_{112}, \dot{x}_2^*)$	$(a_{113}, \dot{x}_2^*)$
(a)	$(+,+)$	(II,SO)	$G_{110} > 0$	(UP,pF)	(DP,pF)
(b)	$(+,+)$	(II,SI)	$G_{110} < 0$	(UP,pF)	(DP,pF)
(c)	$(-,+)$	(DI,SI)	$G_{110} < 0$	(DP,pF)	(UP,pF)
(d)	$(+,-)$	(DI,SO)	$G_{110} > 0$	(DP,nF)	(UP,nF)
(e)	$(+,-)$	(II,SO)	$G_{110} > 0$	(UP,nF)	(DP,nF)
(f)	$(-,-)$	(II,SI)	$G_{110} < 0$	(UP,nF)	(DP,nF)

Table 2.30. Equilibrium flows of $\dot{y}_1 = 0$ for $a_{221} = a_{111}(\alpha_{11} > 0, \alpha_{12} < 0)$.

	$(\dot{x}_1, \dot{x}_2)$	y_1			
Figure 2.26	(a_{110}, a_{220})	$(-\infty, a_{111})$	(a_{111}, a_{112})	(a_{112}, a_{113})	(a_{113}, ∞)
(a)	$(+,+)$	$\dot{y}_1^* = 0$ $(-,-)$	$\dot{y}_1^* = 0$ $(+,+)$	$\dot{y}_1 > 0$ $(-,+)$	$\dot{y}_1^* = 0$ $(+,+)$
(b,c)	$(-,+)$	$\dot{y}_1 > 0$ $(+,-)$	$\dot{y}_1 < 0$ $(-,+)$	$\dot{y}_1^* = 0$ $(+,+)$	$\dot{y}_1 < 0$ $(-,+)$
(d,e)	$(+,-)$	$\dot{y}_1 < 0$ $(-,+)$	$\dot{y}_1 > 0$ $(+,-)$	$\dot{y}_1^* = 0$ $(-,-)$	$\dot{y}_1 > 0$ $(+,-)$
(f)	$(-,-)$	$\dot{y}_1^* = 0$ $(+,+)$	$\dot{y}_1^* = 0$ $(-,-)$	$\dot{y}_1 > 0$ $(+,-)$	$\dot{y}_1^* = 0$ $(-,-)$

Table 2.31. Infinite-equilibrium and parabola flows for $a_{221} = a_{113}$ $(\alpha_{11} > 0, \alpha_{12} > 0)$.

Figure 2.27	(a_{110}, a_{220})	$(a_{111}, \dot{x}_2^*)$	$(a_{112}, \dot{x}_2^*)$	(a_{113}, a_{221})	
(a)	$(+,+)$	(DP,nF)	(UP,nF)	(II,SO)	$G_{110} > 0$
(b)	$(-,+)$	(UP,nF)	(DP,nF)	(DI,SI)	$G_{110} < 0$
(c)	$(-,+)$	(UP,nF)	(DP,nF)	(DI,SO)	$G_{110} > 0$
(d)	$(+,-)$	(UP,pF)	(DP,pF)	(DI,SI)	$G_{110} < 0$
(e)	$(+,-)$	(UP,pF)	(DP,pF)	(DI,SO)	$G_{110} > 0$
(f)	$(-,-)$	(DP,pF)	(UP,pF)	(II,SI)	$G_{110} < 0$

Table 2.32. Equilibrium flows of $\dot{y}_1 = 0$ for $a_{221} = a_{113}$ $(\alpha_{11} > 0, \alpha_{12} > 0)$.

	$(\dot{x}_1, \dot{x}_2)$	y_1			
Figure 2.27	(a_{110}, a_{220})	$(-\infty, a_{111})$	(a_{111}, a_{112})	(a_{12}, a_{113})	(a_{113}, ∞)
(a)	$(+,+)$	$\dot{y}_1 < 0$ $(-,-)$	$\dot{y}_1^* = 0$ $(+,-)$	$\dot{y}_1 < 0$ $(-,-)$	$\dot{y}_1 > 0$ $(+,+)$
(b,c)	$(-,+)$	$\dot{y}_1^* = 0$ $(+,-)$	$\dot{y}_1 < 0$ $(-,-)$	$\dot{y}_1^* = 0$ $(+,-)$	$\dot{y}_1^* = 0$ $(-,+)$
(d,e)	$(+,-)$	$\dot{y}_1^* = 0$ $(-,+)$	$\dot{y}_1 > 0$ $(+,+)$	$\dot{y}_1^* = 0$ $(-,+)$	$\dot{y}_1^* = 0$ $(+,-)$
(f)	$(-,-)$	$\dot{y}_1 > 0$ $(+,+)$	$\dot{y}_1^* = 0$ $(-,+)$	$\dot{y}_1 < 0$ $(-,-)$	$\dot{y}_1 < 0$ $(-,-)$

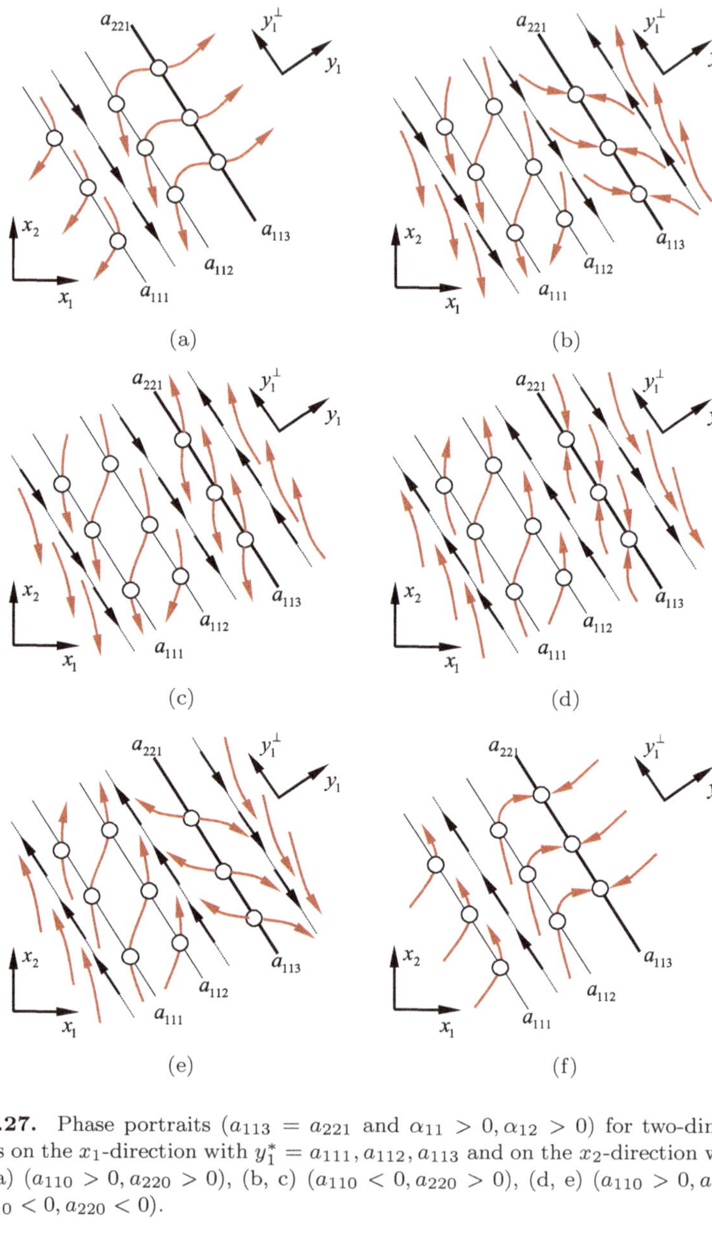

Fig. 2.27. Phase portraits ($a_{113} = a_{221}$ and $\alpha_{11} > 0, \alpha_{12} > 0$) for two-dimensional systems on the x_1-direction with $y_1^* = a_{111}, a_{112}, a_{113}$ and on the x_2-direction with $y_1^* = a_{221}$: (a) ($a_{110} > 0, a_{220} > 0$), (b, c) ($a_{110} < 0, a_{220} > 0$), (d, e) ($a_{110} > 0, a_{220} < 0$), (f) ($a_{110} < 0, a_{220} < 0$).

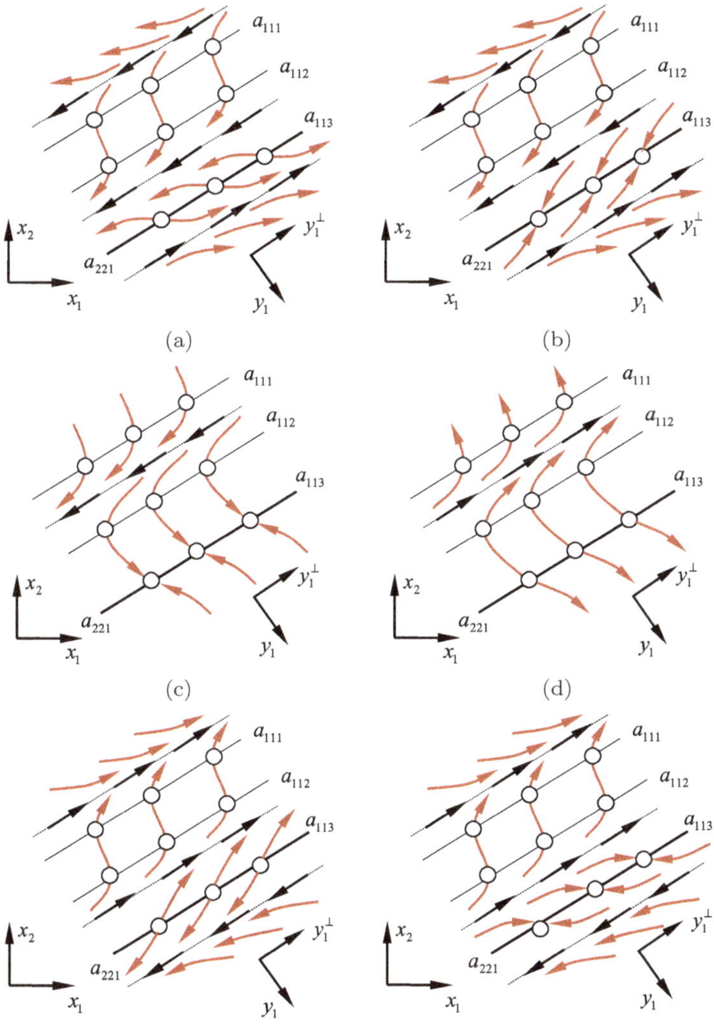

Fig. 2.28. Phase portraits ($a_{113} = a_{221}$ and $\alpha_{11} > 0, \alpha_{12} < 0$) for two-dimensional systems on the x_1-direction with $y_1^* = a_{111}, a_{112}, a_{113}$ and on the x_2-direction with $y_1^* = a_{221}$: (a, b) ($a_{110} > 0, a_{220} > 0$), (c) ($a_{110} < 0, a_{220} > 0$), (d) ($a_{110} > 0, a_{220} < 0$), (e, f) ($a_{110} < 0, a_{220} < 0$).

Table 2.33. Infinite-equilibrium and parabola flows for $a_{221} = a_{113}$ $(\alpha_{11} > 0, \alpha_{12} < 0)$.

Figure 2.28	(a_{110}, a_{220})	$(a_{111}, \dot{x}_2^*)$	$(a_{112}, \dot{x}_2^*)$	(a_{113}, a_{221})	
(a)	$(+,+)$	(UP,nF)	(DP,nF)	(II,SO)	$G_{110} > 0$
(b)	$(+,+)$	(UP,nF)	(DP,nF)	(II,SI)	$G_{110} < 0$
(c)	$(-,+)$	(DP,nF)	(UP,nF)	(DI,SO)	$G_{110} > 0$
(d)	$(+,-)$	(DP,pF)	(UP,pF)	(DI,SI)	$G_{110} < 0$
(e)	$(+,-)$	(UP,pF)	(DP,pF)	(II,SO)	$G_{110} > 0$
(f)	$(-,-)$	(UP,pF)	(DP,pF)	(II,SI)	$G_{110} < 0$

Table 2.34. Equilibrium flows of $\dot{y}_1 = 0$ for $a_{221} = a_{113}$ $(\alpha_{11} > 0, \alpha_{12} < 0)$.

	$(\dot{x}_1, \dot{x}_2)$	y_1			
Figure 2.28	(a_{110}, a_{220})	$(-\infty, a_{111})$	(a_{111}, a_{112})	(a_{112}, a_{113})	(a_{113}, ∞)
(a)	$(+,+)$	$\dot{y}_1^* = 0$ $(-,-)$	$\dot{y}_1 > 0$ $(+,-)$	$\dot{y}_1^* = 0$ $(-,-)$	$\dot{y}_1^* = 0$ $(+,+)$
(b,c)	$(-,+)$	$\dot{y}_1 > 0$ $(+,-)$	$\dot{y}_1^* = 0$ $(-,-)$	$\dot{y}_1 > 0$ $(+,-)$	$\dot{y}_1 < 0$ $(-,+)$
(d,e)	$(+,-)$	$\dot{y}_1 < 0$ $(-,+)$	$\dot{y}_1^* = 0$ $(+,+)$	$\dot{y}_1 < 0$ $(-,+)$	$\dot{y}_1 > 0$ $(+,-)$
(f)	$(-,-)$	$\dot{y}_1^* = 0$ $(+,+)$	$\dot{y}_1 < 0$ $(-,+)$	$\dot{y}_1^* = 0$ $(+,+)$	$\dot{y}_1^* = 0$ $(-,-)$

Table 2.35. Infinite-equilibrium and parabola flows for $a_{221} = a_{112}$ $(\alpha_{11} > 0, \alpha_{12} > 0)$.

Figure 2.29	(a_{110}, a_{220})	$(a_{111}, \dot{x}_2^*)$	(a_{113}, a_{221})		$(a_{112}, \dot{x}_2^*)$
(a)	$(+,+)$	(DP,nF)	(DI,SI)	$G_{110} < 0$	(UP,pF)
(b)	$(+,+)$	(DP,nF)	(DI,SO)	$G_{110} > 0$	(UP,pF)
(c)	$(-,+)$	(UP,nF)	(II,SO)	$G_{110} > 0$	(DP,pF)
(d)	$(+,-)$	(UP,pF)	(II,SI)	$G_{110} < 0$	(DP,nF)
(e)	$(-,-)$	(DP,pF)	(DI,SI)	$G_{110} < 0$	(UP,nF)
(f)	$(-,-)$	(DP,pF)	(DI,SO)	$G_{110} > 0$	(UP,nF)

Table 2.36. Equilibrium flows of $\dot{y}_1 = 0$ for $a_{112} = a_{221}$ $(\alpha_{11} > 0, \alpha_{12} > 0)$.

	$(\dot{x}_1, \dot{x}_2)$	y_1			
Figure 2.29	(a_{110}, a_{220})	$(-\infty, a_{111})$	(a_{111}, a_{112})	(a_{112}, a_{113})	(a_{113}, ∞)
(a,b)	$(+,+)$	$\dot{y}_1 < 0$ $(-,-)$	$\dot{y}_1^* = 0$ $(+,-)$	$\dot{y}_1 < 0$ $(-,-)$	$\dot{y}_1 > 0$ $(+,+)$
(c)	$(-,+)$	$\dot{y}_1^* = 0$ $(+,-)$	$\dot{y}_1 < 0$ $(-,-)$	$\dot{y}_1^* = 0$ $(+,-)$	$\dot{y}_1^* = 0$ $(-,+)$
(d)	$(+,-)$	$\dot{y}_1^* = 0$ $(-,+)$	$\dot{y}_1 > 0$ $(+,+)$	$\dot{y}_1^* = 0$ $(-,+)$	$\dot{y}_1^* = 0$ $(+,-)$
(e,f)	$(-,-)$	$\dot{y}_1 > 0$ $(+,+)$	$\dot{y}_1^* = 0$ $(-,+)$	$\dot{y}_1 < 0$ $(-,-)$	$\dot{y}_1 < 0$ $(-,-)$

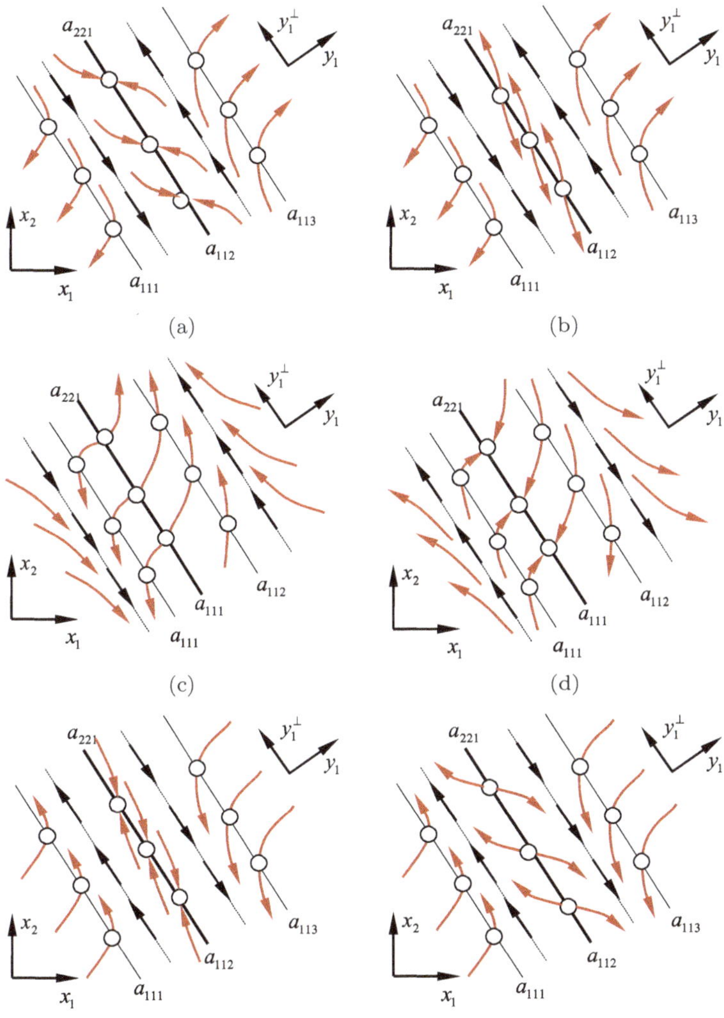

Fig. 2.29. Phase portraits ($a_{112} = a_{221}$ and $\alpha_{11} > 0, \alpha_{12} > 0$) for two-dimensional systems on the x_1-direction with $y_1^* = a_{111}, a_{112}, a_{113}$ and on the x_2-direction with $y_1^* = a_{221}$: (a, b) ($a_{110} > 0, a_{220} > 0$), (c) ($a_{110} < 0, a_{220} > 0$), (d) ($a_{110} > 0, a_{220} < 0$), (e, f) ($a_{110} < 0, a_{220} < 0$).

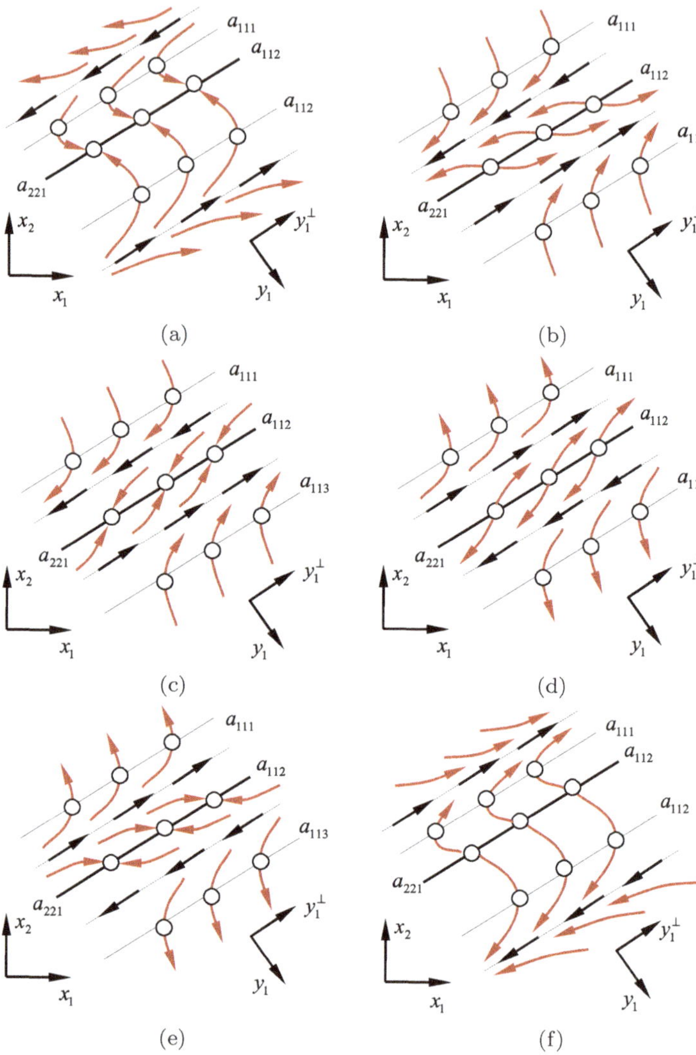

Fig. 2.30. Phase portraits ($a_{112} = a_{221}$ and $\alpha_{11} > 0, \alpha_{12} < 0$) for two-dimensional systems on the x_1-direction with $y_1^* = a_{111}, a_{112}, a_{113}$ and on the x_2-direction with $y_1^* = a_{221}$: (a) ($a_{110} > 0, a_{220} > 0$), (b, c) ($a_{110} < 0, a_{220} > 0$), (d, e) ($a_{110} > 0$, $a_{220} < 0$), (f) ($a_{110} < 0, a_{220} < 0$).

Table 2.37. Infinite-equilibrium and parabola flows for $a_{221} = a_{112}$ ($\alpha_{11} > 0, \alpha_{12} < 0$).

Figure 2.30	(a_{110}, a_{220})	$(a_{111}, \dot{x}_2^*)$	(a_{113}, a_{221})		$(a_{112}, \dot{x}_2^*)$
(a)	(+, +)	(UP,nF)	(DI,SI)	$G_{110} > 0$	(DP,pF)
(b)	(−, +)	(DP,nF)	(II,SO)	$G_{110} > 0$	(UP,pF)
(c)	(−, +)	(DP,nF)	(II,SI)	$G_{110} < 0$	(UP,pF)
(d)	(+, −)	(DP,pF)	(II,SO)	$G_{110} > 0$	(UP,nF)
(e)	(+, −)	(DP,pF)	(II,SI)	$G_{110} < 0$	(UP,nF)
(f)	(−, −)	(UP,pF)	(DI,SI)	$G_{110} < 0$	(DP,pF)

Table 2.38. Equilibrium flows of $\dot{y}_1 = 0$ for $a_{112} = a_{221}$ ($\alpha_{11} > 0, \alpha_{12} < 0$).

	$(\dot{x}_1, \dot{x}_2)$	y_1			
Figure 2.30	(a_{110}, a_{220})	$(-\infty, a_{111})$	(a_{111}, a_{112})	(a_{112}, a_{113})	(a_{113}, ∞)
(a)	(+, +)	$\dot{y}_1^* = 0$ (−, −)	$\dot{y}_1 < 0$ (−, +)	$\dot{y}_1 > 0$ (+, −)	$\dot{y}_1^* = 0$ (+, +)
(b,c)	(−, +)	$\dot{y}_1 > 0$ (+, −)	$\dot{y}_1^* = 0$ (−, −)	$\dot{y}_1^* = 0$ (+, +)	$\dot{y}_1 < 0$ (−, +)
(d,e)	(+, −)	$\dot{y}_1 < 0$ (−, +)	$\dot{y}_1^* = 0$ (+, +)	$\dot{y}_1^* = 0$ (−, −)	$\dot{y}_1 > 0$ (+, −)
(f)	(−, −)	$\dot{y}_1^* = 0$ (+, +)	$\dot{y}_1 > 0$ (+, −)	$\dot{y}_1 < 0$ (−, +)	$\dot{y}_1^* = 0$ (−, −)

Chapter 3

Single-Linear-Bivariate Quadratic and Cubic Systems

In this chapter, two-dimensional nonlinear systems possessing single-linear-bivariate quadratic and cubic vector fields are discussed, and the appearing and switching bifurcations for the one-dimensional flows (i.e., up-parabola and down-parabola flows and inflection flows) are discussed herein. The increasing-inflection and deceasing-inflection flows are the appearing bifurcations for up-parabola and down-parabola flows. The third-order parabola flows are the appearing bifurcations among up-parabola and down-parabola flows, and for parabola and inflection flows. The inflection-sink and source infinite-equilibriums are the switching bifurcations of two parabola flows in the two directions and for two inflections in the two directions. The parabola-sink and source infinite-equilibriums are the switching bifurcations of parabola and inflection flows in the two directions and for inflection and third-order parabola flows in the two directions. The second-order inflection source/sink infinite-equilibriums are the switching bifurcations of parabola and third-order parabola flows in the two directions.

3.1 Single-Linear-Bivariate Quadratic and Cubic Vector Fields

In this section, consider dynamical systems with single-linear-bivariate quadratic and cubic fields. The corresponding stability and bifurcations of equilibriums are presented through the following theorem.

Theorem 3.1. *Consider a two-dimensional, quadratic dynamical system for* $A_{jj0} \neq 0$ $(j = j_1, j_2 \in \{1, 2\}, j_1 \neq j_2)$ *as*

$$\begin{aligned} \dot{x}_{j_1} &= f_{j_1}(\mathbf{x}, \mathbf{p}) = A_{j_1 j_1 3} + A_{j_1 j_1 2} y_{j_1} + A_{j_1 j_1 1} y_{j_1}^2 + A_{j_1 j_1 0} y_{j_1}^3, \\ \dot{x}_{j_2} &= f_{j_2}(\mathbf{x}, \mathbf{p}) = A_{j_2 j_2 3} + A_{j_2 j_2 2} y_{j_1} + A_{j_2 j_2 1} y_{j_1}^2 + A_{j_2 j_2} y_{j_1}^3, \end{aligned} \tag{3.1}$$

with

$$\begin{aligned} y_{j_1} &= \alpha_{j_1 j_1} x_{j_1} + \alpha_{j_1 j_2} x_{j_2}, \quad y_{j_1}^{\perp} = \beta_{j_1 j_1} x_{j_1} + \beta_{j_1 j_2} x_{j_2}; \\ \mathbf{n}_{j_1} &= (\alpha_{j_1 j_1}, \alpha_{j_1 j_2})^{\mathrm{T}}, \quad \mathbf{n}_{j_1}^{\perp} = (\beta_{j_1 j_1}, \beta_{j_1 j_2})^{\mathrm{T}}; \\ \mathbf{n}_{j_1} \cdot \mathbf{n}_{j_1}^{\perp} &= \alpha_{j_1 j_1} \beta_{j_1 j_1} + \alpha_{j_1 j_2} \beta_{j_1 j_2} = 0. \end{aligned} \tag{3.2}$$

For $A_{j_2 j_2 0} = 0$ *with* $(A_{j_2 j_2 1}, A_{j_2 j_2 2}, A_{j_2 j_2 3} \neq 0)$, *the two-dimensional dynamical system is*

$$\begin{aligned} \dot{x}_{j_2} &= a_{j_2 j_2 0}(y_{j_1}^2 + B_{j_2 j_2} y_{j_1} + C_{j_2 j_2}), \\ \dot{x}_{j_1} &= a_{j_1 j_1 0}(y_{j_1} - b_{j_1 j_1 1})(y_{j_1}^2 + B_{j_1 j_1} y_{j_1} + C_{j_1 j_1}), \end{aligned} \tag{3.3}$$

with

$$\begin{aligned} &\Delta_{j_2 j_2} = (B_{j_2 j_2})^2 - 4C_{j_2 j_2}, \quad \Delta_{j_1 j_1} = (B_{j_1 j_1})^2 - 4C_{j_1 j_1}, \\ &a_{j_1 j_1 0} = A_{j_1 j_1 0}, \quad a_{j_2 j_2 0} = A_{j_2 j_2 1}. \end{aligned} \tag{3.4}$$

(i) If

$$\Delta_{j_1 j_1} < 0 \quad \text{and} \quad \Delta_{j_2 j_2} < 0, \tag{3.5}$$

then

$$\begin{aligned} \dot{x}_{j_1} &= a_{j_1 j_1 0}(y_{j_1} - a_{j_1 j_1 1})[(y_{j_1} - a_{j_1 j_1})^2 + b_{j_1 j_1}], \\ \dot{x}_{j_2} &= a_{j_2 j_2 0}[(y_{j_1} - a_{j_2 j_2})^2 + b_{j_2 j_2}], \end{aligned} \tag{3.6}$$

where

$$\begin{aligned} a_{j_2 j_2} &= -\frac{1}{2} B_{j_2 j_2}, \quad b_{j_2 j_2} = -\frac{1}{4} \Delta_{j_2 j_2}, \\ a_{j_1 j_1} &= -\frac{1}{2} B_{j_1 j_1}, \quad b_{j_1 j_1} = -\frac{1}{4} \Delta_{j_2 j_2}, a_{j_1 j_1 1} = b_{j_1 j_1 1}. \end{aligned} \tag{3.7}$$

There is a dynamical system in the y_{j_1}*-direction as*

$$\begin{aligned} \dot{y}_{j_1} = {} & \alpha_{j_1 j_1} a_{j_1 j_1 0}(y_{j_1} - a_{j_1 j_1 1})[(y_{j_1} - a_{j_1 j_1})^2 + b_{j_1 j_1}] \\ & + \alpha_{j_1 j_2} a_{j_2 j_2 0}[(y_{j_1} - a_{j_2 j_2})^2 + b_{j_2 j_2}], \end{aligned} \tag{3.8}$$

with a dynamical system in the $y_{j_1}^{\perp}$*-direction as*

$$\begin{aligned} \dot{y}_{j_1}^{\perp} = {} & \beta_{j_1 j_1} a_{j_1 j_1 0}(y_{j_1} - a_{j_1 j_1 1})[(y_{j_1} - a_{j_1 j_1})^2 + b_{j_1 j_1}] \\ & + \beta_{j_1 j_2} a_{j_2 j_2 0}[(y_{j_1} - a_{j_2 j_2})^2 + b_{j_2 j_2}]. \end{aligned} \tag{3.9}$$

The flow of $y_1^* = a_{j_1 j_1 1}$ *with* $\dot{x}_{j_2}^* = a_{j_2 j_2 0}[(a_{j_1 j_1 1} - a_{j_2 j_2})^2 + b_{j_2 j_2}]$ *possesses the following properties:*

- *For* $a_{j_1j_10}\alpha_{j_1j_2} > 0$ *and* $a_{j_2j_20} > 0$,

$$(a_{j_1j_11}, \dot{x}^*_{j_2}) = \underbrace{(\text{UP}, \text{pF})}_{\text{up-parabola flow } (+)}. \tag{3.10}$$

- *For* $a_{j_1j_10}\alpha_{j_1j_2} < 0$ *and* $a_{j_2j_20} > 0$,

$$(a_{j_1j_1}, \dot{x}^*_{j_2}) = \underbrace{(\text{DP}, \text{pF})}_{\text{down-parabola flow } (+)}. \tag{3.11}$$

- *For* $a_{j_1j_10}\alpha_{j_1j_2} > 0$ *and* $a_{j_2j_20} < 0$,

$$(a_{j_1j_1}, \dot{x}^*_{j_2}) = \underbrace{(\text{DP}, \text{nF})}_{\text{down-parabola flow } (-)}. \tag{3.12}$$

- *For* $a_{j_1j_10}\alpha_{j_1j_2} < 0$ *and* $a_{j_2j_20} < 0$,

$$(a_{j_1j_11}, \dot{x}^*_{j_2}) = \underbrace{(\text{UP}, \text{nF})}_{\text{up-parabola flow } (-)}. \tag{3.13}$$

(ii) If

$$\Delta_{j_1j_1} = 0 \quad \text{and} \quad \Delta_{j_2j_2} < 0, \tag{3.14}$$

then

$$\begin{aligned} \dot{x}_{j_1} &= a_{j_1j_10}(y_{j_1} - a_{j_1j_1s_1})(y_{j_1} - a_{j_1j_1s_2})^2, \\ \dot{x}_{j_2} &= a_{j_2j_20}[(y_{j_1} - a_{j_2j_2})^2 + b_{j_2j_2}], \end{aligned} \tag{3.15}$$

where

$$\begin{aligned} a_{j_2j_2} &= -\frac{1}{2}B_{j_2j_2}, \quad b_{j_2j_2} = -\frac{1}{4}\Delta_{j_2j_2}; \\ a_{j_1j_1s_w} &= -\frac{1}{2}B_{j_1j_1}, \quad b_{j_1j_1} = 0, \quad a_{j_1j_1s_1} = b_{j_1j_1}. \end{aligned} \tag{3.16}$$

(ii_1) For $a_{j_1j_1s_1} \neq a_{j_1j_1s_2}$, *there is a dynamical system in the* y_{j_1}*-direction as*

$$\begin{aligned} \dot{y}_{j_1} &= \alpha_{j_1j_1}a_{j_1j_10}(y_{j_1} - a_{j_1j_1s_1})(y_{j_1} - a_{j_1j_1s_2})^2 \\ &\quad + \alpha_{j_1j_2}a_{j_2j_20}[(y_{j_1} - a_{j_2j_2})^2 + b_{j_2j_2}], \end{aligned} \tag{3.17}$$

with a dynamical system in the $y^{\perp}_{j_1}$*-direction as*

$$\begin{aligned} \dot{y}^{\perp}_{j_1} &= \beta_{j_1j_1}a_{j_1j_10}(y_{j_1} - a_{j_1j_1s_1})(y_{j_1} - a_{j_1j_1s_2})^2 \\ &\quad + \beta_{j_1j_2}a_{j_2j_20}[(y_{j_1} - a_{j_2j_2})^2 + b_{j_2j_2}]. \end{aligned} \tag{3.18}$$

(ii_{1a}) The flow of $y^*_1 = a_{j_1j_1s_1} \neq a_{j_1j_1s_2}$ *with* $\dot{x}^*_{j_2} = a_{j_2j_20}[(a_{j_1j_1s_1} - a_{j_2j_2})^2 + b_{j_2j_2}]$ *possesses the following properties:*

- *For* $a_{j_1j_10}\alpha_{j_1j_2}(a_{j_1j_1s_1}-a_{j_1j_1s_2})^2>0$ *and* $a_{j_2j_20}>0$,

$$(a_{j_1j_1s_1},\dot{x}_{j_2}^*)=\underbrace{(\mathrm{UP},\mathrm{pF})}_{\text{up-parabola flow }(+)}. \tag{3.19}$$

- *For* $a_{j_1j_10}\alpha_{j_1j_2}(a_{j_1j_1s_1}-a_{j_1j_1s_2})^2<0$ *and* $a_{j_2j_20}>0$,

$$(a_{j_1j_1s_1},\dot{x}_{j_2}^*)=\underbrace{(\mathrm{DP},\mathrm{pF})}_{\text{down-parabola flow }(+)}. \tag{3.20}$$

- *For* $a_{j_1j_10}\alpha_{j_1j_2}(a_{j_1j_1s_1}-a_{j_1j_1s_2})^2>0$ *and* $a_{j_2j_20}<0$,

$$(a_{j_1j_1s_1},\dot{x}_{j_2}^*)=\underbrace{(\mathrm{DP},\mathrm{nF})}_{\text{down-parabola flow }(-)}. \tag{3.21}$$

- *For* $a_{j_1j_10}\alpha_{j_1j_2}(a_{j_1j_1s_1}-a_{j_1j_1s_2})^2<0$ *and* $a_{j_2j_20}<0$,

$$(a_{j_1j_1s_1},\dot{x}_{j_2}^*)=\underbrace{(\mathrm{UP},\mathrm{nF})}_{\text{up-parabola flow }(-)}, \tag{3.22}$$

(ii_{1b}) *The flow of* $y_1^*=a_{j_1j_1s_2}\neq a_{j_1j_1s_1}$ *with* $\dot{x}_{j_2}^*=a_{j_2j_20}[(a_{j_1j_1s_2}-a_{j_2j_2})^2+b_{j_2j_2}]$ *possesses the following properties:*

- *For* $a_{j_1j_10}\alpha_{j_1j_2}^2(a_{j_1j_1s_2}-a_{j_1j_1s_1})>0$ *and* $a_{j_2j_20}>0$,

$$(a_{j_1j_1s_2},\dot{x}_{j_2}^*)=\underbrace{(\mathrm{II},\mathrm{pF})}_{\text{increasing-inflection flow }(+)}. \tag{3.23}$$

- *For* $a_{j_1j_10}\alpha_{j_1j_2}^2(a_{j_1j_1s_2}-a_{j_1j_1s_1})<0$ *and* $a_{j_2j_20}>0$,

$$(a_{j_1j_1s_2},\dot{x}_{j_2}^*)=\underbrace{(\mathrm{DI},\mathrm{pF})}_{\text{decreasing-inflection flow }(+)}. \tag{3.24}$$

- *For* $a_{j_1j_10}\alpha_{j_1j_2}^2(a_{j_1j_1s_2}-a_{j_1j_1s_1})>0$ *and* $a_{j_2j_20}<0$,

$$(a_{j_1j_1s_2},\dot{x}_{j_2}^*)=\underbrace{(\mathrm{DI},\mathrm{nF})}_{\text{decreasing-inflection flow }(-)}. \tag{3.25}$$

- *For* $a_{j_1j_10}\alpha_{j_1j_2}^2(a_{j_1j_1s_2}-a_{j_1j_1s_1})<0$ *and* $a_{j_2j_20}<0$,

$$(a_{j_1j_1s_2},\dot{x}_{j_2}^*)=\underbrace{(\mathrm{II},\mathrm{nF})}_{\text{increasing-inflection flow }(-)}. \tag{3.26}$$

– *The inflection flows are the appearing bifurcations of up-parabola and down-parabola flows.*

(*ii*$_2$) *For* $a_{j_1j_1s_1} = a_{j_1j_1s_2} = a_{j_1j_1 1}$,

$$\begin{aligned}\dot{x}_{j_1} &= a_{j_1j_1 0}(y_{j_1} - a_{j_1j_1})^3,\\ \dot{x}_{j_2} &= a_{j_2j_2 0}[(y_{j_1} - a_{j_2j_2})^2 + b_{j_2j_2}].\end{aligned} \tag{3.27}$$

There is a dynamical system in the y_{j_1}*-direction as*

$$\dot{y}_{j_1} = \alpha_{j_1j_1}a_{j_1j_1 0}(y_{j_1} - a_{j_1j_1 1})^3 + \alpha_{j_1j_2}a_{j_2j_2 0}[(y_{j_1} - a_{j_2j_2})^2 + b_{j_2j_2}], \tag{3.28}$$

with a dynamical system in the $y_{j_1}^{\perp}$*-direction as*

$$\dot{y}_{j_1}^{\perp} = \beta_{j_1j_1}a_{j_1j_1 0}(y_{j_1} - a_{j_1j_1 1})^3 + \beta_{j_1j_2}a_{j_2j_2 0}[(y_{j_1} - a_{j_2j_2})^2 + b_{j_2j_2}]. \tag{3.29}$$

The flow of $y_1^* = a_{j_1j_1 1}$ *with* $\dot{x}_{j_2}^* = a_{j_2j_2 0}[(a_{j_1j_1 1} - a_{j_2j_2})^2 + b_{j_2j_2}]$ *possesses the following properties*:

- *For* $a_{j_1j_1 0}\alpha_{j_1j_2}^3 > 0$ *and* $a_{j_2j_2 0} > 0$,

$$(a_{j_1j_1}, \dot{x}_{j_2}^*) = \underbrace{(3^{\text{rd}}\text{UP}, \text{pF})}_{\text{third-order up-parabola flow } (+)}. \tag{3.30}$$

- *For* $a_{j_1j_1 0}\alpha_{j_1j_2}^3 < 0$ *and* $a_{j_2j_2 0} > 0$,

$$(a_{j_1j_1}, \dot{x}_{j_2}^*) = \underbrace{(3^{\text{rd}}\text{DP}, \text{pF})}_{\text{third-order down-parabola flow } (+)}. \tag{3.31}$$

- *For* $a_{j_1j_1 0}\alpha_{j_1j_2}^3 > 0$ *and* $a_{j_2j_2 0} < 0$,

$$(a_{j_1j_1 1}, \dot{x}_{j_2}^*) = \underbrace{(3^{rd}\text{DP}, \text{nF})}_{\text{third-order down-parabola flow } (-)}. \tag{3.32}$$

- *For* $a_{j_1j_1 0}\alpha_{j_1j_2}^3 < 0$ *and* $a_{j_2j_2 0} < 0$,

$$(a_{j_1j_1}, \dot{x}_{j_2}^*) = \underbrace{(3^{\text{rd}}\text{UP}, \text{nF})}_{\text{third-order up-parabola flow } (-)}. \tag{3.33}$$

- *The third-order up-parabola flows are the appearing bifurcations of upparabola, down-parabola, and up-parabola flows.*
- *The third-order down-parabola flows are the appearing bifurcations of down-parabola, up-parabola, and down-parabola flows.*

(iii) If

$$\Delta_{j_1 j_1} > 0 \quad \text{and} \quad \Delta_{j_2 j_2} < 0, \tag{3.34}$$

then

$$\begin{aligned} \dot{x}_{j_1} &= a_{j_1 j_1 0}(y_{j_1} - a_{j_1 j_1 1})(y_{j_1} - a_{j_1 j_1 2})(y_{j_1} - a_{j_1 j_1 3}), \\ \dot{x}_{j_2} &= a_{j_2 j_2 0}[(y_{j_1} - a_{j_2 j_2})^2 + b_{j_2 j_2}], \end{aligned} \tag{3.35}$$

where

$$\begin{aligned} &a_{j_2 j_2} = -\frac{1}{2}B_{j_2 j_2}, b_{j_2 j_2} = -\frac{1}{4}\Delta_{j_2 j_2}; \\ &\{a_{j_1 j_1 1}, a_{j_1 j_1 2}, a_{j_1 j_1 3}\} = \text{sort}\{b_{j_1 j_1 1}, b_{j_1 j_1 2}, b_{j_1 j_1 3}\}, \\ &\quad a_{j_1 j_1 s_1} < a_{j_1 j_1 s_2}, s_1 < s_2; \\ &b_{j_1 j_1 2}, b_{j_1 j_1 3} = -\frac{1}{2}(B_{j_1 j_1} \pm \sqrt{\Delta_{j_1 j_1}}), a_{j_1 j_1 0} = A_{j_1 j_1 0}. \end{aligned} \tag{3.36}$$

There is a dynamical system in the y_{j_1}-direction as

$$\dot{y}_{j_1} = \alpha_{j_1 j_1} a_{j_1 j_1 0} \prod_{s_1=1}^{3} (y_{j_1} - a_{j_1 j_1 s_1}) + \alpha_{j_1 j_2} a_{j_2 j_2 0}[(y_{j_1} - a_{j_2 j_2 1})^2 + b_{j_2 j_2}], \tag{3.37}$$

with a dynamical system in the $y_{j_1}^{\perp}$-direction as

$$\dot{y}_{j_1}^{\perp} = \beta_{j_1 j_1} a_{j_1 j_1 0} \prod_{s_1=1}^{3} (y_{j_1} - a_{j_1 j_1 s_1}) + \beta_{j_1 j_2} a_{j_2 j_2 0}[(y_{j_1} - a_{j_2 j_2 1})^2 + b_{j_2 j_2}]. \tag{3.38}$$

With

$$\dot{x}_{j_2}^{*} = a_{j_2 j_2 0}[(a_{j_1 j_1 s_1} - a_{j_2 j_2})^2 + b_{j_2 j_2}], \tag{3.39}$$

the equilibrium flow of $y_{j_1}^{*} = a_{j_1 j_1 s_1}(s_1, s_2 \in \{1,2,3\}, s_1 \neq s_2)$ in the x_{j_1}-direction possesses the following properties:

- *For $a_{j_1 j_1 0}\alpha_{j_1 j_2} \prod_{s_2=1, s_2 \neq s_1}^{3} (a_{j_1 j_1 s_1} - a_{j_1 j_1 s_2}) > 0$ and $a_{j_2 j_2 0} > 0$,*
$$(a_{j_1 j_1 s_1}, \dot{x}_{j_2}^{*}) = \underbrace{(\text{UP}, \text{pF})}_{\text{up-parabola flow }(+)}. \tag{3.40}$$
- *For $a_{j_1 j_1 0}\alpha_{j_1 j_2} \prod_{s_2=1, s_2 \neq s_1}^{3} (a_{j_1 j_1 s_1} - a_{j_1 j_1 s_2}) < 0$ and $a_{j_2 j_2 0} > 0$,*
$$(a_{j_1 j_1 s_1}, \dot{x}_{j_2}^{*}) = \underbrace{(\text{DP}, \text{pF})}_{\text{down-parabola flow }(+)}. \tag{3.41}$$

- *For* $a_{j_1j_10}\alpha_{j_1j_2}\prod_{s_2=1,s_2\neq s_1}^{3}(a_{j_1j_1s_1}-a_{j_1j_1s_2})>0$ *and* $a_{j_2j_20}<0$,

$$(a_{j_1j_1s_1},\dot{x}_{j_2}^{*})=\underbrace{(\mathrm{DP},\mathrm{nF})}_{\text{down-parabola flow }(-)}. \tag{3.42}$$

- *For* $a_{j_1j_10}\alpha_{j_1j_2}\prod_{s_2=1,s_2\neq s_1}^{3}(a_{j_1j_1s_1}-a_{j_1j_1s_2})<0$ *and* $a_{j_2j_20}<0$,

$$(a_{j_1,j_1s_1},\dot{x}_{j_2}^{*})=\underbrace{(\mathrm{UP},\mathrm{nF})}_{\text{up-parabola flow }(-)}. \tag{3.43}$$

(*iv*) *If*

$$\Delta_{j_1j_1}<0 \quad \text{and} \quad \Delta_{j_2j_2}=0, \tag{3.44}$$

then

$$\begin{aligned}\dot{x}_{j_1}&=a_{j_1j_0}(y_{j_1}-a_{j_1j_11})[(y_{j_1}-a_{j_1j_1})^2+b_{j_1j_1}],\\ \dot{x}_{j_2}&=a_{j_2j_20}(y_{j_1}-a_{j_2j_21})^2,\end{aligned} \tag{3.45}$$

where

$$\begin{aligned}&a_{j_2j_21}=-\frac{1}{2}B_{j_2j_2},\quad b_{j_2j_2}=0,\quad a_{j_2j_20}=A_{j_2j_21};\\ &a_{j_1j_11}=b_{j_1j_11},\quad a_{j_1j_1}=-\frac{1}{2}B_{j_1j_11},\quad b_{j_1j_1}=-\frac{1}{4}\Delta_{j_1j_1},\quad a_{j_1j_10}=A_{j_1j_10}.\end{aligned} \tag{3.46}$$

(*iv*$_1$) *For* $a_{j_1j_11}\neq a_{j_2j_21}$, *there is a dynamical system in the* y_{j_1}*-direction as*

$$\begin{aligned}\dot{y}_{j_1}&=\alpha_{j_1j_1}a_{j_1j_10}(y_{j_1}-a_{j_1j_11})[(y_{j_1}-a_{j_1j_1})^2+b_{j_1j_1}]\\ &\quad+\alpha_{j_1j_2}a_{j_2j_20}(y_{j_1}-a_{j_2j_2})^2,\end{aligned} \tag{3.47}$$

with a dynamical system in the $y_{j_1}^{\perp}$*-direction as*

$$\begin{aligned}\dot{y}_{j_1}^{\perp}&=\beta_{j_1j_1}a_{j_1j_10}(y_{j_1}-a_{j_1j_11})[(y_{j_1}-a_{j_1j_1})^2+b_{j_1j_1}]\\ &\quad+\beta_{j_1j_2}a_{j_2j_20}(y_{j_1}-a_{j_2j_2})^2.\end{aligned} \tag{3.48}$$

(*iv*$_{1a}$) *The flow of* $y_1^*=a_{j_1j_11}$ *with* $\dot{x}_{j_2}^*=a_{j_2j_20}(a_{j_1j_11}-a_{j_2j_21})^2$ *possesses the following properties*:

- *For* $\alpha_{j_1j_2}a_{j_1j_10}>0$ *and* $a_{j_2j_20}(a_{j_1j_11}-a_{j_2j_21})^2>0$,

$$(a_{j_1j_1},\dot{x}_{j_2}^{*})=\underbrace{(\mathrm{UP},\mathrm{pF})}_{\text{up-parabola flow }(+)}. \tag{3.49}$$

- *For* $\alpha_{j_1j_2}a_{j_1j_10} < 0$ *and* $a_{j_2j_20}(a_{j_1j_11} - a_{j_2j_21})^2 > 0$,

$$(a_{j_1j_11}, \dot{x}^*_{j_2}) = \underbrace{(\mathrm{DP}, \mathrm{pF})}_{\text{down-parabola flow } (+)}. \tag{3.50}$$

- *For* $\alpha_{j_1j_2}a_{j_1j_10} > 0$ *and* $a_{j_2j_20}(a_{j_1j_11} - a_{j_2j_21})^2 < 0$,

$$(a_{j_1j_1}, \dot{x}^*_{j_2}) = \underbrace{(\mathrm{DP}, \mathrm{nF})}_{\text{down-parabola flow } (-)}. \tag{3.51}$$

- *For* $\alpha_{j_1j_2}a_{j_1j_10} < 0$ *and* $a_{j_2j_20}(a_{j_1j_11} - a_{j_2j_21})^2 < 0$,

$$(a_{j_1j_11}, \dot{x}^*_{j_2}) = \underbrace{(\mathrm{UP}, \mathrm{nF})}_{\text{up-parabola flow } (-)}. \tag{3.52}$$

(iv_{1b}) *The flow of* $y_1^* = a_{j_2j_21}$ *with* $\dot{x}^*_{j_1} = a_{j_1j_10}(a_{j_2j_21} - a_{j_1j_11})[(a_{j_2j_21} - a_{j_1j_1})^2 + b_{j_1j_1}]$ *possesses the following properties*:

- *For* $a_{j_1j_10}(a_{j_2j_21} - a_{j_1j_11}) > 0$ *and* $a_{j_2j_20}\alpha^2_{j_1j_1} > 0$,

$$(\dot{x}^*_{j_1}, a_{j_2j_21}) = \underbrace{(\mathrm{pF}, \mathrm{II})}_{\text{increasing-inflection flow } (+)}. \tag{3.53}$$

- *For* $a_{j_1j_10}(a_{j_2j_21} - a_{j_1j_11}) < 0$ *and* $a_{j_2j_20}\alpha^2_{j_1j_1} > 0$,

$$(\dot{x}^*_{j_1}, a_{j_2j_21}) = \underbrace{(\mathrm{nF}, \mathrm{DI})}_{\text{decreasing-inflection flow } (-)}. \tag{3.54}$$

- *For* $a_{j_1j_10}(a_{j_2j_21} - a_{j_1j_11}) > 0$ *and* $a_{j_2j_20}\alpha^2_{j_1j_1} < 0$,

$$(\dot{x}^*_{j_1}, a_{j_2j_21}) = \underbrace{(\mathrm{pF}, \mathrm{DI})}_{\text{decreasing-inflection flow } (+)}. \tag{3.55}$$

- *For* $a_{j_1j_10}(a_{j_2j_21} - a_{j_1j_11}) < 0$ *and* $a_{j_2j_20}\alpha^2_{j_1j_1} < 0$,

$$(\dot{x}^*_{j_1}, a_{j_2j_21}) = \underbrace{(\mathrm{nF}, \mathrm{II})}_{\text{increasing-inflection flow } (-)}. \tag{3.56}$$

(iv₂) For $a_{j_1j_11} = a_{j_2j_21}$,

$$\begin{aligned}\dot{x}_{j_1} &= a_{j_1j_10}(y_{j_1} - a_{j_1j_11})[(y_{j_1} - a_{j_1j_1})^2 + b_{j_1j_1}],\\ \dot{x}_{j_2} &= a_{j_2j_20}(y_{j_1} - a_{j_2j_2})^2.\end{aligned} \tag{3.57}$$

There is a dynamical system in the y_{j_1}*-direction as*

$$\begin{aligned}\dot{y}_{j_1} = (y_{j_1} - a_{j_1j_1})\{&\alpha_{j_1j_1}a_{j_1j_10}[(y_{j_1} - a_{j_1j_1})^2\\ &+ b_{j_1j_1}] + \alpha_{j_1j_2}a_{j_2j_20}(y_{j_1} - a_{j_2j_21})\}.\end{aligned} \tag{3.58}$$

with a dynamical system in the $y_{j_1}^{\perp}$*-direction as*

$$\begin{aligned}\dot{y}_{j_1}^{\perp} = (y_{j_1} - a_{j_1j_1})\{&\beta_{j_1j_1}a_{j_1j_10}[(y_{j_1} - a_{j_1j_1})^2 + b_{j_1j_1}]\\ &+\beta_{j_1j_2}a_{j_2j_20}(y_{j_1} - a_{j_2j_21})]\},\end{aligned} \tag{3.59}$$

where

$$\begin{aligned}G_{j_1j_10} &= \frac{\partial}{\partial y_{j_1}}\left[\mathbf{n}^{\mathrm{T}} \cdot \mathbf{F}(y_{j_1}, \mathbf{p})\right]\Big|_{y_{j_1}^* = a_{j_1j_11} = a_{j_2j_21}}\\ &= \alpha_{j_1j_1}a_{j_1j_10}\left[(a_{j_2j_21} - a_{j_1j_1})^2 + b_{j_1j_1}\right].\end{aligned} \tag{3.60}$$

The infinite-equilibrium flow of $y_1^* = a_{j_1j_11} = a_{j_2j_21}$ *possesses the following properties*:

- *For* $\alpha_{j_1j_1}a_{j_1j_10} > 0$ *and* $a_{j_2j_20}\alpha_{j_1j_1}^2 > 0$,

$$(a_{j_1j_11}, a_{j_2j_21}) = \underbrace{(\mathrm{SO}, \mathrm{UP})}_{\text{up-parabola source}}. \tag{3.61}$$

- *For* $\alpha_{j_1j_1}a_{j_1j_10} < 0$ *and* $a_{j_2j_20}\alpha_{j_1j_1}^2 > 0$,

$$(a_{j_1j_11}, a_{j_2j_21}) = \underbrace{(\mathrm{SI}, \mathrm{DP})}_{\text{down-parabola sink}}. \tag{3.62}$$

- *For* $\alpha_{j_1j_1}a_{j_1j_10} > 0$ *and* $a_{j_2j_20}\alpha_{j_1j_1}^2 < 0$,

$$(a_{j_1j_11}, a_{j_2j_21}) = \underbrace{(\mathrm{SO}, \mathrm{DP})}_{\text{down-parabola source}}. \tag{3.63}$$

- *For* $\alpha_{j_1j_1}a_{j_1j_10} < 0$ *and* $a_{j_2j_20}\alpha_{j_1j_1}^2 < 0$,

$$(a_{j_1j_11}, a_{j_2j_21}) = \underbrace{(\mathrm{SI}, \mathrm{UP})}_{\text{up-parabola sink}}. \tag{3.64}$$

(v) If

$$\Delta_{j_1j_1} = 0 \quad \text{and} \quad \Delta_{j_2j_2} = 0, \tag{3.65}$$

then

$$\begin{aligned} \dot{x}_{j_1} &= a_{j_1j_10}(y_{j_1} - a_{j_1j_1s_1})(y_{j_1} - a_{j_1j_1s_2})^2, \\ \dot{x}_{j_2} &= a_{j_2j_20}(y_{j_1} - a_{j_2j_21})^2, \\ &j_1, j_2 \in \{1,2\};\ j_1 \neq j_2, \end{aligned} \tag{3.66}$$

where

$$a_{j_2j_21} = -\frac{1}{2}B_{j_2j_2}, \quad b_{j_2j_2} = 0, \quad a_{j_2j_20} = A_{j_2j_21};$$
$$a_{j_1j_1s_1} = b_{j_1j_11}, \quad a_{j_1j_1s_2} = -\frac{1}{2}B_{j_1j_11}, \quad b_{j_1j_1} = 0, \quad a_{j_1j_10} = A_{j_1j_10}. \tag{3.67}$$

(v_1) For $a_{j_2j_21} \neq a_{j_1j_1s_1}, a_{j_1j_1s_2}$, there is a dynamical system in the y_{j_1}-direction as

$$\dot{y}_{j_1} = \alpha_{j_1j_1}a_{j_1j_10}(y_{j_1} - a_{j_1j_1s_1})(y_{j_1} - a_{j_1j_1s_2})^2 + \alpha_{j_1j_2}a_{j_2j_20}(y_{j_1} - a_{j_2j_21})^2 \tag{3.68}$$

with a dynamical system in the $y_{j_1}^{\perp}$-direction as

$$\dot{y}_{j_1}^{\perp} = \beta_{j_1j_1}a_{j_1j_10}(y_{j_1} - a_{j_1j_1s_1})(y_{j_1} - a_{j_1j_1s_2})^2 + \beta_{j_1j_2}a_{j_2j_20}(y_{j_1} - a_{j_2j_2})^2. \tag{3.69}$$

(v_{1a}) The flow of $y_{j_1}^* = a_{j_1j_1s_1} \neq a_{j_2j_21}$ with $\dot{x}_{j_2}^* = a_{j_2j_20}(a_{j_1j_1s_1} - a_{j_2j_21})^2$ for the x_{j_1}-direction has the following properties:

- *For $a_{j_1j_10}\alpha_{j_1j_2}(a_{j_1j_1s_1} - a_{j_1j_1s_2})^2 > 0$ and $a_{j_2j_20}(a_{j_1j_1s_1} - a_{j_2j_21})^2 > 0$,*

$$(a_{j_1j_1s_1}, \dot{x}_{j_2}^*) = \underbrace{(\mathrm{UP}, \mathrm{pF})}_{\text{up-parabola flow } (+)}. \tag{3.70}$$

- *For $a_{j_1j_10}\alpha_{j_1j_2}(a_{j_1j_1s_1} - a_{j_1j_1s_2})^2 < 0$ and $a_{j_2j_20}(a_{j_1j_1s_1} - a_{j_2j_21})^2 > 0$,*

$$(a_{j_1j_1s_1}, \dot{x}_{j_2}^*) = \underbrace{(\mathrm{DP}, \mathrm{pF})}_{\text{down-parabola flow } (+)}. \tag{3.71}$$

- *For $a_{j_1j_10}\alpha_{j_1j_2}(a_{j_1j_1s_1} - a_{j_1j_1s_2})^2 > 0$ and $a_{j_2j_20}(a_{j_1j_1s_1} - a_{j_2j_21})^2 < 0$,*

$$(a_{j_1j_1s_1}, \dot{x}_{j_2}^*) = \underbrace{(\mathrm{DP}, \mathrm{nF})}_{\text{down-parabola flow } (-)}. \tag{3.72}$$

- *For* $a_{j_1j_10}\alpha_{j_1j_2}(a_{j_1j_1s_1} - a_{j_1j_1s_2})^2 < 0$ *and* $a_{j_2j_20}(a_{j_1j_1s_1} - a_{j_2j_21})^2 < 0$,

$$(a_{j_1j_1s_1}, \dot{x}^*_{j_2}) = \underbrace{(\mathrm{UP}, \mathrm{nF})}_{\text{up-parabola flow } (-)}. \tag{3.73}$$

(v_{1b}) *The flow of* $y^*_{j_1} = a_{j_1j_1s_2} \neq a_{j_2j_21}$ *with* $\dot{x}^*_{j_2} = a_{j_2j_20}(a_{j_1j_1s_1} - a_{j_2j_21})^2$ *for the* x_{j_1}*-direction with* $a_{j_1j_11} \neq a_{j_2j_21}$ *has the following properties:*

- *For* $a_{j_1j_10}\alpha^2_{j_1j_2}(a_{j_1j_1s_2} - a_{j_1j_1s_1}) > 0$ *and* $a_{j_2j_20}(a_{j_1j_1s_2} - a_{j_2j_21})^2 > 0$,

$$(a_{j_1j_1s_2}, \dot{x}^*_{j_2}) = \underbrace{(\mathrm{II}, \mathrm{pF})}_{\text{increasing-inflection flow } (+)}. \tag{3.74}$$

- *For* $a_{j_1j_10}\alpha^2_{j_1j_2}(a_{j_1j_1s_2} - a_{j_1j_1s_1}) > 0$ *and* $a_{j_2j_20}(a_{j_1j_1s_2} - a_{j_2j_21})^2 > 0$,

$$(a_{j_1j_1s_2}, \dot{x}^*_{j_2}) = \underbrace{(\mathrm{DI}, \mathrm{pF})}_{\text{decreasing-inflection flow } (+)}. \tag{3.75}$$

- *For* $a_{j_1j_10}\alpha^2_{j_1j_2}(a_{j_1j_1s_2} - a_{j_1j_1s_1}) > 0$ *and* $a_{j_2j_20}(a_{j_1j_1s_2} - a_{j_2j_2})^2 < 0$,

$$(a_{j_1j_1s_2}, \dot{x}^*_{j_2}) = \underbrace{(\mathrm{DI}, \mathrm{nF})}_{\text{decreasing-inflection flow } (-)}. \tag{3.76}$$

- *For* $a_{j_1j_10}\alpha^2_{j_1j_2}(a_{j_1j_1s_2} - a_{j_1j_1s_1}) < 0$ *and* $a_{j_2j_20}(a_{j_1j_1s_2} - a_{j_2j_21})^2 < 0$,

$$(a_{j_1j_1s_2}, \dot{x}^*_{j_2}) = \underbrace{(\mathrm{II}, \mathrm{nF})}_{\text{increasing-inflection flow } (-)}. \tag{3.77}$$

– *The increasing- and decreasing-inflection flows are the appearing bifurcations for the up-parabola and down-parabola flows.*

(v_{1c}) *The flow of* $y^*_{j_1} = a_{j_2j_21} \neq a_{j_1j_1s_1}, a_{j_1j_1s_2}$ *for the* x_{j_2}*-direction has the following properties with* $\dot{x}^*_{j_1} = a_{j_1j_10}(a_{j_2j_21} - a_{j_1j_1s_1})(a_{j_2j_21} - a_{j_1j_1s_2})^2$:

- *For* $a_{j_1j_10}(a_{j_2j_21} - a_{j_1j_1s_1})(a_{j_2j_21} - a_{j_1j_1s_2})^2 > 0$ *and* $a_{j_2j_20}\ \alpha^2_{j_1j_1} > 0$,

$$(\dot{x}^*_{j_1}, a_{j_2j_21}) = \underbrace{(\mathrm{pF}, \mathrm{II})}_{\text{increasing-inflection flow } (+)}. \tag{3.78}$$

- *For* $a_{j_1j_10}(a_{j_2j_21} - a_{j_1j_1s_1})(a_{j_2j_21} - a_{j_1j_1s_2})^2 < 0$ *and* $a_{j_2j_20}\alpha^2_{j_1j_1} > 0$,

$$(\dot{x}^*_{j_1}, a_{j_2j_21}) = \underbrace{(\mathrm{nF}, \mathrm{DI})}_{\text{decreasing-inflection flow } (-)}. \tag{3.79}$$

- *For* $a_{j_1j_10}(a_{j_2j_21}-a_{j_1j_1s_1})(a_{j_2j_21}-a_{j_1j_1s_2})^2>0$ *and* $a_{j_2j_20}\alpha^2_{j_1j_1}<0$,

$$(\dot{x}^*_{j_1},a_{j_2j_21})=\underbrace{(\mathrm{pF},\mathrm{DI})}_{\text{decreasing-inflection flow }(+)}. \quad (3.80)$$

- *For* $a_{j_1j_10}(a_{j_2j_21}-a_{j_1j_1s_1})(a_{j_2j_21}-a_{j_1j_1s_2})^2<0$ *and* $a_{j_2j_20}\alpha^2_{j_1j_1}<0$

$$(\dot{x}^*_{j_1},a_{j_2j_21})=\underbrace{(\mathrm{nF},\mathrm{II})}_{\text{increasing-inflection flow }(-)}. \quad (3.81)$$

– *The increasing- and decreasing-inflection flows are the appearing bifurcations for the up-parabola and down-parabola flows.*

(v_2) *For* $a_{j_1j_1s_1}=a_{j_1j_1s_2}\equiv a_{j_1j_11}$,

$$\begin{aligned}&\dot{x}_{j_1}=a_{j_1j_10}(y_{j_1}-a_{j_1j_1})^3,\\&\dot{x}_{j_2}=a_{j_2j_20}(y_{j_1}-a_{j_2j_21})^2,\\&j_1,j_2\in\{1,2\};\ j_1\neq j_2.\end{aligned} \quad (3.82)$$

For $a_{j_2j_21}\neq a_{j_1j_11}$, *there is a dynamical system in the* y_{j_1}*-direction as*

$$\dot{y}_{j_1}=\alpha_{j_1j_1}a_{j_1j_10}(y_{j_1}-a_{j_1j_11})^3+\alpha_{j_1j_2}a_{j_2j_20}(y_{j_1}-a_{j_2j_21})^2, \quad (3.83)$$

with a dynamical system in the $y^{\perp}_{j_1}$*-direction as*

$$\dot{y}^{\perp}_{j_1}=\beta_{j_1j_1}a_{j_1j_10}(y_{j_1}-a_{j_1j_11})^3+\beta_{j_1j_2}a_{j_2j_20}(y_{j_1}-a_{j_2j_21})^2. \quad (3.84)$$

(v_{2a}) *The flow of* $y^*_{j_1}=a_{j_1j_11}\neq a_{j_2j_21}$ *with* $\dot{x}^*_{j_2}=a_{j_2j_20}(a_{j_1j_11}-a_{j_2j_21})^2$ *for the* x_{j_1}*-direction has the following properties*:

- *For* $a_{j_1j_10}\alpha^3_{j_1j_2}>0$ *and* $a_{j_2j_20}(a_{j_1j_11}-a_{j_2j_21})^2>0$,

$$(a_{j_1j_1}1,\dot{x}^*_{j_2})=\underbrace{(3^{\mathrm{rd}}\mathrm{UP},\mathrm{pF})}_{\text{third-order up-parabola flow }(+)}. \quad (3.85)$$

- *For* $a_{j_1j_10}\alpha^3_{j_1j_2}<0$ *and* $a_{j_2j_20}(a_{j_1j_11}-a_{j_2j_21})^2>0$,

$$(a_{j_1j_11},\dot{x}^*_{j_2})=\underbrace{(3^{\mathrm{rd}}\mathrm{DP},\mathrm{pF})}_{\text{third-order down-parabola flow }(+)}. \quad (3.86)$$

- *For* $a_{j_1j_10}\alpha^3_{j_1j_2}>0$ *and* $a_{j_2j_20}(a_{j_1j_11}-a_{j_2j_21})^2<0$,

$$(a_{j_1j_11},\dot{x}^*_{j_2})=\underbrace{(3^{\mathrm{rd}}\mathrm{DP},\mathrm{nF})}_{\text{third-order down-parabola flow }(-)}. \quad (3.87)$$

- *For* $a_{j_1j_10}\alpha^3_{j_1j_2} < 0$ *and* $a_{j_2j_20}(a_{j_1j_11} - a_{j_2j_21})^2 < 0$,

$$(a_{j_1j_11}, \dot{x}^*_{j_2}) = \underbrace{(3^{\text{rd}}\text{UP}, \text{nF})}_{\text{third-order up-parabola flow } (-)}. \tag{3.88}$$

- *The third-order up-parabola flows are the appearing bifurcations for up-parabola, down-parabola, and up-parabola flows.*
- *The third-order down-parabola flows are the appearing bifurcations for down-parabola, up-parabola, and down-parabola flows.*

(v_{2b}) *The flow of* $y^*_{j_1} = a_{j_2j_21} \neq a_{j_1j_11}$ *for the* x_{j_2}*-direction has the following properties with* $\dot{x}^*_{j_1} = a_{j_1j_10}(a_{j_2j_21} - a_{j_1j_11})^3$:

- *For* $a_{j_1j_10}(a_{j_2j_21} - a_{j_1j_11})^3 > 0$ *and* $a_{j_2j_20}\alpha^2_{j_1j_1} > 0$,

$$(\dot{x}^*_{j_1}, a_{j_2j_21}) = \underbrace{(\text{pF}, \text{II})}_{\text{increasing-infelction flow } (+)}. \tag{3.89}$$

- *For* $a_{j_1j_10}(a_{j_2j_21} - a_{j_1j_11})^3 < 0$ *and* $a_{j_2j_20}\alpha^2_{j_1j_1} > 0$,

$$(\dot{x}^*_{j_1}, a_{j_2j_21}) = \underbrace{(\text{nF}, \text{DI})}_{\text{decreasing-infelction flow } (-)}. \tag{3.90}$$

- *For* $a_{j_1j_10}(a_{j_2j_21} - a_{j_1j_11})^3 > 0$ *and* $a_{j_2j_20}\alpha^2_{j_1j_1} < 0$,

$$(\dot{x}^*_{j_1}, a_{j_2j_21}) = \underbrace{(\text{pF}, \text{DI})}_{\text{decreasing-infelction flow } (+)}. \tag{3.91}$$

- *For* $a_{j_1j_10}(a_{j_2j_21} - a_{j_1j_11})^3 < 0$ *and* $a_{j_2j_20}\alpha^2_{j_1j_1} < 0$,

$$(\dot{x}^*_{j_1}, a_{j_2j_21}) = \underbrace{(\text{nF}, \text{II})}_{\text{increasing-infelction flow } (-)}. \tag{3.92}$$

- *The increasing- and decreasing-inflection flows are the appearing bifurcations for the up-parabola and down-parabola flows.*

(v_3) *For* $a_{j_1j_1s_1} = a_{j_2j_21}$,

$$\begin{aligned} \dot{x}_{j_1} &= a_{j_1j_10}(y_{j_1} - a_{j_1j_1s_1})(y_{j_1} - a_{j_1j_1s_2})^2, \\ \dot{x}_{j_2} &= a_{j_2j_20}(y_{j_1} - a_{j_2j_21})^2. \end{aligned} \tag{3.93}$$

There is a dynamical system in the y_{j_1}*-direction as*

$$\dot{y}_{j_1} = (y_{j_1} - a_{j_1j_1s_1})[\alpha_{j_1j_1}a_{j_1j_10}(y_{j_1} - a_{j_1j_1s_2})^2 + \alpha_{j_1j_2}a_{j_2j_20}(y_{j_1} - a_{j_2j_21})], \tag{3.94}$$

with a dynamical system in the $y_{j_1}^{\perp}$-direction as

$$\begin{aligned}\dot{y}_{j_1}^{\perp} &= (y_{j_1} - a_{j_1j_1s_1})[\beta_{j_1j_1}a_{j_1j_10}(y_{j_1} - a_{j_1j_1s_2})^2],\\ &\quad + \beta_{j_1j_2}a_{j_2j_20}(y_{j_1} - a_{j_2j_2})\end{aligned} \tag{3.95}$$

where

$$\begin{aligned}G_{j_1j_10} &= \frac{\partial}{\partial y_{j_1}}[\mathbf{n}^{\mathrm{T}} \cdot \mathbf{F}(y_{j_1}, \mathbf{p})]\bigg|_{y_{j_1}^* = a_{j_1j_1s_1} = a_{j_2j_21}}\\ &= \alpha_{j_1j_1}a_{j_1j_10}(a_{j_1j_1s_1} - a_{j_1j_1s_2})^2.\end{aligned} \tag{3.96}$$

The infinite-equilibrium flow of $y_1^* = a_{j_1j_11} = a_{j_2j_21}$ possesses the following properties:

- *For $\alpha_{j_1j_1}a_{j_1j_10}(a_{j_1j_1s_1} - a_{j_1j_1s_2})^2 > 0$ and $a_{j_2j_20}\alpha_{j_1j_1}^2 > 0$,*
$$(a_{j_1j_1s_1}, a_{j_2j_21}) = \underbrace{(\mathrm{SO}, \mathrm{UP})}_{\text{up-parabola source}}. \tag{3.97}$$
- *For $\alpha_{j_1j_1}a_{j_1j_10}(a_{j_1j_1s_1} - a_{j_1j_1s_2})^2 < 0$ and $a_{j_2j_2}\alpha_{j_1j_1}^2 > 0$,*
$$(a_{j_1j_1s_1}, a_{j_2j_21}) = \underbrace{(\mathrm{SI}, \mathrm{DP})}_{\text{down-parabola sink}}. \tag{3.98}$$
- *For $\alpha_{j_1j_1}a_{j_1j_10}(a_{j_1j_1s_1} - a_{j_1j_1s_2})^2 > 0$ and $a_{j_2j_20}\alpha_{j_1j_1}^2 < 0$,*
$$(a_{j_1j_1s_1}, a_{j_2j_21}) = \underbrace{(\mathrm{SO}, \mathrm{DP})}_{\text{down-parabola source}}. \tag{3.99}$$
- *For $\alpha_{j_1j_1}a_{j_1j_10}(a_{j_1j_1s_1} - a_{j_1j_1s_2})^2 < 0$ and $a_{j_2j_20}\alpha_{j_1j_1}^2 < 0$,*
$$(a_{j_1j_1s_1}, a_{j_2j_21}) = \underbrace{(\mathrm{SI}, \mathrm{UP})}_{\text{up-parabola sink}}. \tag{3.100}$$

(v_4) *For $a_{j_1j_1s_2} = a_{j_2j_21}$,*

$$\begin{aligned}\dot{x}_{j_1} &= a_{j_1j_10}(y_{j_1} - a_{j_1j_1s_1})(y_{j_1} - a_{j_1j_1s_2})^2,\\ \dot{x}_{j_2} &= a_{j_2j_20}(y_{j_1} - a_{j_2j_21})^2,\\ j_1, j_2 &\in \{1, 2\};\ j_1 \neq j_2.\end{aligned} \tag{3.101}$$

There is a dynamical system in the y_{j_1}-direction as

$$\dot{y}_{j_1} = (y_{j_1} - a_{j_1j_1s_2})^2[\alpha_{j_1j_1}a_{j_1j_10}(y_{j_1} - a_{j_1j_1s_1}) + \alpha_{j_1j_2}a_{j_2j_20}], \tag{3.102}$$

with a dynamical system in the $y_{j_1}^{\perp}$-direction as

$$\dot{y}_{j_1}^{\perp} = (y_{j_1} - a_{j_1j_1s_2})^2[\beta_{j_1j_1}a_{j_1j_10}(y_{j_1} - a_{j_1j_1s_1}) + \beta_{j_1j_2}a_{j_2j_20}], \tag{3.103}$$

where

$$G_{j_1j_10} = \frac{\partial}{\partial y_{j_1}}[\mathbf{n}^{\mathrm{T}} \cdot \mathbf{F}(y_{j_1}, \mathbf{p})]\bigg|_{y^*_{j_1}=a_{j_1j_1s_2}=a_{j_2j_21}} = 0,$$

$$\begin{aligned} G^{(1)}_{j_1j_10} &= \frac{1}{2!}\frac{\partial^2}{\partial y^2_{j_1}}[\mathbf{n}^{\mathrm{T}} \cdot \mathbf{F}(y_{j_1}, \mathbf{p})]\bigg|_{y^*_{j_1}=a_{j_1j_1s_2}=a_{j_1j_21}} \\ &= \alpha_{j_1j_1}a_{j_1j_10}(a_{j_1j_1s_2} - a_{j_1j_1s_1}) + \alpha_{j_1j_2}a_{j_2j_20}. \end{aligned} \tag{3.104}$$

The infinite-equilibrium of $y^*_{j_1} = a_{j_1j_1s_2} = a_{j_2j_21}$ *has the following properties*:

- *For* $a_{j_1j_10}(a_{j_1j_1s_2} - a_{j_1j_1s_1})/a_{j_2j_20} > 0$ *and* $G^{(1)}_{j_1j_10} > 0$,

$$(a_{j_1j_1s_2}, a_{j_2j_21}) = \underbrace{(\mathrm{II}, \mathrm{US})}_{\text{increasing-inflection upper-saddle}}. \tag{3.105}$$

- *For* $a_{j_1j_10}(a_{j_1j_1s_2} - a_{j_1j_1s_1})/a_{j_2j_20} > 0$ *and* $G^{(1)}_{j_1j_10} < 0$,

$$(a_{j_1j_1s_2}, a_{j_2j_21}) = \underbrace{(\mathrm{II}, \mathrm{LS})}_{\text{increasing-inflection lower-saddle}}. \tag{3.106}$$

- *For* $a_{j_1j_10}(a_{j_1j_1s_2} - a_{j_1j_1s_1})/a_{j_2j_20} < 0$ *and* $G^{(1)}_{j_1j_10} > 0$,

$$(a_{j_1j_1s_2}, a_{j_2j_21}) = \underbrace{(\mathrm{DI}, \mathrm{US})}_{\text{decreasing-inflection upper-saddle}}. \tag{3.107}$$

- *For* $a_{j_1j_10}(a_{j_1j_1s_2} - a_{j_1j_1s_1})/a_{j_2j_20} < 0$ *and* $G^{(1)}_{j_1j_10} < 0$,

$$(a_{j_1j_1s_2}, a_{j_2j_21}) = \underbrace{(\mathrm{DI}, \mathrm{LS})}_{\text{decreasing-inflection lower-saddle}}. \tag{3.108}$$

- *The inflection-saddle infinite-equilibriums are the appearing bifurcations for the up-parabola and down-parabola flows in the* x_{j_1}*- and* x_{j_2}*-directions.*
- *The inflection-saddle infinite-equilibriums are the double-switching bifurcations for the up-parabola and down-parabola flows in the* x_{j_1}*- and* x_{j_2}*-directions.*

(v_5) *For* $a_{j_1 j_1 1} = a_{j_2 j_2 1}$,

$$\begin{aligned} \dot{x}_{j_1} &= a_{j_1 j_1 0}(y_{j_1} - a_{j_1 j_1 1})^3, \\ \dot{x}_{j_2} &= a_{j_2 j_2 0}(y_{j_1} - a_{j_2 j_2 1})^2, \\ j_1, j_2 &\in \{1,2\};\ j_1 \neq j_2. \end{aligned} \tag{3.109}$$

There is a dynamical system in the y_{j_1}*-direction as*

$$\dot{y}_{j_1} = (y_{j_1} - a_{j_1 j_1 1})^2[\alpha_{j_1 j_1} a_{j_1 j_1 0}(y_{j_1} - a_{j_1 j_1 1}) + \alpha_{j_1 j_2} a_{j_2 j_2 0}], \tag{3.110}$$

with a dynamical system in the $y_{j_1}^{\perp}$*-direction as*

$$\dot{y}_{j_1}^{\perp} = (y_{j_1} - a_{j_1 j_1 1})^2[\beta_{j_1 j_1} a_{j_1 j_1 0}(y_{j_1} - a_{j_1 j_1 1}) + \beta_{j_1 j_2} a_{j_2 j_2 0}], \tag{3.111}$$

where

$$\begin{aligned} G_{j_2 j_2 0} &= \frac{\partial}{\partial y_{j_1}}[\mathbf{n}^{\mathrm{T}} \cdot \mathbf{F}(y_{j_1}, \mathbf{p})]\bigg|_{y_{j_1}^* = a_{j_1 j_1 1} = a_{j_2 j_2 1}} = 0, \\ G_{j_1 j_1 0}^{(1)} &= \frac{1}{2!}\frac{\partial^2}{\partial y_{j_1}^2}[\mathbf{n}^{\mathrm{T}} \cdot \mathbf{F}(y_{j_1}, \mathbf{p})]\bigg|_{y_{j_1}^* = a_{j_1 j_1 1} = a_{j_2 j_2 1}} = \alpha_{j_1 j_2} a_{j_2 j_2 0}. \end{aligned} \tag{3.112}$$

The infinite-equilibrium of $y_{j_1}^* = a_{j_1 j_1 1} = a_{j_2 j_2 1}$ *has the following properties:*

- *For* $\alpha_{j_1 j_2}^2 a_{j_1 j_1 0} > 0$ *and* $\alpha_{j_1 j_2} a_{j_2 j_2 0} > 0$,

$$(a_{j_1 j_1 1}, a_{j_2 j_2 1}) = \underbrace{(\mathrm{UP}, \mathrm{US})}_{\text{up-parabola upper-saddle}}. \tag{3.113}$$

- *For* $\alpha_{j_1 j_2}^2 a_{j_1 j_1 0} < 0$ *and* $\alpha_{j_1 j_2} a_{j_2 j_2 0} > 0$,

$$(a_{j_1 j_1 1}, a_{j_2 j_2 1}) = \underbrace{(\mathrm{DP}, \mathrm{US})}_{\text{down-parabola upper-saddle}}. \tag{3.114}$$

- *For* $\alpha_{j_1 j_2}^2 a_{j_1 j_1 0} > 0$ *and* $\alpha_{j_1 j_2} a_{j_2 j_2 0} > 0$,

$$(a_{j_1 j_1 1}, a_{j_2 j_2 1}) = \underbrace{(\mathrm{DP}, \mathrm{LS})}_{\text{down-parabola lower-saddle}}. \tag{3.115}$$

- *For* $\alpha_{j_1 j_2}^2 a_{j_1 j_1 0} < 0$ *and* $\alpha_{j_1 j_2} a_{j_2 j_2 0} < 0$,

$$(a_{j_1 j_1 1}, a_{j_2 j_2 1}) = \underbrace{(\mathrm{UP}, \mathrm{LS})}_{\text{up-parabola lower-saddle}}. \tag{3.116}$$

(vi) If

$$\Delta_{j_1j_1} > 0 \quad \text{and} \quad \Delta_{j_2j_2} = 0, \tag{3.117}$$

then

$$\begin{aligned}\dot{x}_{j_1} &= a_{j_1j_10}(y_{j_1} - a_{j_1j_11})(y_{j_1} - a_{j_1j_12})(y_{j_1} - a_{j_1j_13}),\\ \dot{x}_{j_2} &= a_{j_2j_20}(y_{j_1} - a_{j_2j_21})^2,\end{aligned} \tag{3.118}$$

where

$$\begin{aligned}\{a_{j_1j_11}, a_{j_1j_12}, a_{j_1j_13}\} &= \text{sort}\{b_{j_1j_11}, b_{j_1j_12}, b_{j_1j_13}\}, a_{j_1j_1s_1} < a_{j_1j_1s_2}, s_1 < s_2;\\ b_{j_1j_12}, b_{j_1j_13} &= -\frac{1}{2}(B_{j_1j_11} \pm \sqrt{\Delta_{j_1j_1}}), \quad a_{j_1j_10} = A_{j_1j_10};\\ a_{j_2j_21} &= -\frac{1}{2}B_{j_2j_2},\ j_2 \in \{1,2\}.\end{aligned} \tag{3.119}$$

There is a dynamical system in the y_{j_1}-direction as

$$\dot{y}_{j_1} = \alpha_{j_1j_1}a_{j_1j_10}\prod_{s_1=1}^{3}(y_{j_1} - a_{j_1j_1s_1}) + \alpha_{j_1j_2}a_{j_2j_20}(y_{j_1} - a_{j_2j_21})^2, \tag{3.120}$$

with a dynamical system in the $y_{j_1}^{\perp}$-direction as

$$\dot{y}_{j_1}^{\perp} = \beta_{j_1j_1}a_{j_1j_10}\prod_{s_1=1}^{3}(y_{j_1} - a_{j_1j_1s_1}) + \beta_{j_1j_2}a_{j_2j_20}(y_{j_1} - a_{j_2j_21})^2. \tag{3.121}$$

(vi_{1a}) The equilibrium flows of $y_{j_1}^* = a_{j_1j_1s_1}(s_1, s_2 \in \{1,2,3\}, s_1 \neq s_2)$ for the x_{j_1}-direction have four cases with $\dot{x}_{j_2}^* = a_{j_2j_20}(a_{j_1j_1s_1} - a_{j_2j_2})^2$:

- *For $a_{j_1j_10}\alpha_{j_1j_2}\prod_{s_2=1,s_2\neq s_1}^{3}(a_{j_1j_1s_1} - a_{j_1j_1s_2}) > 0$ and $a_{j_2j_20}(a_{j_1j_1s_1} - a_{j_2j_21})^2 > 0$,*

$$(a_{j_1j_1s_1}, \dot{x}_{j_2}^*) = \underbrace{(\text{UP}, \text{pF})}_{\text{up-parabola flow } (+)}. \tag{3.122}$$

- *For $a_{j_1j_10}\alpha_{j_1j_2}\prod_{s_2=1,s_2\neq s_1}^{3}(a_{j_1j_1s_1} - a_{j_1j_1s_2}) < 0$ and $a_{j_2j_20}(a_{j_1j_1s_1} - a_{j_2j_21})^2 > 0$,*

$$(a_{j_1j_1s_1}, \dot{x}_{j_2}^*) = \underbrace{(\text{DP}, \text{pF})}_{\text{Down-parabola flow } (+)}. \tag{3.123}$$

- *For* $a_{j_1j_10}\alpha_{j_1j_2}\prod_{s_2=1,s_2\neq s_1}^{3}(a_{j_1j_1s_1}-a_{j_1j_1s_2})>0$ *and* $a_{j_2j_20}(a_{j_1j_1s_1}-a_{j_2j_21})^2<0$,

$$(a_{j_1j_1s_1},\dot{x}_{j_2}^{*})=\underbrace{(\mathrm{DP},\mathrm{nF})}_{\text{Down-parabola flow }(-)}. \tag{3.124}$$

- *For* $a_{j_1j_10}\alpha_{j_1j_2}\prod_{s_2=1,s_2\neq s_1}^{3}(a_{j_1j_1s_1}-a_{j_1j_1s_2})<0$ *and* $a_{j_2j_20}(a_{j_1j_1s_1}-a_{j_2j_21})^2<0$,

$$(a_{j_1,j_1s_1},\dot{x}_{j_2}^{*})=\underbrace{(\mathrm{UP},\mathrm{nF})}_{\text{up-parabola flow }(-)}. \tag{3.125}$$

(vi_{1b}) *The equilibrium flow of* $y_{j_1}^{*}=a_{j_2j_21}$ *for the* x_{j_2}*-direction has four cases with* $\dot{x}_{j_1}^{*}=a_{j_1j_10}\prod_{s_1=1}^{3}(a_{j_2j_21}-a_{j_1j_1s_1})$:

- *For* $a_{j_1j_10}\prod_{s_1=1}^{3}(a_{j_2j_21}-a_{j_1j_1s_1})>0$ *and* $a_{j_2j_20}\alpha_{j_1j_1}^{2}>0$,

$$(\dot{x}_{j_1}^{*},a_{j_2j_21})=\underbrace{(\mathrm{pF},\mathrm{II})}_{\text{increasing-inflection flow }(+)}. \tag{3.126}$$

- *For* $a_{j_1j_10}\prod_{s_1=1}^{3}(a_{j_2j_21}-a_{j_1j_1s_1})<0$ *and* $a_{j_2j_20}\alpha_{j_1j_1}^{2}>0$,

$$(\dot{x}_{j_1}^{*},a_{j_2j_21})=\underbrace{(\mathrm{nF},\mathrm{DI})}_{\text{decreasing-inflection flow }(-)}. \tag{3.127}$$

- *For* $a_{j_1j_10}\prod_{s_1=1}^{3}(a_{j_2j_21}-a_{j_1j_1s_1})>0$ *and* $a_{j_2j_20}\alpha_{j_1j_1}^{2}<0$,

$$(\dot{x}_{j_1}^{*},a_{j_2j_21})=\underbrace{(\mathrm{pF},\mathrm{DI})}_{\text{decreasing-inflection flow }(+)}. \tag{3.128}$$

- *For* $a_{j_1j_10}\prod_{s_1=1}^{3}(a_{j_2j_21}-a_{j_1j_1s_1})<0$ *and* $a_{j_2j_20}\alpha_{j_1j_1}^{2}>0$,

$$(\dot{x}_{j_1}^{*},a_{j_2j_21})=\underbrace{(\mathrm{nF},\mathrm{II})}_{\text{increasing-inflection flow }(-)}. \tag{3.129}$$

– *The decreasing- and increasing-inflection flows are the appearing bifurcations for the up-parabola and parabola flows.*

(vi_2) *For* $a_{j_2j_21}=a_{j_1j_1s_1}(s_1,s_2\in\{1,2,3\})$,

$$\begin{aligned}
&\dot{x}_{j_1}=a_{j_1j_10}(y_{j_1}-a_{j_1j_1s_1})\prod_{s_2=1,s_2\neq s_1}^{3}(y_{j_1}-a_{j_1j_1s_2}),\\
&\dot{x}_{j_2}=a_{j_2j_20}(y_{j_1}-a_{j_2j_21})^2,\\
&j_1,j_2\in\{1,2\};\ j_1\neq j_2;\ s_1,s_2\in\{1,2,3\},\ s_1\neq s_2.
\end{aligned} \tag{3.130}$$

There is a dynamical system in the y_{j_1}-direction as

$$\dot{y}_{j_1} = (y_{j_1} - a_{j_1 j_1 s_1})\Big[\alpha_{j_1 j_1} a_{j_1 j_1 0} \prod_{s_2=1, s_2 \neq s_1}^{3} (y_{j_1} - a_{j_1 j_1 s_2}) + \alpha_{j_1 j_2} a_{j_2 j_2 0}(y_{j_1} - a_{j_2 j_2 1})\Big], \tag{3.131}$$

with a dynamical system in the $y_{j_1}^{\perp}$-direction as

$$\dot{y}_{j_1}^{\perp} = (y_{j_1} - a_{j_1 j_1 s_1})\Big[\beta_{j_1 j_1} a_{j_1 j_1 0} \prod_{s_2=1, s_2 \neq s_1}^{3} (y_{j_1} - a_{j_1 j_1 s_2}) + \beta_{j_1 j_2} a_{j_2 j_2 0}(y_{j_1} - a_{j_2 j_2 1})\Big], \tag{3.132}$$

where

$$G_{j_1 j_1 0} = \frac{\partial}{\partial y_{j_1}}[\mathbf{n}^{\mathrm{T}} \cdot \mathbf{F}(y_{j_1}, \mathbf{p})]\Big|_{y_{j_1}^{*} = a_{j j_1 s_1} = a_{j_2 j_2 1}} = \alpha_{j_1 j_1} a_{j_1 j_1 0} \prod_{s_2=1, s_2 \neq s_1}^{3} (a_{j_1 j_1 s_1} - a_{j_1 j_1 s_2}). \tag{3.133}$$

The infinite-equilibriums of $y_{j_1}^{*} = a_{j_1 j_1 s_1}$ for the x_{j_1}-direction and $y_{j_1}^{*} = a_{j_2 j_2 1}$ for the x_{j_2}-direction are four cases:

- *For $\alpha_{j_1 j_1} a_{j_1 j_1 0} \prod_{s_2=1, s_2 \neq s_1}^{3} (a_{j_1 j_1 s_1} - a_{j_1 j_1 s_2}) > 0$ and $a_{j_2 j_2} \alpha_{j_1 j_1}^2 > 0$,*

$$(a_{j_1 j_1 s_1}, a_{j_2 j_2 1}) = \underbrace{(\mathrm{SO}, \mathrm{UP})}_{\text{up-parabola source}}. \tag{3.134}$$

- *For $\alpha_{j_1 j_1} a_{j_1 j_1 0} \prod_{s_2=1, s_2 \neq s_1}^{3} (a_{j_1 j_1 s_1} - a_{j_1 j_1 s_2}) < 0$ and $a_{j_2 j_2} \alpha_{j_1 j_1}^2 > 0$,*

$$(a_{j_1 j_1 s_1}, a_{j_2 j_2 1}) = \underbrace{(\mathrm{SI}, \mathrm{DP})}_{\text{down-parabola sink}}. \tag{3.135}$$

- *For $\alpha_{j_1 j_1} a_{j_1 j_1 0} \prod_{s_2=1, s_2 \neq s_1}^{3} (a_{j_1 j_1 s_1} - a_{j_1 j_1 s_2}) > 0$ and $a_{j_2 j_2} \alpha_{j_1 j_1}^2 < 0$,*

$$(a_{j_1 j_1 s_1}, a_{j_2 j_2 1}) = \underbrace{(\mathrm{SO}, \mathrm{DP})}_{\text{down-parabola source}}. \tag{3.136}$$

- *For* $\alpha_{j_1j_1}a_{j_1j_10}\prod_{s_2=1,s_2\neq s_1}^{3}(a_{j_1j_1s_1}-a_{j_1j_1s_2})<0$ *and* $a_{j_2j_20}\alpha_{j_1j_1}^2<0$,

$$(a_{j_1j_1s_1},a_{j_2j_21})=\underbrace{(\mathrm{SI},\mathrm{UP})}_{\text{up-parabola sink}}. \tag{3.137}$$

(vii) If

$$\Delta_{j_1j_1}<0 \quad \text{and} \quad \Delta_{j_2j_2}>0, \tag{3.138}$$

then

$$\begin{aligned}\dot{x}_{j_1}&=a_{j_1j_10}(y_{j_1}-a_{j_1j_11})[(y_{j_1}-a_{j_1j_1})^2+b_{j_1j_1}],\\ \dot{x}_{j_2}&=a_{j_2j_20}(y_{j_1}-a_{j_2j_1})(y_{j_1}-a_{j_2j_22}),\end{aligned} \tag{3.139}$$

where

$$\begin{aligned}&\{a_{j_2j_21},a_{j_2j_22}\}=\mathrm{sort}\{b_{j_2j_21},b_{j_2j_22}\},\quad a_{j_2j_21}<a_{j_2j_22};\\ &b_{j_2j_21},b_{j_2j_22}=-\frac{1}{2}(B_{j_2j_2}\pm\sqrt{\Delta_{j_2j_2}}),\quad a_{j_2j_20}=A_{j_2j_21};\\ &a_{j_1j_11}=b_{j_1j_11},\quad a_{j_1j_1}=-\frac{1}{2}B_{j_1j_11},\\ &b_{j_1j_1}=-\frac{1}{4}\Delta_{j_1j_1},\quad a_{j_1j_10}=A_{j_1j_10}.\end{aligned} \tag{3.140}$$

(vii₁) For $a_{j_1j_11}\neq a_{j_2j_2l_1}(l_1,l_2\in\{1,2\})$*, there is a dynamical system in the* y_{j_1}*-direction as*

$$\begin{aligned}\dot{y}_{j_1}&=\alpha_{j_1j_1}a_{j_1j_10}(y_{j_1}-a_{j_1j_1})[(y_{j_1}-a_{j_1j_1})^2+b_{j_1j_1}]\\ &\quad+\alpha_{j_1j_2}a_{j_2j_20}(y_{j_1}-a_{j_2j_21})(y_{j_1}-a_{j_2j_22}),\end{aligned} \tag{3.141}$$

with a dynamical system in the $y_{j_1}^{\perp}$*-direction as*

$$\begin{aligned}\dot{y}_{j_1}^{\perp}&=\beta_{j_1j_1}a_{j_1j_10}(y_{j_1}-a_{j_1j_11})[(y_{j_1}-a_{j_1j_1})^2+b_{j_1j_1}]\\ &\quad+\beta_{j_1j_2}a_{j_2j_20}(y_{j_1}-a_{j_2j_21})(y_{j_1}-a_{j_2j_22}),\end{aligned} \tag{3.142}$$

(vii_{1a}) The flow of $y_1^*=a_{j_1j_11}$ *with* $\dot{x}_{j_2}^*=a_{j_2j_20}\prod_{l_1=1}^{2}(a_{j_1j_11}-a_{j_2j_2l_1})$ *possesses the following properties:*

- *For* $\alpha_{j_1j_1}a_{j_1j_10}>0$ *and* $a_{j_2j_20}\prod_{l_1=1}^{2}(a_{j_1j_11}-a_{j_2j_2l_1})>0$,

$$(a_{j_1j_11},\dot{x}_{j_2}^*)=\underbrace{(\mathrm{UP},\mathrm{pF})}_{\text{up-parabola flow }(+)}. \tag{3.143}$$

- *For* $\alpha_{j_1j_1}a_{j_1j_10}<0$ *and* $a_{j_2j_20}\prod_{l_1=1}^{2}(a_{j_1j_11}-a_{j_2j_2l_1})>0$,

$$(a_{j_1j_11},\dot{x}_{j_2}^*)=\underbrace{(\mathrm{DP},\mathrm{pF})}_{\text{down-parabola flow }(+)}. \tag{3.144}$$

- *For* $\alpha_{j_1j_1}a_{j_1j_10} > 0$ *and* $a_{j_2j_20}\prod_{l_1=1}^{2}(a_{j_1j_11} - a_{j_2j_21}) < 0$,

$$(a_{j_1j_11}, \dot{x}^*_{j_2}) = \underbrace{(\mathrm{DP}, \mathrm{nF})}_{\text{down-parabola flow } (-)}. \tag{3.145}$$

- *For* $\alpha_{j_1j_1}a_{j_1j_10} < 0$ *and* $a_{j_2j_20}\prod_{l_1=1}^{2}(a_{j_1j_11} - a_{j_2j_2l_1}) < 0$,

$$(a_{j_1j_11}, \dot{x}^*_{j_2}) = \underbrace{(\mathrm{UP}, \mathrm{nF})}_{\text{up-parabola flow } (-)}. \tag{3.146}$$

(vii_{1b}) *The flow of* $y_1^* = a_{j_2j_2l_1}$ *with* $\dot{x}^*_{j_1} = a_{j_1j_10}(a_{j_2j_2l_1} - a_{j_1j_11})[(a_{j_2j_21} - a_{j_1j_1})^2 + b_{j_1j_1}]$ *possesses the following properties:*

- *For* $a_{j_1j_10}(a_{j_2j_2l_1} - a_{j_1j_11}) > 0$ *and* $a_{j_2j_20}\alpha_{j_1j_1}(a_{j_2j_2l_1} - a_{j_2j_2l_2}) > 0$,

$$(\dot{x}^*_{j_1}, a_{j_2j_2l_1}) = \underbrace{(\mathrm{pF}, \mathrm{UP})}_{\text{up-parabola flow } (+)}. \tag{3.147}$$

- *For* $a_{j_1j_10}(a_{j_2j_2l_1} - a_{j_1j_11}) < 0$ *and* $a_{j_2j_20}\alpha_{j_1j_1}(a_{j_2j_2l_1} - a_{j_2j_2l_2}) > 0$,

$$(\dot{x}^*_{j_1}, a_{j_2j_2l_1}) = \underbrace{(\mathrm{nF}, \mathrm{DP})}_{\text{down-parabola flow } (-)}. \tag{3.148}$$

- *For* $a_{j_1j_10}(a_{j_2j_2l_1} - a_{j_1j_11}) > 0$ *and* $a_{j_2j_20}\alpha_{j_1j_1}(a_{j_2j_2l_1} - a_{j_2j_2l_2}) < 0$,

$$(\dot{x}^*_{j_1}, a_{j_2j_2l_1}) = \underbrace{(\mathrm{pF}, \mathrm{DP})}_{\text{down-parabola flow } (+)}. \tag{3.149}$$

- *For* $a_{j_1j_10}(a_{j_2j_2l_1} - a_{j_1j_11}) < 0$ *and* $a_{j_2j_20}\alpha_{j_1j_1}(a_{j_2j_2l_1} - a_{j_2j_2l_2}) < 0$,

$$(\dot{x}^*_{j_1}, a_{j_2j_2l_1}) = \underbrace{(\mathrm{nF}, \mathrm{UP})}_{\text{up-parabola flow } (-)}. \tag{3.150}$$

(vii_2) *For* $a_{j_1j_11} = a_{j_2j_2l_1}$,

$$\begin{aligned}\dot{x}_{j_1} &= a_{j_1j_10}(y_{j_1} - a_{j_1j_11})[(y_{j_1} - a_{j_1j_1})^2 + b_{j_1j_1}],\\ \dot{x}_{j_2} &= a_{j_2j_20}(y_{j_1} - a_{j_2j_2l_1})(y_{j_1} - a_{j_2j_2l_2}).\end{aligned} \tag{3.151}$$

There is a dynamical system in the y_{j_1}*-direction as*

$$\begin{aligned}\dot{y}_{j_1} = (y_{j_1} - a_{j_1j_11})\{&\alpha_{j_1j_1}a_{j_1j_10}[(y_{j_1} - a_{j_1j_1})^2\\ &+ b_{j_1j_1}] + \alpha_{j_1j_2}a_{j_2j_20}(y_{j_1} - a_{j_2j_2l_2})\},\end{aligned} \tag{3.152}$$

with a dynamical system in the $y^{\perp}_{j_1}$*-direction as*

$$\begin{aligned}\dot{y}^{\perp}_{j_1} = (y_{j_1} - a_{j_1j_11})\{&\beta_{j_1j_1}a_{j_1j_10}[(y_{j_1} - a_{j_1j_1})^2 + b_{j_1j_1}]\\ &+ \beta_{j_1j_2}a_{j_2j_20}(y_{j_1} - a_{j_2j_2l_2})\},\end{aligned} \tag{3.153}$$

where

$$\begin{aligned} G_{j_1 j_1 0} &= \frac{\partial}{\partial y_{j_1}}[\mathbf{n}^{\mathrm{T}} \cdot \mathbf{F}(y_{j_1}, \mathbf{p})]\Big|_{y_{j_1}^* = a_{j_1 j_1 1} = a_{j_2 j_2 l_1}} \\ &= \alpha_{j_1 j_1} a_{j_1 j_1 0}[(a_{j_1 j_1 1} - a_{j_1 j_1})^2 + b_{j_1 j_1}] \\ &\quad + \alpha_{j_1 j_2} a_{j_2 j_2 0}(a_{j_2 j_2 l_1} - a_{j_2 j_2 l_2}). \end{aligned} \tag{3.154}$$

The flow of $y_1^* = a_{j_1 j_1 1} = a_{j_2 j_2 l_1}$ possesses the following properties:

- *For $a_{j_1 j_1 0}/a_{j_2 j_2 0}(a_{j_2 j_2 l_1} - a_{j_2 j_2 l_2}) > 0$ and $G_{j_1 j_1 0} > 0$,*

$$(a_{j_1 j_1 1}, a_{j_2 j_2 l_1}) = \underbrace{(\mathrm{II}, \mathrm{SO})}_{\text{increaung-inflection source}}. \tag{3.155}$$

- *For $a_{j_1 j_1 0}/a_{j_2 j_2 0}(a_{j_2 j_2 l_1} - a_{j_2 j_2 l_2}) > 0$ and $G_{j_1 j_1 0} < 0$,*

$$(a_{j_1 j_1 1}, a_{j_2 j_2 l_1}) = \underbrace{(\mathrm{II}, \mathrm{SI})}_{\text{increaing-inflection sink}}. \tag{3.156}$$

- *For $a_{j_1 j_1 0}/a_{j_2 j_2 0}(a_{j_2 j_2 l_1} - a_{j_2 j_2 l_2}) < 0$ and $G_{j_1 j_1 0} > 0$,*

$$(a_{j_1 j_1 1}, a_{j_2 j_2 l_1}) = \underbrace{(\mathrm{DI}, \mathrm{SO})}_{\text{decreaing-inflection source}}. \tag{3.157}$$

- *For $a_{j_1 j_1 0}/a_{j_2 j_2 0}(a_{j_2 j_2 l_1} - a_{j_2 j_2 l_2}) < 0$ and $G_{j_1 j_1 0} < 0$,*

$$(a_{j_1 j_1 1}, a_{j_2 j_2 l_1}) = \underbrace{(\mathrm{DI}, \mathrm{SI})}_{\text{decreaing-inflection sink}}. \tag{3.158}$$

(viii) If

$$\Delta_{j_1 j_1} = 0 \quad \text{and} \quad \Delta_{j_2 j_2} > 0, \tag{3.159}$$

then

$$\begin{aligned} \dot{x}_{j_1} &= a_{j_1 j_1 0}(y_{j_1} - a_{j_1 j_1 s_1})(y_{j_1} - a_{j_1 j_1 s_2})^2, \\ \dot{x}_{j_2} &= a_{j_2 j_2 0}(y_{j_1} - a_{j_2 j_2 1})(y_{j_1} - a_{j_2 j_2 2}), \\ & j_1,\ j_2 \in \{1, 2\};\ j_1 \neq j_2; \end{aligned} \tag{3.160}$$

where

$$\begin{aligned} &\{a_{j_2 j_2 1}, a_{j_2 j_2 2}\} = \mathrm{sort}\{b_{j_2 j_2 1}, b_{j_2 j_2 2}\}, \quad a_{j_2 j_2 1} < a_{j_2 j_2 2}; \\ &b_{j_2 j_2 1}, b_{j_2 j_2 2} = -\frac{1}{2}(B_{j_2 j_2} \pm \sqrt{\Delta_{j_2 j_2}}), \quad a_{j_2 j_2 0} = A_{j_2 j_2 1}; \\ &a_{j_1 j_1 l_1} = b_{j_1 j_1 1}, \quad a_{j_1 j_1 l_2} = -\frac{1}{2} B_{j_1 j_1}, \quad b_{j_1 j_1} = 0, \quad a_{j_1 j_1 0} = A_{j_1 j_1 0}. \end{aligned} \tag{3.161}$$

($viii_1$) For $a_{j_2j_2l_1} \neq a_{j_1j_1s_1}, a_{j_1j_1s_2}$, there is a dynamical system in the y_{j_1}-direction as

$$\begin{aligned}\dot{y}_{j_1} &= \alpha_{j_1j_1} a_{j_1j_10}(y_{j_1} - a_{j_1j_1s_1})(y_{j_1} - a_{j_1j_1s_2})^2 \\ &\quad + \alpha_{j_1j_2} a_{j_2j_20}(y_{j_1} - a_{j_2j_21})(y_{j_1} - a_{j_2j_22}),\end{aligned} \tag{3.162}$$

with a dynamical system in the $y^{\perp}_{j_1}$-direction as

$$\begin{aligned}\dot{y}^{\perp}_{j_1} &= \beta_{j_1j_1} a_{j_1j_10}(y_{j_1} - a_{j_1j_1s_1})(y_{j_1} - a_{j_1j_1s_2})^2 \\ &\quad + \beta_{j_1j_2} a_{j_2j_20}(y_{j_1} - a_{j_2j_21})(y_{j_1} - a_{j_2j_22}).\end{aligned} \tag{3.163}$$

($viii_{1a}$) The flow of $y^*_{j_1} = a_{j_1j_1s_1} \neq a_{j_2j_2l_1} (l_1 = 1,2)$ for the x_{j_1}-direction has the following properties with $\dot{x}^*_{j_2} = a_{j_2j_20}\prod_{l_1=1}^{2}(a_{j_1j_1s_1} - a_{j_2j_2l_1})^2$:

- *For $a_{j_1j_10}\alpha_{j_1j_2}(a_{j_1j_1s_1} - a_{j_1j_1s_2})^2 > 0$ and $a_{j_2j_20}\prod_{l_1=1}^{2}(a_{j_1j_1s_1} - a_{j_2j_2l_1}) > 0$,*

$$(a_{j_1j_1s_1}, \dot{x}^*_{j_2}) = \underbrace{(\mathrm{UP}, \mathrm{pF})}_{\text{up-parabola flow } (+)}. \tag{3.164}$$

- *For $a_{j_1j_10}\alpha_{j_1j_2}(a_{j_1j_1s_1} - a_{j_1j_1s_2})^2 < 0$ and $a_{j_2j_20}\prod_{l_1=1}^{2}(a_{j_1j_1s_1} - a_{j_2j_2l_1}) > 0$,*

$$(a_{j_1j_1s_1}, \dot{x}^*_{j_2}) = \underbrace{(\mathrm{DP}, \mathrm{pF})}_{\text{down-parabola flow } (+)}. \tag{3.165}$$

- *For $a_{j_1j_10}\alpha_{j_1j_2}(a_{j_1j_1s_1} - a_{j_1j_1s_2})^2 > 0$ and $a_{j_2j_20}\prod_{l_1=1}^{2}(a_{j_1j_1s_1} - a_{j_2j_2l_1}) < 0$,*

$$(a_{j_1j_1s_1}, \dot{x}^*_{j_2}) = \underbrace{(\mathrm{DP}, \mathrm{nF})}_{\text{down-parabola flow } (-)}. \tag{3.166}$$

- *For $a_{j_1j_10}\alpha_{j_1j_2}(a_{j_1j_1s_1} - a_{j_1j_1s_2})^2 < 0$ and $a_{j_2j_20}\prod_{l_1=1}^{2}(a_{j_1j_1s_1} - a_{j_2j_2l_1}) < 0$,*

$$(a_{j_1j_1s_1}, \dot{x}^*_{j_2}) = \underbrace{(\mathrm{UP}, \mathrm{nF})}_{\text{up-parabola flow } (-)}. \tag{3.167}$$

($viii_{1b}$) The flows of $y^*_{j_1} = a_{j_1j_1s_2} \neq a_{j_2j_2l_1} (l_1 = 1,2)$ for the x_{j_1}-direction have the following properties with $\dot{x}^*_{j_2} = a_{j_2j_20}\prod_{l_1=1}^{2}(a_{j_1j_1s_2} - a_{j_2j_2l_1})$:

- *For $a_{j_1j_10}\alpha^2_{j_1j_2}(a_{j_1j_1s_2} - a_{j_1j_1s_1}) > 0$ and $a_{j_2j_20}\prod_{l_1=1}^{2}(a_{j_1j_1s_2} - a_{j_2j_2l_1}) > 0$,*

$$(a_{j_1j_1s_2}, \dot{x}^*_{j_2}) = \underbrace{(\mathrm{II}, \mathrm{pF})}_{\text{increasing-inflection flow } (-)}. \tag{3.168}$$

- *For* $a_{j_1j_10}\alpha^2_{j_1j_2}(a_{j_1j_1s_2} - a_{j_1j_1s_1}) > 0$ *and* $a_{j_2j_20}\prod_{l_1=1}^{2}(a_{j_1j_1s_2} - a_{j_2j_2l_1}) > 0$,

$$(a_{j_1j_1s_2}, \dot{x}^*_{j_2}) = \underbrace{(\mathrm{DI}, \mathrm{pF})}_{\text{decreasing-inflection flow }(+)}. \tag{3.169}$$

- *For* $a_{j_1j_10}\alpha^2_{j_1j_2}(a_{j_1j_1s_2} - a_{j_1j_1s_1}) > 0$ *and* $a_{j_2j_20}\prod_{l_1=1}^{2}(a_{j_1j_1s_2} - a_{j_2j_2l_1}) < 0$,

$$(a_{j_1j_1s_2}, \dot{x}^*_{j_2}) = \underbrace{(\mathrm{DI}, \mathrm{nF})}_{\text{decreasing-inflection flow }(-)}. \tag{3.170}$$

- *For* $a_{j_1j_10}\alpha^2_{j_1j_2}(a_{j_1j_1s_2} - a_{j_1j_1s_1}) < 0$ *and* $a_{j_2j_20}\prod_{l_1=1}^{2}(a_{j_1j_1s_2} - a_{j_2j_2l_1}) < 0$,

$$(a_{j_1,j_1s_2}, \dot{x}^*_{j_2}) = \underbrace{(\mathrm{II}, \mathrm{nF})}_{\text{increasing-inflection flow }(-)}. \tag{3.171}$$

- *The increasing- and decreasing-inflection flows are the appearing bifurcations for the up-parabola and down-parabola flows.*

(viii$_{1c}$*) The flow of* $y^*_{j_1} = a_{j_2j_2l_1} \neq a_{j_1j_1s_1}, a_{j_1j_1s_2}$ *for the* x_{j_2}*-direction has the following properties with* $\dot{x}^*_{j_1} = a_{j_1j_10}(a_{j_2j_2l_1} - a_{j_1j_1s_1})(a_{j_2j_2l_1} - a_{j_1j_1s_2})^2$:

- *For* $a_{j_1j_10}(a_{j_2j_2l_1} - a_{j_1j_1s_1})(a_{j_2j_2l_1} - a_{j_1j_1s_2})^2 > 0$ *and* $a_{j_2j_20}\alpha_{j_1j_1}\ (a_{j_2j_2l_1} - a_{j_2j_2l_2}) > 0$,

$$(\dot{x}^*_{j_1}, a_{j_2j_2l_1}) = \underbrace{(\mathrm{pF}, \mathrm{UP})}_{\text{up-parabola flow }(+)}. \tag{3.172}$$

- *For* $a_{j_1j_10}(a_{j_2j_2l_1} - a_{j_1j_1s_1})(a_{j_2j_2l_1} - a_{j_1j_1s_2})^2 < 0$ *and* $a_{j_2j_20}\alpha_{j_1j_1}\ (a_{j_2j_2l_1} - a_{j_2j_2l_2}) > 0$,

$$(\dot{x}^*_{j_1}, a_{j_2j_2l_1}) = \underbrace{(\mathrm{nF}, \mathrm{DP})}_{\text{down-parabola flow }(-)}. \tag{3.173}$$

- *For* $a_{j_1j_10}(a_{j_2j_2l_1} - a_{j_1j_1s_1})(a_{j_2j_2l_1} - a_{j_1j_1s_2})^2 > 0$ *and* $a_{j_2j_20}\alpha_{j_1j_1}\ (a_{j_2j_2l_1} - a_{j_2j_2l_2}) < 0$,

$$(\dot{x}^*_{j_1}, a_{j_2j_2l_1}) = \underbrace{(\mathrm{pF}, \mathrm{DP})}_{\text{down-parabola flow }(+)}. \tag{3.174}$$

- *For* $a_{j_1j_10}(a_{j_2j_2l_1} - a_{j_1j_1s_1})(a_{j_2j_2l_1} - a_{j_1j_1s_2})^2 < 0$ *and* $a_{j_2j_20}\alpha_{j_1j_1}\ (a_{j_2j_2l_1} - a_{j_2j_2l_2}) < 0$,

$$(\dot{x}^*_{j_1}, a_{j_2j_2l_1}) = \underbrace{(\mathrm{nF}, \mathrm{UP})}_{\text{up-parabola flow }(-)}. \tag{3.175}$$

(viii$_2$) For $a_{j_1j_1s_1} = a_{j_1j_1s_2} \equiv a_{j_1j_1 1}$,

$$\begin{aligned} \dot{x}_{j_1} &= a_{j_1j_1 0}(y_{j_1} - a_{j_1j_1 1})^3, \\ \dot{x}_{j_2} &= a_{j_2j_2 0}(y_{j_1} - a_{j_2j_2 1})(y_{j_1} - a_{j_2j_2 2}), \\ j_1, j_2 &\in \{1,2\};\ j_1 \neq j_2. \end{aligned} \tag{3.176}$$

For $a_{j_2j_2l_1} \neq a_{j_1j_1 1}$, *there is a dynamical system in the* y_{j_1}*-direction as*

$$\begin{aligned} \dot{y}_{j_1} &= \alpha_{j_1j_1} a_{j_1j_1 0}(y_{j_1} - a_{j_1j_1 1})^3 \\ &\quad + \alpha_{j_1j_2} a_{j_2j_2 0} \prod_{l_1=1}^{2} (y_{j_1} - a_{j_2j_2l_1}), \end{aligned} \tag{3.177}$$

with a dynamical system in the $y_{j_1}^{\perp}$*-direction as*

$$\begin{aligned} \dot{y}_{j_1}^{\perp} &= \beta_{j_1j_1} a_{j_1j_1 0}(y_{j_1} - a_{j_1j_1 1})^3 \\ &\quad + \beta_{j_1j_2} a_{j_2j_2 0} \prod_{l_1=1}^{2} (y_{j_1} - a_{j_2j_2l_1}). \end{aligned} \tag{3.178}$$

(viii$_{2a}$) The flow of $y_{j_1}^* = a_{j_1j_1 1} \neq a_{j_2j_2l_1} (l_1 = 1,2)$ *for the* x_{j_1}*-direction has the following properties with* $\dot{x}_{j_2}^* = a_{j_2j_2 0} \prod_{l_1=1}^{2} (a_{j_1j_1 1} - a_{j_2j_2l_1})$:

- *For* $a_{j_1j_1 0}\alpha_{j_1j_2}^3 > 0$ *and* $a_{j_2j_2 0} \prod_{l_1=1}^{2} (a_{j_1j_1 1} - a_{j_2j_2l_1}) > 0$,

$$(a_{j_1j_1 1}, \dot{x}_{j_2}^*) = \underbrace{(3^{\text{rd}}\text{UP}, \text{pF})}_{\text{third-order up-parabola flow } (+)}. \tag{3.179}$$

- *For* $a_{j_1j_1 0}\alpha_{j_1j_2}^3 < 0$ *and* $a_{j_2j_2 0} \prod_{l_1=1}^{2} (a_{j_1j_1 1} - a_{j_2j_2l_1}) > 0$,

$$(a_{j_1,j_1 1}, \dot{x}_{j_2}^*) = \underbrace{(3^{\text{rd}}\text{DP}, \text{pF})}_{\text{third-order down-parabola flow } (+)}. \tag{3.180}$$

- *For* $a_{j_1j_1 0}\alpha_{j_1j_2}^3 > 0$ *and* $a_{j_2j_2 0} \prod_{l_1=1}^{2} (a_{j_1j_1 1} - a_{j_2j_2l_1}) < 0$,

$$(a_{j_1j_1 1}, \dot{x}_{j_2}^*) = \underbrace{(3^{\text{rd}}\text{DP}, \text{nF})}_{\text{third-order down-parabola flow } (-)}. \tag{3.181}$$

- *For* $a_{j_1j_1 0}\alpha_{j_1j_2}^3 < 0$ *and* $a_{j_2j_2 0} \prod_{l_1=1}^{2} (a_{j_1j_1 1} - a_{j_2j_2l_1}) < 0$,

$$(a_{j_1,j_1 1}, \dot{x}_{j_2}^*) = \underbrace{(3^{\text{rd}}\text{UP}, \text{nF})}_{\text{third-order up-parabola flow } (-)}. \tag{3.182}$$

– *The third-order up-parabola flows are the appearing bifurcations for up-parabola, down-parabola, and up-parabola flows.*
– *The third-order down-parabola flows are the appearing bifurcations for down-parabola, up-parabola, and down-parabola flows.*

($viii_{2b}$) *The flow of* $y_{j_1}^* = a_{j_2j_2l_1} \neq a_{j_1j_1 1}$ *for the* x_{j_2}*-direction has the following properties with* $\dot{x}_{j_1}^* = a_{j_1j_1 0}(a_{j_2j_2l_1} - a_{j_1j_1 1})^3$:

- *For* $a_{j_1j_1 0}(a_{j_2j_2 1} - a_{j_1j_1 1})^3 > 0$ *and* $a_{j_2j_2 0}\alpha_{j_1j_1}(a_{j_2j_2l_1} - a_{j_2j_2l_2}) > 0$,

$$(\dot{x}_{j_1}^*, a_{j_2j_2l_1}) = \underbrace{(\mathrm{pF}, \mathrm{UP})}_{\text{up-parabola flow } (+)}. \tag{3.183}$$

- *For* $a_{j_1j_1 0}(a_{j_2j_2 1} - a_{j_1j_1 1})^3 < 0$ *and* $a_{j_2j_2 0}\alpha_{j_1j_1}(a_{j_2j_2l_1} - a_{j_2j_2l_2}) > 0$,

$$(\dot{x}_{j_1}^*, a_{j_2j_2l_1}) = \underbrace{(\mathrm{nF}, \mathrm{DP})}_{\text{down-parabola flow } (-)}. \tag{3.184}$$

- *For* $a_{j_1j_1 0}(a_{j_2j_2 1} - a_{j_1j_1 1})^3 > 0$ *and* $a_{j_2j_2 0}\alpha_{j_1j_1}(a_{j_2j_2l_1} - a_{j_2j_2l_2}) < 0$,

$$(\dot{x}_{j_1}^*, a_{j_2j_2l_1}) = \underbrace{(\mathrm{pF}, \mathrm{DP})}_{\text{down-parabola flow } (+)}. \tag{3.185}$$

- *For* $a_{j_1j_1 0}(a_{j_2j_2 1} - a_{j_1j_1 1})^3 < 0$ *and* $a_{j_2j_2 0}\alpha_{j_1j_1}(a_{j_2j_2l_1} - a_{j_2j_2l_2}) < 0$,

$$(\dot{x}_{j_1}^*, a_{j_2j_2l_1}) = \underbrace{(\mathrm{nF}, \mathrm{UP})}_{\text{up-parabola flow } (-)}. \tag{3.186}$$

($viii_3$) *For* $a_{j_1j_1s_1} = a_{j_2j_2l_1}$,

$$\begin{aligned} \dot{x}_{j_1} &= a_{j_1j_1 0}(y_{j_1} - a_{j_1j_1s_1})(y_{j_1} - a_{j_1j_1s_2})^2, \\ \dot{x}_{j_2} &= a_{j_2j_2 0}(y_{j_1} - a_{j_2j_2l_1})(y_{j_1} - a_{j_2j_2l_2}). \end{aligned} \tag{3.187}$$

There is a dynamical system in the y_{j_1}*-direction as*

$$\begin{aligned} \dot{y}_{j_1} = {} & (y_{j_1} - a_{j_1j_1s_1})[\alpha_{j_1j_1}a_{j_1j_1 0}(y_{j_1} - a_{j_1j_1s_2})^2 \\ & + \alpha_{j_1j_2}a_{j_2j_2 0}(y_{j_1} - a_{j_2j_2l_2})], \end{aligned} \tag{3.188}$$

with a dynamical system in the $y_{j_1}^{\perp}$*-direction as*

$$\begin{aligned} \dot{y}_{j_1}^{\perp} = {} & (y_{j_1} - a_{j_1j_1s_1})[\beta_{j_1j_1}a_{j_1j_1 0}(y_{j_1} - a_{j_1j_1s_2})^2 \\ & + \beta_{j_1j_2}a_{j_2j_2 0}(y_{j_1} - a_{j_2j_2l_2})], \end{aligned} \tag{3.189}$$

where

$$G_{j_1j_10} = \frac{\partial}{\partial y_{j_1}}[\mathbf{n}^{\mathrm{T}} \cdot \mathbf{F}(y_{j_1}, \mathbf{p})]\Big|_{y^*_{j_1}=a_{j_1j_1s_1}=a_{j_2j_2l_1}}$$
$$= \alpha_{j_1j_1}a_{j_1j_10}(a_{j_1j_1s_1} - a_{j_1j_1s_2})^2 + \alpha_{j_1j_2}a_{j_2j_20}(a_{j_2j_2l_1} - a_{j_2j_2l_2}). \tag{3.190}$$

The infinite-equilibrium flows of $y_1^* = a_{j_1j_1s_1} = a_{j_2j_2l_1}$ *possess the flowing properties*:

- *For* $a_{j_1j_10}(a_{j_1j_1s_1} - a_{j_1j_1s_2})^2/a_{j_2j_20}(a_{j_2j_2l_1} - a_{j_2j_2l_2}) > 0$ *and* $G_{j_1j_10} > 0$,
$$(a_{j_1j_1s_1}, a_{22l_1}) = \underbrace{(\mathrm{II}, \mathrm{SO})}_{\text{increasing-inflection source}}. \tag{3.191}$$

- *For* $a_{j_1j_10}(a_{j_1j_1s_1} - a_{j_1j_1s_2})^2/a_{j_2j_20}(a_{j_2j_2l_1} - a_{j_2j_2l_2}) > 0$ *and* $G_{j_1j_10} < 0$,
$$(a_{j_1j_1s_1}, a_{22l_1}) = \underbrace{(\mathrm{II}, \mathrm{SI})}_{\text{increasing-inflection sink}}. \tag{3.192}$$

- *For* $a_{j_1j_10}(a_{j_1j_1s_1} - a_{j_1j_1s_2})^2/a_{j_2j_20}(a_{j_2j_2l_1} - a_{j_2j_2l_2}) < 0$ *and* $G_{j_1j_10} > 0$,
$$(a_{j_1j_1s_1}, a_{22l_1}) = \underbrace{(\mathrm{DI}, \mathrm{SO})}_{\text{decreasing-inflection source}}. \tag{3.193}$$

- *For* $a_{j_1j_10}(a_{j_1j_1s_1} - a_{j_1j_1s_2})^2/a_{j_2j_20}(a_{j_2j_2l_1} - a_{j_2j_2l_2}) < 0$ *and* $G_{j_1j_10} < 0$,
$$(a_{j_1j_1s_1}, a_{22l_1}) = \underbrace{(\mathrm{DI}, \mathrm{SI})}_{\text{increasing-inflection sink}}. \tag{3.194}$$

- *The inflection-source and sink infinite-equilibriums are the switching bifurcations for parabola flows in the* x_{j_1}*- and* x_{j_2}*-directions.*

(viii$_4$) For $a_{j_1j_1s_2} = a_{j_2j_2l_1}$,

$$\begin{aligned} \dot{x}_{j_1} &= a_{j_1j_10}(y_{j_1} - a_{j_1j_1s_1})(y_{j_1} - a_{j_1j_1s_2})^2, \\ \dot{x}_{j_2} &= a_{j_2j_20}(y_{j_1} - a_{j_2j_2l_1})(y_{j_1} - a_{j_2j_2l_2}), \\ &j_1, j_2 \in \{1, 2\}; \ j_1 \neq j_2. \end{aligned} \tag{3.195}$$

There is a dynamical system in the y_{j_1}-direction as

$$\begin{aligned}\dot{y}_{j_1} &= (y_{j_1} - a_{j_1j_1s_2})[\alpha_{j_1j_1}a_{j_1j_10}(y_{j_1} - a_{j_1j_1s_1})(y_{j_1} - a_{j_1j_1s_2})\\ &\quad + \alpha_{j_1j_2}a_{j_2j_20}(y_{j_1} - a_{j_2j_2l_2})],\end{aligned} \tag{3.196}$$

with a dynamical system in the $y_{j_1}^{\perp}$-direction as

$$\begin{aligned}\dot{y}_{j_1}^{\perp} &= (y_{j_1} - a_{j_1j_1s_2})[\beta_{j_1j_1}a_{j_1j_10}(y_{j_1} - a_{j_1j_1s_1})(y_{j_1} - a_{j_1j_1s_2})\\ &\quad + \beta_{j_1j_2}a_{j_2j_20}(y_{j_1} - a_{j_2j_2l_2})],\end{aligned} \tag{3.197}$$

where

$$\begin{aligned}G_{j_1j_10} &= \frac{\partial}{\partial y_{j_1}}[\mathbf{n}^{\mathrm{T}} \cdot \mathbf{F}(y_{j_1}, \mathbf{p})]\Big|_{y_{j_1}^* = a_{j_1j_1s_2} = a_{j_2j_2l_1}}\\ &= \alpha_{j_1j_2}a_{j_2j_20}(a_{j_2j_2l_1} - a_{j_2j_2l_2}).\end{aligned} \tag{3.198}$$

The infinite-equilibrium of $y_{j_1}^* = a_{j_1j_1s_2} = a_{j_2j_2l_1}$ has the following properties:

- *For $a_{j_1j_10}\alpha_{j_1j_2}^2(a_{j_1j_1s_2} - a_{j_1j_1s_1}) > 0$ and $\alpha_{j_1j_2}a_{j_2j_20}(a_{j_2j_2l_1} - a_{j_2j_2l_2}) > 0$,*

$$(a_{j_1j_1s_2}, a_{j_2j_2l_1}) = \underbrace{(\mathrm{UP}, \mathrm{SO})}_{\text{up-parabola source}}. \tag{3.199}$$

- *For $a_{j_1j_10}\alpha_{j_1j_2}^2(a_{j_1j_1s_2} - a_{j_1j_1s_1}) < 0$ and $\alpha_{j_1j_2}a_{j_2j_20}(a_{j_2j_2l_1} - a_{j_2j_2l_2}) > 0$,*

$$(a_{j_1j_1s_2}, a_{j_2j_2l_1}) = \underbrace{(\mathrm{DP}, \mathrm{SO})}_{\text{down-parabola source}}. \tag{3.200}$$

- *For $a_{j_1j_10}\alpha_{j_1j_2}^2(a_{j_1j_1s_2} - a_{j_1j_1s_1}) > 0$ and $\alpha_{j_1j_2}a_{j_2j_20}(a_{j_2j_2l_1} - a_{j_2j_2l_2}) < 0$,*

$$(a_{j_1j_1s_2}, a_{j_2j_2l_1}) = \underbrace{(\mathrm{DP}, \mathrm{SI})}_{\text{down-parabola sink}}. \tag{3.201}$$

- *For $a_{j_1j_10}\alpha_{j_1j_2}^2(a_{j_1j_1s_2} - a_{j_1j_1s_1}) < 0$ and $\alpha_{j_1j_2}a_{j_2j_20}(a_{j_2j_2l_1} - a_{j_2j_2l_2}) < 0$,*

$$(a_{j_1j_1s_2}, a_{j_2j_2l_1}) = \underbrace{(\mathrm{UP}, \mathrm{SI})}_{\text{up-parabola source}}. \tag{3.202}$$

- *The parabola source and sink infinite-equilibriums are the switching bifurcations for the increasing-inflection and parabola flows in the* x_{j_1}*- and* x_{j_2}*-directions.*
- *The inflection-saddle infinite-equilibriums are the appearing bifurcations from a parabola flow and* x_{j_2}*-direction to the up-parabola and down-parabola flows in the* x_{j_1}*-directions and a parabola flow in the* x_{j_2}*-directions.*

(*viii*$_5$) *For* $a_{j_1 j_1 1} = a_{j_2 j_2 l_2}$,

$$\begin{aligned} \dot{x}_{j_1} &= a_{j_1 j_1 0}(y_{j_1} - a_{j_1 j_1 1})^3, \\ \dot{x}_{j_2} &= a_{j_2 j_2 0}(y_{j_1} - a_{j_2 j_2 l_1})(y_{j_1} - a_{j_2 j_2 l_2}), \\ & j_1, j_2 \in \{1, 2\};\ j_1 \neq j_2. \end{aligned} \tag{3.203}$$

There is a dynamical system in the y_{j_1}*-direction as*

$$\begin{aligned} \dot{y}_{j_1} = (y_{j_1} - a_{j_1 1})[&\alpha_{j_1 j_1} a_{j_1 j_1 0}(y_{j_1} - a_{j_1 j_1 1})^2 \\ &+ \alpha_{j_1 j_2} a_{j_2 j_2 0}(y_{j_1} - a_{j_2 j_2 l_2})], \end{aligned} \tag{3.204}$$

with a dynamical system in the $y_{j_1}^{\perp}$*-direction as*

$$\begin{aligned} \dot{y}_{j_1}^{\perp} = (y_{j_1} - a_{j_1 j_1 1})[&\beta_{j_1 j_1} a_{j_1 j_1 0}(y_{j_1} - a_{j_1 j_1 1})^2 \\ &+ \beta_{j_1 j_2} a_{j_2 j_2 0}(y_{j_1} - a_{j_2 j_2 l_2})], \end{aligned} \tag{3.205}$$

where

$$\begin{aligned} G_{j_1 j_1 0} &= \frac{\partial}{\partial y_{j_1}} [\mathbf{n}^{\mathrm{T}} \cdot \mathbf{F}(y_{j_1}, \mathbf{p})] \Big|_{y_{j_1}^* = a_{j_1 j_1 1} = a_{j_2 j_2 l_1}} \\ &= \alpha_{j_1 j_2} a_{j_2 j_2 0}(a_{j_2 j_2 l_1} - a_{j_2 j_2 l_2}). \end{aligned} \tag{3.206}$$

The infinite-equilibrium of $y_{j_1}^* = a_{j_1 j_1 1} = a_{j_2 j_2 l_1}$ *has the following properties*:

- *For* $\alpha_{j_1 j_2}^3 a_{j_1 j_1 0} > 0$ *and* $\alpha_{j_1 j_2} a_{j_2 j_2 0}(a_{j_2 j_2 l_1} - a_{j_2 j_2 l_2}) > 0$,

$$(a_{j_1 j_1}, a_{j_2 j_2 l_1}) = \underbrace{(2^{\text{nd}}\text{II}, \text{SO})}_{\text{second-order increasing-inflection source}}. \tag{3.207}$$

- *For* $\alpha_{j_1 j_2}^3 a_{j_1 j_1 0} < 0$ *and* $\alpha_{j_1 j_2} a_{j_2 j_2 0}(a_{j_2 j_2 l_1} - a_{j_2 j_2 l_2}) > 0$,

$$(a_{j_1 j_1 1}, a_{j_2 j_2 l_1}) = \underbrace{(2^{\text{nd}}\text{DI}, \text{SO})}_{\text{second-order decreasing-inflection source}}. \tag{3.208}$$

- *For* $\alpha^3_{j_1j_2}a_{j_1j_10}>0$ *and* $\alpha_{j_1j_2}a_{j_2j_20}(a_{j_2j_2l_1}-a_{j_2j_2l_2})<0$,

$$(a_{j_1j_11},a_{j_2j_2l_1})=\underbrace{(2^{\rm nd}{\rm DI},{\rm SI})}_{\text{second-order decreasing-inflection sink}}. \tag{3.209}$$

- *For* $\alpha^3_{j_1j_2}a_{j_1j_10}<0$ *and* $\alpha_{j_1j_2}a_{j_2j_20}(a_{j_2j_2l_1}-a_{j_2j_2l_2})<0$,

$$(a_{j_1j_11},a_{j_2j_2l_1})=\underbrace{(2^{\rm nd}{\rm II},{\rm SI})}_{\text{second-order increasing-inflection sink}}. \tag{3.210}$$

- *The second-order inflection-source and sink infinite-equilibriums are the switching bifurcations for third-order parabola and parabola flows in the* x_{j_1}- *and* x_{j_2}-*directions.*
- *The second-order inflection-source and sink infinite-equilibriums are the appearing bifurcations from a parabola flow and a parabola flow in the* x_{j_1}- *and* x_{j_2}-*directions to three parabola flows and a parabola flow in the* x_{j_1}- *and* x_{j_2}-*directions.*

(ix) If

$$\Delta_{j_1j_1}>0 \quad \text{and} \quad \Delta_{j_2j_2}>0, \tag{3.211}$$

then

$$\begin{aligned}\dot{x}_{j_1}&=a_{j_1j_10}(y_{j_1}-a_{j_1j_11})(y_{j_1}-a_{j_1j_12})(y_{j_1}-a_{j_1j_13}),\\ \dot{x}_{j_2}&=a_{j_2j_20}(y_{j_1}-a_{j_2j_21})(y_{j_1}-a_{j_2j_22}),\end{aligned} \tag{3.212}$$

where

$$\begin{aligned}&\{a_{j_1j_11},a_{j_1j_12},a_{j_1j_13}\}={\rm sort}\{b_{j_1j_11},b_{j_1j_12},b_{j_1j_13}\},a_{j_1j_1s_1}<a_{j_1j_1s_2},s_1<s_2;\\ &b_{j_1j_12},b_{j_1j_13}=-\frac{1}{2}(B_{j_1j_11}\pm\sqrt{\Delta_{j_1j_1}}),a_{j_1j_10}=A_{j_1j_10};\\ &\{a_{j_2j_21},a_{j_2j_22}\}={\rm sort}\{b_{j_2j_21},b_{j_2j_22}\},a_{j_2j_21}<a_{j_2j_22};\\ &b_{j_2j_11},b_{j_2j_22}=-\frac{1}{2}(B_{j_2j_21}\pm\sqrt{\Delta_{j_2j_2}}),a_{j_2j_20}=A_{j_2j_21}.\end{aligned} \tag{3.213}$$

There is a dynamical system in the y_{j_1}-*direction as*

$$\dot{y}_{j_1}=\alpha_{j_1j_1}a_{j_1j_10}\prod_{s_1=1}^{3}(y_{j_1}-a_{j_1j_1s_1})+\alpha_{j_1j_2}a_{j_2j_20}\prod_{l_1=1}^{2}(y_{j_1}-a_{j_2j_2l_1}), \tag{3.214}$$

with a dynamical system in the $y^{\perp}_{j_1}$-*direction as*

$$\dot{y}^{\perp}_{j_1}=\beta_{j_1j_1}a_{j_1j_10}\prod_{s_1=1}^{3}(y_{j_1}-a_{j_1j_1s_1})+\beta_{j_1j_2}a_{j_2j_20}\prod_{l_1=1}^{2}(y_{j_1}-a_{j_2j_2l_1}). \tag{3.215}$$

(ix_{1a}) *The equilibrium flows of* $y_{j_1}^* = a_{j_1j_1s_1}(s_1, s_2 \in \{1,2\}, s_1 \neq s_2)$ *for the* x_{j_1}*-direction have the following properties with* $\dot{x}_{j_2}^* = a_{j_2j_20}\prod_{l_1=1}^{2}(a_{j_1j_1s_1} - a_{j_2j_2l_1})$:

- *For* $a_{j_1j_10}\alpha_{j_1j_2}\prod_{s_2=1,s_2\neq s_1}^{3}(a_{j_1j_1s_1} - a_{j_1j_1s_2}) > 0$ *and* $a_{j_2j_20}\prod_{l_1=1}^{2}(a_{j_1j_1s_1} - a_{j_2j_2l_1}) > 0$,

$$(a_{j_1j_1s_1}, \dot{x}_{j_2}^*) = \underbrace{(\text{UP}, \text{pF})}_{\text{up-parabola flow }(+)}. \tag{3.216}$$

- *For* $a_{j_1j_10}\alpha_{j_1j_2}\prod_{s_2=1,s_2\neq s_1}^{3}(a_{j_1j_1s_1} - a_{j_1j_1s_2}) < 0$ *and* $a_{j_2j_20}\prod_{l_1=1}^{2}(a_{j_1j_1s_1} - a_{j_2j_2l_1}) > 0$,

$$(a_{j_1j_1s_1}, \dot{x}_{j_2}^*) = \underbrace{(\text{DP}, \text{pF})}_{\text{Down-parabola flow }(+)}. \tag{3.217}$$

- *For* $a_{j_1j_10}\alpha_{j_1j_2}\prod_{s_2=1,s_2\neq s_1}^{3}(a_{j_1j_1s_1} - a_{j_1j_1s_2}) > 0$ *and* $a_{j_2j_20}\prod_{l_1=1}^{2}(a_{j_1j_1s_1} - a_{j_2j_2j_1}) < 0$,

$$(a_{j_1j_1s_1}, \dot{x}_{j_2}^*) = \underbrace{(\text{DP}, \text{nF})}_{\text{Down-parabola flow }(-)}. \tag{3.218}$$

- *For* $a_{j_1j_10}\alpha_{j_1j_2}\prod_{s_2=1,s_2\neq s_1}^{3}(a_{j_1j_1s_1} - a_{j_1j_1s_2}) < 0$ *and* $a_{j_2j_20}\prod_{l_1=1}^{2}(a_{j_1j_1s_1} - a_{j_2j_2l_1}) < 0$,

$$(a_{j_1j_1s_1}, \dot{x}_{j_2}^*) = \underbrace{(\text{UP}, \text{nF})}_{\text{up-parabola flow }(-)}. \tag{3.219}$$

(ix_{1b}) *The equilibrium flows of* $y_{j_1}^* = a_{j_2j_2l_1}$ *for the* x_{j_2}*-direction have the following properties with* $\dot{x}_{j_1}^* = a_{j_1j_10}\prod_{s_1=1}^{3}(a_{j_2j_2l_1} - a_{j_1j_1s_1})$:

- *For* $a_{j_1j_10}\prod_{s_1=1}^{3}(a_{j_2j_2l_1} - a_{j_1j_1s_1}) > 0$ *and* $a_{j_2j_20}\alpha_{j_1j_1}(a_{j_2j_2l_1} - a_{j_1j_1l_2}) > 0$,

$$(\dot{x}_{j_1}^*, a_{j_2j_2l_1}) = \underbrace{(\text{pF}, \text{UP})}_{\text{up-parabola flow }(+)}. \tag{3.220}$$

- *For* $a_{j_1j_10}\prod_{s_1=1}^{3}(a_{j_2j_2l_1} - a_{j_1j_1s_1}) < 0$ *and* $a_{j_2j_20}\alpha_{j_1j_1}(a_{j_2j_2l_1} - a_{j_1j_1l_2}) > 0$,

$$(\dot{x}_{j_1}^*, a_{j_2j_2l_1}) = \underbrace{(\text{nF}, \text{DP})}_{\text{down-parabola flow }(-)}. \tag{3.221}$$

- *For* $a_{j_1j_10}\prod_{s_1=1}^{3}(a_{j_2j_2l_1} - a_{j_1j_1s_1}) > 0$ *and* $a_{j_2j_20}\alpha_{j_1j_1}(a_{j_2j_2l_1} - a_{j_1j_1l_2}) < 0$,

$$(\dot{x}_{j_1}^*, a_{j_2j_2l_1}) = \underbrace{(\mathrm{pF}, \mathrm{DP})}_{\text{down-parabola flow }(+)}. \tag{3.222}$$

- *For* $a_{j_1j_10}\prod_{s_1=1}^{3}(a_{j_2j_2l_1} - a_{j_1j_1s_1}) < 0$ *and* $a_{j_2j_20}\alpha_{j_1j_1}(a_{j_2j_2l_1} - a_{j_1j_1l_2}) < 0$,

$$(\dot{x}_{j_1}^*, a_{j_2j_2l_1}) = \underbrace{(\mathrm{nF}, \mathrm{UP})}_{\text{up-parabola flow }(-)}. \tag{3.223}$$

(ix_2) *For* $a_{j_2j_2l_1} = a_{j_1j_1s_1} (s_1, s_2 \in \{1,2,3\}, l_1, l_2 \in \{1,2\})$,

$$\begin{aligned}
\dot{x}_{j_1} &= a_{j_1j_10}(y_{j_1} - a_{j_1j_1s_1}) \prod_{s_2=1, s_2\neq s_1}^{3} (y_{j_1} - a_{j_1j_1s_2}),\\
\dot{x}_{j_2} &= a_{j_2j_20}(y_{j_1} - a_{j_2j_2l_1})(y_{j_1} - a_{j_2j_2l_2}),\\
& j_1, j_2 \in \{1,2\};\ j_1 \neq j_2; s_1, s_2 \in \{1,2,3\},\ s_1 \neq s_2.
\end{aligned} \tag{3.224}$$

There is a dynamical system in the y_{j_1}*-direction as*

$$\begin{aligned}
\dot{y}_{j_1} = (y_{j_1} - a_{j_1j_1s_1})\Big[&\alpha_{j_1j_1}a_{j_1j_10} \prod_{s_2=1, s_2\neq s_1}^{3} (y_{j_1} - a_{j_1j_1s_2})\\
&+ \alpha_{j_1j_2}a_{j_2j_20}(y_{j_1} - a_{j_2j_2l_2})\Big],
\end{aligned} \tag{3.225}$$

with a dynamical system in the $y_{j_1}^{\perp}$*-direction as*

$$\begin{aligned}
\dot{y}_{j_1}^{\perp} = (y_{j_1} - a_{j_1j_1s_1})\Big[&\beta_{j_1j_1}a_{j_1j_10} \prod_{s_2=1, s_2\neq s_1}^{3} (y_{j_1} - a_{j_1j_1s_2})\\
&+ \beta_{j_1j_2}a_{j_2j_20}(y_{j_1} - a_{j_2j_2l_2})\Big],
\end{aligned} \tag{3.226}$$

where

$$\begin{aligned}
G_{j_1j_10} &= \frac{\partial}{\partial y_{j_1}}[\mathbf{n}^{\mathrm{T}} \cdot \mathbf{F}(y_{j_1}, \mathbf{p})]\Big|_{y_{j_1}^* = a_{j_1j_1s_1} = a_{j_2j_2l_1}}\\
&= \alpha_{j_1j_1}a_{j_1j_10} \prod_{s_2=1, s_2\neq s_1}^{3} (a_{j_1j_1s_1} - a_{j_1j_1s_2})\\
&\quad + \alpha_{j_1j_2}a_{j_2j_20}(a_{j_2j_2l_1} - a_{j_2j_2l_2}).
\end{aligned} \tag{3.227}$$

The infinite-equilibriums of $y_{j_1}^* = a_{j_2j_2l_1} = a_{j_1j_1s_1}$ *have the following properties:*

- *For* $a_{j_1j_10}\prod_{s_2=1,s_2\neq s_1}^{3}(a_{j_1j_1s_1}-a_{j_1j_1s_2})/a_{j_2j_20}(a_{j_2j_2l_1}-a_{j_2j_2l_2})>0$ *and* $G_{j_1j_10}>0$,

$$(a_{j_1j_1s_1}, a_{j_2j_2l_1}) = \underbrace{(\text{II}, \text{SO})}_{\text{increasing-inflection source}}. \tag{3.228}$$

- *For* $a_{j_1j_10}\prod_{s_2=1,s_2\neq s_1}^{3}(a_{j_1j_1s_1}-a_{j_1j_1s_2})/a_{j_2j_20}(a_{j_2j_2l_1}-a_{j_2j_2l_2})>0$ *and* $G_{j_1j_10}<0$,

$$(a_{j_1j_1s_1}, a_{j_2j_2l_1}) = \underbrace{(\text{II}, \text{SI})}_{\text{increasing-inflection sink}}. \tag{3.229}$$

- *For* $a_{j_1j_10}\prod_{s_2=1,s_2\neq s_1}^{3}(a_{j_1j_1s_1}-a_{j_1j_1s_2})/a_{j_2j_20}(a_{j_2j_2l_1}-a_{j_2j_2l_2})<0$ *and* $G_{j_1j_10}>0$,

$$(a_{j_1j_1s_1}, a_{j_2j_2l_1}) = \underbrace{(\text{DI}, \text{SO})}_{\text{decreasing-inflection source}}. \tag{3.230}$$

- *For* $a_{j_1j_10}\prod_{s_2=1,s_2\neq s_1}^{3}(a_{j_1j_1s_1}-a_{j_1j_1s_2})/a_{j_2j_20}(a_{j_2j_2l_1}-a_{j_2j_2l_2})<0$ *and* $G_{j_1j_10}<0$,

$$(a_{j_1j_1s_1}, a_{j_2j_2l_1}) = \underbrace{(\text{DI}, \text{SI})}_{\text{decreasing-inflection sink}}. \tag{3.231}$$

3.2 Proof of Theorem 3.1

Consider a two-dimensional dynamical system as

$$\begin{aligned}
\dot{x}_{j_2} &= a_{j_2j_20}(y_{j_1}^2 + B_{j_2j_2}y_{j_1} + C_{j_2j_2}),\\
\dot{x}_{j_1} &= a_{j_1j_10}(y_{j_1} - b_{j_1j_11})(y_{j_1}^2 + B_{j_1j_1}y_{j_1} + C_{j_1j_1}),
\end{aligned}$$

with

$$\begin{aligned}
&\Delta_{j_2j_2} = (B_{j_2j_2})^2 - 4C_{j_2j_2}, \quad \Delta_{j_1j_1} = (B_{j_1j_1})^2 - 4C_{j_1j_1},\\
&a_{j_1j_10} = A_{j_1j_10}, \quad a_{j_2j_20} = A_{j_2j_21}.
\end{aligned}$$

(i) If

$$\Delta_{j_1j_1} < 0 \quad \text{and} \quad \Delta_{j_2j_2} < 0,$$

then

$$\begin{aligned}
\dot{x}_{j_1} &= a_{j_1j_10}(y_{j_1} - a_{j_1j_11})[(y_{j_1} - a_{j_1j_1})^2 + b_{j_1j_1}],\\
\dot{x}_{j_2} &= a_{j_2j_20}[(y_{j_1} - a_{j_2j_2})^2 + b_{j_2j_2}],
\end{aligned}$$

where

$$a_{j_2j_2} = -\frac{1}{2}B_{j_2j_2}, \quad b_{j_2j_2} = -\frac{1}{4}\Delta_{j_2j_2},$$
$$a_{j_1j_1} = -\frac{1}{2}B_{j_1j_1}, \quad b_{j_1j_1} = -\frac{1}{4}\Delta_{j_2j_2}, \quad a_{j_1j_11} = b_{j_1j_11}.$$

From the original systems, there is a dynamical system in the y_{j_1}-direction as

$$\begin{aligned}\dot{y}_{j_1} &= \alpha_{j_1j_1}a_{j_1j_10}(y_{j_1} - a_{j_1j_11})[(y_{j_1} - a_{j_1j_1})^2 + b_{j_1j_1}]\\ &\quad + \alpha_{j_1j_2}a_{j_2j_20}[(y_{j_1} - a_{j_2j_2})^2 + b_{j_2j_2}],\end{aligned}$$

with a dynamical system in the $y_{j_1}^{\perp}$-direction as

$$\begin{aligned}\dot{y}_{j_1}^{\perp} &= \beta_{j_1j_1}a_{j_1j_10}(y_{j_1} - a_{j_1j_11})[(y_{j_1} - a_{j_1j_1})^2 + b_{j_1j_1}]\\ &\quad + \beta_{j_1j_2}a_{j_2j_20}[(y_{j_1} - a_{j_2j_2})^2 + b_{j_2j_2}].\end{aligned}$$

In phase space, at $y_1^* = a_{j_1j_11}$,

$$\left.\frac{dx_{j_1}}{dx_{j_2}}\right|_{y_1^*=a_{j_1j_11}} = \left.\frac{a_{j_1j_10}(y_{j_1} - a_{j_1j_11})[(y_{j_1} - a_{j_1j_1})^2 + b_{j_1j_1}]}{a_{j_2j_20}[(y_{j_1} - a_{j_2j_2})^2 + b_{j_2j_2}]}\right|_{y_1^*=a_{j_1j_11}} = 0.$$

If

$$\left.\frac{d^2x_{j_1}}{dx_{j_2}^2}\right|_{y_1^*=a_{j_1j_11}} = \frac{a_{j_1j_10}\alpha_{j_1j_2}[(a_{j_1j_11} - a_{j_1j_1})^2 + b_{j_1j_1}]}{a_{j_2j_20}[(a_{j_1j_11} - a_{j_2j_2})^2 + b_{j_2j_2}]} > 0,$$

there is an up-parabola flow at $y_1^* = a_{j_1j_11}$. If

$$\left.\frac{d^2x_{j_1}}{dx_{j_2}^2}\right|_{y_1^*=a_{j_1j_11}} = \frac{a_{j_1j_10}\alpha_{j_1j_2}[(a_{j_1j_11} - a_{j_1j_1})^2 + b_{j_1j_1}]}{a_{j_2j_20}[(a_{j_1j_11} - a_{j_2j_2})^2 + b_{j_2j_2}]} < 0,$$

there is a down-parabola flow at $y_1^* = a_{j_1j_11}$. With

$$\dot{x}_{j_2}^* = a_{j_2j_20}[(a_{j_1j_11} - a_{j_2j_2})^2 + b_{j_2j_2}],$$

the parabola flows are positive and negative for $\dot{x}_{j_2}^* > 0$ and $\dot{x}_{j_2}^* < 0$, respectively.

Due to

$$(a_{j_1j_11} - a_{j_1j_1})^2 + b_{j_1j_1} > 0 \quad \text{and} \quad (a_{j_1j_11} - a_{j_2j_2})^2 + b_{j_2j_2} > 0,$$

the flow of $y_1^* = a_{j_1j_11}$ with $\dot{x}_{j_2}^* = a_{j_2j_20}[(a_{j_1j_11} - a_{j_2j_2})^2 + b_{j_2j_2}]$ possesses the following properties as in Eqs. (3.10)–(3.13):

- For $a_{j_1j_1 0}\alpha_{j_1j_2} > 0$ and $a_{j_2j_2 0} > 0$,

$$(a_{j_1j_1 1}, \dot{x}^*_{j_2}) = \underbrace{(\mathrm{UP}, \mathrm{pF})}_{\text{up-parabola flow }(+)}.$$

- For $a_{j_1j_1 0}\alpha_{j_1j_2} < 0$ and $a_{j_2j_2 0} > 0$,

$$(a_{j_1j_1 1}, \dot{x}^*_{j_2}) = \underbrace{(\mathrm{DP}, \mathrm{pF})}_{\text{down-parabola flow }(+)}.$$

- For $a_{j_1j_1 0}\alpha_{j_1j_2} > 0$ and $a_{j_2j_2 0} < 0$,

$$(a_{j_1j_1 1}, \dot{x}^*_{j_2}) = \underbrace{(\mathrm{DP}, \mathrm{nF})}_{\text{down-parabola flow }(-)}.$$

- For $a_{j_1j_1 0}\alpha_{j_1j_2} < 0$ and $a_{j_2j_2 0} < 0$,

$$(a_{j_1j_1 1}, \dot{x}^*_{j_2}) = \underbrace{(\mathrm{UP}, \mathrm{nF})}_{\text{up-parabola flow }(-)}.$$

(ii) If

$$\Delta_{j_1j_1} = 0 \quad \text{and} \quad \Delta_{j_2j_2} < 0,$$

then

$$\begin{aligned}
\dot{x}_{j_1} &= a_{j_1j_1 0}(y_{j_1} - a_{j_1j_1 s_1})(y_{j_1} - a_{j_1j_1 s_2})^2,\\
\dot{x}_{j_2} &= a_{j_2j_2 0}[(y_{j_1} - a_{j_2j_2})^2 + b_{j_2j_2}],
\end{aligned}$$

where

$$\begin{aligned}
a_{j_2j_2} &= -\frac{1}{2}B_{j_2j_2}, \quad b_{j_2j_2} = -\frac{1}{4}\Delta_{j_2j_2};\\
a_{j_1j_1 s_w} &= -\frac{1}{2}B_{j_1j_1}, \quad b_{j_1j_1} = 0, \quad a_{j_1j_1 s_1} = b_{j_1j_1}.
\end{aligned}$$

(ii_1) For $a_{j_1j_1 s_1} \neq a_{j_1j_1 s_2}$, a dynamical system in the y_{j_1}-direction can be constructed as

$$\begin{aligned}
\dot{y}_{j_1} = {} & \alpha_{j_1j_1}a_{j_1j_1 0}(y_{j_1} - a_{j_1j_1 s_1})(y_{j_1} - a_{j_1j_1 s_2})^2\\
& + \alpha_{j_1j_2}a_{j_2j_2 0}[(y_{j_1} - a_{j_2j_2})^2 + b_{j_2j_2}],
\end{aligned}$$

and a dynamical system in the $y^{\perp}_{j_1}$-direction is constructed as

$$\begin{aligned}
\dot{y}^{\perp}_{j_1} = {} & \beta_{j_1j_1}a_{j_1j_1 0}(y_{j_1} - a_{j_1j_1 s_1})(y_{j_1} - a_{j_1j_1 s_2})^2\\
& + \beta_{j_1j_2}a_{j_2j_2 0}[(y_{j_1} - a_{j_2j_2})^2 + b_{j_2j_2}].
\end{aligned}$$

(ii$_{1a}$) In phase space, at $y_1^* = a_{j_1j_1s_1}$,

$$\left.\frac{dx_{j_1}}{dx_{j_2}}\right|_{y_1^*=a_{j_1j_1s_1}} = \frac{a_{j_1j_10}(y_{j_1}-a_{j_1j_1s_1})(y_{j_1}-a_{j_1j_1s_2})^2}{a_{j_2j_20}[(y_{j_1}-a_{j_2j_2})^2+b_{j_2j_2}]}|_{y_1^*=a_{j_1j_1s_1}} = 0.$$

If

$$\left.\frac{d^2x_{j_1}}{dx_{j_2}^2}\right|_{y_1^*=a_{j_1j_1s_1}} = \frac{a_{j_1j_10}\alpha_{j_1j_2}(a_{j_1j_1s_1}-a_{j_1j_1s_2})^2}{a_{j_2j_20}[(a_{j_1j_1s_1}-a_{j_2j_2})^2+b_{j_2j_2}]} > 0,$$

there is an up-parabola flow at $y_1^* = a_{j_1j_1s_1}$. If

$$\left.\frac{d^2x_{j_1}}{dx_{j_2}^2}\right|_{y_1^*=a_{j_1j_1s_1}} = \frac{a_{j_1j_10}\alpha_{j_1j_2}(a_{j_1j_1s_1}-a_{j_1j_1s_2})^2}{a_{j_2j_20}[(a_{j_1j_1s_1}-a_{j_2j_2})^2+b_{j_2j_2}]} < 0,$$

there is a down-parabola flow at $y_1^* = a_{j_1j_1s_1}$. With

$$\dot{x}_{j_2}^* = a_{j_2j_20}[(a_{j_1j_1s_1}-a_{j_2j_2})^2+b_{j_2j_2}],$$

the parabola flows are positive and negative for $\dot{x}_{j_2}^* > 0$ and $\dot{x}_{j_2}^* < 0$, respectively.

Due to

$$(a_{j_1j_1s_1}-a_{j_2j_2})^2+b_{j_2j_2} > 0,$$

the flow of $y_1^* = a_{j_1j_1s_1} \neq a_{j_1j_1s_2}$ with $\dot{x}_{j_2}^* = a_{j_2j_20}[(a_{j_1j_1s_1}-a_{j_2j_2})^2+b_{j_2j_2}]$ possesses the following properties as in Eqs. (3.19)–(3.22):

- For $a_{j_1j_10}\alpha_{j_1j_2}(a_{j_1j_1s_1}-a_{j_1j_1s_2})^2 > 0$ and $a_{j_2j_20} > 0$,

$$(a_{j_1j_1s_1}, \dot{x}_{j_2}^*) = \underbrace{(\text{UP}, \text{pF})}_{\text{up-parabola flow } (+)}.$$

- For $a_{j_1j_10}\alpha_{j_1j_2}(a_{j_1j_1s_1}-a_{j_1j_1s_2})^2 < 0$ and $a_{j_2j_20} > 0$,

$$(a_{j_1j_1s_1}, \dot{x}_{j_2}^*) = \underbrace{(\text{DP}, \text{pF})}_{\text{down-parabola flow } (+)}.$$

- For $a_{j_1j_10}\alpha_{j_1j_2}(a_{j_1j_1s_1}-a_{j_1j_1s_2})^2 > 0$ and $a_{j_2j_20} < 0$,

$$(a_{j_1j_1s_1}, \dot{x}_{j_2}^*) = \underbrace{(\text{DP}, \text{nF})}_{\text{down-parabola flow } (-)}.$$

- For $a_{j_1j_10}\alpha_{j_1j_2}(a_{j_1j_1s_1}-a_{j_1j_1s_2})^2 < 0$ and $a_{j_2j_20} < 0$,

$$(a_{j_1j_1s_1}, \dot{x}_{j_2}^*) = \underbrace{(\text{UP}, \text{nF})}_{\text{up-parabola flow } (-)}.$$

(ii_{1b}) In phase space, at $y_1^* = a_{j_1j_1s_2}$,

$$\left.\frac{dx_{j_1}}{dx_{j_2}}\right|_{y_1^*=a_{j_1j_1s_2}} = \left.\frac{a_{j_1j_10}(y_{j_1} - a_{j_1j_1s_1})(y_{j_1} - a_{j_1j_1s_2})^2}{a_{j_2j_20}[(y_{j_1} - a_{j_2j_2})^2 + b_{j_2j_2}]}\right|_{y_1^*=a_{j_1j_1s_2}} = 0,$$

$$\left.\frac{d^2x_{j_1}}{dx_{j_2}^2}\right|_{y_1^*=a_{j_1j_1s_2}} = 0.$$

If

$$\left.\frac{d^3x_{j_1}}{dx_{j_2}^3}\right|_{y_1^*=a_{j_1j_1s_2}} = \frac{2a_{j_1j_10}\alpha_{j_1j_2}^2(a_{j_1j_1s_2} - a_{j_1j_1s_1})}{a_{j_2j_20}[(a_{j_1j_1s_2} - a_{j_2j_2})^2 + b_{j_2j_2}]} > 0,$$

there is an increasing-inflection flow at $y_1^* = a_{j_1j_1s_2}$. If

$$\left.\frac{d^3x_{j_1}}{dx_{j_2}^3}\right|_{y_1^*=a_{j_1j_1s_2}} = \frac{2a_{j_1j_10}\alpha_{j_1j_2}^2(a_{j_1j_1s_2} - a_{j_1j_1s_1})}{a_{j_2j_20}[(a_{j_1j_1s_2} - a_{j_2j_2})^2 + b_{j_2j_2}]} < 0,$$

there is a decreasing-inflection flow at $y_1^* = a_{j_1j_1s_2}$. With

$$\dot{x}_{j_2}^* = a_{j_2j_20}[(a_{j_1j_1s_2} - a_{j_2j_2})^2 + b_{j_2j_2}],$$

the inflection flows are positive and negative for $\dot{x}_{j_2}^* > 0$ and $\dot{x}_{j_2}^* < 0$, respectively.

Due to

$$(a_{j_1j_1s_2} - a_{j_2j_2})^2 + b_{j_2j_2} > 0,$$

the flow of $y_1^* = a_{j_1j_1s_2} \neq a_{j_1j_1s_1}$ with $\dot{x}_{j_2}^* = a_{j_2j_20}[(a_{j_1j_1s_2} - a_{j_2j_2})^2 + b_{j_2j_2}]$ possesses the following properties as in Eqs. (3.23)–(3.26):

- For $a_{j_1j_10}\alpha_{j_1j_2}^2(a_{j_1j_1s_2} - a_{j_1j_1s_1}) > 0$ and $a_{j_2j_20} > 0$,

$$(a_{j_1j_1s_2}, \dot{x}_{j_2}^*) = \underbrace{(\text{II}, \text{pF})}_{\text{increasing-inflection flow } (+)}.$$

- For $a_{j_1j_10}\alpha_{j_1j_2}^2(a_{j_1j_1s_2} - a_{j_1j_1s_1}) < 0$ and $a_{j_2j_20} > 0$,

$$(a_{j_1j_1s_2}, \dot{x}_{j_2}^*) = \underbrace{(\text{DI}, \text{pF})}_{\text{decreasing-inflection flow } (+)}.$$

- For $a_{j_1j_10}\alpha_{j_1j_2}^2(a_{j_1j_1s_2} - a_{j_1j_1s_1}) > 0$ and $a_{j_2j_20} < 0$,

$$(a_{j_1j_1s_2}, \dot{x}_{j_2}^*) = \underbrace{(\text{DI}, \text{nF})}_{\text{decreasing-inflection flow } (-)}.$$

- For $a_{j_1j_10}\alpha^2_{j_1j_2}(a_{j_1j_1s_2} - a_{j_1j_1s_1}) < 0$ and $a_{j_2j_20} < 0$,

$$(a_{j_1j_1s_2}, \dot{x}^*_{j_2}) = \underbrace{(\text{II}, \text{nF})}_{\text{increasing-inflection flow }(-)}.$$

– The inflection flows are the appearing bifurcations of up-parabola and down-parabola flows.

(ii$_2$) For $a_{j_1j_1s_1} = a_{j_1j_1s_2} = a_{j_1j_1 1}$, consider the dynamical system as

$$\dot{x}_{j_1} = a_{j_1j_10}(y_{j_1} - a_{j_1j_11})^3,$$
$$\dot{x}_{j_2} = a_{j_2j_20}[(y_{j_1} - a_{j_2j_2})^2 + b_{j_2j_2}].$$

A dynamical system in the y_{j_1}-direction is constructed as

$$\dot{y}_{j_1} = \alpha_{j_1j_1}a_{j_1j_10}(y_{j_1} - a_{j_1j_11})^3 + \alpha_{j_1j_2}a_{j_2j_20}[(y_{j_1} - a_{j_2j_2})^2 + b_{j_2j_2}],$$

and a dynamical system in the $y^{\perp}_{j_1}$-direction is constructed as

$$\dot{y}^{\perp}_{j_1} = \beta_{j_1j_1}a_{j_1j_10}(y_{j_1} - a_{j_1j_11})^3 + \beta_{j_1j_2}a_{j_2j_20}[(y_{j_1} - a_{j_2j_2})^2 + b_{j_2j_2}].$$

In phase space, at $y^*_1 = a_{j_1j_11}$,

$$\left.\frac{dx_{j_1}}{dx_{j_2}}\right|_{y^*_1=a_{j_1j_11}} = \left.\frac{a_{j_1j_10}(y_{j_1} - a_{j_1j_11})^3}{a_{j_2j_20}[(y_{j_1} - a_{j_2j_2})^2 + b_{j_2j_2}]}\right|_{y^*_1=a_{j_1j_11}} = 0,$$

$$\left.\frac{d^2x_{j_1}}{dx^2_{j_2}}\right|_{y^*_1=a_{j_1j_11}} = \left.\frac{d^3x_{j_1}}{dx^3_{j_2}}\right|_{y^*_1=a_{j_1j_11}} = 0.$$

If

$$\left.\frac{d^4x_{j_1}}{dx^4_{j_2}}\right|_{y^*_1=a_{j_1j_11}} = \frac{6a_{j_1j_10}\alpha^3_{j_1j_2}}{a_{j_2j_20}[(a_{j_1j_11} - a_{j_2j_2})^2 + b_{j_2j_2}]} > 0,$$

there is a third-order up-parabola flow at $y^*_1 = a_{j_1j_11}$. If

$$\left.\frac{d^3x_{j_1}}{dx^3_{j_2}}\right|_{y^*_1=a_{j_1j_11}} = \frac{6a_{j_1j_10}\alpha^3_{j_1j_2}}{a_{j_2j_20}[(a_{j_1j_11} - a_{j_2j_2})^2 + b_{j_2j_2}]} < 0,$$

there is a third-order down-parabola flow at $y^*_1 = a_{j_1j_11}$. With

$$\dot{x}^*_{j_2} = a_{j_2j_20}[(a_{j_1j_11} - a_{j_2j_2})^2 + b_{j_2j_2}],$$

the third-order parabola flows are positive and negative for $\dot{x}^*_{j_2} > 0$ and $\dot{x}^*_{j_2} < 0$, respectively.

Due to

$$(a_{j_1j_1s_2} - a_{j_2j_2})^2 + b_{j_2j_2} > 0,$$

the flow of $y_1^* = a_{j_1j_1 1}$ with $\dot{x}_{j_2}^* = a_{j_2j_2 0}[(a_{j_1j_1 1} - a_{j_2j_2})^2 + b_{j_2j_2}]$ possesses the following properties as in Eqs. (3.30)–(3.33):

- For $a_{j_1j_1 0}\alpha_{j_1j_2}^3 > 0$ and $a_{j_2j_2 0} > 0$,
$$(a_{j_1j_1 1}, \dot{x}_{j_2}^*) = \underbrace{(3^{\text{rd}}\text{UP}, \text{pF})}_{\text{third-order up-parabola flow }(+)}$$
- For $a_{j_1j_1 0}\alpha_{j_1j_2}^3 < 0$ and $a_{j_2j_2 0} > 0$,
$$(a_{j_1j_1 1}, \dot{x}_{j_2}^*) = \underbrace{(3^{\text{rd}}\text{DP}, \text{pF})}_{\text{third-order down-parabola flow }(+)}.$$
- For $a_{j_1j_1 0}\alpha_{j_1j_2}^3 > 0$ and $a_{j_2j_2 0} < 0$,
$$(a_{j_1j_1}, \dot{x}_{j_2}^*) = \underbrace{(3^{\text{rd}}\text{DP}, \text{nF})}_{\text{third-order down-parabola flow }(-)}.$$
- For $a_{j_1j_1 0}\alpha_{j_1j_2}^3 < 0$ and $a_{j_2j_2 0} < 0$,
$$(a_{j_1j_1 1}, \dot{x}_{j_2}^*) = \underbrace{(3^{\text{rd}}\ \text{UP,nF})}_{\text{third-order up-parabola flow }(-)}.$$

– The third-order up-parabola flows are the appearing bifurcations of upparabola, down-parabola, and up-parabola flows.
– The third-order down-parabola flows are the appearing bifurcations of down-parabola, up-parabola, and down-parabola flows.

(iii) If

$$\Delta_{j_1j_1} > 0 \quad \text{and} \quad \Delta_{j_2j_2} < 0$$

then

$$\begin{aligned}
\dot{x}_{j_1} &= a_{j_1j_1 0}(y_{j_1} - a_{j_1j_1 1})(y_{j_1} - a_{j_1j_1 2})(y_{j_1} - a_{j_1j_1 3}),\\
\dot{x}_{j_2} &= a_{j_2j_2 0}[(y_{j_1} - a_{j_2j_2})^2 + b_{j_2j_2}],
\end{aligned}$$

where

$$\begin{aligned}
&a_{j_2j_2} = -\frac{1}{2}B_{j_2j_2}, \quad b_{j_2j_2} = -\frac{1}{4}\Delta_{j_2j_2};\\
&\{a_{j_1j_1 1}, a_{j_1j_1 2}, a_{j_1j_1 3}\} = \text{sort}\{b_{j_1j_1 1}, b_{j_1j_1 2}, b_{j_1j_1 3}\},\\
&\qquad a_{j_1j_1s_1} < a_{j_1j_1s_2}, s_1 < s_2;\\
&b_{j_1j_1 2}, b_{j_1j_1 3} = -\frac{1}{2}(B_{j_1j_1} \pm \sqrt{\Delta_{j_1j_1}}), a_{j_1j_1 0} = A_{j_1j_1 0}.
\end{aligned}$$

A dynamical system in the y_{j_1}-direction is constructed as

$$\dot{y}_{j_1} = \alpha_{j_1 j_1} a_{j_1 j_1 0} \prod_{s_1=1}^{3} (y_{j_1} - a_{j_1 j_1 s_1}) + \alpha_{j_1 j_2} a_{j_2 j_2 0}[(y_{j_1} - a_{j_2 j_2})^2 + b_{j_2 j_2}],$$

and a dynamical system in the $y_{j_1}^{\perp}$-direction is constructed as

$$\dot{y}_{j_1}^{\perp} = \beta_{j_1 j_1} a_{j_1 j_1 0} \prod_{s_1=1}^{3} (y_{j_1} - a_{j_1 j_1 s_1}) + \beta_{j_1 j_2} a_{j_2 j_2 0}[(y_{j_1} - a_{j_2 j_2})^2 + b_{j_2 j_2}].$$

In phase space, at $y_1^* = a_{j_1 j_1 s_1}$,

$$\left.\frac{dx_{j_1}}{dx_{j_2}}\right|_{y_1^* = a_{j_1 j_1 s_1}} = \left.\frac{a_{j_1 j_1 0}(y_{j_1} - a_{j_1 j_1 s_1}) \prod_{s_2=1, s_2 \neq s_1}^{3} (y_{j_1} - a_{j_1 j_1 s_2})}{a_{j_2 j_2 0}[(y_{j_1} - a_{j_2 j_2})^2 + b_{j_2 j_2}]}\right|_{y_1^* = a_{j_1 j_1 s_1}} = 0.$$

If

$$\left.\frac{d^2 x_{j_1}}{dx_{j_2}^2}\right|_{y_1^* = a_{j_1 j_1 s_1}} = \frac{a_{j_1 j_1 0} \prod_{s_2=1, s_2 \neq s_1}^{3} (a_{j_1 j_1 s_1} - a_{j_1 j_1 s_2})}{a_{j_2 j_2 0}[(a_{j_1 j_1 s_1} - a_{j_2 j_2})^2 + b_{j_2 j_2}]} > 0,$$

there is an up-parabola flow at $y_1^* = a_{j_1 j_1 s_1}$. If

$$\left.\frac{d^2 x_{j_1}}{dx_{j_2}^2}\right|_{y_1^* = a_{j_1 j_1 s_1}} = \frac{a_{j_1 j_1 0} \prod_{s_2=1, s_2 \neq s_1}^{3} (a_{j_1 j_1 s_1} - a_{j_1 j_1 s_2})}{a_{j_2 j_2 0}[(a_{j_1 j_1 s_1} - a_{j_2 j_2})^2 + b_{j_2 j_2}]} < 0,$$

there is a down-parabola flow at $y_1^* = a_{j_1 j_1 s_1}$. With

$$\dot{x}_{j_2}^* = a_{j_2 j_2 0}[(a_{j_1 j_1 s_1} - a_{j_2 j_2})^2 + b_{j_2 j_2}],$$

the parabola flows are positive and negative for $\dot{x}_{j_2}^* > 0$ and $\dot{x}_{j_2}^* < 0$, respectively.

With

$$(a_{j_1 j_1 s_1} - a_{j_2 j_2})^2 + b_{j_2 j_2} > 0,$$

the equilibrium flow of $y_{j_1}^* = a_{j_1 j_1 s_1} (s_1, s_2 \in \{1, 2, 3\}, s_1 \neq s_2)$ in the x_{j_1}-direction possesses the following properties as in Eqs. (3.40)–(3.43):

- For $a_{j_1 j_1 0} \alpha_{j_1 j_2} \prod_{s_2=1, s_2 \neq s_1}^{3} (a_{j_1 j_1 s_1} - a_{j_1 j_1 s_2}) > 0$ and $a_{j_2 j_2 0} > 0$,

$$(a_{j_1 j_1 s_1}, \dot{x}_{j_2}^*) = \underbrace{(\mathrm{UP}, \mathrm{pF})}_{\text{up-parabola flow }(+)}.$$

- For $a_{j_1j_1}\alpha_{j_1j_2}\prod_{s_2=1,s_2\neq s_1}^{3}(a_{j_1j_1s_1}-a_{j_1j_1s_2})<0$ and $a_{j_2j_20}>0$,
$$(a_{j_1j_1s_1},\dot{x}_{j_2}^*)=\underbrace{(\text{DP},\text{pF})}_{\text{down-parabola flow }(+)}.$$
- For $a_{j_1j_10}\alpha_{j_1j_2}\prod_{s_2=1,s_2\neq s_1}^{3}(a_{j_1j_1s_1}-a_{j_1j_1s_2})>0$ and $a_{j_2j_20}<0$,
$$(a_{j_1j_1s_1},\dot{x}_{j_2}^*)=\underbrace{(\text{DP},\text{nF})}_{\text{down-parabola flow }(-)}.$$
- For $a_{j_1j_10}\alpha_{j_1j_2}\prod_{s_2=1,s_2\neq s_1}^{3}(a_{j_1j_1s_1}-a_{j_1j_1s_2})<0$ and $a_{j_2j_20}<0$,
$$(a_{j_1j_1s_1},\dot{x}_{j_2}^*)=\underbrace{(\text{UP},\text{nF})}_{\text{up-parabola flow }(-)}.$$

(iv) If
$$\Delta_{j_1j_1}<0\quad\text{and}\quad\Delta_{j_2j_2}=0,$$
then
$$\begin{aligned}\dot{x}_{j_1}&=a_{j_1j_10}(y_{j_1}-a_{j_1j_11})[(y_{j_1}-a_{j_1j_1})^2+b_{j_1j_1}],\\ \dot{x}_{j_2}&=a_{j_2j_20}(y_{j_1}-a_{j_2j_21})^2,\end{aligned}$$
where
$$a_{j_2j_21}=-\frac{1}{2}B_{j_2j_2},\quad b_{j_2j_2}=0,\quad a_{j_2j_20}=A_{j_2j_21};$$
$$a_{j_1j_11}=b_{j_1j_11},\quad a_{j_1j_1}=-\frac{1}{2}B_{j_1j_11},\quad b_{j_1j_1}=-\frac{1}{4}\Delta_{j_1j_1},\quad a_{j_1j_10}=A_{j_1j_10}.$$
(iv$_1$) For $a_{j_1j_11}\neq a_{j_2j_21}$, a dynamical system in the y_{j_1}-direction is constructed as
$$\begin{aligned}\dot{y}_{j_1}&=\alpha_{j_1j_1}a_{j_1j_10}(y_{j_1}-a_{j_1j_11})[(y_{j_1}-a_{j_1j_1})^2+b_{j_1j_1}]\\ &\quad+\alpha_{j_1j_2}a_{j_2j_20}(y_{j_1}-a_{j_2j_21})^2,\end{aligned}$$
and a dynamical system in the $y_{j_1}^{\perp}$-direction is constructed as
$$\begin{aligned}\dot{y}_{j_1}^{\perp}&=\beta_{j_1j_1}a_{j_1j_10}(y_{j_1}-a_{j_1j_11})[(y_{j_1}-a_{j_1j_1})^2+b_{j_1j_1}]\\ &\quad+\beta_{j_1j_2}a_{j_2j_20}(y_{j_1}-a_{j_2j_21})^2.\end{aligned}$$
(iv$_{1a}$) In phase space, at $y_1^*=a_{j_1j_11}$,
$$\left.\frac{dx_{j_1}}{dx_{j_2}}\right|_{y_1^*=a_{j_1j_11}}=\left.\frac{a_{j_1j_10}(y_{j_1}-a_{j_1j_11})[(y_{j_1}-a_{j_1j_1})^2+b_{j_1j_1}]}{a_{j_2j_20}(y_{j_1}-a_{j_2j_21})^2}\right|_{y_1^*=a_{j_1j_11}}=0.$$
If
$$\left.\frac{d^2x_{j_1}}{dx_{j_2}^2}\right|_{y_1^*=a_{j_1j_11}}=\frac{a_{j_1j_10}\alpha_{j_1j_2}[(a_{j_1j_1}-a_{j_1j_1})^2+b_{j_1j_1}]}{a_{j_2j_20}(a_{j_1j_11}-a_{j_2j_21})^2}>0,$$

there is an up-parabola flow at $y_1^* = a_{j_1j_11}$. If

$$\left.\frac{d^2x_{j_1}}{dx_{j_2}^2}\right|_{y_1^*=a_{j_1j_11}} = \frac{a_{j_1j_10}\alpha_{j_1j_2}[(a_{j_1j_11}-a_{j_1j_1})^2+b_{j_1j_1}]}{a_{j_2j_20}(a_{j_1j_11}-a_{j_2j_21})^2} < 0,$$

there is a down-parabola flow at $y_1^* = a_{j_1j_11}$. With

$$\dot{x}_{j_2}^* = a_{j_2j_20}(a_{j_1j_11}-a_{j_2j_21})^2,$$

the parabola flows are positive and negative for $\dot{x}_{j_2}^* > 0$ and $\dot{x}_{j_2}^* < 0$, respectively.

Due to

$$(a_{j_1j_11}-a_{j_1j_1})^2+b_{j_1j_1} > 0,$$

the flow of $y_1^* = a_{j_1j_11}$ with $\dot{x}_{j_2}^* = a_{j_2j_20}(a_{j_1j_11}-a_{j_2j_21})^2$ possesses the following properties as in Eqs. (3.49)–(3.52):

- For $\alpha_{j_1j_2}a_{j_1j_10} > 0$ and $a_{j_2j_20}(a_{j_1j_11}-a_{j_2j_21})^2 > 0$,

$$(a_{j_1j_11}, \dot{x}_{j_2}^*) = \underbrace{(\mathrm{UP}, \mathrm{pF})}_{\text{up-parabola flow } (+)}.$$

- For $\alpha_{j_1j_2}a_{j_1j_10} < 0$ and $a_{j_2j_20}(a_{j_1j_11}-a_{j_2j_21})^2 > 0$,

$$(a_{j_1j_11}, \dot{x}_{j_2}^*) = \underbrace{(\mathrm{DP}, \mathrm{pF})}_{\text{down-parabola flow } (+)}.$$

- For $\alpha_{j_1j_2}a_{j_1j_10} > 0$ and $a_{j_2j_20}(a_{j_1j_11}-a_{j_2j_21})^2 < 0$,

$$(a_{j_1j_11}, \dot{x}_{j_2}^*) = \underbrace{(\mathrm{DP}, \mathrm{nF})}_{\text{down-parabola flow } (-)}.$$

- For $\alpha_{j_1j_2}a_{j_1j_10} < 0$ and $a_{j_2j_20}(a_{j_1j_11}-a_{j_2j_21})^2 < 0$,

$$(a_{j_1j_11}, \dot{x}_{j_2}^*) = \underbrace{(\mathrm{UP}, \mathrm{nF})}_{\text{up-parabola flow } (-)}.$$

(iv$_{1b}$) In phase space, at $y_1^* = a_{j_2 j_2 1}$,

$$\left.\frac{dx_{j_2}}{dx_{j_1}}\right|_{y_1^*=a_{j_2j_21}} = \left.\frac{a_{j_2j_20}(y_{j_1} - a_{j_2j_21})^2}{a_{j_1j_10}(y_{j_1} - a_{j_1j_11})[(y_{j_1} - a_{j_1j_1})^2 + b_{j_1j_1}]}\right|_{y_1^*=a_{j_2j_21}} = 0,$$

$$\left.\frac{d^2x_{j_2}}{dx_{j_1}^2}\right|_{y_1^*=a_{j_2j_21}} = 0.$$

If

$$\left.\frac{d^3x_{j_2}}{dx_{j_1}^3}\right|_{y_1^*=a_{j_2j_21}} = \frac{2a_{j_2j_20}\alpha_{j_1j_1}^2}{a_{j_1j_10}(a_{j_2j_21} - a_{j_1j_11})[(a_{j_2j_21} - a_{j_1j_1})^2 + b_{j_1j_1}]} > 0,$$

there is an increasing-inflection flow at $y_1^* = a_{j_2j_21}$. If

$$\left.\frac{d^3x_{j_2}}{dx_{j_1}^3}\right|_{y_1^*=a_{j_2j_21}} = \frac{2a_{j_2j_20}\alpha_{j_1j_1}^2}{a_{j_1j_10}(a_{j_2j_21} - a_{j_1j_11})[(a_{j_2j_21} - a_{j_1j_1})^2 + b_{j_1j_1}]} > 0,$$

there is a decreasing-inflection flow at $y_1^* = a_{j_2j_21}$. With

$$\dot{x}_{j_1}^* = a_{j_1j_10}(a_{j_2j_21} - a_{j_1j_11})[(a_{j_2j_21} - a_{j_1j_1})^2 + b_{j_1j_1}],$$

the inflection flows are positive and negative for $\dot{x}_{j_1}^* > 0$ and $\dot{x}_{j_1}^* < 0$, respectively.

Due to

$$(a_{j_2j_21} - a_{j_1j_1})^2 + b_{j_1j_1} > 0,$$

the flow of $y_1^* = a_{j_2j_21}$ with $\dot{x}_{j_1}^* = a_{j_1j_10}(a_{j_2j_21} - a_{j_1j_11})[(a_{j_2j_21} - a_{j_1j_1})^2 + b_{j_1j_1}]$ possesses the following properties as in Eqs. (3.53)–(3.56):

- For $a_{j_1j_10}(a_{j_2j_21} - a_{j_1j_11}) > 0$ and $a_{j_2j_20}\alpha_{j_1j_1}^2 > 0$,

$$(\dot{x}_{j_1}^*, a_{j_2j_21}) = \underbrace{(\text{pF}, \text{II})}_{\text{increasing-inflection flow } (+)}.$$

- For $a_{j_1j_10}(a_{j_2j_21} - a_{j_1j_11}) < 0$ and $a_{j_2j_20}\alpha_{j_1j_1}^2 > 0$,

$$(\dot{x}_{j_1}^*, a_{j_2j_21}) = \underbrace{(\text{nF}, \text{DI})}_{\text{decreasing-inflection flow } (-)}.$$

- For $a_{j_1j_10}(a_{j_2j_21} - a_{j_1j_11}) > 0$ and $a_{j_2j_20}\alpha_{j_1j_1}^2 < 0$,

$$(\dot{x}_{j_1}^*, a_{j_2j_21}) = \underbrace{(\text{pF}, \text{DI})}_{\text{decreasing-inflection flow } (+)}.$$

- For $a_{j_1j_10}(a_{j_2j_21} - a_{j_1j_11}) < 0$ and $a_{j_2j_20}\alpha^2_{j_1j_1} < 0$,

$$(\dot{x}^*_{j_1}, a_{j_2j_21}) = \underbrace{(\mathrm{nF}, \mathrm{II})}_{\text{increasing-inflection flow } (-)}.$$

(iv$_2$) For $a_{j_1j_11} = a_{j_2j_21}$,

$$\begin{aligned}\dot{x}_{j_1} &= a_{j_1j_10}(y_{j_1} - a_{j_1j_11})[(y_{j_1} - a_{j_1j_1})^2 + b_{j_1j_1}],\\ \dot{x}_{j_2} &= a_{j_2j_20}(y_{j_1} - a_{j_2j_21})^2.\end{aligned}$$

A dynamical system in the y_{j_1}-direction for $a_{j_1j_11} = a_{j_2j_21}$ is constructed as

$$\begin{aligned}\dot{y}_{j_1} = (y_{j_1} - a_{j_1j_11})\{&\alpha_{j_1j_1}a_{j_1j_10}[(y_{j_1} - a_{j_1j_1})^2 + b_{j_1j_1}]\\ &+ \alpha_{j_1j_2}a_{j_2j_20}(y_{j_1} - a_{j_2j_21})\},\end{aligned}$$

and a dynamical system in the $y^{\perp}_{j_1}$-direction for $a_{j_1j_11} = a_{j_2j_21}$ is constructed as

$$\begin{aligned}\dot{y}^{\perp}_{j_1} = (y_{j_1} - a_{j_1j_11})\{&\beta_{j_1j_1}a_{j_1j_10}[(y_{j_1} - a_{j_1j_1})^2 + b_{j_1j_1}]\\ &+ \beta_{j_1j_2}a_{j_2j_20}(y_{j_1} - a_{j_2j_21})\},\end{aligned}$$

and define a G-function as

$$\begin{aligned}G_{j_1j_10} &= \frac{\partial}{\partial y_{j_1}}[\mathbf{n}^{\mathrm{T}} \cdot \mathbf{F}(y_{j_1}, \mathbf{p})]\Big|_{y^*_{j_1} = a_{j_1j_11} = a_{j_2j_21}}\\ &= \alpha_{j_1j_1}a_{j_1j_10}[(a_{j_2j_21} - a_{j_1j_1})^2 + b_{j_1j_1}].\end{aligned}$$

In phase space, at $y^*_1 = a_{j_1j_11} = a_{j_2j_21}$,

$$\frac{dx_{j_2}}{dx_{j_1}}\Big|_{y^*_1 = a_{j_2j_21}} = \frac{a_{j_2j_20}(y_{j_1} - a_{j_2j_21})}{a_{j_1j_10}[(y_{j_1} - a_{j_1j_1})^2 + b_{j_1j_1}]}\Big|_{y^*_1 = a_{j_2j_21}} = 0.$$

If

$$\frac{d^2x_{j_2}}{dx^2_{j_1}}\Big|_{y^*_1 = a_{j_2j_21}} = \frac{a_{j_2j_20}\alpha^2_{j_1j_1}}{a_{j_1j_10}\alpha_{j_1j_1}[(a_{j_2j_21} - a_{j_1j_1})^2 + b_{j_1j_1}]} > 0,$$

there is an up-parabola flow at $y^*_1 = a_{j_2j_21} = a_{j_1j_11}$. If

$$\frac{d^2x_{j_2}}{dx^2_{j_1}}\Big|_{y^*_1 = a_{j_2j_21}} = \frac{a_{j_2j_20}\alpha^2_{j_1j_1}}{a_{j_1j_10}\alpha_{j_1j_1}[(a_{j_2j_21} - a_{j_1j_11})^2 + b_{j_1j_1}]} < 0,$$

there is a down-parabola flow at $y^*_1 = a_{j_2j_21} = a_{j_1j_11}$.

The variational equation at $y_1^* = a_{j_2j_21} = a_{j_1j_11}$ is

$$\Delta\dot{y}_{j_1} = \alpha_{j_1j_1}a_{j_1j_10}[(a_{j_1j_11} - a_{j_1j_1})^2 + b_{j_1j_1}] = G_{j_1j_10}\Delta y_{j_1}.$$

Due to

$$(a_{j_2j_21} - a_{j_1j_1})^2 + b_{j_1j_1} > 0,$$

the parabola flows for $\alpha_{j_1j_1}a_{j_1j_10} > 0$ and $\alpha_{j_1j_1}a_{j_1j_10} < 0$ are source and sink, respectively.

Therefore, the infinite-equilibrium flow of $y_1^* = a_{j_1j_11} = a_{j_2j_21}$ possesses the following properties as in Eqs. (3.61)–(3.64):

- For $\alpha_{j_1j_1}a_{j_1j_10} > 0$ and $a_{j_2j_20}\alpha_{j_1j_1}^2 > 0$,
$$(a_{j_1j_11}, a_{j_2j_21}) = \underbrace{(\text{SO}, \text{UP})}_{\text{up-parabola source}}.$$
- For $\alpha_{j_1j_1}a_{j_1j_10} < 0$ and $a_{j_2j_20}\alpha_{j_1j_1}^2 > 0$,
$$(a_{j_1j_11}, a_{j_2j_21}) = \underbrace{(\text{SI}, \text{DP})}_{\text{down-parabola sink}}.$$
- For $\alpha_{j_1j_1}a_{j_1j_10} > 0$ and $a_{j_2j_20}\alpha_{j_1j_1}^2 < 0$,
$$(a_{j_1j_1}, a_{j_2j_21}) = \underbrace{(\text{SO}, \text{DP})}_{\text{down-parabola source}}.$$
- For $\alpha_{j_1j_1}a_{j_1j_10} < 0$ and $a_{j_2j_20}\alpha_{j_1j_1}^2 < 0$,
$$(a_{j_1j_11}, a_{j_2j_21}) = \underbrace{(\text{SI}, \text{UP})}_{\text{up-parabola sink}}.$$

(v) For

$$\Delta_{j_1j_1} = 0 \quad \text{and} \quad \Delta_{j_2j_2} = 0,$$

the following equations are obtained:

$$\begin{aligned}
\dot{x}_{j_1} &= a_{j_1j_10}(y_{j_1} - a_{j_1j_1s_1})(y_{j_1} - a_{j_1j_1s_2})^2, \\
\dot{x}_{j_2} &= a_{j_2j_20}(y_{j_1} - a_{j_2j_21})^2, \\
j_1, j_2 &\in \{1, 2\};\ j_1 \neq j_2;
\end{aligned}$$

where

$$a_{j_2j_21} = -\frac{1}{2}B_{j_2j_2}, \quad b_{j_2j_2} = 0, \quad a_{j_2j_20} = A_{j_2j_21};$$

$$a_{j_1j_1s_1} = b_{j_1j_11}, \quad a_{j_1j_1s_2} = -\frac{1}{2}B_{j_1j_11}, \quad b_{j_1j_1} = 0, \quad a_{j_1j_10} = A_{j_1j_10}.$$

(v_1) For $a_{j_2j_21} \neq a_{j_1j_1s_1}, a_{j_1j_1s_2}$, a dynamical system in the y_{j_1}-direction is constructed as

$$\dot{y}_{j_1} = \alpha_{j_1j_1}a_{j_1j_10}(y_{j_1} - a_{j_1j_1s_1})(y_{j_1} - a_{j_1j_1s_2})^2 + \alpha_{j_1j_2}a_{j_2j_20}(y_{j_1} - a_{j_2j_21})^2,$$

and a dynamical system in the $y_{j_1}^{\perp}$-direction is constructed as

$$\dot{y}_{j_1}^{\perp} = \beta_{j_1j_1}a_{j_1j_10}(y_{j_1} - a_{j_1j_1s_1})(y_{j_1} - a_{j_1j_1s_2})^2 + \beta_{j_1j_2}a_{j_2j_20}(y_{j_1} - a_{j_2j_21})^2.$$

($\mathrm{v}_{1\mathrm{a}}$) In phase space, at $y_1^* = a_{j_1j_1s_1}$,

$$\left.\frac{dx_{j_1}}{dx_{j_2}}\right|_{y_1^*=a_{j_1j_1s_1}} = \left.\frac{a_{j_1j_10}(y_{j_1} - a_{j_1j_1s_1})(y_{j_1} - a_{j_1j_1s_2})^2}{a_{j_2j_20}(y_{j_1} - a_{j_2j_21})^2}\right|_{y_1^*=a_{j_1j_1s_1}} = 0.$$

If

$$\left.\frac{d^2x_{j_1}}{dx_{j_2}^2}\right|_{y_1^*=a_{j_1j_1s_1}} = \frac{a_{j_1j_10}\alpha_{j_1j_2}(a_{j_1j_1s_1} - a_{j_1j_1s_2})^2}{a_{j_2j_20}(a_{j_1j_1s_1} - a_{j_2j_21})^2} > 0,$$

there is an up-parabola flow at $y_1^* = a_{j_1j_1s_1}$. If

$$\left.\frac{d^2x_{j_1}}{dx_{j_2}^2}\right|_{y_1^*=a_{j_1j_1s_1}} = \frac{a_{j_1j_10}\alpha_{j_1j_2}(a_{j_1j_1s_1} - a_{j_1j_1s_2})^2}{a_{j_2j_20}(a_{j_1j_1s_1} - a_{j_2j_21})^2} < 0,$$

there is a down-parabola flow at $y_1^* = a_{j_1j_1s_1}$. With

$$\dot{x}_{j_2}^* = a_{j_2j_20}(a_{j_1j_1s_1} - a_{j_2j_21})^2,$$

the parabola flows are positive and negative for $\dot{x}_{j_2}^* > 0$ and $\dot{x}_{j_2}^* < 0$, respectively.

Therefore, the flow of $y_{j_1}^* = a_{j_1j_1s_1} \neq a_{j_2j_21}$ with $\dot{x}_{j_2}^* = a_{j_2j_20}(a_{j_1j_1s_1} - a_{j_2j_21})^2$ for the x_{j_1}-direction has the following properties as in Eqs. (3.70)–(3.73):

- For $a_{j_1j_10}\alpha_{j_1j_2}(a_{j_1j_1s_1} - a_{j_1j_1s_2})^2 > 0$ and $a_{j_2j_20}(a_{j_1j_1s_1} - a_{j_2j_2})^2 > 0$,
$$(a_{j_1j_1s_1}, \dot{x}_{j_2}^*) = \underbrace{(\mathrm{UP}, \mathrm{pF})}_{\text{up-parabola flow } (+)}.$$
- For $a_{j_1j_10}\alpha_{j_1j_2}(a_{j_1j_1s_1} - a_{j_1j_1s_2})^2 < 0$ and $a_{j_2j_20}(a_{j_1j_1s_1} - a_{j_2j_21})^2 > 0$,
$$(a_{j_1j_1s_1}, \dot{x}_{j_2}^*) = \underbrace{(\mathrm{DP}, \mathrm{pF})}_{\text{down-parabola flow } (+)}.$$
- For $a_{j_1j_10}\alpha_{j_1j_2}(a_{j_1j_1s_1} - a_{j_1j_1s_2})^2 > 0$ and $a_{j_2j_20}(a_{j_1j_1s_1} - a_{j_2j_2})^2 < 0$,
$$(a_{j_1j_1s_1}, \dot{x}_{j_2}^*) = \underbrace{(\mathrm{DP}, \mathrm{nF})}_{\text{down-parabola flow } (-)}.$$

- For $a_{j_1j_10}\alpha_{j_1j_2}(a_{j_1j_1s_1}-a_{j_1j_1s_2})^2<0$ and $a_{j_2j_20}(a_{j_1j_1s_1}-a_{j_2j_21})^2<0$,

$$(a_{j_1j_1s_1},\dot{x}^*_{j_2})=\underbrace{(\mathrm{UP},\mathrm{nF})}_{\text{up-parabola flow }(-)}.$$

($\mathrm{v}_{1\mathrm{b}}$) In phase space, at $y_1^*=a_{j_1j_1s_2}$,

$$\left.\frac{dx_{j_1}}{dx_{j_2}}\right|_{y_1^*=a_{j_1j_1s_2}}=\left.\frac{a_{j_1j_10}(y_{j_1}-a_{j_1j_1s_1})(y_{j_1}-a_{j_1j_1s_2})^2}{a_{j_2j_20}(y_{j_1}-a_{j_2j_21})^2}\right|_{y_1^*=a_{j_1j_1s_2}}=0,$$

$$\left.\frac{d^2x_{j_1}}{dx_{j_2}^2}\right|_{y_1^*=a_{j_1j_1s_2}}=0.$$

If

$$\left.\frac{d^3x_{j_1}}{dx_{j_2}^3}\right|_{y_1^*=a_{j_1j_1s_2}}=\frac{2a_{j_1j_10}\alpha^2_{j_1j_2}(a_{j_1j_1s_2}-a_{j_1j_1s_1})}{a_{j_2j_20}(a_{j_1j_1s_2}-a_{j_2j_21})^2}>0,$$

there is an increasing-inflection flow at $y_1^*=a_{j_1j_1s_2}$. If

$$\left.\frac{d^3x_{j_1}}{dx_{j_2}^3}\right|_{y_1^*=a_{j_1j_1s_2}}=\frac{2a_{j_1j_10}\alpha^2_{j_1j_2}(a_{j_1j_1s_2}-a_{j_1j_1s_1})}{a_{j_2j_20}(a_{j_1j_1s_2}-a_{j_2j_21})^2}<0,$$

there is a decreasing-inflection flow at $y_1^*=a_{j_1j_1s_2}$. With

$$\dot{x}^*_{j_2}=a_{j_2j_20}(a_{j_1j_1s_2}-a_{j_2j_21})^2,$$

the inflection flows are positive and negative for $\dot{x}^*_{j_2}>0$ and $\dot{x}^*_{j_2}<0$, respectively.

Therefore, the flows of $y^*_{j_1}=a_{j_1j_1s_2}\neq a_{j_2j_21}$ with $\dot{x}^*_{j_2}=a_{j_2j_20}(a_{j_1j_1s_2}-a_{j_2j_21})^2$ for the x_{j_1}-direction with $a_{j_1j_11}\neq a_{j_2j_21}$ have the following properties as in Eqs. (3.74)–(3.77):

- For $a_{j_1j_10}\alpha^2_{j_1j_2}(a_{j_1j_1s_2}-a_{j_1j_1s_1})>0$ and $a_{j_2j_20}(a_{j_1j_1s_2}-a_{j_2j_21})^2>0$,

$$(a_{j_1j_1s_2},\dot{x}^*_{j_2})=\underbrace{(\mathrm{II},\mathrm{pF})}_{\text{increasing-inflection flow }(+)}.$$

- For $a_{j_1j_10}\alpha^2_{j_1j_2}(a_{j_1j_1s_2}-a_{j_1j_1s_1})>0$ and $a_{j_2j_20}(a_{j_1j_1s_2}-a_{j_2j_21})^2>0$,

$$(a_{j_1j_1s_2},\dot{x}^*_{j_2})=\underbrace{(\mathrm{DI},\mathrm{pF})}_{\text{decreasing-inflection flow }(+)}.$$

- For $a_{j_1j_10}\alpha^2_{j_1j_2}(a_{j_1j_1s_2}-a_{j_1j_1s_1})>0$ and $a_{j_2j_20}(a_{j_1j_1s_2}-a_{j_2j_21})^2<0$,

$$(a_{j_1j_1s_2},\dot{x}^*_{j_2})=\underbrace{(\mathrm{DI},\mathrm{nF})}_{\text{decreasing-inflection flow }(-)}.$$

- For $a_{j_1j_10}\alpha^2_{j_1j_2}(a_{j_1j_1s_2} - a_{j_1j_1s_1}) < 0$ and $a_{j_2j_20}(a_{j_1j_1s_2} - a_{j_2j_21})^2 < 0$,

$$(a_{j_1j_1s_2}, \dot{x}^*_{j_2}) = \underbrace{(\mathrm{II}, \mathrm{nF})}_{\text{increasing-inflection flow }(+)}.$$

– The increasing- and decreasing-inflection flows are the appearing bifurcations for the up-parabola and down-parabola flows.

(v_{1c}) In phase space, at $y^*_1 = a_{j_2j_21}$,

$$\left.\frac{dx_{j_2}}{dx_{j_1}}\right|_{y^*_1=a_{j_2j_21}} = \left.\frac{a_{j_2j_20}(y_{j_1} - a_{j_2j_21})^2}{a_{j_1j_10}(y_{j_1} - a_{j_1j_1s_1})(y_{j_1} - a_{j_1j_1s_2})^2}\right|_{y^*_1=a_{j_2j_21}} = 0,$$

$$\left.\frac{d^2x_{j_2}}{dx^2_{j_1}}\right|_{y^*_1=a_{j_2j_21}} = 0.$$

If

$$\left.\frac{d^3x_{j_2}}{dx^3_{j_1}}\right|_{y^*_1=a_{j_2j_21}} = \frac{2a_{j_2j_20}\alpha^2_{j_1j_1}}{a_{j_1j_10}(a_{j_2j_21} - a_{j_1j_1s_1})(a_{j_2j_21} - a_{j_1j_1s_2})^2} > 0,$$

there is an increasing-inflection flow at $y^*_1 = a_{j_2j_21}$. If

$$\left.\frac{d^3x_{j_2}}{dx^3_{j_1}}\right|_{y^*_1=a_{j_2j_21}} = \frac{2a_{j_2j_20}\alpha^2_{j_1j_1}}{a_{j_1j_10}(a_{j_2j_21} - a_{j_1j_1s_1})(a_{j_2j_21} - a_{j_1j_1s_2})^2} < 0,$$

there is a decreasing-inflection flow at $y^*_1 = a_{j_2j_21}$. With

$$\dot{x}^*_{j_1} = a_{j_1j_10}(a_{j_2j_21} - a_{j_1j_1s_1})(a_{j_2j_21} - a_{j_1j_1s_2})^2,$$

the inflection flows are positive and negative for $\dot{x}^*_{j_1} > 0$ and $\dot{x}^*_{j_1} < 0$, respectively.

Therefore, the flows of $y^*_{j_1} = a_{j_2j_21} \neq a_{j_1j_1s_1}, a_{j_1j_1s_2}$ for the x_{j_2}-direction have the following properties with $\dot{x}^*_{j_1} = a_{j_1j_10}(a_{j_2j_21} - a_{j_1j_1s_1})(a_{j_2j_21} - a_{j_1j_1s_2})^2$ as in Eqs. (3.78)–(3.81):

- For $a_{j_1j_10}(a_{j_2j_21} - a_{j_1j_1s_1})(a_{j_2j_21} - a_{j_1j_1s_2})^2 > 0$ and $a_{j_2j_20}\alpha^2_{j_1j_1} > 0$,

$$(\dot{x}^*_{j_1}, a_{j_2j_21}) = \underbrace{(\mathrm{pF}, \mathrm{II})}_{\text{increasing-inflection flow }(+)}.$$

- For $a_{j_1j_10}(a_{j_2j_21} - a_{j_1j_1s_1})(a_{j_2j_21} - a_{j_1j_1s_2})^2 < 0$ and $a_{j_2j_20}\alpha^2_{j_1j_1} > 0$,

$$(\dot{x}^*_{j_1}, a_{j_2j_21}) = \underbrace{(\mathrm{nF}, \mathrm{DI})}_{\text{decreasing-inflection flow }(-)}.$$

- For $a_{j_1j_10}(a_{j_2j_21} - a_{j_1j_1s_1})(a_{j_2j_21} - a_{j_1j_1s_2})^2 > 0$ and $a_{j_2j_20}\alpha_{j_1j_1}^2 < 0$,

$$(\dot{x}_{j_1}^*, a_{j_2j_21}) = \underbrace{(\text{pF}, \text{DI})}_{\text{decreasing-inflection flow } (+)}.$$

- For $a_{j_1j_10}(a_{j_2j_21} - a_{j_1j_1s_1})(a_{j_2j_21} - a_{j_1j_1s_2})^2 < 0$ and $a_{j_2j_20}\alpha_{j_1j_1}^2 < 0$,

$$(\dot{x}_{j_1}^*, a_{j_2j_21}) = \underbrace{(\text{nF}, \text{II})}_{\text{increasing-inflection flow } (-)}.$$

– The increasing- and decreasing-inflection flows are the appearing bifurcations for the up-parabola and down-parabola flows.

(v_2) For $a_{j_1j_1s_1} = a_{j_1j_1s_2} \equiv a_{j_1j_11}$,

$$\begin{aligned}
\dot{x}_{j_1} &= a_{j_1j_10}(y_{j_1} - a_{j_1j_11})^3, \\
\dot{x}_{j_2} &= a_{j_2j_20}(y_{j_1} - a_{j_2j_21})^2, \\
j_1, j_2 &\in \{1, 2\}; j_1 \neq j_2.
\end{aligned}$$

For $a_{j_2j_21} \neq a_{j_1j_11}$, a dynamical system in the y_{j_1}-direction is constructed as

$$\dot{y}_{j_1} = \alpha_{j_1j_1}a_{j_1j_10}(y_{j_1} - a_{j_1j_11})^3 + \alpha_{j_1j_2}a_{j_2j_20}(y_{j_1} - a_{j_2j_21})^2,$$

and a dynamical system in the $y_{j_1}^{\perp}$-direction is constructed as

$$\dot{y}_{j_1}^{\perp} = \beta_{j_1j_1}a_{j_1j_10}(y_{j_1} - a_{j_1j_1}1)^3 + \beta_{j_1j_2}a_{j_2j_20}(y_{j_1} - a_{j_2j_21})^2.$$

($\text{v}_{2\text{a}}$) In phase space, at $y_1^* = a_{j_1j_11}$,

$$\begin{aligned}
\left.\frac{dx_{j_1}}{dx_{j_2}}\right|_{y_1^*=a_{j_1j_11}} &= \left.\frac{a_{j_1j_00}(y_{j_1} - a_{j_1j_11})^3}{a_{j_2j_20}(y_{j_1} - a_{j_2j_21})^2}\right|_{y_1^*=a_{j_1j_11}} = 0, \\
\left.\frac{d^2x_{j_1}}{dx_{j_2}^2}\right|_{y_1^*=a_{j_1j_11}} &= \left.\frac{d^3x_{j_1}}{dx_{j_2}^3}\right|_{y_1^*=a_{j_1j_11}} = 0.
\end{aligned}$$

If

$$\left.\frac{d^4x_{j_1}}{dx_{j_2}^4}\right|_{y_1^*=a_{j_1j_11}} = \frac{6a_{j_1j_10}\alpha_{j_1j_2}^3}{a_{j_2j_20}(a_{j_1j_11} - a_{j_2j_21})^2} > 0,$$

there is a third-order up-parabola flow at $y_1^* = a_{j_1j_11}$. If

$$\left.\frac{d^4x_{j_1}}{dx_{j_2}^4}\right|_{y_1^*=a_{j_1j_11}} = \frac{6a_{j_1j_10}a_{j_1j_2}^3}{a_{j_2j_20}(a_{j_1j_11} - a_{j_2j_21})^2} < 0,$$

there is a third-order down-parabola flow at $y_1^* = a_{j_1 j_1 1}$. With

$$\dot{x}_{j_2}^* = a_{j_2 j_2 0}(a_{j_1 j_1 1} - a_{j_2 j_2 1})^2,$$

the third-order parabola flows are positive and negative for $\dot{x}_{j_2}^* > 0$ and $\dot{x}_{j_2}^* < 0$, respectively.

Therefore, the flow of $y_{j_1}^* = a_{j_1 j_1 1} \neq a_{j_2 j_2 1}$ with $\dot{x}_{j_2}^* = a_{j_2 j_2 0}(a_{j_1 j_1 1} - a_{j_2 j_2 1})^2$ for the x_{j_1}-direction has the following properties as in Eqs. (3.85)–(3.88):

- For $a_{j_1 j_1 0}\alpha_{j_1 j_2}^3 > 0$ and $a_{j_2 j_2 0}(a_{j_1 j_1 1} - a_{j_2 j_2 1})^2 > 0$,
$$(a_{j_1 j_1 1}, \dot{x}_{j_2}^*) = \underbrace{(3^{\text{rd}}\text{UP}, \text{pF})}_{\text{third-order up-parabola flow }(+)}.$$
- For $a_{j_1 j_1 0}\alpha_{j_1 j_2}^3 < 0$ and $a_{j_2 j_2 0}(a_{j_1 j_1 1} - a_{j_2 j_2 1})^2 > 0$,
$$(a_{j_1 j_1 1}, \dot{x}_{j_2}^*) = \underbrace{(3^{\text{rd}}\text{DP}, \text{pF})}_{\text{third-order down-parabola flow }(+)}.$$
- For $a_{j_1 j_1 0}\alpha_{j_1 j_2}^3 > 0$ and $a_{j_2 j_2 0}(a_{j_1 j_1 1} - a_{j_2 j_2 1})^2 < 0$,
$$(a_{j_1 j_1}, \dot{x}_{j_2}^*) = \underbrace{(3^{\text{rd}}\text{DP}, \text{nF})}_{\text{third-order down-parabola flow }(-)}.$$
- For $a_{j_1 j_1 0}\alpha_{j_1 j_2}^3 < 0$ and $a_{j_2 j_2 0}(a_{j_1 j_1 1} - a_{j_2 j_2 1})^2 < 0$,
$$(a_{j_1 j_1}, \dot{x}_{j_2}^*) = \underbrace{(3^{\text{rd}}\text{UP}, \text{nF})}_{\text{third-order up-parabola flow }(-)}.$$

- The third-order up-parabola flows are the appearing bifurcations for up-parabola, down-parabola, and up-parabola flows.
- The third-order down-parabola flows are the appearing bifurcations for down-parabola, up-parabola, and down-parabola flows.

($\text{v}_{2\text{b}}$) In phase space, at $y_1^* = a_{j_2 j_2 1}$,

$$\left.\frac{dx_{j_2}}{dx_{j_1}}\right|_{y_1^* = a_{j_2 j_2 1}} = \left.\frac{a_{j_2 j_2 0}(y_{j_1} - a_{j_2 j_2 1})^2}{a_{j_1 j_1 0}(y_{j_1} - a_{j_1 j_1 1})^3}\right|_{y_1^* = a_{j_2 j_2 1}} = 0,$$

$$\left.\frac{d^2 x_{j_2}}{dx_{j_1}^2}\right|_{y_1^* = a_{j_2 j_2 1}} = 0.$$

If

$$\left.\frac{d^3 x_{j_2}}{dx_{j_1}^3}\right|_{y_1^* = a_{j_2 j_2 1}} = \frac{2a_{j_2 j_2 0}\alpha_{j_1 j_1}^2}{a_{j_1 j_1 0}(a_{j_2 j_2 1} - a_{j_1 j_1 1})^3} > 0,$$

there is an increasing-inflection flow at $y_1^* = a_{j_2j_21}$. If

$$\left.\frac{d^3 x_{j_2}}{dx_{j_1}^3}\right|_{y_1^*=a_{j_2j_21}} = \frac{2a_{j_2j_20}\alpha_{j_1j_1}^2}{a_{j_1j_10}(a_{j_2j_21}-a_{j_1j_11})^3} < 0,$$

there is a decreasing-inflection flow at $y_1^* = a_{j_2j_21}$. With

$$\dot{x}_{j_1}^* = a_{j_1j_10}(a_{j_2j_21} - a_{j_1j_11})^3,$$

the inflection flows are positive and negative for $\dot{x}_{j_1}^* > 0$ and $\dot{x}_{j_1}^* < 0$, respectively.

Therefore, the flow of $y_{j_1}^* = a_{j_2j_21} \neq a_{j_1j_11}$ for the x_{j_2}-direction has the following properties with $\dot{x}_{j_1}^* = a_{j_1j_10}(a_{j_2j_21} - a_{j_1j_11})^3$ as in Eqs. (3.89)–(3.92):

- For $a_{j_1j_10}(a_{j_2j_21} - a_{j_1j_11})^3 > 0$ and $a_{j_2j_20}\alpha_{j_1j_1}^2 > 0$,

$$(\dot{x}_{j_1}^*, a_{j_2j_21}) = \underbrace{(\text{pF}, \text{II})}_{\text{increasing-infelction flow } (+)}.$$

- For $a_{j_1j_10}(a_{j_2j_21} - a_{j_1j_11})^3 < 0$ and $a_{j_2j_20}\alpha_{j_1j_1}^2 > 0$,

$$(\dot{x}_{j_1}^*, a_{j_2j_21}) = \underbrace{(\text{nF}, \text{DI})}_{\text{decreasing-infelction flow } (-)}.$$

- For $a_{j_1j_10}(a_{j_2j_21} - a_{j_1j_11})^3 > 0$ and $a_{j_2j_20}\alpha_{j_1j_1}^2 < 0$,

$$(\dot{x}_{j_1}^*, a_{j_2j_21}) = \underbrace{(\text{pF}, \text{DI})}_{\text{decreasing-infelction flow } (+)}.$$

- For $a_{j_1j_10}(a_{j_2j_21} - a_{j_1j_11})^3 < 0$ and $a_{j_2j_20}\alpha_{j_1j_1}^2 < 0$,

$$(\dot{x}_{j_1}^*, a_{j_2j_21}) = \underbrace{(\text{nF}, \text{II})}_{\text{increasing-infelction flow } (-)}.$$

– The increasing- and decreasing-inflection flows are the appearing bifurcations for the up-parabola and down-parabola flows.

(v3) For $a_{j_1j_1s_1} = a_{j_2j_21}$, consider a dynamical system as

$$\begin{aligned}
\dot{x}_{j_1} &= a_{j_1j_10}(y_{j_1} - a_{j_1j_1s_1})(y_{j_1} - a_{j_1j_1s_2})^2,\\
\dot{x}_{j_2} &= a_{j_2j_20}(y_{j_1} - a_{j_2j_21})^2.
\end{aligned}$$

A dynamical system in the y_{j_1}-direction is constructed as

$$\dot{y}_{j_1} = (y_{j_1} - a_{j_1j_1s_1})[\alpha_{j_1j_1}a_{j_1j_10}(y_{j_1} - a_{j_1j_1s_2})^2 + \alpha_{j_1j_2}a_{j_2j_20}(y_{j_1} - a_{j_2j_21})],$$

and a dynamical system in the $y_{j_1}^{\perp}$-direction is constructed as

$$\dot{y}_{j_1}^{\perp} = (y_{j_1} - a_{j_1j_1s_1})[\beta_{j_1j_1}a_{j_1j_10}(y_{j_1} - a_{j_1j_1s_2})^2 + \beta_{j_1j_2}a_{j_2j_20}(y_{j_1} - a_{j_2j_21})].$$

Define a G-function as

$$\begin{aligned}
G_{j_1j_10} &= \frac{\partial}{\partial y_{j_1}}[\mathbf{n}^{\mathrm{T}} \cdot \mathbf{F}(y_{j_1}, \mathbf{p})]\bigg|_{y_{j_1}^* = a_{j_1j_1s_1} = a_{j_2j_21}}\\
&= \alpha_{j_1j_1}a_{j_1j_10}(a_{j_1j_1s_1} - a_{j_1j_1s_2})^2.
\end{aligned}$$

In phase space, at $y_1^* = a_{j_1j_1s_1} = a_{j_2j_21}$,

$$\frac{dx_{j_2}}{dx_{j_1}}\bigg|_{y_1^* = a_{j_2j_21}} = \frac{a_{j_2j_20}(y_{j_1} - a_{j_2j_21})}{a_{j_1j_10}(y_{j_1} - a_{j_1j_1s_2})^2}\bigg|_{y_1^* = a_{j_2j_21}} = 0.$$

If

$$\frac{d^2x_{j_2}}{dx_{j_1}^2}\bigg|_{y_1^* = a_{j_2j_21}} = \frac{a_{j_2j_20}\alpha_{j_1j_1}^2}{a_{j_1j_10}\alpha_{j_1j_1}(a_{j_2j_21} - a_{j_1j_1s_2})^2} > 0,$$

there is an up-parabola flow at $y_1^* = a_{j_1j_1s_1} = a_{j_2j_21}$. If

$$\frac{d^2x_{j_2}}{dx_{j_1}^2}\bigg|_{y_1^* = a_{j_2j_21}} = \frac{a_{j_2j_20}\alpha_{j_1j_1}^2}{a_{j_1j_10}\alpha_{j_1j_1}(a_{j_2j_21} - a_{j_1j_1s_2})^2} < 0,$$

there is a down-parabola flow at $y_1^* = a_{j_1j_1s_1} = a_{j_2j_21}$. The variational equation in the y_{j_1}-direction is

$$\Delta\dot{y}_{j_1} = \alpha_{j_1j_1}a_{j_1j_10}(a_{j_1j_1s_1} - a_{j_1j_1s_2})^2\Delta y_{j_1} = G_{j_1j_10}\Delta y_{j_1}.$$

The parabola flows in the y_{j_1}-direction are source and sink for $G_{j_1j_10} > 0$ and $G_{j_1j_10} < 0$, respectively.

Therefore, the infinite-equilibrium flows of $y_1^* = a_{j_1j_1s_1} = a_{j_2j_21}$ possess the following properties as in Eqs. (3.97)–(3.100):

- For $\alpha_{j_1j_1}a_{j_1j_10}(a_{j_1j_1s_1} - a_{j_1j_1s_2})^2 > 0$ and $a_{j_2j_20}\alpha_{j_1j_1}^2 > 0$,
$$(a_{j_1j_1s_1}, a_{j_2j_2}) = \underbrace{(\text{SO}, \text{UP})}_{\text{up-parabola source}}.$$
- For $\alpha_{j_1j_1}a_{j_1j_10}(a_{j_1j_1s_1} - a_{j_1j_1s_2})^2 < 0$ and $a_{j_2j_20}\alpha_{j_1j_1}^2 > 0$,
$$(a_{j_1j_1s_1}, a_{j_2j_21}) = \underbrace{(\text{SI}, \text{DP})}_{\text{down-parabola sink}}.$$
- For $\alpha_{j_1j_1}a_{j_1j_10}(a_{j_1j_1s_1} - a_{j_1j_1s_2})^2 > 0$ and $a_{j_2j_20}\alpha_{j_1j_1}^2 < 0$,
$$(a_{j_1j_1s_1}, a_{j_2j_21}) = \underbrace{(\text{SO}, \text{DP})}_{\text{down-parabola source}}.$$
- For $\alpha_{j_1j_1}a_{j_1j_10}(a_{j_1j_1s_1} - a_{j_1j_1s_2})^2 < 0$ and $a_{j_2j_20}\alpha_{j_1j_1}^2 < 0$,
$$(a_{j_1j_1s_1}, a_{j_2j_21}) = \underbrace{(\text{SI}, \text{UP})}_{\text{up-parabola sink}}.$$

(v$_4$) For $a_{j_1j_1s_2} = a_{j_2j_21}$, consider a dynamical system as

$$\begin{aligned}
\dot{x}_{j_1} &= a_{j_1j_10}(y_{j_1} - a_{j_1j_1s_1})(y_{j_1} - a_{j_1j_1s_2})^2,\\
\dot{x}_{j_2} &= a_{j_2j_20}(y_{j_1} - a_{j_2j_21})^2,\\
&j_1, j_2 \in \{1, 2\}; j_1 \neq j_2.
\end{aligned}$$

A dynamical system in the y_{j_1}-direction is constructed as

$$\dot{y}_{j_1} = (y_{j_1} - a_{j_1j_1s_2})^2[\alpha_{j_1j_1}a_{j_1j_10}(y_{j_1} - a_{j_1j_1s_1}) + \alpha_{j_1j_2}a_{j_2j_20}],$$

and a dynamical system in the $y_{j_1}^{\perp}$-direction is constructed as

$$\dot{y}_{j_1}^{\perp} = (y_{j_1} - a_{j_1j_1s_2})^2[\beta_{j_1j_1}a_{j_1j_10}(y_{j_1} - a_{j_1j_1s_1}) + \beta_{j_1j_2}a_{j_2j_20}].$$

Define a G-function as

$$\begin{aligned}
G_{j_1j_10} &= \frac{\partial}{\partial y_{j_1}}[\mathbf{n}^{\mathrm{T}} \cdot \mathbf{F}(y_{j_1}, \mathbf{p})]\bigg|_{y_{j_1}^* = a_{j_1j_1j_2} = a_{j_2j_21}} = 0,\\
G_{j_1j_10}^{(1)} &= \frac{1}{2!}\frac{\partial^2}{\partial y_{j_1}^2}[\mathbf{n}^{\mathrm{T}} \cdot \mathbf{F}(y_{j_1}, \mathbf{p})]\bigg|_{y_{j_1}^* = a_{j_1j_1j_2} = a_{j_2j_21}}\\
&= \alpha_{j_1j_1}a_{j_1j_10}(a_{j_1j_1s_2} - a_{j_1j_1s_1}) + \alpha_{j_1j_2}a_{j_2j_20}.
\end{aligned}$$

In phase space, at $y_1^* = a_{j_1j_1s_2} = a_{j_2j_21}$, if

$$\left.\frac{dx_{j_2}}{dx_{j_1}}\right|_{y_1^*=a_{j_2j_21}} = \frac{a_{j_2j_20}}{a_{j_1j_10}(a_{j_1j_1s_2} - a_{j_1j_1s_1})} > 0,$$

there is an increasing-inflection flow at $y_1^* = a_{j_1j_1s_2} = a_{j_2j_21}$, and if

$$\left.\frac{dx_{j_2}}{dx_{j_1}}\right|_{y_1^*=a_{j_2j_21}} = \frac{a_{j_2j_20}}{a_{j_1j_10}(a_{j_1j_1s_2} - a_{j_1j_1s_1})} < 0,$$

there is a decreasing-inflection flow at $y_1^* = a_{j_1j_1s_1} = a_{j_2j_21}$. The variational equation in the y_{j_1}-direction is

$$\begin{aligned}\Delta\dot{y}_{j_1} &= [\alpha_{j_1j_1}a_{j_1j_10}(a_{j_1j_1s_2} - a_{j_1j_1s_1}) + \alpha_{j_1j_2}a_{j_2j_20}](\Delta y_{j_1})^2\\ &= G^{(1)}_{j_1j_10}(\Delta y_{j_1})^2.\end{aligned}$$

The inflection flows in the y_{j_1}-direction are upper-saddle and lower-saddle for $G^{(1)}_{j_1j_10} > 0$ and $G^{(1)}_{j_1j_10} < 0$, respectively.

Therefore, the infinite-equilibrium of $y_{j_1}^* = a_{j_1j_11} = a_{j_2j_21}$ has the following properties as in Eqs. (3.105)–(3.108):

- For $a_{j_1j_10}(a_{j_1j_1s_2} - a_{j_1j_1s_1})/a_{j_2j_20} > 0$ and $G^{(1)}_{j_1j_10} > 0$,
$$(a_{j_1j_1s_2}, a_{j_2j_21}) = \underbrace{(\mathrm{II}, \mathrm{US})}_{\text{increasing-inflection upper-saddle}}.$$
- For $a_{j_1j_10}(a_{j_1j_1s_2} - a_{j_1j_1s_1})/a_{j_2j_20} > 0$ and $G^{(1)}_{j_1j_10} < 0$,
$$(a_{j_1j_1s_2}, a_{j_2j_21}) = \underbrace{(\mathrm{II}, \mathrm{LS})}_{\text{increasing-inflection lower-saddle}}.$$
- For $a_{j_1j_10}(a_{j_1j_1s_2} - a_{j_1j_1s_1})/a_{j_2j_20} < 0$ and $G^{(1)}_{j_1j_10} > 0$,
$$(a_{j_1j_1s_2}, a_{j_2j_21}) = \underbrace{(\mathrm{DI}, \mathrm{US})}_{\text{decreasing-inflection upper-saddle}}.$$
- For $a_{j_1j_10}(a_{j_1j_1s_2} - a_{j_1j_1s_1})/a_{j_2j_20} < 0$ and $G^{(1)}_{j_1j_10} < 0$,
$$(a_{j_1j_1s_2}, a_{j_2j_21}) = \underbrace{(\mathrm{DI}, \mathrm{LS})}_{\text{decreasing-inflection lower-saddle}}.$$

– The inflection-saddle infinite-equilibriums are the appearing bifurcations for the up-parabola and down-parabola flows in the x_{j_1}- and x_{j_2}-directions.

– The inflection-saddle infinite-equilibriums are the double-switching bifurcations for the up-parabola and down-parabola flows in the x_{j_1} and x_{j_2}-directions.

(v_5) For $a_{j_1j_1 1} = a_{j_2j_2 1}$, consider a dynamical system as

$$\begin{aligned}\dot{x}_{j_1} &= a_{j_1j_1 0}(y_{j_1} - a_{j_1j_1 1})^3,\\ \dot{x}_{j_2} &= a_{j_2j_2 0}(y_{j_1} - a_{j_2j_2 1})^2,\\ &j_1, j_2 \in \{1,2\};\ j_1 \neq j_2.\end{aligned}$$

A dynamical system in the y_{j_1}-direction is constructed as

$$\dot{y}_{j_1} = (y_{j_1} - a_{j_1j_1})^2[\alpha_{j_1j_1}a_{j_1j_1 0}(y_{j_1} - a_{j_1j_1 1}) + \alpha_{j_1j_2}a_{j_2j_2 0}],$$

and a dynamical system in the $y_{j_1}^{\perp}$-direction is constructed as

$$\dot{y}_{j_1}^{\perp} = (y_{j_1} - a_{j_1j_1 1})^2[\beta_{j_1j_1}a_{j_1j_1 0}(y_{j_1} - a_{j_1j_1 1}) + \beta_{j_1j_2}a_{j_2j_2 0}].$$

Define a G-function as

$$G_{j_1j_1 0} = \frac{\partial}{\partial y_{j_1}}[\mathbf{n}^{\mathrm{T}}\cdot\mathbf{F}(y_{j_1},\mathbf{p})]\Big|_{y_{j_1}^* = a_{j_1j_1 1} = a_{j_2j_2 1}} = 0,$$

$$G_{j_1j_1 0}^{(1)} = \frac{1}{2!}\frac{\partial^2}{\partial y_{j_1}^2}[\mathbf{n}^{\mathrm{T}}\cdot\mathbf{F}(y_{j_1},\mathbf{p})]\Big|_{y_{j_1}^* = a_{j_1j_1 1} = a_{j_2j_2 1}} = \alpha_{j_1j_2}a_{j_2j_2 0}.$$

In phase space, at $y_{j_1}^* = a_{j_1j_1 1} = a_{j_2j_2 1}$,

$$\frac{dx_{j_1}}{dx_{j_2}}\Big|_{y_1^* = a_{j_1j_1 1}} = \frac{a_{j_1j_1 0}(y_{j_1} - a_{j_1j_1 1})}{a_{j_2j_2 0}}\Big|_{y_1^* = a_{j_1j_1 1}} = 0.$$

If

$$\frac{d^2x_{j_1}}{dx_{j_2}^2}\Big|_{y_1^* = a_{j_1j_1 1}} = \frac{a_{j_1j_1 0}\alpha_{j_1j_2}^2}{a_{j_2j_2 0}\alpha_{j_1j_2}} > 0,$$

there is an up-parabola flow at $y_{j_1}^* = a_{j_1j_1 1} = a_{j_2j_2 1}$. If

$$\frac{d^3x_{j_1}}{dx_{j_2}^3}\Big|_{y_1^* = a_{j_1j_1 1}} = \frac{a_{j_1j_1 0}\alpha_{j_1j_2}^2}{a_{j_2j_2 0}\alpha_{j_1j_2}} < 0,$$

there is a down-parabola flow at $y_{j_1}^* = a_{j_1j_1 1} = a_{j_2j_2 1}$. The variational equation in the y_{j_1}-direction is

$$\Delta\dot{y}_{j_1} = \alpha_{j_1j_2}a_{j_2j_2 0}(\Delta y_{j_1})^2 = G_{j_1j_1 0}^{(1)}(\Delta y_{j_1})^2.$$

The parabola flows in the y_{j_1}-direction are upper-saddle and lower-saddle for $G_{j_1j_1 0}^{(1)} > 0$ and $G_{j_1j_1 0}^{(1)} < 0$, respectively.

Therefore, the infinite-equilibrium of $y^*_{j_1} = a_{j_1j_1 1} = a_{j_2j_2 1}$ has the following properties as in Eqs. (3.113)–(3.116):

- For $\alpha^2_{j_1j_2}a_{j_1j_1 0} > 0$ and $\alpha_{j_1j_2}a_{j_2j_2 0} > 0$,
$$(a_{j_1j_1 1}, a_{j_2j_2 1}) = \underbrace{(\mathrm{UP}, \mathrm{US})}_{\text{up-parabola upper-saddle}}.$$
- For $\alpha^2_{j_1j_2}a_{j_1j_1 0} < 0$ and $\alpha_{j_1j_2}a_{j_2j_2 0} > 0$,
$$(a_{j_1j_1 1}, a_{j_2j_2 1}) = \underbrace{(\mathrm{DP}, \mathrm{US})}_{\text{down-parabola upper-saddle}}.$$
- For $\alpha^2_{j_1j_2}a_{j_1j_1 0} > 0$ and $\alpha_{j_1j_2}a_{j_2j_2 0} > 0$,
$$(a_{j_1j_1 1}, a_{j_2j_2 1}) = \underbrace{(\mathrm{DP}, \mathrm{LS})}_{\text{down-parabola lower-saddle}}.$$
- For $\alpha^2_{j_1j_2}a_{j_1j_1 0} < 0$ and $\alpha_{j_1j_2}a_{j_2j_2 0} < 0$,
$$(a_{j_1j_1 1}, a_{j_2j_2 1}) = \underbrace{(\mathrm{UP}, \mathrm{LS})}_{\text{up-parabola lower-saddle}}.$$

(vi) For

$$\Delta_{j_1j_1} > 0 \quad \text{and} \quad \Delta_{j_2j_2} = 0,$$

the dynamical system becomes

$$\begin{aligned}
\dot{x}_{j_1} &= a_{j_1j_1 0}(y_{j_1} - a_{j_1j_1 1})(y_{j_1} - a_{j_1j_1 2})(y_{j_1} - a_{j_1j_1 3}),\\
\dot{x}_{j_2} &= a_{j_2j_2 0}(y_{j_1} - a_{j_2j_2 1})^2,
\end{aligned}$$

where

$$\begin{aligned}
&\{a_{j_1j_1 1}, a_{j_1j_1 2}, a_{j_1j_1 3}\} = \mathrm{sort}\{b_{j_1j_1 1}, b_{j_1j_1 2}, b_{j_1j_1 3}\},\\
&\quad a_{j_1j_1 s_1} < a_{j_1j_1 s_2},\ s_1 < s_2;\\
&b_{j_1j_1 2}, b_{j_1j_1 3} = -\frac{1}{2}(B_{j_1j_1 1} \pm \sqrt{\Delta_{j_1j_1}}), \quad a_{j_1j_1 0} = A_{j_1j_1 0};\\
&a_{j_2j_2 1} = -\frac{1}{2}B_{j_2j_2}, j_2 \in \{1, 2\}.
\end{aligned}$$

A dynamical system in the y_{j_1}-direction is constructed as

$$\dot{y}_{j_1} = \alpha_{j_1j_1}a_{j_1j_1 0}\prod_{s_1=1}^{3}(y_{j_1} - a_{j_1j_1 s_1}) + \alpha_{j_1j_2}a_{j_2j_2 0}(y_{j_1} - a_{j_2j_2 1})^2,$$

and a dynamical system in the $y_{j_1}^{\perp}$-direction is constructed as

$$\dot{y}_{j_1}^{\perp} = \beta_{j_1 j_1} a_{j_1 j_1 0} \prod_{s_1=1}^{3} (y_{j_1} - a_{j_1 j_1 s_1}) + \beta_{j_1 j_2} a_{j_2 j_2 0} (y_{j_1} - a_{j_2 j_2 1})^2.$$

(vi$_{1a}$) In phase space, at $y_1^* = a_{j_1 j_1 s_1}$,

$$\left.\frac{dx_{j_1}}{dx_{j_2}}\right|_{y_1^*=a_{j_1 j_1 s_1}} = \left.\frac{a_{j_1 j_1 0}(y_{j_1} - a_{j_1 j_1 s_1}) \prod_{s_2=1, s_2\neq s_1}^{3} (y_{j_1} - a_{j_1 j_1 s_2})}{a_{j_2 j_2 0}(y_{j_1} - a_{j_2 j_2 1})^2}\right|_{y_1^*=a_{j_1 j_1 s_1}} = 0.$$

If

$$\left.\frac{d^2 x_{j_1}}{dx_{j_2}^2}\right|_{y_1^*=a_{j_1 j_1 s_1}} = \frac{a_{j_1 j_1 0}\alpha_{j_1 j_2} \prod_{s_2=1, s_2\neq s_1}^{3} (a_{j_1 j_1 s_1} - a_{j_1 j_1 s_2})}{a_{j_2 j_2 0}(a_{j_1 j_1 s_1} - a_{j_2 j_2 1})^2} > 0,$$

there is an up-parabola flow at $y_1^* = a_{j_1 j_1 s_1}$. If

$$\left.\frac{d^2 x_{j_1}}{dx_{j_2}^2}\right|_{y_1^*=a_{j_1 j_1 s_1}} = \frac{a_{j_1 j_1 0}\alpha_{j_1 j_2} \prod_{s_2=1, s_2\neq s_1}^{3} (a_{j_1 j_1 s_1} - a_{j_1 j_1 s_2})}{a_{j_2 j_2 0}(a_{j_1 j_1 s_1} - a_{j_2 j_2 1})^2} < 0,$$

there is a down-parabola flow at $y_1^* = a_{j_1 j_1 s_1}$. With

$$\dot{x}_{j_2}^* = a_{j_2 j_2 0}(a_{j_1 j_1 s_1} - a_{j_2 j_2 1})^2,$$

the parabola flows are positive and negative for $\dot{x}_{j_2}^* > 0$ and $\dot{x}_{j_2}^* < 0$, respectively.

Therefore, the simple equilibriums of $y_{j_1}^* = a_{j_1 j_1 s_1} (s_1, s_2 \in \{1,2,3\}, s_1 \neq s_2)$ for the x_{j_1}-direction have four cases with $\dot{x}_{j_2}^* = a_{j_2 j_2 0}(a_{j_1 j_1 s_1} - a_{j_2 j_2 1})^2$ as in Eqs. (3.122)–(3.125):

- For $a_{j_1 j_1 0}\alpha_{j_1 j_2} \prod_{s_2=1, s_2\neq s_1}^{3} (a_{j_1 j_1 s_1} - a_{j_1 j_1 s_2}) > 0$ and $a_{j_2 j_2 0}(a_{j_1 j_1 s_1} - a_{j_2 j_2 1})^2 > 0$,

$$(a_{j_1 j_1 s_1}, \dot{x}_{j_2}^*) = \underbrace{(\mathrm{UP}, \mathrm{pF})}_{\text{up-parabola flow } (+)}.$$

- For $a_{j_1 j_1 0}\alpha_{j_1 j_2} \prod_{s_2=1, s_2\neq s_1}^{3} (a_{j_1 j_1 s_1} - a_{j_1 j_1 s_2}) < 0$ and $a_{j_2 j_2 0}(a_{j_1 j_1 s_1} - a_{j_2 j_2 1})^2 > 0$,

$$(a_{j_1 j_1 s_1}, \dot{x}_{j_2}^*) = \underbrace{(\mathrm{DP}, \mathrm{pF})}_{\text{Down-parabola flow } (+)}.$$

- For $a_{j_1j_10}\alpha_{j_1j_2}\prod_{s_2=1,s_2\neq s_1}^{3}(a_{j_1j_1s_1}-a_{j_1j_1s_2})>0$ and $a_{j_2j_20}(a_{j_1j_1s_1}-a_{j_2j_21})^2<0$,
$$(a_{j_1j_1s_1},\dot{x}^*_{j_2})=\underbrace{(\mathrm{DP},\mathrm{nF})}_{\text{Down-parabola flow }(-)}.$$
- For $a_{j_1j_10}\alpha_{j_1j_2}\prod_{s_2=1,s_2\neq s_1}^{3}(a_{j_1j_1s_1}-a_{j_1j_1s_2})<0$ and $a_{j_2j_20}(a_{j_1j_1s_1}-a_{j_2j_21})^2<0$,
$$(a_{j_1j_1s_1},\dot{x}^*_{j_2})=\underbrace{(\mathrm{UP},\mathrm{nF})}_{\text{up-parabola flow }(-)}.$$

(vi$_{1\mathrm{b}}$) In phase space, at $y_1^*=a_{j_2j_21}$,
$$\left.\frac{dx_{j_2}}{dx_{j_1}}\right|_{y_1^*=a_{j_2j_21}}=\left.\frac{a_{j_2j_20}(y_{j_1}-a_{j_2j_21})^2}{a_{j_1j_10}\prod_{s_1=1}^{3}(y_{j_1}-a_{j_1j_1s_1})}\right|_{y_1^*=a_{j_2j_21}}=0,$$
$$\left.\frac{d^2x_{j_2}}{dx_{j_1}^2}\right|_{y_1^*=a_{j_2j_21}}=0.$$
If
$$\left.\frac{d^3x_{j_2}}{dx_{j_1}^3}\right|_{y_1^*=a_{j_2j_21}}=\frac{2a_{j_2j_20}\alpha_{j_1j_1}^2}{a_{j_1j_10}\prod_{s_1=1}^{3}(a_{j_2j_21}-a_{j_1j_1s_1})}>0,$$
there is an increasing-inflection flow at $y_1^*=a_{j_2j_21}$. If
$$\left.\frac{d^3x_{j_2}}{dx_{j_1}^3}\right|_{y_1^*=a_{j_2j_21}}=\frac{2a_{j_2j_20}\alpha_{j_1j_1}^2}{a_{j_1j_10}\prod_{s_1=1}^{3}(a_{j_2j_21}-a_{j_1j_1s_1})}<0,$$
there is a decreasing-inflection flow at $y_1^*=a_{j_2j_21}$. With
$$\dot{x}^*_{j_1}=a_{j_1j_10}\prod_{s_1=1}^{3}(a_{j_2j_21}-a_{j_1j_1s_1}),$$
the inflection flows are positive and negative for $\dot{x}^*_{j_1}>0$ and $\dot{x}^*_{j_1}<0$, respectively.

Therefore, the repeated equilibrium of $y^*_{j_1}=a_{j_2j_21}$ for the x_{j_2}-direction has four cases with $\dot{x}^*_{j_1}=a_{j_1j_10}\prod_{s_1=1}^{3}(a_{j_2j_21}-a_{j_1j_1s_1})$ as in Eqs. (3.126)–(3.129):

- For $a_{j_1j_10}\prod_{s_1=1}^{3}(a_{j_2j_21}-a_{j_1j_1s_1})>0$ and $a_{j_2j_20}\alpha_{j_1j_1}^2>0$,
$$(\dot{x}^*_{j_1},a_{j_2j_21})=\underbrace{(\mathrm{pF},\mathrm{II})}_{\text{increasing-inflection flow }(+)}.$$

- For $a_{j_1j_10}\prod_{s_1=1}^{3}(a_{j_2j_21}-a_{j_1j_1s_1})<0$ and $a_{j_2j_20}\alpha_{j_1j_1}^2>0$,
$$(\dot{x}_{j_1}^*, a_{j_2j_2}1) = \underbrace{(\text{nF}, \text{DI})}_{\text{decreasing-inflection flow } (-)}.$$
- For $a_{j_1j_10}\prod_{s_1=1}^{3}(a_{j_2j_21}-a_{j_1j_1s_1})>0$ and $a_{j_2j_20}\alpha_{j_1j_1}^2<0$,
$$(\dot{x}_{j_1}^*, a_{j_2j_21}) = \underbrace{(\text{pF}, \text{DI})}_{\text{decreaseng-inflection flow } (+)}.$$
- For $a_{j_1j_10}\prod_{s_1=1}^{3}(a_{j_2j_21}-a_{j_1j_1s_1})<0$ and $a_{j_2j_20}\alpha_{j_1j_1}^2>0$,
$$(\dot{x}_{j_1}^*, a_{j_2j_21}) = \underbrace{(\text{nF}, \text{II})}_{\text{increasing-inflection flow } (-)}.$$

– The decreasing- and increasing-inflection flows are the appearing bifurcations for the up-parabola and parabola flows.

(vi$_2$) For $a_{j_2j_21}=a_{j_1j_1s_1}(s_1,s_2\in\{1,2,3\})$, consider a dynamical system as

$$\begin{aligned}
&\dot{x}_{j_1} = a_{j_1j_10}(y_{j_1}-a_{j_1j_1s_1})\prod_{s_2=1,s_2\neq s_1}^{3}(y_{j_1}-a_{j_1j_1s_2}),\\
&\dot{x}_{j_2} = a_{j_2j_20}(y_{j_1}-a_{j_2j_21})^2,\\
&j_1,j_2\in\{1,2\};\quad j_1\neq j_2;\ s_1,s_2\in\{1,2,3\},\ s_1\neq s_2.
\end{aligned}$$

There is a dynamical system in the y_{j_1}-direction as

$$\begin{aligned}
\dot{y}_{j_1} &= (y_{j_1}-a_{j_1j_1s_1})[\alpha_{j_1j_1}a_{j_1j_10}\prod_{s_2=1,s_2\neq s_1}^{3}(y_{j_1}-a_{j_1j_1s_2})\\
&\quad+\alpha_{j_1j_2}a_{j_2j_20}(y_{j_1}-a_{j_2j_21})],
\end{aligned}$$

with a dynamical system in the $y_{j_1}^{\perp}$-direction as

$$\begin{aligned}
\dot{y}_{j_1}^{\perp} = (y_{j_1}-a_{j_1j_1s_1})\Bigg[&\beta_{j_1j_1}a_{j_1j_10}\prod_{s_2=1,s_2\neq s_1}^{3}(y_{j_1}-a_{j_1j_1s_2})\\
&+\beta_{j_1j_2}a_{j_2j_20}(y_{j_1}-a_{j_2j_21})\Bigg].
\end{aligned}$$

In phase space, at $y_1^*=a_{j_1j_1s_1}=a_{j_2j_21}$,

$$\left.\frac{dx_{j_2}}{dx_{j_1}}\right|_{y_1^*=a_{j_2j_21}} = \left.\frac{a_{j_2j_20}(y_{j_1}-a_{j_2j_21})}{a_{j_1j_10}\prod_{s_2=1,s_2\neq s_1}^{3}(y_{j_1}-a_{j_1j_1s_2})}\right|_{y_1^*=a_{j_2j_21}} = 0.$$

If

$$\left.\frac{d^2x_{j_2}}{dx_{j_1}^2}\right|_{y_1^*=a_{j_2j_2j_1}} = \frac{a_{j_2j_20}\alpha_{j_1j_1}^2}{a_{j_1j_10}\alpha_{j_1j_1}\prod_{s_2=1,s_2\neq s_1}^{3}(a_{j_2j_21}-a_{j_1j_1s_2})} > 0,$$

there is an up-parabola flow at $y_1^* = a_{j_1j_1s_1} = a_{j_2j_21}$. If

$$\left.\frac{d^2x_{j_2}}{dx_{j_1}^2}\right|_{y_1^*=a_{j_2j_21}} = \frac{a_{j_2j_20}\alpha_{j_1j_1}^2}{a_{j_1j_10}\alpha_{j_1j_1}\prod_{s_2=1,s_2\neq s_1}^{3}(a_{j_2j_21}-a_{j_1j_1s_2})} < 0,$$

there is a down-parabola flow at $y_1^* = a_{j_1j_1s_1} = a_{j_2j_21}$. The variational equation in the y_{j_1}-direction is

$$\Delta\dot{y}_{j_1} = \alpha_{j_1j_1}a_{j_1j_10}\prod_{s_2=1,s_2\neq s_1}^{3}(a_{j_1j_1s_1}-a_{j_1j_1s_2})\Delta y_{j_1} = G_{j_1j_10}\Delta y_{j_1},$$

where

$$\begin{aligned} G_{j_1j_10} &= \left.\frac{\partial}{\partial y_{j_1}}\left[\mathbf{n}^{\mathrm{T}}\cdot\mathbf{F}(y_{j_1},\mathbf{p})\right]\right|_{y_{j_1}^*=a_{j_1j_1s_1}=a_{j_2j_21}} \\ &= \alpha_{j_1j_1}a_{j_1j_10}\prod_{s_2=1,s_2\neq s_1}^{3}(a_{j_1j_1s_1}-a_{j_1j_1s_2}). \end{aligned}$$

The parabola flows in the y_{j_1}-direction are source and sink for $G_{j_1j_10} > 0$ and $G_{j_1j_10} < 0$, respectively.

Therefore, the infinite-equilibriums of $y_{j_1}^* = a_{j_1j_1s_1}$ for the x_{j_1}-direction and $y_{j_1}^* = a_{j_2j_21}$ for the x_{j_2}-direction are four cases as in Eqs. (3.134)–(3.137):

- For $\alpha_{j_1j_1}a_{j_1j_10}\prod_{s_2=1,s_2\neq s_1}^{3}(a_{j_1j_1s_1}-a_{j_1j_1s_2}) > 0$ and $a_{j_2j_20}\alpha_{j_1j_1}^2 > 0$,

$$(a_{j_1j_1s_1}, a_{j_2j_21}) = \underbrace{(\mathrm{SO},\mathrm{UP})}_{\text{up-parabola source}}.$$

- For $\alpha_{j_1j_1}a_{j_1j_10}\prod_{s_2=1,s_2\neq s_1}^{3}(a_{j_1j_1s_1}-a_{j_1j_1s_2}) < 0$ and $a_{j_2j_20}\alpha_{j_1j_1}^2 > 0$,

$$(a_{j_1j_1s_1}, a_{j_2j_21}) = \underbrace{(\mathrm{SI},\mathrm{DP})}_{\text{down-parabola sink}}.$$

- For $\alpha_{j_1j_1}a_{j_1j_10}\prod_{s_2=1,s_2\neq s_1}^{3}(a_{j_1j_1s_1}-a_{j_1j_1s_2}) > 0$ and $a_{j_2j_20}\alpha_{j_1j_1}^2 < 0$,

$$(a_{j_1j_1s_1}, a_{j_2j_21}) = \underbrace{(\mathrm{SO},\mathrm{DP})}_{\text{down-parabola source}}.$$

- For $\alpha_{j_1j_1}a_{j_1j_10}\prod_{s_2=1,s_2\neq s_1}^{3}(a_{j_1j_1s_1}-a_{j_1j_1s_2})<0$ and $a_{j_2j_20}\alpha_{j_1j_1}^2<0$,

$$(a_{j_1j_1s_1},a_{j_2j_21})=\underbrace{(\mathrm{SI},\mathrm{UP})}_{\text{up-parabola sink}}.$$

(vii) For

$$\Delta_{j_1j_1}<0\quad\text{and}\quad\Delta_{j_2j_2}>0,$$

the dynamical system becomes

$$\begin{aligned}\dot{x}_{j_1}&=a_{j_1j_10}(y_{j_1}-a_{j_1j_11})[(y_{j_1}-a_{j_1j_1})^2+b_{j_1j_1}],\\ \dot{x}_{j_2}&=a_{j_2j_20}(y_{j_1}-a_{j_2j_2})(y_{j_1}-a_{j_2j_22}),\end{aligned}$$

where

$$\begin{aligned}&\{a_{j_2j_21},a_{j_2j_21}\}=\mathrm{sort}\{b_{j_2j_21},b_{j_2j_22}\},\quad a_{j_2j_21}<a_{j_2j_22};\\ &b_{j_2j_21},b_{j_2j_22}=-\frac{1}{2}(B_{j_2j_2}\pm\sqrt{\Delta_{j_2j_2}}),\quad a_{j_2j_20}=A_{j_2j_21};\\ &a_{j_1j_11}=b_{j_1j_11},\quad a_{j_1j_1}=-\frac{1}{2}B_{j_1j_11},\quad b_{j_1j_1}=-\frac{1}{4}\Delta_{j_1j_1},\quad a_{j_1j_10}=A_{j_1j_10}.\end{aligned}$$

(vii$_1$) For $a_{j_1j_11}\neq a_{j_2j_2l_1}(l_1,l_2\in\{1,2\})$, a dynamical system in the y_{j_1}-direction is constructed as

$$\begin{aligned}\dot{y}_{j_1}&=\alpha_{j_1j_1}a_{j_1j_10}(y_{j_1}-a_{j_1j_11})[(y_{j_1}-a_{j_1j_1})^2+b_{j_1j_1}]\\ &\quad+\alpha_{j_1j_2}a_{j_2j_20}(y_{j_1}-a_{j_2j_21})(y_{j_1}-a_{j_2j_22}),\end{aligned}$$

and a dynamical system in the $y_{j_1}^{\perp}$-direction is constructed as

$$\begin{aligned}\dot{y}_{j_1}^{\perp}&=\beta_{j_1j_1}a_{j_1j_10}(y_{j_1}-a_{j_1j_11})[(y_{j_1}-a_{j_1j_1})^2+b_{j_1j_1}]\\ &\quad+\beta_{j_1j_2}a_{j_2j_20}(y_{j_1}-a_{j_2j_21})(y_{j_1}-a_{j_2j_22}).\end{aligned}$$

(vii$_{1a}$) In phase space, at $y_1^*=a_{j_1j_11}$,

$$\left.\frac{dx_{j_1}}{dx_{j_2}}\right|_{y_1^*=a_{j_1j_11}}=\left.\frac{a_{j_1j_10}(y_{j_1}-a_{j_1j_11})[(y_{j_1}-a_{j_1j_1})^2+b_{j_1j_1}]}{a_{j_2j_20}\prod_{l_1=1}^{2}(y_{j_1}-a_{j_2j_2l_1})}\right|_{y_1^*=a_{j_1j_11}}=0.$$

If

$$\left.\frac{d^2x_{j_1}}{dx_{j_2}^2}\right|_{y_1^*=a_{j_1j_11}}=\frac{a_{j_1j_10}\alpha_{j_1j_2}[(a_{j_1j_11}-a_{j_1j_1})^2+b_{j_1j_1}]}{a_{j_2j_20}\prod_{l_1=1}^{2}(a_{j_1j_11}-a_{j_2j_2l_1})}>0,$$

there is an up-parabola flow at $y_1^* = a_{j_1j_11}$. If

$$\left.\frac{d^2x_{j_1}}{dx_{j_2}^2}\right|_{y_1^*=a_{j_1j_11}} = \frac{a_{j_1j_10}\alpha_{j_1j_2}[(a_{j_1j_11}-a_{j_1j_1})^2+b_{j_1j_1}]}{a_{j_2j_20}\prod_{l_1=1}^{2}(a_{j_1j_11}-a_{j_2j_2l_1})} < 0,$$

there is a down-parabola flow at $y_1^* = a_{j_1j_11}$. With

$$\dot{x}_{j_2}^* = a_{j_2j_20}\prod_{l_1=1}^{2}(a_{j_1j_11}-a_{j_2j_2l_1}),$$

the parabola flows are positive and negative for $\dot{x}_{j_2}^* > 0$ and $\dot{x}_{j_2}^* < 0$, respectively.

Therefore, the flow of $y_1^* = a_{j_1j_11}$ with $\dot{x}_{j_2}^* = a_{j_2j_20}\prod_{l_1=1}^{2}(a_{j_1j_11}-a_{j_2j_2l_1})$ possesses the following properties as in Eqs. (3.143)–(3.146):

- For $\alpha_{j_1j_1}a_{j_1j_10} > 0$ and $a_{j_2j_20}\prod_{l_1=1}^{2}(a_{j_1j_11}-a_{j_2j_2l_1}) > 0$,
$$(a_{j_1j_11},\dot{x}_{j_2}^*) = \underbrace{(\mathrm{UP},\mathrm{pF})}_{\text{up-parabola flow }(+)}.$$
- For $\alpha_{j_1j_1}a_{j_1j_10} < 0$ and $a_{j_2j_20}\prod_{l_1=1}^{2}(a_{j_1j_11}-a_{j_2j_2l_1}) > 0$,
$$(a_{j_1j_11},\dot{x}_{j_2}^*) = \underbrace{(\mathrm{DP},\mathrm{pF})}_{\text{down-parabola flow }(+)}.$$
- For $\alpha_{j_1j_1}a_{j_1j_10} > 0$ and $a_{j_2j_20}\prod_{l_1=1}^{2}(a_{j_1j_11}-a_{j_2j_2l_1}) < 0$,
$$(a_{j_1j_11},\dot{x}_{j_2}^*) = \underbrace{(\mathrm{DP},\mathrm{nF})}_{\text{down-parabola flow }(-)}.$$
- For $\alpha_{j_1j_1}a_{j_1j_10} < 0$ and $a_{j_2j_20}\prod_{l_1=1}^{2}(a_{j_1j_11}-a_{j_2j_2l_1}) < 0$,
$$(a_{j_1j_11},\dot{x}_{j_2}^*) = \underbrace{(\mathrm{UP},\mathrm{nF})}_{\text{up-parabola flow }(-)}.$$

(vii$_{1b}$) In phase space, at $y_1^* = a_{j_2j_2l_1}$,

$$\left.\frac{dx_{j_2}}{dx_{j_1}}\right|_{y_1^*=a_{j_2j_2l_1}} = \left.\frac{a_{j_2j_20}(y_{j_1}-a_{j_2j_2l_1})(y_{j_1}-a_{j_2j_2l_2})}{a_{j_1j_10}(y_{j_1}-a_{j_1j_11})[(y_{j_1}-a_{j_1j_1})^2+b_{j_1j_1}]}\right|_{y_1^*=a_{j_2j_2l_1}} = 0.$$

If

$$\left.\frac{d^2x_{j_2}}{dx_{j_1}^2}\right|_{y_1^*=a_{j_2j_2l_1}} = \frac{a_{j_2j_20}\alpha_{j_1j_1}(a_{j_2j_2l_1}-a_{j_2j_2l_2})}{a_{j_1j_10}(a_{j_2j_2l_1}-a_{j_1j_11})[(a_{j_2j_2l_1}-a_{j_1j_1})^2+b_{j_1j_1}]} > 0,$$

there is an up-parabola flow at $y_1^* = a_{j_2j_2l_1}$. If

$$\left.\frac{d^2x_{j_2}}{dx_{j_1}^2}\right|_{y_1^*=a_{j_2j_2l_1}} = \frac{a_{j_2j_20}\alpha_{j_1j_1}(a_{j_2j_2l_1} - a_{j_2j_2l_2})}{a_{j_1j_10}(a_{j_2j_2l_1} - a_{j_1j_11})[(a_{j_2j_2l_1} - a_{j_1j_1})^2 + b_{j_1j_1}]} < 0,$$

there is a down-parabola flow at $y_1^* = a_{j_2j_2l_1}$. With

$$\dot{x}_{j_1}^* = a_{j_1j_10}(a_{j_2j_2l_1} - a_{j_1j_11})[(a_{j_2j_2l_1} - a_{j_1j_1})^2 + b_{j_1j_1}],$$

the parabola flows are positive and negative for $\dot{x}_{j_1}^* > 0$ and $\dot{x}_{j_1}^* < 0$, respectively.

Due to

$$(a_{j_2j_2l_1} - a_{j_1j_1})^2 + b_{j_1j_1} > 0,$$

the flow of $y_1^* = a_{j_2j_2l_1}$ possesses the following properties as in Eqs. (3.147)–(3.150):

- For $a_{j_1j_10}(a_{j_2j_2l_1} - a_{j_1j_11}) > 0$ and $a_{j_2j_20}\alpha_{j_1j_1}(a_{j_2j_2l_1} - a_{j_2j_2l_2}) > 0$,

$$(\dot{x}_{j_1}^*, a_{j_2j_2l_1}) = \underbrace{(\mathrm{pF}, \mathrm{UP})}_{\text{up-parabola flow}(+)}.$$

- For $a_{j_1j_10}(a_{j_2j_2l_1} - a_{j_1j_11}) < 0$ and $a_{j_2j_20}\alpha_{j_1j_1}(a_{j_2j_2l_1} - a_{j_2j_2l_2}) > 0$,

$$(\dot{x}_{j_1}^*, a_{j_2j_2l_1}) = \underbrace{(\mathrm{nF}, \mathrm{DP})}_{\text{down-parabola flow}(-)}.$$

- For $a_{j_1j_10}(a_{j_2j_2l_1} - a_{j_1j_11}) > 0$ and $a_{j_2j_20}\alpha_{j_1j_1}(a_{j_2j_2l_1} - a_{j_2j_2l_2}) < 0$,

$$(\dot{x}_{j_1}^*, a_{j_2j_2l_1}) = \underbrace{(\mathrm{pF}, \mathrm{DP})}_{\text{down-parabola flow}(+)}.$$

- For $a_{j_1j_10}(a_{j_2j_2l_1} - a_{j_1j_11}) < 0$ and $a_{j_2j_20}\alpha_{j_1j_1}(a_{j_2j_2l_1} - a_{j_2j_2l_2}) < 0$,

$$(\dot{x}_{j_1}^*, a_{j_2j_2l_1}) = \underbrace{(\mathrm{nF}, \mathrm{UP})}_{\text{up-parabola flow}(-)}.$$

(vii$_2$) For $a_{j_1j_11} = a_{j_2j_2l_1}$, consider a dynamical system as

$$\begin{aligned}\dot{x}_{j_1} &= a_{j_1j_10}(y_{j_1} - a_{j_1j_11})[(y_{j_1} - a_{j_1j_1})^2 + b_{j_1j_1}],\\ \dot{x}_{j_2} &= a_{j_2j_20}(y_{j_1} - a_{j_2j_2l_1})(y_{j_1} - a_{j_2j_2l_2}).\end{aligned}$$

A dynamical system in the y_{j_1}-direction is constructed as

$$\begin{aligned}\dot{y}_{j_1} &= (y_{j_1} - a_{j_1j_1 1})\{\alpha_{j_1j_1}a_{j_1j_1 0}[(y_{j_1} - a_{j_1j_1})^2 \\ &\quad + b_{j_1j_1}] + \alpha_{j_1j_2}a_{j_2j_2 0}(y_{j_1} - a_{j_2j_2l_2})\},\end{aligned}$$

and a dynamical system in the $y_{j_1}^{\perp}$-direction is constructed as

$$\begin{aligned}\dot{y}_{j_1}^{\perp} &= (y_{j_1} - a_{j_1j_1 1})\{\beta_{j_1j_1}a_{j_1j_1 0}[(y_{j_1} - a_{j_1j_1})^2 \\ &\quad + b_{j_1j_1}] + \beta_{j_1j_2}a_{j_2j_2 0}(y_{j_1} - a_{j_2j_2l_2})\}.\end{aligned}$$

Define a G-function as

$$\begin{aligned}G_{j_1j_1 0} &= \frac{\partial}{\partial y_{j_1}}[\mathbf{n}^{\mathrm{T}} \cdot \mathbf{F}(y_{j_1}, \mathbf{p})]\Big|_{y_{j_1}^* = a_{j_1j_1 1} = a_{j_2j_2l_1}} \\ &= \alpha_{j_1j_1}a_{j_1j_1 0}[(a_{j_1j_1 1} - a_{j_1j_1})^2 + b_{j_1j_1}] + \alpha_{j_1j_2}a_{j_2j_2 0}(a_{j_2j_2l_1} - a_{j_2j_2l_2}).\end{aligned}$$

In phase space, at $y_1^* = a_{j_1j_1 1} = a_{22l_1}$, if

$$\frac{dx_{j_2}}{dx_{j_1}}\Big|_{y_1^* = a_{j_2j_2l_1}} = \frac{a_{j_2j_2 0}(a_{j_2j_2l_1} - a_{j_2j_2l_2})}{a_{j_1j_1 0}[(a_{j_2j_2l_1} - a_{j_1j_1})^2 + b_{j_1j_1}]} > 0,$$

there is an increasing-inflection flow at $y_1^* = a_{j_2j_2l_1} = a_{j_1j_1 1}$, and if

$$\frac{dx_{j_2}}{dx_{j_1}}\Big|_{y_1^* = a_{j_2j_2l_1}} = \frac{a_{j_2j_2 0}(a_{j_2j_2l_1} - a_{j_2j_2l_2})}{a_{j_1j_1 0}[(a_{j_2j_2l_1} - a_{j_1j_1})^2 + b_{j_1j_1}]} < 0,$$

there is a decreasing-inflection flow at $y_1^* = a_{j_2j_2l_1} = a_{j_1j_1 1}$.

The variational equation at $y_1^* = a_{j_2j_2l_1} = a_{j_1j_1 1}$ is

$$\begin{aligned}\Delta\dot{y}_{j_1} &= \{\alpha_{j_1j_1}a_{j_1j_1 0}[(a_{j_1j_1 1} - a_{j_1j_1})^2 + b_{j_1j_1}] \\ &\quad + \alpha_{j_1j_2}a_{j_2j_2 0}(a_{j_2j_2l_1} - a_{j_2j_2l_2})\}\Delta y_{j_1} \\ &= G_{j_1j_1 0}\Delta y_{j_1}.\end{aligned}$$

The inflection flows are source and sink for $G_{j_1j_1 0} > 0$ and $G_{j_1j_1 0} < 0$, respectively.

Therefore, the infinite-equilibrium flow of $y_1^* = a_{j_2j_2l_1} = a_{j_1j_1 1}$ possesses the following properties as in Eqs. (3.155)–(3.158):

- For $a_{j_1j_1 0}/a_{j_2j_2 0}(a_{j_2j_2l_1} - a_{j_2j_2l_2}) > 0$ and $G_{j_1j_1 0} > 0$,

$$(a_{j_1j_1 1}, a_{j_2j_2l_1}) = \underbrace{(\mathrm{II}, \mathrm{SO})}_{\text{increaing-inflection source}}.$$

- For $a_{j_1j_10}/a_{j_2j_20}(a_{j_2j_2l_1} - a_{j_2j_2l_2}) > 0$ and $G_{j_1j_10} < 0$,

$$(a_{j_1j_11}, a_{j_2j_2l_1}) = \underbrace{(\text{II}, \text{SI})}_{\text{increaing-inflection sink}}.$$

- For $a_{j_1j_10}/a_{j_2j_20}(a_{j_2j_2l_1} - a_{j_2j_2l_2}) < 0$ and $G_{j_1j_10} > 0$,

$$(a_{j_1j_11}, a_{j_2j_2l_1}) = \underbrace{(\text{DI}, \text{SO})}_{\text{decreaing-inflection source}}.$$

- For $a_{j_1j_10}/a_{j_2j_20}(a_{j_2j_2l_1} - a_{j_2j_2l_2}) < 0$ and $G_{j_1j_10} < 0$,

$$(a_{j_1j_11}, a_{j_2j_2l_1}) = \underbrace{(\text{DI}, \text{SI})}_{\text{decreaing-inflection sink}}.$$

(viii) For

$$\Delta_{j_1j_1} = 0 \quad \text{and} \quad \Delta_{j_2j_2} > 0,$$

the dynamical system becomes

$$\begin{aligned}
\dot{x}_{j_1} &= a_{j_1j_10}(y_{j_1} - a_{j_1j_1s_1})(y_{j_1} - a_{j_1j_1s_2})^2, \\
\dot{x}_{j_2} &= a_{j_2j_20}(y_{j_1} - a_{j_2j_21})(y_{j_1} - a_{j_2j_22}), \\
j_1&, j_2 \in \{1, 2\};\ j_1 \neq j_2;
\end{aligned}$$

where

$$\begin{aligned}
&\{a_{j_2j_21}, a_{j_2j_22}\} = \text{sort}\{b_{j_2j_21}, b_{j_2j_22}\}, a_{j_2j_21} < a_{j_2j_22}; \\
&b_{j_2j_21}, b_{j_2j_22} = -\frac{1}{2}(B_{j_2j_2} \pm \sqrt{\Delta_{j_2j_2}}), a_{j_2j_20} = A_{j_2j_21}; \\
&a_{j_1j_1l_1} = b_{j_1j_11}, \quad a_{j_1j_1l_2} = -\frac{1}{2}B_{j_1j_1}, \quad b_{j_1j_1} = 0, \quad a_{j_1j_10} = A_{j_1j_10}.
\end{aligned}$$

(viii$_1$) For $a_{j_2j_21} \neq a_{j_1j_1s_1}, a_{j_1j_1s_2}$, a dynamical system in the y_{j_1}-direction is constructed as

$$\begin{aligned}
\dot{y}_{j_1} &= \alpha_{j_1j_1}a_{j_1j_10}(y_{j_1} - a_{j_1j_1s_1})(y_{j_1} - a_{j_1j_1s_2})^2 \\
&\quad + \alpha_{j_1j_2}a_{j_2j_20}(y_{j_1} - a_{j_2j_21})(y_{j_1} - a_{j_2j_22}),
\end{aligned}$$

and a dynamical system in the $y_{j_1}^{\perp}$-direction is constructed as

$$\begin{aligned}
\dot{y}_{j_1}^{\perp} &= \beta_{j_1j_1}a_{j_1j_10}(y_{j_1} - a_{j_1j_1s_1})(y_{j_1} - a_{j_1j_1s_2})^2 \\
&\quad + \beta_{j_1j_2}a_{j_2j_20}(y_{j_1} - a_{j_2j_21})(y_{j_1} - a_{j_2j_22}).
\end{aligned}$$

(viii$_{1a}$) In phase space, at $y_1^* = a_{j_1 j_1 s_1}$,

$$\left.\frac{dx_{j_1}}{dx_{j_2}}\right|_{y_1^*=a_{j_1 j_1 s_1}} = \left.\frac{a_{j_1 j_1 0}(y_{j_1} - a_{j_1 j_1 s_1})(y_{j_1} - a_{j_1 j_1 s_2})^2}{a_{j_2 j_2 0}\prod_{l_1=1}^{2}(y_{j_1} - a_{j_2 j_2 l_1})}\right|_{y_1^*=a_{j_1 j_1 s_1}} = 0.$$

If

$$\left.\frac{d^2 x_{j_1}}{dx_{j_2}^2}\right|_{y_1^*=a_{j_1 j_1 s_1}} = \frac{a_{j_1 j_1 0}\alpha_{j_1 j_2}(a_{j_1 j_1 s_1} - a_{j_1 j_1 s_2})^2}{a_{j_2 j_2 0}\prod_{l_1=1}^{2}(a_{j_1 j_1 s_1} - a_{j_2 j_2 l_1})} > 0,$$

there is an up-parabola flow at $y_1^* = a_{j_1 j_1 s_1}$. If

$$\left.\frac{d^2 x_{j_1}}{dx_{j_2}^2}\right|_{y_1^*=a_{j_1 j_1 s_1}} = \frac{a_{j_1 j_1 0}\alpha_{j_1 j_2}(a_{j_1 j_1 s_1} - a_{j_1 j_1 s_2})^2}{a_{j_2 j_2 0}\prod_{l_1=1}^{2}(a_{j_1 j_1 s_1} - a_{j_2 j_2 l_1})} < 0,$$

there is a down-parabola flow at $y_1^* = a_{j_1 j_1 s_1}$. With

$$\dot{x}_{j_2}^* = a_{j_2 j_2 0}\prod_{l_1=1}^{2}(a_{j_1 j_1 s_1} - a_{j_2 j_2 l_1}),$$

the parabola flows are positive and negative for $\dot{x}_{j_2}^* > 0$ and $\dot{x}_{j_2}^* < 0$, respectively.

Therefore, the flows of $y_{j_1}^* = a_{j_1 j_1 s_1} \neq a_{j_2 j_2 l_1}$ ($l_1 = 1, 2$) for the x_{j_1}-direction have the following properties as in Eqs. (3.164)–(3.167):

- For $a_{j_1 j_1 0}\alpha_{j_1 j_2}(a_{j_1 j_1 s_1} - a_{j_1 j_1 s_2})^2 > 0$ and $a_{j_2 j_2 0}\prod_{l_1=1}^{2}(a_{j_1 j_1 s_1} - a_{j_2 j_2 l_1}) > 0$,

$$(a_{j_1 j_1 s_1}, \dot{x}_{j_2}^*) = \underbrace{(\mathrm{UP}, \mathrm{pF})}_{\text{up-parabola flow } (+)}.$$

- For $a_{j_1 j_1 0}\alpha_{j_1 j_2}(a_{j_1 j_1 s_1} - a_{j_1 j_1 s_2})^2 < 0$ and $a_{j_2 j_2 0}\prod_{l_1=1}^{2}(a_{j_1 j_1 s_1} - a_{j_2 j_2 l_1}) > 0$,

$$(a_{j_1 j_1 s_1}, \dot{x}_{j_2}^*) = \underbrace{(\mathrm{DP}, \mathrm{pF})}_{\text{down-parabola flow } (+)}.$$

- For $a_{j_1 j_1 0}\alpha_{j_1 j_2}(a_{j_1 j_1 s_1} - a_{j_1 j_1 s_2})^2 > 0$ and $a_{j_2 j_2 0}\prod_{l_1=1}^{2}(a_{j_1 j_1 s_1} - a_{j_2 j_2 l_1}) < 0$,

$$(a_{j_1 j_1 s_1}, \dot{x}_{j_2}^*) = \underbrace{(\mathrm{DP}, \mathrm{nF})}_{\text{down-parabola flow } (-)}.$$

- For $a_{j_1j_10}\alpha_{j_1j_2}(a_{j_1j_1s_1} - a_{j_1j_1s_2})^2 < 0$ and $a_{j_2j_20}\prod_{l_1=1}^{2}(a_{j_1j_1s_1} - a_{j_2j_2l_1}) < 0$,

$$(a_{j_1j_1s_1}, \dot{x}^*_{j_2}) = \underbrace{(\text{UP}, \text{nF})}_{\text{up-parabola flow } (-)}.$$

(viii$_{1b}$) In phase space, at $y_1^* = a_{j_1j_1s_2}$,

$$\left.\frac{dx_{j_1}}{dx_{j_2}}\right|_{y_1^*=a_{j_1j_1s_2}} = \left.\frac{a_{j_1j_10}(y_{j_1} - a_{j_1j_1s_1})(y_{j_1} - a_{j_1j_1s_2})^2}{a_{j_2j_20}\prod_{l_1=1}^{2}(y_{j_1} - a_{j_2j_2l_1})}\right|_{y_1^*=a_{j_1j_1s_2}} = 0,$$

$$\left.\frac{d^2x_{j_1}}{dx_{j_2}^2}\right|_{y_1^*=a_{j_1j_1s_2}} = 0.$$

If

$$\left.\frac{d^3x_{j_1}}{dx_{j_2}^3}\right|_{y_1^*=a_{j_1j_1s_2}} = \frac{2a_{j_1j_10}\alpha_{j_1j_2}^2(a_{j_1j_1s_2} - a_{j_1j_1s_1})}{a_{j_2j_20}\prod_{l_1=1}^{2}(a_{j_1j_1s_1} - a_{j_2j_2l_1})} > 0,$$

there is an increasing-inflection flow at $y_1^* = a_{j_1j_1s_2}$. If

$$\left.\frac{d^3x_{j_1}}{dx_{j_2}^3}\right|_{y_1^*=a_{j_1j_1s_2}} = \frac{2a_{j_1j_10}\alpha_{j_1j_2}^2(a_{j_1j_1s_2} - a_{j_1j_1s_1})}{a_{j_2j_20}\prod_{l_1=1}^{2}(a_{j_1j_1s_1} - a_{j_2j_2l_1})} < 0,$$

there is a decreasing-inflection flow at $y_1^* = a_{j_1j_1s_2}$. With

$$\dot{x}^*_{j_2} = a_{j_2j_20}\prod_{l_1=1}^{2}(a_{j_1j_1s_1} - a_{j_2j_2l_1}),$$

the inflection flows are positive and negative for $\dot{x}^*_{j_2} > 0$ and $\dot{x}^*_{j_2} < 0$, respectively.

Therefore, the flows of $y^*_{j_1} = a_{j_1j_1s_2} \neq a_{j_2j_2l_1}$ $(l_1 = 1, 2)$ for the x_{j_1}-direction have the following properties as in Eqs. (3.168)–(3.171):

- For $a_{j_1j_10}\alpha_{j_1j_2}^2(a_{j_1j_1s_2} - a_{j_1j_1s_1}) > 0$ and $a_{j_2j_20}\prod_{l_1=1}^{2}(a_{j_1j_1s_2} - a_{j_2j_2l_1}) > 0$,

$$(a_{j_1j_1s_2}, \dot{x}^*_{j_2}) = \underbrace{(\text{II}, \text{pF})}_{\text{increasing-inflection flow}(-)}.$$

- For $a_{j_1j_10}\alpha_{j_1j_2}^2(a_{j_1j_1s_2} - a_{j_1j_1s_1}) > 0$ and $a_{j_2j_20}\prod_{l_1=1}^{2}(a_{j_1j_1s_2} - a_{j_2j_2l_1}) > 0$,

$$(a_{j_1j_1s_2}, \dot{x}^*_{j_2}) = \underbrace{(\text{DI}, \text{pF})}_{\text{decreasing-inflection flow } (+)}.$$

- For $a_{j_1j_10}\alpha^2_{j_1j_2}(a_{j_1j_1s_2} - a_{j_1j_1s_1}) > 0$ and $a_{j_2j_20}\prod_{l_1=1}^{2}(a_{j_1j_1s_2} - a_{j_2j_2l_1}) < 0$,

$$(a_{j_1j_1s_2}, \dot{x}^*_{j_2}) = \underbrace{(\mathrm{DI}, \mathrm{nF})}_{\text{decreasing-inflection flow }(-)}.$$

- For $a_{j_1j_10}\alpha^2_{j_1j_2}(a_{j_1j_1s_2} - a_{j_1j_1s_1}) < 0$ and $a_{j_2j_20}\prod_{l_1=1}^{2}(a_{j_1j_1s_2} - a_{j_2j_2l_1}) < 0$,

$$(a_{j_1j_1s_2}, \dot{x}^*_{j_2}) = \underbrace{(\mathrm{II}, \mathrm{nF})}_{\text{increasing-inflection flow }(-)}.$$

- The increasing- and decreasing-inflection flows are the appearing bifurcations for the up-parabola and down-parabola flows.

(viii$_{1c}$) In phase space, at $y^*_1 = a_{j_2j_2l_1}$,

$$\left.\frac{dx_{j_2}}{dx_{j_1}}\right|_{y^*_1=a_{j_2j_2l_1}} = \left.\frac{a_{j_2j_20}(y_{j_1} - a_{j_2j_2l_1})(y_{j_1} - a_{j_2j_2l_2})}{a_{j_1j_10}(y_{j_1} - a_{j_1j_1s_1})(y_{j_1} - a_{j_1j_1s_2})^2}\right|_{y^*_1=a_{j_2j_21}} = 0.$$

If

$$\left.\frac{d^2x_{j_2}}{dx^2_{j_1}}\right|_{y^*_1=a_{j_2j_2l_1}} = \frac{a_{j_2j_20}\alpha_{j_1j_1}(a_{j_2j_2l_1} - a_{j_2j_2l_2})}{a_{j_1j_10}(a_{j_2j_2l_1} - a_{j_1j_1s_1})(a_{j_2j_2l_1} - a_{j_1j_1s_2})^2} > 0,$$

there is an up-parabola flow at $y^*_1 = a_{j_2j_2l_1}$. If

$$\left.\frac{d^2x_{j_2}}{dx^2_{j_1}}\right|_{y^*_1=a_{j_2j_2l_1}} = \frac{a_{j_2j_20}\alpha_{j_1j_1}(a_{j_2j_2l_1} - a_{j_2j_2l_2})}{a_{j_1j_10}(a_{j_2j_2l_1} - a_{j_1j_1s_1})(a_{j_2j_2l_1} - a_{j_1j_1s_2})^2} < 0,$$

there is a down-parabola flow at $y^*_1 = a_{j_2j_2l_1}$. With

$$\dot{x}^*_{j_1} = a_{j_1j_10}(a_{j_2j_2l_1} - a_{j_1j_1s_1})(a_{j_2j_2l_1} - a_{j_1j_1s_2})^2,$$

the parabola flows are positive and negative for $\dot{x}^*_{j_1} > 0$ and $\dot{x}^*_{j_1} < 0$, respectively.

Therefore, the flow of $y^*_{j_1} = a_{j_2j_2l_1} \neq a_{j_1j_1s_1}, a_{j_1j_1s_2}$ for the x_{j_2}-direction has the following properties as in Eqs. (3.172)–(3.175):

- For $a_{j_1j_10}(a_{j_2j_2l_1} - a_{j_1j_1s_1})(a_{j_2j_2l_1} - a_{j_1j_1s_2})^2 > 0$ and $a_{j_2j_20}\alpha_{j_1j_1}(a_{j_2j_2l_1} - a_{j_2j_2l_2}) > 0$,

$$(\dot{x}^*_{j_1}, a_{j_2j_2l_1}) = \underbrace{(\mathrm{pF}, \mathrm{UP})}_{\text{up-parabola flow }(+)}.$$

- For $a_{j_1j_10}(a_{j_2j_2l_1} - a_{j_1j_1s_1})(a_{j_2j_2l_1} - a_{j_1j_1s_2})^2 < 0$ and $a_{j_2j_20}\alpha_{j_1j_1}(a_{j_2j_2l_1} - a_{j_2j_2l_2}) > 0$,

$$(\dot{x}^*_{j_1}, a_{j_2j_2l_1}) = \underbrace{(\text{nF}, \text{DP})}_{\text{down-parabola flow } (-)}.$$

- For $a_{j_1j_10}(a_{j_2j_2l_1} - a_{j_1j_1s_1})(a_{j_2j_2l_1} - a_{j_1j_1s_2})^2 > 0$ and $a_{j_2j_20}\alpha_{j_1j_1}(a_{j_2j_2l_1} - a_{j_2j_2l_2}) < 0$,

$$(\dot{x}^*_{j_1}, a_{j_2j_2l_1}) = \underbrace{(\text{pF}, \text{DP})}_{\text{down-parabola flow } (+)}.$$

- For $a_{j_1j_10}(a_{j_2j_2l_1} - a_{j_1j_1s_1})(a_{j_2j_2l_1} - a_{j_1j_1s_2})^2 < 0$ and $a_{j_2j_20}\alpha_{j_1j_1}(a_{j_2j_2l_1} - a_{j_2j_2l_2}) < 0$,

$$(\dot{x}^*_{j_1}, a_{j_2j_2l_1}) = \underbrace{(\text{nF}, \text{UP})}_{\text{up-parabola flow } (-)}.$$

(viii$_2$) For $a_{j_1j_1s_1} = a_{j_1j_1s_2} \equiv a_{j_1j_11}$, consider a system as

$$\begin{aligned}
&\dot{x}_{j_1} = a_{j_1j_10}(y_{j_1} - a_{j_1j_11})^3,\\
&\dot{x}_{j_2} = a_{j_2j_20}(y_{j_1} - a_{j_2j_21})(y_{j_1} - a_{j_2j_22}),\\
&j_1, j_2 \in \{1, 2\};\ j_1 \neq j_2.
\end{aligned}$$

For $a_{j_2j_2l_1} \neq a_{j_1j_11}$, a dynamical system in the y_{j_1}-direction is constructed as

$$\dot{y}_{j_1} = \alpha_{j_1j_1}a_{j_1j_10}(y_{j_1} - a_{j_1j_1s_1})^3 + \alpha_{j_1j_2}a_{j_2j_20}\prod_{l_1=1}^{2}(y_{j_1} - a_{j_2j_2l_1}),$$

and a dynamical system in the $y^{\perp}_{j_1}$-direction is constructed as

$$\dot{y}^{\perp}_{j_1} = \beta_{j_1j_1}a_{j_1j_10}(y_{j_1} - a_{j_1j_1s_1})^3 + \beta_{j_1j_2}a_{j_2j_20}\prod_{l_1=1}^{2}(y_{j_1} - a_{j_2j_2l_1}).$$

(viii$_{2a}$) In phase space, at $y^*_1 = a_{j_1j_11}$,

$$\begin{aligned}
\left.\frac{dx_{j_1}}{dx_{j_2}}\right|_{y^*_1 = a_{j_1j_11}} &= \left.\frac{a_{j_1j_10}(y_{j_1} - a_{j_1j_11})^3}{a_{j_2j_20}\prod_{l_1=1}^{2}(y_{j_1} - a_{j_2j_2l_1})}\right|_{y^*_1 = a_{j_1j_11}} = 0,\\
\left.\frac{d^2x_{j_1}}{dx^2_{j_2}}\right|_{y^*_1 = a_{j_1j_11}} &= \left.\frac{d^3x_{j_1}}{dx^3_{j_2}}\right|_{y^*_1 = a_{j_1j_11}} = 0.
\end{aligned}$$

If

$$\left.\frac{d^4 x_{j_1}}{dx_{j_2}^4}\right|_{y_1^*=a_{j_1j_1 1}} = \frac{6a_{j_1j_1 0}\alpha_{j_1j_2}^3}{a_{j_2j_2 0}\prod_{l_1=1}^{2}(a_{j_1j_1 1}-a_{j_2j_2l_1})} > 0,$$

there is a third-order up-parabola flow at $y_1^* = a_{j_1j_1 1}$. If

$$\left.\frac{d^3 x_{j_1}}{dx_{j_2}^3}\right|_{y_1^*=a_{j_1j_1 1}} = \frac{6a_{j_1j_1 0}\alpha_{j_1j_2}^3}{a_{j_2j_2 0}\prod_{l_1=1}^{2}(a_{j_1j_1 1}-a_{j_2j_2l_1})} < 0,$$

there is a third-order down-parabola flow at $y_1^* = a_{j_1j_1 1}$. With

$$\dot{x}_{j_2}^* = a_{j_2j_2 0}\prod_{l_1=1}^{2}(a_{j_1j_1 1}-a_{j_2j_2l_1}),$$

the third-order parabola flows are positive and negative for $\dot{x}_{j_2}^* > 0$ and $\dot{x}_{j_2}^* < 0$, respectively.

Therefore, the flow of $y_{j_1}^* = a_{j_1j_1 1} \neq a_{j_2j_2l_1}(l_1 = 1,2)$ for the x_{j_1}-direction has the following properties with $\dot{x}_{j_2}^* = a_{j_2j_2 0}\prod_{l_1=1}^{2}(a_{j_1j_1 1} - a_{j_2j_2l_1})$ as in Eqs. (3.179)–(3.182):

- For $a_{j_1j_1 0}\alpha_{j_1j_2}^3 > 0$ and $a_{j_2j_2 0}\prod_{l_1=1}^{2}(a_{j_1j_1 1}-a_{j_2j_2l_1}) > 0$,

$$(a_{j_1j_1 1}, \dot{x}_{j_2}^*) = \underbrace{(3^{\rm rd}\mathrm{UP}, \mathrm{pF})}_{\text{third-order up-parabola flow } (+)}.$$

- For $a_{j_1j_1 0}\alpha_{j_1j_2}^3 < 0$ and $a_{j_2j_2 0}\prod_{l_1=1}^{2}(a_{j_1j_1 1}-a_{j_2j_2l_1}) > 0$,

$$(a_{j_1j_1 1}, \dot{x}_{j_2}^*) = \underbrace{(3^{\rm rd}\mathrm{DP}, \mathrm{pF})}_{\text{third-order down-parabola flow } (+)}.$$

- For $a_{j_1j_1 0}\alpha_{j_1j_2}^3 > 0$ and $a_{j_2j_2 0}\prod_{l_1=1}^{2}(a_{j_1j_1 1}-a_{j_2j_2l_1}) < 0$,

$$(a_{j_1j_1 1}, \dot{x}_{j_2}^*) = \underbrace{(3^{\rm rd}\mathrm{DP}, \mathrm{nF})}_{\text{third-order down-parabola flow } (-)}.$$

- For $a_{j_1j_1 0}\alpha_{j_1j_2}^3 < 0$ and $a_{j_2j_2 0}\prod_{l_1=1}^{2}(a_{j_1j_1 1}-a_{j_2j_2l_1}) < 0$,

$$(a_{j_1j_1 1}, \dot{x}_{j_2}^*) = \underbrace{(3^{\rm rd}\mathrm{UP}, \mathrm{nF})}_{\text{third-order up-parabola flow } (-)}.$$

– The third-order up-parabola flows are the appearing bifurcations for up-parabola, down-parabola, and up-parabola flows.

– The third-order down-parabola flows are the appearing bifurcations for down-parabola, up-parabola, and down-parabola flows.

(viii$_{2b}$) In phase space, at $y_1^* = a_{j_2j_2l_1}$,

$$\left.\frac{dx_{j_2}}{dx_{j_1}}\right|_{y_1^*=a_{j_2j_2l_1}} = \left.\frac{a_{j_2j_20}(y_{j_1}-a_{j_2j_2l_1})(y_{j_1}-a_{j_2j_2l_2})}{a_{j_1j_10}(y_{j_1}-a_{j_1j_11})^3}\right|_{y_1^*=a_{j_2j_2l_1}} = 0.$$

If

$$\left.\frac{d^2x_{j_2}}{dx_{j_1}^2}\right|_{y_1^*=a_{j_2j_2l_1}} = \frac{a_{j_2j_20}\alpha_{j_1j_1}(a_{j_2j_2l_1}-a_{j_2j_2l_2})}{a_{j_1j_10}(a_{j_2j_2l_1}-a_{j_1j_11})^3} > 0,$$

there is an up-parabola flow at $y_1^* = a_{j_2j_2l_1}$. If

$$\left.\frac{d^2x_{j_2}}{dx_{j_1}^2}\right|_{y_1^*=a_{j_2j_2l_1}} = \frac{a_{j_2j_20}\alpha_{j_1j_1}(a_{j_2j_2l_1}-a_{j_2j_2l_2})}{a_{j_1j_10}(a_{j_2j_2l_1}-a_{j_1j_11})^3} < 0,$$

there is a down-parabola flow at $y_1^* = a_{j_2j_2l_1}$. With

$$\dot{x}_{j_1}^* = a_{j_1j_10}(a_{j_2j_2l_1}-a_{j_1j_11})^3,$$

the inflection flows are positive and negative for $\dot{x}_{j_1}^* > 0$ and $\dot{x}_{j_1}^* < 0$, respectively.

Therefore, the flow of $y_{j_1}^* = a_{j_2j_2l_1} \neq a_{j_1j_11}$ for the x_{j_2}-direction has the following properties as in Eqs. (3.183)–(3.186):

- For $a_{j_1j_10}(a_{j_2j_2l_1}-a_{j_1j_11})^3 > 0$ and $a_{j_2j_20}\alpha_{j_1j_1}(a_{j_2j_2l_1}-a_{j_2j_2l_2}) > 0$,

$$(\dot{x}_{j_1}^*, a_{j_2j_2l_1}) = \underbrace{(\text{pF}, \text{UP})}_{\text{up-parabola flow } (+)}.$$

- For $a_{j_1j_10}(a_{j_2j_2l_1}-a_{j_1j_11})^3 < 0$ and $a_{j_2j_20}\alpha_{j_1j_1}(a_{j_2j_2l_1}-a_{j_2j_2l_2}) > 0$,

$$(\dot{x}_{j_1}^*, a_{j_2j_2l_1}) = \underbrace{(\text{nF}, \text{DP})}_{\text{down-parabola flow } (-)}.$$

- For $a_{j_1j_10}(a_{j_2j_2l_1}-a_{j_1j_11})^3 > 0$ and $a_{j_2j_20}\alpha_{j_1j_1}(a_{j_2j_2l_1}-a_{j_2j_2l_2}) < 0$,

$$(\dot{x}_{j_1}^*, a_{j_2j_2l_1}) = \underbrace{(\text{pF}, \text{DP})}_{\text{down-parabola flow } (+)}.$$

- For $a_{j_1j_10}(a_{j_2j_2l_1}-a_{j_1j_11})^3 < 0$ and $a_{j_2j_20}\alpha_{j_1j_1}(a_{j_2j_2l_1}-a_{j_2j_2l_2}) < 0$,

$$(\dot{x}_{j_1}^*, a_{j_2j_2l_1}) = \underbrace{(\text{nF}, \text{UP})}_{\text{up-parabola flow } (-)}.$$

(viii$_3$) For $a_{j_1j_1s_1} = a_{j_2j_2l_1}$, consider a dynamical system as

$$\dot{x}_{j_1} = a_{j_1j_10}(y_{j_1} - a_{j_1j_1s_1})(y_{j_1} - a_{j_1j_1s_2})^2,$$
$$\dot{x}_{j_2} = a_{j_2j_20}(y_{j_1} - a_{j_2j_2l_1})(y_{j_1} - a_{j_2j_2l_2}).$$

The corresponding dynamical system in the y_{j_1}-direction is constructed as

$$\dot{y}_{j_1} = (y_{j_1} - a_{j_1j_1s_1})[\alpha_{j_1j_1}a_{j_1j_10}(y_{j_1} - a_{j_1j_1s_2})^2 + \alpha_{j_1j_2}a_{j_2j_20}(y_{j_1} - a_{j_2j_2l_2})],$$

and the corresponding dynamical system in the $y_{j_1}^{\perp}$-direction is constructed as

$$\dot{y}_{j_1}^{\perp} = (y_{j_1} - a_{j_1j_1s_1})[\beta_{j_1j_1}a_{j_1j_10}(y_{j_1} - a_{j_1j_1s_2})^2 + \beta_{j_1j_2}a_{j_2j_20}(y_{j_1} - a_{j_2j_2l_2})].$$

Define a G-function as

$$G_{j_1j_10} = \frac{\partial}{\partial y_{j_1}}[\mathbf{n}^{\mathrm{T}} \cdot \mathbf{F}(y_{j_1}, \mathbf{p})]\bigg|_{y_{j_1}^* = a_{j_1j_1s_1} = a_{j_2j_2l_1}}$$
$$= \alpha_{j_1j_1}a_{j_1j_10}(a_{j_1j_1s_1} - a_{j_1j_1s_2})^2 + \alpha_{j_1j_2}a_{j_2j_20}(a_{j_2j_2l_1} - a_{j_2j_2l_2}).$$

In phase space, at $y_1^* = a_{j_1j_1s_1} = a_{j_2j_2l_1}$, if

$$\frac{dx_{j_2}}{dx_{j_1}}\bigg|_{y_1^* = a_{j_2j_2l_1}} = \frac{a_{j_2j_20}(a_{j_2j_2l_1} - a_{j_2j_2l_2})}{a_{j_1j_10}(a_{j_1j_1s_1} - a_{j_1j_1s_2})^2} > 0,$$

there is an increasing-inflection flow at $y_1^* = a_{j_1j_1s_1} = a_{j_2j_2l_1}$, and if

$$\frac{dx_{j_2}}{dx_{j_1}}\bigg|_{y_1^* = a_{j_2j_2l_1}} = \frac{a_{j_2j_20}(a_{j_2j_2l_1} - a_{j_2j_2l_2})}{a_{j_1j_10}(a_{j_1j_1s_1} - a_{j_1j_1s_2})^2} < 0,$$

there is a decreasing-inflection flow at $y_1^* = a_{j_1j_1s_1} = a_{j_2j_2l_1}$. The variational equation in the y_{j_1}-direction is

$$\Delta\dot{y}_{j_1} = [\alpha_{j_1j_1}a_{j_1j_10}(a_{j_1j_1s_1} - a_{j_1j_1s_2})^2 + \alpha_{j_1j_2}a_{j_2j_20}(a_{j_2j_2l_1} - a_{j_2j_2l_2})]\Delta y_{j_1}$$
$$= G_{j_1j_10}\Delta y_{j_1}.$$

The parabola flows in the y_{j_1}-direction are source and sink for in $G_{j_1j_10} > 0$ and $G_{j_1j_10} < 0$, respectively.

Therefore, the infinite-equilibrium flows of $y_1^* = a_{j_1j_1s_1} = a_{j_2j_2l_1}$ possess the flowing properties as in Eqs. (3.191)–(3.194):

- For $a_{j_1j_10}(a_{j_1j_1s_1} - a_{j_1j_1s_2})^2 / a_{j_2j_20}(a_{j_2j_2l_1} - a_{j_2j_2l_2}) > 0$ and $G_{j_1j_10} > 0$,
$$(a_{j_1j_1s_1}, a_{22l_1}) = \underbrace{(\mathrm{II}, \mathrm{SO})}_{\text{increasing-inflection source}}.$$

- For $a_{j_1j_10}(a_{j_1j_1s_1}-a_{j_1j_1s_2})^2/a_{j_2j_20}(a_{j_2j_2l_1}-a_{j_2j_2l_2})>0$ and $G_{j_1j_10}<0$,

$$(a_{j_1j_1s_1},a_{22l_1})=\underbrace{(\mathrm{II},\mathrm{SI})}_{\text{increasing-inflection sink}}.$$

- For $a_{j_1j_10}(a_{j_1j_1s_1}-a_{j_1j_1s_2})^2/a_{j_2j_20}(a_{j_2j_2l_1}-a_{j_2j_2l_2})<0$ and $G_{j_1j_10}>0$,

$$(a_{j_1j_1s_1},a_{22l_1})=\underbrace{(\mathrm{DI},\mathrm{SO})}_{\text{decreasing-inflection source}}.$$

- For $a_{j_1j_10}(a_{j_1j_1s_1}-a_{j_1j_1s_2})^2/a_{j_2j_20}(a_{j_2j_2l_1}-a_{j_2j_2l_2})<0$ and $G_{j_1j_10}<0$,

$$(a_{j_1j_1s_1},a_{22l_1})=\underbrace{(\mathrm{DI},\mathrm{SI})}_{\text{increasing-inflection sink}}.$$

– The inflection-source and sink infinite-equilibriums are the switching bifurcations for parabola flows in the x_{j_1}- and x_{j_2}-directions.

(viii$_4$) For $a_{j_1j_1s_2}=a_{j_2j_2l_1}$, consider a dynamical system as

$$\begin{aligned}
\dot{x}_{j_1}&=a_{j_1j_10}(y_{j_1}-a_{j_1j_1s_1})(y_{j_1}-a_{j_1j_1s_2})^2,\\
\dot{x}_{j_2}&=a_{j_2j_20}(y_{j_1}-a_{j_2j_2l_1})(y_{j_1}-a_{j_2j_2l_2}),\\
&j_1,j_2\in\{1,2\};\ j_1\neq j_2.
\end{aligned}$$

The corresponding dynamical system in the y_{j_1}-direction is constructed as

$$\begin{aligned}
\dot{y}_{j_1}=(y_{j_1}-a_{j_1j_1s_2})[&\alpha_{j_1j_1}a_{j_1j_10}(y_{j_1}-a_{j_1j_1s_1})(y_{j_1}-a_{j_1j_1s_2})\\
&+\alpha_{j_1j_2}a_{j_2j_20}(y_{j_1}-a_{j_2j_2l_2})],
\end{aligned}$$

and the corresponding dynamical system in the $y_{j_1}^{\perp}$-direction is constructed as

$$\begin{aligned}
\dot{y}_{j_1}^{\perp}=(y_{j_1}-a_{j_1j_1s_2})[&\beta_{j_1j_1}a_{j_1j_10}(y_{j_1}-a_{j_1j_1s_1})(y_{j_1}-a_{j_1j_1s_2})\\
&+\beta_{j_1j_2}a_{j_2j_20}(y_{j_1}-a_{j_2j_2l_2})].
\end{aligned}$$

Define a G-function as

$$\begin{aligned}
G_{j_1j_10}&=\frac{\partial}{\partial y_{j_1}}[\mathbf{n}^{\mathrm{T}}\cdot\mathbf{F}(y_{j_1},\mathbf{p})]\Big|_{y_{j_1}^*=a_{j_1j_1s_2}=a_{j_2j_2l_1}}\\
&=\alpha_{j_1j_2}a_{j_2j_20}(a_{j_2j_2l_1}-a_{j_2j_2l_2}).
\end{aligned}$$

In phase space, at $y_1^* = a_{j_1j_1s_2} = a_{j_2j_2l_1}$,

$$\frac{dx_{j_1}}{dx_{j_2}}\bigg|_{y_1^*=a_{j_1j_1s_2}} = \frac{a_{j_1j_10}(y_1 - a_{j_1j_1s_1})(y_1 - a_{j_1j_1s_2})}{a_{j_2j_20}(y_1 - a_{j_2j_2l_2})}\bigg|_{y_1^*=a_{j_1j_1s_2}} = 0.$$

If

$$\frac{d^2x_{j_1}}{dx_{j_2}^2}\bigg|_{y_1^*=a_{j_1j_1s_2}} = \frac{a_{j_1j_10}a_{j_1j_2}(a_{j_1j_1s_2} - a_{j_1j_1s_1})}{a_{j_2j_20}(a_{j_2j_2l_1} - a_{j_2j_2l_2})} > 0,$$

there is an up-parabola flow at $y_1^* = a_{j_1j_1s_2} = a_{j_2j_2l_1}$, and if

$$\frac{d^2x_{j_1}}{dx_{j_2}^2}\bigg|_{y_1^*=a_{j_1j_1s_2}} = \frac{a_{j_1j_10}a_{j_1j_2}(a_{j_1j_1s_2} - a_{j_1j_1s_1})}{a_{j_2j_20}(a_{j_2j_2l_1} - a_{j_2j_2l_2})} > 0,$$

there is a down-parabola flow at $y_1^* = a_{j_1j_1s_2} = a_{j_2j_2l_1}$. The variational equation in the y_{j_1}-direction is

$$\Delta\dot{y}_{j_1} = \alpha_{j_1j_2}a_{j_2j_20}(a_{j_2j_2l_1} - a_{j_2j_2l_2})\Delta y_{j_1} = G_{j_1j_10}\Delta y_{j_1}.$$

The parabola flows in the y_{j_1}-direction are source and sink for $G_{j_1j_10} > 0$ and $G_{j_1j_10} < 0$, respectively.

Therefore, the infinite-equilibrium of $y_{j_1}^* = a_{j_1j_1s_2} = a_{j_2j_2l_1}$ has the following properties as in Eqs. (3.199)–(3.202):

- For $a_{j_1j_10}\alpha_{j_1j_2}^2(a_{j_1j_1s_2} - a_{j_1j_1s_1}) > 0$ and $\alpha_{j_1j_2}a_{j_2j_20}(a_{j_2j_2l_1} - a_{j_2j_2l_2}) > 0$,

$$(a_{j_1j_1s_2}, a_{j_2j_2l_1}) = \underbrace{(\text{UP}, \text{SO})}_{\text{up-parabola source}}.$$

- For $a_{j_1j_10}\alpha_{j_1j_2}^2(a_{j_1j_1s_2} - a_{j_1j_1s_1}) < 0$ and $\alpha_{j_1j_2}a_{j_2j_20}(a_{j_2j_2l_1} - a_{j_2j_2l_2}) > 0$,

$$(a_{j_1j_1s_2}, a_{j_2j_2l_1}) = \underbrace{(\text{DP}, \text{SO})}_{\text{down-parabola source}}.$$

- For $a_{j_1j_10}\alpha_{j_1j_2}^2(a_{j_1j_1s_2} - a_{j_1j_1s_1}) > 0$ and $\alpha_{j_1j_2}a_{j_2j_20}(a_{j_2j_2l_1} - a_{j_2j_2l_2}) < 0$,

$$(a_{j_1j_1s_2}, a_{j_2j_2l_1}) = \underbrace{(\text{DP}, \text{SI})}_{\text{down-parabola sink}}.$$

- For $a_{j_1j_10}\alpha_{j_1j_2}^2(a_{j_1j_1s_2} - a_{j_1j_1s_1}) < 0$ and $\alpha_{j_1j_2}a_{j_2j_20}(a_{j_2j_2l_1} - a_{j_2j_2l_2}) < 0$,

$$(a_{j_1j_1s_2}, a_{j_2j_2l_1}) = \underbrace{(\text{UP}, \text{SI})}_{\text{up-parabola source}}.$$

- The parabola source and sink infinite-equilibriums are the switching bifurcations for the increasing-inflection and parabola flows in the x_{j_1}- and x_{j_2}-directions.
- The inflection-saddle infinite-equilibriums are the appearing bifurcations from a parabola flow and x_{j_2}-direction to the up-parabola and down-parabola flows in the x_{j_1}-directions and a parabola flow in the x_{j_2}-directions.

(viii$_5$) For $a_{j_1j_11} = a_{j_2j_2l_2}$, consider a dynamical system as

$$\begin{aligned}
\dot{x}_{j_1} &= a_{j_1j_10}(y_{j_1} - a_{j_1j_11})^3,\\
\dot{x}_{j_2} &= a_{j_2j_20}(y_{j_1} - a_{j_2j_2l_1})(y_{j_1} - a_{j_2j_2l_2}),\\
&j_1, j_2 \in \{1,2\}; j_1 \neq j_2.
\end{aligned}$$

The corresponding dynamical system in the y_{j_1}-direction is constructed as

$$\begin{aligned}
\dot{y}_{j_1} = (y_{j_1} - a_{j_1j_11})[&\alpha_{j_1j_1}a_{j_1j_10}(y_{j_1} - a_{j_1j_11})^2\\
&+ \alpha_{j_1j_2}a_{j_2j_20}(y_{j_1} - a_{j_2j_2l_2})],
\end{aligned}$$

and a dynamical system in the $y_{j_1}^{\perp}$-direction is constructed as

$$\begin{aligned}
\dot{y}_{j_1}^{\perp} = (y_{j_1} - a_{j_1j_11})[&\beta_{j_1j_1}a_{j_1j_10}(y_{j_1} - a_{j_1j_11})^2\\
&+ \beta_{j_1j_2}a_{j_2j_20}(y_{j_1} - a_{j_2j_2l_2})].
\end{aligned}$$

Define a G-function as

$$\begin{aligned}
G_{j_1j_10} &= \frac{\partial}{\partial y_{j_1}}[\mathbf{n}^{\mathrm{T}} \cdot \mathbf{F}(y_{j_1}, \mathbf{p})]\Big|_{y_{j_1}^* = a_{j_1j_11} = a_{j_2j_2l_1}}\\
&= \alpha_{j_1j_2}a_{j_2j_20}(a_{j_2j_2l_1} - a_{j_2j_2l_2}).
\end{aligned}$$

In phase space, at $y_{j_1}^* = a_{j_1j_11} = a_{j_2j_2l_1}$,

$$\begin{aligned}
\frac{dx_{j_1}}{dx_{j_2}}\Big|_{y_1^* = a_{j_1j_11}} &= \frac{a_{j_1j_10}(y_{j_1} - a_{j_1j_11})^2}{a_{j_2j_20}(y_{j_1} - a_{j_2j_2l_2})}\Big|_{y_1^* = a_{j_1j_11}} = 0,\\
\frac{d^2x_{j_1}}{dx_{j_2}^2}\Big|_{y_1^* = a_{j_1j_11}} &= 0.
\end{aligned}$$

If

$$\frac{d^3x_{j_1}}{dx_{j_2}^3}\Big|_{y_1^* = a_{j_1j_11}} = \frac{2a_{j_1j_10}\alpha_{j_1j_2}^3}{a_{j_2j_20}\alpha_{j_1j_2}(a_{j_2j_2l_1} - a_{j_2j_2l_2})} > 0,$$

there is a third-order increasing-inflection flow at $y_{j_1}^* = a_{j_1j_11} = a_{j_2j_2l_1}$. If

$$\frac{d^3x_{j_1}}{dx_{j_2}^3}\Big|_{y_1^* = a_{j_1j_11}} = \frac{2a_{j_1j_10}\alpha_{j_1j_2}^3}{a_{j_2j_20}\alpha_{j_1j_2}(a_{j_2j_2l_1} - a_{j_2j_2l_2})} < 0,$$

there is a third-order decreasing-inflection flow at $y^*_{j_1} = a_{j_1j_11} = a_{j_2j_2l_1}$. The variational equation in the y_{j_1}-direction is

$$\Delta\dot{y}_{j_1} = \alpha_{j_1j_2}a_{j_2j_20}(a_{j_2j_2l_1} - a_{j_2j_2l_2})\Delta y_{j_1} = G_{j_1j_10}\Delta y_{j_1}.$$

The parabola flows in the y_{j_1}-direction are upper-saddle and lower-saddle for $G_{j_1j_10} > 0$ and $G_{j_1j_10} < 0$, respectively.

Therefore, the infinite-equilibrium of $y^*_{j_1} = a_{j_1j_11} = a_{j_2j_2l_1}$ has the following properties as in Eqs. (3.207)–(3.210):

- For $\alpha^3_{j_1j_2}a_{j_1j_10} > 0$ and $\alpha_{j_1j_2}a_{j_2j_20}(a_{j_2j_2l_1} - a_{j_2j_2l_2}) > 0$,
$$(a_{j_1j_11}, a_{j_2j_2l_1}) = \underbrace{(2^{\text{nd}}\text{II}, \text{SO})}_{\text{second-order increasing-inflection source}}.$$
- For $\alpha^3_{j_1j_2}a_{j_1j_10} < 0$ and $\alpha_{j_1j_2}a_{j_2j_20}(a_{j_2j_2l_1} - a_{j_2j_2l_2}) > 0$,
$$(a_{j_1j_11}, a_{j_2j_2l_1}) = \underbrace{(2^{\text{nd}}\text{DI}, \text{SO})}_{\text{second-order decreasing-inflection source}}.$$
- For $\alpha^3_{j_1j_2}a_{j_1j_10} > 0$ and $\alpha_{j_1j_2}a_{j_2j_20}(a_{j_2j_2l_1} - a_{j_2j_2l_2}) < 0$,
$$(a_{j_1j_11}, a_{j_2j_2l_1}) = \underbrace{(2^{\text{nd}}\text{DI}, \text{SI})}_{\text{second-order decreasing-inflection sink}}.$$
- For $\alpha^3_{j_1j_2}a_{j_1j_10} < 0$ and $\alpha_{j_1j_2}a_{j_2j_20}(a_{j_2j_2l_1} - a_{j_2j_2l_2}) < 0$,
$$(a_{j_1j_11}, a_{j_2j_2l_1}) = \underbrace{(2^{\text{nd}}\text{II}, \text{SI})}_{\text{second-order increasing-inflection sink}}.$$

- The second-order inflection-source and sink infinite-equilibriums are the switching bifurcations for third-order parabola and parabola flows in the x_{j_1}- and x_{j_2}-directions.
- The second-order inflection-source and sink infinite-equilibriums are the appearing bifurcations from a parabola flow and a parabola flow in the x_{j_1}- and x_{j_2}-directions to three parabola flows and a parabola flow in the x_{j_1}- and x_{j_2}-directions.

(ix) For

$$\Delta_{j_1j_1} > 0 \quad \text{and} \quad \Delta_{j_2j_2} > 0,$$

the dynamical system becomes

$$\dot{x}_{j_1} = a_{j_1j_10}(y_{j_1} - a_{j_1j_11})(y_{j_1} - a_{j_1j_12})(y_{j_1} - a_{j_1j_13}),$$
$$\dot{x}_{j_2} = a_{j_2j_20}(y_{j_1} - a_{j_2j_21})(y_{j_1} - a_{j_2j_22}),$$

where

$$\{a_{j_1j_11}, a_{j_1j_12}, a_{j_1j_13}\} = \text{sort}\{b_{j_1j_11}, b_{j_1j_12}, b_{j_1j_13}\}, \quad a_{j_1j_1s_1} < a_{j_1j_1s_2},$$
$$s_1 < s_2;$$
$$b_{j_1j_12}, b_{j_1j_13} = -\frac{1}{2}(B_{j_1j_11} \pm \sqrt{\Delta_{j_1j_1}}), \quad a_{j_1j_10} = A_{j_1j_10}.$$
$$\{a_{j_2j_21}, a_{j_2j_22}\} = \text{sort}\{b_{j_2j_21}, b_{j_2j_22}\}, \quad a_{j_2j_21} < a_{j_2j_22};$$
$$b_{j_2j_21}, \quad b_{j_2j_22} = -\frac{1}{2}(B_{j_2j_21} \pm \sqrt{\Delta_{j_2j_2}}), \quad a_{j_2j_20} = A_{j_2j_21}.$$

The corresponding dynamical system in the y_{j_1}-direction is constructed as

$$\dot{y}_{j_1} = \alpha_{j_1j_1}a_{j_1j_10}\prod_{s_1=1}^{3}(y_{j_1} - a_{j_1j_1s_1}) + \alpha_{j_1j_2}a_{j_2j_20}\prod_{l_1=1}^{2}(y_{j_1} - a_{j_2j_2l_1}),$$

and the corresponding dynamical system in the $y_{j_1}^{\perp}$-direction is constructed as

$$\dot{y}_{j_1}^{\perp} = \beta_{j_1j_1}a_{j_1j_10}\prod_{s_1=1}^{3}(y_{j_1} - a_{j_1j_1s_1}) + \beta_{j_1j_2}a_{j_2j_20}\prod_{l_1=1}^{2}(y_{j_1} - a_{j_2j_2l_1}).$$

($\text{ix}_{1\text{a}}$) In phase space, at $y_1^* = a_{j_1j_1s_1}$,

$$\left.\frac{dx_{j_1}}{dx_{j_2}}\right|_{y_1^*=a_{j_1j_1s_1}} = \left.\frac{a_{j_1j_10}(y_{j_1} - a_{j_1j_1s_1})\prod_{s_2=1,s_2\neq s_1}^{3}(y_{j_1} - a_{j_1j_1s_2})}{a_{j_2j_20}\prod_{l_1=1}^{2}(y_{j_1} - a_{j_2j_2l_1})}\right|_{y_1^*=a_{j_1j_1s_1}} = 0.$$

If

$$\left.\frac{d^2x_{j_1}}{dx_{j_2}^2}\right|_{y_1^*=a_{j_1j_1s_1}} = \frac{a_{j_1j_10}\alpha_{j_1j_2}\prod_{s_2=1,s_2\neq s_1}^{3}(a_{j_1j_1s_1} - a_{j_1j_1s_2})}{a_{j_2j_20}\prod_{l_1=1}^{2}(a_{j_1j_1s_1} - a_{j_2j_2l_1})} > 0,$$

there is an up-parabola flow at $y_1^* = a_{j_1j_1s_1}$. If

$$\left.\frac{d^2x_{j_1}}{dx_{j_2}^2}\right|_{y_1^*=a_{j_1j_1s_1}} = \frac{a_{j_1j_10}\alpha_{j_1j_2}\prod_{s_2=1,s_2\neq s_1}^{3}(a_{j_1j_1s_1} - a_{j_1j_1s_2})}{a_{j_2j_20}\prod_{l_1=1}^{2}(a_{j_1j_1s_1} - a_{j_2j_2l_1})} < 0,$$

there is a down-parabola flow at $y_1^* = a_{j_1j_1s_1}$. With

$$\dot{x}_{j_2}^* = a_{j_2j_20}\prod_{l_1=1}^{2}(a_{j_1j_1s_1} - a_{j_2j_2l_1}),$$

the parabola flows are positive and negative for $\dot{x}_{j_2}^* > 0$ and $\dot{x}_{j_2}^* < 0$, respectively.

Therefore, the equilibrium flows of $y^*_{j_1} = a_{j_1 j_1 s_1}(s_1, s_2 \in \{1,2\}, s_1 \neq s_2)$ for the x_{j_1}-direction have the following properties as in Eqs. (3.216)–(3.219):

- For $a_{j_1 j_1 0}\alpha_{j_1 j_2}\prod_{s_2=1, s_2\neq s_1}^{3}(a_{j_1 j_1 s_1} - a_{j_1 j_1 s_2}) > 0$ and $a_{j_2 j_2 0}\prod_{l_1=1}^{2}(a_{j_1 j_1 s_1} - a_{j_2 j_2 l_1}) > 0$,
$$(a_{j_1 j_1 s_1}, \dot{x}^*_{j_2}) = \underbrace{(\text{UP}, \text{pF})}_{\text{up-parabola flow } (+)}.$$
- For $a_{j_1 j_1 0}\alpha_{j_1 j_2}\prod_{s_2=1, s_2\neq s_1}^{3}(a_{j_1 j_1 s_1} - a_{j_1 j_1 s_2}) < 0$ and $a_{j_2 j_2 0}\prod_{l_1=1}^{2}(a_{j_1 j_1 s_1} - a_{j_2 j_2 l_1}) > 0$,
$$(a_{j_1 j_1 s_1}, \dot{x}^*_{j_2}) = \underbrace{(\text{DP}, \text{pF})}_{\text{Down-parabola flow } (+)}.$$
- For $a_{j_1 j_1 0}\alpha_{j_1 j_2}\prod_{s_2=1, s_2\neq s_1}^{3}(a_{j_1 j_1 s_1} - a_{j_1 j_1 s_2}) > 0$ and $a_{j_2 j_2 0}\prod_{l_1=1}^{2}(a_{j_1 j_1 s_1} - a_{j_2 j_2 l_1}) < 0$,
$$(a_{j_1 j_1 s_1}, \dot{x}^*_{j_2}) = \underbrace{(\text{DP}, \text{nF})}_{\text{Down-parabola flow } (-)}.$$
- For $a_{j_1 j_1 0}\alpha_{j_1 j_2}\prod_{s_2=1, s_2\neq s_1}^{3}(a_{j_1 j_1 s_1} - a_{j_1 j_1 s_2}) < 0$ and $a_{j_2 j_2 0}\prod_{l_1=1}^{2}(a_{j_1 j_1 s_1} - a_{j_2 j_2 l_1}) < 0$,
$$(a_{j_1 j_1 s_1}, \dot{x}^*_{j_2}) = \underbrace{(\text{UP}, \text{nF})}_{\text{up-parabola flow } (-)}.$$

(viii$_{1\text{b}}$) In phase space, at $y^*_1 = a_{j_2 j_2 l_1}$,

$$\left.\frac{dx_{j_2}}{dx_{j_1}}\right|_{y^*_1 = a_{j_2 j_2 l_1}} = \left.\frac{a_{j_2 j_2 0}(y_{j_1} - a_{j_2 j_2 l_1})(y_{j_1} - a_{j_2 j_2 l_2})}{a_{j_1 j_1 0}\prod_{s_1=1}^{3}(y_{j_1} - a_{j_1 j_1 s_1})}\right|_{y^*_1 = a_{j_2 j_2 l_1}} = 0.$$

If

$$\left.\frac{d^2 x_{j_2}}{dx^2_{j_1}}\right|_{y^*_1 = a_{j_2 j_2 l_1}} = \frac{a_{j_2 j_2 0}\alpha_{j_1 j_1}(a_{j_2 j_2 l_1} - a_{j_2 j_2 l_2})}{a_{j_1 j_1 0}\prod_{s_1=1}^{3}(a_{j_2 j_2 l_1} - a_{j_1 j_1 s_1})} > 0,$$

there is an up-parabola flow at $y^*_1 = a_{j_2 j_2 l_1}$. If

$$\left.\frac{d^2 x_{j_2}}{dx^2_{j_1}}\right|_{y^*_1 = a_{j_2 j_2 l_1}} = \frac{a_{j_2 j_2 0}\alpha_{j_1 j_1}(a_{j_2 j_2 l_1} - a_{j_2 j_2 l_2})}{a_{j_1 j_1 0}\prod_{s_1=1}^{3}(a_{j_2 j_2 l_1} - a_{j_1 j_1 s_1})} < 0,$$

there is a down-parabola flow at $y_1^* = a_{j_2j_2l_1}$. With

$$\dot{x}_{j_1}^* = a_{j_1j_10} \prod_{s_1=1}^{3} (a_{j_2j_21} - a_{j_1j_1s_1}),$$

the inflection flows are positive and negative for $\dot{x}_{j_1}^* > 0$ and $\dot{x}_{j_1}^* < 0$, respectively.

Therefore, the equilibrium flow of $y_{j_1}^* = a_{j_2j_2l_1}$ for the x_{j_2}-direction has the following properties as in Eqs. (3.220)–(3.223):

- For $a_{j_1j_10}\prod_{s_1=1}^{3}(a_{j_2j_2l_1} - a_{j_1j_1s_1}) > 0$ and $a_{j_2j_20}\alpha_{j_1j_1}(a_{j_2j_2l_1} - a_{j_1j_1l_2}) > 0$,

$$(\dot{x}_{j_1}^*, a_{j_2j_2l_1}) = \underbrace{(\text{pF,UP})}_{\text{up-parabola flow } (+)}.$$

- For $a_{j_1j_10}\prod_{s_1=1}^{3}(a_{j_2j_2l_1} - a_{j_1j_1s_1}) < 0$ and $a_{j_2j_20}\alpha_{j_1j_1}(a_{j_2j_2l_1} - a_{j_1j_1l_2}) > 0$,

$$(\dot{x}_{j_1}^*, a_{j_2j_2l_1}) = \underbrace{(\text{nF, DP})}_{\text{down-parabola flow } (-)}.$$

- For $a_{j_1j_10}\prod_{s_1=1}^{3}(a_{j_2j_2l_1} - a_{j_1j_1s_1}) > 0$ and $a_{j_2j_20}\alpha_{j_1j_1}(a_{j_2j_2l_1} - a_{j_1j_1l_2}) < 0$,

$$(\dot{x}_{j_1}^*, a_{j_2j_2l_1}) = \underbrace{(\text{pF, DP})}_{\text{down-parabola flow } (+)}.$$

- For $a_{j_1j_10}\prod_{s_1=1}^{3}(a_{j_2j_2l_1} - a_{j_1j_1s_1}) < 0$ and $a_{j_2j_20}\alpha_{j_1j_1}(a_{j_2j_2l_1} - a_{j_1j_1l_2}) < 0$,

$$(\dot{x}_{j_1}^*, a_{j_2j_2l_1}) = \underbrace{(\text{nF, UP})}_{\text{up-parabola flow } (-)}.$$

(ix$_2$) For $a_{j_2j_2l_1} = a_{j_1j_1s_1}$ $(s_1, s_2 \in \{1,2,3\}, l_1, l_2 \in \{1,2\})$, consider a dynamical system as

$$\begin{aligned}
\dot{x}_{j_1} &= a_{j_1j_10}(y_{j_1} - a_{j_1j_1s_1}) \prod_{s_2=1, s_2\neq s_1}^{3} (y_{j_1} - a_{j_1j_1s_2}),\\
\dot{x}_{j_2} &= a_{j_2j_20}(y_{j_1} - a_{j_2j_2l_1})(y_{j_1} - a_{j_2j_2l_2}),\\
&j_1, j_2 \in \{1,2\};\ j_1 \neq j_2;\ s_1, s_2 \in \{1,2,3\}, s_1 \neq s_2.
\end{aligned}$$

The corresponding dynamical system in the y_{j_1}-direction is constructed as

$$\dot{y}_{j_1} = (y_{j_1} - a_{j_1j_1s_1})\Big[\alpha_{j_1j_1}a_{j_1j_10}\prod_{s_2=1,s_2\neq s_1}^{3}(y_{j_1} - a_{j_1j_1s_2}) \\ + \alpha_{j_1j_2}a_{j_2j_20}(y_{j_1} - a_{j_2j_2l_2})\Big],$$

and the corresponding dynamical system in the $y_{j_1}^{\perp}$-direction is constructed as

$$\dot{y}_{j_1}^{\perp} = (y_{j_1} - a_{j_1j_1s_1})\Big[\beta_{j_1j_1}a_{j_1j_10}\prod_{s_2=1,s_2\neq s_1}^{3}(y_{j_1} - a_{j_1j_1s_2}) \\ + \beta_{j_1j_2}a_{j_2j_20}(y_{j_1} - a_{j_2j_2l_2})\Big].$$

Define a G-function as

$$G_{j_1j_10} = \frac{\partial}{\partial y_{j_1}}[\mathbf{n}^{\mathrm{T}}\cdot\mathbf{F}(y_{j_1},\mathbf{p})]\Big|_{y_{j_1}^{*}=a_{j_1j_1s_1}=a_{j_2j_2l_1}} \\ = \alpha_{j_1j_1}a_{j_1j_10}\prod_{s_2=1,s_2\neq s_1}^{3}(a_{j_1j_1s_1} - a_{j_1j_1s_2}) \\ + \alpha_{j_1j_2}a_{j_2j_20}(a_{j_2j_2l_1} - a_{j_2j_2l_2}).$$

In phase space, at $y_1^{*} = a_{j_1j_1s_1} = a_{j_2j_2l_1}$, if

$$\frac{dx_{j_2}}{dx_{j_1}}\Big|_{y_1^{*}=a_{j_2j_2l_1}} = \frac{a_{j_2j_20}(a_{j_2j_2l_1} - a_{j_2j_2l_2})}{a_{j_1j_10}\prod_{s_2=1,s_2\neq s_1}^{3}(a_{j_1j_1s_1} - a_{j_1j_1s_2})} > 0,$$

there is an increasing-inflection flow at $y_1^{*} = a_{j_1j_1s_1} = a_{j_2j_2l_1}$, and if

$$\frac{dx_{j_2}}{dx_{j_1}}\Big|_{y_1^{*}=a_{j_2j_2l_1}} = \frac{a_{j_2j_20}(a_{j_2j_2l_1} - a_{j_2j_2l_2})}{a_{j_1j_10}\prod_{s_2=1,s_2\neq s_1}^{3}(a_{j_1j_1s_1} - a_{j_1j_1s_2})} < 0,$$

there is a down-parabola flow at $y_1^{*} = a_{j_1j_1s_1} = a_{j_2j_2l_1}$. The variational equation in the y_{j_1}-direction is

$$\Delta\dot{y}_{j_1} = \Big[\alpha_{j_1j_1}a_{j_1j_10}\prod_{s_2=1,s_2\neq s_1}^{3}(a_{j_1j_1s_1} - a_{j_1j_1s_2}) \\ + \alpha_{j_1j_2}a_{j_2j_20}(a_{j_2j_2l_1} - a_{j_2j_2l_2})\Big]\Delta y_{j_1} \\ = G_{j_1j_10}\Delta y_{j_1}.$$

The parabola flows in the y_{j_1}-direction are source and sink for $G_{j_1 j_1 0} > 0$ and $G_{j_1 j_1 0} < 0$, respectively.

The infinite-equilibriums of $y_1^* = a_{j_1 j_1 s_1} = a_{j_2 j_2 l_1}$ have the following properties as in Eqs. (3.228)–(3.231):

- For $a_{j_1 j_1 0} \prod_{s_2=1, s_2 \neq s_1}^{3} (a_{j_1 j_1 s_1} - a_{j_1 j_1 s_2}) / a_{j_2 j_2 0}(a_{j_2 j_2 l_1} - a_{j_2 j_2 l_2}) > 0$ and $G_{j_1 j_1 0} > 0$,

$$(a_{j_1 j_1 s_1}, a_{j_2 j_2 l_1}) = \underbrace{(\text{II}, \text{SO})}_{\text{increasing-inflection source}}.$$

- For $a_{j_1 j_1 0} \prod_{s_2=1, s_2 \neq s_1}^{3} (a_{j_1 j_1 s_1} - a_{j_1 j_1 s_2}) / a_{j_2 j_2 0}(a_{j_2 j_2 l_1} - a_{j_2 j_2 l_2}) > 0$ and $G_{j_1 j_1 0} < 0$,

$$(a_{j_1 j_1 s_1}, a_{j_2 j_2 l_1}) = \underbrace{(\text{II}, \text{SI})}_{\text{increasing-inflection sink}}.$$

- For $a_{j_1 j_1 0} \prod_{s_2=1, s_2 \neq s_1}^{3} (a_{j_1 j_1 s_1} - a_{j_1 j_1 s_2}) / a_{j_2 j_2 0}(a_{j_2 j_2 l_1} - a_{j_2 j_2 l_2}) < 0$ and $G_{j_1 j_1 0} > 0$,

$$(a_{j_1 j_1 s_1}, a_{j_2 j_2 l_1}) = \underbrace{(\text{DI}, \text{SO})}_{\text{decreasing-inflection source}}.$$

- For $a_{j_1 j_1 0} \prod_{s_2=1, s_2 \neq s_1}^{3} (a_{j_1 j_1 s_1} - a_{j_1 j_1 s_2}) / a_{j_2 j_2 0}(a_{j_2 j_2 l_1} - a_{j_2 j_2 l_2}) < 0$ and $G_{j_1 j_1 0} < 0$,

$$(a_{j_1 j_1 s_1}, a_{j_2 j_2 l_1}) = \underbrace{(\text{DI}, \text{SI})}_{\text{decreasing-inflection sink}}.$$

From the above local analysis of such a dynamical system, this theorem is proved. □

3.3 Second-Order Singular Quadratic and Cubic Systems

In this section, dynamical systems with single-linear-bivariate quadratic and cubic vector fields are discussed.

3.3.1 *Inflection flows and appearing bifurcations*

Consider

$$\begin{aligned} \dot{x}_1 &= a_{110}(y_1 - a_{111})(y_1 - a_{112})^2, \\ \dot{x}_2 &= a_{220}(y_1 - a_{221})^2, \end{aligned} \tag{3.232}$$

where $a_{111} < a_{112} < a_{221}$ with $a_{211} > a_{112}$,

$$y_1 = \alpha_{11} x_1 + \alpha_{12} x_2. \tag{3.233}$$

There is a dynamical system in the y_1-direction as

$$\dot{y}_1 = \alpha_{11}a_{110}(y_1 - a_{111})(y_1 - a_{112})^2 + \alpha_{12}a_{220}(y_1 - a_{221})^2, \tag{3.234}$$

with a dynamical system in the $y_{j_1}^{\perp}$-direction as

$$\dot{y}_1^{\perp} = \beta_{11}a_{110}(y_1 - a_{111})(y_1 - a_{112})^2 + \beta_{12}a_{220}(y_1 - a_{221})^2, \tag{3.235}$$

where

$$\begin{aligned} y_1^{\perp} &= \beta_{11}x_1 + \beta_{12}x_2, \\ \beta_{11}\alpha_{11} &+ \beta_{12}\alpha_{12} = 0. \end{aligned} \tag{3.236}$$

(i) For $a_{111} > a_{221}$ $(\alpha_{11} > 0, \alpha_{12} > 0)$, presented in Fig. 3.1 are the one-dimensional flows of $(y_1^*, \dot{x}_2) = (a_{111}, \dot{x}_2), (a_{112}, \dot{x}_2)$ and $(\dot{x}_1, y_1^*) = (\dot{x}_1, a_{221})$. The corresponding properties for $\alpha_{11} > 0$ and $\alpha_{12} > 0$ are presented as follows:

$$\left\{\begin{matrix} (\dot{x}_1, a_{221}) \\ (a_{111}, \dot{x}_2) \\ (a_{112}, \dot{x}_2) \end{matrix}\right\} = \left\{\begin{matrix} \underbrace{(\text{nF}, \text{DI})}_{\text{decreasing-inflection flow }(-)} \\ \underbrace{(\text{UP}, \text{pF})}_{\text{up-parabola flow }(+)} \\ \underbrace{(\text{II}, \text{pF})}_{\text{increasing-inflection flow }(+)} \end{matrix}\right\}$$

$$\text{for } a_{110} > 0 \quad \text{and} \quad a_{220} > 0, \tag{3.237}$$

$$\left\{\begin{matrix} (\dot{x}_1, a_{221}) \\ (a_{111}, \dot{x}_2) \\ (a_{112}, \dot{x}_2) \end{matrix}\right\} = \left\{\begin{matrix} \underbrace{(\text{pF}, \text{II})}_{\text{increasing-inflection flow }(+)} \\ \underbrace{(\text{DP}, \text{pF})}_{\text{down-parabola flow }(+)} \\ \underbrace{(\text{DI}, \text{pF})}_{\text{increasing-inflection flow }(+)} \end{matrix}\right\}$$

$$\text{for } a_{110} < 0 \quad \text{and} \quad a_{220} > 0, \tag{3.238}$$

$$\left\{\begin{matrix} (\dot{x}_1, a_{221}) \\ (a_{111}, \dot{x}_2) \\ (a_{112}, \dot{x}_2) \end{matrix}\right\} = \left\{\begin{matrix} \underbrace{(\text{nF}, \text{II})}_{\text{increasing-inflection flow }(-)} \\ \underbrace{(\text{DP}, \text{nF})}_{\text{down-parabola flow }(-)} \\ \underbrace{(\text{DI}, \text{nF})}_{\text{increasing-inflection flow }(-)} \end{matrix}\right\}$$

$$\text{for } a_{110} > 0 \quad \text{and} \quad a_{220} < 0, \tag{3.239}$$

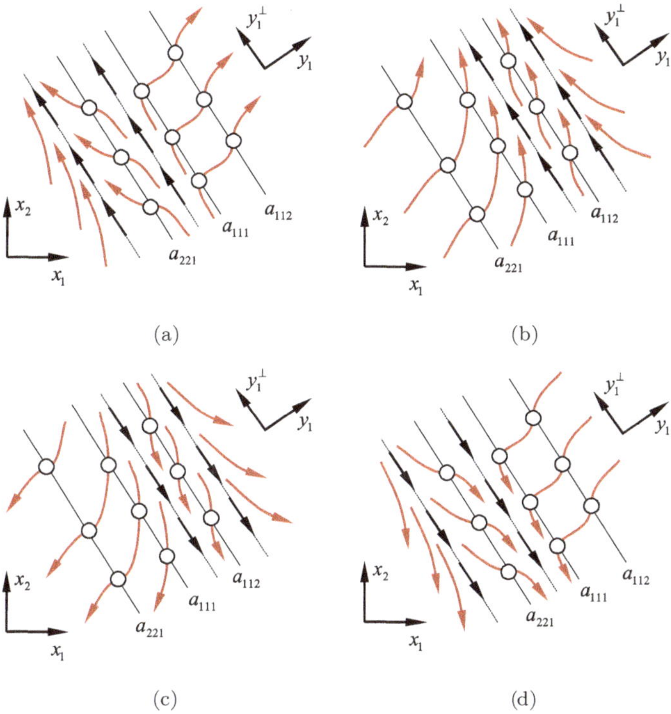

Fig. 3.1. Phase portraits ($a_{111} > a_{221}$ and $\alpha_{11} > 0, \alpha_{12} > 0$) for two-dimensional systems on the x_1-direction with $y_1^* = a_{111}, a_{112}$ and on the x_2-direction with $y_1^* = a_{221}$: (a) ($a_{110} > 0, a_{220} > 0$), (b) ($a_{110} < 0, a_{220} > 0$), (c) ($a_{110} > 0, a_{220} < 0$), (d) ($a_{110} < 0, a_{220} < 0$).

$$\left\{\begin{matrix} (\dot{x}_1, a_{221}) \\ (a_{111}, \dot{x}_2) \\ (a_{112}, \dot{x}_2) \end{matrix}\right\} = \left\{\begin{matrix} \underbrace{(\mathrm{pF}, \mathrm{DI})}_{\text{decreasing-inflection flow } (+)} \\ \underbrace{(\mathrm{UP}, \mathrm{nF})}_{\text{up-parabola flow } (-)} \\ \underbrace{(\mathrm{II}, \mathrm{nF})}_{\text{increasing-inflection flow } (-)} \end{matrix}\right\}$$

$$\text{for } a_{110} < 0 \quad \text{and} \quad a_{220} < 0. \tag{3.240}$$

The equilibrium dynamics on the y_1-direction are listed in Table 3.1.

(ii) For $a_{111} > a_{221}$ ($\alpha_{11} > 0, \alpha_{12} > 0$), presented in Fig. 3.2 are the one-dimensional flows of $(y_1^*, \dot{x}_2) = (a_{111}, \dot{x}_2), (a_{112}, \dot{x}_2)$ and $(\dot{x}_1, y_1^*) = (\dot{x}_1, a_{221})$. The corresponding properties for $\alpha_{11} > 0$ and $\alpha_{12} < 0$ are

Table 3.1. Equilibrium flows of $\dot{y}_1 = 0$ ($\alpha_{11} > 0, \alpha_{12} > 0$).

	$(\dot{x}_1, \dot{x}_2)$	y_1			
Figure 3.1	(a_{110}, a_{220})	$(-\infty, a_{221})$	(a_{221}, a_{111})	(a_{111}, a_{112})	(a_{112}, ∞)
(a)	$(+,+)$	$\dot{y}_1^* = 0$ $(-,+)$	$\dot{y}_1^* = 0$ $(-,+)$	$\dot{y}_1 > 0$ $(+,+)$	$\dot{y}_1 > 0$ $(+,+)$
(b)	$(-,+)$	$\dot{y}_1 > 0$ $(+,+)$	$\dot{y}_1 > 0$ $(+,+)$	$\dot{y}_1^* = 0$ $(-,+)$	$\dot{y}_1^* = 0$ $(-,+)$
(c)	$(+,-)$	$\dot{y}_1 < 0$ $(-,-)$	$\dot{y}_1 < 0$ $(-,-)$	$\dot{y}_1^* = 0$ $(+,-)$	$\dot{y}_1^* = 0$ $(+,-)$
(d)	$(-,-)$	$\dot{y}_1^* = 0$ $(+,-)$	$\dot{y}_1^* = 0$ $(+,-)$	$\dot{y}_1 < 0$ $(-,-)$	$\dot{y}_1 < 0$ $(-,-)$

presented as follows:

$$\left\{ \begin{array}{c} (\dot{x}_1, a_{221}) \\ (a_{111}, \dot{x}_2) \\ (a_{112}, \dot{x}_2) \end{array} \right\} = \left\{ \begin{array}{c} \underbrace{(\mathrm{nF}, \mathrm{DI})}_{\text{decreasing-inflection flow } (-)} \\ \underbrace{(\mathrm{DP}, \mathrm{pF})}_{\text{down-parabola flow } (+)} \\ \underbrace{(\mathrm{II}, \mathrm{pF})}_{\text{increasing-inflection flow } (+)} \end{array} \right\}$$

$$\text{for } a_{110} > 0 \quad \text{and} \quad a_{220} > 0, \tag{3.241}$$

$$\left\{ \begin{array}{c} (\dot{x}_1, a_{221}) \\ (a_{111}, \dot{x}_2) \\ (a_{112}, \dot{x}_2) \end{array} \right\} = \left\{ \begin{array}{c} \underbrace{(\mathrm{pF}, \mathrm{II})}_{\text{increasing-inflection flow } (+)} \\ \underbrace{(\mathrm{UP}, \mathrm{pF})}_{\text{up-parabola flow } (+)} \\ \underbrace{(\mathrm{DI}, \mathrm{pF})}_{\text{increasing-inflection flow } (+)} \end{array} \right\}$$

$$\text{for } a_{110} < 0 \quad \text{and} \quad a_{220} > 0, \tag{3.242}$$

$$\left\{ \begin{array}{c} (\dot{x}_1, a_{221}) \\ (a_{111}, \dot{x}_2) \\ (a_{112}, \dot{x}_2) \end{array} \right\} = \left\{ \begin{array}{c} \underbrace{(\mathrm{nF}, \mathrm{II})}_{\text{increasing-inflection flow } (-)} \\ \underbrace{(\mathrm{UP}, \mathrm{nF})}_{\text{up-parabola flow } (-)} \\ \underbrace{(\mathrm{DI}, \mathrm{nF})}_{\text{increasing-inflection flow } (-)} \end{array} \right\}$$

$$\text{for } a_{110} > 0 \quad \text{and} \quad a_{220} < 0, \tag{3.243}$$

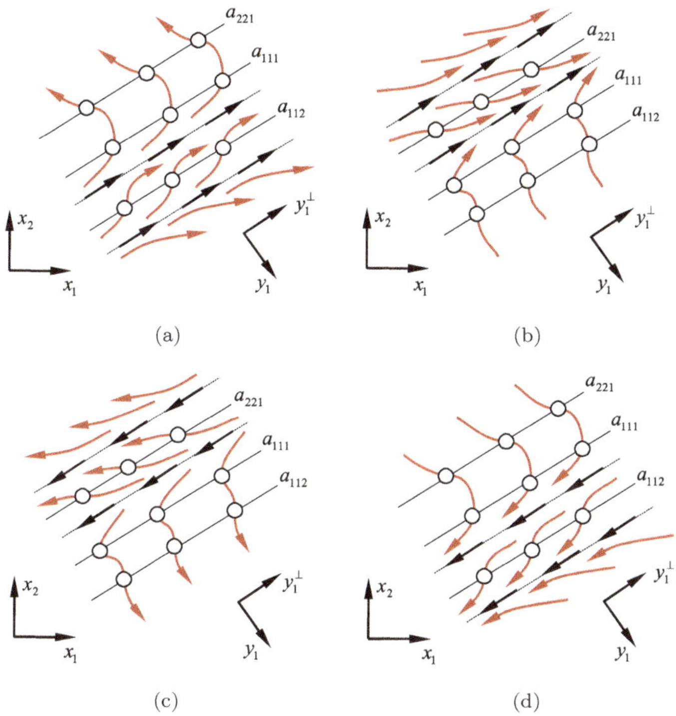

Fig. 3.2. Phase portraits ($a_{111} > a_{221}$ and $\alpha_{11} > 0, \alpha_{12} < 0$) for two-dimensional systems on the x_1-direction with $y_1^* = a_{111}, a_{112}$ and on the x_2-direction with $y_1^* = a_{221}$: (a) ($a_{110} > 0, a_{220} > 0$), (b) ($a_{110} < 0, a_{220} > 0$), (c) ($a_{110} > 0, a_{220} < 0$), (d) ($a_{110} < 0, a_{220} < 0$).

$$\begin{Bmatrix} (\dot{x}_1, a_{221}) \\ (a_{111}, \dot{x}_2) \\ (a_{112}, \dot{x}_2) \end{Bmatrix} = \begin{Bmatrix} \underbrace{(\mathrm{pF}, \mathrm{DI})}_{\text{decreasing-inflection flow }(+)} \\ \underbrace{(\mathrm{DP}, \mathrm{nF})}_{\text{down-parabola flow }(-)} \\ \underbrace{(\mathrm{II}, \mathrm{nF})}_{\text{increasing-inflection flow }(-)} \end{Bmatrix}$$

$$\text{for } a_{110} < 0 \quad \text{and} \quad a_{220} < 0. \tag{3.244}$$

In addition, the equilibrium dynamics on the y_1-direction are listed in Table 3.2.

(iii) For $a_{112} < a_{221}$ ($\alpha_{11} > 0, \alpha_{12} > 0$), presented in Fig. 3.3 are the one-dimensional flows of $(y_1^*, \dot{x}_2) = (a_{111}, \dot{x}_2), (a_{112}, \dot{x}_2)$ and $(\dot{x}_1, y_1^*) =$

Table 3.2. Equilibrium flows of $\dot{y}_1 = 0$ ($\alpha_{11} > 0, \alpha_{12} < 0$).

Figure 3.2	$(\dot{x}_1, \dot{x}_2)$ (a_{110}, a_{220})	y_1 $(-\infty, a_{221})$	(a_{221}, a_{111})	(a_{111}, a_{112})	(a_{112}, ∞)
(a)	$(+,+)$	$\dot{y}_1 < 0$ $(-,+)$	$\dot{y}_1 < 0$ $(-,+)$	$\dot{y}_1^* = 0$ $(+,+)$	$\dot{y}_1^* = 0$ $(+,+)$
(b)	$(-,+)$	$\dot{y}_1^* = 0$ $(+,+)$	$\dot{y}_1^* = 0$ $(+,+)$	$\dot{y}_1 < 0$ $(-,+)$	$\dot{y}_1 < 0$ $(-,+)$
(c)	$(+,-)$	$\dot{y}_1^* = 0$ $(-,-)$	$\dot{y}_1^* = 0$ $(-,-)$	$\dot{y}_1 > 0$ $(+,-)$	$\dot{y}_1 > 0$ $(+,-)$
(d)	$(-,-)$	$\dot{y}_1 > 0$ $(+,-)$	$\dot{y}_1 > 0$ $(+,-)$	$\dot{y}_1^* = 0$ $(-,-)$	$\dot{y}_1^* = 0$ $(-,-)$

$(\dot{x}_1, a_{221})$. The corresponding properties for $\alpha_{11} > 0$ and $\alpha_{12} > 0$ are presented as follows:

$$\left\{\begin{matrix} (a_{111}, \dot{x}_2) \\ (a_{112}, \dot{x}_2) \\ (\dot{x}_1, a_{221}) \end{matrix}\right\} = \left\{\begin{matrix} \underbrace{(\mathrm{UP}, \mathrm{pF})}_{\text{up-parabola flow } (+)} \\ \underbrace{(\mathrm{II}, \mathrm{pF})}_{\text{increasing-inflection flow } (+)} \\ \underbrace{(\mathrm{pF}, \mathrm{II})}_{\text{increasing-inflection flow } (+)} \end{matrix}\right\}$$

$$\text{for } a_{110} > 0 \quad \text{and} \quad a_{220} > 0, \tag{3.245}$$

$$\left\{\begin{matrix} (a_{111}, \dot{x}_2) \\ (a_{112}, \dot{x}_2) \\ (\dot{x}_1, a_{221}) \end{matrix}\right\} = \left\{\begin{matrix} \underbrace{(\mathrm{DP}, \mathrm{pF})}_{\text{down-parabola flow } (+)} \\ \underbrace{(\mathrm{DI}, \mathrm{pF})}_{\text{decreasing-inflection flow } (+)} \\ \underbrace{(\mathrm{nF}, \mathrm{DI})}_{\text{increasing-inflection flow } (-)} \end{matrix}\right\}$$

$$\text{for } a_{110} < 0 \quad \text{and} \quad a_{220} > 0, \tag{3.246}$$

$$\left\{\begin{matrix} (a_{111}, \dot{x}_2) \\ (a_{112}, \dot{x}_2) \\ (\dot{x}_1, a_{221}) \end{matrix}\right\} = \left\{\begin{matrix} \underbrace{(\mathrm{DP}, \mathrm{nF})}_{\text{down-parabola flow } (-)} \\ \underbrace{(\mathrm{DI}, \mathrm{nF})}_{\text{decreasing-inflection flow } (-)} \\ \underbrace{(\mathrm{pF}, \mathrm{DI})}_{\text{increasing-inflection flow } (+)} \end{matrix}\right\}$$

$$\text{for } a_{110} > 0 \quad \text{and} \quad a_{220} < 0, \tag{3.247}$$

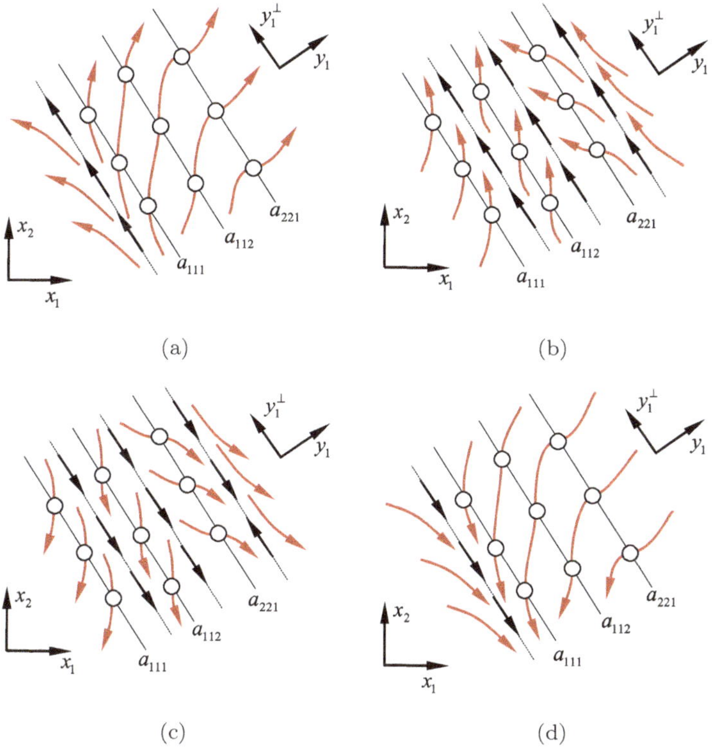

Fig. 3.3. Phase portraits ($a_{112} < a_{221}$ and $\alpha_{11} > 0, \alpha_{12} > 0$) for two-dimensional systems on the x_1-direction with $y_1^* = a_{111}, a_{112}$ and on the x_2-direction with $y_1^* = a_{221}$: (a) ($a_{110} > 0, a_{220} > 0$), (b) ($a_{110} < 0, a_{220} > 0$), (c) ($a_{110} > 0, a_{220} < 0$), (d) ($a_{110} < 0, a_{220} < 0$).

$$\left\{\begin{array}{c}(a_{111}, \dot{x}_2)\\(a_{112}, \dot{x}_2)\\(\dot{x}_1, a_{221})\end{array}\right\} = \left\{\begin{array}{c}\underbrace{(\text{UP}, \text{nF})}_{\text{up-parabola flow }(-)}\\\underbrace{(\text{II}, \text{nF})}_{\text{increasing-inflection flow }(-)}\\\underbrace{(\text{nF}, \text{II})}_{\text{increasing-inflection flow }(-)}\end{array}\right\}$$
$$\text{for } a_{110} < 0 \quad \text{and} \quad a_{220} < 0. \tag{3.248}$$

The equilibrium dynamics on the y_1-direction are listed in Table 3.3.

(iv) For $a_{112} < a_{221}$ ($\alpha_{11} > 0, \alpha_{12} < 0$), presented in Fig. 3.4 are the one-dimensional flows of $(y_1^*, \dot{x}_2) = (a_{111}, \dot{x}_2), (a_{112}, \dot{x}_2)$ and $(\dot{x}_1, y_1^*) =$

Table 3.3. Equilibrium flows of $\dot{y}_1 = 0$ $(\alpha_{11} > 0, \alpha_{12} > 0)$.

Figure 3.3	$(\dot{x}_1, \dot{x}_2)$	y_1			
	(a_{110}, a_{220})	$(-\infty, a_{111})$	(a_{111}, a_{112})	(a_{112}, a_{221})	(a_{221}, ∞)
(a)	$(+,+)$	$\dot{y}_1^* = 0$ $(-,+)$	$\dot{y}_1 > 0$ $(+,+)$	$\dot{y}_1 > 0$ $(+,+)$	$\dot{y}_1 > 0$ $(+,+)$
(b)	$(-,+)$	$\dot{y}_1 > 0$ $(+,+)$	$\dot{y}_1^* = 0$ $(-,+)$	$\dot{y}_1^* = 0$ $(-,+)$	$\dot{y}_1^* = 0$ $(-,+)$
(c)	$(+,-)$	$\dot{y}_1 < 0$ $(-,-)$	$\dot{y}_1^* = 0$ $(+,-)$	$\dot{y}_1^* = 0$ $(+,-)$	$\dot{y}_1^* = 0$ $(+,-)$
(d)	$(-,-)$	$\dot{y}_1^* = 0$ $(-,+)$	$\dot{y}_1 < 0$ $(-,-)$	$\dot{y}_1 < 0$ $(-,-)$	$\dot{y}_1 < 0$ $(-,-)$

$(\dot{x}_1, a_{221})$. The corresponding properties for $\alpha_{11} > 0$ and $\alpha_{12} < 0$ are presented as follows:

$$\left\{\begin{matrix} (a_{111}, \dot{x}_2) \\ (a_{112}, \dot{x}_2) \\ (\dot{x}_1, a_{221}) \end{matrix}\right\} = \left\{\begin{matrix} \underbrace{(\mathrm{DP}, \mathrm{pF})}_{\text{down-parabola flow } (+)} \\ \underbrace{(\mathrm{II}, \mathrm{pF})}_{\text{increasing-inflection flow } (+)} \\ \underbrace{(\mathrm{pF}, \mathrm{II})}_{\text{increasing-inflection flow } (+)} \end{matrix}\right\}$$

$$\text{for } a_{110} > 0 \quad \text{and} \quad a_{220} > 0, \tag{3.249}$$

$$\left\{\begin{matrix} (a_{111}, \dot{x}_2) \\ (a_{112}, \dot{x}_2) \\ (\dot{x}_1, a_{221}) \end{matrix}\right\} = \left\{\begin{matrix} \underbrace{(\mathrm{UP}, \mathrm{pF})}_{\text{up-parabola flow } (+)} \\ \underbrace{(\mathrm{DI}, \mathrm{pF})}_{\text{decreasing-inflection flow } (+)} \\ \underbrace{(\mathrm{nF}, \mathrm{DI})}_{\text{increasing-inflection flow } (-)} \end{matrix}\right\}$$

$$\text{for } a_{110} < 0 \quad \text{and} \quad a_{220} > 0, \tag{3.250}$$

$$\left\{\begin{matrix} (a_{111}, \dot{x}_2) \\ (a_{112}, \dot{x}_2) \\ (\dot{x}_1, a_{221}) \end{matrix}\right\} = \left\{\begin{matrix} \underbrace{(\mathrm{UP}, \mathrm{nF})}_{\text{up-parabola flow } (-)} \\ \underbrace{(\mathrm{DI}, \mathrm{nF})}_{\text{decreasing-inflection flow } (-)} \\ \underbrace{(\mathrm{pF}, \mathrm{DI})}_{\text{increasing-inflection flow } (+)} \end{matrix}\right\}$$

$$\text{for } a_{110} > 0 \quad \text{and} \quad a_{220} < 0, \tag{3.251}$$

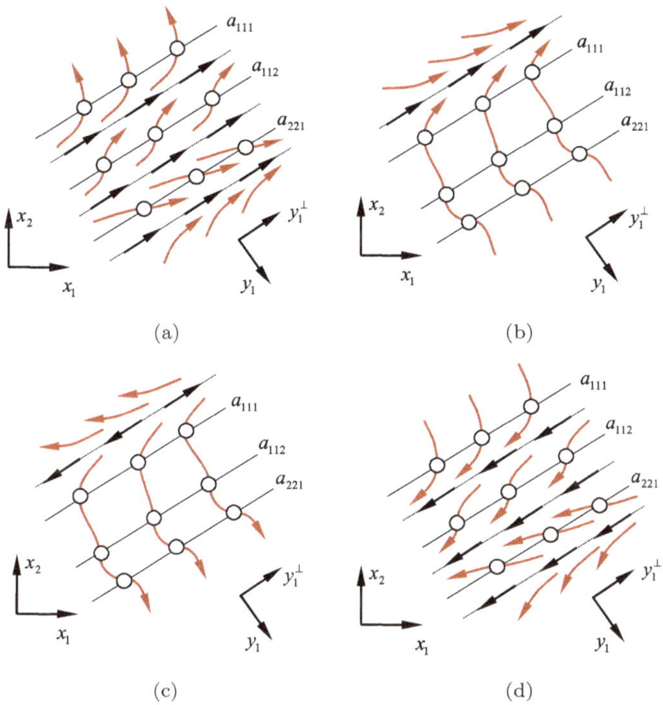

Fig. 3.4. Phase portraits ($a_{112} < a_{221}$ and $\alpha_{11} > 0, \alpha_{12} < 0$) for two-dimensional systems on the x_1-direction with $y_1^* = a_{111}, a_{112}$ and on the x_2-direction with $y_1^* = a_{221}$: (a) ($a_{110} > 0, a_{220} > 0$), (b) ($a_{110} < 0, a_{220} > 0$), (c) ($a_{110} > 0, a_{220} < 0$), (d) ($a_{110} < 0, a_{220} < 0$).

$$\begin{Bmatrix} (a_{111}, \dot{x}_2) \\ (a_{112}, \dot{x}_2) \\ (\dot{x}_1, a_{221}) \end{Bmatrix} = \begin{Bmatrix} \underbrace{(\mathrm{DP}, \mathrm{nF})}_{\text{down-parabola flow } (-)} \\ \underbrace{(\mathrm{II}, \mathrm{nF})}_{\text{increasing-inflection flow } (-)} \\ \underbrace{(\mathrm{nF}, \mathrm{II})}_{\text{increasing-inflection flow } (-)} \end{Bmatrix}$$

$$\text{for } a_{110} < 0 \quad \text{and} \quad a_{220} < 0. \tag{3.252}$$

The equilibrium dynamics on the y_1-direction are listed in Table 3.4.

(v) For $a_{221} \in (a_{111}, a_{112})$ ($\alpha_{11} > 0, \alpha_{12} > 0$), presented in Fig. 3.5 are the one-dimensional flows of $(y_1^*, \dot{x}_2) = (a_{111}, \dot{x}_2), (a_{112}, \dot{x}_2)$ and $(\dot{x}_1, y_1^*) = (\dot{x}_1, a_{221})$. The corresponding properties for $\alpha_{11} > 0$ and $\alpha_{12} > 0$ are

Table 3.4. Equilibrium flows of $\dot{y}_1 = 0$ $(\alpha_{11} > 0, \alpha_{12} < 0)$.

Figure 3.4	$(\dot{x}_1, \dot{x}_2)$ (a_{110}, a_{220})	y_1 $(-\infty, a_{111})$	(a_{111}, a_{112})	(a_{112}, a_{221})	(a_{221}, ∞)
(a)	$(+,+)$	$\dot{y}_1 < 0$ $(-,+)$	$\dot{y}_1^* = 0$ $(+,+)$	$\dot{y}_1^* = 0$ $(+,+)$	$\dot{y}_1^* = 0$ $(+,+)$
(b)	$(-,+)$	$\dot{y}_1^* = 0$ $(+,+)$	$\dot{y}_1 < 0$ $(-,+)$	$\dot{y}_1 < 0$ $(-,+)$	$\dot{y}_1 < 0$ $(-,+)$
(c)	$(+,-)$	$\dot{y}_1^* = 0$ $(-,-)$	$\dot{y}_1 > 0$ $(+,-)$	$\dot{y}_1 > 0$ $(+,-)$	$\dot{y}_1 > 0$ $(+,-)$
(d)	$(-,-)$	$\dot{y}_1 > 0$ $(+,-)$	$\dot{y}_1^* = 0$ $(-,-)$	$\dot{y}_1^* = 0$ $(-,-)$	$\dot{y}_1^* = 0$ $(-,-)$

presented as follows:

$$\left\{\begin{matrix}(a_{111}, \dot{x}_2)\\(\dot{x}_1, a_{221})\\(a_{112}, \dot{x}_2)\end{matrix}\right\} = \left\{\begin{matrix}\underbrace{(\text{UP}, \text{pF})}_{\text{up-parabola flow } (+)}\\\underbrace{(\text{pF}, \text{II})}_{\text{increasing-inflection flow } (+)}\\\underbrace{(\text{II}, \text{pF})}_{\text{increasing-inflection flow } (+)}\end{matrix}\right\}$$

$$\text{for } a_{110} > 0 \quad \text{and} \quad a_{220} > 0, \tag{3.253}$$

$$\left\{\begin{matrix}(a_{111}, \dot{x}_2)\\(\dot{x}_1, a_{221})\\(a_{112}, \dot{x}_2)\end{matrix}\right\} = \left\{\begin{matrix}\underbrace{(\text{DP}, \text{pF})}_{\text{down-parabola flow } (+)}\\\underbrace{(\text{nF}, \text{DI})}_{\text{increasing-inflection flow } (-)}\\\underbrace{(\text{DI}, \text{pF})}_{\text{increasing-inflection flow } (+)}\end{matrix}\right\}$$

$$\text{for } a_{110} < 0 \quad \text{and} \quad a_{220} > 0, \tag{3.254}$$

$$\left\{\begin{matrix}(a_{111}, \dot{x}_2)\\(\dot{x}_1, a_{221})\\(a_{112}, \dot{x}_2)\end{matrix}\right\} = \left\{\begin{matrix}\underbrace{(\text{DP}, \text{nF})}_{\text{down-parabola flow } (-)}\\\underbrace{(\text{nF}, \text{DI})}_{\text{decreasing-inflection flow } (-)}\\\underbrace{(\text{DI}, \text{nF})}_{\text{increasing-inflection flow } (-)}\end{matrix}\right\}$$

$$\text{for } a_{110} > 0 \quad \text{and} \quad a_{220} < 0, \tag{3.255}$$

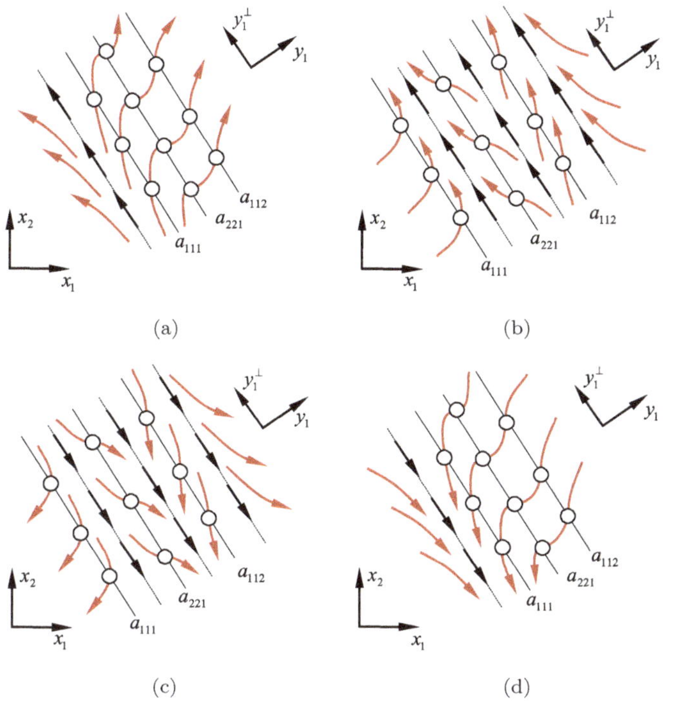

Fig. 3.5. Phase portraits ($a_{111} < a_{221} < a_{112}$ and $\alpha_{11} > 0, \alpha_{12} > 0$) for two-dimensional systems on the x_1-direction with $y_1^* = a_{111}, a_{112}$ and on the x_2-direction with $y_1^* = a_{221}$: (a) ($a_{110} > 0, a_{220} > 0$), (b) ($a_{110} < 0, a_{220} < 0$), (c) ($a_{110} > 0, a_{220} < 0$), (d) ($a_{110} < 0, a_{220} < 0$).

$$\left\{\begin{matrix} (a_{111}, \dot{x}_2) \\ (\dot{x}_1, a_{221}) \\ (a_{112}, \dot{x}_2) \end{matrix}\right\} = \left\{\begin{matrix} \underbrace{(\text{UP}, \text{nF})}_{\text{up-parabola flow } (-)} \\ \underbrace{(\text{nF}, \text{II})}_{\text{increasing-inflection flow } (-)} \\ \underbrace{(\text{II}, \text{nF})}_{\text{increasing-inflection flow } (-)} \end{matrix}\right\}$$

$$\text{for } a_{110} < 0 \quad \text{and} \quad a_{220} < 0. \tag{3.256}$$

The equilibrium dynamics on the y_1-direction are listed in Table 3.5.

(vi) For $a_{221} \in (a_{111}, a_{112})$ ($\alpha_{11} > 0, \alpha_{12} < 0$), presented in Fig. 3.6 are the one-dimensional flows of$(y_1^*, \dot{x}_2) = (a_{111}, \dot{x}_2), (a_{112}, \dot{x}_2)$ and $(\dot{x}_1, y_1^*) = (\dot{x}_1, a_{221})$. The corresponding properties for $\alpha_{11} > 0$ and $\alpha_{12} < 0$ are

Table 3.5. Equilibrium flows of $\dot{y}_1 = 0$ $(\alpha_{11} > 0, \alpha_{12} > 0)$.

	$(\dot{x}_1, \dot{x}_2)$	y_1			
Figure 3.5	(a_{110}, a_{220})	$(-\infty, a_{111})$	(a_{111}, a_{221})	(a_{221}, a_{112})	(a_{112}, ∞)
(a)	$(+,+)$	$\dot{y}_1^* = 0$ $(-,+)$	$\dot{y}_1 > 0$ $(+,+)$	$\dot{y}_1 > 0$ $(+,+)$	$\dot{y}_1 > 0$ $(+,+)$
(b)	$(-,+)$	$\dot{y}_1 > 0$ $(+,+)$	$\dot{y}_1^* = 0$ $(-,+)$	$\dot{y}_1^* = 0$ $(-,+)$	$\dot{y}_1^* = 0$ $(-,+)$
(c)	$(+,-)$	$\dot{y}_1 < 0$ $(-,-)$	$\dot{y}_1^* = 0$ $(+,-)$	$\dot{y}_1^* = 0$ $(+,-)$	$\dot{y}_1^* = 0$ $(+,-)$
(d)	$(-,-)$	$\dot{y}_1^* = 0$ $(-,+)$	$\dot{y}_1 < 0$ $(-,-)$	$\dot{y}_1 < 0$ $(-,-)$	$\dot{y}_1 < 0$ $(-,-)$

presented as follows:

$$\left\{\begin{array}{c}(a_{111}, \dot{x}_2)\\(\dot{x}_1, a_{221})\\(a_{112}, \dot{x}_2)\end{array}\right\} = \left\{\begin{array}{c}\underbrace{(\mathrm{DP}, \mathrm{pF})}_{\text{down-parabola flow }(+)}\\\underbrace{(\mathrm{pF}, \mathrm{II})}_{\text{increasing-inflection flow }(+)}\\\underbrace{(\mathrm{II}, \mathrm{pF})}_{\text{increasing-inflection flow }(+)}\end{array}\right\}$$

$$\text{for } a_{110} > 0 \quad \text{and} \quad a_{220} > 0, \tag{3.257}$$

$$\left\{\begin{array}{c}(a_{111}, \dot{x}_2)\\(\dot{x}_1, a_{221})\\(a_{112}, \dot{x}_2)\end{array}\right\} = \left\{\begin{array}{c}\underbrace{(\mathrm{UP}, \mathrm{pF})}_{\text{up-parabola flow }(+)}\\\underbrace{(\mathrm{nF}, \mathrm{DI})}_{\text{decreasing-inflection flow }(-)}\\\underbrace{(\mathrm{DI}, \mathrm{pF})}_{\text{increasing-inflection flow }(+)}\end{array}\right\}$$

$$\text{for } a_{110} < 0 \quad \text{and} \quad a_{220} > 0, \tag{3.258}$$

$$\left\{\begin{array}{c}(a_{111}, \dot{x}_2)\\(\dot{x}_1, a_{221})\\(a_{112}, \dot{x}_2)\end{array}\right\} = \left\{\begin{array}{c}\underbrace{(\mathrm{UP}, \mathrm{nF})}_{\text{up-parabola flow }(-)}\\\underbrace{(\mathrm{pF}, \mathrm{DI})}_{\text{decreasing-inflection flow }(+)}\\\underbrace{(\mathrm{DI}, \mathrm{nF})}_{\text{increasing-inflection flow }(-)}\end{array}\right\}$$

$$\text{for } a_{110} < 0 \quad \text{and} \quad a_{220} > 0, \tag{3.259}$$

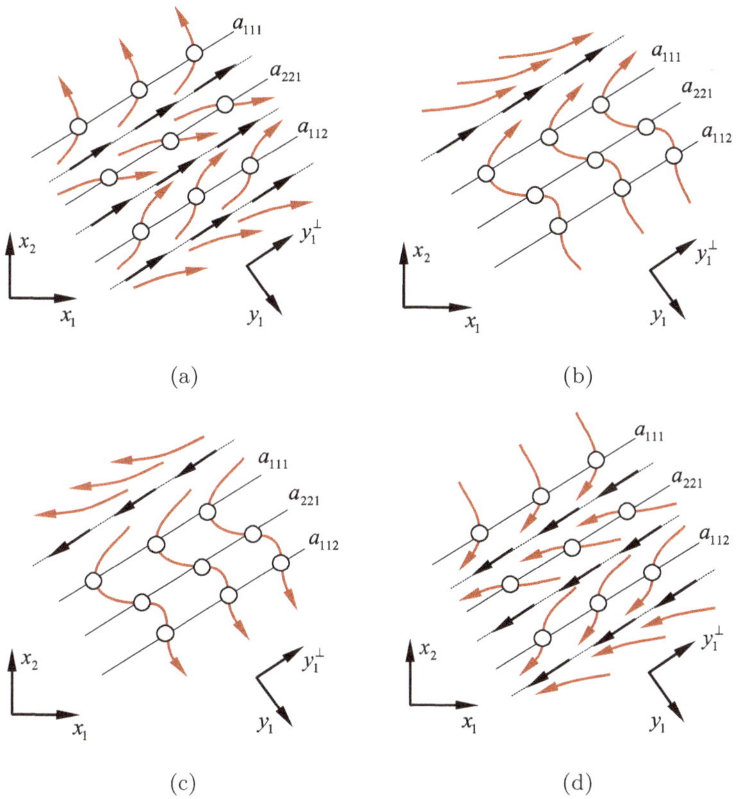

Fig. 3.6. Phase portraits ($a_{111} < a_{221} < a_{112}$ and $\alpha_{11} > 0, \alpha_{12} < 0$) for two-dimensional systems on the x_1-direction with $y_1^* = a_{111}, a_{112}$ and on the x_2-direction with $y_1^* = a_{221}$: (a) ($a_{110} > 0, a_{220} > 0$), (b) ($a_{110} < 0, a_{220} > 0$), (c) ($a_{110} > 0, a_{220} < 0$), (d) ($a_{110} < 0, a_{220} < 0$).

$$\left\{\begin{matrix} (a_{111}, \dot{x}_2) \\ (\dot{x}_1, a_{221}) \\ (a_{112}, \dot{x}_2) \end{matrix}\right\} = \left\{\begin{matrix} \underbrace{(\text{DP}, \text{nF})}_{\text{down-parabola flow } (-)} \\ \underbrace{(\text{nF}, \text{II})}_{\text{increasing-inflection flow } (-)} \\ \underbrace{(\text{II}, \text{nF})}_{\text{increasing-inflection flow } (-)} \end{matrix}\right\}$$

$$\text{for } a_{110} < 0 \quad \text{and} \quad a_{220} < 0. \tag{3.260}$$

The equilibrium dynamics on the y_1-direction are listed in Table 3.6.

Table 3.6. Equilibrium flows of $\dot{y}_1 = 0$ $(\alpha_{11} > 0, \alpha_{12} < 0)$.

	$(\dot{x}_1, \dot{x}_2)$	y_1			
Figure 3.6	(a_{110}, a_{220})	$(-\infty, a_{111})$	(a_{111}, a_{221})	(a_{221}, a_{112})	(a_{112}, ∞)
(a)	$(+,+)$	$\dot{y}_1 < 0$ $(-,+)$	$\dot{y}_1^* = 0$ $(+,+)$	$\dot{y}_1^* = 0$ $(+,+)$	$\dot{y}_1^* = 0$ $(+,+)$
(b)	$(-,+)$	$\dot{y}_1^* = 0$ $(-,-)$	$\dot{y}_1 < 0$ $(-,+)$	$\dot{y}_1 < 0$ $(-,+)$	$\dot{y}_1 < 0$ $(-,+)$
(c)	$(+,-)$	$\dot{y}_1^* = 0$ $(-,-)$	$\dot{y}_1 > 0$ $(+,-)$	$\dot{y}_1 > 0$ $(+,-)$	$\dot{y}_1 > 0$ $(+,-)$
(d)	$(-,-)$	$\dot{y}_1 < 0$ $(+,-)$	$\dot{y}_1^* = 0$ $(-,-)$	$\dot{y}_1^* = 0$ $(-,-)$	$\dot{y}_1^* = 0$ $(-,-)$

3.3.2 *Inflection-saddle and parabola sources and sinks*

(A) For $a_{111} = a_{221} < a_{112}$, consider

$$\begin{aligned}\dot{x}_1 &= a_{110}(y_1 - a_{111})(y_1 - a_{112})^2,\\ \dot{x}_2 &= a_{220}(y_1 - a_{221})^2;\end{aligned} \tag{3.261}$$

where

$$y_1 = \alpha_{11}x_1 + \alpha_{12}x_2. \tag{3.262}$$

There is a dynamical system in the y_1-direction as

$$\dot{y}_1 = \alpha_{11}a_{110}(y_1 - a_{111})(y_1 - a_{112})^2 + \alpha_{12}a_{220}(y_1 - a_{221})^2, \tag{3.263}$$

with a dynamical system in the $y_{j_1}^{\perp}$-direction as

$$\dot{y}_1 = \beta_{11}a_{110}(y_1 - a_{111})(y_1 - a_{112})^2 + \beta_{12}a_{220}(y_1 - a_{221})^2, \tag{3.264}$$

and a G-function is

$$G_{110} = \alpha_{11}a_{110}(a_{111} - a_{112})^2. \tag{3.265}$$

(i) For $a_{111} = a_{221}(\alpha_{11} > 0, \alpha_{12} > 0)$, the phase portraits for the two-dimensional system with the equilibrium of $y_1^* = a_{111}, a_{112}$ on the x_1-direction and with the equilibrium of $y_1^* = a_{221}$ on the x_2-direction are presented in Fig. 3.7. The corresponding properties are given as follows:

- For $a_{110} > 0$ and $a_{220} > 0$,

$$\left\{\begin{matrix}(a_{111}, a_{221})\\(a_{112}, \dot{x}_2^*)\end{matrix}\right\} = \left\{\begin{matrix}\underbrace{(\text{UP}, \text{SO})}_{\text{up-parabola source}}\\ \underbrace{(\text{II}, \text{pF})}_{\text{increasing-inflection flow }(+)}\end{matrix}\right\} \text{ for } G_{110} > 0. \tag{3.266}$$

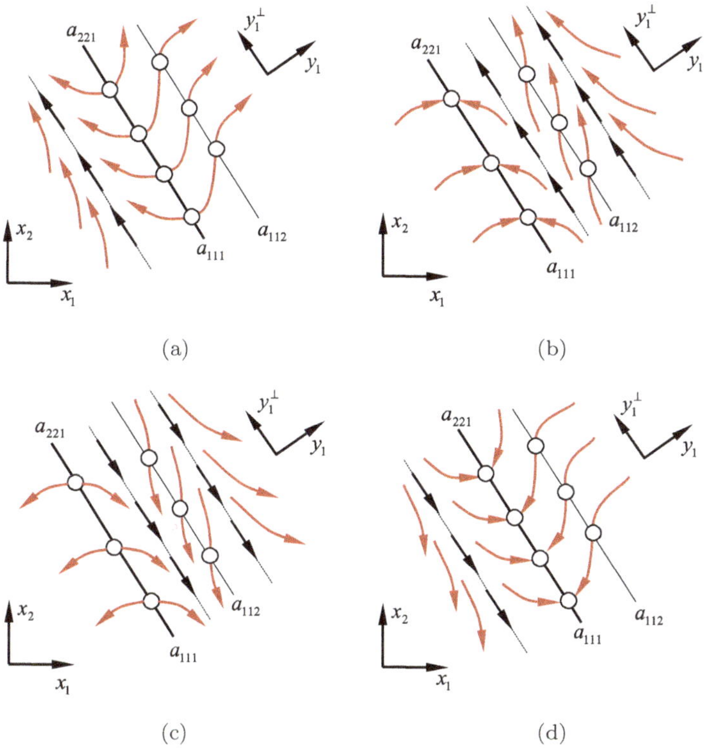

Fig. 3.7. Phase portraits ($a_{111} = a_{221}$ and $\alpha_{11} > 0, \alpha_{12} > 0$) for two-dimensional systems on the x_1-direction with $y_1^* = a_{111}, a_{112}$ and on the x_2-direction with $y_1^* = a_{221}$: (a) ($a_{110} > 0, a_{220} > 0$), (b) ($a_{110} < 0, a_{220} > 0$), (c) ($a_{110} > 0, a_{220} < 0$), (d) ($a_{110} < 0, a_{220} < 0$).

- For $a_{110} < 0$ and $a_{220} > 0$,

$$\left\{\begin{matrix}(a_{111}, a_{221})\\(a_{112}, \dot{x}_2^*)\end{matrix}\right\} = \left\{\begin{matrix}\underbrace{(\mathrm{DP}, \mathrm{SI})}_{\text{down-parabola sink}}\\\underbrace{(\mathrm{II}, \mathrm{pF})}_{\text{increasing-inflection flow }(+)}\end{matrix}\right\} \text{ for } G_{110} < 0. \quad (3.267)$$

- For $a_{110} > 0$ and $a_{220} < 0$,

$$\left\{\begin{matrix}(a_{111}, a_{221})\\(a_{112}, \dot{x}_2^*)\end{matrix}\right\} = \left\{\begin{matrix}\underbrace{(\mathrm{DP}, \mathrm{SO})}_{\text{down-parabola source}}\\\underbrace{(\mathrm{II}, \mathrm{nF})}_{\text{increasing-inflection flow }(-)}\end{matrix}\right\} \text{ for } G_{110} > 0. \quad (3.268)$$

Table 3.7. Equilibrium flows of $\dot{y}_1 = 0$ for $a_{111} = a_{221}$ $(\alpha_{11} > 0, \alpha_{12} > 0)$.

	$(\dot{x}_1, \dot{x}_2)$	y_1		
Figure 3.7	(a_{110}, a_{220})	$(-\infty, a_{111})$	(a_{111}, a_{112})	(a_{112}, ∞)
(a)	$(+,+)$	$\dot{y}_1^* = 0$ $(-,+)$	$\dot{y}_1 > 0$ $(+,+)$	$\dot{y}_1 > 0$ $(+,+)$
(b)	$(-,+)$	$\dot{y}_1 > 0$ $(+,+)$	$\dot{y}_1^* = 0$ $(-,+)$	$\dot{y}_1^* = 0$ $(-,+)$
(c)	$(+,-)$	$\dot{y}_1 < 0$ $(-,-)$	$\dot{y}_1^* = 0$ $(+,-)$	$\dot{y}_1^* = 0$ $(+,-)$
(d)	$(-,-)$	$\dot{y}_1^* = 0$ $(+,-)$	$\dot{y}_1 < 0$ $(-,-)$	$\dot{y}_1 < 0$ $(-,-)$

- For $a_{110} < 0$ and $a_{220} < 0$,

$$\left\{ \begin{array}{c} (a_{111}, a_{221}) \\ (a_{112}, \dot{x}_2^*) \end{array} \right\} = \left\{ \begin{array}{c} \underbrace{(\mathrm{UP}, \mathrm{SI})}_{\text{up-parabola sink}} \\ \underbrace{(\mathrm{II}, \mathrm{nF})}_{\text{increasing-inflection flow } (-)} \end{array} \right\} \quad \text{for } G_{110} < 0. \tag{3.269}$$

The corresponding equilibrium flows of $\dot{y}_1 = 0$ in phase space are listed in Table 3.7.

(ii) For $a_{111} = a_{221} (\alpha_{11} > 0, \alpha_{12} < 0)$, the phase portraits for the two-dimensional system with the equilibrium of $y_1^* = a_{111}, a_{112}$ on the x_1-direction and with the equilibrium of $y_1^* = a_{221}$ on the x_2-direction are presented in Fig. 3.8. The corresponding properties are given as follows:

- For $a_{110} > 0$ and $a_{220} > 0$,

$$\left\{ \begin{array}{c} (a_{111}, a_{221}) \\ (a_{112}, \dot{x}_2^*) \end{array} \right\} = \left\{ \begin{array}{c} \underbrace{(\mathrm{UP}, \mathrm{SO})}_{\text{up-parabola source}} \\ \underbrace{(\mathrm{II}, \mathrm{pF})}_{\text{increasing-inflection flow } (+)} \end{array} \right\} \quad \text{for } G_{110} > 0. \tag{3.270}$$

- For $a_{110} < 0$ and $a_{220} > 0$,

$$\left\{ \begin{array}{c} (a_{111}, a_{221}) \\ (a_{112}, \dot{x}_2^*) \end{array} \right\} = \left\{ \begin{array}{c} \underbrace{(\mathrm{DP}, \mathrm{SI})}_{\text{down-parabola sink}} \\ \underbrace{(\mathrm{DI}, \mathrm{pF})}_{\text{decreasing-inflection flow } (+)} \end{array} \right\} \quad \text{for } G_{110} < 0. \tag{3.271}$$

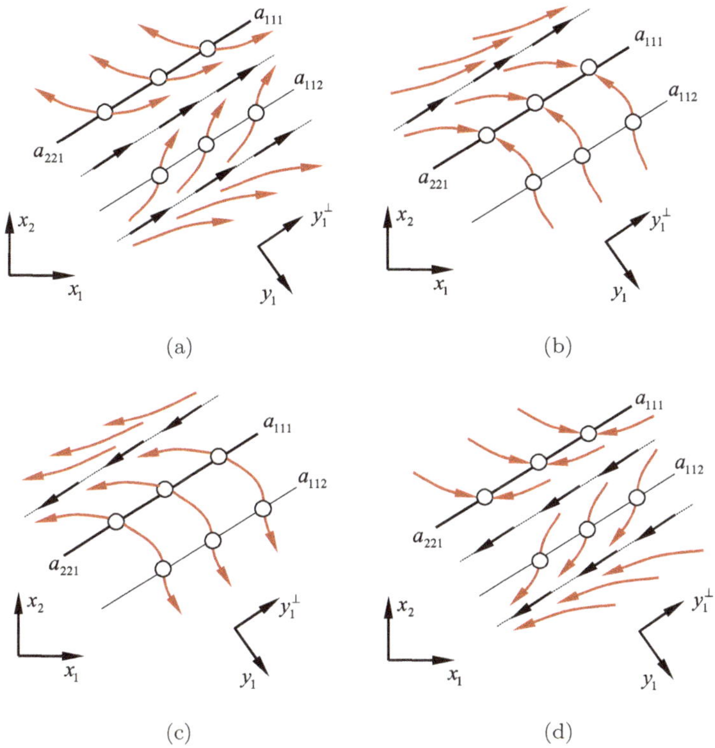

Fig. 3.8. Phase portraits ($a_{111} = a_{221}$ and $\alpha_{11} > 0, \alpha_{12} < 0$) for two-dimensional systems on the x_1-direction with $y_1^* = a_{111}, a_{112}$ and on the x_2-direction with $y_1^* = a_{221}$: (a) ($a_{110} > 0, \quad a_{220} > 0$), (b) ($a_{110} < 0, a_{220} > 0$), (c) ($a_{110} > 0, a_{220} < 0$), (d) ($a_{110} < 0, a_{220} < 0$).

- For $a_{110} > 0$ and $a_{220} < 0$,

$$\left\{ \begin{array}{c} (a_{111}, a_{221}) \\ (a_{112}, \dot{x}_2^*) \end{array} \right\} = \left\{ \begin{array}{c} \underbrace{(\mathrm{DP}, \mathrm{SO})}_{\text{down-parabola source}} \\ \underbrace{(\mathrm{DI}, \mathrm{nF})}_{\text{decreasing-inflection flow } (-)} \end{array} \right\} \quad \text{for } G_{110} > 0. \quad (3.272)$$

- For $a_{110} < 0$ and $a_{220} < 0$,

$$\left\{ \begin{array}{c} (a_{111}, a_{221}) \\ (a_{112}, \dot{x}_2^*) \end{array} \right\} = \left\{ \begin{array}{c} \underbrace{(\mathrm{UP}, \mathrm{SI})}_{\text{up-parabola sink}} \\ \underbrace{(\mathrm{II}, \mathrm{nF})}_{\text{increasing-inflection flow } (-)} \end{array} \right\} \quad \text{for } G_{110} < 0. \quad (3.273)$$

Table 3.8. Equilibrium flows of $\dot{y}_1 = 0$ for $a_{111} = a_{221}$ $(\alpha_{11} > 0, \alpha_{12} < 0)$.

Figure 3.8	$(\dot{x}_1, \dot{x}_2)$ (a_{110}, a_{220})	y_1 $(-\infty, a_{111})$	(a_{111}, a_{112})	(a_{112}, ∞)
(a)	$(+,+)$	$\dot{y}_1 < 0$ $(-,+)$	$\dot{y}_1^* = 0$ $(+,+)$	$\dot{y}_1^* = 0$ $(+,+)$
(b)	$(-,+)$	$\dot{y}_1^* = 0$ $(+,+)$	$\dot{y}_1 < 0$ $(-,+)$	$\dot{y}_1 < 0$ $(-,+)$
(c)	$(+,-)$	$\dot{y}_1^* = 0$ $(-,-)$	$\dot{y}_1 > 0$ $(+,-)$	$\dot{y}_1 > 0$ $(+,-)$
(d)	$(-,-)$	$\dot{y}_1 > 0$ $(+,-)$	$\dot{y}_1^* = 0$ $(-,-)$	$\dot{y}_1^* = 0$ $(-,-)$

The corresponding equilibrium flows of $\dot{y}_1 = 0$ in phase space are listed in Table 3.8.

(B) For $a_{112} = a_{221} < a_{112}$, consider

$$\begin{aligned}\dot{x}_1 &= a_{110}(y_1 - a_{111})(y_1 - a_{112})^2,\\ \dot{x}_2 &= a_{220}(y_1 - a_{221})^2,\end{aligned} \tag{3.274}$$

where

$$y_1 = \alpha_{11}x_1 + \alpha_{12}x_2. \tag{3.275}$$

There is a dynamical system in the y_1-direction as

$$\dot{y}_1 = \alpha_{11}a_{110}(y_1 - a_{111})(y_1 - a_{112})^2 + \alpha_{12}a_{220}(y_1 - a_{221})^2, \tag{3.276}$$

with a dynamical system in the $y_{j_1}^{\perp}$-direction as

$$\dot{y}_1 = \beta_{11}a_{110}(y_1 - a_{111})(y_1 - a_{112})^2 + \beta_{12}a_{220}(y_1 - a_{221})^2, \tag{3.277}$$

and a G-function is

$$G_{110}^{(1)} = \alpha_{11}a_{110}(a_{112} - a_{111}) + \alpha_{12}a_{220}. \tag{3.278}$$

(i) For $a_{112} = a_{221}(\alpha_{11} > 0, \alpha_{12} > 0)$, the phase portraits for the two-dimensional system with the equilibrium of $y_1^* = a_{111}, a_{112}$ on the x_1-direction and with the equilibrium of $y_1^* = a_{221}$ on the x_2-direction are presented in Fig. 3.9. The corresponding properties are given as follows:

- For $a_{110} > 0$ and $a_{220} > 0$,

$$\left\{\begin{matrix}(a_{111}, \dot{x}_2^*)\\(a_{112}, a_{221})\end{matrix}\right\} = \left\{\begin{matrix}\underbrace{(\text{UP}, \text{pF})}_{\text{up-parabola flow }(+)}\\ \underbrace{(\text{II}, \text{US})}_{\text{increasing-inflection upper-saddle}}\end{matrix}\right\} \quad \text{for } G_{110}^{(1)} > 0. \tag{3.279}$$

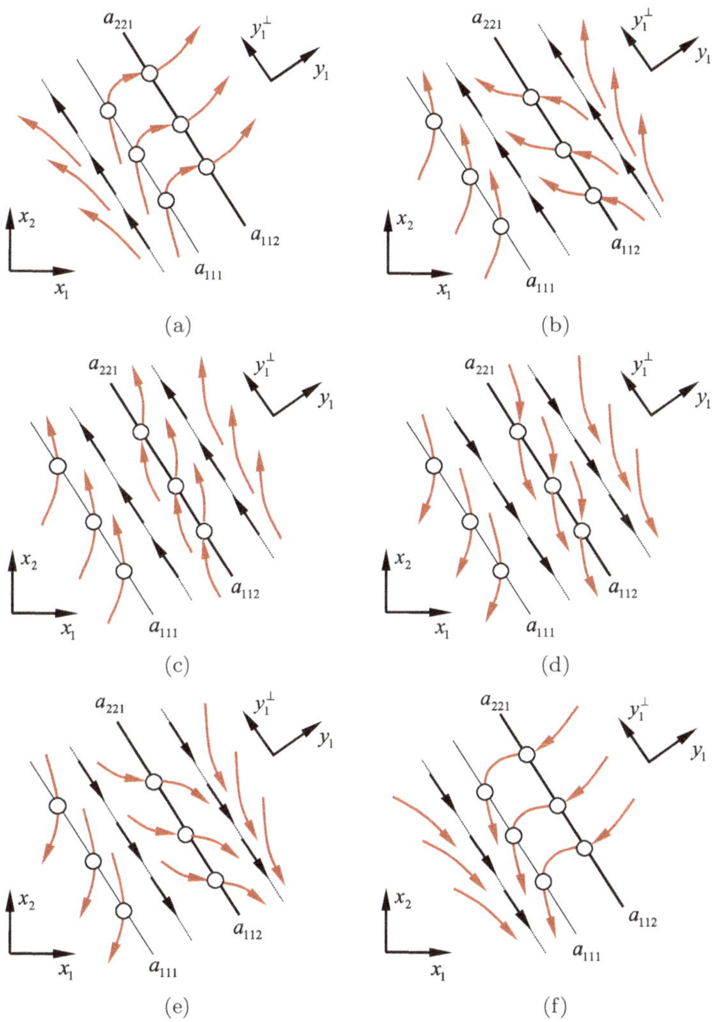

Fig. 3.9. Phase portraits ($a_{112} = a_{221}$ and $\alpha_{11} > 0, \alpha_{12} > 0$) for the two-dimensional system on the x_1-direction with $y_1^* = a_{111}, a_{112}$ and on the x_2-direction with $y_1^* = a_{221}$: (a) ($a_{110} > 0, a_{220} > 0$), (b,c) ($a_{110} < 0, a_{220} > 0$), (d,e) ($a_{110} > 0, a_{220} < 0$), (f) ($a_{110} < 0, a_{220} < 0$).

- For $a_{110} < 0$ and $a_{220} > 0$,

$$\left\{\begin{array}{c}(a_{111}, \dot{x}_2^*)\\(a_{112}, a_{221})\end{array}\right\} = \left\{\begin{array}{c}\underbrace{(\mathrm{UP}, \mathrm{pF})}_{\text{up-parabola flow }(+)}\\\underbrace{(\mathrm{DI}, , \mathrm{LS})}_{\text{decreasing-inflection lower-saddle}}\end{array}\right\} \text{ for } G_{110}^{(1)} < 0;$$

$$\left\{\begin{array}{c}(a_{111}, \dot{x}_2^*)\\(a_{112}, a_{221})\end{array}\right\} = \left\{\begin{array}{c}\underbrace{(\mathrm{UP}, \mathrm{pF})}_{\text{up-parabola flow }(+)}\\\underbrace{(\mathrm{DI}, \mathrm{US})}_{\text{decreasing-inflection upper-saddle}}\end{array}\right\} \text{ for } G_{110}^{(1)} > 0. \tag{3.280}$$

- For $a_{110} > 0$ and $a_{220} < 0$,

$$\left\{\begin{array}{c}(a_{111}, \dot{x}_2^*)\\(a_{112}, a_{221})\end{array}\right\} = \left\{\begin{array}{c}\underbrace{(\mathrm{UP}, \mathrm{nF})}_{\text{up-parabola flow }(-)}\\\underbrace{(\mathrm{DI}, \mathrm{LS})}_{\text{decreasing-inflection lower-saddle}}\end{array}\right\} \text{ for } G_{110}^{(1)} < 0;$$

$$\left\{\begin{array}{c}(a_{111}, \dot{x}_2^*)\\(a_{112}, a_{221})\end{array}\right\} = \left\{\begin{array}{c}\underbrace{(\mathrm{UP}, \mathrm{nF})}_{\text{up-parabola flow }(-)}\\\underbrace{(\mathrm{DI}, \mathrm{US})}_{\text{decreasing-inflection upper-saddle}}\end{array}\right\} \text{ for } G_{110}^{(1)} > 0. \tag{3.281}$$

- For $a_{110} < 0$ and $a_{220} < 0$,

$$\left\{\begin{array}{c}(a_{111}, \dot{x}_2^*)\\(a_{112}, a_{221})\end{array}\right\} = \left\{\begin{array}{c}\underbrace{(\mathrm{UP}, \mathrm{nF})}_{\text{up-parabola flow }(-)}\\\underbrace{(\mathrm{II}, \mathrm{LS})}_{\text{increasing-inflection lower-saddle}}\end{array}\right\} \text{ for } G_{110}^{(1)} < 0. \tag{3.282}$$

The corresponding equilibrium flows of $\dot{y}_1 = 0$ in phase space are in Table 3.9.

(ii) For $a_{112} = a_{221}(\alpha_{11} > 0, \alpha_{12} > 0)$, the phase portraits for the two-dimensional system with the equilibrium of $y_1^* = a_{111}, a_{112}$ on the x_1-direction and with the equilibrium of $y_1^* = a_{221}$ on the x_2-direction are presented in Fig. 3.10. The corresponding properties are given as follows:

Table 3.9. Equilibrium flows of $\dot{y}_1 = 0$ for $a_{111} = a_{221}$ $(\alpha_{11} > 0, \alpha_{12} > 0)$.

Figure 3.9	$(\dot{x}_1, \dot{x}_2)$ (a_{110}, a_{220})	y_1 $(-\infty, a_{111})$	(a_{111}, a_{112})	(a_{112}, ∞)
(a)	$(+,+)$	$\dot{y}_1^* = 0$ $(-,+)$	$\dot{y}_1 > 0$ $(+,+)$	$\dot{y}_1 > 0$ $(+,+)$
(b,c)	$(-,+)$	$\dot{y}_1 > 0$ $(+,+)$	$\dot{y}_1^* = 0$ $(-,+)$	$\dot{y}_1^* = 0$ $(-,+)$
(d,e)	$(+,-)$	$\dot{y}_1 < 0$ $(-,-)$	$\dot{y}_1^* = 0$ $(+,-)$	$\dot{y}_1^* = 0$ $(+,-)$
(f)	$(-,-)$	$\dot{y}_1^* = 0$ $(+,-)$	$\dot{y}_1 < 0$ $(-,-)$	$\dot{y}_1 < 0$ $(-,-)$

- For $a_{110} > 0$ and $a_{220} > 0$,

$$\left\{\begin{matrix}(a_{111}, \dot{x}_2^*)\\(a_{112}, a_{221})\end{matrix}\right\} = \left\{\begin{matrix}\underbrace{(\text{DP}, \text{pF})}_{\text{down-parabola flow } (+)}\\\underbrace{(\text{II}, \text{US})}_{\text{increasing-inflection upper-saddle}}\end{matrix}\right\} \text{ for } G_{110}^{(1)} > 0;$$

$$\left\{\begin{matrix}(a_{111}, \dot{x}_2^*)\\(a_{112}, a_{221})\end{matrix}\right\} = \left\{\begin{matrix}\underbrace{(\text{DP}, \text{pF})}_{\text{down-parabola flow } (+)}\\\underbrace{(\text{II}, \text{LS})}_{\text{increasing-inflection lower-saddle}}\end{matrix}\right\} \text{ for } G_{110}^{(1)} < 0. \tag{3.283}$$

- For $a_{110} < 0$ and $a_{220} > 0$,

$$\left\{\begin{matrix}(a_{111}, \dot{x}_2^*)\\(a_{112}, a_{221})\end{matrix}\right\} = \left\{\begin{matrix}\underbrace{(\text{UP}, \text{pF})}_{\text{up-parabola flow } (+)}\\\underbrace{(\text{DI}, \text{LS})}_{\text{decreasing-inflection lower-saddle}}\end{matrix}\right\} \text{ for } G_{110}^{(1)} < 0. \tag{3.284}$$

- For $a_{110} > 0$ and $a_{220} < 0$,

$$\left\{\begin{matrix}(a_{111}, \dot{x}_2^*)\\(a_{112}, a_{221})\end{matrix}\right\} = \left\{\begin{matrix}\underbrace{(\text{UP}, \text{nF})}_{\text{up-parabola flow } (-)}\\\underbrace{(\text{DI}, \text{US})}_{\text{decreasing-inflection upper-saddle}}\end{matrix}\right\} \text{ for } G_{110}^{(1)} > 0. \tag{3.285}$$

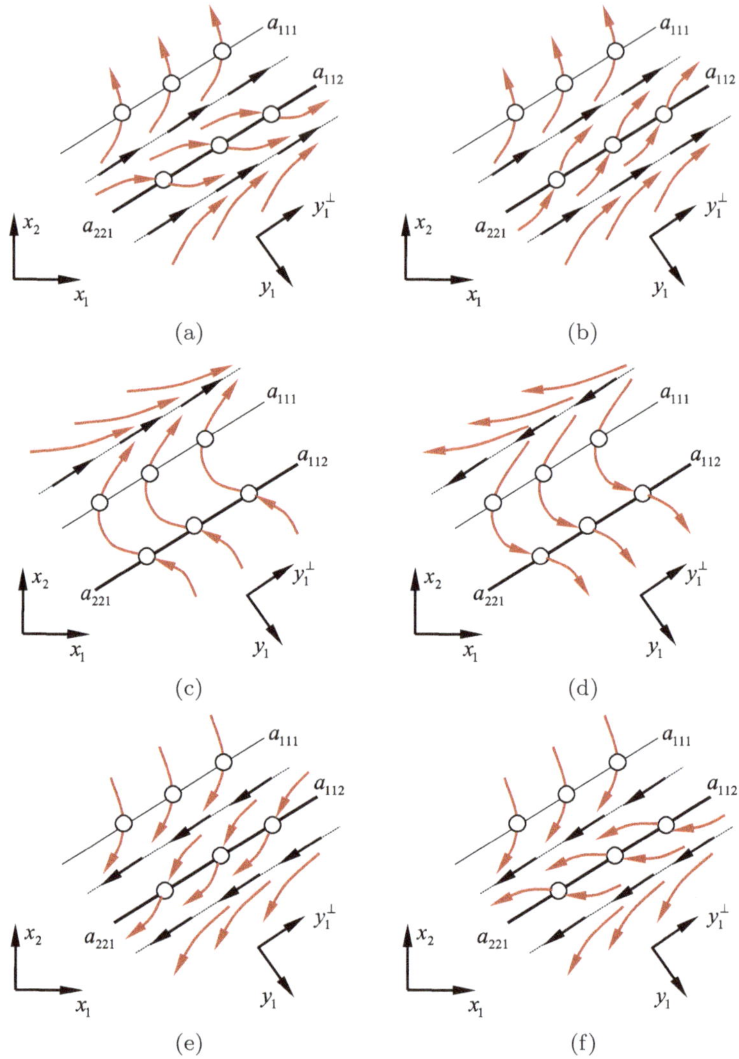

Fig. 3.10. Phase portraits ($a_{112} = a_{221}$ and $\alpha_{11} > 0, \alpha_{12} < 0$) for two-dimensional systems on the x_1-direction with $y_1^* = a_{111}, a_{112}$ and on the x_2-direction with $y_1^* = a_{221}$: (a,b) ($a_{110} > 0, a_{220} > 0$), (c) ($a_{110} < 0, a_{220} > 0$), (d) ($a_{110} > 0, a_{220} < 0$), (e,f) ($a_{110} < 0, a_{220} < 0$).

Table 3.10. Equilibrium flows of $\dot{y}_1 = 0$ for $a_{112} = a_{221}$ $(\alpha_{11} > 0, \alpha_{12} < 0)$.

Figure 3.10	$(\dot{x}_1, \dot{x}_2)$ (a_{110}, a_{220})	y_1 $(-\infty, a_{111})$	(a_{111}, a_{112})	(a_{112}, ∞)
(a)	$(+,+)$	$\dot{y}_1 < 0$ $(-,+)$	$\dot{y}_1^* = 0$ $(+,+)$	$\dot{y}_1^* = 0$ $(+,+)$
(b,c)	$(-,+)$	$\dot{y}_1^* = 0$ $(+,+)$	$\dot{y}_1 < 0$ $(-,+)$	$\dot{y}_1 < 0$ $(-,+)$
(d,e)	$(+,-)$	$\dot{y}_1^* = 0$ $(-,-)$	$\dot{y}_1 > 0$ $(+,-)$	$\dot{y}_1 > 0$ $(+,-)$
(f)	$(-,-)$	$\dot{y}_1 > 0$ $(+,-)$	$\dot{y}_1^* = 0$ $(-,-)$	$\dot{y}_1^* = 0$ $(-,-)$

- For $a_{110} < 0$ and $a_{220} < 0$,

$$
\left\{ \begin{matrix} (a_{111}, \dot{x}_2^*) \\ (a_{112}, a_{221}) \end{matrix} \right\} = \left\{ \begin{matrix} \underbrace{(\text{DP}, \text{nF})}_{\text{down-parabola flow } (-)} \\ \underbrace{(\text{II}, \text{LS})}_{\text{increasing-inflection lower-saddle}} \end{matrix} \right\} \text{ for } G_{110}^{(1)} < 0;
$$

$$
\left\{ \begin{matrix} (a_{111}, \dot{x}_2^*) \\ (a_{112}, a_{221}) \end{matrix} \right\} = \left\{ \begin{matrix} \underbrace{(\text{DP}, \text{nF})}_{\text{down-parabola flow } (-)} \\ \underbrace{(\text{II}, \text{US})}_{\text{increasing-inflection upper-saddle}} \end{matrix} \right\} \text{ for } G_{110}^{(1)} < 0. \tag{3.286}
$$

The corresponding equilibrium flows of $\dot{y}_1 = 0$ in phase space are in Table 3.10.

3.4 Singular Parabola Flows and Infinite-Equilibriums

In this section, the inflection and third-order parabola flows are presented, and the corresponding appearing bifurcations are discussed. The inflection and third-order flow switching dynamics are presented.

3.4.1 *Third-order parabola and inflection flows*

Consider

$$
\begin{aligned}
\dot{x}_1 &= a_{110}(y_1 - a_{111})^3, \\
\dot{x}_2 &= a_{220}(y_1 - a_{221})^2,
\end{aligned} \tag{3.287}
$$

with

$$y_1 = \alpha_{11}x_1 + \alpha_{12}x_2. \tag{3.288}$$

There is a dynamical system in the y_1-direction as

$$\dot{y}_1 = \alpha_{11}a_{110}(y_1 - a_{111})^3 + \alpha_{12}a_{220}(y_1 - a_{221})^2, \tag{3.289}$$

with a dynamical system in the $y_{j_1}^{\perp}$-direction as

$$\dot{y}_1^{\perp} = \beta_{11}a_{110}(y_1 - a_{111})^3 + \beta_{12}a_{220}(y_1 - a_{221})^2. \tag{3.290}$$

For $a_{111} < a_{221}(\alpha_{11} > 0, \alpha_{12} > 0)$, the phase portraits for two-dimensional systems with the equilibrium of $y_1^* = a_{111}$ on the x_1-direction and the equilibrium of $y_1^* = a_{221}$ on the x_2-direction are presented in Fig. 3.11. The corresponding properties are given as follows:

- For $a_{110} > 0$ and $a_{220} > 0$,

$$\begin{Bmatrix} (a_{111}, \dot{x}_2^*) \\ (\dot{x}_1^*, a_{221}) \end{Bmatrix} = \begin{Bmatrix} \underbrace{(3^{\text{rd}}\text{UP}, \text{pF})}_{\text{third-order up-parabola flow } (+)} \\ \underbrace{(\text{pF}, \text{II})}_{\text{increasing-inflection flow } (+)} \end{Bmatrix}. \tag{3.291}$$

- For $a_{110} < 0$ and $a_{220} > 0$,

$$\begin{Bmatrix} (a_{111}, \dot{x}_2^*) \\ (\dot{x}_1^*, a_{221}) \end{Bmatrix} = \begin{Bmatrix} \underbrace{(3^{\text{rd}}\text{DP}, \text{pF})}_{\text{third-order down-parabola flow } (+)} \\ \underbrace{(\text{nF}, \text{DI})}_{\text{increasing-inflection flow } (-)} \end{Bmatrix}. \tag{3.292}$$

- For $a_{110} > 0$ and $a_{220} < 0$,

$$\begin{Bmatrix} (a_{111}, \dot{x}_2^*) \\ (\dot{x}_1^*, a_{221}) \end{Bmatrix} = \begin{Bmatrix} \underbrace{(3^{\text{rd}}\text{DP}, \text{nF})}_{\text{third-order down-parabola flow } (-)} \\ \underbrace{(\text{pF}, \text{DI})}_{\text{increasing-inflection flow } (+)} \end{Bmatrix}. \tag{3.293}$$

- For $a_{110} < 0$ and $a_{220} < 0$,

$$\begin{Bmatrix} (a_{111}, \dot{x}_2^*) \\ (\dot{x}_1^*, a_{221}) \end{Bmatrix} = \begin{Bmatrix} \underbrace{(3^{\text{rd}}\text{UP}, \text{nF})}_{\text{third-order up-parabola flow } (-)} \\ \underbrace{(\text{nF}, \text{II})}_{\text{increasing-inflection flow } (-)} \end{Bmatrix}. \tag{3.294}$$

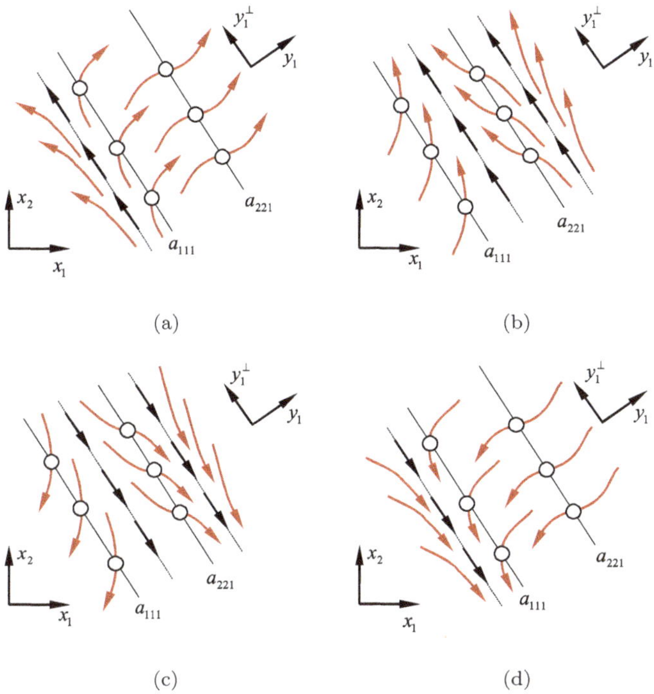

Fig. 3.11. Phase portraits ($a_{111} < a_{221}$ and $\alpha_{11} > 0, \alpha_{12} > 0$) for two-dimensional systems on the x_1-direction with $y_1^* = a_{111}$ and on the x_2-direction with $y_1^* = a_{221}$: (a) (3^{rd}UP,pF)-flow of $y_1^* = a_{111}$ and (pF,II)-flow of $y_1^* = a_{112}$ ($a_{110} > 0, a_{220} > 0$), (b) (3^{rd}DP,pF)-flow of $y_1^* = a_{111}$ and (nF,DI)-flow of $y_1^* = a_{112}$ ($a_{110} < 0, a_{220} > 0$), (c) (3^{rd}DP,nF)-flow of $y_1^* = a_{111}$ and (pF,DI)-flow of $y_1^* = a_{112}$ ($a_{110} > 0, a_{220} < 0$), (d) (3^{rd}UP, nF)-flow of $y_1^* = a_{111}$ and (nF,II)-flow of $y_1^* = a_{112}$ ($a_{110} < 0, a_{220} < 0$).

The corresponding equilibrium flows of $\dot{y}_1 = 0$ in phase space are in Table 3.11.

(ii) For $a_{111} < a_{221}(\alpha_{11} > 0, \alpha_{12} < 0)$, the phase portraits for two-dimensional systems with the equilibrium of $y_1^* = a_{111}$ on the x_1-direction and the equilibrium of $y_1^* = a_{221}$ on the x_2-direction are presented in Fig. 3.12. The corresponding properties are given as follows:

- For $a_{110} > 0$ and $a_{220} > 0$,

$$\left\{ \begin{array}{c} (a_{111}, \dot{x}_2^*) \\ (\dot{x}_1^*, a_{221}) \end{array} \right\} = \left\{ \begin{array}{c} \underbrace{(3^{\text{rd}}\text{DP}, \text{pF})}_{\text{third-order down-parabola flow } (+)} \\ \underbrace{(\text{pF}, \text{II})}_{\text{increasing-inflection flow } (+)} \end{array} \right\}. \tag{3.295}$$

Table 3.11. Equilibrium flows of $\dot{y}_1 = 0$ for $a_{111} = a_{221}$ $(\alpha_{11} > 0, \alpha_{12} > 0)$.

Figure 3.11	$(\alpha_{11}\dot{x}_1, \alpha_{12}\dot{x}_2)$ (a_{110}, a_{220})	y_1 $(-\infty, a_{111})$	(a_{111}, a_{221})	(a_{221}, ∞)
(a)	$(+,+)$	$\dot{y}_1^* = 0$ $(-,+)$	$\dot{y}_1 > 0$ $(+,+)$	$\dot{y}_1 > 0$ $(+,+)$
(b)	$(-,+)$	$\dot{y}_1 > 0$ $(+,+)$	$\dot{y}_1^* = 0$ $(-,+)$	$\dot{y}_1^* = 0$ $(-,+)$
(c)	$(+,-)$	$\dot{y}_1 < 0$ $(-,-)$	$\dot{y}_1^* = 0$ $(+,-)$	$\dot{y}_1^* = 0$ $(+,-)$
(d)	$(-,-)$	$\dot{y}_1^* = 0$ $(+,-)$	$\dot{y}_1 < 0$ $(-,-)$	$\dot{y}_1 < 0$ $(-,-)$

- For $a_{110} < 0$ and $a_{220} > 0$,

$$\begin{Bmatrix} (a_{111}, \dot{x}_2^*) \\ (\dot{x}_1^*, a_{221}) \end{Bmatrix} = \begin{Bmatrix} \underbrace{(3^{\text{rd}}\text{UP}, \text{pF})}_{\text{third-order up-parabola flow } (+)} \\ \underbrace{(\text{nF}, \text{DI})}_{\text{increasing-inflection flow } (-)} \end{Bmatrix}. \tag{3.296}$$

- For $a_{110} > 0$ and $a_{220} < 0$,

$$\begin{Bmatrix} (a_{111}, \dot{x}_2^*) \\ (\dot{x}_1^*, a_{221}) \end{Bmatrix} = \begin{Bmatrix} \underbrace{(3^{\text{rd}}\text{DP}, \text{nF})}_{\text{third-order down-parabola flow } (-)} \\ \underbrace{(\text{pF}, \text{DI})}_{\text{increasing-inflection flow } (+)} \end{Bmatrix}. \tag{3.297}$$

- For $a_{110} < 0$ and $a_{220} < 0$,

$$\begin{Bmatrix} (a_{111}, \dot{x}_2^*) \\ (\dot{x}_1^*, a_{221}) \end{Bmatrix} = \begin{Bmatrix} \underbrace{(3^{\text{rd}}\text{UP}, \text{nF})}_{\text{third-order up-parabola flow } (-)} \\ \underbrace{(\text{nF}, \text{II})}_{\text{increasing-inflection flow } (-)} \end{Bmatrix}. \tag{3.298}$$

The corresponding equilibrium flows of $\dot{y}_1 = 0$ in phase space are in Table 3.12.

The readers can complete a similar discussion for $a_{111} > a_{221}$ with $(\alpha_{11} > 0, \alpha_{12} > 0)$ and $(\alpha_{11} > 0, \alpha_{12} < 0)$.

3.4.2 *Parabola-saddle infinite-equilibriums*

Consider

$$\begin{aligned} \dot{x}_1 &= a_{110}(y_1 - a_{111})^3, \\ \dot{x}_2 &= a_{210}(y_1 - a_{221})^2. \end{aligned} \tag{3.299}$$

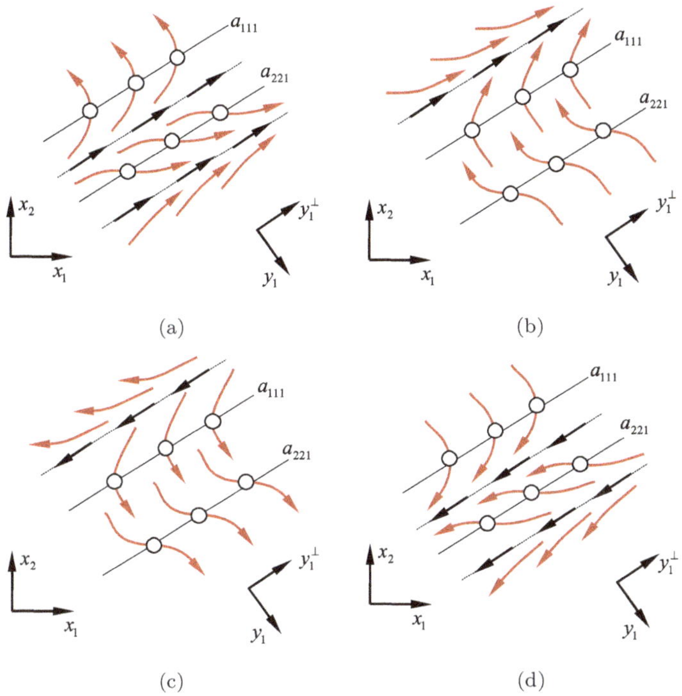

Fig. 3.12. Phase portraits ($a_{111} < a_{221}$ and $\alpha_{11} > 0, \alpha_{12} < 0$) for two-dimensional systems on the x_1-direction with $y_1^* = a_{111}$ and on the x_2-direction with $y_1^* = a_{221}$: (a) (3^{rd}DP,pF)-flow of $y_1^* = a_{111}$ and (pF,II)-flow of $y_1^* = a_{112}$ ($a_{110} > 0, a_{220} > 0$), (b) (3^{rd}UP,pF)-flow of $y_1^* = a_{111}$ and (nF,DI)-flow of $y_1^* = a_{112}$ ($a_{110} < 0, a_{220} > 0$), (c) (3^{rd}UP,nF)-flow of $y_1^* = a_{111}$ and (pF,DI)-flow of $y_1^* = a_{112}$ ($a_{110} > 0, a_{220} < 0$), (d) (3^{rd}DP,nF)-flow of $y_1^* = a_{111}$ and (nF,II)-flow of $y_1^* = a_{112}$ ($a_{110} < 0, a_{220} < 0$).

There is a dynamical system in the y_1-direction as

$$\dot{y}_1 = \alpha_{11}a_{110}(y_1 - a_{111})^3 + \alpha_{12}a_{220}(y_1 - a_{221})^2, \tag{3.300}$$

with a dynamical system in the $y_{j_1}^\perp$-direction as

$$\dot{y}_1^\perp = \beta_{11}a_{110}(y_1 - a_{111})^3 + \beta_{12}a_{220}(y_1 - a_{221})^2, \tag{3.301}$$

and

$$G_{110}^{(1)} = \alpha_{12}a_{220}. \tag{3.302}$$

For $a_{111} = a_{221}(\alpha_{11} > 0, \alpha_{12} > 0)$, the phase portraits for a two-dimensional system with the second-order parabola infinite-equilibrium of $y_1^* = a_{111} = a_{221}$ on the x_1- and x_2-directions are presented in Figs. 3.13(a)–(d) for the (UP, US)-infinite-equilibrium ($a_{110} > 0, a_{220} > 0$),

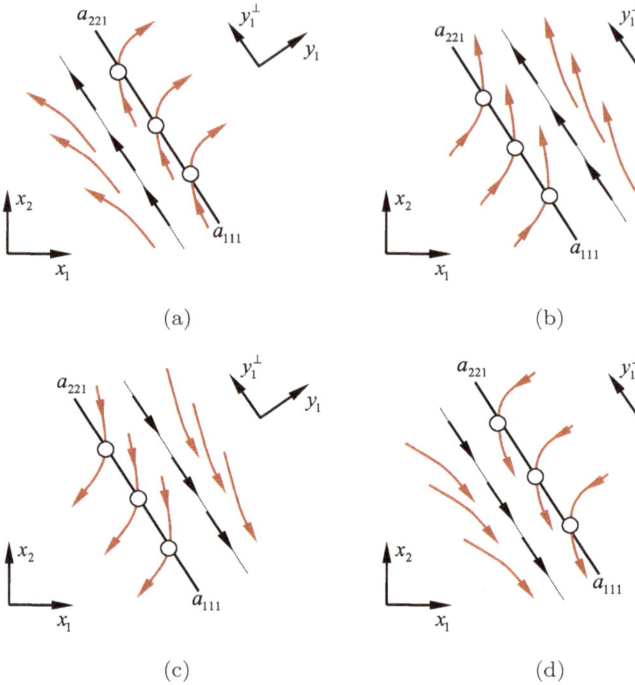

Fig. 3.13. Phase portraits ($a_{111} = a_{221}$ and $\alpha_{11} > 0, \alpha_{12} > 0$) for two-dimensional systems on the x_1-direction with $y_1^* = a_{111}$ and on the x_2-direction with $y_1^* = a_{221}$: (a) (UP,US)-infinite-equilibrium of $y_1^* = a_{111} = a_{221}$ ($a_{110} > 0, a_{220} > 0$), (b) (DP,US)-infinite-equilibrium of $y_1^* = a_{111} = a_{221}$ ($a_{110} < 0, a_{220} > 0$), (c) (DP,LS)-infinite-equilibrium of $y_1^* = a_{111} = a_{221}$ ($a_{110} > 0, a_{220} < 0$), (d) (UP,LS)-infinite equilibrium of $y_1^* = a_{111} = a_{221}$ ($a_{110} < 0, a_{220} < 0$).

Table 3.12. Equilibrium flows of $\dot{y}_1 = 0$ for $a_{111} < a_{221}$ ($\alpha_{11} > 0, \alpha_{12} < 0$).

Figure 3.12	$(\dot{x}_1, \dot{x}_2)$ (a_{110}, a_{220})	y_1 $(-\infty, a_{111})$	(a_{111}, a_{221})	(a_{221}, ∞)
(a)	$(+,+)$	$\dot{y}_1 < 0$ $(-,+)$	$\dot{y}_1^* = 0$ $(+,+)$	$\dot{y}_1^* = 0$ $(+,+)$
(b)	$(-,+)$	$\dot{y}_1^* = 0$ $(+,+)$	$\dot{y}_1 < 0$ $(-,+)$	$\dot{y}_1 < 0$ $(-,+)$
(c)	$(+,-)$	$\dot{y}_1^* = 0$ $(-,-)$	$\dot{y}_1 > 0$ $(+,-)$	$\dot{y}_1 > 0$ $(+,-)$
(d)	$(-,-)$	$\dot{y}_1 > 0$ $(+,-)$	$\dot{y}_1^* = 0$ $(-,-)$	$\dot{y}_1^* = 0$ $(-,-)$

Table 3.13. Equilibrium flows of $\dot{y}_1 = 0$ for $a_{111} < a_{221}$ ($\alpha_{11} > 0, \alpha_{12} > 0$).

Figure 3.13	$(\dot{x}_1, \dot{x}_2)$	y_1	
	(a_{110}, a_{220})	$(-\infty, a_{111})$	(a_{221}, ∞)
(a)	$(+,+)$	$\dot{y}_1^* = 0$ $(-,+)$	$\dot{y}_1 > 0$ $(+,+)$
(b)	$(-,+)$	$\dot{y}_1 > 0$ $(+,+)$	$\dot{y}_1^* = 0$ $(-,+)$
(c)	$(+,-)$	$\dot{y}_1 > 0$ $(-,-)$	$\dot{y}_1^* = 0$ $(+,-)$
(d)	$(-,-)$	$\dot{y}_1^* = 0$ $(+,-)$	$\dot{y}_1 < 0$ $(-,-)$

the (DP, US)-infinite-equilibrium ($a_{110} < 0, a_{220} > 0$), the (DP, LS)-infinite-equilibrium ($a_{110} > 0, a_{220} < 0$), and the (UP, LS)-infinite-equilibrium ($a_{110} < 0, a_{220} < 0$), respectively:

$$(a_{111}, a_{221}) = \begin{cases} \underbrace{(\text{UP}, \text{US})}_{\text{up-parabola upper-saddle}} & \text{for } a_{j_1j_10} > 0 \quad \text{and} \quad a_{j_1j_20} > 0, \\ \underbrace{(\text{DP}, \text{US})}_{\text{down-parabola upper-saddle}} & \text{for } a_{j_1j_10} < 0 \quad \text{and} \quad a_{j_1j_20} > 0, \\ \underbrace{(\text{DP}, \text{LS})}_{\text{down-parabola lower-saddle}} & \text{for } a_{j_1j_10} > 0 \quad \text{and} \quad a_{j_1j_20} < 0, \\ \underbrace{(\text{UP}, \text{LS})}_{\text{up-parabola lower-saddle}} & \text{for } a_{j_1j_10} > 0 \quad \text{and} \quad a_{j_1j_20} < 0. \end{cases} \tag{3.303}$$

The corresponding equilibrium flows of $\dot{y}_1 = 0$ in phase space are in Table 3.13.

(ii) For $a_{111} = a_{221}(\alpha_{11} > 0, \alpha_{12} < 0)$, the phase portraits for a two-dimensional system with the second-order infinite-equilibrium of $y_1^* = a_{111} = a_{221}$ on the x_1-direction and on the x_2-direction are presented in Figs. 3.14(a)–(d) for the (DP, US)-infinite-equilibrium ($a_{110} > 0, a_{220} < 0$), the (UP, US)-infinite-equilibrium ($a_{110} < 0, a_{220} > 0$), the

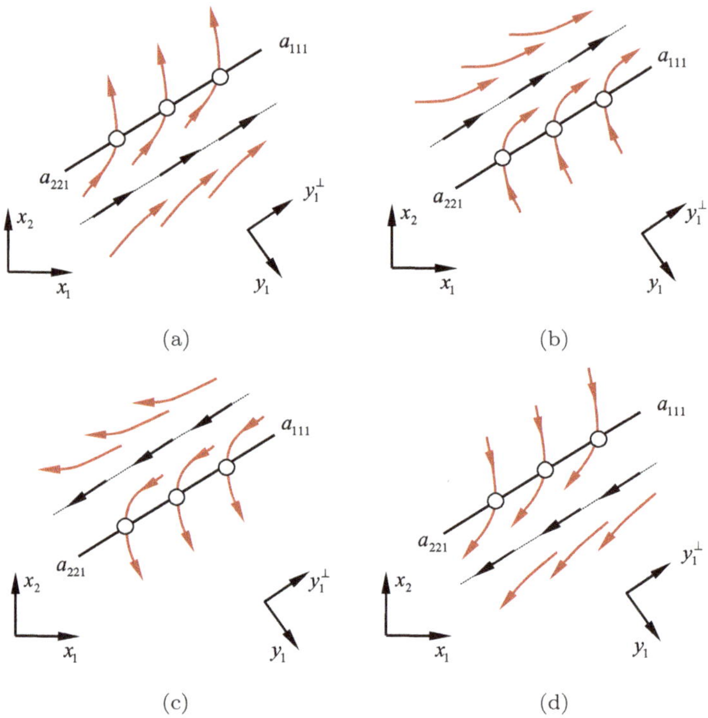

Fig. 3.14. Phase portraits ($a_{111} = a_{221}$ and $\alpha_{11} > 0, \alpha_{12} < 0$) for two-dimensional systems on the x_1-direction with $y_1^* = a_{111}$ and on the x_2-direction with $y_1^* = a_{221}$: (a) (DP, US)-infinite-equilibrium of $y_1^* = a_{111} = a_{221}$ ($a_{110} > 0, a_{220} > 0$), (b) (UP, US)-infinite-equilibrium of $y_1^* = a_{111} = a_{221}$ ($a_{110} < 0, a_{220} > 0$), (c) (UP, LS)-infinite-equilibrium of $y_1^* = a_{111} = a_{221}$ ($a_{110} > 0, a_{220} < 0$), (d) (DP, LS)-infinite-equilibrium of $y_1^* = a_{111} = a_{221}$ ($a_{110} < 0, a_{220} < 0$).

(UP, LS)-infinite-equilibrium ($a_{110} > 0, a_{220} < 0$), and the (DP, LS)-infinite-equilibrium ($a_{110} < 0, a_{220} < 0$), respectively:

$$(a_{111}, a_{221}) = \begin{cases} \underbrace{(\mathrm{DP}, \mathrm{US})}_{\text{down-parabola upper-saddle}} & \text{for } a_{j_1j_10} > 0 \quad \text{and} \quad a_{j_1j_20} > 0, \\ \underbrace{(\mathrm{UP}, \mathrm{US})}_{\text{up-parabola upper-saddle}} & \text{for } a_{j_1j_10} < 0 \quad \text{and} \quad a_{j_1j_20} > 0, \\ \underbrace{(\mathrm{UP}, \mathrm{LS})}_{\text{up-parabola lower-saddle}} & \text{for } a_{j_1j_10} > 0 \quad \text{and} \quad a_{j_1j_20} < 0, \\ \underbrace{(\mathrm{DP}, \mathrm{LS})}_{\text{down-parabola lower-saddle}} & \text{for } a_{j_1j_10} < 0 \quad \text{and} \quad a_{j_1j_20} < 0. \end{cases} \tag{3.304}$$

Table 3.14. Equilibrium flows of $\dot{y}_1 = 0$ for $a_{111} = a_{221}$ $(\alpha_{11} > 0, \alpha_{12} < 0)$.

	$(\dot{x}_1, \dot{x}_2)$	y_1	
Figure 3.14	(a_{110}, a_{220})	$(-\infty, a_{111})$	(a_{221}, ∞)
(a)	$(+,+)$	$\dot{y}_1 > 0$ $(-,+)$	$\dot{y}_1^* = 0$ $(+,+)$
(b)	$(-,+)$	$\dot{y}_1^* = 0$ $(+,+)$	$\dot{y}_1 > 0$ $(-,+)$
(c)	$(+,-)$	$\dot{y}_1^* = 0$ $(-,-)$	$\dot{y}_1 > 0$ $(+,-)$
(d)	$(-,-)$	$\dot{y}_1 > 0$ $(+,-)$	$\dot{y}_1^* = 0$ $(-,-)$

The corresponding equilibrium flows of $\dot{y}_1 = 0$ in phase space are in Table 3.14.

Chapter 4

Single-Linear-Bivariate Double-Cubic Systems

In this chapter, two-dimensional nonlinear systems possessing two single-linear-bivariate cubic vector fields are discussed, and the appearing and switching bifurcations for the one-dimensional flows (i.e., the first-order and third-order up-parabola and down-parabola flows and inflection flows) are discussed herein. The increasing-inflection and deceasing-inflection flows are the appearing bifurcations for up-parabola and down-parabola flows. The third-order parabola flows are the appearing bifurcations among up-parabola and down-parabola flows, and for parabola and inflection flows. The inflection-sink and source infinite-equilibriums are the switching bifurcations of two parabola flows in the two directions and for two inflections in the two directions. The parabola-sink and source infinite-equilibriums are the switching bifurcations for parabola and inflection flows in the two directions and for inflection and third-order parabola flows in the two directions. The second-order inflection-source/sink infinite-equilibriums are the switching bifurcations for parabola and third-order parabola flows in the two directions. The parabola-saddle infinite-equilibriums are the switching bifurcations of the parabola and inflection flows. The third-order inflection-source/sink infinite-equilibriums are the switching bifurcations for the third-order up-parabola flows and the third-order down-parabola flows and for the third-order down-parabola flows and the third-order up-parabola flows.

4.1 Two Single-Linear-Bivariate Cubic Vector Fields

In this section, consider dynamical systems with two single-linear-bivariate cubic fields. The corresponding stability and bifurcations of equilibriums are presented through the following theorem.

Theorem 4.1. *Consider a two-dimensional, cubic dynamical system for* $A_{jj0} \neq 0$ $(j = j_1, j_2 \in \{1,2\}, j_1 \neq j_2)$ *as*

$$\begin{aligned}\dot{x}_{j_1} &= f_{j_1}(\mathbf{x},\mathbf{p}) = A_{j_1j_13} + A_{j_1j_12}y_{j_1} + A_{j_1j_11}y_{j_1}^2 + A_{j_1j_10}y_{j_1}^3,\\ \dot{x}_{j_2} &= f_{j_2}(\mathbf{x},\mathbf{p}) = A_{j_2j_23} + A_{j_2j_22}y_{j_1} + A_{j_2j_21}y_{j_1}^2 + A_{j_2j_20}y_{j_1}^3,\end{aligned} \tag{4.1}$$

with

$$\begin{aligned} y_{j_1} &= \alpha_{j_1j_1}x_{j_1} + \alpha_{j_1j_2}x_{j_2}, \quad y_{j_1}^{\perp} = \beta_{j_1j_1}x_{j_1} + \beta_{j_1j_2}x_{j_2};\\ \mathbf{n}_{j_1} &= (\alpha_{j_1j_1}, \alpha_{j_1j_2})^{\mathrm{T}}, \quad \mathbf{n}_{j_1}^{\perp} = (\beta_{j_1j_1}, \beta_{j_1j_2})^{\mathrm{T}};\\ \mathbf{n}_{j_1}\cdot\mathbf{n}_{j_1}^{\perp} &= \alpha_{j_1j_1}\beta_{j_1j_1} + \alpha_{j_1j_2}\beta_{j_1j_2} = 0.\end{aligned} \tag{4.2}$$

For $A_{j_2j_20} \neq 0$, *the two-dimensional dynamical system is*

$$\begin{aligned}\dot{x}_{j_2} &= a_{j_2j_20}(y_{j_1} - b_{j_2j_21})(y_{j_1}^2 + B_{j_2j_2}y_{j_1} + C_{j_2j_2}),\\ \dot{x}_{j_1} &= a_{j_1j_10}(y_{j_1} - b_{j_1j_11})(y_{j_1}^2 + B_{j_1j_1}y_{j_1} + C_{j_1j_1}),\end{aligned} \tag{4.3}$$

with

$$\begin{aligned}\Delta_{j_2j_2} &= (B_{j_2j_2})^2 - 4C_{j_2j_2}, \quad \Delta_{j_1j_1} = (B_{j_1j_1})^2 - 4C_{j_1j_1},\\ a_{j_1j_10} &= A_{j_1j_10}, \quad a_{j_2j_20} = A_{j_2j_20}.\end{aligned} \tag{4.4}$$

(i) If

$$\Delta_{j_1j_1} < 0 \quad \text{and} \quad \Delta_{j_2j_2} < 0, \tag{4.5}$$

then

$$\begin{aligned}\dot{x}_{j_1} &= a_{j_1j_10}(y_{j_1} - a_{j_1j_11})[(y_{j_1} - a_{j_1j_1})^2 + b_{j_1j_1}],\\ \dot{x}_{j_2} &= a_{j_2j_20}(y_{j_1} - a_{j_2j_21})[(y_{j_1} - a_{j_2j_2})^2 + b_{j_2j_2}],\end{aligned} \tag{4.6}$$

where

$$\begin{aligned} a_{j_2j_2} &= -\frac{1}{2}B_{j_2j_2}, \quad b_{j_2j_2} = -\frac{1}{4}\Delta_{j_2j_2}, \quad a_{j_2j_21} = b_{j_2j_21}\\ a_{j_1j_1} &= -\frac{1}{2}B_{j_1j_1}, \quad b_{j_1j_1} = -\frac{1}{4}\Delta_{j_2j_2}, \quad a_{j_1j_11} = b_{j_1j_11}.\end{aligned} \tag{4.7}$$

(i_1) There is a dynamical system in the y_{j_1}*-direction as*

$$\begin{aligned}\dot{y}_{j_1} = {} & \alpha_{j_1j_1}a_{j_1j_10}(y_{j_1} - a_{j_1j_11})[(y_{j_1} - a_{j_1j_1})^2 + b_{j_1j_1}]\\ & + \alpha_{j_1j_2}a_{j_2j_20}(y_{j_1} - a_{j_2j_21})[(y_{j_1} - a_{j_2j_2})^2 + b_{j_2j_2}],\end{aligned} \tag{4.8}$$

with a dynamical system in the $y_{j_1}^{\perp}$*-direction as*

$$\begin{aligned}\dot{y}_{j_1}^{\perp} = {} & \beta_{j_1j_1}a_{j_1j_10}(y_{j_1} - a_{j_1j_11})[(y_{j_1} - a_{j_1j_1})^2 + b_{j_1j_1}]\\ & + \beta_{j_1j_2}a_{j_2j_20}(y_{j_1} - a_{j_2j_21})[(y_{j_1} - a_{j_2j_2})^2 + b_{j_2j_2}].\end{aligned} \tag{4.9}$$

(i_{1a}) The flow of $y_1^* = a_{j_1j_11}$ *with* $\dot{x}_{j_2}^* = a_{j_2j_20}(a_{j_1j_11} - a_{j_2j_21})[(a_{j_1j_11} - a_{j_2j_2})^2 + b_{j_2j_2}]$ *possesses the following properties:*

- *For* $a_{j_1j_10}\alpha_{j_1j_2} > 0$ *and* $a_{j_2j_20}(a_{j_1j_11} - a_{j_2j_21}) > 0$,

$$(a_{j_1j_11}, \dot{x}^*_{j_2}) = \underbrace{(\mathrm{UP}, \mathrm{pF})}_{\text{up-parabola flow } (+)}. \tag{4.10}$$

- *For* $a_{j_1j_10}\alpha_{j_1j_2} < 0$ *and* $a_{j_2j_20}(a_{j_1j_11} - a_{j_2j_21}) > 0$,

$$(a_{j_1j_11}, \dot{x}^*_{j_2}) = \underbrace{(\mathrm{DP}, \mathrm{pF})}_{\text{down-parabola flow } (+)}. \tag{4.11}$$

- *For* $a_{j_1j_10}\alpha_{j_1j_2} > 0$ *and* $a_{j_2j_20}(a_{j_1j_11} - a_{j_2j_21}) < 0$,

$$(a_{j_1j_11}, \dot{x}^*_{j_2}) = \underbrace{(\mathrm{DP}, \mathrm{nF})}_{\text{down-parabola flow } (-)}. \tag{4.12}$$

- *For* $a_{j_1j_10}\alpha_{j_1j_2} < 0$ *and* $a_{j_2j_20}(a_{j_1j_11} - a_{j_2j_21}) < 0$,

$$(a_{j_1j_11}, \dot{x}^*_{j_2}) = \underbrace{(\mathrm{UP}, \mathrm{nF})}_{\text{up-parabola flow } (-)}. \tag{4.13}$$

(i_{1b}) *The flow of* $y^*_1 = a_{j_2j_21}$ *with* $\dot{x}^*_{j_1} = a_{j_1j_10}(a_{j_2j_21} - a_{j_1j_11})[(a_{j_2j_21} - a_{j_1j_1})^2 + b_{j_1j_1}]$ *possesses the following properties*:

- *For* $a_{j_1j_10}(a_{j_2j_21} - a_{j_1j_11}) > 0$ *and* $a_{j_2j_20}\alpha_{j_1j_1} > 0$,

$$(\dot{x}^*_{j_1}, a_{j_2j_21}) = \underbrace{(\mathrm{pF}, \mathrm{UP})}_{\text{up-parabola flow } (+)}. \tag{4.14}$$

- *For* $a_{j_1j_10}(a_{j_2j_21} - a_{j_1j_11}) < 0$ *and* $a_{j_2j_20}\alpha_{j_1j_1} > 0$,

$$(\dot{x}^*_{j_1}, a_{j_2j_21}) = \underbrace{(\mathrm{nF}, \mathrm{DP})}_{\text{down-parabola flow } (-)}. \tag{4.15}$$

- *For* $a_{j_1j_10}(a_{j_2j_21} - a_{j_1j_11}) > 0$ *and* $a_{j_2j_20}\alpha_{j_1j_1} < 0$,

$$(\dot{x}^*_{j_1}, a_{j_2j_21}) = \underbrace{(\mathrm{pF}, \mathrm{DP})}_{\text{down-parabola flow } (+)}. \tag{4.16}$$

- *For* $a_{j_1j_10}(a_{j_2j_21} - a_{j_1j_11}) < 0$ *and* $a_{j_2j_20}\alpha_{j_1j_1} < 0$,

$$(\dot{x}^*_{j_1}, a_{j_2j_21}) = \underbrace{(\mathrm{nF}, \mathrm{UP})}_{\text{up-parabola flow } (-)}. \tag{4.17}$$

(i_2) *For* $a_{j_1j_11} = a_{j_2j_21}$, *there is a dynamical system in the* y_{j_1}-*direction as*

$$\dot{y}_{j_1} = (y_{j_1} - a_{j_1j_11})\{\alpha_{j_1j_1}a_{j_1j_10}[(y_{j_1} - a_{j_1j_1})^2 + b_{j_1j_1}] + \alpha_{j_1j_2}a_{j_2j_20}[(y_{j_1} - a_{j_2j_2})^2 + b_{j_2j_2}]\}, \tag{4.18}$$

and a dynamical system in the $y_{j_1}^{\perp}$-*direction is defined as*

$$\dot{y}_{j_1}^{\perp} = (y_{j_1} - a_{j_1j_11})\{\beta_{j_1j_1}a_{j_1j_10}[(y_{j_1} - a_{j_1j_1})^2 + b_{j_1j_1}] + \beta_{j_1j_2}a_{j_2j_20}[(y_{j_1} - a_{j_2j_2})^2 + b_{j_2j_2}]\}, \tag{4.19}$$

with a G-function as

$$\begin{aligned} G_{j_1j_10} &= \frac{\partial}{\partial y_{j_1}}[\mathbf{n}^{\mathrm{T}} \cdot \mathbf{F}(y_{j_1}, \mathbf{p})]\bigg|_{y_{j_1}^* = a_{j_1j_11} = a_{j_2j_21}} \\ &= \alpha_{j_1j_1}a_{j_1j_10}[(a_{j_1j_11} - a_{j_1j_1})^2 + b_{j_1j_1}] \\ &\quad + \alpha_{j_1j_2}a_{j_2j_20}[(a_{j_1j_11} - a_{j_2j_2})^2 + b_{j_2j_2}]. \end{aligned} \tag{4.20}$$

The flow of $y_1^* = a_{j_1j_11} = a_{j_2j_21}$ *possesses the following properties:*

- *For* $a_{j_1j_10}/a_{j_2j_20} > 0$ *and* $G_{j_1j_10} > 0$,

$$(a_{j_1j_11}, a_{j_2j_21}) = \underbrace{(\mathrm{II}, \mathrm{SO})}_{\text{increasing-inflection source}}. \tag{4.21}$$

- *For* $a_{j_1j_10}/a_{j_2j_20} > 0$ *and* $G_{j_1j_10} < 0$,

$$(a_{j_1j_11}, a_{j_2j_21}) = \underbrace{(\mathrm{II}, \mathrm{SI})}_{\text{increasing-inflection sink}}. \tag{4.22}$$

- *For* $a_{j_1j_10}/a_{j_2j_20} < 0$ *and* $G_{j_1j_10} > 0$,

$$(a_{j_1j_11}, a_{j_2j_21}) = \underbrace{(\mathrm{DI}, \mathrm{SO})}_{\text{decreasing-inflection source}}. \tag{4.23}$$

- *For* $a_{j_1j_10}/a_{j_2j_20} < 0$ *and* $G_{j_1j_10} < 0$,

$$(a_{j_1j_11}, a_{j_2j_21}) = \underbrace{(\mathrm{DI}, \mathrm{SI})}_{\text{decreasing-inflection sink}}. \tag{4.24}$$

(ii) If

$$\Delta_{j_1j_1} = 0 \quad \text{and} \quad \Delta_{j_2j_2} < 0, \tag{4.25}$$

then

$$\begin{aligned} \dot{x}_{j_1} &= a_{j_1j_10}(y_{j_1} - a_{j_1j_1s_1})(y_{j_1} - a_{j_1j_1s_2})^2, \\ \dot{x}_{j_2} &= a_{j_2j_20}(y_{j_1} - a_{j_2j_21})[(y_{j_1} - a_{j_2j_2})^2 + b_{j_2j_2}], \end{aligned} \tag{4.26}$$

where

$$\begin{aligned} a_{j_2j_2} &= -\frac{1}{2}B_{j_2j_2}, \quad b_{j_2j_2} = -\frac{1}{4}\Delta_{j_2j_2}, \quad a_{j_2j_21} = b_{j_2j_21}; \\ a_{j_1j_1s_w} &= -\frac{1}{2}B_{j_1j_1}, \quad b_{j_1j_1} = 0, \quad a_{j_1j_1s_1} = b_{j_1j_11}. \end{aligned} \tag{4.27}$$

(ii_1) For $a_{j_1j_1s_1} \neq a_{j_1j_1s_2}$, there is a dynamical system in the y_{j_1}-direction as

$$\begin{aligned} \dot{y}_{j_1} = {} & \alpha_{j_1j_1}a_{j_1j_10}(y_{j_1} - a_{j_1j_1s_1})(y_{j_1} - a_{j_1j_1s_2})^2 \\ & + \alpha_{j_1j_2}a_{j_2j_20}(y_{j_1} - a_{j_2j_21})[(y_{j_1} - a_{j_2j_2})^2 + b_{j_2j_2}], \end{aligned} \tag{4.28}$$

with a dynamical system in the $y_{j_1}^{\perp}$-direction as

$$\begin{aligned} \dot{y}_{j_1}^{\perp} = {} & \beta_{j_1j_1}a_{j_1j_10}(y_{j_1} - a_{j_1j_1s_1})(y_{j_1} - a_{j_1j_1s_2})^2 \\ & + \beta_{j_1j_2}a_{j_2j_20}(y_{j_1} - a_{j_2j_21})[(y_{j_1} - a_{j_2j_2})^2 + b_{j_2j_2}]. \end{aligned} \tag{4.29}$$

(ii_{1a}) The flow of $y_1^* = a_{j_1j_1s_1} \neq a_{j_1j_1s_2}$ possesses the following properties with $\dot{x}_{j_2}^* = a_{j_2j_20}(a_{j_1j_1s_1} - a_{j_2j_21})[(a_{j_1j_1s_1} - a_{j_2j_2})^2 + b_{j_2j_2}]$:

- *For $a_{j_1j_10}\alpha_{j_1j_2}(a_{j_1j_1s_1} - a_{j_1j_1s_2})^2 > 0$ and $a_{j_2j_20}(a_{j_1j_1s_1} - a_{j_2j_21}) > 0$,*

$$(a_{j_1j_1s_1}, \dot{x}_{j_2}^*) = \underbrace{(\text{UP}, \text{pF})}_{\text{up-parabola flow } (+)}. \tag{4.30}$$

- *For $a_{j_1j_10}\alpha_{j_1j_2}(a_{j_1j_1s_1} - a_{j_1j_1s_2})^2 < 0$ and $a_{j_2j_20}(a_{j_1j_1s_1} - a_{j_2j_21}) > 0$,*

$$(a_{j_1j_1s_1}, \dot{x}_{j_2}^*) = \underbrace{(\text{DP}, \text{pF})}_{\text{down-parabola flow } (+)}. \tag{4.31}$$

- *For $a_{j_1j_10}\alpha_{j_1j_2}(a_{j_1j_1s_1} - a_{j_1j_1s_2})^2 > 0$ and $a_{j_2j_20}(a_{j_1j_1s_1} - a_{j_2j_21}) < 0$,*

$$(a_{j_1j_1s_1}, \dot{x}_{j_2}^*) = \underbrace{(\text{DP}, \text{nF})}_{\text{down-parabola flow } (-)}. \tag{4.32}$$

- *For* $a_{j_1j_10}\alpha_{j_1j_2}(a_{j_1j_1s_1}-a_{j_1j_1s_2})^2<0$ *and* $a_{j_2j_20}(a_{j_1j_1s_1}-a_{j_2j_21})<0$,

$$(a_{j_1j_1s_1},\dot{x}^*_{j_2})=\underbrace{(\mathrm{UP},\mathrm{nF})}_{\text{up-parabola flow }(-)}. \tag{4.33}$$

(ii_{1b}) *The flow of* $y_1^*=a_{j_1j_1s_2}\neq a_{j_1j_1s_1}$ *possesses the following properties with* $\dot{x}^*_{j_2}=a_{j_2j_20}(a_{j_1j_1s_2}-a_{j_2j_21})[(a_{j_1j_1s_2}-a_{j_2j_2})^2+b_{j_2j_2}]$:

- *For* $a_{j_1j_10}\alpha^2_{j_1j_2}(a_{j_1j_1s_2}-a_{j_1j_1s_1})>0$ *and* $a_{j_2j_20}(a_{j_1j_1s_2}-a_{j_2j_21})>0$,

$$(a_{j_1j_1s_2},\dot{x}^*_{j_2})=\underbrace{(\mathrm{II},\mathrm{pF})}_{\text{increasing-inflection flow }(+)}. \tag{4.34}$$

- *For* $a_{j_1j_10}\alpha^2_{j_1j_2}(a_{j_1j_1s_2}-a_{j_1j_1s_1})<0$ *and* $a_{j_2j_20}(a_{j_1j_1s_2}-a_{j_2j_21})>0$,

$$(a_{j_1j_1s_2},\dot{x}^*_{j_2})=\underbrace{(\mathrm{DI},\mathrm{pF})}_{\text{decreasing-inflection flow }(+)}. \tag{4.35}$$

- *For* $a_{j_1j_10}\alpha^2_{j_1j_2}(a_{j_1j_1s_2}-a_{j_1j_1s_1})>0$ *and* $a_{j_2j_20}(a_{j_1j_1s_2}-a_{j_2j_21})<0$,

$$(a_{j_1j_1s_2},\dot{x}^*_{j_2})=\underbrace{(\mathrm{DI},\mathrm{nF})}_{\text{decreasing-inflection flow }(-)}. \tag{4.36}$$

- *For* $a_{j_1j_10}\alpha^2_{j_1j_2}(a_{j_1j_1s_2}-a_{j_1j_1s_1})<0$ *and* $a_{j_2j_20}(a_{j_1j_1s_2}-a_{j_2j_21})<0$,

$$(a_{j_1j_1s_2},\dot{x}^*_{j_2})=\underbrace{(\mathrm{II},\mathrm{nF})}_{\text{increasing-inflection flow }(-)}. \tag{4.37}$$

- *The inflection flows are the appearing bifurcations of up-parabola and down-parabola flows.*

(ii_{1c}) *The flow of* $y_1^*=a_{j_2j_21}\neq a_{j_1j_1s_1},a_{j_1j_1s_2}$ *possesses the following properties with* $\dot{x}^*_{j_1}=a_{j_1j_10}(a_{j_2j_21}-a_{j_1j_1s_1})(a_{j_2j_21}-a_{j_1j_1s_2})^2$:

- *For* $a_{j_1j_10}(a_{j_2j_21}-a_{j_1j_1s_1})(a_{j_2j_21}-a_{j_1j_1s_2})^2>0$ *and* $a_{j_2j_20}\alpha_{j_1j_1}>0$,

$$(\dot{x}^*_{j_1},a_{j_2j_21})=\underbrace{(\mathrm{pF},\mathrm{UP})}_{\text{up-parabola flow }(+)}. \tag{4.38}$$

- *For* $a_{j_1j_10}(a_{j_2j_21}-a_{j_1j_1s_1})(a_{j_2j_21}-a_{j_1j_1s_2})^2<0$ *and* $a_{j_2j_20}\alpha_{j_1j_1}>0$,

$$(\dot{x}^*_{j_1},a_{j_2j_21})=\underbrace{(\mathrm{nF},\mathrm{DP})}_{\text{down-parabola flow }(-)}. \tag{4.39}$$

- *For* $a_{j_1j_10}(a_{j_2j_21}-a_{j_1j_1s_1})(a_{j_2j_21}-a_{j_1j_1s_2})^2>0$ *and* $a_{j_2j_20}\alpha_{j_1j_1}<0$,

$$(\dot{x}^*_{j_1},a_{j_2j_21})=\underbrace{(\mathrm{pF},\mathrm{DP})}_{\text{down-parabola flow }(+)}. \tag{4.40}$$

- *For* $a_{j_1j_10}(a_{j_2j_21}-a_{j_1j_1s_1})(a_{j_2j_21}-a_{j_1j_1s_2})^2<0$ *and* $a_{j_2j_20}\alpha_{j_1j_1}<0$,

$$(\dot{x}^*_{j_1},a_{j_2j_21})=\underbrace{(\mathrm{nF},\mathrm{UP})}_{\text{up-parabola flow }(-)}. \tag{4.41}$$

(ii_2) *For* $a_{j_1j_1s_1}=a_{j_1j_1s_2}=a_{j_1j_11}$,

$$\begin{aligned}\dot{x}_{j_1}&=a_{j_1j_10}(y_{j_1}-a_{j_1j_11})^3,\\ \dot{x}_{j_2}&=a_{j_2j_20}(y_{j_1}-a_{j_2j_21})[(y_{j_1}-a_{j_2j_2})^2+b_{j_2j_2}].\end{aligned} \tag{4.42}$$

For $a_{j_1j_11}\neq a_{j_2j_21}$, *there is a dynamical system in the* y_{j_1}*-direction as*

$$\begin{aligned}\dot{y}_{j_1}&=\alpha_{j_1j_1}a_{j_1j_10}(y_{j_1}-a_{j_1j_11})^3+\alpha_{j_1j_2}a_{j_2j_20}(y_{j_1}-a_{j_2j_21})\\ &\quad\times[(y_{j_1}-a_{j_2j_2})^2+b_{j_2j_2}],\end{aligned} \tag{4.43}$$

with a dynamical system in the $y^{\perp}_{j_1}$*-direction as*

$$\begin{aligned}\dot{y}^{\perp}_{j_1}&=\beta_{j_1j_1}a_{j_1j_10}(y_{j_1}-a_{j_1j_11})^3+\beta_{j_1j_2}a_{j_2j_20}(y_{j_1}-a_{j_2j_21})\\ &\quad\times[(y_{j_1}-a_{j_2j_2})^2+b_{j_2j_2}].\end{aligned} \tag{4.44}$$

(ii_{2a}) *The flow of* $y^*_1=a_{j_1j_11}$ *with* $\dot{x}^*_{j_2}=a_{j_2j_20}(a_{j_1j_11}-a_{j_2j_21})[(a_{j_1j_11}-a_{j_2j_2})^2+b_{j_2j_2}]$ *possesses the following properties:*

- *For* $a_{j_1j_10}\alpha^3_{j_1j_2}>0$ *and* $a_{j_2j_20}(a_{j_1j_11}-a_{j_2j_21})>0$,

$$(a_{j_1j_11},\dot{x}^*_{j_2})=\underbrace{(3^{\mathrm{rd}}\mathrm{UP},\mathrm{pF})}_{\text{third-order up-parabola flow }(+)}. \tag{4.45}$$

- *For* $a_{j_1j_10}\alpha^3_{j_1j_2}<0$ *and* $a_{j_2j_20}(a_{j_1j_11}-a_{j_2j_21})>0$,

$$(a_{j_1j_11},\dot{x}^*_{j_2})=\underbrace{(3^{\mathrm{rd}}\mathrm{DP},\mathrm{pF})}_{\text{third-order down-parabola flow }(+)}. \tag{4.46}$$

- *For* $a_{j_1j_10}\alpha^3_{j_1j_2}>0$ *and* $a_{j_2j_20}(a_{j_1j_11}-a_{j_2j_21})<0$,

$$(a_{j_1j_11},\dot{x}^*_{j_2})=\underbrace{(3^{\mathrm{rd}}\mathrm{DP},\mathrm{nF})}_{\text{third-order down-parabola flow }(-)}. \tag{4.47}$$

- *For* $a_{j_1j_10}\alpha^3_{j_1j_2}<0$ *and* $a_{j_2j_20}(a_{j_1j_11}-a_{j_2j_21})<0$,

$$(a_{j_1j_11},\dot{x}^*_{j_2})=\underbrace{(3^{\mathrm{rd}}\mathrm{UP},\mathrm{nF})}_{\text{third-order up-parabola flow }(-)}. \tag{4.48}$$

- *The third-order up-parabola flows are the appearing bifurcations of up-parabola, down-parabola, and up-parabola flows.*
- *The third-order down-parabola flows are the appearing bifurcations of down-parabola, up-parabola, and down-parabola flows.*

(ii_{2b}) The flow of $y_1^* = a_{j_2j_2 1}$ with $\dot{x}_{j_1}^* = a_{j_1j_1 0}(a_{j_2j_2 1} - a_{j_1j_1 1})^3$ possesses the following properties:

- *For $a_{j_1j_1 0}(a_{j_2j_2 1} - a_{j_1j_1 1})^3 > 0$ and $a_{j_2j_2 0}\alpha_{j_1j_1} > 0$,*
$$(\dot{x}_{j_1}^*, a_{j_2j_2 1}) = \underbrace{(\mathrm{pF}, \mathrm{UP})}_{\text{down-parabola flow }(+)}. \tag{4.49}$$
- *For $a_{j_1j_1 0}(a_{j_2j_2 1} - a_{j_1j_1 1})^3 < 0$ and $a_{j_2j_2 0}\alpha_{j_1j_1} > 0$,*
$$(\dot{x}_{j_1}^*, a_{j_2j_2 1}) = \underbrace{(\mathrm{nF}, \mathrm{DP})}_{\text{down-parabola flow }(-)}. \tag{4.50}$$
- *For $a_{j_1j_1 0}(a_{j_2j_2 1} - a_{j_1j_1 1})^3 > 0$ and $a_{j_2j_2 0}\alpha_{j_1j_1} < 0$,*
$$(\dot{x}_{j_1}^*, a_{j_2j_2 1}) = \underbrace{(\mathrm{pF}, \mathrm{DP})}_{\text{down-parabola flow }(+)}. \tag{4.51}$$
- *For $a_{j_1j_1 0}(a_{j_2j_2 1} - a_{j_1j_1 1})^3 < 0$ and $a_{j_2j_2 0}\alpha_{j_1j_1} < 0$,*
$$(\dot{x}_{j_1}^*, a_{j_2j_2 1}) = \underbrace{(\mathrm{nF}, \mathrm{UP})}_{\text{up-parabola flow }(-)}. \tag{4.52}$$

(ii_3) For $a_{j_1j_1 s_1} = a_{j_2j_2 1}$, there is a dynamical system in the y_{j_1}-direction as

$$\begin{aligned}\dot{y}_{j_1} &= (y_{j_1} - a_{j_1j_1 s_1})\{\alpha_{j_1j_1}a_{j_1j_1 0}(y_{j_1} - a_{j_1j_1 s_2})^2 \\ &\quad + \alpha_{j_1j_2}a_{j_2j_2 0}[(y_{j_1} - a_{j_2j_2})^2 + b_{j_2j_2}]\},\end{aligned} \tag{4.53}$$

and a dynamical system in the $y_{j_1}^{\perp}$-direction is defined as

$$\begin{aligned}\dot{y}_{j_1}^{\perp} &= (y_{j_1} - a_{j_1j_1 s_1})\{\beta_{j_1j_1}a_{j_1j_1 0}(y_{j_1} - a_{j_1j_1 s_2})^2 \\ &\quad + \beta_{j_1j_2}a_{j_2j_2 0}[(y_{j_1} - a_{j_2j_2})^2 + b_{j_2j_2}]\},\end{aligned} \tag{4.54}$$

with a G-function as

$$\begin{aligned}G_{j_1j_1 0} &= \frac{\partial}{\partial y_{j_1}}[\mathbf{n}^{\mathrm{T}} \cdot \mathbf{F}(y_{j_1}, \mathbf{p})]\Big|_{y_{j_1}^* = a_{j_1j_1 s_1} = a_{j_2j_2 1}} \\ &= \alpha_{j_1j_1}a_{j_1j_1 0}(a_{j_1j_1 s_1} - a_{j_1j_1 s_2})^2 + \alpha_{j_1j_2}a_{j_2j_2 0} \\ &\quad \times [(a_{j_1j_1 s_1} - a_{j_2j_2})^2 + b_{j_2j_2}].\end{aligned} \tag{4.55}$$

The infinite-equilibrium flow of $y_1^* = a_{j_1j_1s_1} = a_{j_2j_21}$ *possesses the following properties:*

- *For* $a_{j_1j_10}(a_{j_1j_1s_1} - a_{j_1j_1s_2})^2/a_{j_2j_20} > 0$ *and* $G_{j_1j_10} > 0$,

$$(a_{j_1j_1s_1}, a_{j_2j_21}) = \underbrace{(\text{II}, \text{SO})}_{\text{increasing-inflection source}}. \tag{4.56}$$

- *For* $a_{j_1j_10}(a_{j_1j_1s_1} - a_{j_1j_1s_2})^2/a_{j_2j_20} > 0$ *and* $G_{j_1j_10} < 0$,

$$(a_{j_1j_1s_1}, a_{j_2j_21}) = \underbrace{(\text{II}, \text{SI})}_{\text{increasing-inflection sink}}. \tag{4.57}$$

- *For* $a_{j_1j_10}(a_{j_1j_1s_1} - a_{j_1j_1s_2})^2/a_{j_2j_20} < 0$ *and* $G_{j_1j_10} > 0$,

$$(a_{j_1j_1s_1}, a_{j_2j_21}) = \underbrace{(\text{DI}, \text{SO})}_{\text{decreasing-inflection source}}. \tag{4.58}$$

- *For* $a_{j_1j_10}(a_{j_1j_1s_1} - a_{j_1j_1s_2})^2/a_{j_2j_20} < 0$ *and* $G_{j_1j_10} < 0$,

$$(a_{j_1j_1s_1}, a_{j_2j_21}) = \underbrace{(\text{DI}, \text{SI})}_{\text{decreasing-inflection sink}}. \tag{4.59}$$

(ii_4) *For* $a_{j_1j_1s_2} = a_{j_2j_21}$, *there is a dynamical system in the* y_{j_1}*-direction as*

$$\begin{aligned}\dot{y}_{j_1} &= (y_{j_1} - a_{j_1j_1s_2})\{\alpha_{j_1j_1}a_{j_1j_10}(y_{j_1} - a_{j_1j_1s_1})(y_{j_1} - a_{j_1j_1s_2}) \\ &\quad + \alpha_{j_1j_2}a_{j_2j_20}[(y_{j_1} - a_{j_2j_2})^2 + b_{j_2j_2}]\},\end{aligned} \tag{4.60}$$

and a dynamical system in the $y_{j_1}^{\perp}$*-direction is defined as*

$$\begin{aligned}\dot{y}_{j_1}^{\perp} &= (y_{j_1} - a_{j_1j_1s_2})\{\beta_{j_1j_1}a_{j_1j_10}(y_{j_1} - a_{j_1j_1s_1})(y_{j_1} - a_{j_1j_1s_2}) \\ &\quad + \beta_{j_1j_2}a_{j_2j_20}[(y_{j_1} - a_{j_2j_2})^2 + b_{j_2j_2}]\},\end{aligned} \tag{4.61}$$

with a G-function as

$$\begin{aligned}G_{j_1j_10} &= \frac{\partial}{\partial y_{j_1}}[\mathbf{n}^{\mathrm{T}} \cdot \mathbf{F}(y_{j_1}, \mathbf{p})]\bigg|_{y_{j_1}^* = a_{j_1j_1s_2} = a_{j_2j_21}} \\ &= \alpha_{j_1j_2}a_{j_2j_20}[(a_{j_1j_1s_2} - a_{j_2j_2})^2 + b_{j_2j_2}].\end{aligned} \tag{4.62}$$

The infinite-equilibrium flow of $y_1^* = a_{j_1j_1s_1} = a_{j_2j_21}$ *possesses the following properties*:

- *For* $\alpha_{j_1j_2}^2 a_{j_1j_10}(a_{j_1j_1s_2} - a_{j_1j_1s_1}) > 0$ *and* $\alpha_{j_1j_2}a_{j_2j_20} > 0$,
$$(a_{j_1j_1s_2}, a_{j_2j_21}) = \underbrace{(\text{UP}, \text{SO})}_{\text{up-parabola source}}. \tag{4.63}$$
- *For* $\alpha_{j_1j_2}^2 a_{j_1j_10}(a_{j_1j_1s_2} - a_{j_1j_1s_1}) < 0$ *and* $\alpha_{j_1j_2}a_{j_2j_20} > 0$,
$$(a_{j_1j_1s_2}, a_{j_2j_21}) = \underbrace{(\text{DP}, \text{SO})}_{\text{down-parabola source}}. \tag{4.64}$$
- $\alpha_{j_1j_2}^2 a_{j_1j_10}(a_{j_1j_1s_2} - a_{j_1j_1s_1}) > 0$ *and* $\alpha_{j_1j_2}a_{j_2j_20} < 0$,
$$(a_{j_1j_1s_2}, a_{j_2j_21}) = \underbrace{(\text{DP}, \text{SI})}_{\text{down-parabola sink}}. \tag{4.65}$$
- $\alpha_{j_1j_2}^2 a_{j_1j_10}(a_{j_1j_1s_2} - a_{j_1j_1s_1}) < 0$ *and* $\alpha_{j_1j_2}a_{j_2j_20} < 0$,
$$(a_{j_1j_1s_2}, a_{j_2j_21}) = \underbrace{(\text{UP}, \text{SI})}_{\text{up-parabola sink}}. \tag{4.66}$$

(ii_5) *For* $a_{j_1j_1s_2} = a_{j_2j_21}$,
$$\begin{aligned}\dot{x}_{j_1} &= a_{j_1j_10}(y_{j_1} - a_{j_1j_11})^3,\\ \dot{x}_{j_2} &= a_{j_2j_20}(y_{j_1} - a_{j_2j_21})[(y_{j_1} - a_{j_2j_2})^2 + b_{j_2j_2}],\end{aligned} \tag{4.67}$$
there is a dynamical system in the y_{j_1}*-direction as*
$$\begin{aligned}\dot{y}_{j_1} = (y_{j_1} - a_{j_1j_11})\{&\alpha_{j_1j_1}a_{j_1j_10}(y_{j_1} - a_{j_1j_11})^2\\ &+ \alpha_{j_1j_2}a_{j_2j_20}[(y_{j_1} - a_{j_2j_2})^2 + b_{j_2j_2}]\},\end{aligned} \tag{4.68}$$
and a dynamical system in the $y_{j_1}^{\perp}$*-direction is defined as*
$$\begin{aligned}\dot{y}_{j_1}^{\perp} = (y_{j_1} - a_{j_1j_11})\{&\beta_{j_1j_1}a_{j_1j_10}(y_{j_1} - a_{j_1j_11})^2\\ &+ \beta_{j_1j_2}a_{j_2j_20}[(y_{j_1} - a_{j_2j_2})^2 + b_{j_2j_2}]\},\end{aligned} \tag{4.69}$$
with a G-function as
$$\begin{aligned}G_{j_1j_10} &= \frac{\partial}{\partial y_{j_1}}[\mathbf{n}^{\mathrm{T}} \cdot \mathbf{F}(y_{j_1}, \mathbf{p})]\bigg|_{y_{j_1}^* = a_{j_1j_11} = a_{j_2j_21}}\\ &= \alpha_{j_1j_2}a_{j_2j_20}[(a_{j_1j_11} - a_{j_2j_2})^2 + b_{j_2j_2}].\end{aligned} \tag{4.70}$$
The infinite-equilibrium flow of $y_1^* = a_{j_1j_11} = a_{j_2j_21}$ *possesses the following properties*:

- *For* $\alpha^3_{j_1j_2}a_{j_1j_10} > 0$ *and* $\alpha_{j_1j_2}a_{j_2j_20} > 0$,

$$(a_{j_1j_11}, a_{j_2j_21}) = \underbrace{(2^{\text{nd}}\text{II}, \text{SO})}_{\text{second-order incerasing-inflection source}}. \tag{4.71}$$

- *For* $\alpha^3_{j_1j_2}a_{j_1j_10} < 0$ *and* $\alpha_{j_1j_2}a_{j_2j_20} > 0$,

$$(a_{j_1j_11}, a_{j_2j_21}) = \underbrace{(2^{\text{nd}}\text{DI}, \text{SO})}_{\text{second-order decerasing-inflection source}}. \tag{4.72}$$

- *For* $\alpha^3_{j_1j_2}a_{j_1j_10} > 0$ *and* $\alpha_{j_1j_2}a_{j_2j_20} < 0$,

$$(a_{j_1j_11}, a_{j_2j_21}) = \underbrace{(2^{\text{nd}}\text{DI}, \text{SI})}_{\text{second-order decerasing-inflection sink}}. \tag{4.73}$$

- *For* $\alpha^3_{j_1j_2}a_{j_1j_10} < 0$ *and* $\alpha_{j_1j_2}a_{j_2j_20} < 0$,

$$(a_{j_1j_11}, a_{j_2j_21}) = \underbrace{(2^{\text{nd}}\text{II}, \text{SI})}_{\text{second-order incerasing-inflection sink}}. \tag{4.74}$$

(iii) If

$$\Delta_{j_1j_1} > 0 \quad \text{and} \quad \Delta_{j_2j_2} < 0, \tag{4.75}$$

then

$$\begin{aligned} \dot{x}_{j_1} &= a_{j_1j_10}(y_{j_1} - a_{j_1j_11})(y_{j_1} - a_{j_1j_12})(y_{j_1} - a_{j_1j_13}), \\ \dot{x}_{j_2} &= a_{j_2j_20}(y_{j_1} - a_{j_2j_21})[(y_{j_1} - a_{j_2j_2})^2 + b_{j_2j_2}], \end{aligned} \tag{4.76}$$

where

$$\begin{aligned} &a_{j_2j_2} = -\frac{1}{2}B_{j_2j_2}, \quad b_{j_2j_2} = -\frac{1}{4}\Delta_{j_2j_2}, \quad a_{j_2j_21} = b_{j_2j_21}; \\ &\{a_{j_1j_11}, a_{j_1j_12}, a_{j_1j_13}\} = \text{sort}\{b_{j_1j_11}, b_{j_1j_12}, b_{j_1j_13}\}, \\ &\quad a_{j_1j_1s_1} < a_{j_1j_1s_2}, \quad s_1 < s_2; \\ &b_{j_1j_12}, b_{j_1j_13} = -\frac{1}{2}(B_{j_1j_1} \pm \sqrt{\Delta_{j_1j_1}}), \quad a_{j_1j_10} = A_{j_1j_10}. \end{aligned} \tag{4.77}$$

There is a dynamical system in the y_{j_1}*-direction as*

$$\begin{aligned} \dot{y}_{j_1} &= \alpha_{j_1j_1}a_{j_1j_10}\prod_{s_1=1}^{3}(y_{j_1} - a_{j_1j_1s_1}) \\ &\quad + \alpha_{j_1j_2}a_{j_2j_20}(y_{j_1} - a_{j_2j_21})[(y_{j_1} - a_{j_2j_2})^2 + b_{j_2j_2}], \end{aligned} \tag{4.78}$$

with a dynamical system in the $y_{j_1}^{\perp}$*-direction as*

$$\dot{y}_{j_1}^{\perp} = \beta_{j_1 j_1} a_{j_1 j_1 0} \prod_{s_1=1}^{3} (y_{j_1} - a_{j_1 j_1 s_1}) + \beta_{j_1 j_2} a_{j_2 j_2 0} (y_{j_1} - a_{j_2 j_2 1})[(y_{j_1} - a_{j_2 j_2})^2 + b_{j_2 j_2}]. \tag{4.79}$$

With

$$\dot{x}_{j_2}^{*} = a_{j_2 j_2 0}(a_{j_1 j_1 s_1} - a_{j_2 j_2 1})[(a_{j_1 j_1 s_1} - a_{j_2 j_2})^2 + b_{j_2 j_2}], \tag{4.80}$$

the equilibrium flow of $y_{j_1}^{*} = a_{j_1 j_1 s_1}(s_1, s_2 \in \{1,2,3\}, s_1 \neq s_2)$ *in the* x_{j_1}*-direction possesses the following properties:*

- *For* $a_{j_1 j_1 0}\alpha_{j_1 j_2} \prod_{s_2=1, s_2 \neq s_1}^{3} (a_{j_1 j_1 s_1} - a_{j_1 j_1 s_2}) > 0$ *and* $a_{j_2 j_2 0}(a_{j_1 j_1 s_1} - a_{j_2 j_2 1}) > 0$,

$$(a_{j_1 j_1 s_1}, \dot{x}_{j_2}^{*}) = \underbrace{(\mathrm{UP}, \mathrm{pF})}_{\text{up-parabola flow } (+)}. \tag{4.81}$$

- *For* $a_{j_1 j_1 0}\alpha_{j_1 j_2} \prod_{s_2=1, s_2 \neq s_1}^{3} (a_{j_1 j_1 s_1} - a_{j_1 j_1 s_2}) < 0$ *and* $a_{j_2 j_2 0}(a_{j_1 j_1 s_1} - a_{j_2 j_2 1}) > 0$,

$$(a_{j_1 j_1 s_1}, \dot{x}_{j_2}^{*}) = \underbrace{(\mathrm{DP}, \mathrm{pF})}_{\text{down-parabola flow } (+)}. \tag{4.82}$$

- *For* $a_{j_1 j_1 0}\alpha_{j_1 j_2} \prod_{s_2=1, s_2 \neq s_1}^{3} (a_{j_1 j_1 s_1} - a_{j_1 j_1 s_2}) > 0$ *and* $a_{j_2 j_2 0}(a_{j_1 j_1 s_1} - a_{j_2 j_2 1}) < 0$,

$$(a_{j_1 j_1 s_1}, \dot{x}_{j_2}^{*}) = \underbrace{(\mathrm{DP}, \mathrm{nF})}_{\text{down-parabola flow } (-)}. \tag{4.83}$$

- *For* $a_{j_1 j_1 0}\alpha_{j_1 j_2} \prod_{s_2=1, s_2 \neq s_1}^{3} (a_{j_1 j_1 s_1} - a_{j_1 j_1 s_2}) < 0$ *and* $a_{j_2 j_2 0}(a_{j_1 j_1 s_1} - a_{j_2 j_2 1}) < 0$,

$$(a_{j_1 j_1 s_1}, \dot{x}_{j_2}^{*}) = \underbrace{(\mathrm{UP}, \mathrm{nF})}_{\text{up-parabola flow } (-)}. \tag{4.84}$$

(iii$_{1b}$*) The flow of* $y_1^{*} = a_{j_2 j_2 1} \neq a_{j_1 j_1 s_1}$ *possesses the following properties with* $\dot{x}_{j_1}^{*} = a_{j_1 j_1 0} \prod_{s_1=1}^{3} (a_{j_2 j_2 1} - a_{j_1 j_1 s_1})$:

- *For* $a_{j_1 j_1 0} \prod_{s_1=1}^{3} (a_{j_2 j_2 1} - a_{j_1 j_1 s_1}) > 0$ *and* $a_{j_2 j_2 0}\alpha_{j_1 j_1} > 0$,

$$(\dot{x}_{j_1}^{*}, a_{j_2 j_2 1}) = \underbrace{(\mathrm{pF}, \mathrm{UP})}_{\text{up-parabola flow } (+)}. \tag{4.85}$$

- *For* $a_{j_1j_10}\prod_{s_1=1}^{3}(a_{j_2j_21}-a_{j_1j_1s_1})<0$ *and* $a_{j_2j_20}\alpha_{j_1j_1}>0$,

$$(\dot{x}_{j_1}^{*},a_{j_2j_21})=\underbrace{(\text{nF},\text{DP})}_{\text{down-parabola flow }(-)}. \tag{4.86}$$

- *For* $a_{j_1j_10}\prod_{s_1=1}^{3}(a_{j_2j_21}-a_{j_1j_1s_1})>0$ *and* $a_{j_2j_20}\alpha_{j_1j_1}<0$,

$$(\dot{x}_{j_1}^{*},a_{j_2j_21})=\underbrace{(\text{pF},\text{DP})}_{\text{down-parabola flow }(+)}. \tag{4.87}$$

- *For* $a_{j_1j_10}\prod_{s_1=1}^{3}(a_{j_2j_21}-a_{j_1j_1s_1})<0$ *and* $a_{j_2j_20}\alpha_{j_1j_1}<0$,

$$(\dot{x}_{j_1}^{*},a_{j_2j_21})=\underbrace{(\text{nF},\text{UP})}_{\text{up-parabola flow }(-)}. \tag{4.88}$$

(iii_2) *For* $a_{j_2j_21}=a_{j_1j_1s_1}$,

$$\begin{aligned}\dot{x}_{j_1}&=a_{j_1j_10}(y_{j_1}-a_{j_1j_11})(y_{j_1}-a_{j_1j_12})(y_{j_1}-a_{j_1j_13}),\\ \dot{x}_{j_2}&=a_{j_2j_20}(y_{j_1}-a_{j_2j_21})[(y_{j_1}-a_{j_2j_2})^2+b_{j_2j_2}].\end{aligned} \tag{4.89}$$

There is a dynamical system in the y_{j_1}*-direction as*

$$\begin{aligned}\dot{y}_{j_1}=(y_{j_1}-a_{j_1j_1s_1})\Big\{&\alpha_{j_1j_1}a_{j_1j_10}\prod_{s_2=1,s_2\neq s_1}^{3}(y_{j_1}-a_{j_1j_1s_2})\\ &+\alpha_{j_1j_2}a_{j_2j_20}[(y_{j_1}-a_{j_2j_2})^2+b_{j_2j_2}]\Big\},\end{aligned} \tag{4.90}$$

with a dynamical system in the $y_{j_1}^{\perp}$*-direction as*

$$\begin{aligned}\dot{y}_{j_1}^{\perp}=(y_{j_1}-a_{j_1j_1s_1})\Big\{&\beta_{j_1j_1}a_{j_1j_10}\prod_{s_2=1,s_2\neq s_1}^{3}(y_{j_1}-a_{j_1j_1s_2})\\ &+\beta_{j_1j_2}a_{j_2j_20}[(y_{j_1}-a_{j_2j_2})^2+b_{j_2j_2}]\Big\},\end{aligned} \tag{4.91}$$

and a G-function is

$$\begin{aligned}G_{j_1j_10}&=\frac{\partial}{\partial y_{j_1}}[\mathbf{n}^{\mathrm{T}}\cdot\mathbf{F}(y_{j_1},\mathbf{p})]\Big|_{y_{j_1}^{*}=a_{j_1j_1s_1}=a_{j_2j_21}}\\ &=\alpha_{j_1j_1}a_{j_1j_10}\prod_{s_2=1,s_2\neq s_1}^{3}(a_{j_1j_1s_1}-a_{j_1j_1s_2})\\ &\quad+\alpha_{j_1j_2}a_{j_2j_20}[(a_{j_2j_21}-a_{j_2j_2})^2+b_{j_2j_2}].\end{aligned} \tag{4.92}$$

The flow of $y_1^* = a_{j_1j_1s_1} = a_{j_2j_21}$ *possesses the following properties:*

- *For* $a_{j_1j_10}\prod_{s_2=1,s_2\neq s_1}^{3}(a_{j_1j_1s_1} - a_{j_1j_1s_2})/a_{j_2j_20} > 0$ *and* $G_{j_1j_10} > 0$,
$$(a_{j_1j_1s_1}, a_{j_2j_21}) = \underbrace{(\text{II}, \text{SO})}_{\text{increasing-inflection source}}. \tag{4.93}$$
- *For* $a_{j_1j_10}\prod_{s_2=1,s_2\neq s_1}^{3}(a_{j_1j_1s_1} - a_{j_1j_1s_2})/a_{j_2j_20} > 0$ *and* $G_{j_1j_10} < 0$,
$$(a_{j_1j_1s_1}, a_{j_2j_21}) = \underbrace{(\text{II}, \text{SI})}_{\text{increasing-inflection sink}}. \tag{4.94}$$
- *For* $a_{j_1j_10}\prod_{s_2=1,s_2\neq s_1}^{3}(a_{j_1j_1s_1} - a_{j_1j_1s_2})/a_{j_2j_20} < 0$ *and* $G_{j_1j_10} > 0$,
$$(a_{j_1j_1s_1}, a_{j_2j_21}) = \underbrace{(\text{DI}, \text{SO})}_{\text{decreasing-inflection source}}. \tag{4.95}$$
- *For* $a_{j_1j_10}\prod_{s_2=1,s_2\neq s_1}^{3}(a_{j_1j_1s_1} - a_{j_1j_1s_2})/a_{j_2j_20} < 0$ *and* $G_{j_1j_10} < 0$,
$$(a_{j_1j_1s_1}, a_{j_2j_21}) = \underbrace{(\text{DI}, \text{SI})}_{\text{decreasing-inflection sink}}. \tag{4.96}$$

(iv) If

$$\Delta_{j_1j_1} < 0 \quad \text{and} \quad \Delta_{j_2j_2} = 0, \tag{4.97}$$

then

$$\begin{aligned}\dot{x}_{j_1} &= a_{j_1j_10}(y_{j_1} - a_{j_1j_11})[(y_{j_1} - a_{j_1j_1})^2 + b_{j_1j_1}],\\ \dot{x}_{j_2} &= a_{j_2j_20}(y_{j_1} - a_{j_2j_2l_1})(y_{j_1} - a_{j_2j_2l_2})^2,\end{aligned} \tag{4.98}$$

where

$$\begin{aligned}&a_{j_2j_2l_2} = -\frac{1}{2}B_{j_2j_2}, \quad b_{j_2j_2} = 0, \quad a_{j_2j_20} = A_{j_2j_21}, \quad a_{j_2j_2l_1} = b_{j_2j_21};\\ &a_{j_1j_11} = b_{j_1j_11}, \quad a_{j_1j_1} = -\frac{1}{2}B_{j_1j_11}, \quad b_{j_1j_1} = -\frac{1}{4}\Delta_{j_1j_1}, \quad a_{j_1j_10} = A_{j_1j_10}.\end{aligned} \tag{4.99}$$

(iv₁) For $a_{j_1j_11} \neq a_{j_2j_21}$, *there is a dynamical system in the* y_{j_1}*-direction as*

$$\begin{aligned}\dot{y}_{j_1} &= \alpha_{j_1j_1}a_{j_1j_10}(y_{j_1} - a_{j_1j_11})[(y_{j_1} - a_{j_1j_1})^2 + b_{j_1j_1}]\\ &\quad + \alpha_{j_1j_2}a_{j_2j_20}(y_{j_1} - a_{j_2j_2l_1})(y_{j_1} - a_{j_2j_2l_2})^2,\end{aligned} \tag{4.100}$$

with a dynamical system in the $y_{j_1}^{\perp}$-direction as

$$\dot{y}_{j_1}^{\perp} = \beta_{j_1j_1}a_{j_1j_1 0}(y_{j_1} - a_{j_1j_1 1})[(y_{j_1} - a_{j_1j_1})^2 + b_{j_1j_1}] + \beta_{j_1j_2}a_{j_2j_2 0}(y_{j_1} - a_{j_2j_2l_1})(y_{j_1} - a_{j_2j_2l_2})^2. \tag{4.101}$$

(iv_{1a}) *The flow of $y_1^* = a_{j_1j_1 1}$ with $\dot{x}_{j_2}^* = a_{j_2j_2 0}(a_{j_1j_1 1} - a_{j_2j_2l_1})(a_{j_1j_1 1} - a_{j_2j_2l_2})^2$ possesses the following properties:*

- *For $\alpha_{j_1j_2}a_{j_1j_1 0} > 0$ and $a_{j_2j_2 0}(a_{j_1j_1 1} - a_{j_2j_2l_1})(a_{j_1j_1 1} - a_{j_2j_2l_2})^2 > 0$,*

$$(a_{j_1j_1 1}, \dot{x}_{j_2}^*) = \underbrace{(\mathrm{UP}, \mathrm{pF})}_{\text{up-parabola flow }(+)}. \tag{4.102}$$

- *For $\alpha_{j_1j_2}a_{j_1j_1 0} < 0$ and $a_{j_2j_2 0}(a_{j_1j_1 1} - a_{j_2j_2l_1})(a_{j_1j_1 1} - a_{j_2j_2l_2})^2 > 0$,*

$$(a_{j_1j_1 1}, \dot{x}_{j_2}^*) = \underbrace{(\mathrm{DP}, \mathrm{pF})}_{\text{down-parabola flow }(+)}. \tag{4.103}$$

- *For $\alpha_{j_1j_2}a_{j_1j_1 0} > 0$ and $a_{j_2j_2 0}(a_{j_1j_1 1} - a_{j_2j_2l_1})(a_{j_1j_1 1} - a_{j_2j_2l_2})^2 < 0$,*

$$(a_{j_1j_1 1}, \dot{x}_{j_2}^*) = \underbrace{(\mathrm{DP}, \mathrm{nF})}_{\text{down-parabola flow }(-)}. \tag{4.104}$$

- *For $\alpha_{j_1j_2}a_{j_1j_1 0} < 0$ and $a_{j_2j_2 0}(a_{j_1j_1 1} - a_{j_2j_2l_1})(a_{j_1j_1 1} - a_{j_2j_2l_2})^2 < 0$,*

$$(a_{j_1j_1 1}, \dot{x}_{j_2}^*) = \underbrace{(\mathrm{UP}, \mathrm{nF})}_{\text{up-parabola flow }(-)}. \tag{4.105}$$

(iv_{1b}) *The flow of $y_1^* = a_{j_2j_2l_1}$ with $\dot{x}_{j_1}^* = a_{j_1j_1 0}(a_{j_2j_2l_1} - a_{j_1j_1 1})[(a_{j_2j_2l_1} - a_{j_1j_1})^2 + b_{j_1j_1}]$ possesses the following properties:*

- *For $a_{j_1j_1 0}(a_{j_2j_2l_1} - a_{j_1j_1 1}) > 0$ and $a_{j_2j_2 0}\alpha_{j_1j_1}(a_{j_2j_2l_1} - a_{j_2j_2l_2})^2 > 0$,*

$$(\dot{x}_{j_1}^*, a_{j_2j_2l_1}) = \underbrace{(\mathrm{pF}, \mathrm{UP})}_{\text{up-parabola flow }(+)}. \tag{4.106}$$

- *For $a_{j_1j_1 0}(a_{j_2j_2l_1} - a_{j_1j_1 1}) < 0$ and $a_{j_2j_2 0}\alpha_{j_1j_1}(a_{j_2j_2l_1} - a_{j_2j_2l_2})^2 > 0$,*

$$(\dot{x}_{j_1}^*, a_{j_2j_2l_1}) = \underbrace{(\mathrm{nF}, \mathrm{DP})}_{\text{down-parabola flow }(-)}. \tag{4.107}$$

- *For $a_{j_1j_1 0}(a_{j_2j_2l_1} - a_{j_1j_1 1}) > 0$ and $a_{j_2j_2 0}\alpha_{j_1j_1}(a_{j_2j_2l_1} - a_{j_2j_2l_2})^2 < 0$,*

$$(\dot{x}_{j_1}^*, a_{j_2j_2l_1}) = \underbrace{(\mathrm{pF}, \mathrm{DP})}_{\text{down-parabola flow }(+)}. \tag{4.108}$$

- *For* $a_{j_1j_10}(a_{j_2j_2l_1} - a_{j_1j_11}) < 0$ *and* $a_{j_2j_20}\alpha_{j_1j_1}(a_{j_2j_2l_1} - a_{j_2j_2l_2})^2 < 0$,

$$(\dot{x}^*_{j_1}, a_{j_2j_2l_1}) = \underbrace{(\mathrm{nF}, \mathrm{UP})}_{\text{up-parabola flow } (-)}. \tag{4.109}$$

(iv$_{1c}$) The flow of $y_1^* = a_{j_2j_2l_2}$ *with* $\dot{x}^*_{j_1} = a_{j_1j_10}(a_{j_2j_2l_2} - a_{j_1j_11})[(a_{j_2j_2l_2} - a_{j_1j_1})^2 + b_{j_1j_1}]$ *possesses the following properties:*

- *For* $a_{j_1j_10}(a_{j_2j_2l_2} - a_{j_1j_11}) > 0$ *and* $a_{j_2j_20}\alpha^2_{j_1j_1}(a_{j_2j_2l_2} - a_{j_2j_2l_1}) > 0$,

$$(\dot{x}^*_{j_1}, a_{j_2j_2l_2}) = \underbrace{(\mathrm{pF}, \mathrm{II})}_{\text{increasing-inflection flow } (+)}. \tag{4.110}$$

- *For* $a_{j_1j_10}(a_{j_2j_2l_2} - a_{j_1j_11}) < 0$ *and* $a_{j_2j_20}\alpha^2_{j_1j_1}(a_{j_2j_2l_2} - a_{j_2j_2l_1}) > 0$,

$$(\dot{x}^*_{j_1}, a_{j_2j_2l_2}) = \underbrace{(\mathrm{nF}, \mathrm{DI})}_{\text{decreasing-inflection flow } (-)}. \tag{4.111}$$

- *For* $a_{j_1j_10}(a_{j_2j_2l_2} - a_{j_1j_11}) > 0$ *and* $a_{j_2j_20}\alpha^2_{j_1j_1}(a_{j_2j_2l_2} - a_{j_2j_2l_1}) < 0$,

$$(\dot{x}^*_{j_1}, a_{j_2j_2l_2}) = \underbrace{(\mathrm{pF}, \mathrm{DI})}_{\text{decreasing-inflection flow } (+)}. \tag{4.112}$$

- *For* $a_{j_1j_10}(a_{j_2j_2l_2} - a_{j_1j_11}) < 0$ *and* $a_{j_2j_20}\alpha^2_{j_1j_1}(a_{j_2j_2l_2} - a_{j_2j_2l_1}) < 0$,

$$(\dot{x}^*_{j_1}, a_{j_2j_2l_2}) = \underbrace{(\mathrm{nF}, \mathrm{II})}_{\text{increasing-inflection flow } (-)}. \tag{4.113}$$

(iv$_2$) For $a_{j_2j_2l_1} = a_{j_2j_2l_2} = a_{j_2j_21}$,

$$\begin{aligned}\dot{x}_{j_1} &= a_{j_1j_10}(y_{j_1} - a_{j_1j_11})[(y_{j_1} - a_{j_1j_1})^2 + b_{j_1j_1}],\\ \dot{x}_{j_2} &= a_{j_2j_20}(y_{j_1} - a_{j_2j_21})^3.\end{aligned} \tag{4.114}$$

For $a_{j_1j_11} \neq a_{j_2j_21}$, *there is a dynamical system in the* y_{j_1}*-direction as*

$$\begin{aligned}\dot{y}_{j_1} = {} & \alpha_{j_1j_1}a_{j_1j_10}(y_{j_1} - a_{j_1j_11})[(y_{j_1} - a_{j_1j_1})^2 + b_{j_1j_1}]\\ & + \alpha_{j_1j_2}a_{j_2j_20}(y_{j_1} - a_{j_2j_21})^3,\end{aligned} \tag{4.115}$$

with a dynamical system in the $y^{\perp}_{j_1}$*-direction as*

$$\begin{aligned}\dot{y}^{\perp}_{j_1} = {} & \beta_{j_1j_1}a_{j_1j_10}(y_{j_1} - a_{j_1j_11})[(y_{j_1} - a_{j_1j_1})^2 + b_{j_1j_1}]\\ & + \beta_{j_1j_2}a_{j_2j_20}(y_{j_1} - a_{j_2j_21})^3.\end{aligned} \tag{4.116}$$

(iv_{2a}) The flow of $y_1^* = a_{j_1j_11}$ with $\dot{x}_{j_2}^* = a_{j_2j_20}(a_{j_1j_11} - a_{j_2j_21})^3$ possesses the following properties:

- *For $\alpha_{j_1j_1}a_{j_1j_10} > 0$ and $a_{j_2j_20}(a_{j_1j_11} - a_{j_2j_21})^3 > 0$,*

$$(a_{j_1j_11}, \dot{x}_{j_2}^*) = \underbrace{(\mathrm{UP}, \mathrm{pF})}_{\text{up-parabola flow } (+)}. \tag{4.117}$$

- *For $\alpha_{j_1j_1}a_{j_1j_10} < 0$ and $a_{j_2j_20}(a_{j_1j_11} - a_{j_2j_21})^3 > 0$,*

$$(a_{j_1j_11}, \dot{x}_{j_2}^*) = \underbrace{(\mathrm{DP}, \mathrm{pF})}_{\text{down-parabola flow } (+)}. \tag{4.118}$$

- *For $\alpha_{j_1j_1}a_{j_1j_10} > 0$ and $a_{j_2j_20}(a_{j_1j_11} - a_{j_2j_21})^3 < 0$,*

$$(a_{j_1j_11}, \dot{x}_{j_2}^*) = \underbrace{(\mathrm{DP}, \mathrm{nF})}_{\text{down-parabola flow } (-)}. \tag{4.119}$$

- *For $\alpha_{j_1j_1}a_{j_1j_10} < 0$ and $a_{j_2j_20}(a_{j_1j_11} - a_{j_2j_21})^3 < 0$,*

$$(a_{j_1j_11}, \dot{x}_{j_2}^*) = \underbrace{(\mathrm{UP}, \mathrm{nF})}_{\text{up-parabola flow } (-)}. \tag{4.120}$$

(iv_{2b}) The flow of $y_1^* = a_{j_2j_21}$ with $\dot{x}_{j_1}^* = a_{j_1j_10}(a_{j_2j_21} - a_{j_1j_11})[(a_{j_2j_21} - a_{j_1j_1})^2 + b_{j_1j_1}]$ possesses the following properties:

- *For $a_{j_1j_10}(a_{j_2j_21} - a_{j_1j_11}) > 0$ and $a_{j_2j_20}\alpha_{j_1j_1}^3 > 0$,*

$$(\dot{x}_{j_1}^*, a_{j_2j_21}) = \underbrace{(\mathrm{pF}, 3^{\mathrm{rd}}\mathrm{UP})}_{\text{third-order up-parabola flow } (+)}. \tag{4.121}$$

- *For $a_{j_1j_10}(a_{j_2j_21} - a_{j_1j_11}) < 0$ and $a_{j_2j_20}\alpha_{j_1j_1}^3 > 0$,*

$$(\dot{x}_{j_1}^*, a_{j_2j_21}) = \underbrace{(\mathrm{nF}, 3^{\mathrm{rd}}\mathrm{DP})}_{\text{third-order down-parabola flow } (-)}. \tag{4.122}$$

- *For $a_{j_1j_10}(a_{j_2j_21} - a_{j_1j_11}) > 0$ and $a_{j_2j_20}\alpha_{j_1j_1}^3 < 0$,*

$$(\dot{x}_{j_1}^*, a_{j_2j_21}) = \underbrace{(\mathrm{pF}, 3^{\mathrm{rd}}\mathrm{DP})}_{\text{third-order down-parabola flow } (+)}. \tag{4.123}$$

- *For $a_{j_1j_10}(a_{j_2j_21} - a_{j_1j_11}) < 0$ and $a_{j_2j_20}\alpha_{j_1j_1}^3 < 0$,*

$$(\dot{x}_{j_1}^*, a_{j_2j_21}) = \underbrace{(\mathrm{nF}, 3^{\mathrm{rd}}\mathrm{UP})}_{\text{third-order up-parabola flow } (-)}. \tag{4.124}$$

(iv_3) For $a_{j_1j_1 1} = a_{j_2j_2l_1}$,

$$\begin{aligned}\dot{x}_{j_1} &= a_{j_1j_1 0}(y_{j_1} - a_{j_1j_1 1})[(y_{j_1} - a_{j_1j_1})^2 + b_{j_1j_1}],\\ \dot{x}_{j_2} &= a_{j_2j_2 0}(y_{j_1} - a_{j_2j_2l_1})(y_{j_1} - a_{j_2j_2l_2})^2.\end{aligned} \tag{4.125}$$

There is a dynamical system in the y_{j_1}-direction as

$$\begin{aligned}\dot{y}_{j_1} = (y_{j_1} - a_{j_1j_1 1})\{&\alpha_{j_1j_1}a_{j_1j_1 0}[(y_{j_1} - a_{j_1j_1})^2 + b_{j_1j_1}]\\ &+ \alpha_{j_1j_2}a_{j_2j_2 0}(y_{j_1} - a_{j_2j_2l_2})^2\},\end{aligned} \tag{4.126}$$

with a dynamical system in the $y_{j_1}^{\perp}$-direction as

$$\begin{aligned}\dot{y}_{j_1}^{\perp} = (y_{j_1} - a_{j_1j_1 1})\{&\beta_{j_1j_1}a_{j_1j_1 0}[(y_{j_1} - a_{j_1j_1})^2 + b_{j_1j_1}]\\ &+ \beta_{j_1j_2}a_{j_2j_2 0}(y_{j_1} - a_{j_2j_2l_2})^2]\},\end{aligned} \tag{4.127}$$

where

$$\begin{aligned}G_{j_1j_1 0} &= \frac{\partial}{\partial y_{j_1}}[\mathbf{n}^{\mathrm{T}} \cdot \mathbf{F}(y_{j_1}, \mathbf{p})]\Big|_{y_{j_1}^* = a_{j_1j_1 1} = a_{j_2j_2l_1}}\\ &= \alpha_{j_1j_1}a_{j_1j_1 0}[(a_{j_2j_2 1} - a_{j_1j_1})^2 + b_{j_1j_1}]\\ &\quad + \alpha_{j_1j_2}a_{j_2j_2 0}(a_{j_2j_2l_1} - a_{j_2j_2l_2})^2.\end{aligned} \tag{4.128}$$

The infinite-equilibrium flow of $y_1^* = a_{j_1j_1 1} = a_{j_2j_2 1}$ possesses the following properties:

- *For $G_{j_1j_1 0} > 0$ and $a_{j_2j_2 0}(a_{j_2j_2l_1} - a_{j_2j_2l_2})^2/a_{j_1j_1 0} > 0$,*

$$(a_{j_1j_1 1}, a_{j_2j_2l_1}) = \underbrace{(\mathrm{SO}, \mathrm{II})}_{\text{increasing-inflection source}}. \tag{4.129}$$

- *For $G_{j_1j_1 0} < 0$ and $a_{j_2j_2 0}(a_{j_2j_2l_1} - a_{j_2j_2l_2})^2/a_{j_1j_1 0} > 0$,*

$$(a_{j_1j_1 1}, a_{j_2j_2l_1}) = \underbrace{(\mathrm{SI}, \mathrm{II})}_{\text{increasing-inflection sink}}. \tag{4.130}$$

- *For $G_{j_1j_1 0} > 0$ and $a_{j_2j_2 0}(a_{j_2j_2l_1} - a_{j_2j_2l_2})^2/a_{j_1j_1 0} < 0$,*

$$(a_{j_1j_1 1}, a_{j_2j_2l_1}) = \underbrace{(\mathrm{SO}, \mathrm{DI})}_{\text{decreasing-inflection source}}. \tag{4.131}$$

- *For $G_{j_1j_1 0} < 0$ and $a_{j_2j_2 0}(a_{j_2j_2l_1} - a_{j_2j_2l_2})^2/a_{j_1j_1 0} < 0$,*

$$(a_{j_1j_1 1}, a_{j_2j_2l_1}) = \underbrace{(\mathrm{SI}, \mathrm{DI})}_{\text{decreasing-inflection sink}}. \tag{4.132}$$

(iv_4) For $a_{j_1j_1 1} = a_{j_2j_2l_2}$,

$$\begin{aligned}\dot{x}_{j_1} &= a_{j_1j_1 0}(y_{j_1} - a_{j_1j_1 1})[(y_{j_1} - a_{j_1j_1})^2 + b_{j_1j_1}],\\ \dot{x}_{j_2} &= a_{j_2j_2 0}(y_{j_1} - a_{j_2j_2l_1})(y_{j_1} - a_{j_2j_2l_2})^2.\end{aligned} \tag{4.133}$$

There is a dynamical system in the y_{j_1}-direction as

$$\begin{aligned}\dot{y}_{j_1} = (y_{j_1} - a_{j_1j_1 1})\{&\alpha_{j_1j_1}a_{j_1j_1 0}[(y_{j_1} - a_{j_1j_1})^2 + b_{j_1j_1}]\\ &+ \alpha_{j_1j_2}a_{j_2j_2 0}(y_{j_1} - a_{j_2j_2l_1})(y_{j_1} - a_{j_2j_2l_2})\},\end{aligned} \tag{4.134}$$

with a dynamical system in the $y_{j_1}^{\perp}$-direction as

$$\begin{aligned}\dot{y}_{j_1}^{\perp} = (y_{j_1} - a_{j_1j_1 1})\{&\beta_{j_1j_1}a_{j_1j_1 0}[(y_{j_1} - a_{j_1j_1})^2 + b_{j_1j_1}]\\ &+ \beta_{j_1j_2}a_{j_2j_2 0}(y_{j_1} - a_{j_2j_2l_1})(y_{j_1} - a_{j_2j_2l_2})]\},\end{aligned} \tag{4.135}$$

where

$$\begin{aligned}G_{j_1j_1 0} &= \frac{\partial}{\partial y_{j_1}}[\mathbf{n}^{\mathrm{T}} \cdot \mathbf{F}(y_{j_1}, \mathbf{p})]\bigg|_{y_{j_1}^* = a_{j_1j_1 1} = a_{j_2j_2 1}}\\ &= \alpha_{j_1j_1}a_{j_1j_1 0}[(a_{j_1j_1 1} - a_{j_1j_1})^2 + b_{j_1j_1}].\end{aligned} \tag{4.136}$$

The infinite-equilibrium flow of $y_1^* = a_{j_1j_1 1} = a_{j_2j_2l_2}$ possesses the following properties:

- *For $\alpha_{j_1j_1}a_{j_1j_1 0} > 0$ and $a_{j_2j_2 0}\alpha_{j_1j_1}^2(a_{j_2j_2l_2} - a_{j_2j_2l_1}) > 0$,*

$$(a_{j_1j_1 1}, a_{j_2j_2l_2}) = \underbrace{(\mathrm{SO}, \mathrm{UP})}_{\text{up-parabola source}}. \tag{4.137}$$

- *For $\alpha_{j_1j_1}a_{j_1j_1 0} < 0$ and $a_{j_2j_2 0}\alpha_{j_1j_1}^2(a_{j_2j_2l_2} - a_{j_2j_2l_1}) > 0$,*

$$(a_{j_1j_1 1}, a_{j_2j_2l_2}) = \underbrace{(\mathrm{SI}, \mathrm{DP})}_{\text{down-parabola sink}}. \tag{4.138}$$

- *For $\alpha_{j_1j_1}a_{j_1j_1 0} > 0$ and $a_{j_2j_2 0}\alpha_{j_1j_1}^2(a_{j_2j_2l_2} - a_{j_2j_2l_1}) < 0$,*

$$(a_{j_1j_1 1}, a_{j_2j_2l_2}) = \underbrace{(\mathrm{SO}, \mathrm{DP})}_{\text{down-parabola source}}. \tag{4.139}$$

- *For $\alpha_{j_1j_1}a_{j_1j_1 0} < 0$ and $a_{j_2j_2 0}\alpha_{j_1j_1}^2(a_{j_2j_2l_2} - a_{j_2j_2l_1}) < 0$,*

$$(a_{j_1j_1 1}, a_{j_2j_2l_2}) = \underbrace{(\mathrm{SI}, \mathrm{UP})}_{\text{up-parabola sink}}. \tag{4.140}$$

(iv_5) For $a_{j_1j_11} = a_{j_2j_21}$,

$$\begin{aligned}\dot{x}_{j_1} &= a_{j_1j_10}(y_{j_1} - a_{j_1j_11})[(y_{j_1} - a_{j_1j_1})^2 + b_{j_1j_1}],\\ \dot{x}_{j_2} &= a_{j_2j_20}(y_{j_1} - a_{j_2j_21})^3.\end{aligned} \tag{4.141}$$

There is a dynamical system in the y_{j_1}-direction as

$$\begin{aligned}\dot{y}_{j_1} = (y_{j_1} - a_{j_1j_11})\{&\alpha_{j_1j_1}a_{j_1j_10}[(y_{j_1} - a_{j_1j_1})^2 + b_{j_1j_1}]\\ &+ \alpha_{j_1j_2}a_{j_2j_20}(y_{j_1} - a_{j_2j_21})^2\},\end{aligned} \tag{4.142}$$

with a dynamical system in the $y_{j_1}^{\perp}$-direction as

$$\begin{aligned}\dot{y}_{j_1}^{\perp} = (y_{j_1} - a_{j_1j_11})\{&\beta_{j_1j_1}a_{j_1j_10}[(y_{j_1} - a_{j_1j_1})^2 + b_{j_1j_1}]\\ &+ \beta_{j_1j_2}a_{j_2j_20}(y_{j_1} - a_{j_2j_21})^2]\},\end{aligned} \tag{4.143}$$

where

$$\begin{aligned}G_{j_1j_10} &= \frac{\partial}{\partial y_{j_1}}[\mathbf{n}^{\mathrm{T}} \cdot \mathbf{F}(y_{j_1}, \mathbf{p})]\bigg|_{y_{j_1}^* = a_{j_1j_11} = a_{j_2j_21}}\\ &= \alpha_{j_1j_1}a_{j_1j_10}[(a_{j_2j_21} - a_{j_1j_1})^2 + b_{j_1j_1}].\end{aligned} \tag{4.144}$$

The infinite-equilibrium flow of $y_1^* = a_{j_1j_11} = a_{j_2j_2l_2}$ possesses the following properties:

- *For $\alpha_{j_1j_1}a_{j_1j_10} > 0$ and $a_{j_2j_20}\alpha_{j_1j_1}^3 > 0$,*

$$(a_{j_1j_11}, a_{j_2j_21}) = \underbrace{(\mathrm{SO}, 2^{\mathrm{nd}}\mathrm{II})}_{\text{second-order increasing-infelction source}}. \tag{4.145}$$

- *For $\alpha_{j_1j_1}a_{j_1j_10} < 0$ and $a_{j_2j_20}\alpha_{j_1j_1}^3 > 0$,*

$$(a_{j_1j_11}, a_{j_2j_21}) = \underbrace{(\mathrm{SI}, 2^{\mathrm{nd}}\mathrm{DI})}_{\text{second-order decreasing-infelction sink}}. \tag{4.146}$$

- *For $\alpha_{j_1j_1}a_{j_1j_10} > 0$ and $a_{j_2j_20}\alpha_{j_1j_1}^3 < 0$,*

$$(a_{j_1j_11}, a_{j_2j_21}) = \underbrace{(\mathrm{SO}, 2^{\mathrm{nd}}\mathrm{DI})}_{\text{second-order decreasing-infelction source}}. \tag{4.147}$$

- *For $\alpha_{j_1j_1}a_{j_1j_10} < 0$ and $a_{j_2j_20}\alpha_{j_1j_1}^3 < 0$,*

$$(a_{j_1j_11}, a_{j_2j_21}) = \underbrace{(\mathrm{SI}, 2^{\mathrm{nd}}\mathrm{II})}_{\text{second-order increasing-infelction sink}}. \tag{4.148}$$

(*v*) *If*

$$\Delta_{j_1j_1} = 0 \quad \text{and} \quad \Delta_{j_2j_2} = 0, \tag{4.149}$$

then

$$\begin{aligned}
\dot{x}_{j_1} &= a_{j_1j_10}(y_{j_1} - a_{j_1j_1s_1})(y_{j_1} - a_{j_1j_1s_2})^2,\\
\dot{x}_{j_2} &= a_{j_2j_20}(y_{j_1} - a_{j_2j_2l_1})(y_{j_1} - a_{j_2j_2l_2})^2,\\
&j_1, j_2 \in \{1,2\};\ j_1 \neq j_2;
\end{aligned} \tag{4.150}$$

where

$$\begin{aligned}
&a_{j_1j_1l_1} = b_{j_1j_11}, \quad a_{j_2j_2l_2} = -\frac{1}{2}B_{j_2j_2}, \quad b_{j_2j_2} = 0, \quad a_{j_2j_20} = A_{j_2j_20};\\
&a_{j_1j_1s_1} = b_{j_1j_11}, \quad a_{j_1j_1s_2} = -\frac{1}{2}B_{j_1j_11}, \quad b_{j_1j_1} = 0, \quad a_{j_1j_10} = A_{j_1j_10}.
\end{aligned} \tag{4.151}$$

(v_1) *For* $a_{j_2j_2l_1}, a_{j_2j_2l_2} \neq a_{j_1j_1s_1}, a_{j_1j_1s_2}$, *there is a dynamical system in the* y_{j_1}*-direction as*

$$\begin{aligned}
\dot{y}_{j_1} = {}&\alpha_{j_1j_1}a_{j_1j_10}(y_{j_1} - a_{j_1j_1s_1})(y_{j_1} - a_{j_1j_1s_2})^2\\
&+ \alpha_{j_1j_2}a_{j_2j_20}(y_{j_1} - a_{j_2j_2l_1})(y_{j_1} - a_{j_2j_2l_2})^2,
\end{aligned} \tag{4.152}$$

with a dynamical system in the $y_{j_1}^{\perp}$*-direction as*

$$\begin{aligned}
\dot{y}_{j_1}^{\perp} = {}&\beta_{j_1j_1}a_{j_1j_10}(y_{j_1} - a_{j_1j_1s_1})(y_{j_1} - a_{j_1j_1s_2})^2\\
&+ \beta_{j_1j_2}a_{j_2j_20}(y_{j_1} - a_{j_2j_2l_1})(y_{j_1} - a_{j_2j_2l_2})^2.
\end{aligned} \tag{4.153}$$

(v_{1a}) *The flow of* $y_{j_1}^{*} = a_{j_1j_1s_1} \neq a_{j_2j_2l_1}, a_{j_2j_2l_2}$ *for the* x_{j_1}*-direction has the following properties with* $\dot{x}_{j_2}^{*} = a_{j_2j_20}(a_{j_1j_1s_1} - a_{j_2j_2l_1})(a_{j_1j_1s_1} - a_{j_2j_2l_2})^2$:

- *For* $a_{j_1j_10}\alpha_{j_1j_2}(a_{j_1j_1s_1} - a_{j_1j_1s_2})^2 > 0$ *and* $a_{j_2j_20}(a_{j_1j_1s_1} - a_{j_2j_2l_1}) \times (a_{j_1j_1s_1} - a_{j_2j_2l_2})^2 > 0$,

$$(a_{j_1j_1s_1}, \dot{x}_{j_2}^{*}) = \underbrace{(\mathrm{UP}, \mathrm{pF})}_{\text{up-parabola flow } (+)}. \tag{4.154}$$

- *For* $a_{j_1j_10}\alpha_{j_1j_2}(a_{j_1j_1s_1} - a_{j_1j_1s_2})^2 < 0$ *and* $a_{j_2j_20}(a_{j_1j_1s_1} - a_{j_2j_2l_1}) \times (a_{j_1j_1s_1} - a_{j_2j_2l_2})^2 > 0$,

$$(a_{j_1j_1s_1}, \dot{x}_{j_2}^{*}) = \underbrace{(\mathrm{DP}, \mathrm{pF})}_{\text{down-parabola flow } (+)}. \tag{4.155}$$

- *For* $a_{j_1j_10}\alpha_{j_1j_2}(a_{j_1j_1s_1} - a_{j_1j_1s_2})^2 > 0$ *and* $a_{j_2j_20}(a_{j_1j_1s_1} - a_{j_2j_2l_1}) \times (a_{j_1j_1s_1} - a_{j_2j_2l_2})^2 < 0$,

$$(a_{j_1j_1s_1}, \dot{x}^*_{j_2}) = \underbrace{(\mathrm{DP}, \mathrm{nF})}_{\text{down-parabola flow } (-)}. \tag{4.156}$$

- *For* $a_{j_1j_10}\alpha_{j_1j_2}(a_{j_1j_1s_1} - a_{j_1j_1s_2})^2 < 0$ *and* $a_{j_2j_20}(a_{j_1j_1s_1} - a_{j_2j_2l_1}) \times (a_{j_1j_1s_1} - a_{j_2j_2l_2})^2 > 0$,

$$(a_{j_1j_1s_1}, \dot{x}^*_{j_2}) = \underbrace{(\mathrm{UP}, \mathrm{nF})}_{\text{up-parabola flow } (-)}. \tag{4.157}$$

(v_{1b}) *The flow of* $y^*_{j_1} = a_{j_1j_1s_2} \neq a_{j_2j_2l_1}, a_{j_2j_2l_2}$ *for the* x_{j_1}*-direction with* $a_{j_1j_11} \neq a_{j_2j_21}$ *has the following properties with* $\dot{x}^*_{j_2} = a_{j_2j_20}(a_{j_1j_1s_2} - a_{j_2j_2l_1})(a_{j_1j_1s_2} - a_{j_2j_2l_2})^2$:

- *For* $a_{j_1j_10}\alpha^2_{j_1j_2}(a_{j_1j_1s_2} - a_{j_1j_1s_1}) > 0$ *and* $a_{j_2j_20}(a_{j_1j_1s_2} - a_{j_2j_2l_1}) \times (a_{j_1j_1s_2} - a_{j_2j_2l_2})^2 > 0$,

$$(a_{j_1j_1s_2}, \dot{x}^*_{j_2}) = \underbrace{(\mathrm{II}, \mathrm{pF})}_{\text{increasing-inflection flow}(+)}. \tag{4.158}$$

- *For* $a_{j_1j_10}\alpha^2_{j_1j_2}(a_{j_1j_1s_2} - a_{j_1j_1s_1}) > 0$ *and* $a_{j_2j_20}(a_{j_1j_1s_2} - a_{j_2j_2l_1}) \times (a_{j_1j_1s_2} - a_{j_2j_2l_2})^2 > 0$,

$$(a_{j_1j_1s_2}, \dot{x}^*_{j_2}) = \underbrace{(\mathrm{DI}, \mathrm{pF})}_{\text{decreasing-inflection flow } (+)}. \tag{4.159}$$

- *For* $a_{j_1j_10}\alpha^2_{j_1j_2}(a_{j_1j_1s_2} - a_{j_1j_1s_1}) > 0$ *and* $a_{j_2j_20}(a_{j_1j_1s_2} - a_{j_2j_2l_1}) \times (a_{j_1j_1s_2} - a_{j_2j_2l_2})^2 < 0$,

$$(a_{j_1j_1s_2}, \dot{x}^*_{j_2}) = \underbrace{(\mathrm{DI}, \mathrm{nF})}_{\text{decreasing-inflection flow } (-)}. \tag{4.160}$$

- *For* $a_{j_1j_10}\alpha^2_{j_1j_2}(a_{j_1j_1s_2} - a_{j_1j_1s_1}) < 0$ *and* $a_{j_2j_20}(a_{j_1j_1s_2} - a_{j_2j_2l_1}) \times (a_{j_1j_1s_2} - a_{j_2j_2l_2})^2 < 0$,

$$(a_{j_1j_1s_2}, \dot{x}^*_{j_2}) = \underbrace{(\mathrm{II}, \mathrm{nF})}_{\text{increasing-inflection flow } (-)}. \tag{4.161}$$

The increasing- and decreasing-inflection flows are the appearing bifurcations for the up-parabola and down-parabola flows.

(v_{1c}) *The flow of* $y_{j_1}^* = a_{j_2j_2l_1} \neq a_{j_1j_1s_1}, a_{j_1j_1s_2}$ *for the* x_{j_2}*-direction has the following properties with* $\dot{x}_{j_1}^* = a_{j_1j_10}(a_{j_2j_2l_1} - a_{j_1j_1s_1})(a_{j_2j_2l_1} - a_{j_1j_1s_2})^2$:

- *For* $a_{j_1j_10}(a_{j_2j_2l_1} - a_{j_1j_1s_1})(a_{j_2j_2l_1} - a_{j_1j_1s_2})^2 > 0$ *and* $a_{j_2j_20}\alpha_{j_1j_1} \times (a_{j_2j_2l_1} - a_{j_1j_1l_2})^2 > 0$,

$$(\dot{x}_{j_1}^*, a_{j_2j_2l_1}) = \underbrace{(\mathrm{pF}, \mathrm{UP})}_{\text{up-parabola flow } (+)}. \tag{4.162}$$

- *For* $a_{j_1j_10}(a_{j_2j_2l_1} - a_{j_1j_1s_1})(a_{j_2j_2l_1} - a_{j_1j_1s_2})^2 < 0$ *and* $a_{j_2j_20}\alpha_{j_1j_1} \times (a_{j_2j_2l_1} - a_{j_1j_1l_2})^2 > 0$,

$$(\dot{x}_{j_1}^*, a_{j_2j_2l_1}) = \underbrace{(\mathrm{nF}, \mathrm{DP})}_{\text{down-parabola flow } (-)}. \tag{4.163}$$

- *For* $a_{j_1j_10}(a_{j_2j_2l_1} - a_{j_1j_1s_1})(a_{j_2j_2l_1} - a_{j_1j_1s_2})^2 > 0$ *and* $a_{j_2j_20}\alpha_{j_1j_1} \times (a_{j_2j_2l_1} - a_{j_1j_1l_2})^2 < 0$,

$$(\dot{x}_{j_1}^*, a_{j_2j_2l_1}) = \underbrace{(\mathrm{pF}, \mathrm{DP})}_{\text{down-parabola flow } (+)}. \tag{4.164}$$

- *For* $a_{j_1j_10}(a_{j_2j_2l_1} - a_{j_1j_1s_1})(a_{j_2j_2l_1} - a_{j_1j_1s_2})^2 < 0$ *and* $a_{j_2j_20}\alpha_{j_1j_1} \times (a_{j_2j_2l_1} - a_{j_1j_1l_2})^2 < 0$,

$$(\dot{x}_{j_1}^*, a_{j_2j_2l_1}) = \underbrace{(\mathrm{nF}, \mathrm{UP})}_{\text{up-parabola flow } (-)}. \tag{4.165}$$

(v_{1d}) *The flow of* $y_{j_1}^* = a_{j_2j_2l_2} \neq a_{j_1j_1s_1}, a_{j_1j_1s_2}$ *for the* x_{j_2}*-direction has the following properties with* $\dot{x}_{j_1}^* = a_{j_1j_10}(a_{j_2j_2l_2} - a_{j_1j_1s_1})(a_{j_2j_2l_2} - a_{j_1j_1s_2})^2$:

- *For* $a_{j_1j_10}(a_{j_2j_2l_2} - a_{j_1j_1s_1})(a_{j_2j_2l_2} - a_{j_1j_1s_2})^2 > 0$ *and* $a_{j_2j_20}\alpha_{j_1j_1}^2 \times (a_{j_2j_2l_2} - a_{j_1j_1l_1}) > 0$,

$$(\dot{x}_{j_1}^*, a_{j_2j_21}) = \underbrace{(\mathrm{pF}, \mathrm{II})}_{\text{increasing-infelction flow } (+)}. \tag{4.166}$$

- *For* $a_{j_1j_10}(a_{j_2j_2l_2} - a_{j_1j_1s_1})(a_{j_2j_2l_2} - a_{j_1j_1s_2})^2 < 0$ *and* $a_{j_2j_20}\alpha_{j_1j_1}^2 \times (a_{j_1j_1l_2} - a_{j_2j_2l_1}) > 0$,

$$(\dot{x}_{j_1}^*, a_{j_2j_21}) = \underbrace{(\mathrm{nF}, \mathrm{DI})}_{\text{decreasing-infelction flow}(-)}. \tag{4.167}$$

- *For* $a_{j_1j_10}(a_{j_2j_2l_1} - a_{j_1j_1s_1})(a_{j_2j_2l_2} - a_{j_1j_1s_2})^2 > 0$ *and* $a_{j_2j_20}\alpha_{j_1j_1}^2 \times (a_{j_1j_1l_2} - a_{j_2j_2l_1}) < 0$,

$$(\dot{x}_{j_1}^*, a_{j_2j_21}) = \underbrace{(\mathrm{pF}, \mathrm{DI})}_{\text{decreasing-infelction flow } (+)}. \tag{4.168}$$

- *For* $a_{j_1j_10}(a_{j_2j_2l_1} - a_{j_1j_1s_1})(a_{j_2j_2l_2} - a_{j_1j_1s_2})^2 < 0$ *and* $a_{j_2j_20}\alpha_{j_1j_1}^2 \times (a_{j_1j_1l_2} - a_{j_2j_2l_1}) < 0$,

$$(\dot{x}_{j_1}^*, a_{j_2j_21}) = \underbrace{(\text{nF}, \text{II})}_{\text{increasing-infelction flow } (-)}. \tag{4.169}$$

– *The increasing- and decreasing-inflection flows are the appearing bifurcations for the up-parabola and down-parabola flows.*

(v_2) *For* $a_{j_1j_1s_1} = a_{j_1j_1s_2} \equiv a_{j_1j_11}$,

$$\begin{aligned} \dot{x}_{j_1} &= a_{j_1j_10}(y_{j_1} - a_{j_1j_11})^3, \\ \dot{x}_{j_2} &= a_{j_2j_20}(y_{j_1} - a_{j_2j_2l_1})(y_{j_1} - a_{j_2j_2l_2})^2, \\ &j_1, j_2 \in \{1, 2\}; j_1 \neq j_2. \end{aligned} \tag{4.170}$$

For $a_{j_2j_2l_1}, a_{j_2j_2l_2} \neq a_{j_1j_11}$, *there is a dynamical system in the* y_{j_1}*-direction as*

$$\begin{aligned} \dot{y}_{j_1} = {}& \alpha_{j_1j_1}a_{j_1j_10}(y_{j_1} - a_{j_1j_11})^3 + \alpha_{j_1j_2}a_{j_2j_20} \\ & \times (y_{j_1} - a_{j_2j_2l_1})(y_{j_1} - a_{j_2j_2l_2})^2, \end{aligned} \tag{4.171}$$

with a dynamical system in the $y_{j_1}^{\perp}$*-direction as*

$$\begin{aligned} \dot{y}_{j_1}^{\perp} = {}& \beta_{j_1j_1}a_{j_1j_10}(y_{j_1} - a_{j_1j_11})^3 + \beta_{j_1j_2}a_{j_2j_20} \\ & \times (y_{j_1} - a_{j_2j_2l_1})(y_{j_1} - a_{j_2j_2l_2})^2. \end{aligned} \tag{4.172}$$

(v_{2a}) *The flow of* $y_{j_1}^* = a_{j_1j_11} \neq a_{j_2j_2l_1}, a_{j_2j_2l_2}$ *for the* x_{j_1}*-direction has the following properties with* $\dot{x}_{j_2}^* = a_{j_2j_20}(a_{j_1j_11} - a_{j_2j_2l_1})(a_{j_1j_11} - a_{j_2j_2l_2})^2$:

- *For* $a_{j_1j_10}\alpha_{j_1j_2}^3 > 0$ *and* $a_{j_2j_20}(a_{j_1j_11} - a_{j_2j_2l_1})(a_{j_1j_11} - a_{j_2j_2l_2})^2 > 0$,

$$(a_{j_1j_11}, \dot{x}_{j_2}^*) = \underbrace{(3^{\text{rd}}\text{UP}, \text{pF})}_{\text{third-order up-parabola flow } (+)}. \tag{4.173}$$

- *For* $a_{j_1j_10}\alpha_{j_1j_2}^3 < 0$ *and* $a_{j_2j_20}(a_{j_1j_11} - a_{j_2j_2l_1})(a_{j_1j_11} - a_{j_2j_2l_2})^2 > 0$,

$$(a_{j_1j_11}, \dot{x}_{j_2}^*) = \underbrace{(3^{\text{rd}}\text{DP}, \text{pF})}_{\text{third-order down-parabola flow } (+)}. \tag{4.174}$$

- *For* $a_{j_1j_10}\alpha_{j_1j_2}^3 > 0$ *and* $a_{j_2j_20}(a_{j_1j_11} - a_{j_2j_2l_1})(a_{j_1j_11} - a_{j_2j_2l_2})^2 < 0$,

$$(a_{j_1j_11}, \dot{x}_{j_2}^*) = \underbrace{(3^{\text{rd}}\text{DP}, \text{nF})}_{\text{third-order down-parabola flow } (-)}. \tag{4.175}$$

- *For* $a_{j_1j_10}\alpha^3_{j_1j_2} < 0$ *and* $a_{j_2j_20}(a_{j_1j_11} - a_{j_2j_2l_1})(a_{j_1j_11} - a_{j_2j_2l_2})^2 < 0$,

$$(a_{j_1j_11}, \dot{x}^*_{j_2}) = \underbrace{(3^{\text{rd}}\text{UP}, \text{nF})}_{\text{third-order up-parabola flow }(-)}. \tag{4.176}$$

– *The third-order up-parabola flows are the appearing bifurcations for up-parabola, down-parabola, and up-parabola flows.*
– *The third-order down-parabola flows are the appearing bifurcations for down-parabola, up-parabola, and down-parabola flows.*

(v_{2b}) *The flow of* $y^*_{j_1} = a_{j_2j_2l_1} \neq a_{j_1j_11}, a_{j_2j_2l_2}$ *for the* x_{j_2}*-direction has the following properties with* $\dot{x}^*_{j_1} = a_{j_1j_10}(a_{j_2j_2l_1} - a_{j_1j_11})^3$:

- *For* $a_{j_1j_10}(a_{j_2j_2l_1} - a_{j_1j_11})^3 > 0$ *and* $a_{j_2j_20}\alpha_{j_1j_1}(a_{j_2j_2l_1} - a_{j_2j_2l_2})^2 > 0$,

$$(\dot{x}^*_{j_1}, a_{j_2j_2l_1}) = \underbrace{(\text{pF}, \text{UP})}_{\text{up-parabola flow }(+)}. \tag{4.177}$$

- *For* $a_{j_1j_10}(a_{j_2j_2l_1} - a_{j_1j_11})^3 < 0$ *and* $a_{j_2j_20}\alpha_{j_1j_1}(a_{j_2j_2l_1} - a_{j_2j_2l_2})^2 > 0$,

$$(\dot{x}^*_{j_1}, a_{j_2j_2l_1}) = \underbrace{(\text{nF}, \text{DP})}_{\text{down-parabola flow }(-)}. \tag{4.178}$$

- *For* $a_{j_1j_10}(a_{j_2j_2l_1} - a_{j_1j_11})^3 > 0$ *and* $a_{j_2j_20}\alpha_{j_1j_1}(a_{j_2j_2l_1} - a_{j_2j_2l_2})^2 < 0$,

$$(\dot{x}^*_{j_1}, a_{j_2j_2l_1}) = \underbrace{(\text{pF}, \text{DP})}_{\text{down-parabola flow }(+)}. \tag{4.179}$$

- *For* $a_{j_1j_10}(a_{j_2j_2l_1} - a_{j_1j_11})^3 < 0$ *and* $a_{j_2j_20}\alpha_{j_1j_1}(a_{j_2j_2l_1} - a_{j_2j_2l_2})^2 < 0$,

$$(\dot{x}^*_{j_1}, a_{j_2j_2l_1}) = \underbrace{(\text{nF}, \text{UP})}_{\text{up-parabola flow }(+)}. \tag{4.180}$$

(v_{2c}) *The flow of* $y^*_{j_1} = a_{j_2j_2l_2} \neq a_{j_1j_11}$ *for the* x_{j_2}*-direction has the following properties with* $\dot{x}^*_{j_1} = a_{j_1j_10}(a_{j_2j_2l_2} - a_{j_1j_11})^3$:

- *For* $a_{j_1j_10}(a_{j_2j_2l_2} - a_{j_1j_11})^3 > 0$ *and* $a_{j_2j_20}\alpha^2_{j_1j_1}(a_{j_2j_2l_2} - a_{j_2j_2l_1}) > 0$,

$$(\dot{x}^*_{j_1}, a_{j_2j_2l_2}) = \underbrace{(\text{pF}, \text{II})}_{\text{increasing-inflection flow }(+)}. \tag{4.181}$$

- *For* $a_{j_1j_10}(a_{j_2j_2l_2} - a_{j_1j_11})^3 < 0$ *and* $a_{j_2j_20}\alpha^2_{j_1j_1}(a_{j_2j_2l_2} - a_{j_2j_2l_1}) > 0$,

$$(\dot{x}^*_{j_1}, a_{j_2j_2l_2}) = \underbrace{(\text{nF}, \text{DI})}_{\text{decreasing-inflection flow }(-)}. \tag{4.182}$$

- *For* $a_{j_1j_10}(a_{j_2j_2l_2}-a_{j_1j_11})^3>0$ *and* $a_{j_2j_20}\alpha^2_{j_1j_1}(a_{j_2j_2l_2}-a_{j_2j_2l_1})<0$,

$$(\dot{x}^*_{j_1},a_{j_2j_2l_2})=\underbrace{(\mathrm{pF},\mathrm{DI})}_{\text{decreasing-inflection flow }(+)}. \tag{4.183}$$

- *For* $a_{j_1j_10}(a_{j_2j_2l_2}-a_{j_1j_11})^3<0$ *and* $a_{j_2j_20}\alpha^2_{j_1j_1}(a_{j_2j_2l_2}-a_{j_2j_2l_1})<0$,

$$(\dot{x}^*_{j_1},a_{j_2j_2l_2})=\underbrace{(\mathrm{nF},\mathrm{II})}_{\text{increasing-inflection flow }(-)}. \tag{4.184}$$

– *The increasing- and decreasing-inflection flows are the appearing bifurcations for the up-parabola and down-parabola flows.*

(v_3) *For* $a_{j_2j_2l_1}=a_{j_2j_2l_2}\equiv a_{j_2j_21}$,

$$\begin{aligned}
&\dot{x}_{j_1}=a_{j_1j_10}(y_{j_1}-a_{j_1j_1s_1})(y_{j_1}-a_{j_1j_1s_2})^2,\\
&\dot{x}_{j_2}=a_{j_2j_20}(y_{j_1}-a_{j_2j_21})^3,\\
&j_1,j_2\in\{1,2\};\ j_1\neq j_2.
\end{aligned} \tag{4.185}$$

For $a_{j_1j_1s_1},a_{j_1j_1s_2}\neq a_{j_2j_21}$, *there is a dynamical system in the* y_{j_1}*-direction as*

$$\begin{aligned}
\dot{y}_{j_1}&=\alpha_{j_1j_1}a_{j_1j_10}(y_{j_1}-a_{j_1j_1s_1})(y_{j_1}-a_{j_1j_1s_2})^2\\
&\quad+\alpha_{j_1j_2}a_{j_2j_20}(y_{j_1}-a_{j_2j_21})^3,
\end{aligned} \tag{4.186}$$

with a dynamical system in the $y^{\perp}_{j_1}$ *-direction as*

$$\begin{aligned}
\dot{y}^{\perp}_{j_1}&=\beta_{j_1j_1}a_{j_1j_10}(y_{j_1}-a_{j_1j_1s_1})(y_{j_1}-a_{j_1j_1s_2})^2\\
&\quad+\beta_{j_1j_2}a_{j_2j_20}(y_{j_1}-a_{j_2j_21})^3.
\end{aligned} \tag{4.187}$$

(v_{3a}) *The flow of* $y^*_{j_1}=a_{j_1j_1s_1}\neq a_{j_2j_21}$ *with* $\dot{x}^*_{j_2}=a_{j_2j_20}(a_{j_1j_1s_1}-a_{j_2j_21})^3$ *for the* x_{j_1}*-direction has the following properties*:

- *For* $a_{j_1j_10}\alpha_{j_1j_2}(a_{j_1j_1s_1}-a_{j_1j_1s_2})^2>0$ *and* $a_{j_2j_20}(a_{j_1j_1s_1}-a_{j_2j_21})^3>0$,

$$(a_{j_1j_1s_1},\dot{x}^*_{j_2})=\underbrace{(\mathrm{UP},\mathrm{pF})}_{\text{up-parabola flow }(+)}. \tag{4.188}$$

- *For* $a_{j_1j_10}\alpha_{j_1j_2}(a_{j_1j_1s_1}-a_{j_1j_1s_2})^2<0$ *and* $a_{j_2j_20}(a_{j_1j_1s_1}-a_{j_2j_21})^3>0$,

$$(a_{j_1j_1s_1},\dot{x}^*_{j_2})=\underbrace{(\mathrm{DP},\mathrm{pF})}_{\text{down-parabola flow }(+)}. \tag{4.189}$$

- *For* $a_{j_1j_10}\alpha_{j_1j_2}(a_{j_1j_1s_1} - a_{j_1j_1s_2})^2 > 0$ *and* $a_{j_2j_20}(a_{j_1j_1s_1} - a_{j_2j_21})^3 < 0$,

$$(a_{j_1j_1s_1}, \dot{x}^*_{j_2}) = \underbrace{(\mathrm{DP}, \mathrm{nF})}_{\text{down-parabola flow } (-)}. \tag{4.190}$$

- *For* $a_{j_1j_10}\alpha_{j_1j_2}(a_{j_1j_1s_1} - a_{j_1j_1s_2})^2 < 0$ *and* $a_{j_2j_20}(a_{j_1j_1s_1} - a_{j_2j_21})^3 < 0$,

$$(a_{j_1j_1s_1}, \dot{x}^*_{j_2}) = \underbrace{(\mathrm{UP}, \mathrm{nF})}_{\text{third-order up-parabola flow } (-)}. \tag{4.191}$$

(v_{3b}) *The flow of* $y^*_{j_1} = a_{j_1j_1s_2} \neq a_{j_2j_21}$ *for the* x_{j_2}*-direction has the following properties with* $\dot{x}^*_{j_2} = a_{j_2j_20}(a_{j_1j_1s_1} - a_{j_2j_21})^3$:

- *For* $a_{j_1j_10}\alpha^2_{j_1j_2}(a_{j_1j_1s_2} - a_{j_1j_1s_1}) > 0$ *and* $a_{j_2j_20}(a_{j_1j_1s_2} - a_{j_2j_21})^3 > 0$,

$$(a_{j_1j_1s_2}, \dot{x}^*_{j_2}) = \underbrace{(\mathrm{II}, \mathrm{pF})}_{\text{increasing-inflection flow } (+)}. \tag{4.192}$$

- *For* $a_{j_1j_10}\alpha^2_{j_1j_2}(a_{j_1j_1s_2} - a_{j_1j_1s_1}) < 0$ *and* $a_{j_2j_20}(a_{j_1j_1s_2} - a_{j_2j_21})^3 > 0$,

$$(a_{j_1j_1s_2}, \dot{x}^*_{j_2}) = \underbrace{(\mathrm{DI}, \mathrm{pF})}_{\text{decreasing-inflection flow } (+)}. \tag{4.193}$$

- *For* $a_{j_1j_10}\alpha^2_{j_1j_2}(a_{j_1j_1s_2} - a_{j_1j_1s_1}) > 0$ *and* $a_{j_2j_20}(a_{j_1j_1s_2} - a_{j_2j_21})^3 < 0$,

$$(a_{j_1j_1s_2}, \dot{x}^*_{j_2}) = \underbrace{(\mathrm{DI}, \mathrm{nF})}_{\text{decreasing-inflection flow } (-)}. \tag{4.194}$$

- *For* $a_{j_1j_10}\alpha^2_{j_1j_2}(a_{j_1j_1s_2} - a_{j_1j_1s_1}) < 0$ *and* $a_{j_2j_20}(a_{j_1j_1s_2} - a_{j_2j_21})^3 < 0$,

$$(a_{j_1j_1s_2}, \dot{x}^*_{j_2}) = \underbrace{(\mathrm{II}, \mathrm{nF})}_{\text{increasing-inflection flow } (-)}. \tag{4.195}$$

- *The inflection flows are the appearing bifurcations for up-parabola and down-parabola flows.*

(v_{3c}) *The flow of* $y^*_{j_1} = a_{j_2j_21} \neq a_{j_1j_1s_1}, a_{j_1j_1s_2}$ *for the* x_{j_2}*-direction has the following properties with* $\dot{x}^*_{j_1} = a_{j_1j_10}(a_{j_2j_21} - a_{j_1j_1s_1})(a_{j_2j_21} - a_{j_1j_1s_2})^2$:

- *For* $a_{j_1j_10}(a_{j_2j_21} - a_{j_1j_1s_1})(a_{j_2j_21} - a_{j_1j_1s_2})^2 > 0$ *and* $a_{j_2j_20}\alpha^3_{j_1j_1} > 0$,

$$(\dot{x}^*_{j_1}, a_{j_2j_21}) = \underbrace{(\mathrm{pF}, 3^{\mathrm{rd}}\mathrm{UP})}_{\text{third-order up-parabola flow } (+)}. \tag{4.196}$$

- *For* $a_{j_1j_10}(a_{j_2j_21} - a_{j_1j_1s_1})(a_{j_2j_21} - a_{j_1j_1s_2})^2 < 0$ *and* $a_{j_2j_20}\alpha^3_{j_1j_1} > 0$,

$$(\dot{x}^*_{j_1}, a_{j_2j_21}) = \underbrace{(\mathrm{nF}, 3^{\mathrm{rd}}\mathrm{DP})}_{\text{third-order down-parabola flow } (-)}. \tag{4.197}$$

- *For* $a_{j_1j_10}(a_{j_2j_21} - a_{j_1j_1s_1})(a_{j_2j_21} - a_{j_1j_1s_2})^2 > 0$ *and* $a_{j_2j_20}\alpha^3_{j_1j_1} < 0$,

$$(\dot{x}^*_{j_1}, a_{j_2j_2l_1}) = \underbrace{(\mathrm{pF}, 3^{\mathrm{rd}}\mathrm{DP})}_{\text{third-order down-parabola flow } (+)}. \tag{4.198}$$

- *For* $a_{j_1j_10}(a_{j_2j_21} - a_{j_1j_1s_1})(a_{j_2j_21} - a_{j_1j_1s_2})^2 < 0$ *and* $a_{j_2j_20}\alpha^3_{j_1j_1} < 0$,

$$(\dot{x}^*_{j_1}, a_{j_2j_21}) = \underbrace{(\mathrm{nF}, 3^{\mathrm{rd}}\mathrm{UP})}_{\text{third-order up-parabola flow } (+)}. \tag{4.199}$$

– *The third-order up-parabola flows are the appearing bifurcations from an up-parabola flow to up-parabola, down-parabola, and up-parabola flows.*
– *The third-order down-parabola flows are the appearing bifurcations from a down-parabola flow to down-parabola, up-parabola, and down-parabola flows.*

(v_4) *For* $a_{j_1j_1s_1} = a_{j_1j_1s_2} \equiv a_{j_1j_11}$ *and* $a_{j_2j_2l_1} = a_{j_2j_2l_2} \equiv a_{j_2j_21}$,

$$\begin{aligned} &\dot{x}_{j_1} = a_{j_1j_10}(y_{j_1} - a_{j_1j_11})^3, \\ &\dot{x}_{j_2} = a_{j_2j_20}(y_{j_1} - a_{j_2j_21})^3, \\ &j_1, j_2 \in \{1, 2\};\ j_1 \neq j_2. \end{aligned} \tag{4.200}$$

For $a_{j_2j_21} \neq a_{j_1j_11}$, *there is a dynamical system in the* y_{j_1}*-direction as*

$$\dot{y}_{j_1} = \alpha_{j_1j_1}a_{j_1j_10}(y_{j_1} - a_{j_1j_11})^3 + \alpha_{j_1j_2}a_{j_2j_20}(y_{j_1} - a_{j_2j_21})^3, \tag{4.201}$$

with a dynamical system in the $y^{\perp}_{j_1}$*-direction as*

$$\dot{y}^{\perp}_{j_1} = \beta_{j_1j_1}a_{j_1j_10}(y_{j_1} - a_{j_1j_11})^3 + \beta_{j_1j_2}a_{j_2j_20}(y_{j_1} - a_{j_2j_21})^3. \tag{4.202}$$

(v_{4a}) *The flow of* $y^*_{j_1} = a_{j_1j_11} \neq a_{j_2j_21}$ *with* $\dot{x}^*_{j_2} = a_{j_2j_20}(a_{j_1j_11} - a_{j_2j_21})^2$ *for the* x_{j_1}*-direction has the following properties*:

- *For* $a_{j_1j_10}\alpha^3_{j_1j_2} > 0$ *and* $a_{j_2j_20}(a_{j_1j_11} - a_{j_2j_2l_2})^3 > 0$,

$$(a_{j_1j_11}, \dot{x}^*_{j_2}) = \underbrace{(3^{\mathrm{rd}}\mathrm{UP}, \mathrm{pF})}_{\text{third-order up-parabola flow } (+)}. \tag{4.203}$$

- *For* $a_{j_1j_10}\alpha^3_{j_1j_2} < 0$ *and* $a_{j_2j_20}(a_{j_1j_11} - a_{j_2j_21})^3 > 0$,

$$(a_{j_1j_11}, \dot{x}^*_{j_2}) = \underbrace{(3^{\mathrm{rd}}\mathrm{DP}, \mathrm{pF})}_{\text{third-order down-parabola flow } (+)}. \tag{4.204}$$

- *For* $a_{j_1j_10}\alpha^3_{j_1j_2} > 0$ *and* $a_{j_2j_20}(a_{j_1j_11} - a_{j_2j_21})^3 < 0$,

$$(a_{j_1j_11}, \dot{x}^*_{j_2}) = \underbrace{(3^{\text{rd}}\text{DP}, \text{nF})}_{\text{third-order down-parabola flow }(-)}. \quad (4.205)$$

- *For* $a_{j_1j_10}\alpha^3_{j_1j_2} < 0$ *and* $a_{j_2j_20}(a_{j_1j_11} - a_{j_2j_21})^3 < 0$,

$$(a_{j_1j_11}, \dot{x}^*_{j_2}) = \underbrace{(3^{\text{rd}}\text{UP}, \text{nF})}_{\text{third-order up-parabola flow }(-)}. \quad (4.206)$$

– *The third-order up-parabola flows are the appearing bifurcations for up-parabola, down-parabola, and up-parabola flows.*
– *The third-order down-parabola flows are the appearing bifurcations for down-parabola, up-parabola, and down-parabola flows.*

(v_{4b}) *The flow of* $y^*_{j_1} = a_{j_2j_21} \neq a_{j_1j_11}$ *for the* x_{j_2}*-direction has the following properties with* $\dot{x}^*_{j_1} = a_{j_1j_10}(a_{j_2j_21} - a_{j_1j_11})^3$:

- *For* $a_{j_1j_10}(a_{j_2j_21} - a_{j_1j_11})^3 > 0$ *and* $a_{j_2j_20}\alpha^3_{j_1j_1} > 0$,

$$(\dot{x}^*_{j_1}, a_{j_2j_21}) = \underbrace{(\text{pF}, 3^{\text{rd}}\text{UP})}_{\text{third-order up-parabola flow }(+)}. \quad (4.207)$$

- *For* $a_{j_1j_10}(a_{j_2j_21} - a_{j_1j_11})^3 < 0$ *and* $a_{j_2j_20}\alpha^3_{j_1j_1} > 0$,

$$(\dot{x}^*_{j_1}, a_{j_2j_21}) = \underbrace{(\text{nF}, 3^{\text{rd}}\text{DP})}_{\text{third-order down-parabola flow }(-)}. \quad (4.208)$$

- *For* $a_{j_1j_10}(a_{j_2j_21} - a_{j_1j_11})^3 > 0$ *and* $a_{j_2j_20}\alpha^3_{j_1j_1} < 0$,

$$(\dot{x}^*_{j_1}, a_{j_2j_2l_1}) = \underbrace{(\text{pF}, 3^{\text{rd}}\text{DP})}_{\text{third-order down-parabola flow }(+)}. \quad (4.209)$$

- *For* $a_{j_1j_10}(a_{j_2j_21} - a_{j_1j_11})^3 < 0$ *and* $a_{j_2j_20}\alpha^3_{j_1j_1} < 0$,

$$(\dot{x}^*_{j_1}, a_{j_2j_21}) = \underbrace{(\text{nF}, 3^{\text{rd}}\text{UP})}_{\text{third-order up-parabola flow }(+)}. \quad (4.210)$$

– *The third-order up-parabola flows are the appearing bifurcations from an up-parabola flow to up-parabola, down-parabola, and up-parabola flows.*
– *The third-order down-parabola flows are the appearing bifurcations from a down-parabola flow to down-parabola, up-parabola, and down-parabola flows.*

(v_5) *For* $a_{j_1j_1s_1} = a_{j_2j_2l_1}$,

$$\begin{aligned}\dot{x}_{j_1} &= a_{j_1j_10}(y_{j_1} - a_{j_1j_1s_1})(y_{j_1} - a_{j_1j_1s_2})^2,\\ \dot{x}_{j_2} &= a_{j_2j_20}(y_{j_1} - a_{j_2j_2l_1})(y_{j_1} - a_{j_2j_2l_2})^2.\end{aligned} \tag{4.211}$$

There is a dynamical system in the y_{j_1}*-direction as*

$$\begin{aligned}\dot{y}_{j_1} = (y_{j_1} - a_{j_1j_1s_1})[&\alpha_{j_1j_1}a_{j_1j_10}(y_{j_1} - a_{j_1j_1s_2})^2\\ &+ \alpha_{j_1j_2}a_{j_2j_20}(y_{j_1} - a_{j_2j_2l_2})^2],\end{aligned} \tag{4.212}$$

with a dynamical system in the $y_{j_1}^{\perp}$*-direction as*

$$\begin{aligned}\dot{y}_{j_1}^{\perp} = (y_{j_1} - a_{j_1j_1s_1})[&\beta_{j_1j_1}a_{j_1j_10}(y_{j_1} - a_{j_1j_1s_2})^2\\ &+ \beta_{j_1j_2}a_{j_2j_20}(y_{j_1} - a_{j_2j_2l_2})^2],\end{aligned} \tag{4.213}$$

where

$$\begin{aligned}G_{j_1j_10} &= \frac{\partial}{\partial y_{j_1}}[\mathbf{n}^{\mathrm{T}} \cdot \mathbf{F}(y_{j_1}, \mathbf{p})]\Big|_{y_{j_1}^* = a_{j_1j_1s_1} = a_{j_2j_2l_1}}\\ &= \alpha_{j_1j_1}a_{j_1j_10}(a_{j_1j_1s_1} - a_{j_1j_1s_2})^2 + \alpha_{j_1j_2}a_{j_2j_20}(a_{j_2j_2l_1} - a_{j_2j_2l_2})^2.\end{aligned} \tag{4.214}$$

The infinite-equilibrium flow of $y_1^* = a_{j_1j_1s_1} = a_{j_2j_2l_1}$ *possesses the following properties*:

- *For* $a_{j_1j_10}(a_{j_1j_1s_1} - a_{j_1j_1s_2})^2/a_{j_2j_20}(a_{j_2j_2l_1} - a_{j_2j_2l_2})^2 > 0$ *and* $G_{j_2j_20} > 0$,

$$(a_{j_1j_1s_1}, a_{j_2j_2l_1}) = \underbrace{(\mathrm{II}, \mathrm{SO})}_{\text{increasing-inflection source}}. \tag{4.215}$$

- *For* $a_{j_1j_10}(a_{j_1j_1s_1} - a_{j_1j_1s_2})^2/a_{j_2j_20}(a_{j_2j_2l_1} - a_{j_2j_2l_2})^2 > 0$ *and* $G_{j_2j_20} < 0$

$$(a_{j_1j_1s_1}, a_{j_2j_2l_1}) = \underbrace{(\mathrm{II}, \mathrm{SI})}_{\text{increasing-inflection sink}}. \tag{4.216}$$

- *For* $a_{j_1j_10}(a_{j_1j_1s_1} - a_{j_1j_1s_2})^2/a_{j_2j_20}(a_{j_2j_2l_1} - a_{j_2j_2l_2})^2 < 0$ *and* $G_{j_2j_20} > 0$,

$$(a_{j_1j_1s_1}, a_{j_2j_2l_1}) = \underbrace{(\mathrm{DI}, \mathrm{SO})}_{\text{decreasing-inflection source}}. \tag{4.217}$$

- *For* $a_{j_1j_10}(a_{j_1j_1s_1}-a_{j_1j_1s_2})^2/a_{j_2j_20}(a_{j_2j_2l_1}-a_{j_2j_2l_2})^2<0$ *and* $G_{j_2j_20}$ <0,

$$(a_{j_1j_1s_1},a_{j_2j_2l_1})=\underbrace{(\mathrm{DI},\mathrm{SI})}_{\text{decreasing-inflection sink}}. \tag{4.218}$$

(v_6) *For* $a_{j_1j_1s_2}=a_{j_2j_2l_1}$,

$$\begin{aligned}\dot{x}_{j_1}&=a_{j_1j_10}(y_{j_1}-a_{j_1j_1s_1})(y_{j_1}-a_{j_1j_1s_2})^2,\\ \dot{x}_{j_2}&=a_{j_2j_20}(y_{j_1}-a_{j_2j_2l_1})(y_{j_1}-a_{j_2j_2l_2})^2.\end{aligned} \tag{4.219}$$

There is a dynamical system in the y_{j_1}*-direction as*

$$\begin{aligned}\dot{y}_{j_1}=&(y_{j_1}-a_{j_1j_1s_2})[\alpha_{j_1j_1}a_{j_1j_10}(y_{j_1}-a_{j_1j_1s_1})(y_{j_1}-a_{j_1j_1s_2})\\ &+\alpha_{j_1j_2}a_{j_2j_20}(y_{j_1}-a_{j_2j_2l_2})^2],\end{aligned} \tag{4.220}$$

with a dynamical system in the $y_{j_1}^{\perp}$*-direction as*

$$\begin{aligned}\dot{y}_{j_1}^{\perp}=&(y_{j_1}-a_{j_1j_1s_2})[\beta_{j_1j_1}a_{j_1j_10}(y_{j_1}-a_{j_1j_1s_1})(y_{j_1}-a_{j_1j_1s_2})\\ &+\beta_{j_1j_2}a_{j_2j_20}(y_{j_1}-a_{j_2j_2l_2})^2],\end{aligned} \tag{4.221}$$

where

$$\begin{aligned}G_{j_1j_10}&=\frac{\partial}{\partial y_{j_1}}[\mathbf{n}^{\mathrm{T}}\cdot\mathbf{F}(y_{j_1},\mathbf{p})]\Big|_{y_{j_1}^*=a_{j_1j_1s_2}=a_{j_2j_2l_1}}\\ &=\alpha_{j_1j_2}a_{j_2j_20}(a_{j_2j_2l_1}-a_{j_2j_2l_2})^2.\end{aligned} \tag{4.222}$$

The infinite-equilibrium of $y_1^*=a_{j_1j_1s_2}=a_{j_2j_2l_1}$ *possesses the following properties*:

- *For* $\alpha_{j_1j_2}^2a_{j_1j_10}(a_{j_1j_1s_1}-a_{j_1j_1s_2})>0$ *and* $\alpha_{j_1j_2}a_{j_2j_20}(a_{j_2j_2l_1}-a_{j_2j_2l_2})^2$ >0,

$$(a_{j_1j_1s_1},a_{j_2j_2l_1})=\underbrace{(\mathrm{UP},\mathrm{SO})}_{\text{up-parabola source}}. \tag{4.223}$$

- *For* $\alpha_{j_1j_2}^2a_{j_1j_10}(a_{j_1j_1s_1}-a_{j_1j_1s_2})<0$ *and* $\alpha_{j_1j_2}a_{j_2j_20}(a_{j_2j_2l_1}-a_{j_2j_2l_2})^2$ >0,

$$(a_{j_1j_1s_1},a_{j_2j_2l_1})=\underbrace{(\mathrm{DP},\mathrm{SO})}_{\text{down-parabola source}}. \tag{4.224}$$

- *For* $\alpha_{j_1j_2}^2 a_{j_1j_10}(a_{j_1j_1s_1} - a_{j_1j_1s_2}) > 0$ *and* $\alpha_{j_1j_2}a_{j_2j_20}(a_{j_2j_2l_1} - a_{j_2j_2l_2})^2 < 0$,

$$(a_{j_1j_1s_1}, a_{j_2j_2l_1}) = \underbrace{(\mathrm{DP}, \mathrm{SI})}_{\text{down-parabola sink}}. \tag{4.225}$$

- *For* $\alpha_{j_1j_2}^2 a_{j_1j_10}(a_{j_1j_1s_1} - a_{j_1j_1s_2}) < 0$ *and* $\alpha_{j_1j_2}a_{j_2j_20}(a_{j_2j_2l_1} - a_{j_2j_2l_2})^2 < 0$,

$$(a_{j_1j_1s_1}, a_{j_2j_2l_1}) = \underbrace{(\mathrm{UP}, \mathrm{SI})}_{\text{up-parabola sink}}. \tag{4.226}$$

(v_7) *For* $a_{j_1j_1s_1} = a_{j_2j_2l_2}$,

$$\begin{aligned} \dot{x}_{j_1} &= a_{j_1j_10}(y_{j_1} - a_{j_1j_1s_1})(y_{j_1} - a_{j_1j_1s_2})^2, \\ \dot{x}_{j_2} &= a_{j_2j_20}(y_{j_1} - a_{j_2j_2l_1})(y_{j_1} - a_{j_2j_2l_2})^2. \end{aligned} \tag{4.227}$$

There is a dynamical system in the y_{j_1}*-direction as*

$$\begin{aligned} \dot{y}_{j_1} = (y_{j_1} - a_{j_1j_1s_1})[&\alpha_{j_1j_1}a_{j_1j_10}(y_{j_1} - a_{j_1j_1s_2})^2 \\ &+ \alpha_{j_1j_2}a_{j_2j_20}(y_{j_1} - a_{j_2j_2l_1})(y_{j_1} - a_{j_2j_2l_2})], \end{aligned} \tag{4.228}$$

with a dynamical system in the $y_{j_1}^{\perp}$*-direction as*

$$\begin{aligned} \dot{y}_{j_1}^{\perp} = (y_{j_1} - a_{j_1j_1s_1})[&\beta_{j_1j_1}a_{j_1j_10}(y_{j_1} - a_{j_1j_1s_2})^2 \\ &+ \beta_{j_1j_2}a_{j_2j_20}(y_{j_1} - a_{j_2j_2l_1})(y_{j_1} - a_{j_2j_2l_2})], \end{aligned} \tag{4.229}$$

where

$$\begin{aligned} G_{j_1j_10} &= \frac{\partial}{\partial y_{j_1}}[\mathbf{n}^{\mathrm{T}} \cdot \mathbf{F}(y_{j_1}, \mathbf{p})]\Big|_{y_{j_1}^* = a_{j_1j_1s_1} = a_{j_2j_2l_2}} \\ &= \alpha_{j_1j_1}a_{j_1j_10}(a_{j_1j_1s_1} - a_{j_1j_1s_2})^2. \end{aligned} \tag{4.230}$$

The infinite-equilibrium flow of $y_1^* = a_{j_1j_1s_1} = a_{j_2j_2l_2}$ *possesses the following properties*:

- *For* $\alpha_{j_1j_1}a_{j_1j_10}(a_{j_1j_1s_1} - a_{j_1j_1s_2})^2 > 0$ *and* $a_{j_2j_20}\alpha_{j_1j_1}^2(a_{j_2j_2l_2} - a_{j_2j_2l_1}) > 0$,

$$(a_{j_1j_1s_1}, a_{j_2j_2l_2}) = \underbrace{(\mathrm{SO}, \mathrm{UP})}_{\text{up-parabola source}}. \tag{4.231}$$

- *For* $\alpha_{j_1j_1}a_{j_1j_10}(a_{j_1j_1s_1}-a_{j_1j_1s_2})^2<0$ *and* $a_{j_2j_20}\alpha_{j_1j_1}^2(a_{j_2j_2l_2}-a_{j_2j_2l_1})>0$,

$$(a_{j_1j_1s_1},a_{j_2j_2l_2})=\underbrace{(\mathrm{SI},\mathrm{DP})}_{\text{down-parabola sink}}. \tag{4.232}$$

- *For* $\alpha_{j_1j_1}a_{j_1j_10}(a_{j_1j_1s_1}-a_{j_1j_1s_2})^2>0$ *and* $a_{j_2j_20}\alpha_{j_1j_1}^2(a_{j_2j_2l_2}-a_{j_2j_2l_1})<0$,

$$(a_{j_1j_1s_1},a_{j_2j_2l_2})=\underbrace{(\mathrm{SO},\mathrm{DP})}_{\text{down-parabola source}}. \tag{4.233}$$

- *For* $\alpha_{j_1j_1}a_{j_1j_10}(a_{j_1j_1s_1}-a_{j_1j_1s_2})^2<0$ *and* $a_{j_2j_20}\alpha_{j_1j_1}^2(a_{j_2j_2l_2}-a_{j_2j_2l_1})<0$,

$$(a_{j_1j_1s_1},a_{j_2j_2l_2})=\underbrace{(\mathrm{SI},\mathrm{UP})}_{\text{up-parabola sink}}. \tag{4.234}$$

(v_8) *For* $a_{j_1j_1s_2}=a_{j_2j_2l_2}$,

$$\begin{aligned}\dot{x}_{j_1}&=a_{j_1j_10}(y_{j_1}-a_{j_1j_1s_1})(y_{j_1}-a_{j_1j_1s_2})^2,\\ \dot{x}_{j_2}&=a_{j_2j_20}(y_{j_1}-a_{j_2j_2l_1})(y_{j_1}-a_{j_2j_2l_2})^2,\\ &j_1,j_2\in\{1,2\};\ j_1\neq j_2.\end{aligned} \tag{4.235}$$

There is a dynamical system in the y_{j_1}*-direction as*

$$\begin{aligned}\dot{y}_{j_1}&=(y_{j_1}-a_{j_1j_1s_2})^2[\alpha_{j_1j_1}a_{j_1j_10}(y_{j_1}-a_{j_1j_1s_1})\\ &\quad+\alpha_{j_1j_2}a_{j_2j_20}(y_{j_1}-a_{j_2j_2l_1})],\end{aligned} \tag{4.236}$$

with a dynamical system in the $y_{j_1}^{\perp}$*-direction as*

$$\begin{aligned}\dot{y}_{j_1}^{\perp}&=(y_{j_1}-a_{j_1j_1s_2})^2[\beta_{j_1j_1}a_{j_1j_10}(y_{j_1}-a_{j_1j_1s_1})\\ &\quad+\beta_{j_1j_2}a_{j_2j_20}(y_{j_1}-a_{j_2j_2l_1})],\end{aligned} \tag{4.237}$$

where

$$\begin{aligned}G_{j_1j_10}&=\frac{\partial}{\partial y_{j_1}}[\mathbf{n}^{\mathrm{T}}\cdot\mathbf{F}(y_{j_1},\mathbf{p})]\Big|_{y_{j_1}^*=a_{j_1j_1s_2}=a_{j_2j_2l_2}}=0,\\ G_{j_1j_10}^{(1)}&=\frac{1}{2!}\,\frac{\partial^2}{\partial y_{j_1}^2}[\mathbf{n}^{\mathrm{T}}\cdot\mathbf{F}(y_{j_1},\mathbf{p})]\Big|_{y_{j_1}^*=a_{j_1j_1s_2}=a_{j_2j_2l_2}}\\ &=\alpha_{j_1j_1}a_{j_1j_10}(a_{j_1j_1s_2}-a_{j_1j_1s_1})+\alpha_{j_1j_2}a_{j_2j_20}(a_{j_2j_2l_2}-a_{j_2j_2l_1}).\end{aligned} \tag{4.238}$$

The infinite-equilibrium of $y^*_{j_1} = a_{j_1 j_1 s_2} = a_{j_2 j_2 l_2}$ *has the following properties*:

- *For* $a_{j_1 j_1 0}(a_{j_1 j_1 s_2} - a_{j_1 j_1 s_1})/a_{j_2 j_2 0}(a_{j_2 j_2 l_2} - a_{j_2 j_2 l_1}) > 0$ *and* $G^{(1)}_{j_1 j_1 0} > 0$,

$$(a_{j_1 j_1 s_2}, a_{j_2 j_2 1}) = \underbrace{(\text{II}, \text{US})}_{\text{increasing-inflection upper-saddle}} . \tag{4.239}$$

- *For* $a_{j_1 j_1 0}(a_{j_1 j_1 s_2} - a_{j_1 j_1 s_1})/a_{j_2 j_2 0}(a_{j_2 j_2 l_2} - a_{j_2 j_2 l_1}) > 0$ *and* $G^{(1)}_{j_1 j_1 0} < 0$,

$$(a_{j_1 j_1 s_2}, a_{j_2 j_2 1}) = \underbrace{(\text{II}, \text{LS})}_{\text{increasing-inflection lower-saddle}} . \tag{4.240}$$

- *For* $a_{j_1 j_1 0}(a_{j_1 j_1 s_2} - a_{j_1 j_1 s_1})/a_{j_2 j_2 0}(a_{j_2 j_2 l_2} - a_{j_2 j_2 l_1}) < 0$ *and* $G^{(1)}_{j_1 j_1 0} > 0$,

$$(a_{j_1 j_1 s_2}, a_{j_2 j_2 1}) = \underbrace{(\text{DI}, \text{US})}_{\text{decreasing-inflection upper-saddle}} . \tag{4.241}$$

- *For* $a_{j_1 j_1 0}(a_{j_1 j_1 s_2} - a_{j_1 j_1 s_1})/a_{j_2 j_2 0}(a_{j_2 j_2 l_2} - a_{j_2 j_2 l_1}) < 0$ *and* $G^{(1)}_{j_1 j_1 0} < 0$,

$$(a_{j_1 j_1 s_2}, a_{j_2 j_2 1}) = \underbrace{(\text{DI}, \text{LS})}_{\text{decreasing-inflection lower-saddle}} . \tag{4.242}$$

- *The inflection-saddle infinite-equilibriums are the appearing bifurcations for the up-parabola and down-parabola flow in the* x_{j_1}*- and* x_{j_2}*-directions.*
- *The inflection-saddle infinite-equilibriums are the double-switching bifurcations for the up-parabola and down-parabola flows in the* x_{j_1}*- and* x_{j_2}*-directions.*

(v_9) *For* $a_{j_1 j_1 s_1} = a_{j_1 j_1 s_2} \equiv a_{j_1 j_1 1}$ *and* $a_{j_1 j_1 1} = a_{j_2 j_2 l_1}$,

$$\begin{aligned} \dot{x}_{j_1} &= a_{j_1 j_1 0}(y_{j_1} - a_{j_1 j_1 1})^3, \\ \dot{x}_{j_2} &= a_{j_2 j_2 0}(y_{j_1} - a_{j_2 j_2 l_1})(y_{j_1} - a_{j_2 j_2 l_2})^2, \\ & j_1, j_2 \in \{1, 2\};\ j_1 \neq j_2. \end{aligned} \tag{4.243}$$

For $a_{j_1 j_1 1} = a_{j_2 j_2 l_1}$, *there is a dynamical system in the* y_{j_1}*-direction as*

$$\begin{aligned} \dot{y}_{j_1} = (y_{j_1} - a_{j_2 j_2 l_1})[&\alpha_{j_1 j_1} a_{j_1 j_1 0}(y_{j_1} - a_{j_1 j_1 1})^2 \\ &+ \alpha_{j_1 j_2} a_{j_2 j_2 0}(y_{j_1} - a_{j_2 j_2 l_2})^2], \end{aligned} \tag{4.244}$$

with a dynamical system in the $y_{j_1}^{\perp}$-direction as

$$\dot{y}_{j_1}^{\perp} = (y_{j_1} - a_{j_2j_2l_1})[\beta_{j_1j_1}a_{j_1j_10}(y_{j_1} - a_{j_1j_11})^2 + \beta_{j_1j_2}a_{j_2j_20}(y_{j_1} - a_{j_2j_2l_2})^2], \tag{4.245}$$

and a G-function as

$$\begin{aligned} G_{j_0j_01} &= \frac{\partial}{\partial y_{j_1}}[\mathbf{n}^{\mathrm{T}} \cdot \mathbf{F}(y_{j_1}, \mathbf{p})]\bigg|_{y_{j_1}^* = a_{j_1j_11} = a_{j_2j_2l_1}} \\ &= \alpha_{j_1j_2}a_{j_2j_20}(a_{j_2j_2l_1} - a_{j_2j_2l_2})^2. \end{aligned} \tag{4.246}$$

The infinite-equilibrium flow of $y_{j_1}^* = a_{j_1j_11} = a_{j_2j_2l_1}$ has the following properties:

- *For $a_{j_1j_10}\alpha_{j_1j_2}^3 > 0$ and $\alpha_{j_1j_2}a_{j_2j_20}(a_{j_2j_2l_1} - a_{j_2j_2l_2})^2 > 0$,*

$$(a_{j_1j_11}, a_{j_2j_2l_1}) = \underbrace{(2^{\mathrm{nd}}\mathrm{II}, \mathrm{SO})}_{\text{second-order increasing-inflection source}}. \tag{4.247}$$

- *For $a_{j_1j_10}\alpha_{j_1j_2}^3 < 0$ and $\alpha_{j_1j_2}a_{j_2j_20}(a_{j_2j_2l_1} - a_{j_2j_2l_2})^2 > 0$,*

$$(a_{j_1j_11}, a_{j_2j_2l_1}) = \underbrace{(2^{\mathrm{nd}}\mathrm{DI}, \mathrm{SO})}_{\text{second-order decreasing-inflection source}}. \tag{4.248}$$

- *For $a_{j_1j_10}\alpha_{j_1j_2}^3 > 0$ and $\alpha_{j_1j_2}a_{j_2j_20}(a_{j_2j_2l_1} - a_{j_2j_2l_2})^2 < 0$,*

$$(a_{j_1j_11}, a_{j_2j_2l_1}) = \underbrace{(2^{\mathrm{nd}}\mathrm{DI}, \mathrm{SI})}_{\text{second-order decreasing-inflection sink}}. \tag{4.249}$$

- *For $a_{j_1j_10}\alpha_{j_1j_2}^3 < 0$ and $\alpha_{j_1j_2}a_{j_2j_20}(a_{j_2j_2l_1} - a_{j_2j_2l_2})^2 < 0$,*

$$(a_{j_1j_11}, a_{j_2j_2l_1}) = \underbrace{(2^{\mathrm{nd}}\mathrm{II}, \mathrm{SI})}_{\text{second-order increasing-inflection sink}}. \tag{4.250}$$

– *The second-order inflection infinite-equilibrium flows are the switching bifurcations for up-parabola, down-parabola, and up-parabola flows.*

(v_{10}) *For* $a_{j_1j_1s_1} = a_{j_1j_1s_2} \equiv a_{j_1j_11}$ *and* $a_{j_1j_11} = a_{j_2j_2l_2}$,

$$\begin{aligned} \dot{x}_{j_1} &= a_{j_1j_10}(y_{j_1} - a_{j_1j_11})^3, \\ \dot{x}_{j_2} &= a_{j_2j_20}(y_{j_1} - a_{j_2j_2l_1})(y_{j_1} - a_{j_2j_2l_2})^2, \\ & j_1, j_2 \in \{1,2\};\ j_1 \neq j_2. \end{aligned} \tag{4.251}$$

For $a_{j_1j_11} = a_{j_2j_2l_2}$, *there is a dynamical system in the* y_{j_1}*-direction as*

$$\begin{aligned} \dot{y}_{j_1} &= (y_{j_1} - a_{j_2j_2l_2})^2[\alpha_{j_1j_1}a_{j_1j_10}(y_{j_1} - a_{j_1j_11}) \\ &\quad + \alpha_{j_1j_2}a_{j_2j_20}(y_{j_1} - a_{j_2j_2l_1})], \end{aligned} \tag{4.252}$$

with a dynamical system in the $y_{j_1}^{\perp}$*-direction as*

$$\begin{aligned} \dot{y}_{j_1}^{\perp} &= (y_{j_1} - a_{j_2j_2l_2})^2[\beta_{j_1j_1}a_{j_1j_10}(y_{j_1} - a_{j_1j_11}) \\ &\quad + \beta_{j_1j_2}a_{j_2j_20}(y_{j_1} - a_{j_2j_2l_1})], \end{aligned} \tag{4.253}$$

and a G-function as

$$\begin{aligned} G_{j_1j_11} &= \frac{\partial}{\partial y_{j_1}}[\mathbf{n}^{\mathrm{T}} \cdot \mathbf{F}(y_{j_1}, \mathbf{p})]\Big|_{y_{j_1}^* = a_{j_1j_11} = a_{j_2j_2l_2}} = 0, \\ G_{j_1j_11}^{(1)} &= \frac{1}{2!}\frac{\partial^2}{\partial y_{j_1}^2}[\mathbf{n}^{\mathrm{T}} \cdot \mathbf{F}(y_{j_1}, \mathbf{p})]\Big|_{y_{j_1}^* = a_{j_1j_11} = a_{j_2j_2l_2}} \\ &= \alpha_{j_1j_2}a_{j_2j_20}(a_{j_2j_2l_2} - a_{j_2j_2l_1}). \end{aligned} \tag{4.254}$$

The infinite-equilibrium flow of $y_{j_1}^* = a_{j_1j_11} = a_{j_2j_2l_2}$ *has the following properties*:

- *For* $a_{j_1j_10}\alpha_{j_1j_2}^2 > 0$ *and* $\alpha_{j_1j_2}a_{j_2j_20}(a_{j_2j_2l_2} - a_{j_2j_2l_1}) > 0$,

$$(a_{j_1j_11}, a_{j_2j_2l_1}) = \underbrace{(\mathrm{UP}, \mathrm{US})}_{\text{up-parabola upper-saddle}}. \tag{4.255}$$

- *For* $a_{j_1j_10}\alpha_{j_1j_2}^2 < 0$ *and* $\alpha_{j_1j_2}a_{j_2j_20}(a_{j_2j_2l_2} - a_{j_2j_2l_1}) > 0$,

$$(a_{j_1j_11}, a_{j_2j_2l_1}) = \underbrace{(\mathrm{DP}, \mathrm{US})}_{\text{down-parabola upper-saddle}}. \tag{4.256}$$

- *For* $a_{j_1j_10}\alpha_{j_1j_2}^2 > 0$ *and* $\alpha_{j_1j_2}a_{j_2j_20}(a_{j_2j_2l_2} - a_{j_2j_2l_1}) < 0$,

$$(a_{j_1j_11}, a_{j_2j_2l_1}) = \underbrace{(\mathrm{DP}, \mathrm{LS})}_{\text{down-parabola lower-saddle}}. \tag{4.257}$$

- *For* $a_{j_1j_10}\alpha^2_{j_1j_2} < 0$ *and* $\alpha_{j_1j_2}a_{j_2j_20}(a_{j_2j_2l_2} - a_{j_2j_2l_1}) < 0$,

$$(a_{j_1j_11}, a_{j_2j_2l_1}) = \underbrace{(\mathrm{UP}, \mathrm{LS})}_{\text{up-parabola lower-saddle}}. \tag{4.258}$$

– *The parabola-saddle infinite-equilibrium flows are the switching bifurcations for up-parabola, down-parabola, and up-parabola flows.*

(v_{11}) *For* $a_{j_2j_2l_1} = a_{j_2j_2l_{\partial 2}} = a_{j_2j_21}$ *and* $a_{j_1j_1s_1} = a_{j_2j_21}$,

$$\begin{aligned}
&\dot{x}_{j_1} = a_{j_1j_10}(y_{j_1} - a_{j_1j_1s_1})(y_{j_1} - a_{j_1j_1s_2})^2,\\
&\dot{x}_{j_2} = a_{j_2j_20}(y_{j_1} - a_{j_2j_21})^3,\\
&j_1, j_2 \in \{1, 2\};\ j_1 \neq j_2.
\end{aligned} \tag{4.259}$$

For $a_{j_1j_1s_1} = a_{j_2j_21}$, *there is a dynamical system in the* y_{j_1}*-direction as*

$$\begin{aligned}
\dot{y}_{j_1} = {} & (y_{j_1} - a_{j_1j_1s_1})[\alpha_{j_1j_1}a_{j_1j_10}(y_{j_1} - a_{j_1j_1s_2})^2\\
& + \alpha_{j_1j_2}a_{j_2j_20}(y_{j_1} - a_{j_2j_21})^2],
\end{aligned} \tag{4.260}$$

with a dynamical system in the $y^{\perp}_{j_1}$*-direction as*

$$\begin{aligned}
\dot{y}^{\perp}_{j_1} = {} & (y_{j_1} - a_{j_1j_1s_1})[\beta_{j_1j_1}a_{j_1j_10}(y_{j_1} - a_{j_1j_1s_2})^2\\
& + \beta_{j_1j_2}a_{j_2j_20}(y_{j_1} - a_{j_2j_21})^2],
\end{aligned} \tag{4.261}$$

and a G-function is

$$\begin{aligned}
G_{j_1j_10} &= \frac{\partial}{\partial y_{j_1}}[\mathbf{n}^{\mathrm{T}} \cdot \mathbf{F}(y_{j_1}, \mathbf{p})]\bigg|_{y^*_{j_1} = a_{j_1j_1s_1} = a_{j_2j_21}}\\
&= \alpha_{j_1j_1}a_{j_1j_10}(a_{j_1j_1s_1} - a_{j_1j_1s_2})^2.
\end{aligned} \tag{4.262}$$

The infinite-equilibrium flow of $y^*_{j_1} = a_{j_1j_1s_1} = a_{j_2j_21}$ *has the following properties*:

- *For* $\alpha_{j_1j_1}a_{j_1j_10}(a_{j_1j_1s_1} - a_{j_1j_1s_2})^2 > 0$ *and* $a_{j_2j_20}\alpha^3_{j_1j_1} > 0$,

$$(a_{j_1j_1s_1}, a_{j_2j_21}) = \underbrace{(\mathrm{SO}, 2^{\mathrm{nd}}\mathrm{II})}_{\text{second-order increasing-inflection source}}. \tag{4.263}$$

- *For* $a_{j_1j_10}\alpha_{j_1j_2}(a_{j_1j_1s_1} - a_{j_1j_1s_2})^2 < 0$ *and* $a_{j_2j_20}\alpha^3_{j_1j_1} > 0$,

$$(a_{j_1j_1s_1}, a_{j_2j_21}) = \underbrace{(\mathrm{SI}, 2^{\mathrm{nd}}\mathrm{DI})}_{\text{second-order decreasing-inflection sink}}. \tag{4.264}$$

- *For* $a_{j_1j_10}\alpha_{j_1j_2}(a_{j_1j_1s_1}-a_{j_1j_1s_2})^2>0$ *and* $a_{j_2j_20}\alpha^3_{j_1j_1}<0$,

$$(a_{j_1j_1s_1},a_{j_2j_21})=\underbrace{(\mathrm{SO},2^{\mathrm{nd}}\mathrm{DI})}_{\text{second-order decreasing-inflection source}}. \tag{4.265}$$

- *For* $a_{j_1j_10}\alpha_{j_1j_2}(a_{j_1j_1s_1}-a_{j_1j_1s_2})^2<0$ *and* $a_{j_2j_20}\alpha^3_{j_1j_1}<0$,

$$(a_{j_1j_1s_1},a_{j_2j_21})=\underbrace{(\mathrm{SI},2^{\mathrm{nd}}\mathrm{II})}_{\text{second-order increasing-inflection sink}}. \tag{4.266}$$

– *The second-order inflection infinite-equilibrium flows are the switching bifurcations for up-parabola, down-parabola, and up-parabola flows.*

(v_{12}) *For* $a_{j_2j_2l_1}=a_{j_2j_2l_{\theta 2}}=a_{j_2j_21}$ *and* $a_{j_1j_1s_2}=a_{j_2j_21}$,

$$\begin{aligned}
\dot{x}_{j_1}&=a_{j_1j_10}(y_{j_1}-a_{j_1j_1s_1})(y_{j_1}-a_{j_1j_1s_2})^2,\\
\dot{x}_{j_2}&=a_{j_2j_20}(y_{j_1}-a_{j_2j_21})^3,\\
&\quad j_1,j_2\in\{1,2\};j_1\neq j_2.
\end{aligned} \tag{4.267}$$

For $a_{j_1j_1s_2}=a_{j_2j_21}$, *there is a dynamical system in the* y_{j_1}*-direction as*

$$\begin{aligned}
\dot{y}_{j_1}&=(y_{j_1}-a_{j_1j_1s_2})^2[\alpha_{j_1j_1}a_{j_1j_10}(y_{j_1}-a_{j_1j_1s_1})\\
&\quad+\alpha_{j_1j_2}a_{j_2j_20}(y_{j_1}-a_{j_2j_21})],
\end{aligned} \tag{4.268}$$

with a dynamical system in the $y^{\perp}_{j_1}$*-direction as*

$$\begin{aligned}
\dot{y}^{\perp}_{j_1}&=(y_{j_1}-a_{j_1j_1s_2})^2[\beta_{j_1j_1}a_{j_1j_10}(y_{j_1}-a_{j_1j_1s_1})\\
&\quad+\beta_{j_1j_2}a_{j_2j_20}(y_{j_1}-a_{j_2j_21})],
\end{aligned} \tag{4.269}$$

and a G-function is

$$\begin{aligned}
G_{j_1j_10}&=\frac{\partial}{\partial y_{j_1}}[\mathbf{n}^{\mathrm{T}}\cdot\mathbf{F}(y_{j_1},\mathbf{p})]\Big|_{y^*_{j_1}=a_{j_1j_1s_2}=a_{j_2j_21}}=0,\\
G^{(1)}_{j_1j_10}&=\frac{1}{2!}\frac{\partial^2}{\partial y^2_{j_1}}[\mathbf{n}^{\mathrm{T}}\cdot\mathbf{F}(y_{j_1},\mathbf{p})]\Big|_{y^*_{j_1}=a_{j_1j_1s_2}=a_{j_2j_21}}\\
&=\alpha_{j_1j_1}a_{j_1j_10}(a_{j_1j_1s_2}-a_{j_1j_1s_1}).
\end{aligned} \tag{4.270}$$

The infinite-equilibrium flow of $y^*_{j_1} = a_{j_1 j_1 s_1} = a_{j_2 j_2 1}$ *has the following properties*:

- *For* $\alpha_{j_1 j_1} a_{j_1 j_1 0}(a_{j_1 j_1 s_2} - a_{j_1 j_1 s_1}) > 0$ *and* $a_{j_2 j_2 0}\alpha^2_{j_1 j_1} > 0$,

$$(a_{j_1 j_1 s_2}, a_{j_2 j_2 1}) = \underbrace{(\mathrm{US}, \mathrm{UP})}_{\text{up-parabola upper-saddle}}. \tag{4.271}$$

- *For* $\alpha_{j_1 j_1} a_{j_1 j_1 0}(a_{j_1 j_1 s_2} - a_{j_1 j_1 s_1}) < 0$ *and* $a_{j_2 j_2 0}\alpha^2_{j_1 j_1} > 0$,

$$(a_{j_1 j_1 s_2}, a_{j_2 j_2 1}) = \underbrace{(\mathrm{LS}, \mathrm{DP})}_{\text{down-parabola lower-saddle}}. \tag{4.272}$$

- *For* $\alpha_{j_1 j_1} a_{j_1 j_1 0}(a_{j_1 j_1 s_2} - a_{j_1 j_1 s_1}) > 0$ *and* $a_{j_2 j_2 0}\alpha^2_{j_1 j_1} < 0$,

$$(a_{j_1 j_1 s_2}, a_{j_2 j_2 1}) = \underbrace{(\mathrm{US}, \mathrm{DP})}_{\text{down-parabola upper-saddle}}. \tag{4.273}$$

- *For* $\alpha_{j_1 j_1} a_{j_1 j_1 0}(a_{j_1 j_1 s_2} - a_{j_1 j_1 s_1}) < 0$ *and* $a_{j_2 j_2 0}\alpha^2_{j_1 j_1} < 0$,

$$(a_{j_1 j_1 s_2}, a_{j_2 j_2 1}) = \underbrace{(\mathrm{LS}, \mathrm{UP})}_{\text{up-parabola lower-saddle}}. \tag{4.274}$$

– *The parabola-saddle infinite-equilibrium flows are the switching bifurcations for up-parabola, down-parabola, and up-parabola flows.*

(v_{13}) *For* $a_{j_1 j_1 1} = a_{j_2 j_2 1}$,

$$\begin{aligned}
\dot{x}_{j_1} &= a_{j_1 j_1 0}(y_{j_1} - a_{j_1 j_1 1})^3, \\
\dot{x}_{j_2} &= a_{j_2 j_2 0}(y_{j_1} - a_{j_2 j_2 1})^3, \\
& j_1, j_2 \in \{1, 2\}; j_1 \neq j_2.
\end{aligned} \tag{4.275}$$

There is a dynamical system in the y_{j_1}*-direction as*

$$\dot{y}_{j_1} = (y_{j_1} - a_{j_1 j_1 1})^3(\alpha_{j_1 j_1} a_{j_1 j_1 0} + \alpha_{j_1 j_2} a_{j_2 j_2 0}), \tag{4.276}$$

with a dynamical system in the $y^{\perp}_{j_1}$*-direction as*

$$\dot{y}^{\perp}_{j_1} = (y_{j_1} - a_{j_1 j_1 1})^3(\beta_{j_1 j_1} a_{j_1 j_1 0} + \beta_{j_1 j_2} a_{j_2 j_2 0}), \tag{4.277}$$

where

$$G_{j_1j_10} = \frac{\partial}{\partial y_{j_1}}[\mathbf{n}^{\mathrm{T}} \cdot \mathbf{F}(y_{j_1}, \mathbf{p})]\bigg|_{y^*_{j_1}=a_{j_1j_11}=a_{j_2j_21}} = 0,$$

$$G^{(1)}_{j_1j_10} = \frac{1}{2!}\frac{\partial^2}{\partial y^2_{j_1}}[\mathbf{n}^{\mathrm{T}} \cdot \mathbf{F}(y_{j_1}, \mathbf{p})]\bigg|_{y^*_{j_1}=a_{j_1j_11}=a_{j_2j_21}} = 0,$$

$$G^{(2)}_{j_1j_10} = \frac{1}{3!}\frac{\partial^3}{\partial y^3_{j_1}}[\mathbf{n}^{\mathrm{T}} \cdot \mathbf{F}(y_{j_1}, \mathbf{p})]\bigg|_{y^*_{j_1}=a_{j_1j_11}=a_{j_2j_21}} = \alpha_{j_1j_1}a_{j_1j_10} + \alpha_{j_1j_2}a_{j_2j_20}. \tag{4.278}$$

The infinite-equilibrium of $y^*_{j_1} = a_{j_1j_11} = a_{j_2j_21}$ *has the following properties:*

- *For* $a_{j_1j_10}/a_{j_2j_20} > 0$ *and* $G^{(2)}_{j_1j_10} > 0$,

$$(a_{j_1j_11}, a_{j_2j_21}) = \underbrace{(\mathrm{II}, 3^{\mathrm{rd}}\mathrm{SO})}_{\text{increasing-inflection third-order source}}. \tag{4.279}$$

- *For* $a_{j_1j_10}/a_{j_2j_20} > 0$ *and* $G^{(2)}_{j_1j_10} < 0$,

$$(a_{j_1j_11}, a_{j_2j_21}) = \underbrace{(\mathrm{II}, 3^{\mathrm{rd}}\mathrm{SI})}_{\text{increasing-inflection third-order sink}}. \tag{4.280}$$

- *For* $a_{j_1j_10}/a_{j_2j_20} < 0$ *and* $G^{(2)}_{j_1j_10} > 0$,

$$(a_{j_1j_11}, a_{j_2j_21}) = \underbrace{(\mathrm{DI}, 3^{\mathrm{rd}}\mathrm{SO})}_{\text{decreasing-inflection third-order source}}. \tag{4.281}$$

- *For* $a_{j_1j_10}/a_{j_2j_20} < 0$ *and* $G^{(2)}_{j_1j_10} < 0$,

$$(a_{j_1j_11}, a_{j_2j_21}) = \underbrace{(\mathrm{DI}, 3^{\mathrm{rd}}\mathrm{SI})}_{\text{decreasing-inflection third-order sink}}. \tag{4.282}$$

- *The inflection-third-order source (sink) infinite-equilibriums are the appearing bifurcations for the up-parabola and down-parabola flows in the* x_{j_1}*- and* x_{j_2}*-directions.*
- *The inflection-third-order source (sink) infinite-equilibriums are the double-switching bifurcations for the up-parabola and down-parabola flows in the* x_{j_1}*- and* x_{j_2}*-directions.*

(vi) If

$$\Delta_{j_1 j_1} > 0 \quad \text{and} \quad \Delta_{j_2 j_2} = 0, \tag{4.283}$$

then

$$\begin{aligned} \dot{x}_{j_1} &= a_{j_1 j_1 0}(y_{j_1} - a_{j_1 j_1 1})(y_{j_1} - a_{j_1 j_1 2})(y_{j_1} - a_{j_1 j_1 3}), \\ \dot{x}_{j_2} &= a_{j_2 j_2 0}(y_{j_1} - a_{j_2 j_2 l_1})(y_{j_1} - a_{j_2 j_2 l_2})^2, \end{aligned} \tag{4.284}$$

where

$$\begin{gathered} \{a_{j_1 j_1 1}, a_{j_1 j_1 2}, a_{j_1 j_1 3}\} = \text{sort}\{b_{j_1 j_1 1}, b_{j_1 j_1 2}, b_{j_1 j_1 3}\}, \;\; a_{j_1 j_1 s_1} < a_{j_1 j_1 s_2}, s_1 < s_2; \\ b_{j_1 j_1 2}, b_{j_1 j_1 3} = -\frac{1}{2}(B_{j_1 j_1 1} \pm \sqrt{\Delta_{j_1 j_1}}), a_{j_1 j_1 0} = A_{j_1 j_1 0}; \\ a_{j_2 j_2 l_1} = b_{j_2 j_2 1}, a_{j_2 j_2 l_2} = -\frac{1}{2} B_{j_2 j_2}, a_{j_2 j_2 0} = A_{j_2 j_2 0}; \quad j_1, j_2 \in \{1, 2\}. \end{gathered} \tag{4.285}$$

There is a dynamical system in the y_{j_1}-direction as

$$\begin{aligned} \dot{y}_{j_1} = {} & \alpha_{j_1 j_1} a_{j_1 j_1 0} \prod_{s_1=1}^{3} (y_{j_1} - a_{j_1 j_1 s_1}) \\ & + \alpha_{j_1 j_2} a_{j_2 j_2 0}(y_{j_1} - a_{j_2 j_2 l_1})(y_{j_1} - a_{j_2 j_2 l_2})^2, \end{aligned} \tag{4.286}$$

with a dynamical system in the $y_{j_1}^{\perp}$-direction as

$$\begin{aligned} \dot{y}_{j_1}^{\perp} = {} & \beta_{j_1 j_1} a_{j_1 j_1 0} \prod_{s_1=1}^{3} (y_{j_1} - a_{j_1 j_1 s_1}) \\ & + \beta_{j_1 j_2} a_{j_2 j_2 0}(y_{j_1} - a_{j_2 j_2 l_1})(y_{j_1} - a_{j_2 j_2 l_2})^2. \end{aligned} \tag{4.287}$$

(vi_{1a}) The equilibrium flow of $y_{j_1}^* = a_{j_1 j_1 s_1} (s_1, s_2 \in \{1, 2, 3\}, s_1 \neq s_2)$ for the x_{j_1}-direction with $\dot{x}_{j_2}^* = a_{j_2 j_2 0}(a_{j_1 j_1 s_1} - a_{j_2 j_2 l_1})(a_{j_1 j_1 s_1} - a_{j_2 j_2 l_1})^2$ has the following properties:

- *For* $a_{j_1 j_1 0} \alpha_{j_1 j_2} \prod_{s_2=1, s_2 \neq s_1}^{3} (a_{j_1 j_1 s_1} - a_{j_1 j_1 s_2}) > 0, a_{j_2 j_2 0}(a_{j_1 j_1 s_1} - a_{j_2 j_2 l_1})(a_{j_1 j_1 s_1} - a_{j_2 j_2 l_1})^2 > 0$,

$$(a_{j_1 j_1 s_1}, \dot{x}_{j_2}^*) = \underbrace{(\text{UP}, \text{pF})}_{\text{up-parabola flow } (+)}. \tag{4.288}$$

- *For* $a_{j_1 j_1 0} \alpha_{j_1 j_2} \prod_{s_2=1, s_2 \neq s_1}^{3} (a_{j_1 j_1 s_1} - a_{j_1 j_1 s_2}) < 0, a_{j_2 j_2 0}(a_{j_1 j_1 s_1} - a_{j_2 j_2 l_1})(a_{j_1 j_1 s_1} - a_{j_2 j_2 l_1})^2 > 0$,

$$(a_{j_1 j_1 s_1}, \dot{x}_{j_2}^*) = \underbrace{(\text{DP}, \text{pF})}_{\text{Down-parabola flow } (+)}. \tag{4.289}$$

- *For* $a_{j_1j_10}\alpha_{j_1j_2}\prod_{s_2=1,s_2\neq s_1}^{3}(a_{j_1j_1s_1} - a_{j_1j_1s_2}) > 0, a_{j_2j_20}(a_{j_1j_1s_1} - a_{j_2j_2l_1})(a_{j_1j_1s_1} - a_{j_2j_2l_1})^2 < 0$,

$$(a_{j_1j_1s_1}, \dot{x}^*_{j_2}) = \underbrace{(\mathrm{DP}, \mathrm{nF})}_{\text{Down-parabola flow }(-)}. \tag{4.290}$$

- *For* $a_{j_1j_10}\alpha_{j_1j_2}\prod_{s_2=1,s_2\neq s_1}^{3}(a_{j_1j_1s_1} - a_{j_1j_1s_2}) < 0, a_{j_2j_20}(a_{j_1j_1s_1} - a_{j_2j_2l_1})(a_{j_1j_1s_1} - a_{j_2j_2l_1})^2 < 0$,

$$(a_{j_1j_1s_1}, \dot{x}^*_{j_2}) = \underbrace{(\mathrm{UP}, \mathrm{nF})}_{\text{up-parabola flow }(-)}. \tag{4.291}$$

(vi_{1b}) The equilibrium flow of $y^*_{j_1} = a_{j_2j_2l_1}$ *for the* x_{j_2}*-direction has the following properties with* $\dot{x}^*_{j_1} = a_{j_1j_10}\prod_{s_1=1}^{3}(a_{j_2j_2l_1} - a_{j_1j_1s_1})$:

- *For* $a_{j_1j_10}\prod_{s_1=1}^{3}(a_{j_2j_2l_1} - a_{j_1j_1s_1}) > 0$ *and* $a_{j_2j_20}\alpha_{j_1j_1}(a_{j_2j_2l_1} - a_{j_2j_2l_2})^2 > 0$,

$$(\dot{x}^*_{j_1}, a_{j_2j_2l_1}) = \underbrace{(\mathrm{pF}, \mathrm{UP})}_{\text{up-parabola flow }(+)}. \tag{4.292}$$

- *For* $a_{j_1j_10}\prod_{s_1=1}^{3}(a_{j_2j_2l_1} - a_{j_1j_1s_1}) < 0$ *and* $a_{j_2j_20}\alpha_{j_1j_1}(a_{j_2j_2l_1} - a_{j_2j_2l_2})^2 > 0$,

$$(\dot{x}^*_{j_1}, a_{j_2j_2l_1}) = \underbrace{(\mathrm{nF}, \mathrm{DP})}_{\text{down-parabola flow}(-)}. \tag{4.293}$$

- *For* $a_{j_1j_10}\prod_{s_1=1}^{3}(a_{j_2j_2l_1} - a_{j_1j_1s_1}) > 0$ *and* $a_{j_2j_20}\alpha_{j_1j_1}(a_{j_2j_2l_1} - a_{j_2j_2l_2})^2 < 0$,

$$(\dot{x}^*_{j_1}, a_{j_2j_2l_1}) = \underbrace{(\mathrm{pF}, \mathrm{DP})}_{\text{down-parabola flow}(+)}. \tag{4.294}$$

- *For* $a_{j_1j_10}\prod_{s_1=1}^{3}(a_{j_2j_2l_1} - a_{j_1j_1s_1}) < 0$ *and* $a_{j_2j_20}\alpha_{j_1j_1}(a_{j_2j_2l_1} - a_{j_2j_2l_2})^2 > 0$,

$$(\dot{x}^*_{j_1}, a_{j_2j_2l_1}) = \underbrace{(\mathrm{nF}, \mathrm{UP})}_{\text{up-parabola flow }(-)}. \tag{4.295}$$

(vi_{1c}) *The equilibrium flow of* $y^*_{j_1} = a_{j_2j_2l_2}$ *for the* x_{j_2}*-direction has the following properties with* $\dot{x}^*_{j_1} = a_{j_1j_10}\prod_{s_1=1}^{3}(a_{j_2j_2l_2} - a_{j_1j_1s_1})$:

- *For* $a_{j_1j_10}\prod_{s_1=1}^{3}(a_{j_2j_2l_2} - a_{j_1j_1s_1}) > 0$ *and* $a_{j_2j_20}\alpha^2_{j_1j_1}(a_{j_2j_2l_2} - a_{j_2j_2l_1}) > 0$,

$$(\dot{x}^*_{j_1}, a_{j_2j_2l_2}) = \underbrace{(\mathrm{pF}, \mathrm{II})}_{\text{increasing-inflection flow}(+)}. \tag{4.296}$$

- *For* $a_{j_1j_10}\prod_{s_1=1}^{3}(a_{j_2j_2l_2} - a_{j_1j_1s_1}) < 0$ *and* $a_{j_2j_20}\alpha^2_{j_1j_1}(a_{j_2j_2l_2} - a_{j_2j_2l_1}) > 0$,

$$(\dot{x}^*_{j_1}, a_{j_2j_2l_2}) = \underbrace{(\mathrm{nF}, \mathrm{DI})}_{\text{decreasing-inflection flow }(-)}. \tag{4.297}$$

- *For* $a_{j_1j_10}\prod_{s_1=1}^{3}(a_{j_2j_2l_2} - a_{j_1j_1s_1}) > 0$ *and* $a_{j_2j_20}\alpha^2_{j_1j_1}(a_{j_2j_2l_2} - a_{j_2j_2l_1}) < 0$,

$$(\dot{x}^*_{j_1}, a_{j_2j_2l_2}) = \underbrace{(\mathrm{pF}, \mathrm{DI})}_{\text{decreasing-inflection flow }(+)}. \tag{4.298}$$

- *For* $a_{j_1j_10}\prod_{s_1=1}^{3}(a_{j_2j_2l_2} - a_{j_1j_1s_1}) < 0$ *and* $a_{j_2j_20}\alpha^2_{j_1j_1}(a_{j_2j_2l_2} - a_{j_2j_2l_1}) > 0$,

$$(\dot{x}^*_{j_1}, a_{j_2j_2l_2}) = \underbrace{(\mathrm{nF}, \mathrm{II})}_{\text{increasing-inflection flow }(-)}. \tag{4.299}$$

– *The decreasing- and increasing-inflection flows are the appearing bifurcations for the up-parabola and parabola flows.*

(vi_2) *For* $a_{j_2j_2l_2} = a_{j_2j_2l_1} = a_{j_2j_21}$,

$$\begin{aligned}\dot{x}_{j_1} &= a_{j_1j_10}(y_{j_1} - a_{j_1j_11})(y_{j_1} - a_{j_1j_12})(y_{j_1} - a_{j_1j_13}),\\ \dot{x}_{j_2} &= a_{j_2j_20}(y_{j_1} - a_{j_2j_21})^3.\end{aligned} \tag{4.300}$$

There is a dynamical system in the y_{j_1}*-direction as*

$$\dot{y}_{j_1} = \alpha_{j_1j_1}a_{j_1j_10}\prod_{s_1=1}^{3}(y_{j_1} - a_{j_1j_1s_1}) + \alpha_{j_1j_2}a_{j_2j_20}(y_{j_1} - a_{j_2j_21})^3, \tag{4.301}$$

with a dynamical system in the $y^{\perp}_{j_1}$*-direction as*

$$\dot{y}^{\perp}_{j_1} = \beta_{j_1j_1}a_{j_1j_10}\prod_{s_1=1}^{3}(y_{j_1} - a_{j_1j_1s_1}) + \beta_{j_1j_2}a_{j_2j_20}(y_{j_1} - a_{j_2j_21})^3. \tag{4.302}$$

(vi_{2a}) *The equilibrium flows of* $y_{j_1}^* = a_{j_1 j_1 s_1} (s_1, s_2 \in \{1, 2, 3\}, s_1 \neq s_2)$ *for the* x_{j_1}*-direction have four cases with* $\dot{x}_{j_2}^* = a_{j_2 j_2 0}(a_{j_1 j_1 s_1} - a_{j_2 j_2 1})^3$:

- *For* $a_{j_1 j_1 0} \alpha_{j_1 j_2} \prod_{s_2=1, s_2 \neq s_1}^{3} (a_{j_1 j_1 s_1} - a_{j_1 j_1 s_2}) > 0$ *and* $a_{j_2 j_2 0}(a_{j_1 j_1 s_1} - a_{j_2 j_2 1})^3 > 0$,

$$(a_{j_1 j_1 s_1}, \dot{x}_{j_2}^*) = \underbrace{(\mathrm{UP}, \mathrm{pF})}_{\text{up-parabola flow } (+)} . \tag{4.303}$$

- *For* $a_{j_1 j_1 0} \alpha_{j_1 j_2} \prod_{s_2=1, s_2 \neq s_1}^{3} (a_{j_1 j_1 s_1} - a_{j_1 j_1 s_2}) < 0$ *and* $a_{j_2 j_2 0}(a_{j_1 j_1 s_1} - a_{j_2 j_2 1})^3 > 0$,

$$(a_{j_1 j_1 s_1}, \dot{x}_{j_2}^*) = \underbrace{(\mathrm{DP}, \mathrm{pF})}_{\text{Down-parabola flow } (+)} . \tag{4.304}$$

- *For* $a_{j_1 j_1 0} \alpha_{j_1 j_2} \prod_{s_2=1, s_2 \neq s_1}^{3} (a_{j_1 j_1 s_1} - a_{j_1 j_1 s_2}) > 0$ *and* $a_{j_2 j_2 0}(a_{j_1 j_1 s_1} - a_{j_2 j_2 1})^3 < 0$,

$$(a_{j_1 j_1 s_1}, \dot{x}_{j_2}^*) = \underbrace{(\mathrm{DP}, \mathrm{nF})}_{\text{Down-parabola flow } (-)} . \tag{4.305}$$

- *For* $a_{j_1 j_1 0} \alpha_{j_1 j_2} \prod_{s_2=1, s_2 \neq s_1}^{3} (a_{j_1 j_1 s_1} - a_{j_1 j_1 s_2}) < 0$ *and* $a_{j_2 j_2 0}(a_{j_1 j_1 s_1} - a_{j_2 j_2 1})^3 < 0$,

$$(a_{j_1 j_1 s_1}, \dot{x}_{j_2}^*) = \underbrace{(\mathrm{UP}, \mathrm{nF})}_{\text{up-parabola flow } (-)} . \tag{4.306}$$

(vi_{2b}) *The equilibrium flow of* $y_{j_1}^* = a_{j_2 j_2 1}$ *for the* x_{j_2}*-direction has the following properties with* $\dot{x}_{j_1}^* = a_{j_1 j_1 0} \prod_{s_1=1}^{3} (a_{j_2 j_2 1} - a_{j_1 j_1 s_1})$:

- *For* $a_{j_1 j_1 0} \prod_{s_1=1}^{3} (a_{j_2 j_2 1} - a_{j_1 j_1 s_1}) > 0$ *and* $a_{j_2 j_2 0} \alpha_{j_1 j_1}^3 > 0$,

$$(\dot{x}_{j_1}^*, a_{j_2 j_2 1}) = \underbrace{(\mathrm{pF}, 3^{\mathrm{rd}}\mathrm{UP})}_{\text{third-order up-parabola flow } (+)} . \tag{4.307}$$

- *For* $a_{j_1 j_1 0} \prod_{s_1=1}^{3} (a_{j_2 j_2 1} - a_{j_1 j_1 s_1}) < 0$ *and* $a_{j_2 j_2 0} \alpha_{j_1 j_1}^3 > 0$,

$$(\dot{x}_{j_1}^*, a_{j_2 j_2 1}) = \underbrace{(\mathrm{nF}, 3^{\mathrm{rd}}\mathrm{DP})}_{\text{third-order down-parabola flow } (-)} . \tag{4.308}$$

- *For* $a_{j_1 j_1 0} \prod_{s_1=1}^{3} (a_{j_2 j_2 1} - a_{j_1 j_1 s_1}) > 0$ *and* $a_{j_2 j_2 0} \alpha_{j_1 j_1}^3 < 0$,

$$(\dot{x}_{j_1}^*, a_{j_2 j_2 1}) = \underbrace{(\mathrm{pF}, 3^{\mathrm{rd}}\mathrm{DP})}_{\text{third-order down-parabola flow } (+)} . \tag{4.309}$$

- *For* $a_{j_1j_10}\prod_{s_1=1}^{3}(a_{j_2j_21}-a_{j_1j_1s_1})<0$ *and* $a_{j_2j_20}\alpha_{j_1j_1}^{3}<0$,

$$(\dot{x}_{j_1}^{*},a_{j_2j_21})=\underbrace{(\mathrm{nF},3^{\mathrm{rd}}\mathrm{UP})}_{\text{third-order up-parabola flow }(-)}. \tag{4.310}$$

(vi_3) *For* $a_{j_2j_2l_1}=a_{j_1j_1s_1}(s_1,s_2\in\{1,2,3\})$,

$$\begin{aligned}
\dot{x}_{j_1}&=a_{j_1j_10}(y_{j_1}-a_{j_1j_1s_1})\prod_{s_2=1,s_2\neq s_1}^{3}(y_{j_1}-a_{j_1j_1s_2}),\\
\dot{x}_{j_2}&=a_{j_2j_20}(y_{j_1}-a_{j_2j_2l_1})(y_{j_1}-a_{j_2j_2l_2})^2,\\
&j_1,j_2\in\{1,2\};\ j_1\neq j_2;\ s_1,s_2\in\{1,2,3\},s_1\neq s_2.
\end{aligned} \tag{4.311}$$

For $a_{j_2j_2l_1}=a_{j_1j_1s_1}$, *there is a dynamical system in the* y_{j_1}*-direction as*

$$\begin{aligned}
\dot{y}_{j_1}&=(y_{j_1}-a_{j_1j_1s_1})\Big[\alpha_{j_1j_1}a_{j_1j_10}\prod_{s_2=1,s_2\neq s_1}^{3}(y_{j_1}-a_{j_1j_1s_2})\\
&\quad+\alpha_{j_1j_2}a_{j_2j_20}(y_{j_1}-a_{j_2j_2l_2})^2\Big],
\end{aligned} \tag{4.312}$$

with a dynamical system in the $y_{j_1}^{\perp}$*-direction as*

$$\begin{aligned}
\dot{y}_{j_1}^{\perp}&=(y_{j_1}-a_{j_1j_1s_1})\Big[\beta_{j_1j_1}a_{j_1j_10}\prod_{s_2=1,s_2\neq s_1}^{3}(y_{j_1}-a_{j_1j_1s_2})\\
&\quad+\beta_{j_1j_2}a_{j_2j_20}(y_{j_1}-a_{j_2j_2l_2})^2\Big],
\end{aligned} \tag{4.313}$$

where

$$\begin{aligned}
G_{j_1j_10}&=\frac{\partial}{\partial y_{j_1}}[\mathbf{n}^{\mathrm{T}}\cdot\mathbf{F}(y_{j_1},\mathbf{p})]\Big|_{y_{j_1}^{*}=a_{j_1j_1s_1}=a_{j_2j_2l_1}}\\
&=\alpha_{j_1j_1}a_{j_1j_10}\prod_{s_2=1,s_2\neq s_1}^{3}(a_{j_1j_1s_1}-a_{j_1j_1s_2})\\
&\quad+\alpha_{j_1j_2}a_{j_2j_20}(a_{j_2j_2l_1}-a_{j_2j_2l_2})^2.
\end{aligned} \tag{4.314}$$

The infinite-equilibrium flow of $y^*_{j_1} = a_{j_1 j_1 s_1} = a_{j_2 j_2 l_1}$ *has the following properties*:

- *For* $a_{j_1 j_1 0}\prod_{s_2=1,s_2\neq s_1}^{3}(a_{j_1 j_1 s_1}-a_{j_1 j_1 s_2})/a_{j_2 j_2 0}(a_{j_2 j_2 l_1}-a_{j_2 j_2 l_2})>0$ *and* $G_{j_1 j_1 0}>0$,

$$(a_{j_1 j_1 s_1}, a_{j_2 j_2 l_1}) = \underbrace{(\mathrm{II},\mathrm{SO})}_{\text{increasing-inflection source}}. \tag{4.315}$$

- *For* $a_{j_1 j_1 0}\prod_{s_2=1,s_2\neq s_1}^{3}(a_{j_1 j_1 s_1}-a_{j_1 j_1 s_2})/a_{j_2 j_2 0}(a_{j_2 j_2 l_1}-a_{j_2 j_2 l_2})>0$ *and* $G_{j_1 j_1 0}<0$,

$$(a_{j_1 j_1 s_1}, a_{j_2 j_2 l_1}) = \underbrace{(\mathrm{II},\mathrm{SI})}_{\text{increasing-inflection sink}}. \tag{4.316}$$

- *For* $a_{j_1 j_1 0}\prod_{s_2=1,s_2\neq s_1}^{3}(a_{j_1 j_1 s_1}-a_{j_1 j_1 s_2})/a_{j_2 j_2 0}(a_{j_2 j_2 l_1}-a_{j_2 j_2 l_2})<0$ *and* $G_{j_1 j_1 0}>0$,

$$(a_{j_1 j_1 s_1}, a_{j_2 j_2 l_1}) = \underbrace{(\mathrm{DI},\mathrm{SO})}_{\text{decreasing-inflection source}}. \tag{4.317}$$

- *For* $a_{j_1 j_1 0}\prod_{s_2=1,s_2\neq s_1}^{3}(a_{j_1 j_1 s_1}-a_{j_1 j_1 s_2})/a_{j_2 j_2 0}(a_{j_2 j_2 l_1}-a_{j_2 j_2 l_2})<0$ *and* $G_{j_1 j_1 0}<0$,

$$(a_{j_1 j_1 s_1}, a_{j_2 j_2 l_1}) = \underbrace{(\mathrm{DI},\mathrm{SI})}_{\text{decreasing-inflection sink}}. \tag{4.318}$$

(vi_4) *For* $a_{j_2 j_2 l_2} = a_{j_1 j_1 s_1} (s_1, s_2 \in \{1,2,3\})$,

$$\begin{aligned}
\dot{x}_{j_1} &= a_{j_1 j_1 0}(y_{j_1}-a_{j_1 j_1 s_1})\prod_{s_2=1,s_2\neq s_1}^{3}(y_{j_1}-a_{j_1 j_1 s_2}),\\
\dot{x}_{j_2} &= a_{j_2 j_2 0}(y_{j_1}-a_{j_2 j_2 l_1})(y_{j_1}-a_{j_2 j_2 l_2})^2,\\
&j_1, j_2 \in \{1,2\};\ j_1\neq j_2;\ s_1,s_2\in\{1,2,3\},\ s_1\neq s_2.
\end{aligned} \tag{4.319}$$

There is a dynamical system in the y_{j_1}*-direction as*

$$\begin{aligned}
\dot{y}_{j_1} = (y_{j_1}-a_{j_1 j_1 s_1})\Big[&\alpha_{j_1 j_1} a_{j_1 j_1 0}\prod_{s_2=1,s_2\neq s_1}^{3}(y_{j_1}-a_{j_1 j_1 s_2})\\
&+\alpha_{j_1 j_1} a_{j_2 j_2 0}(y_{j_1}-a_{j_2 j_2 l_1})(y_{j_1}-a_{j_2 j_2 l_2})\Big],
\end{aligned} \tag{4.320}$$

with a dynamical system in the $y_{j_1}^{\perp}$-direction as

$$\dot{y}_{j_1}^{\perp} = (y_{j_1} - a_{j_1 j_1 s_1})\Big[\beta_{j_1 j_1} a_{j_1 j_1 0} \prod_{s_2=1, s_2 \neq s_1}^{3} (y_{j_1} - a_{j_1 j_1 s_2}) + \beta_{j_1 j_1} a_{j_2 j_2 0}(y_{j_1} - a_{j_2 j_2 l_1})(y_{j_1} - a_{j_2 j_2 l_2})\Big], \tag{4.321}$$

where

$$G_{j_1 j_1 0} = \frac{\partial}{\partial y_{j_1}}[\mathbf{n}^{\mathrm{T}} \cdot \mathbf{F}(y_{j_1}, \mathbf{p})]\Big|_{y_{j_1}^* = a_{j_1 j_1 s_1} = a_{j_2 j_2 1}} = \alpha_{j_1 j_1} a_{j_1 j_1 0} \prod_{s_2=1, s_2 \neq s_1}^{3} (a_{j_1 j_1 s_1} - a_{j_1 j_1 s_2}). \tag{4.322}$$

The infinite-equilibrium of $y_{j_1}^* = a_{j_2 j_2 l_2} = a_{j_1 j_1 s_1}$ has the following properties:

- *For* $\alpha_{j_1 j_1} a_{j_1 j_1 0} \prod_{s_2=1, s_2 \neq s_1}^{3} (a_{j_1 j_1 s_1} - a_{j_1 j_1 s_2}) > 0$ *and* $a_{j_2 j_2 0} \alpha_{j_1 j_1}^2 \times (a_{j_2 j_2 l_2} - a_{j_2 j_2 l_1}) > 0$,

$$(a_{j_1 j_1 s_1}, a_{j_2 j_2 l_2}) = \underbrace{(\mathrm{SO}, \mathrm{UP})}_{\text{up-parabola source}}. \tag{4.323}$$

- *For* $\alpha_{j_1 j_1} a_{j_1 j_1 0} \prod_{s_2=1, s_2 \neq s_1}^{3} (a_{j_1 j_1 s_1} - a_{j_1 j_1 s_2}) < 0$ *and* $a_{j_2 j_2 0} \alpha_{j_1 j_1}^2 \times (a_{j_2 j_2 l_2} - a_{j_2 j_2 l_1}) > 0$,

$$(a_{j_1 j_1 s_1}, a_{j_2 j_2 l_2}) = \underbrace{(\mathrm{SI}, \mathrm{DP})}_{\text{down-parabola sink}}. \tag{4.324}$$

- *For* $\alpha_{j_1 j_1} a_{j_1 j_1 0} \prod_{s_2=1, s_2 \neq s_1}^{3} (a_{j_1 j_1 s_1} - a_{j_1 j_1 s_2}) > 0$ *and* $a_{j_2 j_2 0} \alpha_{j_1 j_1}^2 \times (a_{j_2 j_2 l_2} - a_{j_2 j_2 l_1}) < 0$,

$$(a_{j_1 j_1 s_1}, a_{j_2 j_2 l_2}) = \underbrace{(\mathrm{SO}, \mathrm{DP})}_{\text{down-parabola source}}. \tag{4.325}$$

- *For* $\alpha_{j_1 j_1} a_{j_1 j_1 0} \prod_{s_2=1, s_2 \neq s_1}^{3} (a_{j_1 j_1 s_1} - a_{j_1 j_1 s_2}) < 0$ *and* $a_{j_2 j_2 0} \alpha_{j_1 j_1}^2 \times (a_{j_2 j_2 l_2} - a_{j_2 j_2 l_1}) < 0$,

$$(a_{j_1 j_1 s_1}, a_{j_2 j_2 l_2}) = \underbrace{(\mathrm{SI}, \mathrm{UP})}_{\text{up-parabola sink}}. \tag{4.326}$$

(vi_5) For $a_{j_2j_21} = a_{j_1j_1s_1}(s_1, s_2 \in \{1,2,3\})$,

$$\begin{aligned}
\dot{x}_{j_1} &= a_{j_1j_10}(y_{j_1} - a_{j_1j_1s_1}) \prod_{s_2=1, s_2\neq s_1}^{3} (y_{j_1} - a_{j_1j_1s_2}),\\
\dot{x}_{j_2} &= a_{j_2j_20}(y_{j_1} - a_{j_2j_21})^3,\\
& j_1, j_2 \in \{1,2\};\ j_1 \neq j_2;\ s_1, s_2 \in \{1,2,3\}, s_1 \neq s_2.
\end{aligned} \tag{4.327}$$

For $a_{j_2j_21} = a_{j_1j_1s_1}$, there is a dynamical system in the y_{j_1}-direction as

$$\begin{aligned}
\dot{y}_{j_1} = (y_{j_1} - a_{j_1j_1s_1})\Big[&\alpha_{j_1j_1}a_{j_1j_10} \prod_{s_2=1, s_2\neq s_1}^{3} (y_{j_1} - a_{j_1j_1s_2})\\
&+ \alpha_{j_1j_1}a_{j_2j_20}(y_{j_1} - a_{j_2j_21})^2\Big],
\end{aligned} \tag{4.328}$$

with a dynamical system in the $y_{j_1}^{\perp}$-direction as

$$\begin{aligned}
\dot{y}_{j_1}^{\perp} = (y_{j_1} - a_{j_1j_1s_1})\Big[&\beta_{j_1j_1}a_{j_1j_10} \prod_{s_2=1, s_2\neq s_1}^{3} (y_{j_1} - a_{j_1j_1s_2})\\
&+ \beta_{j_1j_1}a_{j_2j_20}(y_{j_1} - a_{j_2j_21})^2\Big],
\end{aligned} \tag{4.329}$$

where

$$\begin{aligned}
G_{j_1j_10} &= \frac{\partial}{\partial y_{j_1}}[\mathbf{n}^{\mathrm{T}} \cdot \mathbf{F}(y_{j_1}, \mathbf{p})]\Big|_{y_{j_1}^* = a_{j_1j_1s_1} = a_{j_2j_21}}\\
&= \alpha_{j_1j_1}a_{j_1j_10} \prod_{s_2=1, s_2\neq s_1}^{3} (a_{j_1j_1s_1} - a_{j_1j_1s_2}).
\end{aligned} \tag{4.330}$$

The infinite-equilibrium flow of $y_{j_1}^* = a_{j_2j_21} = a_{j_1j_1s_1}$ has the following properties:

- *For $\alpha_{j_1j_1}a_{j_1j_10}\prod_{s_2=1,s_2\neq s_1}^{3}(a_{j_1j_1s_1} - a_{j_1j_1s_2}) > 0$ and $a_{j_2j_20}\alpha_{j_1j_1}^3 > 0$,*

$$(a_{j_1j_1s_1}, a_{j_2j_21}) = \underbrace{(\mathrm{SO}, 2^{\mathrm{nd}}\mathrm{II})}_{\text{second-order increasing-inflection source}}. \tag{4.331}$$

- *For $\alpha_{j_1j_1}a_{j_1j_10}\prod_{s_2=1,s_2\neq s_1}^{3}(a_{j_1j_1s_1} - a_{j_1j_1s_2}) < 0$ and $a_{j_2j_20}\alpha_{j_1j_1}^3 > 0$,*

$$(a_{j_1j_1s_1}, a_{j_2j_21}) = \underbrace{(\mathrm{SI}, 2^{\mathrm{nd}}\mathrm{DI})}_{\text{second-order decreasing-inflection sink}}. \tag{4.332}$$

- *For* $\alpha_{j_1j_1}a_{j_1j_10}\prod_{s_2=1,s_2\neq s_1}^{3}(a_{j_1j_1s_1}-a_{j_1j_1s_2})>0$ *and* $a_{j_2j_20}\alpha_{j_1j_1}^3<0$,

$$(a_{j_1j_1s_1},a_{j_2j_21})=\underbrace{(\mathrm{SO},2^{\mathrm{nd}}\mathrm{DI})}_{\text{second-order decreasing-inflection source}}. \tag{4.333}$$

- *For* $\alpha_{j_1j_1}a_{j_1j_10}\prod_{s_2=1,s_2\neq s_1}^{3}(a_{j_1j_1s_1}-a_{j_1j_1s_2})<0$ *and* $a_{j_2j_20}\alpha_{j_1j_1}^3<0$,

$$(a_{j_1j_1s_1},a_{j_2j_21})=\underbrace{(\mathrm{SI},2^{\mathrm{nd}}\mathrm{II})}_{\text{second-order increasing-inflection sink}}. \tag{4.334}$$

(vii) If

$$\Delta_{j_1j_1}<0 \quad \text{and} \quad \Delta_{j_2j_2}>0, \tag{4.335}$$

then

$$\begin{aligned}\dot{x}_{j_1}&=a_{j_1j_10}(y_{j_1}-a_{j_1j_11})[(y_{j_1}-a_{j_1j_1})^2+b_{j_1j_1}],\\ \dot{x}_{j_2}&=a_{j_2j_20}(y_{j_1}-a_{j_2j_21})(y_{j_1}-a_{j_2j_22})(y_{j_1}-a_{j_2j_23}),\end{aligned} \tag{4.336}$$

where

$$\begin{aligned}&\{a_{j_2j_21},a_{j_2j_22},a_{j_2j_23}\}=\mathrm{sort}\{b_{j_2j_21},b_{j_2j_22},b_{j_2j_23}\},a_{j_2j_2l_1}<a_{j_2j_2l_2},l_1<l_2;\\ &b_{j_2j_22},b_{j_2j_23}=-\frac{1}{2}(B_{j_2j_2}\pm\sqrt{\Delta_{j_2j_2}}),a_{j_2j_20}=A_{j_2j_20};\\ &a_{j_1j_11}=b_{j_1j_11},a_{j_1j_1}=-\frac{1}{2}B_{j_1j_11},b_{j_1j_1}=-\frac{1}{4}\Delta_{j_1j_1},a_{j_1j_10}=A_{j_1j_10}.\end{aligned} \tag{4.337}$$

(vii₁) For $a_{j_1j_11}\neq a_{j_2j_2l_1}(l_1,l_2\in\{1,2\})$, *there is a dynamical system in the* y_{j_1}*-direction as*

$$\begin{aligned}\dot{y}_{j_1}&=\alpha_{j_1j_1}a_{j_1j_10}(y_{j_1}-a_{j_1j_11})[(y_{j_1}-a_{j_1j_1})^2+b_{j_1j_1}]\\ &\quad+\alpha_{j_1j_2}a_{j_2j_20}(y_{j_1}-a_{j_2j_21})(y_{j_1}-a_{j_2j_22})(y_{j_1}-a_{j_2j_23}),\end{aligned} \tag{4.338}$$

with a dynamical system in the $y_{j_1}^{\perp}$*-direction as*

$$\begin{aligned}\dot{y}_{j_1}^{\perp}&=\beta_{j_1j_1}a_{j_1j_10}(y_{j_1}-a_{j_1j_11})[(y_{j_1}-a_{j_1j_1})^2+b_{j_1j_1}]\\ &\quad+\beta_{j_1j_2}a_{j_2j_20}(y_{j_1}-a_{j_2j_21})(y_{j_1}-a_{j_2j_22})(y_{j_1}-a_{j_2j_23}),\end{aligned} \tag{4.339}$$

(vii$_{1a}$) The flow of $y_1^*=a_{j_1j_11}$ *with* $\dot{x}_{j_2}^*=a_{j_2j_20}\prod_{l_1=1}^{3}(a_{j_1j_11}-a_{j_2j_2l_1})$ *possesses the following properties*:

- *For* $\alpha_{j_1j_1}a_{j_1j_10}>0$ *and* $a_{j_2j_20}\prod_{l_1=1}^{3}(a_{j_1j_11}-a_{j_2j_2l_1})>0$,

$$(a_{j_1j_11},\dot{x}_{j_2}^*)=\underbrace{(\mathrm{UP},\mathrm{pF})}_{\text{up-parabola flow }(+)}. \tag{4.340}$$

- *For* $\alpha_{j_1j_1}a_{j_1j_10} < 0$ *and* $a_{j_2j_20}\prod_{l_1=1}^{3}(a_{j_1j_11} - a_{j_2j_2l_1}) > 0$,

$$(a_{j_1j_11}, \dot{x}^*_{j_2}) = \underbrace{(\text{DP}, \text{pF})}_{\text{down-parabola flow } (+)}. \tag{4.341}$$

- *For* $\alpha_{j_1j_1}a_{j_1j_10} > 0$ *and* $a_{j_2j_20}\prod_{l_1=1}^{3}(a_{j_1j_11} - a_{j_2j_2l_1}) < 0$,

$$(a_{j_1j_11}, \dot{x}^*_{j_2}) = \underbrace{(\text{DP}, \text{nF})}_{\text{down-parabola flow } (-)}. \tag{4.342}$$

- *For* $\alpha_{j_1j_1}a_{j_1j_10} < 0$ *and* $a_{j_2j_20}\prod_{l_1=1}^{3}(a_{j_1j_11} - a_{j_2j_2l_1}) < 0$,

$$(a_{j_1j_11}, \dot{x}^*_{j_2}) = \underbrace{(\text{UP}, \text{nF})}_{\text{up-parabola flow } (-)}. \tag{4.343}$$

(vii_{1b}) *The flow of* $y_1^* = a_{j_2j_2l_1}$ *with* $\dot{x}^*_{j_1} = a_{j_1j_10}(a_{j_2j_2l_1} - a_{j_1j_11})[(a_{j_2j_21} - a_{j_1j_1})^2 + b_{j_1j_1}]$ *possesses the following properties*:

- *For* $a_{j_1j_10}(a_{j_2j_2l_1} - a_{j_1j_11}) > 0$ *and* $a_{j_2j_20}\alpha_{j_1j_1}\prod_{l_2=1,l_2\neq l_1}^{3}(a_{j_2j_2l_1} - a_{j_2j_2l_2}) > 0$,

$$(\dot{x}^*_{j_1}, a_{j_2j_2l_1}) = \underbrace{(\text{pF}, \text{UP})}_{\text{up-parabola flow}(+)}. \tag{4.344}$$

- *For* $a_{j_1j_10}(a_{j_2j_2l_1} - a_{j_1j_11}) < 0$ *and* $a_{j_2j_20}\alpha_{j_1j_1}\prod_{l_2=1,l_2\neq l_1}^{3}(a_{j_2j_2l_1} - a_{j_2j_2l_2}) > 0$,

$$(\dot{x}^*_{j_1}, a_{j_2j_2l_1}) = \underbrace{(\text{nF}, \text{DP})}_{\text{down-parabola flow } (-)}. \tag{4.345}$$

- *For* $a_{j_1j_10}(a_{j_2j_2l_1} - a_{j_1j_11}) > 0$ *and* $a_{j_2j_20}\alpha_{j_1j_1}\prod_{l_2=1,l_2\neq l_1}^{3}(a_{j_2j_2l_1} - a_{j_2j_2l_2}) < 0$,

$$(\dot{x}^*_{j_1}, a_{j_2j_2l_1}) = \underbrace{(\text{pF}, \text{DP})}_{\text{down-parabola flow } (+)}. \tag{4.346}$$

- *For* $a_{j_1j_10}(a_{j_2j_2l_1} - a_{j_1j_11}) < 0$ *and* $a_{j_2j_20}\alpha_{j_1j_1}\prod_{l_2=1,l_2\neq l_1}^{3}(a_{j_2j_2l_1} - a_{j_2j_2l_2}) < 0$,

$$(\dot{x}^*_{j_1}, a_{j_2j_2l_1}) = \underbrace{(\text{nF}, \text{UP})}_{\text{up-parabola flow}(-)}. \tag{4.347}$$

(vii$_2$) For $a_{j_1j_11} = a_{j_2j_2l_1}$,

$$\begin{aligned}\dot{x}_{j_1} &= a_{j_1j_10}(y_{j_1} - a_{j_1j_11})[(y_{j_1} - a_{j_1j_1})^2 + b_{j_1j_1}],\\ \dot{x}_{j_2} &= a_{j_2j_20}(y_{j_1} - a_{j_2j_2l_1}) \prod_{l_2=1,l_2\neq l_1}^{3} (y_{j_1} - a_{j_2j_2l_2}).\end{aligned} \tag{4.348}$$

There is a dynamical system in the y_{j_1}-direction as

$$\begin{aligned}\dot{y}_{j_1} = (y_{j_1} - a_{j_1j_11})\Big\{&\alpha_{j_1j_1}a_{j_1j_10}[(y_{j_1} - a_{j_1j_1})^2 + b_{j_1j_1}]\\ &+ \alpha_{j_1j_2}a_{j_2j_20} \prod_{l_2=1,l_2\neq l_1}^{3} (y_{j_1} - a_{j_2j_2l_2})\Big\},\end{aligned} \tag{4.349}$$

with a dynamical system in the $y_{j_1}^{\perp}$-direction as

$$\begin{aligned}\dot{y}_{j_1}^{\perp} = (y_{j_1} - a_{j_1j_11})\Big\{&\beta_{j_1j_1}a_{j_1j_10}[(y_{j_1} - a_{j_1j_1})^2 + b_{j_1j_1}]\\ &+ \beta_{j_1j_2}a_{j_2j_20} \prod_{l_2=1,l_2\neq l_1}^{3} (y_{j_1} - a_{j_2j_2l_2})\Big\},\end{aligned} \tag{4.350}$$

where

$$\begin{aligned}G_{j_1j_10} &= \frac{\partial}{\partial y_{j_1}}[\mathbf{n}^{\mathrm{T}} \cdot \mathbf{F}(y_{j_1}, \mathbf{p})]\Big|_{y_{j_1}^{*}=a_{j_1j_11}=a_{j_2j_2l_1}}\\ &= \alpha_{j_1j_1}a_{j_1j_10}[(a_{j_1j_11} - a_{j_1j_1})^2 + b_{j_1j_1}]\\ &\quad + \alpha_{j_1j_2}a_{j_2j_20} \prod_{l_2=1,l_2\neq l_1}^{3} (a_{j_2j_2l_1} - a_{j_2j_2l_2}).\end{aligned} \tag{4.351}$$

The infinite-equilibrium flow of $y_1^{*} = a_{j_1j_11}$ possesses the following properties:

- *For $a_{j_1j_10}/a_{j_2j_20}(a_{j_2j_2l_1} - a_{j_2j_2l_2}) > 0$ and $G_{j_1j_10} > 0$,*

$$(a_{j_1j_11}, a_{j_2j_2l_1}) = \underbrace{(\mathrm{II}, \mathrm{SO})}_{\text{increaing-inflection source}}. \tag{4.352}$$

- *For $a_{j_1j_10}/a_{j_2j_20}(a_{j_2j_2l_1} - a_{j_2j_2l_2}) > 0$ and $G_{j_1j_10} < 0$,*

$$(a_{j_1j_11}, a_{j_2j_2l_1}) = \underbrace{(\mathrm{II}, \mathrm{SI})}_{\text{increaing-inflection sink}}. \tag{4.353}$$

- *For* $a_{j_1j_10}/a_{j_2j_20}(a_{j_2j_2l_1} - a_{j_2j_2l_2}) < 0$ *and* $G_{j_1j_10} > 0$,

$$(a_{j_1j_11}, a_{j_2j_2l_1}) = \underbrace{(\mathrm{DI}, \mathrm{SO})}_{\text{decreaing-inflection source}}. \tag{4.354}$$

- *For* $a_{j_1j_10}/a_{j_2j_20}(a_{j_2j_2l_1} - a_{j_2j_2l_2}) < 0$ *and* $G_{j_1j_10} < 0$,

$$(a_{j_1j_11}, a_{j_2j_2l_1}) = \underbrace{(\mathrm{DI}, \mathrm{SI})}_{\text{decreaing-inflection sink}}. \tag{4.355}$$

(viii) If

$$\Delta_{j_1j_1} = 0 \quad \text{and} \quad \Delta_{j_2j_2} > 0, \tag{4.356}$$

then

$$\begin{aligned}
\dot{x}_{j_1} &= a_{j_1j_10}(y_{j_1} - a_{j_1j_1s_1})(y_{j_1} - a_{j_1j_1s_2})^2, \\
\dot{x}_{j_2} &= a_{j_2j_20}(y_{j_1} - a_{j_2j_21})(y_{j_1} - a_{j_2j_22})(y_{j_1} - a_{j_2j_23}), \\
j_1, j_2 &\in \{1, 2\};\ j_1 \neq j_2;
\end{aligned} \tag{4.357}$$

where

$$\begin{aligned}
\{a_{j_2j_21}, a_{j_2j_22}, a_{j_2j_23}\} &= \mathrm{sort}\{b_{j_2j_21}, b_{j_2j_22}, b_{j_2j_23}\}, a_{j_2j_2l_1} < a_{j_2j_2l_2}, l_1 < l_2; \\
b_{j_2j_22}, b_{j_2j_23} &= -\frac{1}{2}(B_{j_2j_2} \pm \sqrt{\Delta_{j_2j_2}}), a_{j_2j_20} = A_{j_2j_20}; \\
a_{j_1j_1s_1} = b_{j_1j_11}, a_{j_1j_1s_2} &= -\frac{1}{2}B_{j_1j_1}, b_{j_1j_1} = 0, a_{j_1j_10} = A_{j_1j_10}.
\end{aligned} \tag{4.358}$$

(viii$_1$) or $a_{j_2j_21} \neq a_{j_1j_1s_1}, a_{j_1j_1s_2}$, *there is a dynamical system in the* y_{j_1}-*direction as*

$$\begin{aligned}
\dot{y}_{j_1} &= \alpha_{j_1j_1}a_{j_1j_10}(y_{j_1} - a_{j_1j_1s_1})(y_{j_1} - a_{j_1j_1s_2})^2 \\
&\quad + \alpha_{j_1j_2}a_{j_2j_20}\prod_{l_1=1}^{3}(y_{j_1} - a_{j_2j_2l_1}),
\end{aligned} \tag{4.359}$$

with a dynamical system in the $y_{j_1}^{\perp}$-*direction as*

$$\begin{aligned}
\dot{y}_{j_1}^{\perp} &= \beta_{j_1j_1}a_{j_1j_10}(y_{j_1} - a_{j_1j_1s_1})(y_{j_1} - a_{j_1j_1s_2})^2 \\
&\quad + \beta_{j_1j_2}a_{j_2j_20}\prod_{l_1=1}^{3}(y_{j_1} - a_{j_2j_2l_1}).
\end{aligned} \tag{4.360}$$

(viii$_{1a}$) The flow of $y_{j_1}^* = a_{j_1j_1s_1} \neq a_{j_2j_21}$ *with* $\dot{x}_{j_2}^* = a_{j_2j_20}\prod_{l_1=1}^{3}(a_{j_1j_1s_1} - a_{j_2j_2l_1})^2$ *for the* x_{j_1}-*direction has the following properties*:

- *For* $a_{j_1j_10}\alpha_{j_1j_2}(a_{j_1j_1s_1} - a_{j_1j_1s_2})^2 > 0$ *and* $a_{j_2j_20}\prod_{l_1=1}^{3}(a_{j_1j_1s_1} - a_{j_2j_2l_1}) > 0$,

$$(a_{j_1j_1s_1}, \dot{x}_{j_2}^*) = \underbrace{(\mathrm{UP}, \mathrm{pF})}_{\text{up-parabola flow }(+)}. \tag{4.361}$$

- For $a_{j_1j_10}\alpha_{j_1j_2}(a_{j_1j_1s_1} - a_{j_1j_1s_2})^2 < 0$ and $a_{j_2j_20}\prod_{l_1=1}^{3}(a_{j_1j_1s_1} - a_{j_2j_2l_1}) > 0$,

$$(a_{j_1j_1s_1}, \dot{x}^*_{j_2}) = \underbrace{(\text{DP}, \text{pF})}_{\text{down-parabola flow } (+)}. \tag{4.362}$$

- For $a_{j_1j_10}\alpha_{j_1j_2}(a_{j_1j_1s_1} - a_{j_1j_1s_2})^2 > 0$ and $a_{j_2j_20}\prod_{l_1=1}^{3}(a_{j_1j_1s_1} - a_{j_2j_2l_1}) < 0$,

$$(a_{j_1j_1s_1}, \dot{x}^*_{j_2}) = \underbrace{(\text{DP}, \text{nF})}_{\text{down-parabola flow } (-)}. \tag{4.363}$$

- For $a_{j_1j_10}\alpha_{j_1j_2}(a_{j_1j_1s_1} - a_{j_1j_1s_2})^2 < 0$ and $a_{j_2j_20}\prod_{l_1=1}^{3}(a_{j_1j_1s_1} - a_{j_2j_2l_1}) < 0$,

$$(a_{j_1j_1s_1}, \dot{x}^*_{j_2}) = \underbrace{(\text{UP}, \text{nF})}_{\text{up-parabola flow } (-)}. \tag{4.364}$$

($viii_{1b}$) The flows of $y^*_{j_1} = a_{j_1j_1s_2} \neq a_{j_2j_2l_1}(l_1 = 1, 2, 3)$ for the x_{j_1}-direction with $a_{j_1j_11} \neq a_{j_2j_21}$ have the following properties with $\dot{x}^*_{j_2} = a_{j_2j_20}\prod_{l_1=1}^{3}(a_{j_1j_1s_2} - a_{j_2j_2l_1})$:

- For $a_{j_1j_10}\alpha^2_{j_1j_2}(a_{j_1j_1s_2} - a_{j_1j_1s_1}) > 0$ and $a_{j_2j_20}\prod_{l_1=1}^{3}(a_{j_1j_1s_2} - a_{j_2j_2l_1}) > 0$,

$$(a_{j_1j_11}, \dot{x}^*_{j_2}) = \underbrace{(\text{II}, \text{pF})}_{\text{increasing-inflection flow } (+)}. \tag{4.365}$$

- For $a_{j_1j_10}\alpha^2_{j_1j_2}(a_{j_1j_1s_2} - a_{j_1j_1s_1}) > 0$ and $a_{j_2j_20}\prod_{l_1=1}^{3}(a_{j_1j_1s_2} - a_{j_2j_2l_1}) > 0$,

$$(a_{j_1j_11}, \dot{x}^*_{j_2}) = \underbrace{(\text{DI}, \text{pF})}_{\text{decreasing-inflection flow } (+)}. \tag{4.366}$$

- For $a_{j_1j_10}\alpha^2_{j_1j_2}(a_{j_1j_1s_2} - a_{j_1j_1s_1}) > 0$ and $a_{j_2j_20}\prod_{l_1=1}^{3}(a_{j_1j_1s_2} - a_{j_2j_2l_1}) < 0$,

$$(a_{j_1j_11}, \dot{x}^*_{j_2}) = \underbrace{(\text{DI}, \text{nF})}_{\text{decreasing-inflection flow } (-)}. \tag{4.367}$$

- For $a_{j_1j_10}\alpha^2_{j_1j_2}(a_{j_1j_1s_2} - a_{j_1j_1s_1}) < 0$ and $a_{j_2j_20}\prod_{l_1=1}^{3}(a_{j_1j_1s_2} - a_{j_2j_2l_1}) < 0$,

$$(a_{j_1j_11}, \dot{x}^*_{j_2}) = \underbrace{(\text{II}, \text{nF})}_{\text{increasing-inflection flow } (-)}. \tag{4.368}$$

– *The increasing- and decreasing-inflection flows are the appearing bifurcations for the up-parabola and down-parabola flows.*

(viii$_{1c}$) The flow of $y_{j_1}^* = a_{j_2j_2l_1} \neq a_{j_1j_1s_1}, a_{j_1j_1s_2}$ for the x_{j_2}-direction has the following properties with $\dot{x}_{j_1}^* = a_{j_1j_10}(a_{j_2j_2l_1} - a_{j_1j_1s_1})(a_{j_2j_2l_1} - a_{j_1j_1s_2})^2$:

- *For $\dot{x}_{j_1}^* > 0$ and $a_{j_2j_20}\alpha_{j_1j_1}\prod_{l_2=1,l_2\neq l_1}^{3}(a_{j_2j_2l_1} - a_{j_2j_2l_2}) > 0$,*

$$(\dot{x}_{j_1}^*, a_{j_2j_2l_1}) = \underbrace{(\text{pF}, \text{UP})}_{\text{up-parabola flow }(+)}. \tag{4.369}$$

- *For $\dot{x}_{j_1}^* < 0$ and $a_{j_2j_20}\alpha_{j_1j_1}\prod_{l_2=1,l_2\neq l_1}^{3}(a_{j_2j_2l_1} - a_{j_2j_2l_2}) > 0$,*

$$(\dot{x}_{j_1}^*, a_{j_2j_2l_1}) = \underbrace{(\text{nF}, \text{DP})}_{\text{down-parabola flow }(-)}. \tag{4.370}$$

- *For $\dot{x}_{j_1}^* > 0$ and $a_{j_2j_20}\alpha_{j_1j_1}\prod_{l_2=1,l_2\neq l_1}^{3}(a_{j_2j_2l_1} - a_{j_2j_2l_2}) < 0$,*

$$(\dot{x}_{j_1}^*, a_{j_2j_2l_1}) = \underbrace{(\text{pF}, \text{DP})}_{\text{down-parabola flow }(+)}. \tag{4.371}$$

- *For $\dot{x}_{j_1}^* < 0$ and $a_{j_2j_20}\alpha_{j_1j_1}\prod_{l_2=1,l_2\neq l_1}^{3}(a_{j_2j_2l_1} - a_{j_2j_2l_2}) < 0$,*

$$(\dot{x}_{j_1}^*, a_{j_2j_2l_1}) = \underbrace{(\text{nF}, \text{UP})}_{\text{up-parabola flow }(-)}. \tag{4.372}$$

(viii$_2$) For $a_{j_1j_1s_1} = a_{j_1j_1s_2} \equiv a_{j_1j_11}$,

$$\begin{aligned}
\dot{x}_{j_1} &= a_{j_1j_10}(y_{j_1} - a_{j_1j_11})^3,\\
\dot{x}_{j_2} &= a_{j_2j_20}\prod_{l_1=1}^{3}(y_{j_1} - a_{j_2j_2l_1}),\\
& j_1, j_2 \in \{1,2\};\ j_1 \neq j_2.
\end{aligned} \tag{4.373}$$

For $a_{j_2j_2l_1} \neq a_{j_1j_11}$, there is a dynamical system in the y_{j_1}-direction as

$$\begin{aligned}
\dot{y}_{j_1} &= \alpha_{j_1j_1}a_{j_1j_10}(y_{j_1} - a_{j_1j_1s_1})(y_{j_1} - a_{j_1j_1s_2})^2\\
&\quad + \alpha_{j_1j_2}a_{j_2j_20}\prod_{l_1=1}^{3}(y_{j_1} - a_{j_2j_2l_1}),
\end{aligned} \tag{4.374}$$

with a dynamical system in the $y_{j_1}^{\perp}$-direction as

$$\begin{aligned}
\dot{y}_{j_1}^{\perp} &= \beta_{j_1j_1}a_{j_1j_10}(y_{j_1} - a_{j_1j_1s_1})(y_{j_1} - a_{j_1j_1s_2})^2\\
&\quad + \beta_{j_1j_2}a_{j_2j_20}\prod_{l_1=1}^{3}(y_{j_1} - a_{j_2j_2l_1}).
\end{aligned} \tag{4.375}$$

($viii_{2a}$) The flow of $y^*_{j_1} = a_{j_1j_11} \neq a_{j_2j_2l_1}(l_1 = 1, 2, 3)$ for the x_{j_1}-direction has the following properties with $\dot{x}^*_{j_2} = a_{j_2j_20}\prod_{l_1=1}^{3}(y_{j_1} - a_{j_2j_2l_1})$:

- *For $a_{j_1j_10}\alpha^3_{j_1j_2} > 0$ and $a_{j_2j_20}\prod_{l_1=1}^{3}(y_{j_1} - a_{j_2j_2l_1}) > 0$,*

$$(a_{j_1j_1s_1}, \dot{x}^*_{j_2}) = \underbrace{(3^{\text{rd}}\text{UP}, \text{pF})}_{\text{third-order up-parabola flow } (+)}. \tag{4.376}$$

- *For $a_{j_1j_10}\alpha^3_{j_1j_2} < 0$ and $a_{j_2j_20}\prod_{l_1=1}^{3}(y_{j_1} - a_{j_2j_2l_1}) > 0$,*

$$(a_{j_1j_1s_1}, \dot{x}^*_{j_2}) = \underbrace{(3^{\text{rd}}\text{DP}, \text{pF})}_{\text{third-order down-parabola flow } (+)}. \tag{4.377}$$

- *For $a_{j_1j_10}\alpha^3_{j_1j_2} > 0$ and $a_{j_2j_20}\prod_{l_1=1}^{3}(y_{j_1} - a_{j_2j_2l_1}) < 0$,*

$$(a_{j_1j_1s_1}, \dot{x}^*_{j_2}) = \underbrace{(3^{\text{rd}}\text{DP}, \text{nF})}_{\text{third-order down-parabola flow } (-)}. \tag{4.378}$$

- *For $a_{j_1j_10}\alpha^3_{j_1j_2} < 0$ and $a_{j_2j_20}\prod_{l_1=1}^{3}(y_{j_1} - a_{j_2j_2l_1}) < 0$,*

$$(a_{j_1j_1s_1}, \dot{x}^*_{j_2}) = \underbrace{(3^{\text{rd}}\text{UP}, \text{nF})}_{\text{third-order up-parabola flow } (-)}. \tag{4.379}$$

- *The third-order up-parabola flows are the appearing bifurcations for up-parabola, down-parabola, and up-parabola flows.*
- *The third-order down-parabola flows are the appearing bifurcations for down-parabola, up-parabola, and down-parabola flows.*

($viii_{2b}$) The flow of $y^*_{j_1} = a_{j_2j_2l_1} \neq a_{j_1j_11}$ for the x_{j_2}-direction has the following properties with $\dot{x}^*_{j_1} = a_{j_1j_10}(a_{j_2j_2l_1} - a_{j_1j_11})^3$:

- *For $a_{j_1j_10}(a_{j_2j_21} - a_{j_1j_11})^3 > 0$ and $a_{j_2j_20}\alpha_{j_1j_1}\prod_{l_2=1,l_2\neq l_1}^{3}(a_{j_2j_2l_1} - a_{j_2j_2l_2}) > 0$,*

$$(\dot{x}^*_{j_1}, a_{j_2j_2l_1}) = \underbrace{(\text{pF}, \text{UP})}_{\text{up-parabola flow } (+)}. \tag{4.380}$$

- *For $a_{j_1j_10}(a_{j_2j_21} - a_{j_1j_11})^3 < 0$ and $a_{j_2j_20}\alpha_{j_1j_1}\prod_{l_2=1,l_2\neq l_1}^{3}(a_{j_2j_2l_1} - a_{j_2j_2l_2}) > 0$,*

$$(\dot{x}^*_{j_1}, a_{j_2j_2l_1}) = \underbrace{(\text{nF}, \text{DP})}_{\text{down-parabola flow } (-)}. \tag{4.381}$$

- *For* $a_{j_1j_10}(a_{j_2j_21} - a_{j_1j_11})^3 > 0$ *and* $a_{j_2j_20}\alpha_{j_1j_1}\prod_{l_2=1,l_2\neq l_1}^{3}(a_{j_2j_2l_1} - a_{j_2j_2l_2}) < 0$,

$$(\dot{x}^*_{j_1}, a_{j_2j_2l_1}) = \underbrace{(\mathrm{pF}, \mathrm{DP})}_{\text{down-parabola flow } (+)}. \tag{4.382}$$

- *For* $a_{j_1j_10}(a_{j_2j_21} - a_{j_1j_11})^3 < 0$ *and* $a_{j_2j_20}\alpha_{j_1j_1}\prod_{l_2=1,l_2\neq l_1}^{3}(a_{j_2j_2l_1} - a_{j_2j_2l_2}) < 0$,

$$(\dot{x}^*_{j_1}, a_{j_2j_2l_1}) = \underbrace{(\mathrm{nF}, \mathrm{UP})}_{\text{up-parabola flow } (-)}. \tag{4.383}$$

$(viii_3)$ *For* $a_{j_1j_1s_1} = a_{j_2j_2l_1}$,

$$\begin{aligned}\dot{x}_{j_1} &= a_{j_1j_10}(y_{j_1} - a_{j_1j_1s_1})(y_{j_1} - a_{j_1j_1s_2})^2,\\ \dot{x}_{j_2} &= a_{j_2j_20}(y_{j_1} - a_{j_2j_2l_1})\prod_{l_2=1,l_2\neq l_1}^{3}(y_{j_1} - a_{j_2j_2l_1}).\end{aligned} \tag{4.384}$$

There is a dynamical system in the y_{j_1}*-direction as*

$$\begin{aligned}\dot{y}_{j_1} = (y_{j_1} - a_{j_1j_1s_1})\Big[&\alpha_{j_1j_1}a_{j_1j_10}(y_{j_1} - a_{j_1j_1s_2})^2\\ &+ \alpha_{j_1j_2}a_{j_2j_20}\prod_{l_2=1,l_2\neq l_1}^{3}(y_{j_1} - a_{j_2j_2l_1})\Big],\end{aligned} \tag{4.385}$$

with a dynamical system in the $y^{\perp}_{j_1}$*-direction as*

$$\begin{aligned}\dot{y}^{\perp}_{j_1} = (y_{j_1} - a_{j_1j_1s_1})\Big[&\beta_{j_1j_1}a_{j_1j_10}(y_{j_1} - a_{j_1j_1s_2})^2\\ &+ \beta_{j_1j_2}a_{j_2j_20}\prod_{l_2=1,l_2\neq l_1}^{3}(y_{j_1} - a_{j_2j_2l_1})\Big],\end{aligned} \tag{4.386}$$

where

$$\begin{aligned}G_{j_1j_10} &= \frac{\partial}{\partial y_{j_1}}[\mathbf{n}^{\mathrm{T}} \cdot \mathbf{F}(y_{j_1}, \mathbf{p})]\Big|_{y^*_{j_1}=a_{j_1j_1s_1}=a_{j_2j_2l_1}}\\ &= \alpha_{j_1j_1}a_{j_1j_10}(a_{j_1j_1s_1} - a_{j_1j_1s_2})^2\\ &\quad + \alpha_{j_1j_2}a_{j_2j_20}\prod_{l_2=1,l_2\neq l_1}^{3}(a_{j_2j_2l_1} - a_{j_2j_2l_2}).\end{aligned} \tag{4.387}$$

The infinite-equilibrium flows of $y_1^* = a_{j_1j_1s_1} = a_{j_2j_2l_1}$ possess the flowing properties:

- *For* $a_{j_1j_10}(a_{j_1j_1s_1} - a_{j_1j_1s_2})^2/a_{j_2j_20}\prod_{l_2=1,l_2\neq l_1}^{3}(a_{j_2j_2l_1} - a_{j_2j_2l_2}) > 0$ *and* $G_{j_1j_10} > 0$,

$$(a_{j_1j_1s_1}, a_{j_2j_2l_1}) = \underbrace{(\mathrm{II}, \mathrm{SO})}_{\text{increasing-inflection source}}. \tag{4.388}$$

- *For* $a_{j_1j_10}(a_{j_1j_1s_1} - a_{j_1j_1s_2})^2/a_{j_2j_20}\prod_{l_2=1,l_2\neq l_1}^{3}(a_{j_2j_2l_1} - a_{j_2j_2l_2}) > 0$ *and* $G_{j_1j_10} < 0$,

$$(a_{j_1j_1s_1}, a_{j_2j_2l_1}) = \underbrace{(\mathrm{II}, \mathrm{SI})}_{\text{increasing-inflection sink}}. \tag{4.389}$$

- *For* $a_{j_1j_10}(a_{j_1j_1s_1} - a_{j_1j_1s_2})^2/a_{j_2j_20}\prod_{l_2=1,l_2\neq l_1}^{3}(a_{j_2j_2l_1} - a_{j_2j_2l_2}) < 0$ *and* $G_{j_1j_10} > 0$,

$$(a_{j_1j_1s_1}, a_{j_2j_2l_1}) = \underbrace{(\mathrm{DI}, \mathrm{SO})}_{\text{decreasing-inflection source}}. \tag{4.390}$$

- *For* $a_{j_1j_10}(a_{j_1j_1s_1} - a_{j_1j_1s_2})^2/a_{j_2j_20}\prod_{l_2=1,l_2\neq l_1}^{3}(a_{j_2j_2l_1} - a_{j_2j_2l_2}) < 0$ *and* $G_{j_1j_10} < 0$,

$$(a_{j_1j_1s_1}, a_{j_2j_2l_1}) = \underbrace{(\mathrm{DI}, \mathrm{SI})}_{\text{increasing-inflection sink}}. \tag{4.391}$$

– *The inflection-source and sink infinite-equilibriums are the switching bifurcations for parabola flows in the x_{j_1}- and x_{j_2}-directions.*

($viii_4$) *For* $a_{j_1j_1s_2} = a_{j_2j_2l_1}$,

$$\begin{aligned}\dot{x}_{j_1} &= a_{j_1j_10}(y_{j_1} - a_{j_1j_1s_1})(y_{j_1} - a_{j_1j_1s_2})^2,\\ \dot{x}_{j_2} &= a_{j_2j_20}(y_{j_1} - a_{j_2j_2l_1})\prod_{l_2=1,l_2\neq l_1}^{3}(y_{j_1} - a_{j_2j_2l_2}),\end{aligned} \tag{4.392}$$

$j_1, j_2 \in \{1,2\};\ j_1 \neq j_2.$

There is a dynamical system in the y_{j_1}-direction as

$$\begin{aligned}\dot{y}_{j_1} = (y_{j_1} - a_{j_1j_1s_2})\Big[&\alpha_{j_1j_1}a_{j_1j_10}(y_{j_1} - a_{j_1j_1s_1})(y_{j_1} - a_{j_1j_1s_2})\\ &+ \alpha_{j_1j_2}a_{j_2j_20}\prod_{l_2=1,l_2\neq l_1}^{3}(y_{j_1} - a_{j_2j_2l_2})\Big].\end{aligned} \tag{4.393}$$

With a dynamical system in the $y_{j_1}^{\perp}$-direction as

$$\dot{y}_{j_1}^{\perp} = (y_{j_1} - a_{j_1 j_1 s_2})\Big[\beta_{j_1 j_1} a_{j_1 j_1 0}(y_{j_1} - a_{j_1 j_1 s_1})(y_{j_1} - a_{j_1 j_1 s_2}) + \beta_{j_1 j_2} a_{j_2 j_2 0} \prod_{l_2=1, l_2 \neq l_1}^{3} (y_{j_1} - a_{j_2 j_2 l_2})\Big], \tag{4.394}$$

where

$$\begin{aligned} G_{j_1 j_1 0} &= \frac{\partial}{\partial y_{j_1}}[\mathbf{n}^{\mathrm{T}} \cdot \mathbf{F}(y_{j_1}, \mathbf{p})]\Big|_{y_{j_1}^* = a_{j_1 j_1 s_2} = a_{j_2 j_2 l_1}} \\ &= \alpha_{j_1 j_2} a_{j_2 j_2 0} \prod_{l_2=1, l_2 \neq l_1}^{3} (a_{j_2 j_2 l_1} - a_{j_2 j_2 l_2}). \end{aligned} \tag{4.395}$$

The infinite-equilibrium of $y_{j_1}^* = a_{j_1 j_1 s_2} = a_{j_2 j_2 l_1}$ has the following properties:

- *For $a_{j_1 j_1 0}\alpha_{j_1 j_2}^2 (a_{j_1 j_1 s_2} - a_{j_1 j_1 s_1}) > 0$ and $\alpha_{j_1 j_2} a_{j_2 j_2 0} \prod_{l_2=1, l_2 \neq l_1}^{3} (a_{j_2 j_2 l_1} - a_{j_2 j_2 l_2}) > 0$,*

$$(a_{j_1 j_1 s_2}, a_{j_2 j_2 l_1}) = \underbrace{(\mathrm{UP}, \mathrm{SO})}_{\text{up-parabola source}}. \tag{4.396}$$

- *For $a_{j_1 j_1 0}\alpha_{j_1 j_2}^2 (a_{j_1 j_1 s_2} - a_{j_1 j_1 s_1}) < 0$ and $\alpha_{j_1 j_2} a_{j_2 j_2 0} \prod_{l_2=1, l_2 \neq l_1}^{3} (a_{j_2 j_2 l_1} - a_{j_2 j_2 l_2}) > 0$,*

$$(a_{j_1 j_1 s_2}, a_{j_2 j_2 l_1}) = \underbrace{(\mathrm{DP}, \mathrm{SO})}_{\text{down-parabola source}}. \tag{4.397}$$

- *For $a_{j_1 j_1 0}\alpha_{j_1 j_2}^2 (a_{j_1 j_1 s_2} - a_{j_1 j_1 s_1}) > 0$ and $\alpha_{j_1 j_2} a_{j_2 j_2 0} \prod_{l_2=1, l_2 \neq l_1}^{3} (a_{j_2 j_2 l_1} - a_{j_2 j_2 l_2}) < 0$,*

$$(a_{j_1 j_1 s_2}, a_{j_2 j_2 l_1}) = \underbrace{(\mathrm{DP}, \mathrm{SI})}_{\text{down-parabola sink}}. \tag{4.398}$$

- *For $a_{j_1 j_1 0}\alpha_{j_1 j_2}^2 (a_{j_1 j_1 s_2} - a_{j_1 j_1 s_1}) < 0$ and $\alpha_{j_1 j_2} a_{j_2 j_2 0} \prod_{l_2=1, l_2 \neq l_1}^{3} (a_{j_2 j_2 l_1} - a_{j_2 j_2 l_2}) < 0$,*

$$(a_{j_1 j_1 s_2}, a_{j_2 j_2 l_1}) = \underbrace{(\mathrm{UP}, \mathrm{SI})}_{\text{up-parabola source}}. \tag{4.399}$$

- *The parabola source and sink infinite-equilibriums are the switching bifurcations for the increasing-inflection and parabola flows in the* x_{j_1}*- and* x_{j_2}*-directions.*
- *The inflection-saddle infinite-equilibriums are the appearing bifurcations from a parabola flow and* x_{j_2}*-direction to the up-parabola and down-parabola flows in the* x_{j_1}*-direction and a parabola flow in the* x_{j_2}*-direction.*

(viii$_5$*) For* $a_{j_1j_11} = a_{j_2j_2l_2}$,

$$\begin{aligned}
&\dot{x}_{j_1} = a_{j_1j_10}(y_{j_1} - a_{j_1j_11})^3,\\
&\dot{x}_{j_2} = a_{j_2j_20}(y_{j_1} - a_{j_2j_2l_1}) \prod_{l_2=1,l_2\neq l_1}^{3} (y_{j_1} - a_{j_2j_2l_2}),\\
&j_1, j_2 \in \{1,2\};\ j_1 \neq j_2.
\end{aligned} \tag{4.400}$$

There is a dynamical system in the y_{j_1}*-direction as*

$$\begin{aligned}
\dot{y}_{j_1} = (y_{j_1} - a_{j_1j_11})\Big[&\alpha_{j_1j_1}a_{j_1j_10}(y_{j_1} - a_{j_1j_11})^2\\
&+ \alpha_{j_1j_2}a_{j_2j_20} \prod_{l_2=1,l_2\neq l_1}^{3} (y_{j_1} - a_{j_2j_2l_2})\Big],
\end{aligned} \tag{4.401}$$

with a dynamical system in the $y_{j_1}^{\perp}$*-direction as*

$$\begin{aligned}
\dot{y}_{j_1}^{\perp} = (y_{j_1} - a_{j_1j_11})\Big[&\beta_{j_1j_1}a_{j_1j_10}(y_{j_1} - a_{j_1j_11})^2\\
&+ \beta_{j_1j_2}a_{j_2j_20} \prod_{l_2=1,l_2\neq l_1}^{3} (y_{j_1} - a_{j_2j_2l_2})\Big],
\end{aligned} \tag{4.402}$$

where

$$\begin{aligned}
G_{j_1j_10} &= \frac{\partial}{\partial y_{j_1}}[\mathbf{n}^{\mathrm{T}} \cdot \mathbf{F}(y_{j_1}, \mathbf{p})]\Big|_{y_{j_1}^* = a_{j_1j_11} = a_{j_2j_2l_1}}\\
&= \alpha_{j_1j_2}a_{j_2j_20} \prod_{l_2=1,l_2\neq l_1}^{3} (a_{j_2j_2l_1} - a_{j_2j_2l_2}).
\end{aligned} \tag{4.403}$$

The infinite-equilibrium of $y^*_{j_1} = a_{j_1 j_1 1} = a_{j_2 j_2 l_1}$ *has the following properties*:

- *For* $\alpha^3_{j_1 j_2} a_{j_1 j_1 0} > 0$ *and* $\alpha_{j_1 j_2} a_{j_2 j_2 0} \prod_{l_2=1, l_2 \neq l_1}^{3} (a_{j_2 j_2 l_1} - a_{j_2 j_2 l_2}) > 0$,

$$(a_{j_1 j_1 1}, a_{j_2 j_2 l_1}) = \underbrace{(2^{\text{nd}}\text{II}, \text{SO})}_{\text{second-order increasing-inflection source}}. \tag{4.404}$$

- *For* $\alpha^3_{j_1 j_2} a_{j_1 j_1 0} < 0$ *and* $\alpha_{j_1 j_2} a_{j_2 j_2 0} \prod_{l_2=1, l_2 \neq l_1}^{3} (a_{j_2 j_2 l_1} - a_{j_2 j_2 l_2}) > 0$,

$$(a_{j_1 j_1 1}, a_{j_2 j_2 l_1}) = \underbrace{(2^{\text{nd}}\text{DI}, \text{SO})}_{\text{second-order decreasing-inflection source}}. \tag{4.405}$$

- *For* $\alpha^3_{j_1 j_2} a_{j_1 j_1 0} > 0$ *and* $\alpha_{j_1 j_2} a_{j_2 j_2 0} \prod_{l_2=1, l_2 \neq l_1}^{3} (a_{j_2 j_2 l_1} - a_{j_2 j_2 l_2}) < 0$,

$$(a_{j_1 j_1 1}, a_{j_2 j_2 l_1}) = \underbrace{(2^{\text{nd}}\text{DI}, \text{SI})}_{\text{second-order decreasing-inflection sink}}. \tag{4.406}$$

- *For* $\alpha^3_{j_1 j_2} a_{j_1 j_1 0} < 0$ *and* $\alpha_{j_1 j_2} a_{j_2 j_2 0} \prod_{l_2=1, l_2 \neq l_1}^{3} (a_{j_2 j_2 l_1} - a_{j_2 j_2 l_2}) < 0$,

$$(a_{j_1 j_1 1}, a_{j_2 j_2 l_1}) = \underbrace{(2^{\text{nd}}\text{II}, \text{SI})}_{\text{second-order increasing-inflection sink}}. \tag{4.407}$$

- *The second-order inflection-source and sink infinite-equilibriums are the switching bifurcations for third-order parabola and parabola flows in the* x_{j_1} *and* x_{j_2}*-directions.*
- *The second-order inflection-source and sink infinite-equilibriums are the appearing bifurcations from a parabola flow and a parabola flow in the* x_{j_1}*- and* x_{j_2}*-directions to three parabola flows and a parabola flow in the* x_{j_1}*- and* x_{j_2}*-directions.*

(ix) If

$$\Delta_{j_1 j_1} > 0 \quad \text{and} \quad \Delta_{j_2 j_2} > 0, \tag{4.408}$$

then

$$\begin{aligned} \dot{x}_{j_1} &= a_{j_1 j_1 0}(y_{j_1} - a_{j_1 j_1 1})(y_{j_1} - a_{j_1 j_1 2})(y_{j_1} - a_{j_1 j_1 3}), \\ \dot{x}_{j_2} &= a_{j_2 j_2 0}(y_{j_1} - a_{j_2 j_2 1})(y_{j_1} - a_{j_2 j_2 2})(y_{j_1} - a_{j_2 j_2 3}), \end{aligned} \tag{4.409}$$

where

$$\{a_{j_1j_1 1}, a_{j_1j_1 2}, a_{j_1j_1 3}\} = \text{sort}\{b_{j_1j_1 1}, b_{j_1j_1 2}, b_{j_1j_1 3}\}, \quad a_{j_1j_1s_1} < a_{j_1j_1s_2}, s_1 < s_2;$$

$$b_{j_1j_1 2}, b_{j_1j_1 3} = -\frac{1}{2}(B_{j_1j_1 1} \pm \sqrt{\Delta_{j_1j_1}}), a_{j_1j_1 0} = A_{j_1j_1 0}.$$

$$\{a_{j_2j_2 1}, a_{j_2j_2 2}a_{j_2j_2 3}\} = \text{sort}\{b_{j_2j_2 1}, b_{j_2j_2 2}, b_{j_2j_2 3}\} \quad a_{j_2j_2 1} < a_{j_2j_2 2};$$

$$b_{j_2j_2 2}, b_{j_2j_2 3} = -\frac{1}{2}(B_{j_2j_2 1} \pm \sqrt{\Delta_{j_2j_2}}), \quad a_{j_2j_2 0} = A_{j_2j_2 0}. \tag{4.410}$$

There is a dynamical system in the y_{j_1}-direction as

$$\dot{y}_{j_1} = \alpha_{j_1j_1} a_{j_1j_1 0} \prod_{s_1=1}^{3} (y_{j_1} - a_{j_1j_1s_1}) + \alpha_{j_1j_2} a_{j_2j_2 0} \prod_{l_1=1}^{3} (y_{j_1} - a_{j_2j_2l_1}), \tag{4.411}$$

with a dynamical system in the $y_{j_1}^{\perp}$-direction as

$$\dot{y}_{j_1}^{\perp} = \beta_{j_1j_1} a_{j_1j_1 0} \prod_{s_1=1}^{3} (y_{j_1} - a_{j_1j_1s_1}) + \beta_{j_1j_2} a_{j_2j_2 0} \prod_{l_1=1}^{3} (y_{j_1} - a_{j_2j_2l_1}). \tag{4.412}$$

(ix_{1a}) The equilibrium flows of $y_{j_1}^* = a_{j_1j_1s_1} (s_1, s_2 \in \{1,2\}, s_1 \neq s_2)$ for the x_{j_1}-direction have the following properties with $\dot{x}_{j_2}^* = a_{220} \prod_{l_1=1}^{3} (a_{j_1j_1s_1} - a_{j_2j_2l_1})$:

- *For $a_{j_1j_1 0}\alpha_{j_1j_2} \prod_{s_2=1, s_2\neq s_1}^{3} (a_{j_1j_1s_1} - a_{j_1j_1s_2}) > 0$ and $a_{220} \prod_{l_1=1}^{3} (a_{j_1j_1s_1} - a_{j_2j_2l_1}) > 0$,*

$$(a_{j_1j_1s_1}, \dot{x}_{j_2}^*) = \underbrace{(\text{UP}, \text{pF})}_{\text{up-parabola flow } (+)}. \tag{4.413}$$

- *For $a_{j_1j_1 0}\alpha_{j_1j_2} \prod_{s_2=1, s_2\neq s_1}^{3} (a_{j_1j_1s_1} - a_{j_1j_1s_2}) < 0$ and $a_{220} \prod_{l_1=1}^{3} (a_{j_1j_1s_1} - a_{j_2j_2l_1}) > 0$,*

$$(a_{j_1j_1s_1}, \dot{x}_{j_2}^*) = \underbrace{(\text{DP}, \text{pF})}_{\text{Down-parabola flow } (+)}. \tag{4.414}$$

- *For $a_{j_1j_1 0}\alpha_{j_1j_2} \prod_{s_2=1, s_2\neq s_1}^{3} (a_{j_1j_1s_1} - a_{j_1j_1s_2}) > 0$ and $a_{220} \prod_{l_1=1}^{3} (a_{j_1j_1s_1} - a_{j_2j_2l_1}) < 0$,*

$$(a_{j_1j_1s_1}, \dot{x}_{j_2}^*) = \underbrace{(\text{DP}, \text{nF})}_{\text{Down-parabola flow } (-)}. \tag{4.415}$$

- *For* $a_{j_1j_10}\alpha_{j_1j_2}\prod_{s_2=1,s_2\neq s_1}^{3}(a_{j_1j_1s_1}-a_{j_1j_1s_2})<0$ *and* $a_{220}\prod_{l_1=1}^{3}(a_{j_1j_1s_1}-a_{j_2j_2l_1})<0$,

$$(a_{j_1j_1s_1},\dot{x}^*_{j_2})=\underbrace{(\mathrm{UP},\mathrm{nF})}_{\text{up-parabola flow }(-)}. \tag{4.416}$$

(ix$_{1b}$) The equilibrium flows of $y^*_{j_1}=a_{j_2j_2l_1}$ *for the* x_{j_2}*-direction have the following properties with* $\dot{x}^*_{j_1}=a_{j_1j_10}\prod_{s_1=1}^{3}(a_{j_2j_2l_1}-a_{j_1j_1s_1})$:

- *For* $a_{j_1j_10}\prod_{s_1=1}^{3}(a_{j_2j_2l_1}-a_{j_1j_1s_1})>0$ *and* $a_{j_2j_20}\alpha_{j_1j_1}\prod_{l_2=1,l_2\neq l_1}^{3}(a_{j_2j_2l_1}-a_{j_1j_1l_2})>0$,

$$(\dot{x}^*_{j_1},a_{j_2j_2l_1})=\underbrace{(\mathrm{pF},\mathrm{UP})}_{\text{up-parabola flow }(+)}. \tag{4.417}$$

- *For* $a_{j_1j_10}\prod_{s_1=1}^{3}(a_{j_2j_2l_1}-a_{j_1j_1s_1})<0$ *and* $a_{j_2j_20}\alpha_{j_1j_1}\prod_{l_2=1,l_2\neq l_1}^{3}(a_{j_2j_2l_1}-a_{j_1j_1l_2})>0$,

$$(\dot{x}^*_{j_1},a_{j_2j_2l_1})=\underbrace{(\mathrm{nF},\mathrm{DP})}_{\text{down-parabola flow }(-)}. \tag{4.418}$$

- *For* $a_{j_1j_10}\prod_{s_1=1}^{3}(a_{j_2j_2l_1}-a_{j_1j_1s_1})>0$ *and* $a_{j_2j_20}\alpha_{j_1j_1}\prod_{l_2=1,l_2\neq l_1}^{3}(a_{j_2j_2l_1}-a_{j_1j_1l_2})<0$,

$$(\dot{x}^*_{j_1},a_{j_2j_2l_1})=\underbrace{(\mathrm{pF},\mathrm{DP})}_{\text{down-parabola flow }(+)}. \tag{4.419}$$

- *For* $a_{j_1j_10}\prod_{s_1=1}^{3}(a_{j_2j_2l_1}-a_{j_1j_1s_1})<0$ *and* $a_{j_2j_20}\alpha_{j_1j_1}\prod_{l_2=1,l_2\neq l_1}^{3}(a_{j_2j_2l_1}-a_{j_1j_1l_2})<0$,

$$(\dot{x}^*_{j_1},a_{j_2j_2l_1})=\underbrace{(\mathrm{nF},\mathrm{UP})}_{\text{up-parabola flow }(-)}. \tag{4.420}$$

(ix$_2$) For $a_{j_2j_2l_1}=a_{j_1j_1s_1}(s_1,s_2,l_1,l_2\in\{1,2,3\})$,

$$\begin{aligned}
\dot{x}_{j_1}&=a_{j_1j_10}(y_{j_1}-a_{j_1j_1s_1})\prod_{s_2=1,s_2\neq s_1}^{3}(y_{j_1}-a_{j_1j_1s_2}),\\
\dot{x}_{j_2}&=a_{j_2j_20}(y_{j_1}-a_{j_2j_2l_1})\prod_{l_2=1,l_2\neq l_1}^{3}(y_{j_1}-a_{j_2j_2l_2}),\\
&j_1,j_2\in\{1,2\};\ j_1\neq j_2;\ s_1,s_2\in\{1,2,3\},\ s_1\neq s_2.
\end{aligned}\tag{4.421}$$

There is a dynamical system in the y_{j_1}*-direction as*

$$\dot{y}_{j_1} = (y_{j_1} - a_{j_1 j_1 s_1})\Big[\alpha_{j_1 j_1} a_{j_1 j_1 0} \prod_{s_2=1, s_2\neq s_1}^{3} (y_{j_1} - a_{j_1 j_1 s_2}) + \alpha_{j_1 j_2} a_{j_2 j_2 0} \prod_{l_2=1, l_2\neq l_1}^{3} (y_{j_1} - a_{j_2 j_2 l_2})\Big], \tag{4.422}$$

with a dynamical system in the $y_{j_1}^{\perp}$*-direction as*

$$\dot{y}_{j_1}^{\perp} = (y_{j_1} - a_{j_1 j_1 s_1})\Big[\beta_{j_1 j_1} a_{j_1 j_1 0} \prod_{s_2=1, s_2\neq s_1}^{3} (y_{j_1} - a_{j_1 j_1 s_2}) + \beta_{j_1 j_2} a_{j_2 j_2 0} \prod_{l_2=1, l_2\neq l_1}^{3} (y_{j_1} - a_{j_2 j_2 l_2})\Big], \tag{4.423}$$

where

$$\begin{aligned} G_{j_1 j_1 0} &= \frac{\partial}{\partial y_{j_1}}[\mathbf{n}^{\mathrm{T}} \cdot \mathbf{F}(y_{j_1}, \mathbf{p})]\bigg|_{y_{j_1}^{*} = a_{j_1 j_1 s_1} = a_{j_2 j_2 l_1}} \\ &= \alpha_{j_1 j_1} a_{j_1 j_1 0} \prod_{s_2=1, s_2\neq s_1}^{3} (a_{j_1 j_1 s_1} - a_{j_1 j_1 s_2}) \\ &\quad + \alpha_{j_1 j_2} a_{j_2 j_2 0} \prod_{l_2=1, l_2\neq l_1}^{3} (a_{j_2 j_2 l_1} - a_{j_2 j_2 l_2}). \end{aligned} \tag{4.424}$$

The infinite-equilibriums of $y_{j_1}^{*} = a_{j_2 j_2 l_1} = a_{j_1 j_1 s_1}$ *have the following properties:*

- *For* $a_{j_1 j_1 0} \prod_{s_2=1, s_2\neq s_1}^{3} (a_{j_1 j_1 s_1} - a_{j_1 j_1 s_2}) / a_{j_2 j_2 0} \prod_{l_2=1, l_2\neq l_1}^{3} (a_{j_2 j_2 l_1} - a_{j_2 j_2 l_2}) > 0, G_{j_1 j_1 0} > 0$,

$$(a_{j_1 j_1 s_1}, a_{j_2 j_2 l_1}) = \underbrace{(\mathrm{II}, \mathrm{SO})}_{\text{increasing-inflection source}}. \tag{4.425}$$

- *For* $a_{j_1 j_1 0} \prod_{s_2=1, s_2\neq s_1}^{3} (a_{j_1 j_1 s_1} - a_{j_1 j_1 s_2}) / a_{j_2 j_2 0} \prod_{l_2=1, l_2\neq l_1}^{3} (a_{j_2 j_2 l_1} - a_{j_2 j_2 l_2}) > 0, G_{j_1 j_1 0} < 0$,

$$(a_{j_1 j_1 s_1}, a_{j_2 j_2 l_1}) = \underbrace{(\mathrm{II}, \mathrm{SI})}_{\text{increasing-inflection sink}}. \tag{4.426}$$

- *For* $a_{j_1j_10}\prod_{s_2=1,s_2\neq s_1}^{3}(a_{j_1j_1s_1} - a_{j_1j_1s_2})/a_{j_2j_20}\prod_{l_2=1,l_2\neq l_1}^{3}(a_{j_2j_2l_1} - a_{j_2j_2l_2}) < 0, G_{j_1j_10} > 0,$

$$(a_{j_1j_1s_1}, a_{j_2j_2l_1}) = \underbrace{(\mathrm{DI}, \mathrm{SO})}_{\text{decreasing-inflection source}}. \tag{4.427}$$

- *For* $a_{j_1j_10}\prod_{s_2=1,s_2\neq s_1}^{3}(a_{j_1j_1s_1} - a_{j_1j_1s_2})/a_{j_2j_20}\prod_{l_2=1,l_2\neq l_1}^{3}(a_{j_2j_2l_1} - a_{j_2j_2l_2}) < 0, G_{j_1j_10} < 0,$

$$(a_{j_1j_1s_1}, a_{j_2j_2l_1}) = \underbrace{(\mathrm{DI}, \mathrm{SI})}_{\text{decreasing-inflection sink}}. \tag{4.428}$$

4.2 Proof of Theorem 4.1

Consider a two-dimensional dynamical system as

$$\begin{aligned}\dot{x}_{j_2} &= a_{j_2j_20}(y_{j_1} - b_{j_2j_21})(y_{j_1}^2 + B_{j_2j_2}y_{j_1} + C_{j_2j_2}),\\ \dot{x}_{j_1} &= a_{j_1j_10}(y_{j_1} - b_{j_1j_11})(y_{j_1}^2 + B_{j_1j_1}y_{j_1} + C_{j_1j_1}),\end{aligned}$$

with

$$\begin{aligned}&\Delta_{j_2j_2} = (B_{j_2j_2})^2 - 4C_{j_2j_2}, \quad \Delta_{j_1j_1} = (B_{j_1j_1})^2 - 4C_{j_1j_1},\\ &a_{j_1j_10} = A_{j_1j_10}, \quad a_{j_2j_20} = A_{j_2j_20}.\end{aligned}$$

(i) If

$$\Delta_{j_1j_1} < 0 \quad \text{and} \quad \Delta_{j_2j_2} < 0,$$

then

$$\begin{aligned}\dot{x}_{j_1} &= a_{j_1j_10}(y_{j_1} - a_{j_1j_11})[(y_{j_1} - a_{j_1j_1})^2 + b_{j_1j_1}],\\ \dot{x}_{j_2} &= a_{j_2j_20}(y_{j_1} - a_{j_2j_21})[(y_{j_1} - a_{j_2j_2})^2 + b_{j_2j_2}],\end{aligned}$$

where

$$\begin{aligned}a_{j_2j_2} &= -\frac{1}{2}B_{j_2j_2}, \quad b_{j_2j_2} = -\frac{1}{4}\Delta_{j_2j_2}, \quad a_{j_2j_21} = b_{j_2j_21},\\ a_{j_1j_1} &= -\frac{1}{2}B_{j_1j_1}, \quad b_{j_1j_1} = -\frac{1}{4}\Delta_{j_2j_2}, \quad a_{j_1j_11} = b_{j_1j_11}.\end{aligned}$$

From the original systems, there is a dynamical system in the y_{j_1}-direction as

$$\begin{aligned}\dot{y}_{j_1} = {}&\alpha_{j_1j_1}a_{j_1j_10}(y_{j_1} - a_{j_1j_11})[(y_{j_1} - a_{j_1j_1})^2 + b_{j_1j_1}]\\ &+ \alpha_{j_1j_2}a_{j_2j_20}(y_{j_1} - a_{j_2j_21})[(y_{j_1} - a_{j_2j_2})^2 + b_{j_2j_2}],\end{aligned}$$

with a dynamical system in the $y_{j_1}^{\perp}$-direction as

$$\begin{aligned}\dot{y}_{j_1}^{\perp} &= \beta_{j_1j_1}a_{j_1j_10}(y_{j_1}-a_{j_1j_11})[(y_{j_1}-a_{j_1j_1})^2+b_{j_1j_1}]\\&\quad+\beta_{j_1j_2}a_{j_2j_20}(y_{j_1}-a_{j_2j_21})[(y_{j_1}-a_{j_2j_2})^2+b_{j_2j_2}].\end{aligned}$$

($\mathrm{i}_{1\mathrm{a}}$) In phase space, at $y_1^* = a_{j_1j_11}$,

$$\left.\frac{dx_{j_1}}{dx_{j_2}}\right|_{y_1^*=a_{j_1j_11}} = \left.\frac{a_{j_1j_10}(y_{j_1}-a_{j_1j_11})[(y_{j_1}-a_{j_1j_1})^2+b_{j_1j_1}]}{a_{j_2j_20}(y_{j_1}-a_{j_2j_21})[(y_{j_1}-a_{j_2j_2})^2+b_{j_2j_2}]}\right|_{y_1^*=a_{j_1j_11}} = 0.$$

If

$$\left.\frac{d^2x_{j_1}}{dx_{j_2}^2}\right|_{y_1^*=a_{j_1j_11}} = \frac{a_{j_1j_10}\alpha_{j_1j_2}[(a_{j_1j_11}-a_{j_1j_1})^2+b_{j_1j_1}]}{a_{j_2j_20}(a_{j_1j_11}-a_{j_2j_21})[(a_{j_1j_11}-a_{j_2j_2})^2+b_{j_2j_2}]} > 0,$$

there is an up-parabola flow at $y_1^* = a_{j_1j_11}$. If

$$\left.\frac{d^2x_{j_1}}{dx_{j_2}^2}\right|_{y_1^*=a_{j_1j_11}} = \frac{a_{j_1j_10}\alpha_{j_1j_2}[(a_{j_1j_11}-a_{j_1j_1})^2+b_{j_1j_1}]}{a_{j_2j_20}(a_{j_1j_11}-a_{j_2j_21})[(a_{j_1j_11}-a_{j_2j_2})^2+b_{j_2j_2}]} < 0,$$

there is a down-parabola flow at $y_1^* = a_{j_1j_11}$. With

$$\dot{x}_{j_2}^* = a_{j_2j_20}(a_{j_1j_11}-a_{j_2j_21})[(a_{j_1j_11}-a_{j_2j_2})^2+b_{j_2j_2}],$$

the parabola flows are positive and negative for $\dot{x}_{j_2}^* > 0$ and $\dot{x}_{j_2}^* < 0$, respectively.

Due to

$$(a_{j_1j_11}-a_{j_1j_1})^2+b_{j_1j_1} > 0 \quad \text{and} \quad (a_{j_1j_11}-a_{j_2j_2})^2+b_{j_2j_2} > 0,$$

the flow of $y_1^* = a_{j_1j_11}$ with $\dot{x}_{j_2}^* = a_{j_2j_20}[(a_{j_1j_11}-a_{j_2j_2})^2+b_{j_2j_2}]$ possesses the following properties as in Eqs. (4.14)–(4.17):

- For $a_{j_1j_10}\alpha_{j_1j_2} > 0$ and $a_{j_2j_20}(a_{j_1j_11}-a_{j_2j_21}) > 0$,
$$(a_{j_1j_11}, \dot{x}_{j_2}^*) = \underbrace{(\mathrm{UP}, \mathrm{pF})}_{\text{up-parabola flow }(+)}.$$
- For $a_{j_1j_10}\alpha_{j_1j_2} < 0$ and $a_{j_2j_20}(a_{j_1j_11}-a_{j_2j_21}) > 0$,
$$(a_{j_1j_11}, \dot{x}_{j_2}^*) = \underbrace{(\mathrm{DP}, \mathrm{pF})}_{\text{down-parabola flow }(+)}.$$
- For $a_{j_1j_10}\alpha_{j_1j_2} > 0$ and $a_{j_2j_20}(a_{j_1j_11}-a_{j_2j_21}) < 0$,
$$(a_{j_1j_11}, \dot{x}_{j_2}^*) = \underbrace{(\mathrm{DP}, \mathrm{nF})}_{\text{down-parabola flow }(-)}.$$

- For $a_{j_1j_10}\alpha_{j_1j_2} < 0$ and $a_{j_2j_20}(a_{j_1j_11} - a_{j_2j_21}) < 0$,

$$(a_{j_1j_11}, \dot{x}^*_{j_2}) = \underbrace{(\text{UP}, \text{nF})}_{\text{up-parabola flow } (-)}.$$

($\text{i}_{1\text{b}}$) For $y^*_1 = a_{j_2j_21}$, in phase space,

$$\left.\frac{dx_{j_2}}{dx_{j_1}}\right|_{y^*_1=a_{j_2j_21}} = \left.\frac{a_{j_2j_20}(y_{j_1} - a_{j_2j_21})[(y_{j_1} - a_{j_2j_2})^2 + b_{j_2j_2}]}{a_{j_1j_10}(y_{j_1} - a_{j_1j_11})[(y_{j_1} - a_{j_1j_1})^2 + b_{j_1j_1}]}\right|_{y^*_1=a_{j_2j_21}} = 0.$$

If

$$\left.\frac{d^2x_{j_2}}{dx^2_{j_1}}\right|_{y^*_1=a_{j_2j_21}} = \frac{a_{j_2j_20}\alpha_{j_1j_1}[(a_{j_2j_21} - a_{j_2j_2})^2 + b_{j_2j_2}]}{a_{j_1j_10}(a_{j_2j_21} - a_{j_1j_11})[(a_{j_2j_21} - a_{j_1j_1})^2 + b_{j_1j_1}]} > 0,$$

there is an up-parabola flow at $y^*_1 = a_{j_2j_21}$. If

$$\left.\frac{d^2x_{j_2}}{dx^2_{j_1}}\right|_{y^*_1=a_{j_2j_21}} = \frac{a_{j_2j_20}\alpha_{j_1j_1}[(a_{j_2j_21} - a_{j_2j_2})^2 + b_{j_2j_2}]}{a_{j_1j_10}(a_{j_2j_21} - a_{j_1j_11})[(a_{j_2j_21} - a_{j_1j_1})^2 + b_{j_1j_1}]} < 0,$$

there is a down-parabola flow at $y^*_1 = a_{j_2j_21}$. With

$$\dot{x}^*_{j_1} = a_{j_1j_10}(a_{j_2j_21} - a_{j_1j_11})[(a_{j_2j_21} - a_{j_1j_1})^2 + b_{j_1j_1}],$$

the parabola flows are positive and negative for $\dot{x}^*_{j_1} > 0$ and $\dot{x}^*_{j_1} < 0$, respectively.

Due to

$$(a_{j_2j_21} - a_{j_1j_1})^2 + b_{j_1j_1} > 0 \quad \text{and} \quad (a_{j_2j_21} - a_{j_2j_2})^2 + b_{j_2j_2} > 0,$$

the flow of $y^*_1 = a_{j_2j_21}$ with $\dot{x}^*_{j_1} = a_{j_1j_10}(a_{j_2j_21} - a_{j_1j_11})[(a_{j_2j_21} - a_{j_1j_1})^2 + b_{j_1j_1}]$ possesses the following properties as in Eqs. (4.14)–(4.17):

- For $a_{j_1j_10}(a_{j_2j_21} - a_{j_1j_11}) > 0$ and $a_{j_2j_20}\alpha_{j_1j_1} > 0$,

$$(\dot{x}^*_{j_1}, a_{j_2j_21}) = \underbrace{(\text{pF}, \text{UP})}_{\text{up-parabola flow } (+)}.$$

- For $a_{j_1j_10}(a_{j_2j_21} - a_{j_1j_11}) < 0$ and $a_{j_2j_20}\alpha_{j_1j_1} > 0$,

$$(\dot{x}^*_{j_1}, a_{j_2j_21}) = \underbrace{(\text{nF}, \text{DP})}_{\text{down-parabola flow } (-)}.$$

- For $a_{j_1j_10}(a_{j_2j_21} - a_{j_1j_11}) > 0$ and $a_{j_2j_20}\alpha_{j_1j_1} < 0$,

$$(\dot{x}^*_{j_1}, a_{j_2j_21}) = \underbrace{(\text{pF}, \text{DP})}_{\text{down-parabola flow } (+)}.$$

- For $a_{j_1j_10}(a_{j_2j_21} - a_{j_1j_11}) < 0$ and $a_{j_2j_20}\alpha_{j_1j_1} < 0$,

$$(\dot{x}^*_{j_1}, a_{j_2j_21}) = \underbrace{(\text{nF}, \text{UP})}_{\text{up-parabola flow }(-)}.$$

(i$_2$) For $a_{j_1j_11} = a_{j_2j_21}$, a dynamical system in the y_{j_1}-direction is constructed as

$$\begin{aligned}\dot{y}_{j_1} &= (y_{j_1} - a_{j_1j_11})\{\alpha_{j_1j_1}a_{j_1j_10}[(y_{j_1} - a_{j_1j_1})^2 + b_{j_1j_1}]\\ &\quad + \alpha_{j_1j_2}a_{j_2j_20}[(y_{j_1} - a_{j_2j_2})^2 + b_{j_2j_2}]\},\end{aligned}$$

and a dynamical system in the $y^{\perp}_{j_1}$-direction is constructed as

$$\begin{aligned}\dot{y}^{\perp}_{j_1} &= (y_{j_1} - a_{j_1j_11})\{\beta_{j_1j_1}a_{j_1j_10}[(y_{j_1} - a_{j_1j_1})^2 + b_{j_1j_1}]\\ &\quad + \beta_{j_1j_2}a_{j_2j_20}[(y_{j_1} - a_{j_2j_2})^2 + b_{j_2j_2}]\},\end{aligned}$$

and a G-function is defined as

$$\begin{aligned}G_{j_1j_10} &= \frac{\partial}{\partial y_{j_1}}[\mathbf{n}^{\mathrm{T}} \cdot \mathbf{F}(y_{j_1}, \mathbf{p})]\Big|_{y^*_{j_1}=a_{j_1j_11}=a_{j_2j_21}}\\ &= \alpha_{j_1j_1}a_{j_1j_10}[(a_{j_1j_11} - a_{j_1j_1})^2\\ &\quad + b_{j_1j_1}] + \alpha_{j_1j_2}a_{j_2j_20}[(a_{j_1j_11} - a_{j_2j_2})^2 + b_{j_2j_2}].\end{aligned}$$

For $y^*_1 = a_{j_1j_11} = a_{j_2j_21}$, in phase space, if

$$\frac{dx_{j_2}}{dx_{j_1}}\Big|_{y^*_1=a_{j_2j_21}} = \frac{a_{j_2j_20}[(a_{j_2j_21} - a_{j_2j_2})^2 + b_{j_2j_2}]}{a_{j_1j_10}[(a_{j_2j_21} - a_{j_1j_1})^2 + b_{j_1j_1}]}\Big|_{y^*_1=a_{j_2j_21}} > 0,$$

there is an increasing-inflection flow at $y^*_1 = a_{j_1j_11} = a_{j_2j_21}$. If

$$\frac{dx_{j_2}}{dx_{j_1}}\Big|_{y^*_1=a_{j_2j_21}} = \frac{a_{j_2j_20}[(a_{j_2j_21} - a_{j_2j_2})^2 + b_{j_2j_2}]}{a_{j_1j_10}[(a_{j_2j_21} - a_{j_1j_1})^2 + b_{j_1j_1}]}\Big|_{y^*_1=a_{j_2j_21}} < 0,$$

there is a down-parabola flow at $y^*_1 = a_{j_1j_11} = a_{j_2j_21}$. The corresponding variational equation is

$$\begin{aligned}\Delta\dot{y}_{j_1} &= \{\alpha_{j_1j_1}a_{j_1j_10}[(a_{j_1j_11} - a_{j_1j_1})^2 + b_{j_1j_1}]\\ &\quad + \alpha_{j_1j_2}a_{j_2j_20}[(a_{j_2j_21} - a_{j_2j_2})^2 + b_{j_2j_2}]\}\Delta y_{j_1}\\ &= G_{j_2j_20}\Delta y_{j_1}.\end{aligned}$$

The infinite-equilibrium is source and sink for $G_{j_2j_20} > 0$ and $G_{j_2j_20} < 0$.

With

$$(a_{j_2j_21} - a_{j_1j_1})^2 + b_{j_1j_1} > 0 \quad \text{and} \quad (a_{j_2j_21} - a_{j_2j_2})^2 + b_{j_2j_2} > 0,$$

thus, the flow of $y_1^* = a_{j_1j_11} = a_{j_2j_21}$ possesses the following properties as in Eqs. (4.21)–(4.24):

- For $a_{j_1j_10}/a_{j_2j_20} > 0$ and $G_{j_1j_10} > 0$,

$$(a_{j_1j_11}, a_{j_2j_21}) = \underbrace{(\text{II}, \text{SO})}_{\text{increasing-inflection source}}.$$

- For $a_{j_1j_10}/a_{j_2j_20} > 0$ and $G_{j_1j_10} < 0$,

$$(a_{j_1j_11}, a_{j_2j_21}) = \underbrace{(\text{II}, \text{SI})}_{\text{increasing-inflection sink}}.$$

- For $a_{j_1j_10}/a_{j_2j_20} < 0$ and $G_{j_1j_10} > 0$,

$$(a_{j_1j_11}, a_{j_2j_21}) = \underbrace{(\text{DI}, \text{SO})}_{\text{decreasing-inflection source}}.$$

- For $a_{j_1j_10}/a_{j_2j_20} < 0$ and $G_{j_1j_10} < 0$,

$$(a_{j_1j_11}, a_{j_2j_21}) = \underbrace{(\text{DI}, \text{SI})}_{\text{decreasing-inflection sink}}.$$

(ii) If

$$\Delta_{j_1j_1} = 0 \quad \text{and} \quad \Delta_{j_2j_2} < 0,$$

then

$$\begin{aligned} \dot{x}_{j_1} &= a_{j_1j_10}(y_{j_1} - a_{j_1j_1s_1})(y_{j_1} - a_{j_1j_1s_2})^2, \\ \dot{x}_{j_2} &= a_{j_2j_20}(y_{j_1} - a_{j_2j_21})[(y_{j_1} - a_{j_2j_2})^2 + b_{j_2j_2}], \end{aligned}$$

where

$$\begin{aligned} a_{j_2j_2} &= -\frac{1}{2}B_{j_2j_2}, \quad b_{j_2j_2} = -\frac{1}{4}\Delta_{j_2j_2}; \\ a_{j_1j_1s_w} &= -\frac{1}{2}B_{j_1j_1}, \quad b_{j_1j_1} = 0, \quad a_{j_1j_1s_1} = b_{j_1j_11}. \end{aligned}$$

(ii$_1$) For $a_{j_1j_1s_1} \neq a_{j_1j_1s_2}$, a dynamical system in the y_{j_1}-direction is constructed as

$$\begin{aligned} \dot{y}_{j_1} = {} & \alpha_{j_1j_1}a_{j_1j_10}(y_{j_1} - a_{j_1j_1s_1})(y_{j_1} - a_{j_1j_1s_2})^2 \\ & + \alpha_{j_1j_2}a_{j_2j_20}(y_{j_1} - a_{j_2j_21})[(y_{j_1} - a_{j_2j_2})^2 + b_{j_2j_2}], \end{aligned}$$

and a dynamical system in the $y_{j_1}^{\perp}$-direction is constructed as

$$\begin{aligned}\dot{y}_{j_1}^{\perp} &= \beta_{j_1j_1}a_{j_1j_10}(y_{j_1} - a_{j_1j_1s_1})(y_{j_1} - a_{j_1j_1s_2})^2 \\ &\quad + \beta_{j_1j_2}a_{j_2j_20}(y_{j_1} - a_{j_2j_21})[(y_{j_1} - a_{j_2j_2})^2 + b_{j_2j_2}].\end{aligned}$$

(ii_{1a}) In phase space, at $y_1^* = a_{j_1j_1s_1}$,

$$\left.\frac{dx_{j_1}}{dx_{j_2}}\right|_{y_1^*=a_{j_1j_1s_1}} = \left.\frac{a_{j_1j_10}(y_{j_1} - a_{j_1j_1s_1})(y_{j_1} - a_{j_1j_1s_2})^2}{a_{j_2j_20}(y_{j_1} - a_{j_2j_21})[(y_{j_1} - a_{j_2j_2})^2 + b_{j_2j_2}]}\right|_{y_1^*=a_{j_1j_1s_1}} = 0.$$

If

$$\left.\frac{d^2x_{j_1}}{dx_{j_2}^2}\right|_{y_1^*=a_{j_1j_1s_1}} = \frac{a_{j_1j_10}\alpha_{j_1j_2}(a_{j_1j_1s_1} - a_{j_1j_1s_2})^2}{a_{j_2j_20}(a_{j_1j_1s_1} - a_{j_2j_21})[(a_{j_1j_1s_1} - a_{j_2j_2})^2 + b_{j_2j_2}]} > 0,$$

there is an up-parabola flow at $y_1^* = a_{j_1j_1s_1}$. If

$$\left.\frac{d^2x_{j_1}}{dx_{j_2}^2}\right|_{y_1^*=a_{j_1j_1s_1}} = \frac{a_{j_1j_10}\alpha_{j_1j_2}(a_{j_1j_1s_1} - a_{j_1j_1s_2})^2}{a_{j_2j_20}(a_{j_1j_1s_1} - a_{j_2j_21})[(a_{j_1j_1s_1} - a_{j_2j_2})^2 + b_{j_2j_2}]} < 0,$$

there is a down-parabola flow at $y_1^* = a_{j_1j_1s_1}$. With

$$\dot{x}_{j_2}^* = a_{j_2j_20}(a_{j_1j_1s_1} - a_{j_2j_21})[(a_{j_1j_1s_1} - a_{j_2j_2})^2 + b_{j_2j_2}],$$

the parabola flows are positive and negative for $\dot{x}_{j_2}^* > 0$ and $\dot{x}_{j_2}^* < 0$, respectively.

Due to

$$(a_{j_1j_1s_1} - a_{j_2j_2})^2 + b_{j_2j_2} > 0,$$

the flow of $y_1^* = a_{j_1j_1s_1} \neq a_{j_1j_1s_2}$ with $\dot{x}_{j_2}^* = a_{j_2j_20}[(a_{j_1j_1s_1} - a_{j_2j_2})^2 + b_{j_2j_2}]$ possesses the following properties as in Eqs. (4.30)–(4.33):

- For $a_{j_1j_10}\alpha_{j_1j_2}(a_{j_1j_1s_1} - a_{j_1j_1s_2})^2 > 0$ and $a_{j_2j_20}(a_{j_1j_1s_1} - a_{j_2j_21}) > 0$,
$$(a_{j_1j_1s_1}, \dot{x}_{j_2}^*) = \underbrace{(\mathrm{UP}, \mathrm{pF})}_{\text{up-parabola flow } (+)}.$$
- For $a_{j_1j_10}\alpha_{j_1j_2}(a_{j_1j_1s_1} - a_{j_1j_1s_2})^2 < 0$ and $a_{j_2j_20}(a_{j_1j_1s_1} - a_{j_2j_21}) > 0$,
$$(a_{j_1j_1s_1}, \dot{x}_{j_2}^*) = \underbrace{(\mathrm{DP}, \mathrm{pF})}_{\text{down-parabola flow } (+)}.$$
- For $a_{j_1j_10}\alpha_{j_1j_2}(a_{j_1j_1s_1} - a_{j_1j_1s_2})^2 > 0$ and $a_{j_2j_20}(a_{j_1j_1s_1} - a_{j_2j_21}) < 0$,
$$(a_{j_1j_1s_1}, \dot{x}_{j_2}^*) = \underbrace{(\mathrm{DP}, \mathrm{nF})}_{\text{down-parabola flow } (-)}.$$

- For $a_{j_1j_10}\alpha_{j_1j_2}(a_{j_1j_1s_1}-a_{j_1j_1s_2})^2<0$ and $a_{j_2j_20}(a_{j_1j_1s_1}-a_{j_2j_21})<0$,

$$(a_{j_1j_1s_1},\dot{x}^*_{j_2})=\underbrace{(\mathrm{UP},\mathrm{nF})}_{\text{up-parabola flow }(-)}.$$

(ii$_{1b}$) In phase space, at $y_1^*=a_{j_1j_1s_2}$,

$$\left.\frac{dx_{j_1}}{dx_{j_2}}\right|_{y_1^*=a_{j_1j_1s_2}}=\left.\frac{a_{j_1j_10}(y_{j_1}-a_{j_1j_1s_1})(y_{j_1}-a_{j_1j_1s_2})^2}{a_{j_2j_20}(y_{j_1}-a_{j_2j_21})[(y_{j_1}-a_{j_2j_2})^2+b_{j_2j_2}]}\right|_{y_1^*=a_{j_1j_1s_2}}=0,$$

$$\left.\frac{d^2x_{j_1}}{dx_{j_2}^2}\right|_{y_1^*=a_{j_1j_1s_2}}=0.$$

If

$$\left.\frac{d^3x_{j_1}}{dx_{j_2}^3}\right|_{y_1^*=a_{j_1j_1s_2}}=\frac{2a_{j_1j_10}\alpha^2_{j_1j_2}(a_{j_1j_1s_2}-a_{j_1j_1s_1})}{a_{j_2j_20}(a_{j_1j_1s_2}-a_{j_2j_21})[(a_{j_1j_1s_2}-a_{j_2j_2})^2+b_{j_2j_2}]}>0,$$

there is an increasing-inflection flow at $y_1^*=a_{j_1j_1s_2}$. If

$$\left.\frac{d^3x_{j_1}}{dx_{j_2}^3}\right|_{y_1^*=a_{j_1j_1s_2}}=\frac{2a_{j_1j_10}\alpha^2_{j_1j_2}(a_{j_1j_1s_2}-a_{j_1j_1s_1})}{a_{j_2j_20}(a_{j_1j_1s_2}-a_{j_2j_21})[(a_{j_1j_1s_2}-a_{j_2j_2})^2+b_{j_2j_2}]}<0,$$

there is a decreasing-inflection flow at $y_1^*=a_{j_1j_1s_2}$. With

$$\dot{x}^*_{j_2}=a_{j_2j_20}(a_{j_1j_1s_2}-a_{j_2j_21})[(a_{j_1j_1s_2}-a_{j_2j_2})^2+b_{j_2j_2}],$$

the inflection flows are positive and negative for $\dot{x}^*_{j_2}>0$ and $\dot{x}^*_{j_2}<0$, respectively.

Due to

$$(a_{j_1j_1s_2}-a_{j_2j_2})^2+b_{j_2j_2}>0,$$

the flow of $y_1^*=a_{j_1j_1s_2}\neq a_{j_1j_1s_1}$ with $\dot{x}^*_{j_2}=a_{j_2j_20}[(a_{j_1j_1s_2}-a_{j_2j_2})^2+b_{j_2j_2}]$ possesses the following properties as in Eqs. (4.34)–(4.37):

- For $a_{j_1j_10}\alpha^2_{j_1j_2}(a_{j_1j_1s_2}-a_{j_1j_1s_1})>0$ and $a_{j_2j_20}(a_{j_1j_1s_2}-a_{j_2j_21})>0$,

$$(a_{j_1j_1s_2},\dot{x}^*_{j_2})=\underbrace{(\mathrm{II},\mathrm{pF})}_{\text{increasing-inflection flow }(+)}.$$

- For $a_{j_1j_10}\alpha^2_{j_1j_2}(a_{j_1j_1s_2}-a_{j_1j_1s_1})<0$ and $a_{j_2j_20}(a_{j_1j_1s_2}-a_{j_2j_21})>0$,

$$(a_{j_1j_1s_2},\dot{x}^*_{j_2})=\underbrace{(\mathrm{DI},\mathrm{pF})}_{\text{decreasing-inflection flow }(+)}.$$

- For $a_{j_1j_10}\alpha^2_{j_1j_2}(a_{j_1j_1s_2} - a_{j_1j_1s_1}) > 0$ and $a_{j_2j_20}(a_{j_1j_1s_2} - a_{j_2j_21}) < 0$,

$$(a_{j_1j_1s_2}, \dot{x}^*_{j_2}) = \underbrace{(\mathrm{DI}, \mathrm{nF})}_{\text{decreasing-inflection flow } (-)}.$$

- For $a_{j_1j_10}\alpha^2_{j_1j_2}(a_{j_1j_1s_2} - a_{j_1j_1s_1}) < 0$ and $a_{j_2j_20}(a_{j_1j_1s_2} - a_{j_2j_21}) < 0$,

$$(a_{j_1j_1s_2}, \dot{x}^*_{j_2}) = \underbrace{(\mathrm{II}, \mathrm{nF})}_{\text{increasing-inflection flow } (-)}.$$

- The inflection flows are the appearing bifurcations of up-parabola and down-parabola flows.

(ii_{1c}) For $y_1^* = a_{j_2j_21}$, in phase space,

$$\left.\frac{dx_{j_2}}{dx_{j_1}}\right|_{y_1^*=a_{j_2j_21}} = \left.\frac{a_{j_2j_20}(y_{j_1} - a_{j_2j_21})[(y_{j_1} - a_{j_2j_2})^2 + b_{j_2j_2}]}{a_{j_1j_10}(y_{j_1} - a_{j_1j_1s_1})(y_{j_1} - a_{j_1j_1s_2})^2}\right|_{y_1^*=a_{j_2j_21}} = 0.$$

If

$$\left.\frac{d^2x_{j_2}}{dx^2_{j_1}}\right|_{y_1^*=a_{j_2j_21}} = \frac{a_{j_2j_20}\alpha_{j_1j_1}[(a_{j_2j_21} - a_{j_2j_2})^2 + b_{j_2j_2}]}{a_{j_1j_10}(a_{j_2j_21} - a_{j_1j_1s_1})(a_{j_2j_21} - a_{j_1j_1s_2})^2} > 0,$$

there is an up-parabola flow at $y_1^* = a_{j_2j_21}$. If

$$\left.\frac{d^2x_{j_2}}{dx^2_{j_1}}\right|_{y_1^*=a_{j_2j_21}} = \frac{a_{j_2j_20}\alpha_{j_1j_1}[(a_{j_2j_21} - a_{j_2j_2})^2 + b_{j_2j_2}]}{a_{j_1j_10}(a_{j_2j_21} - a_{j_1j_1s_1})(a_{j_2j_21} - a_{j_1j_1s_2})^2} < 0,$$

there is a down-parabola flow at $y_1^* = a_{j_2j_21}$. With

$$\dot{x}^*_{j_1} = a_{j_1j_10}(a_{j_2j_21} - a_{j_1j_1s_1})(a_{j_2j_21} - a_{j_1j_1s_2})^2,$$

the parabola flows are positive and negative for $\dot{x}^*_{j_2} > 0$ and $\dot{x}^*_{j_2} < 0$, respectively.

Due to

$$(a_{j_2j_21} - a_{j_2j_2})^2 + b_{j_2j_2} > 0,$$

the flow of $y_1^* = a_{j_2j_21} \neq a_{j_1j_1s_1}, a_{j_1j_1s_2}$ possesses the following properties with $\dot{x}^*_{j_1} = a_{j_1j_10}(a_{j_2j_21} - a_{j_1j_1s_1})(a_{j_2j_21} - a_{j_1j_1s_2})^2$ as in Eqs. (4.38)–(4.41):

- For $a_{j_1j_10}(a_{j_2j_21} - a_{j_1j_1s_1})(a_{j_2j_21} - a_{j_1j_1s_2})^2 > 0$ and $a_{j_2j_20}\alpha_{j_1j_1} > 0$,

$$(\dot{x}^*_{j_1}, a_{j_2j_21}) = \underbrace{(\mathrm{pF}, \mathrm{UP})}_{\text{up-parabola flow } (+)}.$$

- For $a_{j_1j_10}(a_{j_2j_21}-a_{j_1j_1s_1})(a_{j_2j_21}-a_{j_1j_1s_2})^2<0$ and $a_{j_2j_20}\alpha_{j_1j_1}>0$,

$$(\dot{x}^*_{j_1},a_{j_2j_21})=\underbrace{(\text{nF},\text{DP})}_{\text{down-parabola flow }(-)}.$$

- For $a_{j_1j_10}(a_{j_2j_21}-a_{j_1j_1s_1})(a_{j_2j_21}-a_{j_1j_1s_2})^2>0$ and $a_{j_2j_20}\alpha_{j_1j_1}<0$,

$$(\dot{x}^*_{j_1},a_{j_2j_21})=\underbrace{(\text{pF},\text{DP})}_{\text{down-parabola flow }(+)}.$$

- For $a_{j_1j_10}(a_{j_2j_21}-a_{j_1j_1s_1})(a_{j_2j_21}-a_{j_1j_1s_2})^2<0$ and $a_{j_2j_20}\alpha_{j_1j_1}<0$,

$$(\dot{x}^*_{j_1},a_{j_2j_21})=\underbrace{(\text{nF},\text{UP})}_{\text{up-parabola flow }(-)}.$$

(ii$_2$) For $a_{j_1j_1s_1}=a_{j_1j_1s_2}=a_{j_1j_11}$,

$$\begin{aligned}\dot{x}_{j_1}&=a_{j_1j_10}(y_{j_1}-a_{j_1j_11})^3,\\ \dot{x}_{j_2}&=a_{j_2j_20}(y_{j_1}-a_{j_2j_21})[(y_{j_1}-a_{j_2j_2})^2+b_{j_2j_2}].\end{aligned}$$

For $a_{j_1j_11}\neq a_{j_2j_21}$, a dynamical system in the y_{j_1}-direction is constructed as

$$\begin{aligned}\dot{y}_{j_1}&=\alpha_{j_1j_1}a_{j_1j_10}(y_{j_1}-a_{j_1j_11})^3+\alpha_{j_1j_2}a_{j_2j_20}(y_{j_1}-a_{j_2j_21})\\ &\quad\times[(y_{j_1}-a_{j_2j_2})^2+b_{j_2j_2}],\end{aligned}$$

and a dynamical system in the $y^{\perp}_{j_1}$-direction is constructed as

$$\begin{aligned}\dot{y}^{\perp}_{j_1}&=\beta_{j_1j_1}a_{j_1j_10}(y_{j_1}-a_{j_1j_11})^3+\beta_{j_1j_2}a_{j_2j_20}(y_{j_1}-a_{j_2j_21})\\ &\quad\times[(y_{j_1}-a_{j_2j_2})^2+b_{j_2j_2}].\end{aligned}$$

(ii$_{2a}$) In phase space, at $y^*_1=a_{j_1j_11}$,

$$\left.\frac{dx_{j_1}}{dx_{j_2}}\right|_{y^*_1=a_{j_1j_11}}=\left.\frac{a_{j_1j_10}(y_{j_1}-a_{j_1j_1s_1})(y_{j_1}-a_{j_1j_1s_2})^2}{a_{j_2j_20}(y_{j_1}-a_{j_2j_21})[(y_{j_1}-a_{j_2j_2})^2+b_{j_2j_2}]}\right|_{y^*_1=a_{j_1j_11}}=0,$$

$$\left.\frac{d^2x_{j_1}}{dx^2_{j_2}}\right|_{y^*_1=a_{j_1j_11}}=\left.\frac{d^3x_{j_1}}{dx^3_{j_2}}\right|_{y^*_1=a_{j_1j_11}}=0.$$

If

$$\left.\frac{d^4x_{j_1}}{dx^4_{j_2}}\right|_{y^*_1=a_{j_1j_11}}=\frac{6a_{j_1j_10}\alpha^3_{j_1j_2}}{a_{j_2j_20}(a_{j_1j_11}-a_{j_2j_21})[(a_{j_1j_11}-a_{j_2j_2})^2+b_{j_2j_2}]}>0,$$

there is a third-order up-parabola flow at $y_1^* = a_{j_1j_11}$. If

$$\left.\frac{d^4x_{j_1}}{dx_{j_2}^4}\right|_{y_1^*=a_{j_1j_11}} = \frac{6a_{j_1j_10}\alpha_{j_1j_2}^3}{a_{j_2j_20}(a_{j_1j_11}-a_{j_2j_21})[(a_{j_1j_11}-a_{j_2j_2})^2+b_{j_2j_2}]} < 0,$$

there is a third-order down-parabola flow at $y_1^* = a_{j_1j_1s_2}$. With

$$\dot{x}_{j_2}^* = a_{j_2j_20}(a_{j_1j_11}-a_{j_2j_21})[(a_{j_1j_11}-a_{j_2j_2})^2+b_{j_2j_2}],$$

the third-order parabola flows are positive and negative for $\dot{x}_{j_2}^* > 0$ and $\dot{x}_{j_2}^* < 0$, respectively.

Due to

$$(a_{j_1j_1s_2}-a_{j_2j_2})^2+b_{j_2j_2} > 0,$$

the flow of $y_1^* = a_{j_1j_11}$ with $\dot{x}_{j_2}^* = a_{j_2j_20}(a_{j_1j_11}-a_{j_2j_21})[(a_{j_1j_11}-a_{j_2j_2})^2+b_{j_2j_2}]$ possesses the following properties as in Eqs. (4.45)–(4.48):

- For $a_{j_1j_10}\alpha_{j_1j_2}^3 > 0$ and $a_{j_2j_20}(a_{j_1j_11}-a_{j_2j_21}) > 0$,

$$(a_{j_1j_11}, \dot{x}_{j_2}^*) = \underbrace{(3^{\text{rd}}\text{UP}, \text{pF})}_{\text{third-order up-parabola flow }(+)}.$$

- For $a_{j_1j_10}\alpha_{j_1j_2}^3 < 0$ and $a_{j_2j_20}(a_{j_1j_11}-a_{j_2j_21}) > 0$,

$$(a_{j_1j_11}, \dot{x}_{j_2}^*) = \underbrace{(3^{\text{rd}}\text{DP}, \text{pF})}_{\text{third-order down-parabola flow }(+)}.$$

- For $a_{j_1j_10}\alpha_{j_1j_2}^3 > 0$ and $a_{j_2j_20}(a_{j_1j_11}-a_{j_2j_21}) < 0$,

$$(a_{j_1j_11}, \dot{x}_{j_2}^*) = \underbrace{(3^{\text{rd}}\text{DP}, \text{nF})}_{\text{third-order down-parabola flow }(-)}.$$

- For $a_{j_1j_10}\alpha_{j_1j_2}^3 < 0$ and $a_{j_2j_20}(a_{j_1j_11}-a_{j_2j_21}) < 0$,

$$(a_{j_1j_11}, \dot{x}_{j_2}^*) = \underbrace{(3^{\text{rd}}\text{UP}, \text{nF})}_{\text{third-order up-parabola flow }(-)}.$$

– The third-order up-parabola flows are the appearing bifurcations of up-parabola, down-parabola, and up-parabola flows.
– The third-order down-parabola flows are the appearing bifurcations of down-parabola, up-parabola, and down-parabola flows.

(ii$_{2b}$) For $y_1^* = a_{j_2j_2 1}$, in phase space,

$$\left.\frac{dx_{j_2}}{dx_{j_1}}\right|_{y_1^*=a_{j_2j_21}} = \left.\frac{a_{j_2j_20}(y_{j_1}-a_{j_2j_21})[(y_{j_1}-a_{j_2j_2})^2+b_{j_2j_2}]}{a_{j_1j_10}(y_{j_1}-a_{j_1j_11})^3}\right|_{y_1^*=a_{j_2j_21}} = 0.$$

If

$$\left.\frac{d^2x_{j_2}}{dx_{j_1}^2}\right|_{y_1^*=a_{j_2j_21}} = \frac{a_{j_2j_20}\alpha_{j_1j_1}[(a_{j_2j_21}-a_{j_2j_2})^2+b_{j_2j_2}]}{a_{j_1j_10}(a_{j_2j_21}-a_{j_1j_11})^3} > 0,$$

there is an up-parabola flow at $y_1^* = a_{j_2j_21}$. If

$$\left.\frac{d^2x_{j_2}}{dx_{j_1}^2}\right|_{y_1^*=a_{j_2j_21}} = \frac{a_{j_2j_20}\alpha_{j_1j_1}[(a_{j_2j_21}-a_{j_2j_2})^2+b_{j_2j_2}]}{a_{j_1j_10}(a_{j_2j_21}-a_{j_1j_11})^3} < 0,$$

there is a down-parabola flow at $y_1^* = a_{j_2j_21}$. With

$$\dot{x}_{j_1}^* = a_{j_1j_10}(a_{j_2j_21}-a_{j_1j_11})^3,$$

the parabola flows are positive and negative for $\dot{x}_{j_2}^* > 0$ and $\dot{x}_{j_2}^* < 0$, respectively.

Due to

$$(a_{j_2j_21}-a_{j_2j_2})^2+b_{j_2j_2} > 0,$$

the flow of $y_1^* = a_{j_2j_21}$ with $\dot{x}_{j_1}^* = a_{j_1j_10}(a_{j_2j_21}-a_{j_1j_11})^3$ possesses the following properties as in Eqs. (4.49)–(4.52):

- For $a_{j_1j_10}(a_{j_2j_21}-a_{j_1j_11})^3 > 0$ and $a_{j_2j_20}\alpha_{j_1j_1} > 0$,
$$(\dot{x}_{j_1}^*, a_{j_2j_21}) = \underbrace{(\text{pF}, \text{UP})}_{\text{up-parabola flow }(+)}.$$
- For $a_{j_1j_10}(a_{j_2j_21}-a_{j_1j_11})^3 < 0$ and $a_{j_2j_20}\alpha_{j_1j_1} > 0$,
$$(\dot{x}_{j_1}^*, a_{j_2j_21}) = \underbrace{(\text{nF}, \text{DP})}_{\text{down-parabola flow }(-)}.$$
- For $a_{j_1j_10}(a_{j_2j_21}-a_{j_1j_11})^3 > 0$ and $a_{j_2j_20}\alpha_{j_1j_1} < 0$,
$$(\dot{x}_{j_1}^*, a_{j_2j_21}) = \underbrace{(\text{pF}, \text{DP})}_{\text{down-parabola flow }(+)}.$$
- For $a_{j_1j_10}(a_{j_2j_21}-a_{j_1j_11})^3 < 0$ and $a_{j_2j_20}\alpha_{j_1j_1} < 0$,
$$(\dot{x}_{j_1}^*, a_{j_2j_21}) = \underbrace{(\text{nF}, \text{UP})}_{\text{up-parabola flow }(-)}.$$

(ii$_3$) For $a_{j_1j_1s_1} = a_{j_2j_21}$, a dynamical system in the y_{j_1}-direction is constructed as

$$\begin{aligned}\dot{y}_{j_1} &= (y_{j_1} - a_{j_1j_1s_1})\{\alpha_{j_1j_1}a_{j_1j_10}(y_{j_1} - a_{j_1j_1s_2})^2\\ &\quad + \alpha_{j_1j_2}a_{j_2j_20}[(y_{j_1} - a_{j_2j_2})^2 + b_{j_2j_2}]\},\end{aligned}$$

and a dynamical system in the $y_{j_1}^{\perp}$-direction is defined as

$$\begin{aligned}\dot{y}_{j_1}^{\perp} &= (y_{j_1} - a_{j_1j_1s_1})\{\beta_{j_1j_1}a_{j_1j_10}(y_{j_1} - a_{j_1j_1s_2})^2\\ &\quad + \beta_{j_1j_2}a_{j_2j_20}[(y_{j_1} - a_{j_2j_2})^2 + b_{j_2j_2}]\},\end{aligned}$$

and a G-function is defined as

$$\begin{aligned}G_{j_1j_10} &= \frac{\partial}{\partial y_{j_1}}[\mathbf{n}^{\mathrm{T}} \cdot \mathbf{F}(y_{j_1}, \mathbf{p})]\bigg|_{y_{j_1}^* = a_{j_1j_1s_1} = a_{j_2j_21}}\\ &= \alpha_{j_1j_1}a_{j_1j_10}(a_{j_1j_1s_1} - a_{j_1j_1s_2})^2\\ &\quad + \alpha_{j_1j_2}a_{j_2j_20}[(a_{j_2j_21} - a_{j_2j_2})^2 + b_{j_2j_2}].\end{aligned}$$

In phase space, at $y_1^* = a_{j_1j_1s_1} = a_{j_2j_21}$, if

$$\frac{dx_{j_1}}{dx_{j_2}}\bigg|_{y_1^* = a_{j_1j_1s_1}} = \frac{a_{j_1j_10}(a_{j_1j_1s_1} - a_{j_1j_1s_2})^2}{a_{j_2j_20}[(a_{j_2j_21} - a_{j_2j_2})^2 + b_{j_2j_2}]} > 0,$$

there is an increasing-inflection infinite-equilibrium flow, and if

$$\frac{dx_{j_1}}{dx_{j_2}}\bigg|_{y_1^* = a_{j_1j_1s_1}} = \frac{a_{j_1j_10}(a_{j_1j_1s_1} - a_{j_1j_1s_2})^2}{a_{j_2j_20}[(a_{j_2j_21} - a_{j_2j_2})^2 + b_{j_2j_2}]} < 0,$$

there is a decreasing-inflection infinite-equilibrium flow. The corresponding variational equation is

$$\begin{aligned}\Delta\dot{y}_{j_1} &= \{\alpha_{j_1j_1}a_{j_1j_10}(a_{j_1j_1s_1} - a_{j_1j_1s_2})^2 + \alpha_{j_1j_2}a_{j_2j_20}\\ &\quad \times [(a_{j_2j_21} - a_{j_2j_2})^2 + b_{j_2j_2}]\}\Delta y_{j_1}\\ &= G_{j_1j_10}\Delta y_{j_1}.\end{aligned}$$

The inflection infinite-equilibrium flows are source and sink for $G_{j_1j_10} > 0$ and $G_{j_1j_10} < 0$, respectively.

Therefore, the flow of $y_1^* = a_{j_1j_1s_1} = a_{j_2j_21}$ possesses the following properties as in Eqs. (4.56)–(4.59):

- For $a_{j_1j_10}(a_{j_1j_1s_1} - a_{j_1j_1s_2})^2/a_{j_2j_20} > 0$ and $G_{j_1j_10} > 0$,

$$(a_{j_1j_1s_1}, a_{j_2j_21}) = \underbrace{(\mathrm{II}, \mathrm{SO})}_{\text{increasing-inflection source}}.$$

- For $a_{j_1j_10}(a_{j_1j_1s_1} - a_{j_1j_1s_2})^2/a_{j_2j_20} > 0$ and $G_{j_1j_10} < 0$,

$$(a_{j_1j_1s_1}, a_{j_2j_21}) = \underbrace{(\text{II}, \text{SI})}_{\text{increasing-inflection sink}}.$$

- For $a_{j_1j_10}(a_{j_1j_1s_1} - a_{j_1j_1s_2})^2/a_{j_2j_20} < 0$ and $G_{j_1j_10} > 0$,

$$(a_{j_1j_1s_1}, a_{j_2j_21}) = \underbrace{(\text{DI}, \text{SO})}_{\text{decreasing-inflection source}}.$$

- For $a_{j_1j_10}(a_{j_1j_1s_1} - a_{j_1j_1s_2})^2/a_{j_2j_20} < 0$ and $G_{j_1j_10} < 0$,

$$(a_{j_1j_1s_1}, a_{j_2j_21}) = \underbrace{(\text{DI}, \text{SI})}_{\text{decreasing-inflection sink}}.$$

(ii$_4$) For $a_{j_1j_1s_2} = a_{j_2j_21}$, a dynamical system in the y_{j_1}-direction is constructed as

$$\begin{aligned}\dot{y}_{j_1} = (y_{j_1} - a_{j_1j_1s_2})\{&\alpha_{j_1j_1}a_{j_1j_10}(y_{j_1} - a_{j_1j_1s_1})(y_{j_1} - a_{j_1j_1s_2})\\ &+ \alpha_{j_1j_2}a_{j_2j_20}[(y_{j_1} - a_{j_2j_2})^2 + b_{j_2j_2}]\},\end{aligned}$$

and a dynamical system in the $y_{j_1}^{\perp}$-direction is constructed as

$$\begin{aligned}\dot{y}_{j_1}^{\perp} = (y_{j_1} - a_{j_1j_1s_2})\{&\beta_{j_1j_1}a_{j_1j_10}(y_{j_1} - a_{j_1j_1s_1})(y_{j_1} - a_{j_1j_1s_2})\\ &+ \beta_{j_1j_2}a_{j_2j_20}[(y_{j_1} - a_{j_2j_2})^2 + b_{j_2j_2}]\},\end{aligned}$$

and a G-function is defined as

$$\begin{aligned}G_{j_1j_10} &= \frac{\partial}{\partial y_{j_1}}[\mathbf{n}^{\mathrm{T}} \cdot \mathbf{F}(y_{j_1}, \mathbf{p})]\Big|_{y_{j_1}^* = a_{j_1j_1s_2} = a_{j_2j_21}}\\ &= \alpha_{j_1j_2}a_{j_2j_20}[(a_{j_1j_1s_2} - a_{j_2j_2})^2 + b_{j_2j_2}].\end{aligned}$$

In phase space, at $y_1^* = a_{j_1j_1s_2} = a_{j_2j_21}$,

$$\frac{dx_{j_1}}{dx_{j_2}}\Big|_{y_1^* = a_{j_1j_1s_2}} = \frac{a_{j_1j_10}(y_{j_1} - a_{j_1j_1s_1})(y_{j_1} - a_{j_1j_1s_2})}{a_{j_2j_20}[(y_{j_1} - a_{j_2j_2})^2 + b_{j_2j_2}]}\Big|_{y_1^* = a_{j_1j_1s_2}} = 0.$$

If

$$\frac{d^2x_{j_1}}{dx_{j_2}^2}\Big|_{y_1^* = a_{j_1j_1s_2}} = \frac{a_{j_1j_10}\alpha_{j_1j_2}^2(a_{j_1j_1s_2} - a_{j_1j_1s_1})}{a_{j_2j_20}\alpha_{j_1j_2}[(a_{j_1j_1s_2} - a_{j_2j_2})^2 + b_{j_2j_2}]} > 0,$$

there is an up-parabola infinite-equilibrium flow at $y_1^* = a_{j_1j_1s_2} = a_{j_2j_21}$. If

$$\left.\frac{d^3x_{j_1}}{dx_{j_2}^3}\right|_{y_1^*=a_{j_1j_1s_2}} = \frac{a_{j_1j_10}\alpha_{j_1j_2}^2(a_{j_1j_1s_2} - a_{j_1j_1s_1})}{a_{j_2j_20}\alpha_{j_1j_2}[(a_{j_1j_1s_2} - a_{j_2j_2})^2 + b_{j_2j_2}]} < 0,$$

there is a down-parabola infinite-equilibrium flow at $y_1^* = a_{j_1j_1s_2} = a_{j_2j_21}$. The corresponding variational equation is

$$\begin{aligned}\Delta\dot{y}_{j_1} &= \alpha_{j_1j_2}a_{j_2j_20}[(a_{j_1j_1s_2} - a_{j_2j_2})^2 + b_{j_2j_2}]\Delta y_{j_1}\\ &= G_{j_2j_20}\Delta y_{j_1},\end{aligned}$$

the parabola infinite-equilibrium flows are source and sink for $G_{j_2j_20} > 0$ and $G_{j_2j_20} < 0$, respectively.

Due to

$$(a_{j_1j_1s_2} - a_{j_2j_2})^2 + b_{j_2j_2} > 0,$$

the infinite-equilibrium flow of $y_1^* = a_{j_1j_1s_1} = a_{j_2j_21}$ possesses the following properties as Eqs. (4.63)–(4.66):

- For $\alpha_{j_1j_2}^2 a_{j_1j_10}(a_{j_1j_1s_2} - a_{j_1j_1s_1}) > 0$ and $\alpha_{j_1j_2}a_{j_2j_20} > 0$,

$$(a_{j_1j_1s_2}, a_{j_2j_21}) = \underbrace{(\text{UP}, \text{SO})}_{\text{up-parabola source}}.$$

- For $\alpha_{j_1j_2}^2 a_{j_1j_10}(a_{j_1j_1s_2} - a_{j_1j_1s_1}) < 0$ and $\alpha_{j_1j_2}a_{j_2j_20} > 0$,

$$(a_{j_1j_1s_2}, a_{j_2j_21}) = \underbrace{(\text{DP}, \text{SO})}_{\text{down-parabola source}}.$$

- $\alpha_{j_1j_2}^2 a_{j_1j_10}(a_{j_1j_1s_2} - a_{j_1j_1s_1}) > 0$ and $\alpha_{j_1j_2}a_{j_2j_20} < 0$,

$$(a_{j_1j_1s_2}, a_{j_2j_21}) = \underbrace{(\text{DP}, \text{SI})}_{\text{down-parabola sink}}.$$

- $\alpha_{j_1j_2}^2 a_{j_1j_10}(a_{j_1j_1s_2} - a_{j_1j_1s_1}) < 0$ and $\alpha_{j_1j_2}a_{j_2j_20} < 0$,

$$(a_{j_1j_1s_2}, a_{j_2j_21}) = \underbrace{(\text{UP}, \text{SI})}_{\text{up-parabola sink}}.$$

(ii$_5$) For $a_{j_1j_1s_2} = a_{j_2j_21}$, the dynamical system becomes

$$\dot{x}_{j_1} = a_{j_1j_10}(y_{j_1} - a_{j_1j_11})^3,$$
$$\dot{x}_{j_2} = a_{j_2j_20}(y_{j_1} - a_{j_2j_21})[(y_{j_1} - a_{j_2j_2})^2 + b_{j_2j_2}],$$

and a dynamical system in the y_{j_1}-direction is constructed as

$$\dot{y}_{j_1} = (y_{j_1} - a_{j_1j_11})\{\alpha_{j_1j_1}a_{j_1j_10}(y_{j_1} - a_{j_1j_11})^2 + \alpha_{j_1j_2}a_{j_2j_20}[(y_{j_1} - a_{j_2j_2})^2 + b_{j_2j_2}]\},$$

and a dynamical system in the $y_{j_1}^{\perp}$-direction is constructed as

$$\dot{y}_{j_1}^{\perp} = (y_{j_1} - a_{j_1j_11})\{\beta_{j_1j_1}a_{j_1j_10}(y_{j_1} - a_{j_1j_11})^2 + \beta_{j_1j_2}a_{j_2j_20}[(y_{j_1} - a_{j_2j_2})^2 + b_{j_2j_2}]\},$$

and a G-function is defined as

$$G_{j_1j_10} = \frac{\partial}{\partial y_{j_1}}[\mathbf{n}^{\mathrm{T}} \cdot \mathbf{F}(y_{j_1}, \mathbf{p})]\Big|_{y_{j_1}^* = a_{j_1j_11} = a_{j_2j_21}} = \alpha_{j_1j_2}a_{j_2j_20}[(a_{j_2j_21} - a_{j_2j_2})^2 + b_{j_2j_2}].$$

For $y_1^* = a_{j_1j_11} = a_{j_2j_21}$, in phase space,

$$\frac{dx_{j_1}}{dx_{j_2}}\Big|_{y_1^* = a_{j_1j_11}} = \frac{a_{j_1j_10}(y_{j_1} - a_{j_1j_11})^2}{a_{j_2j_20}[(y_{j_1} - a_{j_2j_2})^2 + b_{j_2j_2}]}\Big|_{y_1^* = a_{j_1j_11}} = 0,$$

$$\frac{d^2x_{j_1}}{dx_{j_2}^2}\Big|_{y_1^* = a_{j_1j_11}} = 0.$$

If

$$\frac{d^3x_{j_1}}{dx_{j_2}^3}\Big|_{y_1^* = a_{j_1j_11}} = \frac{2a_{j_1j_10}\alpha_{j_1j_2}^3}{a_{j_2j_20}\alpha_{j_1j_2}[(a_{j_1j_11} - a_{j_2j_2})^2 + b_{j_2j_2}]} > 0,$$

there is a third-order increasing-inflection flow at $y_1^* = a_{j_1j_11} = a_{j_2j_21}$. If

$$\frac{d^3x_{j_1}}{dx_{j_2}^3}\Big|_{y_1^* = a_{j_1j_11}} = \frac{2a_{j_1j_10}\alpha_{j_1j_2}^3}{a_{j_2j_20}\alpha_{j_1j_2}[(a_{j_1j_11} - a_{j_2j_2})^2 + b_{j_2j_2}]} < 0,$$

there is a third-order decreasing-inflection flow at $y_1^* = a_{j_1j_11} = a_{j_2j_21}$. The corresponding variational equation is

$$\Delta\dot{y}_{j_1} = \alpha_{j_1j_2}a_{j_2j_20}[(a_{j_2j_21} - a_{j_2j_2})^2 + b_{j_2j_2}]\Delta y_{j_1} = G_{j_1j_10}\Delta y_{j_1},$$

the third-order inflection infinite-equilibrium flows are source and sink for $G_{j_1j_10} > 0$ and $G_{j_1j_10} < 0$, respectively. With

$$(a_{j_2j_21} - a_{j_2j_2})^2 + b_{j_2j_2} > 0,$$

thus, the infinite-equilibrium flow of $y_1^* = a_{j_1j_11} = a_{j_2j_21}$ possesses the following properties as in Eqs. (4.71)–(4.74):

- For $\alpha_{j_1j_2}^3 a_{j_1j_10} > 0$ and $\alpha_{j_1j_2} a_{j_2j_20} > 0$,
$$(a_{j_1j_11}, a_{j_2j_21}) = \underbrace{(2^{\text{nd}}\text{II}, \text{SO})}_{\text{second-order incerasing-inflection source}}.$$
- For $\alpha_{j_1j_2}^3 a_{j_1j_10} < 0$ and $\alpha_{j_1j_2} a_{j_2j_20} > 0$,
$$(a_{j_1j_11}, a_{j_2j_21}) = \underbrace{(2^{\text{nd}}\text{DI}, \text{SO})}_{\text{second-order decerasing-inflection source}}.$$
- For $\alpha_{j_1j_2}^3 a_{j_1j_10} > 0$ and $\alpha_{j_1j_2} a_{j_2j_20} < 0$,
$$(a_{j_1j_11}, a_{j_2j_21}) = \underbrace{(2^{\text{nd}}\text{DI}, \text{SI})}_{\text{second-order decerasing-inflection sink}}.$$
- For $\alpha_{j_1j_2}^3 a_{j_1j_10} < 0$ and $\alpha_{j_1j_2} a_{j_2j_20} < 0$,
$$(a_{j_1j_11}, a_{j_2j_21}) = \underbrace{(2^{\text{nd}}\text{II}, \text{SI})}_{\text{second-order incerasing-inflection sink}}.$$

(iii) If

$$\Delta_{j_1j_1} > 0 \quad \text{and} \quad \Delta_{j_2j_2} < 0,$$

then

$$\begin{aligned}
\dot{x}_{j_1} &= a_{j_1j_10}(y_{j_1} - a_{j_1j_11})(y_{j_1} - a_{j_1j_12})(y_{j_1} - a_{j_1j_13}),\\
\dot{x}_{j_2} &= a_{j_2j_20}(y_{j_1} - a_{j_2j_21})[(y_{j_1} - a_{j_2j_2})^2 + b_{j_2j_2}],
\end{aligned}$$

where

$$\begin{gathered}
a_{j_2j_2} = -\frac{1}{2}B_{j_2j_2}, \quad b_{j_2j_2} = -\frac{1}{4}\Delta_{j_2j_2};\\
\{a_{j_1j_11}, a_{j_1j_12}, a_{j_1j_13}\} = \text{sort}\{b_{j_1j_11}, b_{j_1j_12}, b_{j_1j_13}\},\\
a_{j_1j_1s_1} < a_{j_1j_1s_2}, s_1 < s_2;\\
b_{j_1j_12}, b_{j_1j_13} = -\frac{1}{2}(B_{j_1j_1} \pm \sqrt{\Delta_{j_1j_1}}), \quad a_{j_1j_10} = A_{j_1j_10}.
\end{gathered}$$

A dynamical system in the y_{j_1}-direction is constructed as

$$\dot{y}_{j_1} = \alpha_{j_1 j_1} a_{j_1 j_1 0} \prod_{s_1=1}^{3} (y_{j_1} - a_{j_1 j_1 s_1}) \\ + \alpha_{j_1 j_2} a_{j_2 j_2 0}(y_{j_1} - a_{j_2 j_2 1})[(y_{j_1} - a_{j_2 j_2})^2 + b_{j_2 j_2}],$$

and a dynamical system in the $y_{j_1}^{\perp}$-direction is constructed as

$$\dot{y}_{j_1}^{\perp} = \beta_{j_1 j_1} a_{j_1 j_1 0} \prod_{s_1=1}^{3} (y_{j_1} - a_{j_1 j_1 s_1}) \\ + \beta_{j_1 j_2} a_{j_2 j_2 0}(y_{j_1} - a_{j_2 j_2 1})[(y_{j_1} - a_{j_2 j_2})^2 + b_{j_2 j_2}].$$

In phase space, at $y_1^* = a_{j_1 j_1 s_1}$,

$$\left.\frac{dx_{j_1}}{dx_{j_2}}\right|_{y_1^*=a_{j_1 j_1 s_1}} = \left.\frac{a_{j_1 j_1 0}(y_{j_1} - a_{j_1 j_1 s_1}) \prod_{s_2=1, s_2 \neq s_1}^{3} (y_{j_1} - a_{j_1 j_1 s_2})}{a_{j_2 j_2 0}(y_{j_1} - a_{j_2 j_2 1})[(y_{j_1} - a_{j_2 j_2})^2 + b_{j_2 j_2}]}\right|_{y_1^*=a_{j_1 j_1 s_1}} = 0.$$

If

$$\left.\frac{d^2 x_{j_1}}{dx_{j_2}^2}\right|_{y_1^*=a_{j_1 j_1 s_1}} = \frac{a_{j_1 j_1 0} \prod_{s_2=1, s_2 \neq s_1}^{3} (a_{j_1 j_1 s_1} - a_{j_1 j_1 s_2})}{a_{j_2 j_2 0}(a_{j_1 j_1 s_1} - a_{j_2 j_2 1})[(a_{j_1 j_1 s_1} - a_{j_2 j_2})^2 + b_{j_2 j_2}]} > 0,$$

there is an up-parabola flow at $y_1^* = a_{j_1 j_1 s_1}$. If

$$\left.\frac{d^2 x_{j_1}}{dx_{j_2}^2}\right|_{y_1^*=a_{j_1 j_1 s_1}} = \frac{a_{j_1 j_1 0} \prod_{s_2=1, s_2 \neq s_1}^{3} (a_{j_1 j_1 s_1} - a_{j_1 j_1 s_2})}{a_{j_2 j_2 0}(a_{j_1 j_1 s_1} - a_{j_2 j_2 1})[(a_{j_1 j_1 s_1} - a_{j_2 j_2})^2 + b_{j_2 j_2}]} < 0,$$

there is a down-parabola flow at $y_1^* = a_{j_1 j_1 s_1}$. With

$$\dot{x}_{j_2}^* = a_{j_2 j_2 0}(a_{j_1 j_1 s_1} - a_{j_2 j_2 1})[(a_{j_1 j_1 s_1} - a_{j_2 j_2})^2 + b_{j_2 j_2}],$$

the parabola flows are positive and negative for $\dot{x}_{j_2}^* > 0$ and $\dot{x}_{j_2}^* < 0$, respectively.

With

$$(a_{j_1 j_1 s_1} - a_{j_2 j_2})^2 + b_{j_2 j_2} > 0,$$

the equilibrium flow of $y_{j_1}^* = a_{j_1 j_1 s_1} (s_1, s_2 \in \{1, 2, 3\}, s_1 \neq s_2)$ in the x_{j_1}-direction possesses the following properties as in Eqs. (4.81)–(4.84):

- For $a_{j_1j_10}\alpha_{j_1j_2}\prod_{s_2=1,s_2\neq s_1}^{3}(a_{j_1j_1s_1}-a_{j_1j_1s_2})>0$ and $a_{j_2j_20}(a_{j_1j_1s_1}-a_{j_2j_21})>0$,

$$(a_{j_1j_1s_1},\dot{x}_{j_2}^*)=\underbrace{(\text{UP},\text{pF})}_{\text{up-parabola flow }(+)}.$$

- For $a_{j_1j_10}\alpha_{j_1j_2}\prod_{s_2=1,s_2\neq s_1}^{3}(a_{j_1j_1s_1}-a_{j_1j_1s_2})<0$ and $a_{j_2j_20}(a_{j_1j_1s_1}-a_{j_2j_21})>0$,

$$(a_{j_1j_1s_1},\dot{x}_{j_2}^*)=\underbrace{(\text{DP},\text{pF})}_{\text{down-parabola flow }(+)}.$$

- For $a_{j_1j_10}\alpha_{j_1j_2}\prod_{s_2=1,s_2\neq s_1}^{3}(a_{j_1j_1s_1}-a_{j_1j_1s_2})>0$ and $a_{j_2j_20}(a_{j_1j_1s_1}-a_{j_2j_21})<0$,

$$(a_{j_1j_1s_1},\dot{x}_{j_2}^*)=\underbrace{(\text{DP},\text{nF})}_{\text{down-parabola flow }(-)}.$$

- For $a_{j_1j_10}\alpha_{j_1j_2}\prod_{s_2=1,s_2\neq s_1}^{3}(a_{j_1j_1s_1}-a_{j_1j_1s_2})<0$ and $a_{j_2j_20}(a_{j_1j_1s_1}-a_{j_2j_21})<0$,

$$(a_{j_1j_1s_1},\dot{x}_{j_2}^*)=\underbrace{(\text{UP},\text{nF})}_{\text{up-parabola flow }(-)}.$$

($\text{iii}_{1\text{b}}$) For $y_1^*=a_{j_2j_21}\neq a_{j_1j_1s_1}$, in phase space,

$$\left.\frac{dx_{j_2}}{dx_{j_1}}\right|_{y_1^*=a_{j_2j_21}}=\left.\frac{a_{j_2j_20}(y_{j_1}-a_{j_2j_21})[(y_{j_1}-a_{j_2j_2})^2+b_{j_2j_2}]}{a_{j_1j_10}\prod_{s_1=1}^{3}(y_{j_1}-a_{j_1j_1s_1})}\right|_{y_1^*=a_{j_2j_21}}=0.$$

If

$$\left.\frac{d^2x_{j_2}}{dx_{j_2}^2}\right|_{y_1^*=a_{j_2j_21}}=\frac{a_{j_2j_20}\alpha_{j_1j_1}[(a_{j_2j_21}-a_{j_2j_2})^2+b_{j_2j_2}]}{a_{j_1j_10}\prod_{s_1=1}^{3}(a_{j_2j_21}-a_{j_1j_1s_1})}>0,$$

there is an up-parabola flow at $y_1^*=a_{j_2j_21}$. If

$$\left.\frac{d^2x_{j_2}}{dx_{j_2}^2}\right|_{y_1^*=a_{j_2j_21}}=\frac{a_{j_2j_20}\alpha_{j_1j_1}[(a_{j_2j_21}-a_{j_2j_2})^2+b_{j_2j_2}]}{a_{j_1j_10}\prod_{s_1=1}^{3}(a_{j_2j_21}-a_{j_1j_1s_1})}<0,$$

there is a down-parabola flow at $y_1^*=a_{j_2j_21}$. With

$$\dot{x}_{j_1}^*=a_{j_1j_10}\prod_{s_1=1}^{3}(a_{j_2j_21}-a_{j_1j_1s_1}),$$

the parabola flows are positive and negative for $\dot{x}_{j_2}^*>0$ and $\dot{x}_{j_2}^*<0$, respectively.

Due to

$$(a_{j_2j_21} - a_{j_2j_2})^2 + b_{j_2j_2} > 0,$$

the flow of $y_1^* = a_{j_2j_21} \neq a_{j_1j_1s_1}$ possesses the following properties with $\dot{x}_{j_1}^* = a_{j_1j_10}\prod_{s_1=1}^{3}(a_{j_2j_21} - a_{j_1j_1s_1})$ as in Eqs. (4.85)–(4.88):

- For $a_{j_1j_10}\prod_{s_1=1}^{3}(a_{j_2j_21} - a_{j_1j_1s_1}) > 0$ and $a_{j_2j_20}\alpha_{j_1j_1} > 0$,

$$(\dot{x}_{j_1}^*, a_{j_2j_21}) = \underbrace{(\mathrm{pF}, \mathrm{UP})}_{\text{up-parabola flow } (+)}.$$

- For $a_{j_1j_10}\prod_{s_1=1}^{3}(a_{j_2j_21} - a_{j_1j_1s_1}) < 0$ and $a_{j_2j_20}\alpha_{j_1j_1} > 0$,

$$(\dot{x}_{j_1}^*, a_{j_2j_21}) = \underbrace{(\mathrm{nF}, \mathrm{DP})}_{\text{down-parabola flow } (-)}.$$

- For $a_{j_1j_10}\prod_{s_1=1}^{3}(a_{j_2j_21} - a_{j_1j_1s_1}) > 0$ and $a_{j_2j_20}\alpha_{j_1j_1} < 0$,

$$(\dot{x}_{j_1}^*, a_{j_2j_21}) = \underbrace{(\mathrm{pF}, \mathrm{DP})}_{\text{down-parabola flow } (+)}.$$

- For $a_{j_1j_10}\prod_{s_1=1}^{3}(a_{j_2j_21} - a_{j_1j_1s_1}) < 0$ and $a_{j_2j_20}\alpha_{j_1j_1} < 0$,

$$(\dot{x}_{j_1}^*, a_{j_2j_21}) = \underbrace{(\mathrm{nF}, \mathrm{UP})}_{\text{up-parabola flow } (-)}.$$

(iii$_2$) For $a_{j_2j_21} = a_{j_1j_1s_1}$, the dynamical system becomes

$$\begin{aligned}\dot{x}_{j_1} &= a_{j_1j_10}(y_{j_1} - a_{j_1j_11})(y_{j_1} - a_{j_1j_12})(y_{j_1} - a_{j_1j_13}),\\ \dot{x}_{j_2} &= a_{j_2j_20}(y_{j_1} - a_{j_2j_21})[(y_{j_1} - a_{j_2j_2})^2 + b_{j_2j_2}].\end{aligned}$$

A dynamical system in the y_{j_1}-direction is constructed as

$$\begin{aligned}\dot{y}_{j_1} = (y_{j_1} - a_{j_1j_1s_1})\{&\alpha_{j_1j_1}a_{j_1j_10}\prod_{s_2=1, s_2\neq s_1}^{3}(y_{j_1} - a_{j_1j_1s_2})\\ &+ \alpha_{j_1j_2}a_{j_2j_20}[(y_{j_1} - a_{j_2j_2})^2 + b_{j_2j_2}]\},\end{aligned}$$

and a dynamical system in the $y_{j_1}^{\perp}$-direction is constructed as

$$\begin{aligned}\dot{y}_{j_1}^{\perp} = (y_{j_1} - a_{j_1j_1s_1})\{&\beta_{j_1j_1}a_{j_1j_10}\prod_{s_2=1, s_2\neq s_1}^{3}(y_{j_1} - a_{j_1j_1s_2})\\ &+ \beta_{j_1j_2}a_{j_2j_20}[(y_{j_1} - a_{j_2j_2})^2 + b_{j_2j_2}]\},\end{aligned}$$

and G-function is defined as

$$G_{j_1j_10} = \frac{\partial}{\partial y_{j_1}}[\mathbf{n}^{\mathrm{T}} \cdot \mathbf{F}(y_{j_1}, \mathbf{p})]\Big|_{y_{j_1}^* = a_{j_1j_11} = a_{j_2j_21}}$$

$$= \alpha_{j_1j_1} a_{j_1j_10} \prod_{s_2=1, s_2\neq s_1}^{3} (a_{j_1j_1s_1} - a_{j_1j_1s_2})$$

$$+ \alpha_{j_1j_2} a_{j_2j_20}[(a_{j_2j_21} - a_{j_2j_2})^2 + b_{j_2j_2}].$$

In phase space, at $y_1^* = a_{j_1j_11} = a_{j_2j_21}$,

$$\frac{dx_{j_1}}{dx_{j_2}}\Big|_{y_1^* = a_{j_1j_11}} = \frac{a_{j_1j_10}(y_{j_1} - a_{j_1j_11})^2}{a_{j_2j_20}[(y_{j_1} - a_{j_2j_2})^2 + b_{j_2j_2}]}\Big|_{y_1^* = a_{j_1j_11}} = 0.$$

If

$$\frac{d^2x_{j_1}}{dx_{j_2}^2}\Big|_{y_1^* = a_{j_1j_1s_1}} = \frac{a_{j_1j_10}\alpha_{j_1j_2}^2 \prod_{s_2=1,s_2\neq s_1}^{3} (a_{j_1j_1s_1} - a_{j_1j_1s_2})}{a_{j_2j_20}\alpha_{j_1j_2}[(a_{j_1j_11} - a_{j_2j_2})^2 + b_{j_2j_2}]} > 0,$$

there is a third-order increasing-inflection infinite-equilibrium flow at $y_1^* = a_{j_1j_1s_2}$.

If

$$\frac{d^3x_{j_1}}{dx_{j_2}^3}\Big|_{y_1^* = a_{j_1j_1s_2}} = \frac{a_{j_1j_10}\alpha_{j_1j_2}^2 \prod_{s_2=1,s_2\neq s_1}^{3} (a_{j_1j_1s_1} - a_{j_1j_1s_2})}{a_{j_2j_20}\alpha_{j_1j_2}[(a_{j_1j_11} - a_{j_2j_2})^2 + b_{j_2j_2}]} < 0,$$

there is a third-order decreasing-inflection at $y_1^* = a_{j_1j_1s_2}$. The corresponding variational equation is

$$\Delta\dot{y}_{j_1} = \Big\{\alpha_{j_1j_1} a_{j_1j_10} \prod_{s_2=1,s_2\neq s_1}^{3} (a_{j_1j_1s_1} - a_{j_1j_1s_2})$$

$$+ \alpha_{j_1j_2} a_{j_2j_20}[(a_{j_2j_21} - a_{j_2j_2})^2 + b_{j_2j_2}]\Big\}\Delta y_{j_1}$$

$$= G_{j_2j_20}\Delta y_{j_1},$$

the inflection parabola infinite-equilibrium flows are source and sink for $G_{j_2j_20} > 0$ and $G_{j_2j_20} < 0$, respectively.

Due to

$$(a_{j_1j_1s_2} - a_{j_2j_2})^2 + b_{j_2j_2} > 0,$$

The infinite-equilibrium flow of $y_1^* = a_{j_1j_1s_1} = a_{j_2j_21}$ possesses the following properties as in Eqs. (4.93)–(4.96):

- For $a_{j_1j_10}\prod_{s_2=1,s_2\neq s_1}^{3}(a_{j_1j_1s_1}-a_{j_1j_1s_2})/a_{j_2j_20}>0$ and $G_{j_1j_10}>0$,
$$(a_{j_1j_1s_1},a_{j_2j_21})=\underbrace{(\mathrm{II},\mathrm{SO})}_{\text{increasing-inflection source}}.$$
- For $a_{j_1j_10}\prod_{s_2=1,s_2\neq s_1}^{3}(a_{j_1j_1s_1}-a_{j_1j_1s_2})/a_{j_2j_20}>0$ and $G_{j_1j_10}<0$,
$$(a_{j_1j_1s_1},a_{j_2j_21})=\underbrace{(\mathrm{II},\mathrm{SI})}_{\text{increasing-inflection sink}}.$$
- For $a_{j_1j_10}\prod_{s_2=1,s_2\neq s_1}^{3}(a_{j_1j_1s_1}-a_{j_1j_1s_2})/a_{j_2j_20}<0$ and $G_{j_1j_10}>0$,
$$(a_{j_1j_1s_1},a_{j_2j_21})=\underbrace{(\mathrm{DI},\mathrm{SO})}_{\text{decreasing-inflection source}}.$$
- For $a_{j_1j_10}\prod_{s_2=1,s_2\neq s_1}^{3}(a_{j_1j_1s_1}-a_{j_1j_1s_2})/a_{j_2j_20}<0$ and $G_{j_1j_10}<0$,
$$(a_{j_1j_1s_1},a_{j_2j_21})=\underbrace{(\mathrm{DI},\mathrm{SI})}_{\text{decreasing-inflection sink}}.$$

(iv) If
$$\Delta_{j_1j_1}<0\quad\text{and}\quad\Delta_{j_2j_2}=0,$$
then
$$\begin{aligned}\dot{x}_{j_1}&=a_{j_1j_10}(y_{j_1}-a_{j_1j_11})[(y_{j_1}-a_{j_1j_1})^2+b_{j_1j_1}],\\ \dot{x}_{j_2}&=a_{j_2j_20}(y_{j_1}-a_{j_2j_2l_1})(y_{j_1}-a_{j_2j_2l_2})^2,\end{aligned}$$
where
$$a_{j_2j_2l_1}=b_{j_2j_21},\quad a_{j_2j_2s_2}=-\frac{1}{2}B_{j_2j_2},\quad b_{j_2j_2}=0,\quad a_{j_2j_20}=A_{j_2j_20};$$
$$a_{j_1j_11}=b_{j_1j_11},\quad a_{j_1j_1}=-\frac{1}{2}B_{j_1j_11},\quad b_{j_1j_1}=-\frac{1}{4}\Delta_{j_1j_1},\quad a_{j_1j_10}=A_{j_1j_10}.$$
(iv$_1$) For $a_{j_1j_11}\neq a_{j_2j_21}$, a dynamical system in the y_{j_1}-direction is constructed as
$$\begin{aligned}\dot{y}_{j_1}&=\alpha_{j_1j_1}a_{j_1j_10}(y_{j_1}-a_{j_1j_11})[(y_{j_1}-a_{j_1j_1})^2+b_{j_1j_1}]\\ &\quad+\alpha_{j_1j_2}a_{j_2j_20}(y_{j_1}-a_{j_2j_2l_1})(y_{j_1}-a_{j_2j_2l_2})^2,\end{aligned}$$
and a dynamical system in the $y_{j_1}^{\perp}$-direction is constructed as
$$\begin{aligned}\dot{y}_{j_1}^{\perp}&=\beta_{j_1j_1}a_{j_1j_10}(y_{j_1}-a_{j_1j_11})[(y_{j_1}-a_{j_1j_1})^2+b_{j_1j_1}]\\ &\quad+\beta_{j_1j_2}a_{j_2j_20}(y_{j_1}-a_{j_2j_2l_1})(y_{j_1}-a_{j_2j_2l_2})^2.\end{aligned}$$

($\text{iv}_{1\text{a}}$) In phase space, at $y_1^* = a_{j_1j_1 1}$,

$$\left.\frac{dx_{j_1}}{dx_{j_2}}\right|_{y_1^*=a_{j_1j_11}} = \left.\frac{a_{j_1j_10}(y_{j_1} - a_{j_1j_11})[(y_{j_1} - a_{j_1j_1})^2 + b_{j_1j_1}]}{a_{j_2j_20}(y_{j_1} - a_{j_2j_2l_1})(y_{j_1} - a_{j_2j_2l_2})^2}\right|_{y_1^*=a_{j_1j_11}} = 0.$$

If

$$\left.\frac{d^2x_{j_1}}{dx_{j_2}^2}\right|_{y_1^*=a_{j_1j_11}} = \frac{a_{j_1j_10}\alpha_{j_1j_2}[(a_{j_1j_11} - a_{j_1j_1})^2 + b_{j_1j_1}]}{a_{j_2j_20}(a_{j_1j_11} - a_{j_2j_2l_1})(a_{j_1j_11} - a_{j_2j_2l_2})^2} > 0,$$

there is an up-parabola flow at $y_1^* = a_{j_1j_11}$. If

$$\left.\frac{d^2x_{j_1}}{dx_{j_2}^2}\right|_{y_1^*=a_{j_1j_11}} = \frac{a_{j_1j_10}\alpha_{j_1j_2}[(a_{j_1j_11} - a_{j_1j_1})^2 + b_{j_1j_1}]}{a_{j_2j_20}(a_{j_1j_11} - a_{j_2j_2l_1})(a_{j_1j_11} - a_{j_2j_2l_2})^2} < 0,$$

there is a down-parabola flow at $y_1^* = a_{j_1j_11}$. With

$$\dot{x}_{j_2}^* = a_{j_2j_20}(a_{j_1j_11} - a_{j_2j_2l_1})(a_{j_1j_11} - a_{j_2j_2l_2})^2,$$

the parabola flows are positive and negative for $\dot{x}_{j_2}^* > 0$ and $\dot{x}_{j_2}^* < 0$, respectively.

Due to

$$(a_{j_1j_11} - a_{j_1j_1})^2 + b_{j_1j_1} > 0,$$

the flow of $y_1^* = a_{j_1j_11}$ with $\dot{x}_{j_2}^* = a_{j_2j_20}(a_{j_1j_11} - a_{j_2j_2l_1})(a_{j_1j_11} - a_{j_2j_2l_2})^2$ possesses the following properties as in Eqs. (4.102)–(4.105):

- For $\alpha_{j_1j_2}a_{j_1j_10} > 0$ and $a_{j_2j_20}(a_{j_1j_11} - a_{j_2j_2l_1})(a_{j_1j_11} - a_{j_2j_2l_2})^2 > 0$,
$$(a_{j_1j_11}, \dot{x}_{j_2}^*) = \underbrace{(\text{UP}, \text{pF})}_{\text{up-parabola flow } (+)}.$$
- For $\alpha_{j_1j_2}a_{j_1j_10} < 0$ and $a_{j_2j_20}(a_{j_1j_11} - a_{j_2j_2l_1})(a_{j_1j_11} - a_{j_2j_2l_2})^2 > 0$,
$$(a_{j_1j_11}, \dot{x}_{j_2}^*) = \underbrace{(\text{DP}, \text{pF})}_{\text{down-parabola flow } (+)}.$$
- For $\alpha_{j_1j_2}a_{j_1j_10} > 0$ and $a_{j_2j_20}(a_{j_1j_11} - a_{j_2j_2l_1})(a_{j_1j_11} - a_{j_2j_2l_2})^2 < 0$,
$$(a_{j_1j_11}, \dot{x}_{j_2}^*) = \underbrace{(\text{DP}, \text{nF})}_{\text{down-parabola flow } (-)}.$$
- For $\alpha_{j_1j_2}a_{j_1j_10} < 0$ and $a_{j_2j_20}(a_{j_1j_11} - a_{j_2j_2l_1})(a_{j_1j_11} - a_{j_2j_2l_2})^2 < 0$,
$$(a_{j_1j_11}, \dot{x}_{j_2}^*) = \underbrace{(\text{UP}, \text{nF})}_{\text{up-parabola flow } (-)}.$$

($\mathrm{iv_{1b}}$) For $y_1^* = a_{j_2j_2l_1}$, in phase space,

$$\left.\frac{dx_{j_2}}{dx_{j_1}}\right|_{y_1^*=a_{j_2j_2l_1}} = \left.\frac{a_{j_2j_20}(y_{j_1} - a_{j_2j_2l_1})(y_{j_1} - a_{j_2j_2l_2})^2}{a_{j_1j_10}(y_{j_1} - a_{j_1j_11})[(y_{j_1} - a_{j_1j_1})^2 + b_{j_1j_1}]}\right|_{y_1^*=a_{j_2j_2l_1}} = 0.$$

If

$$\left.\frac{d^2x_{j_2}}{dx_{j_1}^2}\right|_{y_1^*=a_{j_2j_2l_1}} = \frac{a_{j_2j_20}\alpha_{j_1j_1}(a_{j_2j_2l_1} - a_{j_2j_2l_2})^2}{a_{j_1j_10}(a_{j_2j_2l_1} - a_{j_1j_11})[(a_{j_2j_2l_1} - a_{j_1j_1})^2 + b_{j_1j_1}]} > 0,$$

there is an up-parabola flow at $y_1^* = a_{j_2j_2l_1}$. If

$$\left.\frac{d^2x_{j_2}}{dx_{j_2}^2}\right|_{y_1^*=a_{j_2j_2l_1}} = \frac{a_{j_2j_20}\alpha_{j_1j_1}(a_{j_2j_2l_1} - a_{j_2j_2l_2})^2}{a_{j_1j_10}(a_{j_2j_2l_1} - a_{j_1j_11})[(a_{j_2j_2l_1} - a_{j_1j_1})^2 + b_{j_1j_1}]} < 0,$$

there is a down-parabola flow at $y_1^* = a_{j_2j_2l_1}$. With

$$\dot{x}_{j_1}^* = a_{j_1j_10}(a_{j_2j_2l_1} - a_{j_1j_11})[(a_{j_2j_2l_1} - a_{j_1j_1})^2 + b_{j_1j_1}],$$

the parabola flows are positive and negative for $\dot{x}_{j_1}^* > 0$ and $\dot{x}_{j_1}^* < 0$, respectively.

Due to

$$(a_{j_2j_2l_1} - a_{j_1j_1})^2 + b_{j_1j_1} > 0,$$

the flow of $y_1^* = a_{j_2j_2l_1}$ with $\dot{x}_{j_1}^* = a_{j_1j_10}(a_{j_2j_2l_1} - a_{j_1j_11})[(a_{j_2j_2l_1} - a_{j_1j_1})^2 + b_{j_1j_1}]$ possesses the following properties as in Eqs. (4.106)–(4.109):

- For $a_{j_1j_10}(a_{j_2j_2l_1} - a_{j_1j_11}) > 0$ and $a_{j_2j_20}\alpha_{j_1j_1}(a_{j_2j_2l_1} - a_{j_2j_2l_2})^2 > 0$,
$$(\dot{x}_{j_1}^*, a_{j_2j_2l_1}) = \underbrace{(\mathrm{pF}, \mathrm{UP})}_{\text{up-parabola flow }(+)}.$$
- For $a_{j_1j_10}(a_{j_2j_2l_1} - a_{j_1j_11}) < 0$ and $a_{j_2j_20}\alpha_{j_1j_1}(a_{j_2j_2l_1} - a_{j_2j_2l_2})^2 > 0$,
$$(\dot{x}_{j_1}^*, a_{j_2j_2l_1}) = \underbrace{(\mathrm{nF}, \mathrm{DP})}_{\text{down-parabola flow }(-)}.$$
- For $a_{j_1j_10}(a_{j_2j_2l_1} - a_{j_1j_11}) > 0$ and $a_{j_2j_20}\alpha_{j_1j_1}(a_{j_2j_2l_1} - a_{j_2j_2l_2})^2 < 0$,
$$(\dot{x}_{j_1}^*, a_{j_2j_2l_1}) = \underbrace{(\mathrm{pF}, \mathrm{DP})}_{\text{down-parabola flow }(+)}.$$
- For $a_{j_1j_10}(a_{j_2j_2l_1} - a_{j_1j_11}) < 0$ and $a_{j_2j_20}\alpha_{j_1j_1}(a_{j_2j_2l_1} - a_{j_2j_2l_2})^2 < 0$,
$$(\dot{x}_{j_1}^*, a_{j_2j_2l_1}) = \underbrace{(\mathrm{nF}, \mathrm{UP})}_{\text{up-parabola flow }(-)}.$$

(iv_{1c}) In phase space, at $y_1^* = a_{j_2j_2l_2}$,

$$\left.\frac{dx_{j_2}}{dx_{j_1}}\right|_{y_1^*=a_{j_2j_2l_2}} = \left.\frac{a_{j_2j_20}(y_{j_1}-a_{j_2j_2l_1})(y_{j_1}-a_{j_2j_2l_2})^2}{a_{j_1j_10}(y_{j_1}-a_{j_1j_11})[(y_{j_1}-a_{j_1j_1})^2+b_{j_1j_1}]}\right|_{y_1^*=a_{j_2j_2l_2}} = 0,$$

$$\left.\frac{d^2x_{j_2}}{dx_{j_1}^2}\right|_{y_1^*=a_{j_2j_2l_2}} = 0.$$

If

$$\left.\frac{d^3x_{j_2}}{dx_{j_1}^3}\right|_{y_1^*=a_{j_2j_2l_2}} = \frac{2a_{j_2j_20}\alpha_{j_1j_1}^2(a_{j_2j_2l_2}-a_{j_2j_2l_1})}{a_{j_1j_10}(a_{j_2j_2l_2}-a_{j_1j_11})[(a_{j_2j_2l_2}-a_{j_1j_1})^2+b_{j_1j_1}]} > 0,$$

there is an increasing-inflection flow at $y_1^* = a_{j_2j_2l_2}$. If

$$\left.\frac{d^3x_{j_2}}{dx_{j_1}^3}\right|_{y_1^*=a_{j_2j_2l_2}} = \frac{2a_{j_2j_20}\alpha_{j_1j_1}^2(a_{j_2j_2l_2}-a_{j_2j_2l_1})}{a_{j_1j_10}(a_{j_2j_2l_2}-a_{j_1j_11})[(a_{j_2j_2l_2}-a_{j_1j_1})^2+b_{j_1j_1}]} > 0,$$

there is a decreasing-inflection flow at $y_1^* = a_{j_2j_2l_2}$. With

$$\dot{x}_{j_1}^* = a_{j_1j_10}(a_{j_2j_2l_2}-a_{j_1j_11})[(a_{j_2j_2l_2}-a_{j_1j_1})^2+b_{j_1j_1}],$$

the inflection flows are positive and negative for $\dot{x}_{j_1}^* > 0$ and $\dot{x}_{j_1}^* < 0$, respectively.

Due to

$$(a_{j_2j_2l_2}-a_{j_1j_1})^2+b_{j_1j_1} > 0,$$

the flow of $y_1^* = a_{j_2j_2l_2}$ with $\dot{x}_{j_1}^* = a_{j_1j_10}(a_{j_2j_2l_2}-a_{j_1j_11})[(a_{j_2j_2l_2}-a_{j_1j_1})^2+b_{j_1j_1}]$ possesses the following properties as in Eqs. (4.110)–(4.113):

- For $a_{j_1j_10}(a_{j_2j_2l_2}-a_{j_1j_11}) > 0$ and $a_{j_2j_20}\alpha_{j_1j_1}^2(a_{j_2j_2l_2}-a_{j_2j_2l_1}) > 0$,

$$(\dot{x}_{j_1}^*, a_{j_2j_2l_2}) = \underbrace{(\mathrm{pF},\mathrm{II})}_{\text{increasing-inflection flow }(+)}.$$

- For $a_{j_1j_10}(a_{j_2j_2l_2}-a_{j_1j_11}) < 0$ and $a_{j_2j_20}\alpha_{j_1j_1}^2(a_{j_2j_2l_2}-a_{j_2j_2l_1}) > 0$,

$$(\dot{x}_{j_1}^*, a_{j_2j_2l_2}) = \underbrace{(\mathrm{nF},\mathrm{DI})}_{\text{decreasing-inflection flow }(-)}.$$

- For $a_{j_1j_10}(a_{j_2j_2l_2}-a_{j_1j_11}) > 0$ and $a_{j_2j_20}\alpha_{j_1j_1}^2(a_{j_2j_2l_2}-a_{j_2j_2l_1}) < 0$,

$$(\dot{x}_{j_1}^*, a_{j_2j_2l_2}) = \underbrace{(\mathrm{pF},\mathrm{DI})}_{\text{decreasing-inflection flow }(+)}.$$

- For $a_{j_1j_1 0}(a_{j_2j_2l_2}-a_{j_1j_1 1})<0$ and $a_{j_2j_2 0}\alpha_{j_1j_1}^2(a_{j_2j_2l_2}-a_{j_2j_2l_1})<0$,

$$(\dot{x}_{j_1}^*,a_{j_2j_2l_2})=\underbrace{(\text{nF},\text{II})}_{\text{increasing-inflection flow }(-)}.$$

(iv$_2$) For $a_{j_2j_2l_1}=a_{j_2j_2l_2}=a_{j_2j_2 1}$, the dynamical systems become

$$\begin{aligned}\dot{x}_{j_1}&=a_{j_1j_1 0}(y_{j_1}-a_{j_1j_1 1})[(y_{j_1}-a_{j_1j_1})^2+b_{j_1j_1}],\\ \dot{x}_{j_2}&=a_{j_2j_2 0}(y_{j_1}-a_{j_2j_2 1})^3.\end{aligned}$$

For $a_{j_1j_1 1}\neq a_{j_2j_2 1}$, a dynamical system in the y_{j_1}-direction is constructed as

$$\begin{aligned}\dot{y}_{j_1}&=\alpha_{j_1j_1}a_{j_1j_1 0}(y_{j_1}-a_{j_1j_1 1})[(y_{j_1}-a_{j_1j_1})^2+b_{j_1j_1}]\\ &\quad+\alpha_{j_1j_2}a_{j_2j_2 0}(y_{j_1}-a_{j_2j_2 1})^3,\end{aligned}$$

and a dynamical system in the $y_{j_1}^{\perp}$-direction is constructed as

$$\begin{aligned}\dot{y}_{j_1}^{\perp}&=\beta_{j_1j_1}a_{j_1j_1 0}(y_{j_1}-a_{j_1j_1 1})[(y_{j_1}-a_{j_1j_1})^2+b_{j_1j_1}]\\ &\quad+\beta_{j_1j_2}a_{j_2j_2 0}(y_{j_1}-a_{j_2j_2 1})^3.\end{aligned}$$

(iv$_{2\text{a}}$) For $y_1^*=a_{j_1j_1 1}$, in phase space,

$$\left.\frac{dx_{j_1}}{dx_{j_2}}\right|_{y_1^*=a_{j_1j_1 1}}=\left.\frac{a_{j_1j_1 0}(y_{j_1}-a_{j_1j_1 1})[(y_{j_1}-a_{j_1j_1})^2+b_{j_1j_1}]}{a_{j_2j_2 0}(y_{j_1}-a_{j_2j_2 1})^3}\right|_{y_1^*=a_{j_1j_1 1}}=0.$$

If

$$\left.\frac{d^2x_{j_1}}{dx_{j_2}^2}\right|_{y_1^*=a_{j_1j_1 1}}=\frac{a_{j_1j_1 0}\alpha_{j_1j_2}[(a_{j_1j_1 1}-a_{j_1j_1})^2+b_{j_1j_1}]}{a_{j_2j_2 0}(a_{j_1j_1 1}-a_{j_2j_2 1})^3}>0,$$

there is an up-parabola flow at $y_1^*=a_{j_1j_1 1}$. If

$$\left.\frac{d^2x_{j_1}}{dx_{j_2}^2}\right|_{y_1^*=a_{j_1j_1 1}}=\frac{a_{j_1j_1 0}\alpha_{j_1j_2}[(a_{j_1j_1 1}-a_{j_1j_1})^2+b_{j_1j_1}]}{a_{j_2j_2 0}(a_{j_1j_1 1}-a_{j_2j_2 1})^3}<0,$$

there is a down-parabola flow at $y_1^*=a_{j_1j_1 1}$. With

$$\dot{x}_{j_2}^*=a_{j_2j_2 0}(a_{j_1j_1 1}-a_{j_2j_2 1})^3,$$

the parabola flows are positive and negative for $\dot{x}_{j_2}^*>0$ and $\dot{x}_{j_2}^*<0$, respectively.

Due to

$$(a_{j_1j_1 1}-a_{j_1j_1})^2+b_{j_1j_1}>0,$$

the flow of $y_1^* = a_{j_1j_1 1}$ with $\dot{x}_{j_2}^* = a_{j_2j_2 0}(a_{j_1j_1 1} - a_{j_2j_2 1})^3$ possesses the following properties as in Eqs. (4.117)–(4.120):

- For $\alpha_{j_1j_1}a_{j_1j_1 0} > 0$ and $a_{j_2j_2 0}(a_{j_1j_1 1} - a_{j_2j_2 1})^3 > 0$,
$$(a_{j_1j_1 1}, \dot{x}_{j_2}^*) = \underbrace{(\mathrm{UP}, \mathrm{pF})}_{\text{up-parabola flow } (+)}.$$
- For $\alpha_{j_1j_1}a_{j_1j_1 0} < 0$ and $a_{j_2j_2 0}(a_{j_1j_1 1} - a_{j_2j_2 1})^3 > 0$,
$$(a_{j_1j_1 1}, \dot{x}_{j_2}^*) = \underbrace{(\mathrm{DP}, \mathrm{pF})}_{\text{down-parabola flow } (+)}.$$
- For $\alpha_{j_1j_1}a_{j_1j_1 0} > 0$ and $a_{j_2j_2 0}(a_{j_1j_1 1} - a_{j_2j_2 1})^3 < 0$,
$$(a_{j_1j_1 1}, \dot{x}_{j_2}^*) = \underbrace{(\mathrm{DP}, \mathrm{nF})}_{\text{down-parabola flow } (-)}.$$
- For $\alpha_{j_1j_1}a_{j_1j_1 0} < 0$ and $a_{j_2j_2 0}(a_{j_1j_1 1} - a_{j_2j_2 1})^3 < 0$,
$$(a_{j_1j_1 1}, \dot{x}_{j_2}^*) = \underbrace{(\mathrm{UP}, \mathrm{nF})}_{\text{up-parabola flow } (-)}.$$

($\mathrm{iv}_{2\mathrm{b}}$) For $y_1^* = a_{j_2j_2 1}$, in phase space,

$$\left.\frac{dx_{j_2}}{dx_{j_1}}\right|_{y_1^*=a_{j_2j_2 1}} = \left.\frac{a_{j_2j_2 0}(y_{j_1} - a_{j_2j_2 1})^3}{a_{j_1j_1 0}(y_{j_1} - a_{j_1j_1 1})[(y_{j_1} - a_{j_1j_1})^2 + b_{j_1j_1}]}\right|_{y_1^*=a_{j_2j_2 1}} = 0,$$

$$\left.\frac{d^2x_{j_2}}{dx_{j_1}^2}\right|_{y_1^*=a_{j_2j_2 1}} = \left.\frac{d^3x_{j_2}}{dx_{j_1}^3}\right|_{y_1^*=a_{j_2j_2 1}} = 0.$$

If

$$\left.\frac{d^4x_{j_2}}{dx_{j_1}^4}\right|_{y_1^*=a_{j_2j_2 1}} = \frac{6a_{j_2j_2 0}\alpha_{j_1j_1}^3}{a_{j_1j_1 0}(a_{j_2j_2 1} - a_{j_1j_1 1})[(a_{j_2j_2 1} - a_{j_1j_1})^2 + b_{j_1j_1}]} > 0,$$

there is a third-order up-parabola flow at $y_1^* = a_{j_2j_2 1}$. If

$$\left.\frac{d^4x_{j_2}}{dx_{j_1}^4}\right|_{y_1^*=a_{j_2j_2 1}} = \frac{6a_{j_2j_2 0}\alpha_{j_1j_1}^3}{a_{j_1j_1 0}(a_{j_2j_2 1} - a_{j_1j_1 1})[(a_{j_2j_2 1} - a_{j_1j_1})^2 + b_{j_1j_1}]} > 0,$$

there is a third-order up-parabola flow at $y_1^* = a_{j_2j_2 1}$. With

$$\dot{x}_{j_1}^* = a_{j_1j_1 0}(a_{j_2j_2 1} - a_{j_1j_1 1})[(a_{j_2j_2 1} - a_{j_1j_1})^2 + b_{j_1j_1}],$$

the inflection flows are positive and negative for $\dot{x}_{j_1}^* > 0$ and $\dot{x}_{j_1}^* < 0$, respectively.

Due to

$$(a_{j_2j_2 1} - a_{j_1j_1})^2 + b_{j_1j_1} > 0,$$

the flow of $y_1^* = a_{j_2j_21}$ with $\dot{x}_{j_1}^* = a_{j_1j_10}(a_{j_2j_21} - a_{j_1j_11})[(a_{j_2j_21} - a_{j_1j_1})^2 + b_{j_1j_1}]$ possesses the following properties as in Eqs. (4.121)–(4.124):

- For $a_{j_1j_10}(a_{j_2j_21} - a_{j_1j_11}) > 0$ and $a_{j_2j_20}\alpha_{j_1j_1}^3 > 0$,
$$(\dot{x}_{j_1}^*, a_{j_2j_21}) = \underbrace{(\text{pF}, 3^{\text{rd}}\text{UP})}_{\text{third-order up-parabola flow }(+)}.$$
- For $a_{j_1j_10}(a_{j_2j_21} - a_{j_1j_11}) < 0$ and $a_{j_2j_20}\alpha_{j_1j_1}^3 > 0$,
$$(\dot{x}_{j_1}^*, a_{j_2j_21}) = \underbrace{(\text{nF}, 3^{\text{rd}}\text{DP})}_{\text{third-order down-parabola flow }(-)}.$$
- For $a_{j_1j_10}(a_{j_2j_21} - a_{j_1j_11}) > 0$ and $a_{j_2j_20}\alpha_{j_1j_1}^3 < 0$,
$$(\dot{x}_{j_1}^*, a_{j_2j_21}) = \underbrace{(\text{pF}, 3^{\text{rd}}\text{DP})}_{\text{third-order down-parabola flow }(+)}.$$
- For $a_{j_1j_10}(a_{j_2j_21} - a_{j_1j_11}) < 0$ and $a_{j_2j_20}\alpha_{j_1j_1}^3 < 0$,
$$(\dot{x}_{j_1}^*, a_{j_2j_21}) = \underbrace{(\text{nF}, 3^{\text{rd}}\text{UP})}_{\text{third-order up-parabola flow }(-)}.$$

(iv$_3$) For $a_{j_1j_11} = a_{j_2j_2l_1}$, the cubic system becomes

$$\begin{aligned}\dot{x}_{j_1} &= a_{j_1j_10}(y_{j_1} - a_{j_1j_11})[(y_{j_1} - a_{j_1j_1})^2 + b_{j_1j_1}],\\ \dot{x}_{j_2} &= a_{j_2j_20}(y_{j_1} - a_{j_2j_2l_1})(y_{j_1} - a_{j_2j_2l_2})^2.\end{aligned}$$

A corresponding dynamical system in the y_{j_1}-direction is constructed as

$$\begin{aligned}\dot{y}_{j_1} &= (y_{j_1} - a_{j_1j_11})\{\alpha_{j_1j_1}a_{j_1j_10}[(y_{j_1} - a_{j_1j_1})^2 + b_{j_1j_1}]\\ &\quad + \alpha_{j_1j_2}a_{j_2j_20}(y_{j_1} - a_{j_2j_2l_2})^2\},\end{aligned}$$

and a dynamical system in the $y_{j_1}^{\perp}$-direction is constructed as

$$\begin{aligned}\dot{y}_{j_1}^{\perp} &= (y_{j_1} - a_{j_1j_11})\{\beta_{j_1j_1}a_{j_1j_10}[(y_{j_1} - a_{j_1j_1})^2 + b_{j_1j_1}]\\ &\quad + \beta_{j_1j_2}a_{j_2j_20}(y_{j_1} - a_{j_2j_2l_2})^2]\},\end{aligned}$$

and a G-function is defined as

$$\begin{aligned}G_{j_1j_10} &= \frac{\partial}{\partial y_{j_1}}[\mathbf{n}^{\mathrm{T}} \cdot \mathbf{F}(y_{j_1}, \mathbf{p})]\bigg|_{y_{j_1}^* = a_{j_1j_11} = a_{j_2j_2l_1}}\\ &= \alpha_{j_1j_1}a_{j_1j_10}[(a_{j_2j_21} - a_{j_1j_1})^2 + b_{j_1j_1}] + \alpha_{j_1j_2}a_{j_2j_20}(a_{j_2j_2l_1} - a_{j_2j_2l_2})^2.\end{aligned}$$

In phase space, at $y_1^* = a_{j_2j_2l_1} = a_{j_1j_11}$, if

$$\left.\frac{dx_{j_2}}{dx_{j_1}}\right|_{y_1^*=a_{j_2j_2l_1}} = \left.\frac{a_{j_2j_20}(a_{j_2j_2l_1} - a_{j_2j_2l_2})^2}{a_{j_1j_10}[(a_{j_1j_11} - a_{j_1j_1})^2 + b_{j_1j_1}]}\right|_{y_1^*=a_{j_2j_2l_1}} > 0,$$

there is an increasing-inflection flow at $y_1^* = a_{j_2j_2l_1} = a_{j_1j_11}$, and if

$$\left.\frac{dx_{j_2}}{dx_{j_1}}\right|_{y_1^*=a_{j_2j_2l_1}} = \left.\frac{a_{j_2j_20}(a_{j_2j_2l_1} - a_{j_2j_2l_2})^2}{a_{j_1j_10}[(a_{j_1j_11} - a_{j_1j_1})^2 + b_{j_1j_1}]}\right|_{y_1^*=a_{j_2j_2l_1}} < 0,$$

there is a decreasing-inflection flow at $y_1^* = a_{j_2j_2l_1} = a_{j_1j_11}$. The variational equation is

$$\begin{aligned}\Delta\dot{y}_{j_1} &= \{\alpha_{j_1j_1}a_{j_1j_10}[(a_{j_1j_11} - a_{j_1j_1})^2 + b_{j_1j_1}] \\ &\quad + \alpha_{j_1j_2}a_{j_2j_20}(a_{j_2j_2l_1} - a_{j_2j_2l_2})^2\}\Delta y_{j_1} \\ &= G_{j_1j_10}\Delta y_{j_1},\end{aligned}$$

the inflection infinite-equilibrium flows are source and sink for $G_{j_1j_10} > 0$ and $G_{j_1j_10} < 0$, respectively.

Due to

$$(a_{j_1j_11} - a_{j_1j_1})^2 + b_{j_1j_1} > 0,$$

the infinite-equilibrium flow of $y_1^* = a_{j_1j_11} = a_{j_2j_21}$ possesses the following properties as in Eqs. (4.129)–(4.132):

- For $G_{j_1j_10} > 0$ and $a_{j_2j_20}(a_{j_2j_2l_1} - a_{j_2j_2l_2})^2/a_{j_1j_10} > 0$,

$$(a_{j_1j_11}, a_{j_2j_2l_1}) = \underbrace{(\mathrm{SO}, \mathrm{II})}_{\text{increasing-inflection source}}.$$

- For $G_{j_1j_10} < 0$ and $a_{j_2j_20}(a_{j_2j_2l_1} - a_{j_2j_2l_2})^2/a_{j_1j_10} > 0$,

$$(a_{j_1j_11}, a_{j_2j_2l_1}) = \underbrace{(\mathrm{SI}, \mathrm{II})}_{\text{increasing-inflection sink}}.$$

- For $G_{j_1j_10} > 0$ and $a_{j_2j_20}(a_{j_2j_2l_1} - a_{j_2j_2l_2})^2/a_{j_1j_10} < 0$,

$$(a_{j_1j_11}, a_{j_2j_2l_1}) = \underbrace{(\mathrm{SO}, \mathrm{DI})}_{\text{decreasing-inflection source}}.$$

- For $G_{j_1j_10} < 0$ and $a_{j_2j_20}(a_{j_2j_2l_1} - a_{j_2j_2l_2})^2/a_{j_1j_10} < 0$,

$$(a_{j_1j_11}, a_{j_2j_2l_1}) = \underbrace{(\mathrm{SI}, \mathrm{DI})}_{\text{decreasing-inflection sink}}.$$

(iv$_4$) For $a_{j_1j_11} = a_{j_2j_2l_2}$, the cubic dynamical system becomes

$$\dot{x}_{j_1} = a_{j_1j_10}(y_{j_1} - a_{j_1j_11})[(y_{j_1} - a_{j_1j_1})^2 + b_{j_1j_1}],$$
$$\dot{x}_{j_2} = a_{j_2j_20}(y_{j_1} - a_{j_2j_2l_1})(y_{j_1} - a_{j_2j_2l_2})^2.$$

A dynamical system in the y_{j_1}-direction is constructed as

$$\dot{y}_{j_1} = (y_{j_1} - a_{j_1j_11})\{\alpha_{j_1j_1}a_{j_1j_10}[(y_{j_1} - a_{j_1j_1})^2 + b_{j_1j_1}] + \alpha_{j_1j_2}a_{j_2j_20}(y_{j_1} - a_{j_2j_2l_1})(y_{j_1} - a_{j_2j_2l_2})\}$$

and a dynamical system in the $y_{j_1}^{\perp}$-direction is constructed as

$$\dot{y}_{j_1}^{\perp} = (y_{j_1} - a_{j_1j_11})\{\beta_{j_1j_1}a_{j_1j_10}[(y_{j_1} - a_{j_1j_1})^2 + b_{j_1j_1}] + \beta_{j_1j_2}a_{j_2j_20}(y_{j_1} - a_{j_2j_2l_1})(y_{j_1} - a_{j_2j_2l_2})]\}$$

and a G-function is defined as

$$G_{j_1j_10} = \frac{\partial}{\partial y_{j_1}}[\mathbf{n}^{\mathrm{T}} \cdot \mathbf{F}(y_{j_1}, \mathbf{p})]\bigg|_{y_{j_1}^* = a_{j_1j_11} = a_{j_2j_2l_2}}$$
$$= \alpha_{j_1j_1}a_{j_1j_10}[(a_{j_2j_2l_2} - a_{j_1j_1})^2 + b_{j_1j_1}].$$

In phase space, at $y_1^* = a_{j_1j_11} = a_{j_2j_2l_2}$,

$$\frac{dx_{j_2}}{dx_{j_1}}\bigg|_{y_1^* = a_{j_2j_2l_2}} = \frac{a_{j_2j_20}(y_{j_1} - a_{j_2j_2l_2})}{a_{j_1j_10}[(y_{j_1} - a_{j_1j_1})^2 + b_{j_1j_1}]}\bigg|_{y_1^* = a_{j_2j_2l_2}} = 0.$$

If

$$\frac{d^2x_{j_2}}{dx_{j_1}^2}\bigg|_{y_1^* = a_{j_2j_2l_2}} = \frac{a_{j_2j_20}\alpha_{j_1j_1}^2}{a_{j_1j_10}\alpha_{j_1j_1}[(a_{j_2j_2l_2} - a_{j_1j_1})^2 + b_{j_1j_1}]} > 0,$$

there is an up-parabola flow at $y_1^* = a_{j_2j_21} = a_{j_1j_11}$. If

$$\frac{d^2x_{j_2}}{dx_{j_1}^2}\bigg|_{y_1^* = a_{j_2j_2l_2}} = \frac{a_{j_2j_20}\alpha_{j_1j_1}^2}{a_{j_1j_10}\alpha_{j_1j_1}[(a_{j_2j_2l_2} - a_{j_1j_1})^2 + b_{j_1j_1}]} < 0,$$

there is a down-parabola flow at $y_1^* = a_{j_2j_2l_2} = a_{j_1j_11}$.

The variational equation at $y_1^* = a_{j_2j_21} = a_{j_1j_11}$ is

$$\Delta\dot{y}_{j_1} = \alpha_{j_1j_1}a_{j_1j_10}[(a_{j_2j_2l_2} - a_{j_1j_1})^2 + b_{j_1j_1}]\Delta y_{j_1}.$$

Due to

$$(a_{j_2j_2l_2} - a_{j_1j_1})^2 + b_{j_1j_1} > 0,$$

the parabola flows are source and sink for $\alpha_{j_1j_1}a_{j_1j_10} > 0$ and $\alpha_{j_1j_1}a_{j_1j_10} < 0$, respectively.

Therefore, the infinite-equilibrium flow of $y_1^* = a_{j_1j_1 1} = a_{j_2j_2l_2}$ possesses the following properties as in Eqs. (4.137)–(4.140):

- For $\alpha_{j_1j_1}a_{j_1j_1 0} > 0$ and $a_{j_2j_2 0}\alpha_{j_1j_1}^2(a_{j_2j_2l_2} - a_{j_2j_2l_1}) > 0$,

$$(a_{j_1j_1 1}, a_{j_2j_2l_2}) = \underbrace{(\mathrm{SO}, \mathrm{UP})}_{\text{up-parabola source}}.$$

- For $\alpha_{j_1j_1}a_{j_1j_1 0} < 0$ and $a_{j_2j_2 0}\alpha_{j_1j_1}^2(a_{j_2j_2l_2} - a_{j_2j_2l_1}) > 0$,

$$(a_{j_1j_1 1}, a_{j_2j_2l_2}) = \underbrace{(\mathrm{SI}, \mathrm{DP})}_{\text{down-parabola sink}}.$$

- For $\alpha_{j_1j_1}a_{j_1j_1 0} > 0$ and $a_{j_2j_2 0}\alpha_{j_1j_1}^2(a_{j_2j_2l_2} - a_{j_2j_2l_1}) < 0$,

$$(a_{j_1j_1 1}, a_{j_2j_2l_2}) = \underbrace{(\mathrm{SO}, \mathrm{DP})}_{\text{down-parabola source}}.$$

- For $\alpha_{j_1j_1}a_{j_1j_1 0} < 0$ and $a_{j_2j_2 0}\alpha_{j_1j_1}^2(a_{j_2j_2l_2} - a_{j_2j_2l_1}) < 0$,

$$(a_{j_1j_1 1}, a_{j_2j_2l_2}) = \underbrace{(\mathrm{SI}, \mathrm{UP})}_{\text{up-parabola sink}}.$$

(iv_5) For $a_{j_1j_1 1} = a_{j_2j_2 1}$, the cubic dynamical system becomes

$$\begin{aligned}\dot{x}_{j_1} &= a_{j_1j_1 0}(y_{j_1} - a_{j_1j_1 1})[(y_{j_1} - a_{j_1j_1})^2 + b_{j_1j_1}],\\ \dot{x}_{j_2} &= a_{j_2j_2 0}(y_{j_1} - a_{j_2j_2 1})^3.\end{aligned}$$

A dynamical system in the y_{j_1}-direction is constructed as

$$\begin{aligned}\dot{y}_{j_1} = (y_{j_1} - a_{j_1j_1 1})\{&\alpha_{j_1j_1}a_{j_1j_1 0}[(y_{j_1} - a_{j_1j_1})^2 + b_{j_1j_1}]\\ &+ \alpha_{j_1j_2}a_{j_2j_2 0}(y_{j_1} - a_{j_2j_2 1})^2\},\end{aligned}$$

and a dynamical system in the $y_{j_1}^{\perp}$-direction is constructed as

$$\begin{aligned}\dot{y}_{j_1}^{\perp} = (y_{j_1} - a_{j_1j_1 1})\{&\beta_{j_1j_1}a_{j_1j_1 0}[(y_{j_1} - a_{j_1j_1})^2 + b_{j_1j_1}]\\ &+ \beta_{j_1j_2}a_{j_2j_2 0}(y_{j_1} - a_{j_2j_2 1})^2]\},\end{aligned}$$

and a G-function is defined as

$$\begin{aligned}G_{j_1j_1 0} &= \frac{\partial}{\partial y_{j_1}}[\mathbf{n}^{\mathrm{T}} \cdot \mathbf{F}(y_{j_1}, \mathbf{p})]\bigg|_{y_{j_1}^* = a_{j_1j_1 1} = a_{j_2j_2 1}}\\ &= \alpha_{j_1j_1}a_{j_1j_1 0}[(a_{j_2j_2 1} - a_{j_1j_1})^2 + b_{j_1j_1}].\end{aligned}$$

In phase space, at $y_1^* = a_{j_1j_1 1} = a_{j_2j_2 1}$,

$$\left.\frac{dx_{j_2}}{dx_{j_1}}\right|_{y_1^*=a_{j_2j_21}} = \left.\frac{a_{j_2j_20}(y_{j_1} - a_{j_2j_21})^2}{a_{j_1j_10}[(y_{j_1} - a_{j_1j_1})^2 + b_{j_1j_1}]}\right|_{y_1^*=a_{j_2j_21}} = 0,$$

$$\left.\frac{d^2x_{j_2}}{dx_{j_1}^2}\right|_{y_1^*=a_{j_2j_21}} = 0.$$

If

$$\left.\frac{d^3x_{j_2}}{dx_{j_1}^3}\right|_{y_1^*=a_{j_2j_21}} = \frac{a_{j_2j_20}\alpha_{j_1j_1}^3}{a_{j_1j_10}\alpha_{j_1j_1}[(a_{j_2j_21} - a_{j_1j_1})^2 + b_{j_1j_1}]} > 0,$$

there is a second-order increasing-inflection flow at $y_1^* = a_{j_2j_21} = a_{j_1j_11}$. If

$$\left.\frac{d^3x_{j_2}}{dx_{j_1}^3}\right|_{y_1^*=a_{j_2j_21}} = \frac{a_{j_2j_20}\alpha_{j_1j_1}^3}{a_{j_1j_10}\alpha_{j_1j_1}[(a_{j_2j_21} - a_{j_1j_1})^2 + b_{j_1j_1}]} < 0,$$

there is a second-order increasing-inflection flow at $y_1^* = a_{j_2j_21} = a_{j_1j_11}$.

The variational equation at $y_1^* = a_{j_2j_21} = a_{j_1j_11}$ is

$$\Delta\dot{y}_{j_1} = \alpha_{j_1j_1}a_{j_1j_10}[(a_{j_1j_11} - a_{j_1j_1})^2 + b_{j_1j_1}]\Delta y_{j_1}.$$

Due to

$$(a_{j_2j_21} - a_{j_1j_1})^2 + b_{j_1j_1} > 0,$$

the second-order inflection infinite-equilibrium flows are source and sink for $\alpha_{j_1j_1}a_{j_1j_10} > 0$ and $\alpha_{j_1j_1}a_{j_1j_10} < 0$, respectively.

Therefore, the infinite-equilibrium flow of $y_1^* = a_{j_1j_11} = a_{j_2j_2l_2}$ possesses the following properties as in Eqs. (4.145)–(4.148):

- For $\alpha_{j_1j_1}a_{j_1j_10} > 0$ and $a_{j_2j_20}\alpha_{j_1j_1}^3 > 0$,

$$(a_{j_1j_11}, a_{j_2j_21}) = \underbrace{(\mathrm{SO}, 2^{\mathrm{nd}}\mathrm{II})}_{\text{second-order increasing-infelction source}}.$$

- For $\alpha_{j_1j_1}a_{j_1j_10} < 0$ and $a_{j_2j_20}\alpha_{j_1j_1}^3 > 0$,

$$(a_{j_1j_11}, a_{j_2j_21}) = \underbrace{(\mathrm{SI}, 2^{\mathrm{nd}}\mathrm{DI})}_{\text{second-order decreasing-infelction sink}}.$$

- For $\alpha_{j_1j_1}a_{j_1j_10} > 0$ and $a_{j_2j_20}\alpha_{j_1j_1}^3 < 0$,

$$(a_{j_1j_11}, a_{j_2j_21}) = \underbrace{(\mathrm{SO}, 2^{\mathrm{nd}}\mathrm{DI})}_{\text{second-order decreasing-infelction source}}.$$

- For $\alpha_{j_1j_1}a_{j_1j_10} < 0$ and $a_{j_2j_20}\alpha^3_{j_1j_1} < 0$,

$$(a_{j_1j_11}, a_{j_2j_21}) = \underbrace{(\text{SI}, 2^{\text{nd}}\text{II})}_{\text{second-order increasing-infelction sink}}.$$

(v) For

$$\Delta_{j_1j_1} = 0 \quad \text{and} \quad \Delta_{j_2j_2} = 0,$$

the following equations are obtained:

$$\begin{aligned}
\dot{x}_{j_1} &= a_{j_1j_10}(y_{j_1} - a_{j_1j_1s_1})(y_{j_1} - a_{j_1j_1s_2})^2, \\
\dot{x}_{j_2} &= a_{j_2j_20}(y_{j_1} - a_{j_2j_2l_1})(y_{j_1} - a_{j_2j_2l_2})^2, \\
& j_1, j_2 \in \{1, 2\};\ j_1 \neq j_2;
\end{aligned}$$

where

$$a_{j_2j_2l_2} = b_{j_2j_21}, \quad a_{j_2j_2l_2} = -\frac{1}{2}B_{j_2j_2}, \quad b_{j_2j_2} = 0, \quad a_{j_2j_20} = A_{j_2j_20};$$

$$a_{j_1j_1s_1} = b_{j_1j_11}, \quad a_{j_1j_1s_2} = -\frac{1}{2}B_{j_1j_11}, \quad b_{j_1j_1} = 0, \quad a_{j_1j_10} = A_{j_1j_10}.$$

(v_1) For $a_{j_2j_2l_1}, a_{j_2j_2l_2} \neq a_{j_1j_1s_1}, a_{j_1j_1s_2}$, a dynamical system in the y_{j_1}-direction is constructed as

$$\dot{y}_{j_1} = \alpha_{j_1j_1}a_{j_1j_10}(y_{j_1} - a_{j_1j_1s_1})(y_{j_1} - a_{j_1j_1s_2})^2 + \alpha_{j_1j_2}a_{j_2j_20}(y_{j_1} - a_{j_2j_21})^2.$$

and a dynamical system in the $y^{\perp}_{j_1}$-direction is constructed as

$$\dot{y}^{\perp}_{j_1} = \beta_{j_1j_1}a_{j_1j_10}(y_{j_1} - a_{j_1j_1s_1})(y_{j_1} - a_{j_1j_1s_2})^2 + \beta_{j_1j_2}a_{j_2j_20}(y_{j_1} - a_{j_2j_2})^2.$$

($\text{v}_{1\text{a}}$) In phase space, at $y^*_1 = a_{j_1j_1s_1}$,

$$\left.\frac{dx_{j_1}}{dx_{j_2}}\right|_{y^*_1 = a_{j_1j_1s_1}} = \left.\frac{a_{j_1j_10}(y_{j_1} - a_{j_1j_1s_1})(y_{j_1} - a_{j_1j_1s_2})^2}{a_{j_2j_20}(y_{j_1} - a_{j_2j_2l_1})(y_{j_1} - a_{j_2j_2l_2})^2}\right|_{y^*_1 = a_{j_1j_1s_1}} = 0.$$

If

$$\left.\frac{d^2x_{j_1}}{dx^2_{j_2}}\right|_{y^*_1 = a_{j_1j_1s_1}} = \frac{a_{j_1j_10}\alpha_{j_1j_2}(a_{j_1j_1s_1} - a_{j_1j_1s_2})^2}{a_{j_2j_20}(a_{j_1j_1s_1} - a_{j_2j_2l_1})(a_{j_1j_1s_1} - a_{j_2j_2l_2})^2} > 0,$$

there is an up-parabola flow at $y^*_1 = a_{j_1j_1s_1}$. If

$$\left.\frac{d^2x_{j_1}}{dx^2_{j_2}}\right|_{y^*_1 = a_{j_1j_1s_1}} = \frac{a_{j_1j_10}\alpha_{j_1j_2}(a_{j_1j_1s_1} - a_{j_1j_1s_2})^2}{a_{j_2j_20}(a_{j_1j_1s_1} - a_{j_2j_2l_1})(a_{j_1j_1s_1} - a_{j_2j_2l_2})^2} < 0,$$

there is a down-parabola flow at $y^*_1 = a_{j_1j_1s_1}$. With

$$\dot{x}^*_{j_2} = a_{j_2j_20}(a_{j_1j_1s_1} - a_{j_2j_2l_1})(a_{j_1j_1s_1} - a_{j_2j_2l_2})^2,$$

the parabola flows are positive and negative for $\dot{x}^*_{j_2} > 0$ and $\dot{x}^*_{j_2} < 0$, respectively.

Therefore, the flow of $y_{j_1}^* = a_{j_1j_1s_1} \neq a_{j_2j_21}$ for the x_{j_1}-direction has the following properties with $\dot{x}_{j_2}^* = a_{j_2j_20}(a_{j_1j_1s_1} - a_{j_2j_2l_1})(a_{j_1j_1s_1} - a_{j_2j_2l_2})^2$ as in Eqs. (4.154)–(4.157):

- For $a_{j_1j_10}\alpha_{j_1j_2}(a_{j_1j_1s_1} - a_{j_1j_1s_2})^2 > 0$ and $a_{j_2j_20}(a_{j_1j_1s_1} - a_{j_2j_2l_1}) \times (a_{j_1j_1s_1} - a_{j_2j_2l_2})^2 > 0$,
$$(a_{j_1j_1s_1}, \dot{x}_{j_2}^*) = \underbrace{(\text{UP}, \text{pF})}_{\text{up-parabola flow } (+)}.$$
- For $a_{j_1j_10}\alpha_{j_1j_2}(a_{j_1j_1s_1} - a_{j_1j_1s_2})^2 < 0$ and $a_{j_2j_20}(a_{j_1j_1s_1} - a_{j_2j_2l_1}) \times (a_{j_1j_1s_1} - a_{j_2j_2l_2})^2 > 0$,
$$(a_{j_1j_1s_1}, \dot{x}_{j_2}^*) = \underbrace{(\text{DP}, \text{pF})}_{\text{down-parabola flow } (+)}.$$
- For $a_{j_1j_10}\alpha_{j_1j_2}(a_{j_1j_1s_1} - a_{j_1j_1s_2})^2 > 0$ and $a_{j_2j_20}(a_{j_1j_1s_1} - a_{j_2j_2l_1}) \times (a_{j_1j_1s_1} - a_{j_2j_2l_2})^2 < 0$,
$$(a_{j_1j_1s_1}, \dot{x}_{j_2}^*) = \underbrace{(\text{DP}, \text{nF})}_{\text{down-parabola flow } (-)}.$$
- For $a_{j_1j_10}\alpha_{j_1j_2}(a_{j_1j_1s_1} - a_{j_1j_1s_2})^2 < 0$ and $a_{j_2j_20}(a_{j_1j_1s_1} - a_{j_2j_2l_1}) \times (a_{j_1j_1s_1} - a_{j_2j_2l_2})^2 < 0$,
$$(a_{j_1j_1s_1}, \dot{x}_{j_2}^*) = \underbrace{(\text{UP}, \text{nF})}_{\text{up-parabola flow } (-)}.$$

($\text{v}_{1\text{b}}$) In phase space, at $y_1^* = a_{j_1j_1s_2}$,

$$\left.\frac{dx_{j_1}}{dx_{j_2}}\right|_{y_1^*=a_{j_1j_1s_2}} = \left.\frac{a_{j_1j_10}(y_{j_1} - a_{j_1j_1s_1})(y_{j_1} - a_{j_1j_1s_2})^2}{a_{j_2j_20}(y_{j_1} - a_{j_2j_2l_1})(y_{j_1} - a_{j_2j_2l_2})^2}\right|_{y_1^*=a_{j_1j_1s_2}} = 0,$$

$$\left.\frac{d^2x_{j_1}}{dx_{j_2}^2}\right|_{y_1^*=a_{j_1j_1s_2}} = 0.$$

If

$$\left.\frac{d^3x_{j_1}}{dx_{j_2}^3}\right|_{y_1^*=a_{j_1j_1s_2}} = \frac{2a_{j_1j_10}\alpha_{j_1j_2}^2(a_{j_1j_1s_2} - a_{j_1j_1s_1})}{a_{j_2j_20}(a_{j_1j_1s_2} - a_{j_2j_2l_1})(a_{j_1j_1s_2} - a_{j_2j_2l_2})^2} > 0,$$

there is an increasing-inflection flow at $y_1^* = a_{j_1j_1s_2}$. If

$$\left.\frac{d^3x_{j_1}}{dx_{j_2}^3}\right|_{y_1^*=a_{j_1j_1s_2}} = \frac{2a_{j_1j_10}\alpha_{j_1j_2}^2(a_{j_1j_1s_2} - a_{j_1j_1s_1})}{a_{j_2j_20}(a_{j_1j_1s_2} - a_{j_2j_2l_1})(a_{j_1j_1s_2} - a_{j_2j_2l_2})^2} < 0,$$

there is a decreasing-inflection flow at $y_1^* = a_{j_1j_1s_2}$. With

$$\dot{x}_{j2}^* = a_{j_2j_20}(a_{j_1j_1s_2} - a_{j_2j_2l_1})(a_{j_1j_1s_2} - a_{j_2j_2l_2})^2,$$

the inflection flows are positive and negative for $\dot{x}_{j2}^* > 0$ and $\dot{x}_{j2}^* < 0$, respectively.

Thus, the flow of $y_{j_1}^* = a_{j_1j_1s_2} \neq a_{j_2j_21}$ for the x_{j_1}-direction with $a_{j_1j_11} \neq a_{j_2j_21}$ has the following properties with $\dot{x}_{j2}^* = a_{j_2j_20}(a_{j_1j_1s_2} - a_{j_2j_2l_1})(a_{j_1j_1s_2} - a_{j_2j_2l_2})^2$ as in Eqs. (4.158)–(4.161):

- For $a_{j_1j_10}\alpha_{j_1j_2}^2(a_{j_1j_1s_2} - a_{j_1j_1s_1}) > 0$ and $a_{j_2j_20}(a_{j_1j_1s_2} - a_{j_2j_2l_1})(a_{j_1j_1s_2} - a_{j_2j_2l_2})^2 > 0$

$$(a_{j_1j_1s_2}, \dot{x}_{j2}^*) = \underbrace{(\text{II}, \text{pF})}_{\text{increasing-inflection flow } (+)}.$$

- For $a_{j_1j_10}\alpha_{j_1j_2}^2(a_{j_1j_1s_2} - a_{j_1j_1s_1}) > 0$ and $a_{j_2j_20}(a_{j_1j_1s_2} - a_{j_2j_2l_1})(a_{j_1j_1s_2} - a_{j_2j_2l_2})^2 > 0$

$$(a_{j_1j_1s_2}, \dot{x}_{j2}^*) = \underbrace{(\text{DI}, \text{pF})}_{\text{decreasing-inflection flow } (+)}.$$

- For $a_{j_1j_10}\alpha_{j_1j_2}^2(a_{j_1j_1s_2} - a_{j_1j_1s_1}) > 0$ and $a_{j_2j_20}(a_{j_1j_1s_2} - a_{j_2j_2l_1})(a_{j_1j_1s_2} - a_{j_2j_2l_2})^2 < 0$,

$$(a_{j_1j_1s_2}, \dot{x}_{j2}^*) = \underbrace{(\text{DI}, \text{nF})}_{\text{decreasing-inflection flow } (-)}.$$

- For $a_{j_1j_10}\alpha_{j_1j_2}^2(a_{j_1j_1s_2} - a_{j_1j_1s_1}) < 0$ and $a_{j_2j_20}(a_{j_1j_1s_2} - a_{j_2j_2l_1})(a_{j_1j_1s_2} - a_{j_2j_2l_2})^2 < 0$,

$$(a_{j_1j_1s_2}, \dot{x}_{j2}^*) = \underbrace{(\text{II}, \text{nF})}_{\text{increasing-inflection flow } (-)}.$$

– The increasing- and decreasing-inflection flows are the appearing bifurcations for the up-parabola and down-parabola flows.

($\mathrm{v_{1c}}$) For $y_{j_1}^* = a_{j_2j_2l_1} \neq a_{j_1j_1s_1}, a_{j_1j_1s_2}$,

$$\left.\frac{dx_{j_2}}{dx_{j_1}}\right|_{y_1^*=a_{j_2j_2l_1}} = \left.\frac{a_{j_2j_20}(y_{j_1}-a_{j_2j_2l_1})(y_{j_1}-a_{j_2j_2l_2})^2}{a_{j_1j_10}(y_{j_1}-a_{j_1j_1s_1})(y_{j_1}-a_{j_1j_1s_2})^2}\right|_{y_1^*=a_{j_2j_2l_1}} = 0.$$

If

$$\left.\frac{d^2x_{j_2}}{dx_{j_1}^2}\right|_{y_1^*=a_{j_2j_2l_1}} = \frac{a_{j_2j_20}\alpha_{j_1j_1}^2(a_{j_2j_2l_1}-a_{j_2j_2l_2})^2}{a_{j_1j_10}(a_{j_2j_2l_1}-a_{j_1j_1s_1})(a_{j_2j_2l_1}-a_{j_1j_1s_2})^2} > 0,$$

there is an up-parabola flow at $y_1^* = a_{j_2j_2l_1}$. If

$$\left.\frac{d^2x_{j_2}}{dx_{j_1}^2}\right|_{y_1^*=a_{j_2j_2l_1}} = \frac{a_{j_2j_20}\alpha_{j_1j_1}^2(a_{j_2j_2l_1}-a_{j_2j_2l_2})^2}{a_{j_1j_10}(a_{j_2j_2l_1}-a_{j_1j_1s_1})(a_{j_2j_2l_1}-a_{j_1j_1s_2})^2} < 0,$$

there is a down-parabola at $y_1^* = a_{j_2j_2l_1}$. With

$$\dot{x}_{j_1}^* = a_{j_1j_10}(a_{j_2j_2l_1}-a_{j_1j_1s_1})(a_{j_2j_2l_1}-a_{j_1j_1s_2})^2,$$

the inflection flows are positive and negative for $\dot{x}_{j_1}^* > 0$ and $\dot{x}_{j_1}^* < 0$, respectively.

Therefore, the flow of $y_{j_1}^* = a_{j_2j_2l_1}$ for the x_{j_2}-direction has the following properties with $\dot{x}_{j_1}^* = a_{j_1j_10}(a_{j_2j_2l_1}-a_{j_1j_1s_1})(a_{j_2j_2l_1}-a_{j_1j_1s_2})^2$ as in Eqs. (4.162)–(4.165):

- For $a_{j_1j_10}(a_{j_2j_2l_1}-a_{j_1j_1s_1})(a_{j_2j_2l_1}-a_{j_1j_1s_2})^2 > 0$ and $a_{j_2j_20}\alpha_{j_1j_1} \times (a_{j_2j_2l_1}-a_{j_1j_1l_2})^2 > 0$,

$$(\dot{x}_{j_1}^*, a_{j_2j_2l_1}) = \underbrace{(\mathrm{pF}, \mathrm{UP})}_{\text{up-parabola flow } (+)}.$$

- For $a_{j_1j_10}(a_{j_2j_2l_1}-a_{j_1j_1s_1})(a_{j_2j_2l_1}-a_{j_1j_1s_2})^2 < 0$ and $a_{j_2j_20}\alpha_{j_1j_1} \times (a_{j_2j_2l_1}-a_{j_1j_1l_2})^2 > 0$,

$$(\dot{x}_{j_1}^*, a_{j_2j_2l_1}) = \underbrace{(\mathrm{nF}, \mathrm{DP})}_{\text{down-parabola flow } (-)}.$$

- For $a_{j_1j_10}(a_{j_2j_2l_1}-a_{j_1j_1s_1})(a_{j_2j_2l_1}-a_{j_1j_1s_2})^2 > 0$ and $a_{j_2j_20}\alpha_{j_1j_1} \times (a_{j_2j_2l_1}-a_{j_1j_1l_2})^2 < 0$,

$$(\dot{x}_{j_1}^*, a_{j_2j_2l_1}) = \underbrace{(\mathrm{pF}, \mathrm{DP})}_{\text{down-parabola flow } (+)}.$$

- For $a_{j_1j_10}(a_{j_2j_2l_1} - a_{j_1j_1s_1})(a_{j_2j_2l_1} - a_{j_1j_1s_2})^2 < 0$ and $a_{j_2j_20}\alpha_{j_1j_1} \times (a_{j_2j_2l_1} - a_{j_1j_1l_2})^2 < 0$,

$$(\dot{x}^*_{j_1}, a_{j_2j_2l_1}) = \underbrace{(\mathrm{nF}, \mathrm{UP})}_{\text{up-parabola flow }(-)}.$$

(v$_{1d}$) In phase space, at $y^*_1 = a_{j_2j_2l_2}$,

$$\left.\frac{dx_{j_2}}{dx_{j_1}}\right|_{y^*_1=a_{j_2j_2l_2}} = \left.\frac{a_{j_2j_20}(y_{j_1} - a_{j_2j_2l_1})(y_{j_1} - a_{j_2j_2l_2})^2}{a_{j_1j_10}(y_{j_1} - a_{j_1j_1s_1})(y_{j_1} - a_{j_1j_1s_2})^2}\right|_{y^*_1=a_{j_2j_2l_2}} = 0,$$

$$\left.\frac{d^2x_{j_2}}{dx^2_{j_1}}\right|_{y^*_1=a_{j_2j_21}} = 0.$$

If

$$\left.\frac{d^3x_{j_2}}{dx^3_{j_1}}\right|_{y^*_1=a_{j_2j_2l_2}} = \frac{2a_{j_2j_20}\alpha^2_{j_1j_1}(a_{j_2j_2l_2} - a_{j_2j_2l_1})}{a_{j_1j_10}(a_{j_2j_2l_2} - a_{j_1j_1s_1})(a_{j_2j_2l_2} - a_{j_1j_1s_2})^2} > 0,$$

there is an increasing-inflection flow at $y^*_1 = a_{j_2j_2l_2}$. If

$$\left.\frac{d^3x_{j_2}}{dx^3_{j_1}}\right|_{y^*_1=a_{j_2j_2l_2}} = \frac{2a_{j_2j_20}\alpha^2_{j_1j_1}(a_{j_2j_2l_2} - a_{j_2j_2l_1})}{a_{j_1j_10}(a_{j_2j_2l_2} - a_{j_1j_1s_1})(a_{j_2j_2l_2} - a_{j_1j_1s_2})^2} < 0,$$

there is a decreasing-inflection flow at $y^*_1 = a_{j_2j_2l_2}$. With

$$\dot{x}^*_{j_1} = a_{j_1j_10}(a_{j_2j_21} - a_{j_1j_1s_1})(a_{j_2j_21} - a_{j_1j_1s_2})^2,$$

the inflection flows are positive and negative for $\dot{x}^*_{j_1} > 0$ and $\dot{x}^*_{j_1} < 0$, respectively.

Therefore, the flow of $y^*_{j_1} = a_{j_2j_2l_2}$ for the x_{j_2}-direction has the following properties with $\dot{x}^*_{j_1} = a_{j_1j_10}(a_{j_2j_2l_2} - a_{j_1j_1s_1})(a_{j_2j_2l_2} - a_{j_1j_1s_2})^2$ as in Eqs. (4.166)–(4.169):

- For $a_{j_1j_10}(a_{j_2j_2l_2} - a_{j_1j_1s_1})(a_{j_2j_2l_2} - a_{j_1j_1s_2})^2 > 0$ and $a_{j_2j_20}\alpha^2_{j_1j_1} \times (a_{j_2j_2l_2} - a_{j_1j_1l_1}) > 0$,

$$(\dot{x}^*_{j_1}, a_{j_2j_21}) = \underbrace{(\mathrm{pF}, \mathrm{II})}_{\text{increasing-infelction flow }(+)}.$$

- For $a_{j_1j_10}(a_{j_2j_2l_2} - a_{j_1j_1s_1})(a_{j_2j_2l_2} - a_{j_1j_1s_2})^2 < 0$ and $a_{j_2j_20}\alpha^2_{j_1j_1} \times (a_{j_1j_1l_2} - a_{j_2j_2l_1}) > 0$,

$$(\dot{x}^*_{j_1}, a_{j_2j_21}) = \underbrace{(\mathrm{nF}, \mathrm{DI})}_{\text{decreasing-infelction flow }(-)}$$

- For $a_{j_1j_10}(a_{j_2j_2l_1} - a_{j_1j_1s_1})(a_{j_2j_2l_2} - a_{j_1j_1s_2})^2 > 0$ and $a_{j_2j_20}\alpha_{j_1j_1}^2 \times (a_{j_1j_1l_2} - a_{j_2j_2l_1}) < 0$,

$$(\dot{x}_{j_1}^*, a_{j_2j_21}) = \underbrace{(\mathrm{pF}, \mathrm{DI})}_{\text{decreasing-infelction flow } (+)}.$$

- For $a_{j_1j_10}(a_{j_2j_2l_1} - a_{j_1j_1s_1})(a_{j_2j_2l_2} - a_{j_1j_1s_2})^2 < 0$ and $a_{j_2j_20}\alpha_{j_1j_1}^2 \times (a_{j_1j_1l_2} - a_{j_2j_2l_1}) < 0$,

$$(\dot{x}_{j_1}^*, a_{j_2j_21}) = \underbrace{(\mathrm{nF}, \mathrm{II})}_{\text{increasing-infelction flow } (-)}.$$

– The increasing- and decreasing-inflection flows are the appearing bifurcations for the up-parabola and down-parabola flows.

(v_2) For $a_{j_1j_1s_1} = a_{j_1j_1s_2} \equiv a_{j_1j_11}$,

$$\begin{aligned}
\dot{x}_{j_1} &= a_{j_1j_10}(y_{j_1} - a_{j_1j_11})^3,\\
\dot{x}_{j_2} &= a_{j_2j_20}(y_{j_1} - a_{j_2j_2l_1})(y_{j_1} - a_{j_2j_2l_2})^2,\\
&j_1, j_2 \in \{1,2\};\ j_1 \neq j_2.
\end{aligned}$$

For $a_{j_2j_21} \neq a_{j_1j_1s_1}, a_{j_1j_1s_2}$, a dynamical system in the y_{j_1}-direction is constructed as

$$\begin{aligned}
\dot{y}_{j_1} &= \alpha_{j_1j_1}a_{j_1j_10}(y_{j_1} - a_{j_1j_11})^3\\
&\quad + \alpha_{j_1j_2}a_{j_2j_20}(y_{j_1} - a_{j_2j_2l_1})(y_{j_1} - a_{j_2j_2l_2})^2,
\end{aligned}$$

and a dynamical system in the $y_{j_1}^{\perp}$-direction is constructed as

$$\begin{aligned}
\dot{y}_{j_1}^{\perp} &= \beta_{j_1j_1}a_{j_1j_10}(y_{j_1} - a_{j_1j_11})^3\\
&\quad + \beta_{j_1j_2}a_{j_2j_20}(y_{j_1} - a_{j_2j_2l_1})(y_{j_1} - a_{j_2j_2l_2})^2.
\end{aligned}$$

(v_{2a}) In phase space, at $y_1^* = a_{j_1j_11}$,

$$\left.\frac{dx_{j_1}}{dx_{j_2}}\right|_{y_1^*=a_{j_1j_11}} = \left.\frac{a_{j_1j_10}(y_{j_1} - a_{j_1j_11})^3}{a_{j_2j_20}(y_{j_1} - a_{j_2j_2l_1})(y_{j_1} - a_{j_2j_2l_2})^2}\right|_{y_1^*=a_{j_1j_11}} = 0,$$

$$\left.\frac{d^2x_{j_1}}{dx_{j_2}^2}\right|_{y_1^*=a_{j_1j_11}} = \left.\frac{d^3x_{j_1}}{dx_{j_2}^3}\right|_{y_1^*=a_{j_1j_11}} = 0.$$

If

$$\left.\frac{d^4x_{j_1}}{dx_{j_2}^4}\right|_{y_1^*=a_{j_1j_11}} = \frac{6a_{j_1j_10}\alpha_{j_1j_2}^3}{a_{j_2j_20}(a_{j_1j_11} - a_{j_2j_2l_1})(a_{j_1j_11} - a_{j_2j_2l_2})^2} > 0,$$

there is a third-order up-parabola flow at $y_1^* = a_{j_1 j_1 1}$. If

$$\left.\frac{d^4 x_{j_1}}{dx_{j_2}^4}\right|_{y_1^* = a_{j_1 j_1 1}} = \frac{6 a_{j_1 j_1 0} \alpha_{j_1 j_2}^3}{a_{j_2 j_2 0}(a_{j_1 j_1 1} - a_{j_2 j_2 l_1})(a_{j_1 j_1 1} - a_{j_2 j_2 l_2})^2} < 0,$$

there is a third-order down-parabola flow at $y_1^* = a_{j_1 j_1 1}$. With

$$\dot{x}_{j_2}^* = a_{j_2 j_2 0}(a_{j_1 j_1 1} - a_{j_2 j_2 l_1})(a_{j_1 j_1 1} - a_{j_2 j_2 l_2})^2,$$

the third-order parabola flows are positive and negative for $\dot{x}_{j_2}^* > 0$ and $\dot{x}_{j_2}^* < 0$, respectively.

Therefore, the flow of $y_{j_1}^* = a_{j_1 j_1 1} \neq a_{j_2 j_2 1}$ for the x_{j_1}-direction has the following properties with $\dot{x}_{j_2}^* = a_{j_2 j_2 0}(a_{j_1 j_1 1} - a_{j_2 j_2 l_1})(a_{j_1 j_1 1} - a_{j_2 j_2 l_2})^2$ as in Eqs. (3.173)–(3.176):

- For $a_{j_1 j_1 0} \alpha_{j_1 j_2}^3 > 0$ and $a_{j_2 j_2 0}(a_{j_1 j_1 1} - a_{j_2 j_2 l_1})(a_{j_1 j_1 1} - a_{j_2 j_2 l_2})^2 > 0$,

$$(a_{j_1 j_1 1}, \dot{x}_{j_2}^*) = \underbrace{(3^{\text{rd}}\text{UP}, \text{pF})}_{\text{third-order up-parabola flow } (+)}.$$

- For $a_{j_1 j_1 0} \alpha_{j_1 j_2}^3 < 0$ and $a_{j_2 j_2 0}(a_{j_1 j_1 1} - a_{j_2 j_2 l_1})(a_{j_1 j_1 1} - a_{j_2 j_2 l_2})^2 > 0$,

$$(a_{j_1 j_1 1}, \dot{x}_{j_2}^*) = \underbrace{(3^{\text{rd}}\text{DP}, \text{pF})}_{\text{third-order down-parabola flow } (+)}.$$

- For $a_{j_1 j_1 0} \alpha_{j_1 j_2}^3 > 0$ and $a_{j_2 j_2 0}(a_{j_1 j_1 1} - a_{j_2 j_2 l_1})(a_{j_1 j_1 1} - a_{j_2 j_2 l_2})^2 < 0$,

$$(a_{j_1 j_1 1}, \dot{x}_{j_2}^*) = \underbrace{(3^{\text{rd}}\text{DP}, \text{nF})}_{\text{third-order down-parabola flow } (-)}.$$

- For $a_{j_1 j_1 0} \alpha_{j_1 j_2}^3 < 0$ and $a_{j_2 j_2 0}(a_{j_1 j_1 1} - a_{j_2 j_2 l_1})(a_{j_1 j_1 1} - a_{j_2 j_2 l_2})^2 < 0$,

$$(a_{j_1 j_1 1}, \dot{x}_{j_2}^*) = \underbrace{(3^{\text{rd}}\text{UP}, \text{nF})}_{\text{third-order up-parabola flow } (-)}.$$

- The third-order up-parabola flows are the appearing bifurcations for up-parabola, down-parabola, and up-parabola flows.
- The third-order down-parabola flows are the appearing bifurcations for down-parabola, up-parabola, and down-parabola flows.

(v_{2b}) For $y^*_{j_1} = a_{j_2j_2l_1} \neq a_{j_1j_11}$,

$$\left.\frac{dx_{j_2}}{dx_{j_1}}\right|_{y^*_1=a_{j_2j_2l_1}} = \left.\frac{a_{j_2j_20}(y_{j_1} - a_{j_2j_2l_1})(y_{j_1} - a_{j_2j_2l_2})^2}{a_{j_1j_10}(y_{j_1} - a_{j_1j_11})^3}\right|_{y^*_1=a_{j_2j_2l_1}} = 0.$$

If

$$\left.\frac{d^2x_{j_2}}{dx^2_{j_1}}\right|_{y^*_1=a_{j_2j_2l_1}} = \frac{a_{j_2j_20}\alpha^2_{j_1j_1}(a_{j_2j_2l_1} - a_{j_2j_2l_2})^2}{a_{j_1j_10}(a_{j_2j_2l_1} - a_{j_1j_11})^3} > 0,$$

there is an up-parabola flow at $y^*_1 = a_{j_2j_2l_1}$. If

$$\left.\frac{d^2x_{j_2}}{dx^2_{j_1}}\right|_{y^*_1=a_{j_2j_2l_1}} = \frac{a_{j_2j_20}\alpha^2_{j_1j_1}(a_{j_2j_2l_1} - a_{j_2j_2l_2})^2}{a_{j_1j_10}(a_{j_2j_2l_1} - a_{j_1j_11})^3} < 0,$$

there is a down-parabola at $y^*_1 = a_{j_2j_2l_1}$. With

$$\dot{x}^*_{j_1} = a_{j_1j_10}(a_{j_2j_2l_1} - a_{j_1j_11})^3,$$

the inflection flows are positive and negative for $\dot{x}^*_{j_1} > 0$ and $\dot{x}^*_{j_1} < 0$, respectively.

Therefore, the flow of $y^*_{j_1} = a_{j_2j_2l_1} \neq a_{j_1j_1s_1}, a_{j_1j_1s_2}$ for the x_{j_2}-direction has the following properties with $\dot{x}^*_{j_1} = a_{j_1j_10}(a_{j_2j_2l_1} - a_{j_1j_11})^3$ as in Eqs. (4.177)–(4.180):

- For $a_{j_1j_10}(a_{j_2j_2l_1} - a_{j_1j_11})^3 > 0$ and $a_{j_2j_20}\alpha_{j_1j_1}(a_{j_2j_2l_1} - a_{j_2j_2l_2})^2 > 0$,

$$(\dot{x}^*_{j_1}, a_{j_2j_2l_1}) = \underbrace{(\text{pF}, \text{UP})}_{\text{up-parabola flow } (+)}.$$

- For $a_{j_1j_10}(a_{j_2j_2l_1} - a_{j_1j_11})^3 < 0$ and $a_{j_2j_20}\alpha_{j_1j_1}(a_{j_2j_2l_1} - a_{j_2j_2l_2})^2 > 0$,

$$(\dot{x}^*_{j_1}, a_{j_2j_2l_1}) = \underbrace{(\text{nF}, \text{DP})}_{\text{down-parabola flow } (-)}.$$

- For $a_{j_1j_10}(a_{j_2j_2l_1} - a_{j_1j_11})^3 > 0$ and $a_{j_2j_20}\alpha_{j_1j_1}(a_{j_2j_2l_1} - a_{j_2j_2l_2})^2 < 0$,

$$(\dot{x}^*_{j_1}, a_{j_2j_2l_1}) = \underbrace{(\text{pF}, \text{DP})}_{\text{down-parabola flow } (+)}.$$

- For $a_{j_1j_10}(a_{j_2j_2l_1} - a_{j_1j_11})^3 < 0$ and $a_{j_2j_20}\alpha_{j_1j_1}(a_{j_2j_2l_1} - a_{j_2j_2l_2})^2 < 0$,

$$(\dot{x}^*_{j_1}, a_{j_2j_2l_1}) = \underbrace{(\text{nF}, \text{UP})}_{\text{up-parabola flow } (+)}.$$

(v_{2c}) In phase space, at $y_1^* = a_{j_2 j_2 l_2}$,

$$\left.\frac{dx_{j_2}}{dx_{j_1}}\right|_{y_1^*=a_{j_2j_2l_2}} = \left.\frac{a_{j_2j_20}(y_{j_1}-a_{j_2j_2l_1})(y_{j_1}-a_{j_2j_2l_2})^2}{a_{j_1j_10}(y_{j_1}-a_{j_1j_11})^3}\right|_{y_1^*=a_{j_2j_2l_2}} = 0,$$

$$\left.\frac{d^2x_{j_2}}{dx_{j_1}^2}\right|_{y_1^*=a_{j_2j_2l_2}} = 0.$$

If

$$\left.\frac{d^3x_{j_2}}{dx_{j_1}^3}\right|_{y_1^*=a_{j_2j_2l_2}} = \frac{2a_{j_2j_20}\alpha_{j_1j_1}^2(a_{j_2j_2l_2}-a_{j_2j_2l_1})}{a_{j_1j_10}(a_{j_2j_2l_2}-a_{j_1j_11})^3} > 0,$$

there is an increasing-inflection flow at $y_1^* = a_{j_2j_2l_2}$. If

$$\left.\frac{d^3x_{j_2}}{dx_{j_1}^3}\right|_{y_1^*=a_{j_2j_2l_2}} = \frac{2a_{j_2j_20}\alpha_{j_1j_1}^2(a_{j_2j_2l_2}-a_{j_2j_2l_1})}{a_{j_1j_10}(a_{j_2j_2l_2}-a_{j_1j_11})^3} < 0,$$

there is a decreasing-inflection flow at $y_1^* = a_{j_2j_2l_2}$. With

$$\dot{x}_{j_1}^* = a_{j_1j_10}(a_{j_2j_2l_2}-a_{j_1j_11})^3,$$

the inflection flows are positive and negative for $\dot{x}_{j_1}^* > 0$ and $\dot{x}_{j_1}^* < 0$, respectively.

Therefore, the flow of $y_{j_1}^* = a_{j_2j_2l_2} \neq a_{j_1j_11}$ for the x_{j_2}-direction has the following properties with $\dot{x}_{j_1}^* = a_{j_1j_10}(a_{j_2j_2l_2}-a_{j_1j_11})^3$ as in Eqs. (4.181)–(4.184):

- For $a_{j_1j_10}(a_{j_2j_2l_2}-a_{j_1j_11})^3 > 0$ and $a_{j_2j_20}\alpha_{j_1j_1}^2(a_{j_2j_2l_2}-a_{j_2j_2l_1}) > 0$,
$$(\dot{x}_{j_1}^*, a_{j_2j_2l_2}) = \underbrace{(\mathrm{pF}, \mathrm{II})}_{\text{increasing-inflection flow } (+)}.$$
- For $a_{j_1j_10}(a_{j_2j_2l_2}-a_{j_1j_11})^3 < 0$ and $a_{j_2j_20}\alpha_{j_1j_1}^2(a_{j_2j_2l_2}-a_{j_2j_2l_1}) > 0$,
$$(\dot{x}_{j_1}^*, a_{j_2j_2l_2}) = \underbrace{(\mathrm{nF}, \mathrm{DI})}_{\text{decreasing-inflection flow } (-)}.$$
- For $a_{j_1j_10}(a_{j_2j_2l_2}-a_{j_1j_11})^3 > 0$ and $a_{j_2j_20}\alpha_{j_1j_1}^2(a_{j_2j_2l_2}-a_{j_2j_2l_1}) < 0$,
$$(\dot{x}_{j_1}^*, a_{j_2j_2l_2}) = \underbrace{(\mathrm{pF}, \mathrm{DI})}_{\text{decreasing-inflection flow } (+)}.$$
- For $a_{j_1j_10}(a_{j_2j_2l_2}-a_{j_1j_11})^3 < 0$ and $a_{j_2j_20}\alpha_{j_1j_1}^2(a_{j_2j_2l_2}-a_{j_2j_2l_1}) < 0$,
$$(\dot{x}_{j_1}^*, a_{j_2j_2l_2}) = \underbrace{(\mathrm{nF}, \mathrm{II})}_{\text{increasing-inflection flow } (-)}.$$

– The increasing- and decreasing-inflection flows are the appearing bifurcations for the up-parabola and down-parabola flows.

(v$_3$) For $a_{j_2 j_2 l_1} = a_{j_2 j_2 l_2} \equiv a_{j_2 j_2 1}$, the cubic system becomes

$$\begin{aligned}
\dot{x}_{j_1} &= a_{j_1 j_1 0}(y_{j_1} - a_{j_1 j_1 s_1})(y_{j_1} - a_{j_1 j_1 s_2})^2,\\
\dot{x}_{j_2} &= a_{j_2 j_2 0}(y_{j_1} - a_{j_2 j_2 1})^3,\\
&j_1, j_2 \in \{1,2\};\ j_1 \neq j_2.
\end{aligned}$$

For $a_{j_2 j_2 l_1}, a_{j_2 j_2 l_2} \neq a_{j_1 j_1 1}$, a dynamical system in the y_{j_1}-direction is constructed as

$$\begin{aligned}
\dot{y}_{j_1} &= \alpha_{j_1 j_1} a_{j_1 j_1 0}(y_{j_1} - a_{j_1 j_1 s_1})(y_{j_1} - a_{j_1 j_1 s_2})^2\\
&\quad + \alpha_{j_1 j_2} a_{j_2 j_2 0}(y_{j_1} - a_{j_2 j_2 1})^3,
\end{aligned}$$

and a dynamical system in the $y_{j_1}^{\perp}$-direction is constructed as

$$\begin{aligned}
\dot{y}_{j_1}^{\perp} &= \beta_{j_1 j_1} a_{j_1 j_1 0}(y_{j_1} - a_{j_1 j_1 s_1})(y_{j_1} - a_{j_1 j_1 s_2})^2\\
&\quad + \beta_{j_1 j_2} a_{j_2 j_2 0}(y_{j_1} - a_{j_2 j_2 1})^3.
\end{aligned}$$

(v$_{3a}$) For $y_{j_1}^* = a_{j_1 j_1 s_1} \neq a_{j_2 j_2 1}$, in phase space,

$$\left.\frac{dx_{j_1}}{dx_{j_2}}\right|_{y_1^* = a_{j_1 j_1 s_1}} = \left.\frac{a_{j_1 j_1 0}(y_{j_1} - a_{j_1 j_1 s_1})(y_{j_1} - a_{j_1 j_1 s_2})^2}{a_{j_2 j_2 0}(y_{j_1} - a_{j_2 j_2 1})^3}\right|_{y_1^* = a_{j_1 j_1 s_1}} = 0.$$

If

$$\left.\frac{d^2 x_{j_1}}{dx_{j_2}^2}\right|_{y_1^* = a_{j_1 j_1 s_1}} = \frac{a_{j_1 j_1 0}\alpha_{j_1 j_2}(a_{j_1 j_1 s_1} - a_{j_1 j_1 s_2})^2}{a_{j_2 j_2 0}(a_{j_1 j_1 s_1} - a_{j_2 j_2 l_2})^3} > 0,$$

there is an up-parabola flow at $y_1^* = a_{j_1 j_1 s_1}$. If

$$\left.\frac{d^2 x_{j_1}}{dx_{j_2}^2}\right|_{y_1^* = a_{j_1 j_1 s_1}} = \frac{a_{j_1 j_1 0}\alpha_{j_1 j_2}(a_{j_1 j_1 s_1} - a_{j_1 j_1 s_2})^2}{a_{j_2 j_2 0}(a_{j_1 j_1 s_1} - a_{j_2 j_2 l_2})^3} < 0,$$

there is a down-parabola flow at $y_1^* = a_{j_1 j_1 s_1}$. With

$$\dot{x}_{j_2}^* = a_{j_2 j_2 0}(a_{j_1 j_1 s_1} - a_{j_2 j_2 l_2})^3,$$

the parabola flows are positive and negative for $\dot{x}_{j_2}^* > 0$ and $\dot{x}_{j_2}^* < 0$, respectively.

Therefore, the flow of $y_{j_1}^* = a_{j_1 j_1 s_1} \neq a_{j_2 j_2 1}$ with $\dot{x}_{j_2}^* = a_{j_2 j_2 0}(a_{j_1 j_1 s_1} - a_{j_2 j_2 1})^3$ for the x_{j_1}-direction has the following properties as in Eqs. (4.188)–(4.191):

- For $a_{j_1 j_1 0}\alpha_{j_1 j_2}(a_{j_1 j_1 s_1} - a_{j_1 j_1 s_2})^2 > 0$ and $a_{j_2 j_2 0}(a_{j_1 j_1 s_1} - a_{j_2 j_2 1})^3 > 0$,

$$(a_{j_1 j_1 s_1}, \dot{x}_{j_2}^*) = \underbrace{(\mathrm{UP}, \mathrm{pF})}_{\text{up-parabola flow } (+)}.$$

- For $a_{j_1j_10}\alpha_{j_1j_2}(a_{j_1j_1s_1}-a_{j_1j_1s_2})^2<0$ and $a_{j_2j_20}(a_{j_1j_1s_1}-a_{j_2j_21})^3>0$,

$$(a_{j_1j_1s_1},\dot{x}^*_{j_2}) = \underbrace{(\text{DP},\text{pF})}_{\text{down-parabola flow }(+)}.$$

- For $a_{j_1j_10}\alpha_{j_1j_2}(a_{j_1j_1s_1}-a_{j_1j_1s_2})^2>0$ and $a_{j_2j_20}(a_{j_1j_1s_1}-a_{j_2j_21})^3<0$,

$$(a_{j_1j_1s_1},\dot{x}^*_{j_2}) = \underbrace{(\text{DP},\text{nF})}_{\text{down-parabola flow }(-)}.$$

- For $a_{j_1j_10}\alpha_{j_1j_2}(a_{j_1j_1s_1}-a_{j_1j_1s_2})^2<0$ and $a_{j_2j_20}(a_{j_1j_1s_1}-a_{j_2j_21})^3<0$,

$$(a_{j_1j_1s_1},\dot{x}^*_{j_2}) = \underbrace{(\text{UP},\text{nF})}_{\text{third-order up-parabola flow }(-)}.$$

($\text{v}_{3\text{b}}$) In phase space, at $y_1^*=a_{j_1j_1s_2}$,

$$\left.\frac{dx_{j_1}}{dx_{j_2}}\right|_{y_1^*=a_{j_1j_1s_2}} = \left.\frac{a_{j_1j_10}(y_{j_1}-a_{j_1j_1s_1})(y_{j_1}-a_{j_1j_1s_2})^2}{a_{j_2j_20}(y_{j_1}-a_{j_2j_21})^3}\right|_{y_1^*=a_{j_1j_1s_2}} = 0,$$

$$\left.\frac{d^2x_{j_1}}{dx_{j_2}^2}\right|_{y_1^*=a_{j_1j_1s_2}} = 0.$$

If

$$\left.\frac{d^3x_{j_1}}{dx_{j_2}^3}\right|_{y_1^*=a_{j_1j_1s_2}} = \frac{2a_{j_1j_10}\alpha^2_{j_1j_2}(a_{j_1j_1s_2}-a_{j_1j_1s_1})}{a_{j_2j_20}(a_{j_1j_1s_2}-a_{j_2j_21})^3} > 0,$$

there is an increasing-inflection flow at $y_1^*=a_{j_1j_1s_2}$. If

$$\left.\frac{d^3x_{j_1}}{dx_{j_2}^3}\right|_{y_1^*=a_{j_1j_1s_2}} = \frac{2a_{j_1j_10}\alpha^2_{j_1j_2}(a_{j_1j_1s_2}-a_{j_1j_1s_1})}{a_{j_2j_20}(a_{j_1j_1s_2}-a_{j_2j_21})^3} < 0,$$

there is a decreasing-inflection flow at $y_1^*=a_{j_1j_1s_2}$. With

$$\dot{x}^*_{j_2} = a_{j_2j_20}(a_{j_1j_1s_2}-a_{j_2j_21})^3,$$

the inflection flows are positive and negative for $\dot{x}^*_{j_2}>0$ and $\dot{x}^*_{j_2}<0$, respectively.

Therefore, the flow of $y^*_{j_1}=a_{j_1j_1s_2}\neq a_{j_2j_21}$ for the x_{j_2}-direction has the following properties with $\dot{x}^*_{j_2}=a_{j_2j_20}(a_{j_1j_1s_1}-a_{j_2j_21})^3$ as in Eqs. (4.192)–(4.195):

- For $a_{j_1j_10}\alpha^2_{j_1j_2}(a_{j_1j_1s_2}-a_{j_1j_1s_1})>0$ and $a_{j_2j_20}(a_{j_1j_1s_2}-a_{j_2j_21})^3>0$,

$$(a_{j_1j_1s_2},\dot{x}^*_{j_2}) = \underbrace{(\text{II},\text{pF})}_{\text{increasing-inflection flow }(+)}.$$

- For $a_{j_1j_10}\alpha^2_{j_1j_2}(a_{j_1j_1s_2} - a_{j_1j_1s_1}) < 0$ and $a_{j_2j_20}(a_{j_1j_1s_2} - a_{j_2j_21})^3 > 0$,
$$(a_{j_1j_1s_2}, \dot{x}^*_{j_2}) = \underbrace{(\text{DI}, \text{pF})}_{\text{decreasing-inflection flow } (+)}.$$

- For $a_{j_1j_10}\alpha^2_{j_1j_2}(a_{j_1j_1s_2} - a_{j_1j_1s_1}) > 0$ and $a_{j_2j_20}(a_{j_1j_1s_2} - a_{j_2j_21})^3 < 0$,
$$(a_{j_1j_1s_2}, \dot{x}^*_{j_2}) = \underbrace{(\text{DI}, \text{nF})}_{\text{decreasing-inflection flow } (-)}.$$

- For $a_{j_1j_10}\alpha^2_{j_1j_2}(a_{j_1j_1s_2} - a_{j_1j_1s_1}) < 0$ and $a_{j_2j_20}(a_{j_1j_1s_2} - a_{j_2j_21})^3 < 0$,
$$(a_{j_1j_1s_2}, \dot{x}^*_{j_2}) = \underbrace{(\text{II}, \text{nF})}_{\text{increasing-inflection flow } (-)}.$$

- The inflection flows are the appearing bifurcations for up-parabola and down-parabola flows.

($\text{v}_{3\text{c}}$) For $y^*_{j_1} = a_{j_2j_21} \neq a_{j_1j_1s_1}, a_{j_1j_1s_2}$,

$$\left.\frac{dx_{j_2}}{dx_{j_1}}\right|_{y^*_1=a_{j_2j_21}} = \left.\frac{a_{j_2j_20}(y_{j_1} - a_{j_2j_21})^3}{a_{j_1j_10}(y_{j_1} - a_{j_1j_1s_1})(y_{j_1} - a_{j_1j_1s_2})^2}\right|_{y^*_1=a_{j_2j_21}} = 0,$$
$$\left.\frac{d^2x_{j_2}}{dx^2_{j_1}}\right|_{y^*_1=a_{j_2j_21}} = \left.\frac{d^3x_{j_2}}{dx^3_{j_1}}\right|_{y^*_1=a_{j_2j_21}} = 0.$$

If

$$\left.\frac{d^4x_{j_2}}{dx^4_{j_1}}\right|_{y^*_1=a_{j_2j_21}} = \frac{6a_{j_2j_20}\alpha^3_{j_1j_1}}{a_{j_1j_10}(a_{j_2j_21} - a_{j_1j_1s_1})(a_{j_2j_21} - a_{j_1j_1s_2})^2} > 0,$$

there is a third-order up-parabola flow at $y^*_1 = a_{j_2j_21}$. If

$$\left.\frac{d^4x_{j_2}}{dx^4_{j_1}}\right|_{y^*_1=a_{j_2j_21}} = \frac{6a_{j_2j_20}\alpha^3_{j_1j_1}}{a_{j_1j_10}(a_{j_2j_21} - a_{j_1j_1s_1})(a_{j_2j_21} - a_{j_1j_1s_2})^2} < 0,$$

there is a third-order down-parabola flow at $y^*_1 = a_{j_2j_21}$. With

$$\dot{x}^*_{j_1} = a_{j_1j_10}(a_{j_2j_21} - a_{j_1j_1s_1})(a_{j_2j_21} - a_{j_1j_1s_2})^2,$$

the third-order parabola flows are positive and negative for $\dot{x}^*_{j_1} > 0$ and $\dot{x}^*_{j_1} < 0$, respectively.

Therefore, the flow of $y^*_{j_1} = a_{j_2j_21} \neq a_{j_1j_1s_1}, a_{j_1j_1s_2}$ for the x_{j_2}-direction has the following properties with $\dot{x}^*_{j_1} = a_{j_1j_10}(a_{j_2j_21} - a_{j_1j_1s_1})(a_{j_2j_21} - a_{j_1j_1s_2})^2$ as in Eqs. (4.196)–(4.199):

- For $a_{j_1j_10}(a_{j_2j_21} - a_{j_1j_1s_1})(a_{j_2j_21} - a_{j_1j_1s_2})^2 > 0$ and $a_{j_2j_20}\alpha^3_{j_1j_1} > 0$,

$$(\dot{x}^*_{j_1}, a_{j_2j_21}) = \underbrace{(\text{pF}, 3^{\text{rd}}\text{UP})}_{\text{third-order up-parabola flow }(+)}.$$

- For $a_{j_1j_10}(a_{j_2j_21} - a_{j_1j_1s_1})(a_{j_2j_21} - a_{j_1j_1s_2})^2 < 0$ and $a_{j_2j_20}\alpha^3_{j_1j_1} > 0$,

$$(\dot{x}^*_{j_1}, a_{j_2j_21}) = \underbrace{(\text{nF}, 3^{\text{rd}}\text{DP})}_{\text{third-order down-parabola flow }(-)}.$$

- For $a_{j_1j_10}(a_{j_2j_21} - a_{j_1j_1s_1})(a_{j_2j_21} - a_{j_1j_1s_2})^2 > 0$ and $a_{j_2j_20}\alpha^3_{j_1j_1} < 0$,

$$(\dot{x}^*_{j_1}, a_{j_2j_2l_1}) = \underbrace{(\text{pF}, 3^{\text{rd}}\text{DP})}_{\text{third-order down-parabola flow }(+)}.$$

- For $a_{j_1j_10}(a_{j_2j_21} - a_{j_1j_1s_1})(a_{j_2j_21} - a_{j_1j_1s_2})^2 < 0$ and $a_{j_2j_20}\alpha^3_{j_1j_1} < 0$,

$$(\dot{x}^*_{j_1}, a_{j_2j_21}) = \underbrace{(\text{nF}, 3^{\text{rd}}\text{UP})}_{\text{third-order up-parabola flow }(+)}.$$

- The third-order up-parabola flows are the appearing bifurcations from an up-parabola flow to up-parabola, down-parabola, and up-parabola flows.
- The third-order down-parabola flows are the appearing bifurcations from a down-parabola flow to down-parabola, up-parabola, and down-parabola flows.

(v$_4$) For $a_{j_1j_1s_1} = a_{j_1j_1s_2} \equiv a_{j_1j_11}$ and $a_{j_2j_2l_1} = a_{j_2j_2l_2} \equiv a_{j_2j_21}$, the dynamical system is

$$\begin{aligned}
&\dot{x}_{j_1} = a_{j_1j_10}(y_{j_1} - a_{j_1j_11})^3,\\
&\dot{x}_{j_2} = a_{j_2j_20}(y_{j_1} - a_{j_2j_21})^3,\\
&j_1, j_2 \in \{1,2\};\ j_1 \neq j_2.
\end{aligned}$$

For $a_{j_2j_21} \neq a_{j_1j_11}$, a dynamical system in the y_{j_1}-direction is constructed as

$$\dot{y}_{j_1} = \alpha_{j_1j_1}a_{j_1j_10}(y_{j_1} - a_{j_1j_11})^3 + \alpha_{j_1j_2}a_{j_2j_20}(y_{j_1} - a_{j_2j_21})^3,$$

and a dynamical system in the $y_{j_1}^{\perp}$-direction is constructed as

$$\dot{y}_{j_1}^{\perp} = \beta_{j_1 j_1} a_{j_1 j_1 0}(y_{j_1} - a_{j_1 j_1 1})^3 + \beta_{j_1 j_2} a_{j_2 j_2 0}(y_{j_1} - a_{j_2 j_2 1})^3.$$

($\mathrm{v}_{4\mathrm{a}}$) In phase space, at $y_1^* = a_{j_1 j_1 1}$,

$$\left.\frac{dx_{j_1}}{dx_{j_2}}\right|_{y_1^*=a_{j_1 j_1 1}} = \left.\frac{a_{j_1 j_1 0}(y_{j_1} - a_{j_1 j_1 1})^3}{a_{j_2 j_2 0}(y_{j_1} - a_{j_2 j_2 1})^3}\right|_{y_1^*=a_{j_1 j_1 1}} = 0,$$

$$\left.\frac{d^2 x_{j_1}}{dx_{j_2}^2}\right|_{y_1^*=a_{j_1 j_1 1}} = \left.\frac{d^3 x_{j_1}}{dx_{j_2}^3}\right|_{y_1^*=a_{j_1 j_1 1}} = 0.$$

If

$$\left.\frac{d^4 x_{j_1}}{dx_{j_2}^4}\right|_{y_1^*=a_{j_1 j_1 1}} = \frac{6 a_{j_1 j_1 0}\alpha_{j_1 j_2}^3}{a_{j_2 j_2 0}(a_{j_1 j_1 1} - a_{j_2 j_2 1})^3} > 0,$$

there is a third-order up-parabola flow at $y_1^* = a_{j_1 j_1 1}$. If

$$\left.\frac{d^4 x_{j_1}}{dx_{j_2}^4}\right|_{y_1^*=a_{j_1 j_1 1}} = \frac{6 a_{j_1 j_1 0}\alpha_{j_1 j_2}^3}{a_{j_2 j_2 0}(a_{j_1 j_1 1} - a_{j_2 j_2 1})^3} < 0,$$

there is a third-order down-parabola flow at $y_1^* = a_{j_1 j_1 1}$. With

$$\dot{x}_{j_2}^* = a_{j_2 j_2 0}(a_{j_1 j_1 1} - a_{j_2 j_2 1})^3,$$

the third-order parabola flows are positive and negative for $\dot{x}_{j_2}^* > 0$ and $\dot{x}_{j_2}^* < 0$, respectively.

Therefore, the flow of $y_{j_1}^* = a_{j_1 j_1 1} \neq a_{j_2 j_2 1}$ with $\dot{x}_{j_2}^* = a_{j_2 j_2 0}(a_{j_1 j_1 1} - a_{j_2 j_2 1})^2$ for the x_{j_1}-direction has the following properties as in Eqs. (4.203)–(4.206):

- For $a_{j_1 j_1 0}\alpha_{j_1 j_2}^3 > 0$ and $a_{j_2 j_2 0}(a_{j_1 j_1 1} - a_{j_2 j_2 l_2})^3 > 0$,
$$(a_{j_1 j_1 1}, \dot{x}_{j_2}^*) = \underbrace{(3^{\mathrm{rd}}\mathrm{UP}, \mathrm{pF})}_{\text{third-order up-parabola flow }(+)}.$$
- For $a_{j_1 j_1 0}\alpha_{j_1 j_2}^3 < 0$ and $a_{j_2 j_2 0}(a_{j_1 j_1 1} - a_{j_2 j_2 1})^3 > 0$,
$$(a_{j_1 j_1 1}, \dot{x}_{j_2}^*) = \underbrace{(3^{\mathrm{rd}}\mathrm{DP}, \mathrm{pF})}_{\text{third-order down-parabola flow }(+)}.$$
- For $a_{j_1 j_1 0}\alpha_{j_1 j_2}^3 > 0$ and $a_{j_2 j_2 0}(a_{j_1 j_1 1} - a_{j_2 j_2 1})^3 < 0$,
$$(a_{j_1 j_1 1}, \dot{x}_{j_2}^*) = \underbrace{(3^{\mathrm{rd}}\mathrm{DP}, \mathrm{nF})}_{\text{third-order down-parabola flow }(-)}.$$

- For $a_{j_1j_10}\alpha^3_{j_1j_2} < 0$ and $a_{j_2j_20}(a_{j_1j_11} - a_{j_2j_21})^3 < 0$,

$$(a_{j_1j_11}, \dot{x}^*_{j_2}) = \underbrace{(3^{\text{rd}}\text{UP}, \text{nF})}_{\text{third-order up-parabola flow } (-)}.$$

- The third-order up-parabola flows are the appearing bifurcations for up-parabola, down-parabola, and up-parabola flows.
- The third-order down-parabola flows are the appearing bifurcations for down-parabola, up-parabola, and down-parabola flows.

($\text{v}_{4\text{b}}$) For $y^*_{j_1} = a_{j_2j_21} \neq a_{j_1j_11}$,

$$\left.\frac{dx_{j_2}}{dx_{j_1}}\right|_{y^*_1 = a_{j_2j_21}} = \left.\frac{a_{j_2j_20}(y_{j_1} - a_{j_2j_21})^3}{a_{j_1j_10}(y_{j_1} - a_{j_1j_11})^3}\right|_{y^*_1 = a_{j_2j_21}} = 0,$$

$$\left.\frac{d^2x_{j_2}}{dx^2_{j_1}}\right|_{y^*_1 = a_{j_2j_21}} = \left.\frac{d^3x_{j_2}}{dx^3_{j_1}}\right|_{y^*_1 = a_{j_2j_21}} = 0.$$

If

$$\left.\frac{d^4x_{j_2}}{dx^4_{j_1}}\right|_{y^*_1 = a_{j_2j_21}} = \frac{6a_{j_2j_20}\alpha^3_{j_1j_1}}{a_{j_1j_10}(a_{j_2j_21} - a_{j_1j_11})^3} > 0,$$

there is a third-order up-parabola flow at $y^*_1 = a_{j_2j_21}$. If

$$\left.\frac{d^4x_{j_2}}{dx^4_{j_1}}\right|_{y^*_1 = a_{j_2j_21}} = \frac{6a_{j_2j_20}\alpha^3_{j_1j_1}}{a_{j_1j_10}(a_{j_2j_21} - a_{j_1j_11})^3} < 0,$$

there is a third-order down-parabola flow at $y^*_1 = a_{j_2j_21}$. With

$$\dot{x}^*_{j_1} = a_{j_1j_10}(a_{j_2j_21} - a_{j_1j_11})^3,$$

the third-order parabola flows are positive and negative for $\dot{x}^*_{j_1} > 0$ and $\dot{x}^*_{j_1} < 0$, respectively.

Therefore, the flow of $y^*_{j_1} = a_{j_2j_21} \neq a_{j_1j_11}$ for the x_{j_2}-direction has the following properties with $\dot{x}^*_{j_1} = a_{j_1j_10}(a_{j_2j_21} - a_{j_1j_11})^3$ as in Eqs. (4.207)–(4.210):

- For $a_{j_1j_10}(a_{j_2j_21} - a_{j_1j_11})^3 > 0$ and $a_{j_2j_20}\alpha^3_{j_1j_1} > 0$,

$$(\dot{x}^*_{j_1}, a_{j_2j_21}) = \underbrace{(\text{pF}, 3^{\text{rd}}\text{UP})}_{\text{third-order up-parabola flow } (+)}.$$

- For $a_{j_1j_10}(a_{j_2j_21}-a_{j_1j_11})^3<0$ and $a_{j_2j_20}\alpha^3_{j_1j_1}>0$,

$$(\dot{x}^*_{j_1}, a_{j_2j_21}) = \underbrace{(\text{nF}, 3^{\text{rd}}\text{DP})}_{\text{third-order down-parabola flow } (-)}.$$

- For $a_{j_1j_10}(a_{j_2j_21}-a_{j_1j_11})^3>0$ and $a_{j_2j_20}\alpha^3_{j_1j_1}<0$,

$$(\dot{x}^*_{j_1}, a_{j_2j_2l_1}) = \underbrace{(\text{pF}, 3^{\text{rd}}\text{DP})}_{\text{third-order down-parabola flow } (+)}.$$

- For $a_{j_1j_10}(a_{j_2j_21}-a_{j_1j_11})^3<0$ and $a_{j_2j_20}\alpha^3_{j_1j_1}<0$,

$$(\dot{x}^*_{j_1}, a_{j_2j_21}) = \underbrace{(\text{nF}, 3^{\text{rd}}\text{UP})}_{\text{third-order up-parabola flow } (+)}.$$

- The third-order up-parabola flows are the appearing bifurcations from an up-parabola flow to up-parabola, down-parabola, and up-parabola flows.
- The third-order down-parabola flows are the appearing bifurcations from a down-parabola flow to down-parabola, up-parabola, and down-parabola flows.

(v_5) For $a_{j_1j_1s_1}=a_{j_2j_2l_1}$, the dynamical system becomes

$$\begin{aligned}
\dot{x}_{j_1} &= a_{j_1j_10}(y_{j_1}-a_{j_1j_1s_1})(y_{j_1}-a_{j_1j_1s_2})^2,\\
\dot{x}_{j_2} &= a_{j_2j_20}(y_{j_1}-a_{j_2j_2l_1})(y_{j_1}-a_{j_2j_2l_2})^2.
\end{aligned}$$

A dynamical system in the y_{j_1}-direction is constructed as

$$\dot{y}_{j_1} = (y_{j_1}-a_{j_1j_1s_1})[\alpha_{j_1j_1}a_{j_1j_10}(y_{j_1}-a_{j_1j_1s_2})^2+\alpha_{j_1j_2}a_{j_2j_20}(y_{j_1}-a_{j_2j_2l_2})^2],$$

and a dynamical system in the $y^{\perp}_{j_1}$-direction is constructed as

$$\dot{y}^{\perp}_{j_1} = (y_{j_1}-a_{j_1j_1s_1})[\beta_{j_1j_1}a_{j_1j_10}(y_{j_1}-a_{j_1j_1s_2})^2+\beta_{j_1j_2}a_{j_2j_20}(y_{j_1}-a_{j_2j_2l_2})^2],$$

where

$$\begin{aligned}
G_{j_1j_10} &= \frac{\partial}{\partial y_{j_1}}[\mathbf{n}^{\text{T}}\cdot\mathbf{F}(y_{j_1},\mathbf{p})]\Big|_{y^*_{j_1}=a_{j_1j_1s_1}=a_{j_2j_2l_1}}\\
&= \alpha_{j_1j_1}a_{j_1j_10}(a_{j_1j_1s_1}-a_{j_1j_1s_2})^2+\alpha_{j_1j_2}a_{j_2j_20}(a_{j_2j_2l_1}-a_{j_2j_2l_2})^2.
\end{aligned}$$

In phase space, at $y_1^* = a_{j_1j_1s_1} = a_{j_2j_2l_1}$, if

$$\left.\frac{dx_{j_1}}{dx_{j_2}}\right|_{y_1^*=a_{j_1j_1s_1}} = \left.\frac{a_{j_1j_10}(a_{j_1j_1s_1} - a_{j_1j_1s_2})^2}{a_{j_2j_20}(a_{j_2j_2l_1} - a_{j_2j_2l_2})^2}\right|_{y_1^*=a_{j_1j_1s_1}} > 0,$$

there is an increasing-inflection equilibrium flow at $y_1^* = a_{j_1j_1s_1} = a_{j_2j_2l_1}$, and if

$$\left.\frac{dx_{j_1}}{dx_{j_2}}\right|_{y_1^*=a_{j_1j_1s_1}} = \left.\frac{a_{j_1j_10}(a_{j_1j_1s_1} - a_{j_1j_1s_2})^2}{a_{j_2j_20}(a_{j_2j_2l_1} - a_{j_2j_2l_2})^2}\right|_{y_1^*=a_{j_1j_1s_1}} < 0,$$

there is an increasing-inflection equilibrium flow at $y_1^* = a_{j_1j_1s_1} = a_{j_2j_2l_1}$. Since the corresponding variational equation is

$$\begin{aligned}\Delta\dot{y}_{j_1} &= [\alpha_{j_1j_1}a_{j_1j_10}(a_{j_1j_1s_1} - a_{j_1j_1s_2})^2 + \alpha_{j_1j_2}a_{j_2j_20}(a_{j_2j_2l_1} - a_{j_2j_2l_2})^2]\Delta y_{j_1}\\ &= G_{j_1j_10}\Delta y_{j_1},\end{aligned}$$

the inflection flows are source and sink $\dot{x}_{j_2}^* > 0$ and $\dot{x}_{j_2}^* < 0$, respectively.

Therefore, the infinite-equilibrium flow of $y_1^* = a_{j_1j_1s_1} = a_{j_2j_2l_1}$ possesses the following properties as in Eqs. (4.215)–(4.218):

- For $a_{j_1j_10}(a_{j_1j_1s_1} - a_{j_1j_1s_2})^2/a_{j_2j_20}(a_{j_2j_2l_1} - a_{j_2j_2l_2})^2 > 0$ and $G_{j_1j_10} > 0$,

$$(a_{j_1j_1s_1}, a_{j_2j_2l_1}) = \underbrace{(\text{II}, \text{SO})}_{\text{increasing-inflection source}}.$$

- For $a_{j_1j_10}(a_{j_1j_1s_1} - a_{j_1j_1s_2})^2/a_{j_2j_20}(a_{j_2j_2l_1} - a_{j_2j_2l_2})^2 > 0$ and $G_{j_1j_10} < 0$,

$$(a_{j_1j_1s_1}, a_{j_2j_2l_1}) = \underbrace{(\text{II}, \text{SI})}_{\text{increasing-inflection sink}}.$$

- For $a_{j_1j_10}(a_{j_1j_1s_1} - a_{j_1j_1s_2})^2/a_{j_2j_20}(a_{j_2j_2l_1} - a_{j_2j_2l_2})^2 < 0$ and $G_{j_1j_10} > 0$,

$$(a_{j_1j_1s_1}, a_{j_2j_2l_1}) = \underbrace{(\text{DI}, \text{SO})}_{\text{decreasing-inflection source}}.$$

- For $a_{j_1j_10}(a_{j_1j_1s_1} - a_{j_1j_1s_2})^2/a_{j_2j_20}(a_{j_2j_2l_1} - a_{j_2j_2l_2})^2 < 0$ and $G_{j_1j_10} < 0$,

$$(a_{j_1j_1s_1}, a_{j_2j_2l_1}) = \underbrace{(\text{DI}, \text{SI})}_{\text{decreasing-inflection sink}}.$$

(v$_6$) For $a_{j_1j_1s_2} = a_{j_2j_2l_1}$, the dynamical system becomes

$$\dot{x}_{j_1} = a_{j_1j_10}(y_{j_1} - a_{j_1j_1s_1})(y_{j_1} - a_{j_1j_1s_2})^2,$$
$$\dot{x}_{j_2} = a_{j_2j_20}(y_{j_1} - a_{j_2j_2l_1})(y_{j_1} - a_{j_2j_2l_2})^2.$$

There is a dynamical system in the y_{j_1}-direction as

$$\dot{y}_{j_1} = (y_{j_1} - a_{j_1j_1s_2})[\alpha_{j_1j_1}a_{j_1j_10}(y_{j_1} - a_{j_1j_1s_1})(y_{j_1} - a_{j_1j_1s_2}) + \alpha_{j_1j_2}a_{j_2j_20}(y_{j_1} - a_{j_2j_2l_2})^2],$$

with a dynamical system in the $y_{j_1}^{\perp}$-direction as

$$\dot{y}_{j_1}^{\perp} = (y_{j_1} - a_{j_1j_1s_2})[\beta_{j_1j_1}a_{j_1j_10}(y_{j_1} - a_{j_1j_1s_1})(y_{j_1} - a_{j_1j_1s_2}) + \beta_{j_1j_2}a_{j_2j_20}(y_{j_1} - a_{j_2j_2l_2})^2],$$

where

$$G_{j_1j_10} = \frac{\partial}{\partial y_{j_1}}[\mathbf{n}^{\mathrm{T}} \cdot \mathbf{F}(y_{j_1}, \mathbf{p})]\Big|_{y_{j_1}^* = a_{j_1j_1s_2} = a_{j_2j_2l_1}} = \alpha_{j_1j_2}a_{j_2j_20}(a_{j_2j_2l_1} - a_{j_2j_2l_2})^2.$$

In phase space, at $y_1^* = a_{j_1j_1s_2} = a_{j_2j_2l_1}$,

$$\frac{dx_{j_1}}{dx_{j_2}}\Big|_{y_1^* = a_{j_1j_1s_2}} = \frac{a_{j_1j_10}(y_{j_1} - a_{j_1j_1s_1})(y_{j_1} - a_{j_1j_1s_2})}{a_{j_2j_20}(y_{j_1} - a_{j_2j_2l_2})^2}\Big|_{y_1^* = a_{j_1j_1s_2}} = 0.$$

If

$$\frac{d^2x_{j_1}}{dx_{j_2}^2}\Big|_{y_1^* = a_{j_1j_1s_2}} = \frac{2a_{j_1j_10}\alpha_{j_1j_2}^2(a_{j_1j_1s_2} - a_{j_1j_1s_1})}{a_{j_2j_20}\alpha_{j_1j_2}(a_{j_1j_1s_2} - a_{j_2j_2l_2})^2} > 0,$$

there is an up-parabola flow at $y_1^* = a_{j_1j_1s_2}$. If

$$\frac{d^2x_{j_1}}{dx_{j_2}^2}\Big|_{y_1^* = a_{j_1j_1s_2}} = \frac{2a_{j_1j_10}\alpha_{j_1j_2}^2(a_{j_1j_1s_2} - a_{j_1j_1s_1})}{a_{j_2j_20}\alpha_{j_1j_2}(a_{j_1j_1s_2} - a_{j_2j_2l_2})^2} < 0,$$

there is a down-parabola flow at $y_1^* = a_{j_1j_1s_2}$. Since a corresponding variational equation is

$$\Delta\dot{y}_{j_1} = \alpha_{j_1j_2}a_{j_2j_20}(a_{j_2j_2l_1} - a_{j_2j_2l_2})^2\Delta y_{j_1} = G_{j_1j_10}\Delta y_{j_1},$$

the parabola flows are source and sink for $G_{j_1j_10} > 0$ and $G_{j_1j_10} < 0$, respectively.

Therefore, the infinite-equilibrium flow of $y_1^* = a_{j_1j_1s_2} = a_{j_2j_2l_1}$ possesses the following properties as in Eqs. (4.223)–(4.226):

- For $\alpha_{j_1j_2}^2 a_{j_1j_10}(a_{j_1j_1s_1} - a_{j_1j_1s_2}) > 0$ and $\alpha_{j_1j_2} a_{j_2j_20}(a_{j_2j_2l_1} - a_{j_2j_2l_2})^2 > 0$,

$$(a_{j_1j_1s_1}, a_{j_2j_2l_1}) = \underbrace{(\text{UP}, \text{SO})}_{\text{up-parabola source}}.$$

- For $\alpha_{j_1j_2}^2 a_{j_1j_10}(a_{j_1j_1s_1} - a_{j_1j_1s_2}) < 0$ and $\alpha_{j_1j_2} a_{j_2j_20}(a_{j_2j_2l_1} - a_{j_2j_2l_2})^2 > 0$,

$$(a_{j_1j_1s_1}, a_{j_2j_2l_1}) = \underbrace{(\text{DP}, \text{SO})}_{\text{down-parabola source}}.$$

- For $\alpha_{j_1j_2}^2 a_{j_1j_10}(a_{j_1j_1s_1} - a_{j_1j_1s_2}) > 0$ and $\alpha_{j_1j_2} a_{j_2j_20}(a_{j_2j_2l_1} - a_{j_2j_2l_2})^2 < 0$,

$$(a_{j_1j_1s_1}, a_{j_2j_2l_1}) = \underbrace{(\text{DP}, \text{SI})}_{\text{down-parabola sink}}.$$

- For $\alpha_{j_1j_2}^2 a_{j_1j_10}(a_{j_1j_1s_1} - a_{j_1j_1s_2}) < 0$ and $\alpha_{j_1j_2} a_{j_2j_20}(a_{j_2j_2l_1} - a_{j_2j_2l_2})^2 < 0$,

$$(a_{j_1j_1s_1}, a_{j_2j_2l_1}) = \underbrace{(\text{UP}, \text{SI})}_{\text{up-parabola sink}}.$$

(v_7) For $a_{j_1j_1s_1} = a_{j_2j_2l_2}$, the dynamical system becomes

$$\begin{aligned}\dot{x}_{j_1} &= a_{j_1j_10}(y_{j_1} - a_{j_1j_1s_1})(y_{j_1} - a_{j_1j_1s_2})^2,\\ \dot{x}_{j_2} &= a_{j_2j_20}(y_{j_1} - a_{j_2j_2l_1})(y_{j_1} - a_{j_2j_2l_2})^2.\end{aligned}$$

A dynamical system in the y_{j_1}-direction is constructed as

$$\begin{aligned}\dot{y}_{j_1} = (y_{j_1} - a_{j_1j_1s_1})[&\alpha_{j_1j_1} a_{j_1j_10}(y_{j_1} - a_{j_1j_1s_2})^2\\ &+ \alpha_{j_1j_2} a_{j_2j_20}(y_{j_1} - a_{j_2j_2l_1})(y_{j_1} - a_{j_2j_2l_2})],\end{aligned}$$

and a dynamical system in the $y_{j_1}^{\perp}$-direction is constructed as

$$\begin{aligned}\dot{y}_{j_1}^{\perp} = (y_{j_1} - a_{j_1j_1s_1})[&\beta_{j_1j_1} a_{j_1j_10}(y_{j_1} - a_{j_1j_1s_2})^2\\ &+ \beta_{j_1j_2} a_{j_2j_20}(y_{j_1} - a_{j_2j_2l_1})(y_{j_1} - a_{j_2j_2l_2})],\end{aligned}$$

and a G-function is defined as

$$G_{j_1j_10} = \frac{\partial}{\partial y_{j_1}}[\mathbf{n}^{\mathrm{T}} \cdot \mathbf{F}(y_{j_1}, \mathbf{p})]\Big|_{y_{j_1}^* = a_{j_1j_1s_1} = a_{j_2j_2l_2}}$$
$$= \alpha_{j_1j_1}a_{j_1j_10}(a_{j_1j_1s_1} - a_{j_1j_1s_2})^2.$$

In phase space, at $y_1^* = a_{j_1j_1s_1} = a_{j_2j_2l_2}$,

$$\frac{dx_{j_2}}{dx_{j_1}}\Big|_{y_1^* = a_{j_2j_2l_2}} = \frac{a_{j_2j_20}(y_{j_1} - a_{j_2j_2l_1})(y_{j_1} - a_{j_2j_2l_2})}{a_{j_1j_10}(y_{j_1} - a_{j_1j_1s_2})^2}\Big|_{y_1^* = a_{j_2j_2l_2}} = 0.$$

If

$$\frac{d^2x_{j_2}}{dx_{j_1}^2}\Big|_{y_1^* = a_{j_2j_2l_2}} = \frac{2a_{j_2j_20}\alpha_{j_1j_1}^2(a_{j_2j_2l_2} - a_{j_2j_2l_1})}{a_{j_1j_10}\alpha_{j_1j_1}a_{j_2j_2l_2} - a_{j_1j_1s_1})(a_{j_2j_2l_2} - a_{j_1j_1s_2})^2} > 0,$$

there is an up-parabola flow at $y_1^* = a_{j_1j_1s_1} = a_{j_2j_2l_2}$. If

$$\frac{d^2x_{j_2}}{dx_{j_1}^2}\Big|_{y_1^* = a_{j_2j_2l_2}} = \frac{2a_{j_2j_20}\alpha_{j_1j_1}^2(a_{j_2j_2l_2} - a_{j_2j_2l_1})}{a_{j_1j_10}\alpha_{j_1j_1}a_{j_2j_2l_2} - a_{j_1j_1s_1})(a_{j_2j_2l_2} - a_{j_1j_1s_2})^2} < 0,$$

there is a down-parabola flow at $y_1^* = a_{j_1j_1s_1} = a_{j_2j_2l_2}$. Since the corresponding variational equation is

$$\Delta\dot{y}_{j_1} = \alpha_{j_1j_1}a_{j_1j_10}(a_{j_1j_1s_1} - a_{j_1j_1s_2})^2\Delta y_{j_1} = G_{j_1j_10}\Delta y_{j_1},$$

the parabola flows are source and sink for $G_{j_1j_10} > 0$ and $G_{j_1j_10} < 0$, respectively.

Therefore, the infinite-equilibrium flow of $y_1^* = a_{j_1j_1s_1} = a_{j_2j_2l_2}$ possesses the following properties as in Eqs. (4.231)–(4.234):

- For $\alpha_{j_1j_1}a_{j_1j_10}(a_{j_1j_1s_1} - a_{j_1j_1s_2})^2 > 0$ and $a_{j_2j_20}\alpha_{j_1j_1}^2(a_{j_2j_2l_2} - a_{j_2j_2l_1}) > 0$,

$$(a_{j_1j_1s_1}, a_{j_2j_2l_2}) = \underbrace{(\mathrm{SO}, \mathrm{UP})}_{\text{up-parabola source}}.$$

- For $\alpha_{j_1j_1}a_{j_1j_10}(a_{j_1j_1s_1} - a_{j_1j_1s_2})^2 < 0$ and $a_{j_2j_20}\alpha_{j_1j_1}^2(a_{j_2j_2l_2} - a_{j_2j_2l_1}) > 0$,

$$(a_{j_1j_1s_1}, a_{j_2j_2l_2}) = \underbrace{(\mathrm{SI}, \mathrm{DP})}_{\text{down-parabola sink}}.$$

- For $\alpha_{j_1j_1}a_{j_1j_10}(a_{j_1j_1s_1}-a_{j_1j_1s_2})^2>0$ and $a_{j_2j_20}\alpha_{j_1j_1}^2(a_{j_2j_2l_2}-a_{j_2j_2l_1})<0$,

$$(a_{j_1j_1s_1},a_{j_2j_2l_2})=\underbrace{(\mathrm{SO},\mathrm{DP})}_{\text{down-parabola source}}.$$

- For $\alpha_{j_1j_1}a_{j_1j_10}(a_{j_1j_1s_1}-a_{j_1j_1s_2})^2<0$ and $a_{j_2j_20}\alpha_{j_1j_1}^2(a_{j_2j_2l_2}-a_{j_2j_2l_1})<0$,

$$(a_{j_1j_1s_1},a_{j_2j_2l_2})=\underbrace{(\mathrm{SI},\mathrm{UP})}_{\text{up-parabola sink}}.$$

(v_8) For $a_{j_1j_1s_2}=a_{j_2j_2l_2}$, the dynamical system becomes

$$\begin{aligned}
\dot{x}_{j_1}&=a_{j_1j_10}(y_{j_1}-a_{j_1j_1s_1})(y_{j_1}-a_{j_1j_1s_2})^2,\\
\dot{x}_{j_2}&=a_{j_2j_20}(y_{j_1}-a_{j_2j_2l_1})(y_{j_1}-a_{j_2j_2l_2})^2,\\
&j_1,j_2\in\{1,2\};\ j_1\neq j_2.
\end{aligned}$$

A dynamical system in the y_{j_1}-direction is constructed as

$$\begin{aligned}
\dot{y}_{j_1}=(y_{j_1}-a_{j_1j_1s_2})^2[&\alpha_{j_1j_1}a_{j_1j_10}(y_{j_1}-a_{j_1j_1s_1})\\
&+\alpha_{j_1j_2}a_{j_2j_20}(y_{j_1}-a_{j_2j_2l_1})],
\end{aligned}$$

and a dynamical system in the $y_{j_1}^{\perp}$-direction is constructed as

$$\begin{aligned}
\dot{y}_{j_1}^{\perp}=(y_{j_1}-a_{j_1j_1s_2})^2[&\beta_{j_1j_1}a_{j_1j_10}(y_{j_1}-a_{j_1j_1s_1})\\
&+\beta_{j_1j_2}a_{j_2j_20}(y_{j_1}-a_{j_2j_2l_1})],
\end{aligned}$$

and a G-function is defined as

$$\begin{aligned}
G_{j_1j_10}&=\frac{\partial}{\partial y_{j_1}}[\mathbf{n}^{\mathrm{T}}\cdot\mathbf{F}(y_{j_1},\mathbf{p})]\bigg|_{y_{j_1}^*=a_{j_1j_1s_2}=a_{j_2j_2l_2}}=0,\\
G_{j_1j_10}^{(1)}&=\frac{1}{2!}\frac{\partial^2}{\partial y_{j_1}^2}[\mathbf{n}^{\mathrm{T}}\cdot\mathbf{F}(y_{j_1},\mathbf{p})]\bigg|_{y_{j_1}^*=a_{j_1j_1s_2}=a_{j_2j_2l_2}}\\
&=\alpha_{j_1j_1}a_{j_1j_10}(a_{j_1j_1s_2}-a_{j_1j_1s_1})+\alpha_{j_1j_2}a_{j_2j_20}(a_{j_2j_2l_2}-a_{j_2j_2l_1}).
\end{aligned}$$

In phase space, at $y_{j_1}^*=a_{j_1j_1s_2}=a_{j_2j_2l_2}$, if

$$\frac{dx_{j_1}}{dx_{j_2}}\bigg|_{y_1^*=a_{j_1j_1s_2}}=\frac{a_{j_1j_10}(a_{j_1j_1s_2}-a_{j_1j_1s_1})}{a_{j_2j_20}(a_{j_2j_2l_2}-a_{j_2j_2l_1})}>0,$$

there is an increasing-inflection flow at $y_{j_1}^*=a_{j_1j_1s_2}=a_{j_2j_2l_2}$, and if

$$\frac{dx_{j_1}}{dx_{j_2}}\bigg|_{y_1^*=a_{j_1j_1s_2}}=\frac{a_{j_1j_10}(a_{j_1j_1s_2}-a_{j_1j_1s_1})}{a_{j_2j_20}(a_{j_2j_2l_2}-a_{j_2j_2l_1})}<0,$$

there is a decreasing-inflection flow at $y_{j_1}^* = a_{j_1j_1s_2} = a_{j_2j_2l_2}$. Since the corresponding variational equation is

$$\begin{aligned}\Delta\dot{y}_{j_1} &= [\alpha_{j_1j_1}a_{j_1j_10}(a_{j_1j_1s_2} - a_{j_1j_1s_1}) + \alpha_{j_1j_2}a_{j_2j_20}(a_{j_2j_2l_2} - a_{j_2j_2l_1})](\Delta y_{j_1})^2 \\ &= G^{(1)}_{j_1j_10}(\Delta y_{j_1})^2,\end{aligned}$$

the parabola flows are upper-saddle and lower-saddle for $G_{j_1j_10} > 0$ and $G_{j_1j_10} < 0$, respectively.

Therefore, the infinite-equilibrium of $y_{j_1}^* = a_{j_1j_1s_2} = a_{j_2j_2l_2}$ has the following properties as in Eqs. (4.239)–(4.242):

- For $a_{j_1j_10}(a_{j_1j_1s_2} - a_{j_1j_1s_1})/a_{j_2j_20}(a_{j_2j_2l_2} - a_{j_2j_2l_1}) > 0$ and $G^{(1)}_{j_1j_10} > 0$,
$$(a_{j_1j_1s_2}, a_{j_2j_21}) = \underbrace{(\text{II}, \text{US})}_{\text{increasing-inflection upper-saddle}}.$$
- For $a_{j_1j_10}(a_{j_1j_1s_2} - a_{j_1j_1s_1})/a_{j_2j_20}(a_{j_2j_2l_2} - a_{j_2j_2l_1}) > 0$ and $G^{(1)}_{j_1j_10} < 0$,
$$(a_{j_1j_1s_2}, a_{j_2j_21}) = \underbrace{(\text{II}, \text{LS})}_{\text{increasing-inflection lower-saddle}}.$$
- For $a_{j_1j_10}(a_{j_1j_1s_2} - a_{j_1j_1s_1})/a_{j_2j_20}(a_{j_2j_2l_2} - a_{j_2j_2l_1}) < 0$ and $G^{(1)}_{j_1j_10} > 0$,
$$(a_{j_1j_1s_2}, a_{j_2j_21}) = \underbrace{(\text{DI}, \text{US})}_{\text{decreasing-inflection upper-saddle}}.$$
- For $a_{j_1j_10}(a_{j_1j_1s_2} - a_{j_1j_1s_1})/a_{j_2j_20}(a_{j_2j_2l_2} - a_{j_2j_2l_1}) < 0$ and $G^{(1)}_{j_1j_10} < 0$,
$$(a_{j_1j_1s_2}, a_{j_2j_21}) = \underbrace{(\text{DI}, \text{LS})}_{\text{decreasing-inflection lower-saddle}}.$$

– The inflection-saddle infinite-equilibriums are the appearing bifurcations for the up-parabola and down-parabola flows in the x_{j_1}- and x_{j_2}-directions.
– The inflection-saddle infinite-equilibriums are the double-switching bifurcations for the up-parabola and down-parabola flows in the x_{j_1}- and x_{j_2}-directions.

(v_9) For $a_{j_1j_1s_1} = a_{j_1j_1s_2} \equiv a_{j_1j_11}$ and $a_{j_1j_11} = a_{j_2j_2l_1}$, the dynamical system becomes

$$\begin{aligned}\dot{x}_{j_1} &= a_{j_1j_10}(y_{j_1} - a_{j_1j_11})^3, \\ \dot{x}_{j_2} &= a_{j_2j_20}(y_{j_1} - a_{j_2j_2l_1})(y_{j_1} - a_{j_2j_2l_2})^2, \\ j_1, j_2 &\in \{1,2\};\ j_1 \neq j_2.\end{aligned}$$

For $a_{j_1j_11} = a_{j_2j_2l_1}$, a dynamical system in the y_{j_1}-direction is constructed as

$$\dot{y}_{j_1} = (y_{j_1} - a_{j_2j_2l_1})[\alpha_{j_1j_1}a_{j_1j_10}(y_{j_1} - a_{j_1j_11})^2 + \alpha_{j_1j_2}a_{j_2j_20}(y_{j_1} - a_{j_2j_2l_2})^2],$$

and a dynamical system in the $y_{j_1}^{\perp}$-direction is constructed as

$$\dot{y}_{j_1}^{\perp} = (y_{j_1} - a_{j_2j_2l_1})[\beta_{j_1j_1}a_{j_1j_10}(y_{j_1} - a_{j_1j_11})^2 + \beta_{j_1j_2}a_{j_2j_20}(y_{j_1} - a_{j_2j_2l_2})^2],$$

and a G-function is defined as

$$\begin{aligned} G_{j_1j_10} &= \frac{\partial}{\partial y_{j_1}}[\mathbf{n}^{\mathrm{T}} \cdot \mathbf{F}(y_{j_1}, \mathbf{p})]\bigg|_{y_{j_1}^* = a_{j_1j_11} = a_{j_2j_2l_1}} \\ &= \alpha_{j_1j_2}a_{j_2j_20}(a_{j_2j_2l_1} - a_{j_2j_2l_2})^2. \end{aligned}$$

In phase space, at $y_1^* = a_{j_1j_11} = a_{j_2j_2l_1}$,

$$\frac{dx_{j_1}}{dx_{j_2}}\bigg|_{y_1^* = a_{j_1j_11}} = \frac{a_{j_1j_10}(y_{j_1} - a_{j_1j_11})^2}{a_{j_2j_20}(y_{j_1} - a_{j_2j_2l_2})^2}\bigg|_{y_1^* = a_{j_1j_11}} = 0,$$

$$\frac{d^2x_{j_1}}{dx_{j_2}^2}\bigg|_{y_1^* = a_{j_1j_11}} = 0.$$

If

$$\frac{d^3x_{j_1}}{dx_{j_2}^3}\bigg|_{y_1^* = a_{j_1j_11}} = \frac{2a_{j_1j_10}\alpha_{j_1j_2}^3}{a_{j_2j_20}\alpha_{j_1j_2}(a_{j_1j_11} - a_{j_2j_2l_2})^2} > 0,$$

there is a second-order increasing-inflection flow at $y_1^* = a_{j_1j_11} = a_{j_2j_2l_1}$. If

$$\frac{d^3x_{j_1}}{dx_{j_2}^3}\bigg|_{y_1^* = a_{j_1j_11}} = \frac{2a_{j_1j_10}\alpha_{j_1j_2}^3}{a_{j_2j_20}\alpha_{j_1j_2}(a_{j_1j_11} - a_{j_2j_2l_2})^2} < 0,$$

there is a second-order inflection flow at $y_1^* = a_{j_1j_11} = a_{j_2j_2l_1}$. Since the corresponding variational equation is

$$\Delta\dot{y}_{j_1} = \alpha_{j_1j_2}a_{j_2j_20}(a_{j_2j_2l_1} - a_{j_2j_2l_2})^2\Delta y_{j_1} = G_{j_1j_10}\Delta y_{j_1},$$

the second-order inflection flows are source and sink for $G_{j_1j_10} > 0$ and $G_{j_1j_10} < 0$, respectively.

Therefore, the infinite-equilibrium flow of $y_{j_1}^* = a_{j_1j_11} = a_{j_2j_2l_1}$ has the following properties as in Eqs. (4.247)–(4.250):

- For $a_{j_1j_10}\alpha_{j_1j_2}^3 > 0$ and $\alpha_{j_1j_2}a_{j_2j_20}(a_{j_2j_2l_1} - a_{j_2j_2l_2})^2 > 0$,

$$(a_{j_1j_11}, a_{j_2j_2l_1}) = \underbrace{(2^{\mathrm{nd}}\mathrm{II}, \mathrm{SO})}_{\text{second-order increasing-inflection source}}.$$

- For $a_{j_1j_10}\alpha^3_{j_1j_2} < 0$ and $\alpha_{j_1j_2}a_{j_2j_20}(a_{j_2j_2l_1} - a_{j_2j_2l_2})^2 > 0$,

$$(a_{j_1j_11}, a_{j_2j_2l_1}) = \underbrace{(2^{\text{nd}}\text{DI}, \text{SO})}_{\text{second-order decreasing-inflection source}}.$$

- For $a_{j_1j_10}\alpha^3_{j_1j_2} > 0$ and $\alpha_{j_1j_2}a_{j_2j_20}(a_{j_2j_2l_1} - a_{j_2j_2l_2})^2 < 0$,

$$(a_{j_1j_11}, a_{j_2j_2l_1}) = \underbrace{(2^{\text{nd}}\text{DI}, \text{SI})}_{\text{second-order decreasing-inflection sink}}.$$

- For $a_{j_1j_10}\alpha^3_{j_1j_2} < 0$ and $\alpha_{j_1j_2}a_{j_2j_20}(a_{j_2j_2l_1} - a_{j_2j_2l_2})^2 < 0$,

$$(a_{j_1j_11}, a_{j_2j_2l_1}) = \underbrace{(2^{\text{nd}}\text{II}, \text{SI})}_{\text{second-order increasing-inflection sink}}.$$

The third-order inflection infinite-equilibrium flows are the switching bifurcations for up-parabola, down-parabola, and up-parabola flows.

(v_{10}) For $a_{j_1j_1s_1} = a_{j_1j_1s_2} \equiv a_{j_1j_11}$ and $a_{j_1j_11} = a_{j_2j_2l_2}$,

$$\begin{aligned}
&\dot{x}_{j_1} = a_{j_1j_10}(y_{j_1} - a_{j_1j_11})^3,\\
&\dot{x}_{j_2} = a_{j_2j_20}(y_{j_1} - a_{j_2j_2l_1})(y_{j_1} - a_{j_2j_2l_2})^2,\\
&j_1, j_2 \in \{1,2\}; j_1 \neq j_2.
\end{aligned}$$

For $a_{j_1j_11} = a_{j_2j_2l_2}$, a dynamical system in the y_{j_1}-direction is constructed as

$$\begin{aligned}
\dot{y}_{j_1} = (y_{j_1} - a_{j_2j_2l_2})^2[&\alpha_{j_1j_1}a_{j_1j_10}(y_{j_1} - a_{j_1j_11})\\
&+ \alpha_{j_1j_2}a_{j_2j_20}(y_{j_1} - a_{j_2j_2l_1})],
\end{aligned}$$

and a dynamical system in the $y^{\perp}_{j_1}$-direction is constructed as

$$\begin{aligned}
\dot{y}^{\perp}_{j_1} = (y_{j_1} - a_{j_2j_2l_2})^2[&\beta_{j_1j_1}a_{j_1j_10}(y_{j_1} - a_{j_1j_11})\\
&+ \beta_{j_1j_2}a_{j_2j_20}(y_{j_1} - a_{j_2j_2l_1})],
\end{aligned}$$

and a G-function is defined as

$$\begin{aligned}
G_{j_1j_10} &= \frac{\partial}{\partial y_{j_1}}[\mathbf{n}^{\mathrm{T}} \cdot \mathbf{F}(y_{j_1}, \mathbf{p})]\bigg|_{y^*_{j_1} = a_{j_1j_11} = a_{j_2j_2l_2}} = 0,\\
G^{(1)}_{j_1j_11} &= \frac{1}{2!}\frac{\partial^2}{\partial y^2_{j_1}}[\mathbf{n}^{\mathrm{T}} \cdot \mathbf{F}(y_{j_1}, \mathbf{p})]\bigg|_{y^*_{j_1} = a_{j_1j_11} = a_{j_2j_2l_2}}\\
&= \alpha_{j_1j_2}a_{j_2j_20}(a_{j_2j_2l_2} - a_{j_2j_2l_1}).
\end{aligned}$$

In phase space, at $y_1^* = a_{j_1j_11} = a_{j_2j_2l_2}$,

$$\left.\frac{dx_{j_1}}{dx_{j_2}}\right|_{y_1^*=a_{j_1j_11}} = \left.\frac{a_{j_1j_10}(y_{j_1} - a_{j_1j_11})}{a_{j_2j_20}(y_{j_1} - a_{j_2j_2l_1})}\right|_{y_1^*=a_{j_1j_11}} = 0.$$

If

$$\left.\frac{d^2x_{j_1}}{dx_{j_2}^2}\right|_{y_1^*=a_{j_1j_11}} = \frac{6a_{j_1j_10}\alpha_{j_1j_2}^2}{a_{j_2j_20}\alpha_{j_1j_2}(a_{j_1j_11} - a_{j_2j_2l_1})} > 0,$$

there is an up-parabola flow at $y_1^* = a_{j_1j_11} = a_{j_2j_2l_2}$. If

$$\left.\frac{d^2x_{j_1}}{dx_{j_2}^2}\right|_{y_1^*=a_{j_1j_11}} = \frac{6a_{j_1j_10}\alpha_{j_1j_2}^2}{a_{j_2j_20}\alpha_{j_1j_2}(a_{j_1j_11} - a_{j_2j_2l_1})} < 0,$$

there is a down-parabola flow at $y_1^* = a_{j_1j_11} = a_{j_2j_2l_2}$. Since the corresponding variational equation is

$$\dot{y}_{j_1} = \alpha_{j_1j_2}a_{j_2j_20}(a_{j_2j_2l_2} - a_{j_2j_2l_1})(\Delta y_{j_1})^2 = G_{j_1j_10}^{(1)}(\Delta y_{j_1})^2,$$

the parabola flows are upper-saddle and lower-saddle for $G_{j_1j_10}^{(1)} > 0$ and $G_{j_1j_10}^{(1)} < 0$, respectively.

Therefore, the infinite-equilibrium flow of $y_{j_1}^* = a_{j_1j_11} = a_{j_2j_2l_2}$ has the following properties as in Eqs. (4.255)–(4.258):

- For $a_{j_1j_10}\alpha_{j_1j_2}^2 > 0$ and $\alpha_{j_1j_2}a_{j_2j_20}(a_{j_2j_2l_2} - a_{j_2j_2l_1}) > 0$,

$$(a_{j_1j_11}, a_{j_2j_2l_1}) = \underbrace{(\mathrm{UP}, \mathrm{US})}_{\text{up-parabola upper-saddle}}.$$

- For $a_{j_1j_10}\alpha_{j_1j_2}^2 < 0$ and $\alpha_{j_1j_2}a_{j_2j_20}(a_{j_2j_2l_2} - a_{j_2j_2l_1}) > 0$,

$$(a_{j_1j_11}, a_{j_2j_2l_1}) = \underbrace{(\mathrm{DP}, \mathrm{US})}_{\text{down-parabola upper-saddle}}.$$

- For $a_{j_1j_10}\alpha_{j_1j_2}^2 > 0$ and $\alpha_{j_1j_2}a_{j_2j_20}(a_{j_2j_2l_2} - a_{j_2j_2l_1}) < 0$,

$$(a_{j_1j_11}, a_{j_2j_2l_1}) = \underbrace{(\mathrm{DP}, \mathrm{LS})}_{\text{down-parabola lower-saddle}}.$$

- For $a_{j_1j_10}\alpha_{j_1j_2}^2 < 0$ and $\alpha_{j_1j_2}a_{j_2j_20}(a_{j_2j_2l_2} - a_{j_2j_2l_1}) < 0$,

$$(a_{j_1j_11}, a_{j_2j_2l_1}) = \underbrace{(\mathrm{UP}, \mathrm{LS})}_{\text{up-parabola lower-saddle}}.$$

– The parabola-saddle infinite-equilibrium flows are the switching bifurcations for up-parabola, down-parabola, and up-parabola flows.

(v$_{11}$) For $a_{j_2j_2l_1} = a_{j_2j_2l_{\theta 2}} = a_{j_2j_21}$ and $a_{j_1j_1s_1} = a_{j_2j_21}$, the dynamical system becomes

$$\begin{aligned} \dot{x}_{j_1} &= a_{j_1j_10}(y_{j_1} - a_{j_1j_1s_1})(y_{j_1} - a_{j_1j_1s_2})^2, \\ \dot{x}_{j_2} &= a_{j_2j_20}(y_{j_1} - a_{j_2j_21})^3, \\ & j_1, j_2 \in \{1,2\};\ j_1 \neq j_2. \end{aligned}$$

For $a_{j_1j_1s_1} = a_{j_2j_21}$, the dynamical system in the y_{j_1}-direction is constructed as

$$\begin{aligned} \dot{y}_{j_1} &= (y_{j_1} - a_{j_1j_1s_1})\big[\alpha_{j_1j_1}a_{j_1j_10}(y_{j_1} - a_{j_1j_1s_2})^2 \\ &\quad + \alpha_{j_1j_2}a_{j_2j_20}(y_{j_1} - a_{j_2j_21})^2\big], \end{aligned}$$

and a dynamical system in the $y_{j_1}^{\perp}$-direction is constructed as

$$\begin{aligned} \dot{y}_{j_1}^{\perp} &= (y_{j_1} - a_{j_1j_1s_1})\big[\beta_{j_1j_1}a_{j_1j_10}(y_{j_1} - a_{j_1j_1s_2})^2 \\ &\quad + \beta_{j_1j_2}a_{j_2j_20}(y_{j_1} - a_{j_2j_21})^2\big], \end{aligned}$$

and a G-function is defined as

$$\begin{aligned} G_{j_1j_10} &= \frac{\partial}{\partial y_{j_1}}[\mathbf{n}^{\mathrm{T}} \cdot \mathbf{F}(y_{j_1}, \mathbf{p})]\bigg|_{y_{j_1}^* = a_{j_1j_1s_1} = a_{j_2j_21}} \\ &= \alpha_{j_1j_1}a_{j_1j_10}(a_{j_1j_1s_1} - a_{j_1j_1s_2})^2. \end{aligned}$$

For $y_{j_1}^* = a_{j_2j_21} = a_{j_1j_1s_1}$,

$$\begin{aligned} \frac{dx_{j_2}}{dx_{j_1}}\bigg|_{y_1^* = a_{j_2j_21}} &= \frac{a_{j_2j_20}(y_{j_1} - a_{j_2j_21})^2}{a_{j_1j_10}(y_{j_1} - a_{j_1j_1s_2})^2}\bigg|_{y_1^* = a_{j_2j_21}} = 0, \\ \frac{d^2x_{j_2}}{dx_{j_1}^2}\bigg|_{y_1^* = a_{j_2j_21}} &= 0. \end{aligned}$$

If

$$\frac{d^3x_{j_2}}{dx_{j_1}^3}\bigg|_{y_1^* = a_{j_2j_21}} = \frac{2a_{j_2j_20}\alpha_{j_1j_1}^3}{a_{j_1j_10}\alpha_{j_1j_1}(a_{j_2j_21} - a_{j_1j_1s_2})^2} > 0,$$

there is a second-order increasing-inflection flow at $y_{j_1}^* = a_{j_2j_21} = a_{j_1j_1s_1}$. If

$$\frac{d^3x_{j_2}}{dx_{j_1}^3}\bigg|_{y_1^* = a_{j_2j_21}} = \frac{2a_{j_2j_20}\alpha_{j_1j_1}^3}{a_{j_1j_10}\alpha_{j_1j_1}(a_{j_2j_21} - a_{j_1j_1s_2})^2} < 0,$$

there is a second-order increasing-inflection flow at $y^*_{j_1} = a_{j_2j_21} = a_{j_1j_1s_1}$. Since the corresponding variational equation is

$$\Delta\dot{y}_{j_1} = \alpha_{j_1j_1}a_{j_1j_10}(a_{j_1j_1s_1} - a_{j_1j_1s_2})^2\Delta y_{j_1} = G_{j_1j_10}\Delta y_{j_1},$$

the second-order inflection flows are source and sink for $\dot{x}^*_{j_1} > 0$ and $\dot{x}^*_{j_1} < 0$, respectively.

Therefore, the infinite-equilibrium flow of $y^*_{j_1} = a_{j_1j_1s_1} = a_{j_2j_21}$ has the following properties as in Eqs. (4.263)–(4.266):

- For $\alpha_{j_1j_1}a_{j_1j_10}(a_{j_1j_1s_1} - a_{j_1j_1s_2})^2 > 0$ and $a_{j_2j_20}\alpha^3_{j_1j_1} > 0$,
$$(a_{j_1j_1s_1}, a_{j_2j_21}) = \underbrace{(\text{SO}, 2^{\text{nd}}\text{II})}_{\text{second-order increasing-inflection source}}.$$
- For $a_{j_1j_10}\alpha_{j_1j_2}(a_{j_1j_1s_1} - a_{j_1j_1s_2})^2 < 0$ and $a_{j_2j_20}\alpha^3_{j_1j_1} > 0$,
$$(a_{j_1j_1s_1}, a_{j_2j_21}) = \underbrace{(\text{SI}, 2^{\text{nd}}\text{DI})}_{\text{second-order decreasing-inflection sink}}.$$
- For $a_{j_1j_10}\alpha_{j_1j_2}(a_{j_1j_1s_1} - a_{j_1j_1s_2})^2 > 0$ and $a_{j_2j_20}\alpha^3_{j_1j_1} < 0$,
$$(a_{j_1j_1s_1}, a_{j_2j_21}) = \underbrace{(\text{SO}, 2^{\text{nd}}\text{DI})}_{\text{second-order decreasing-inflection source}}.$$
- For $a_{j_1j_10}\alpha_{j_1j_2}(a_{j_1j_1s_1} - a_{j_1j_1s_2})^2 < 0$ and $a_{j_2j_20}\alpha^3_{j_1j_1} < 0$,
$$(a_{j_1j_1s_1}, a_{j_2j_21}) = \underbrace{(\text{SI}, 2^{\text{nd}}\text{II})}_{\text{second-order increasing-inflection sink}}.$$

– The second-order inflection infinite-equilibrium flows are the switching bifurcations for up-parabola, down-parabola, and up-parabola flows.

(v_{12}) For $a_{j_2j_2l_1} = a_{j_2j_2l_{\sigma 2}} = a_{j_2j_21}$ and $a_{j_1j_1s_2} = a_{j_2j_21}$, the dynamical system becomes

$$\begin{aligned}
&\dot{x}_{j_1} = a_{j_1j_10}(y_{j_1} - a_{j_1j_1s_1})(y_{j_1} - a_{j_1j_1s_2})^2,\\
&\dot{x}_{j_2} = a_{j_2j_20}(y_{j_1} - a_{j_2j_21})^3,\\
&j_1, j_2 \in \{1, 2\};\ j_1 \neq j_2.
\end{aligned}$$

For $a_{j_1j_1s_2} = a_{j_2j_21}$, a dynamical system in the y_{j_1}-direction is constructed as

$$\dot{y}_{j_1} = (y_{j_1} - a_{j_1j_1s_2})^2[\alpha_{j_1j_1}a_{j_1j_10}(y_{j_1} - a_{j_1j_1s_1}) + \alpha_{j_1j_2}a_{j_2j_20}(y_{j_1} - a_{j_2j_21})],$$

and a dynamical system in the $y_{j_1}^{\perp}$-direction is constructed as

$$\dot{y}_{j_1}^{\perp} = (y_{j_1} - a_{j_1 j_1 s_2})^2[\beta_{j_1 j_1} a_{j_1 j_1 0}(y_{j_1} - a_{j_1 j_1 s_1}) + \beta_{j_1 j_2} a_{j_2 j_2 0}(y_{j_1} - a_{j_2 j_2 1})],$$

and a G-function is defined as

$$G_{j_1 j_1 0} = \frac{\partial}{\partial y_{j_1}}[\mathbf{n}^{\mathrm{T}} \cdot \mathbf{F}(y_{j_1}, \mathbf{p})]\bigg|_{y_{j_1}^* = a_{j_1 j_1 s_2} = a_{j_2 j_2 1}} = 0,$$

$$\begin{aligned} G_{j_1 j_1 0}^{(1)} &= \frac{1}{2!}\frac{\partial^2}{\partial y_{j_1}^2}[\mathbf{n}^{\mathrm{T}} \cdot \mathbf{F}(y_{j_1}, \mathbf{p})]\bigg|_{y_{j_1}^* = a_{j_1 j_1 s_2} = a_{j_2 j_2 1}} \\ &= \alpha_{j_1 j_1} a_{j_1 j_1 0}(a_{j_1 j_1 s_2} - a_{j_1 j_1 s_1}). \end{aligned}$$

For $y_{j_1}^* = a_{j_2 j_2 1} = a_{j_1 j_1 s_2}$, in phase space,

$$\frac{dx_{j_2}}{dx_{j_1}}\bigg|_{y_1^* = a_{j_2 j_2 1}} = \frac{a_{j_2 j_2 0}(y_{j_1} - a_{j_2 j_2 1})}{a_{j_1 j_1 0}(y_{j_1} - a_{j_1 j_1 s_1})}\bigg|_{y_1^* = a_{j_2 j_2 1}} = 0.$$

If

$$\frac{d^2 x_{j_2}}{dx_{j_1}^2}\bigg|_{y_1^* = a_{j_2 j_2 1}} = \frac{6 a_{j_2 j_2 0} \alpha_{j_1 j_1}^2}{a_{j_1 j_1 0} \alpha_{j_1 j_1}(a_{j_2 j_2 s_2} - a_{j_1 j_1 s_1})} > 0,$$

there is an up-parabola flow at $y_{j_1}^* = a_{j_2 j_2 1} = a_{j_1 j_1 s_2}$. If

$$\frac{d^2 x_{j_2}}{dx_{j_1}^2}\bigg|_{y_1^* = a_{j_2 j_2 1}} = \frac{6 a_{j_2 j_2 0} \alpha_{j_1 j_1}^2}{a_{j_1 j_1 0} \alpha_{j_1 j_1}(a_{j_2 j_2 s_2} - a_{j_1 j_1 s_1})} < 0,$$

there is a down-parabola flow at $y_{j_1}^* = a_{j_2 j_2 1} = a_{j_1 j_1 s_2}$. Since the corresponding variational equation is

$$\Delta \dot{y}_{j_1} = \alpha_{j_1 j_1} a_{j_1 j_1 0}(a_{j_1 j_1 s_2} - a_{j_1 j_1 s_1})(\Delta y_{j_1})^2 = G_{j_1 j_1 0}^{(1)}(\Delta y_{j_1})^2,$$

the parabola flows are upper-saddle and lower-saddle for $G_{j_1 j_1 0}^{(1)} > 0$ and $G_{j_1 j_1 0}^{(1)} < 0$, respectively.

Therefore, the infinite-equilibrium flow of $y_{j_1}^* = a_{j_1 j_1 s_1} = a_{j_2 j_2 1}$ has the following properties as in Eqs. (4.271)–(4.274):

- For $\alpha_{j_1 j_1} a_{j_1 j_1 0}(a_{j_1 j_1 s_2} - a_{j_1 j_1 s_1}) > 0$ and $a_{j_2 j_2 0} \alpha_{j_1 j_1}^2 > 0$,

$$(a_{j_1 j_1 s_2}, a_{j_2 j_2 1}) = \underbrace{(\mathrm{UP}, \mathrm{US})}_{\text{up-parabola upper-saddle}}.$$

- For $\alpha_{j_1j_1}a_{j_1j_10}(a_{j_1j_1s_2} - a_{j_1j_1s_1}) < 0$ and $a_{j_2j_20}\alpha^2_{j_1j_1} > 0$,

$$(a_{j_1j_1s_2}, a_{j_2j_21}) = \underbrace{(\text{DP}, \text{US})}_{\text{down-parabola upper-saddle}}.$$

- For $\alpha_{j_1j_1}a_{j_1j_10}(a_{j_1j_1s_2} - a_{j_1j_1s_1}) > 0$ and $a_{j_2j_20}\alpha^2_{j_1j_1} < 0$,

$$(a_{j_1j_1s_2}, a_{j_2j_21}) = \underbrace{(\text{DP}, \text{LS})}_{\text{down-parabola lower-saddle}}$$

- For $\alpha_{j_1j_1}a_{j_1j_10}(a_{j_1j_1s_2} - a_{j_1j_1s_1}) < 0$ and $a_{j_2j_20}\alpha^2_{j_1j_1} < 0$,

$$(a_{j_1j_1s_2}, a_{j_2j_21}) = \underbrace{(\text{UP}, \text{LS})}_{\text{up-parabola lower-saddle}}.$$

– The parabola-saddle infinite-equilibrium flows are the switching bifurcations for up-parabola, down-parabola, and up-parabola flows.

(v_{13}) For $a_{j_1j_11} = a_{j_2j_21}$, the dynamical system becomes

$$\begin{aligned}
\dot{x}_{j_1} &= a_{j_1j_10}(y_{j_1} - a_{j_1j_11})^3, \\
\dot{x}_{j_2} &= a_{j_2j_20}(y_{j_1} - a_{j_2j_21})^3, \\
& j_1, j_2 \in \{1, 2\};\ j_1 \neq j_2.
\end{aligned}$$

A dynamical system in the y_{j_1}-direction is constructed as

$$\dot{y}_{j_1} = (y_{j_1} - a_{j_1j_11})^3(\alpha_{j_1j_1}a_{j_1j_10} + \alpha_{j_1j_2}a_{j_2j_20}),$$

and a dynamical system in the $y^{\perp}_{j_1}$-direction is constructed as

$$\dot{y}^{\perp}_{j_1} = (y_{j_1} - a_{j_1j_11})^3(\beta_{j_1j_1}a_{j_1j_10} + \beta_{j_1j_2}a_{j_2j_20}),$$

where

$$G_{j_1j_10} = \frac{\partial}{\partial y_{j_1}}[\mathbf{n}^{\mathrm{T}} \cdot \mathbf{F}(y_{j_1}, \mathbf{p})]\Big|_{y^*_{j_1} = a_{j_1j_11} = a_{j_2j_21}} = 0,$$

$$G^{(1)}_{j_1j_10} = \frac{1}{2!}\frac{\partial^2}{\partial y^2_{j_1}}[\mathbf{n}^{\mathrm{T}} \cdot \mathbf{F}(y_{j_1}, \mathbf{p})]\Big|_{y^*_{j_1} = a_{j_1j_11} = a_{j_2j_21}} = 0,$$

$$G^{(2)}_{j_1j_10} = \frac{1}{3!}\frac{\partial^3}{\partial y^3_{j_1}}[\mathbf{n}^{\mathrm{T}} \cdot \mathbf{F}(y_{j_1}, \mathbf{p})]\Big|_{y^*_{j_1} = a_{j_1j_11} = a_{j_2j_21}} = \alpha_{j_1j_1}a_{j_1j_10} + \alpha_{j_1j_2}a_{j_2j_20}.$$

For $y^*_{j_1} = a_{j_2j_21} = a_{j_1j_11}$, in phase space, if

$$\left.\frac{dx_{j_2}}{dx_{j_1}}\right|_{y_1^*=a_{j_2j_21}} = \frac{a_{j_2j_20}}{a_{j_1j_10}} > 0,$$

there is an increasing-inflection flow at $y^*_{j_1} = a_{j_2j_21} = a_{j_1j_11}$. If

$$\left.\frac{dx_{j_2}}{dx_{j_1}}\right|_{y_1^*=a_{j_2j_21}} = \frac{a_{j_2j_20}}{a_{j_1j_10}} < 0,$$

there is a decreasing inflection flow at $y^*_{j_1} = a_{j_2j_21} = a_{j_1j_11}$. Since the corresponding variational equation is

$$\Delta\dot{y}_{j_1} = (\alpha_{j_1j_1}a_{j_1j_10} + \alpha_{j_1j_2}a_{j_2j_20})(\Delta y_{j_1})^3 = G^{(2)}_{j_1j_10}(\Delta y_{j_1})^3,$$

the parabola flows are third-order source and sink for $G^{(2)}_{j_1j_10} > 0$ and $G^{(2)}_{j_1j_10} < 0$, respectively.

Therefore, the infinite-equilibrium of $y^*_{j_1} = a_{j_1j_11} = a_{j_2j_21}$ has the following properties as in Eqs. (4.279)–(4.282):

- For $a_{j_1j_10}/a_{j_2j_20} > 0$ and $G^{(2)}_{j_1j_10} > 0$,

$$(a_{j_1j_11}, a_{j_2j_21}) = \underbrace{(\text{II}, 3^{\text{rd}}\text{SO})}_{\text{increasing-inflection third-order source}}.$$

- For $a_{j_1j_10}/a_{j_2j_20} > 0$ and $G^{(2)}_{j_1j_10} < 0$,

$$(a_{j_1j_11}, a_{j_2j_21}) = \underbrace{(\text{II}, 3^{\text{rd}}\text{SI})}_{\text{increasing-inflection third-order sink}}.$$

- For $a_{j_1j_10}/a_{j_2j_20} < 0$ and $G^{(2)}_{j_1j_10} > 0$,

$$(a_{j_1j_11}, a_{j_2j_21}) = \underbrace{(\text{DI}, 3^{\text{rd}}\text{SO})}_{\text{decreasing-inflection third-order source}}.$$

- For $a_{j_1j_10}/a_{j_2j_20} < 0$ and $G^{(2)}_{j_1j_10} < 0$,

$$(a_{j_1j_11}, a_{j_2j_21}) = \underbrace{(\text{DI}, 3^{\text{rd}}\text{SI})}_{\text{decreasing-inflection third-order sink}}.$$

– The inflection-third-order source (sink) infinite-equilibriums are the appearing bifurcations for the up-parabola and down-parabola flows in the x_{j_1}- and x_{j_2}-directions.

– The inflection-third-order source (sink) infinite-equilibriums are the double-switching bifurcations for the up-parabola and down-parabola flows in the x_{j_1}- and x_{j_2}-directions.

(vi) If

$$\Delta_{j_1j_1} > 0 \quad \text{and} \quad \Delta_{j_2j_2} = 0,$$

the dynamical system becomes

$$\begin{aligned}\dot{x}_{j_1} &= a_{j_1j_10}(y_{j_1} - a_{j_1j_11})(y_{j_1} - a_{j_1j_12})(y_{j_1} - a_{j_1j_13}),\\ \dot{x}_{j_2} &= a_{j_2j_20}(y_{j_1} - a_{j_2j_2l_1})(y_{j_1} - a_{j_2j_2l_2})^2,\end{aligned}$$

where

$$\begin{aligned}&\{a_{j_1j_11}, a_{j_1j_12}, a_{j_1j_13}\} = \text{sort}\{b_{j_1j_11}, b_{j_1j_12}, b_{j_1j_13}\},\\ &\qquad a_{j_1j_1s_1} < a_{j_1j_1s_2}, \quad s_1 < s_2;\\ &\qquad b_{j_1j_12}, b_{j_1j_13} = -\frac{1}{2}(B_{j_1j_11} \pm \sqrt{\Delta_{j_1j_1}}), \quad a_{j_1j_10} = A_{j_1j_10}.\\ &a_{j_2j_2l_1} = b_{j_2j_21}, \quad a_{j_2j_2l_2} = -\frac{1}{2}B_{j_2j_2}, \quad a_{j_2j_20} = A_{j_2j_20};\ j_1, j_2 \in \{1, 2\}.\end{aligned}$$

A dynamical system in the y_{j_1}-direction is constructed as

$$\begin{aligned}\dot{y}_{j_1} = \alpha_{j_1j_1}a_{j_1j_10}&\prod_{s_1=1}^{3}(y_{j_1} - a_{j_1j_1s_1})\\ &+ \alpha_{j_1j_2}a_{j_2j_20}(y_{j_1} - a_{j_2j_2l_1})(y_{j_1} - a_{j_2j_2l_2})^2,\end{aligned}$$

and a dynamical system in the $y_{j_1}^{\perp}$-direction is constructed as

$$\begin{aligned}\dot{y}_{j_1}^{\perp} = \beta_{j_1j_1}a_{j_1j_10}&\prod_{s_1=1}^{3}(y_{j_1} - a_{j_1j_1s_1})\\ &+ \beta_{j_1j_2}a_{j_2j_20}(y_{j_1} - a_{j_2j_2l_1})(y_{j_1} - a_{j_2j_2l_2})^2.\end{aligned}$$

(vi$_{1a}$) In phase space, for $y_{j_1}^* = a_{j_1j_1s_1}(s_1, s_2 \in \{1, 2, 3\}, s_1 \neq s_2)$,

$$\begin{aligned}&\left.\frac{dx_{j_1}}{dx_{j_2}}\right|_{y_1^*=a_{j_1j_1s_1}}\\ &= \left.\frac{a_{j_1j_10}(y_{j_1} - a_{j_1j_1s_1})\prod_{s_2=1,s_2\neq s_1}^{3}(y_{j_1} - a_{j_1j_1s_2})}{a_{j_2j_20}(y_{j_1} - a_{j_2j_2l_1})(y_{j_1} - a_{j_2j_2l_2})^2}\right|_{y_1^*=a_{j_1j_1s_1}} = 0.\end{aligned}$$

If

$$\left.\frac{d^2x_{j_1}}{dx_{j_2}^2}\right|_{y_1^*=a_{j_1j_1s_1}} = \frac{a_{j_1j_10}\alpha_{j_1j_2}\prod_{s_2=1,s_2\neq s_1}^{3}(a_{j_1j_1s_1}-a_{j_1j_1s_2})}{a_{j_2j_20}(a_{j_1j_1s_1}-a_{j_2j_2l_1})(a_{j_1j_1s_1}-a_{j_2j_2l_2})^2} > 0,$$

there is an up-parabola flow at $y_1^* = a_{j_1j_1s_1}$. If

$$\left.\frac{d^2x_{j_1}}{dx_{j_2}^2}\right|_{y_1^*=a_{j_1j_1s_1}} = \frac{a_{j_1j_10}\alpha_{j_1j_2}\prod_{s_2=1,s_2\neq s_1}^{3}(a_{j_1j_1s_1}-a_{j_1j_1s_2})}{a_{j_2j_20}(a_{j_1j_1s_1}-a_{j_2j_2l_1})(a_{j_1j_1s_1}-a_{j_2j_2l_2})^2} < 0,$$

there is a down-parabola flow at $y_1^* = a_{j_1j_1s_1}$. With

$$\dot{x}_{j_2}^* = a_{j_2j_20}(a_{j_1j_1s_1}-a_{j_2j_2l_1})(a_{j_1j_1s_1}-a_{j_2j_2l_2})^2,$$

the parabola flows are positive and negative for $\dot{x}_{j_2}^* > 0$ and $\dot{x}_{j_2}^* < 0$, respectively.

Therefore, the equilibrium flows of $y_{j_1}^* = a_{j_1j_1s_1}(s_1, s_2 \in \{1,2,3\}, s_1 \neq s_2)$ for the x_{j_1}-direction have four cases with $\dot{x}_{j_2}^* = a_{j_2j_20}(a_{j_1j_1s_1} - a_{j_2j_2l_1})(a_{j_1j_1s_1} - a_{j_2j_2l_1})^2$ as in Eqs. (4.288)–(4.291):

- For $a_{j_1j_10}\alpha_{j_1j_2}\prod_{s_2=1,s_2\neq s_1}^{3}(a_{j_1j_1s_1}-a_{j_1j_1s_2}) > 0$ and $a_{j_2j_20}(a_{j_1j_1s_1} - a_{j_2j_2l_1})(a_{j_1j_1s_1} - a_{j_2j_2l_1})^2 > 0$,

$$(a_{j_1j_1s_1}, \dot{x}_{j_2}^*) = \underbrace{(\mathrm{UP}, \mathrm{pF})}_{\text{up-parabola flow }(+)}.$$

- For $a_{j_1j_10}\alpha_{j_1j_2}\prod_{s_2=1,s_2\neq s_1}^{3}(a_{j_1j_1s_1}-a_{j_1j_1s_2}) < 0$ and $a_{j_2j_20}(a_{j_1j_1s_1} - a_{j_2j_2l_1})(a_{j_1j_1s_1} - a_{j_2j_2l_1})^2 > 0$,

$$(a_{j_1j_1s_1}, \dot{x}_{j_2}^*) = \underbrace{(\mathrm{DP}, \mathrm{pF})}_{\text{Down-parabola flow }(+)}.$$

- For $a_{j_1j_10}\alpha_{j_1j_2}\prod_{s_2=1,s_2\neq s_1}^{3}(a_{j_1j_1s_1}-a_{j_1j_1s_2}) > 0$ and $a_{j_2j_20}(a_{j_1j_1s_1} - a_{j_2j_2l_1})(a_{j_1j_1s_1} - a_{j_2j_2l_1})^2 < 0$,

$$(a_{j_1j_1s_1}, \dot{x}_{j_2}^*) = \underbrace{(\mathrm{DP}, \mathrm{nF})}_{\text{Down-parabola flow }(-)}.$$

- For $a_{j_1j_10}\alpha_{j_1j_2}\prod_{s_2=1,s_2\neq s_1}^{3}(a_{j_1j_1s_1}-a_{j_1j_1s_2}) < 0$ and $a_{j_2j_20}(a_{j_1j_1s_1} - a_{j_2j_2l_1})(a_{j_1j_1s_1} - a_{j_2j_2l_1})^2 < 0$,

$$(a_{j_1j_1s_1}, \dot{x}_{j_2}^*) = \underbrace{(\mathrm{UP}, \mathrm{nF})}_{\text{up-parabola flow }(-)}.$$

(vi$_{1b}$) For $y_{j_1}^* = a_{j_2j_2l_1} \neq a_{j_1j_1s_1}, a_{j_1j_1s_2}$,

$$\left.\frac{dx_{j_2}}{dx_{j_1}}\right|_{y_1^*=a_{j_2j_2l_1}} = \left.\frac{a_{j_2j_20}(y_{j_1} - a_{j_2j_2l_1})(y_{j_1} - a_{j_2j_2l_2})^2}{a_{j_1j_10}\prod_{s_1=1}^{3}(y_{j_1} - a_{j_1j_1s_1})}\right|_{y_1^*=a_{j_2j_2l_1}} = 0.$$

If

$$\left.\frac{d^2x_{j_2}}{dx_{j_1}^2}\right|_{y_1^*=a_{j_2j_2l_1}} = \frac{a_{j_2j_20}\alpha_{j_1j_1}^2(a_{j_2j_2l_1} - a_{j_2j_2l_2})^2}{a_{j_1j_10}\prod_{s_1=1}^{3}(a_{j_2j_2l_1} - a_{j_1j_1s_1})} > 0,$$

there is an up-parabola flow at $y_1^* = a_{j_2j_2l_1}$. If

$$\left.\frac{d^2x_{j_2}}{dx_{j_1}^2}\right|_{y_1^*=a_{j_2j_2l_1}} = \frac{a_{j_2j_20}\alpha_{j_1j_1}^2(a_{j_2j_2l_1} - a_{j_2j_2l_2})^2}{a_{j_1j_10}\prod_{s_1=1}^{3}(a_{j_2j_2l_1} - a_{j_1j_1s_1})} < 0,$$

there is a down-parabola at $y_1^* = a_{j_2j_2l_1}$. With

$$\dot{x}_{j_1}^* = a_{j_1j_10}\prod_{s_1=1}^{3}(a_{j_2j_2l_1} - a_{j_1j_1s_1}),$$

the parabola flows are positive and negative for $\dot{x}_{j_1}^* > 0$ and $\dot{x}_{j_1}^* < 0$, respectively.

Thus, the equilibrium flow of $y_{j_1}^* = a_{j_2j_2l_1}$ for the x_{j_2}-direction has four cases with $\dot{x}_{j_1}^* = a_{j_1j_10}\prod_{s_1=1}^{3}(a_{j_2j_2l_1} - a_{j_1j_1s_1})$ as in Eqs. (4.292)–(4.295):

- For $a_{j_1j_10}\prod_{s_1=1}^{3}(a_{j_2j_2l_1} - a_{j_1j_1s_1}) > 0$ and $a_{j_2j_20}\alpha_{j_1j_1}(a_{j_2j_2l_1} - a_{j_2j_2l_2})^2 > 0$,

$$(\dot{x}_{j_1}^*, a_{j_2j_2l_1}) = \underbrace{(\text{pF}, \text{UP})}_{\text{up-parabola flow } (+)}.$$

- For $a_{j_1j_10}\prod_{s_1=1}^{3}(a_{j_2j_2l_1} - a_{j_1j_1s_1}) < 0$ and $a_{j_2j_20}\alpha_{j_1j_1}(a_{j_2j_2l_1} - a_{j_2j_2l_2})^2 > 0$,

$$(\dot{x}_{j_1}^*, a_{j_2j_2l_1}) = \underbrace{(\text{nF}, \text{DP})}_{\text{down-parabola flow } (-)}.$$

- For $a_{j_1j_10}\prod_{s_1=1}^{3}(a_{j_2j_2l_1} - a_{j_1j_1s_1}) > 0$ and $a_{j_2j_20}\alpha_{j_1j_1}(a_{j_2j_2l_1} - a_{j_2j_2l_2})^2 < 0$,

$$(\dot{x}_{j_1}^*, a_{j_2j_2l_1}) = \underbrace{(\text{pF}, \text{DP})}_{\text{down-parabola flow } (+)}.$$

- For $a_{j_1 j_1 0}\prod_{s_1=1}^{3}(a_{j_2 j_2 l_1} - a_{j_1 j_1 s_1}) < 0$ and $a_{j_2 j_2 0}\alpha_{j_1 j_1}(a_{j_2 j_2 l_1} - a_{j_2 j_2 l_2})^2 > 0$,

$$(\dot{x}_{j_1}^*, a_{j_2 j_2 l_1}) = \underbrace{(\text{nF}, \text{UP})}_{\text{up-parabola flow }(-)}.$$

(vi$_{1c}$) In phase space, at $y_1^* = a_{j_2 j_2 l_2}$,

$$\left.\frac{dx_{j_2}}{dx_{j_1}}\right|_{y_1^*=a_{j_2 j_2 l_2}} = \left.\frac{a_{j_2 j_2 0}(y_{j_1} - a_{j_2 j_2 l_1})(y_{j_1} - a_{j_2 j_2 l_2})^2}{a_{j_1 j_1 0}\prod_{s_1=1}^{3}(y_{j_1} - a_{j_1 j_1 s_1})}\right|_{y_1^*=a_{j_2 j_2 l_2}} = 0,$$

$$\left.\frac{d^2 x_{j_2}}{dx_{j_1}^2}\right|_{y_1^*=a_{j_2 j_2 1}} = 0.$$

If

$$\left.\frac{d^3 x_{j_2}}{dx_{j_1}^3}\right|_{y_1^*=a_{j_2 j_2 l_2}} = \frac{2a_{j_2 j_2 0}\alpha_{j_1 j_1}^2(a_{j_2 j_2 l_2} - a_{j_2 j_2 l_1})}{a_{j_1 j_1 0}\prod_{s_1=1}^{3}(a_{j_2 j_2 l_1} - a_{j_1 j_1 s_1})} > 0,$$

there is an increasing-inflection flow at $y_1^* = a_{j_2 j_2 l_2}$. If

$$\left.\frac{d^3 x_{j_2}}{dx_{j_1}^3}\right|_{y_1^*=a_{j_2 j_2 l_2}} = \frac{2a_{j_2 j_2 0}\alpha_{j_1 j_1}^2(a_{j_2 j_2 l_2} - a_{j_2 j_2 l_1})}{a_{j_1 j_1 0}\prod_{s_1=1}^{3}(a_{j_2 j_2 l_1} - a_{j_1 j_1 s_1})} < 0,$$

there is a decreasing-inflection flow at $y_1^* = a_{j_2 j_2 l_2}$. With

$$\dot{x}_{j_1}^* = a_{j_1 j_1 0}(a_{j_2 j_2 1} - a_{j_1 j_1 s_1})(a_{j_2 j_2 1} - a_{j_1 j_1 s_2})^2,$$

the inflection flows are positive and negative for $\dot{x}_{j_1}^* > 0$ and $\dot{x}_{j_1}^* < 0$, respectively.

Thus, the equilibrium flow of $y_{j_1}^* = a_{j_2 j_2 l_2}$ for the x_{j_2}-direction has four cases with $\dot{x}_{j_1}^* = a_{j_1 j_1 0}\prod_{s_1=1}^{3}(a_{j_2 j_2 l_2} - a_{j_1 j_1 s_1})$ as in Eqs. (4.296)–(4.299):

- For $a_{j_1 j_1 0}\prod_{s_1=1}^{3}(a_{j_2 j_2 l_2} - a_{j_1 j_1 s_1}) > 0$ and $a_{j_2 j_2 0}\alpha_{j_1 j_1}^2(a_{j_2 j_2 l_2} - a_{j_2 j_2 l_1}) > 0$,

$$(\dot{x}_{j_1}^*, a_{j_2 j_2 l_2}) = \underbrace{(\text{pF}, \text{II})}_{\text{increasing-inflection flow }(+)}.$$

- For $a_{j_1 j_1 0}\prod_{s_1=1}^{3}(a_{j_2 j_2 l_2} - a_{j_1 j_1 s_1}) < 0$ and $a_{j_2 j_2 0}\alpha_{j_1 j_1}^2(a_{j_2 j_2 l_2} - a_{j_2 j_2 l_1}) > 0$,

$$(\dot{x}_{j_1}^*, a_{j_2 j_2 l_2}) = \underbrace{(\text{nF}, \text{DI})}_{\text{decreasing-inflection flow }(-)}.$$

- For $a_{j_1j_10}\prod_{s_1=1}^{3}(a_{j_2j_2l_2} - a_{j_1j_1s_1}) > 0$ and $a_{j_2j_20}\alpha_{j_1j_1}^2(a_{j_2j_2l_2} - a_{j_2j_2l_1}) < 0$,

$$(\dot{x}_{j_1}^*, a_{j_2j_2l_2}) = \underbrace{(\text{pF}, \text{DI})}_{\text{decreasing-inflection flow }(+)}.$$

- For $a_{j_1j_10}\prod_{s_1=1}^{3}(a_{j_2j_2l_2} - a_{j_1j_1s_1}) < 0$ and $a_{j_2j_20}\alpha_{j_1j_1}^2(a_{j_2j_2l_2} - a_{j_2j_2l_1}) > 0$,

$$(\dot{x}_{j_1}^*, a_{j_2j_2l_2}) = \underbrace{(\text{nF}, \text{II})}_{\text{increasing-inflection flow }(-)}.$$

– The decreasing- and increasing-inflection flows are the appearing bifurcations for the up-parabola and parabola flows.

(vi$_2$) For $a_{j_2j_2l_2} = a_{j_2j_2l_1} = a_{j_2j_21}$, the dynamical system becomes

$$\begin{aligned}\dot{x}_{j_1} &= a_{j_1j_10}(y_{j_1} - a_{j_1j_11})(y_{j_1} - a_{j_1j_12})(y_{j_1} - a_{j_1j_13}),\\ \dot{x}_{j_2} &= a_{j_2j_20}(y_{j_1} - a_{j_2j_21})^3.\end{aligned}$$

A dynamical system in the y_{j_1}-direction is constructed as

$$\dot{y}_{j_1} = \alpha_{j_1j_1}a_{j_1j_10}\prod_{s_1=1}^{3}(y_{j_1} - a_{j_1j_1s_1}) + \alpha_{j_1j_2}a_{j_2j_20}(y_{j_1} - a_{j_2j_21})^3,$$

and a dynamical system in the $y_{j_1}^{\perp}$-direction is constructed as

$$\dot{y}_{j_1}^{\perp} = \beta_{j_1j_1}a_{j_1j_10}\prod_{s_1=1}^{3}(y_{j_1} - a_{j_1j_1s_1}) + \beta_{j_1j_2}a_{j_2j_20}(y_{j_1} - a_{j_2j_21})^3.$$

(vi$_{2a}$) In phase space, $y_{j_1}^* = a_{j_1j_1s_1}(s_1, s_2 \in \{1,2,3\}, s_1 \neq s_2)$,

$$\begin{aligned}&\left.\frac{dx_{j_1}}{dx_{j_2}}\right|_{y_1^*=a_{j_1j_1s_1}}\\ &\quad = \left.\frac{a_{j_1j_10}(y_{j_1} - a_{j_1j_1s_1})\prod_{s_2=1,s_2\neq s_1}^{3}(y_{j_1} - a_{j_1j_1s_2})}{a_{j_2j_20}(y_{j_1} - a_{j_2j_21})^3}\right|_{y_1^*=a_{j_1j_1s_1}} = 0.\end{aligned}$$

If

$$\left.\frac{d^2x_{j_1}}{dx_{j_2}^2}\right|_{y_1^*=a_{j_1j_1s_1}} = \frac{a_{j_1j_10}\alpha_{j_1j_2}\prod_{s_2=1,s_2\neq s_1}^{3}(a_{j_1j_1s_1} - a_{j_1j_1s_2})}{a_{j_2j_20}(a_{j_1j_1s_1} - a_{j_2j_2l_2})^3} > 0,$$

there is an up-parabola flow at $y_1^* = a_{j_1 j_1 s_1}$. If

$$\left.\frac{d^2 x_{j_1}}{dx_{j_2}^2}\right|_{y_1^* = a_{j_1 j_1 s_1}} = \frac{a_{j_1 j_1 0}\alpha_{j_1 j_2}\prod_{s_2=1, s_2\neq s_1}^{3}(a_{j_1 j_1 s_1} - a_{j_1 j_1 s_2})}{a_{j_2 j_2 0}(a_{j_1 j_1 s_1} - a_{j_2 j_2 1})^3} < 0,$$

there is a down-parabola flow at $y_1^* = a_{j_1 j_1 s_1}$. With

$$\dot{x}_{j_2}^* = a_{j_2 j_2 0}(a_{j_1 j_1 s_1} - a_{j_2 j_2 1})^3,$$

the parabola flows are positive and negative for $\dot{x}_{j_2}^* > 0$ and $\dot{x}_{j_2}^* < 0$, respectively.

Therefore, the equilibrium flows of $y_{j_1}^* = a_{j_1 j_1 s_1}(s_1, s_2 \in \{1,2,3\}, s_1 \neq s_2)$ for the x_{j_1}-direction have four cases with $\dot{x}_{j_2}^* = a_{j_2 j_2 0}(a_{j_1 j_1 s_1} - a_{j_2 j_2 1})^3$ as in Eqs. (4.303)–(4.306).

- For $a_{j_1 j_1 0}\alpha_{j_1 j_2}\prod_{s_2=1, s_2\neq s_1}^{3}(a_{j_1 j_1 s_1} - a_{j_1 j_1 s_2}) > 0$ and $a_{j_2 j_2 0}(a_{j_1 j_1 s_1} - a_{j_2 j_2 1})^3 > 0$,

$$(a_{j_1 j_1 s_1}, \dot{x}_{j_2}^*) = \underbrace{(\text{UP}, \text{pF})}_{\text{up-parabola flow } (+)}.$$

- For $a_{j_1 j_1 0}\alpha_{j_1 j_2}\prod_{s_2=1, s_2\neq s_1}^{3}(a_{j_1 j_1 s_1} - a_{j_1 j_1 s_2}) < 0$ and $a_{j_2 j_2 0}(a_{j_1 j_1 s_1} - a_{j_2 j_2 1})^3 > 0$,

$$(a_{j_1 j_1 s_1}, \dot{x}_{j_2}^*) = \underbrace{(\text{DP}, \text{pF})}_{\text{Down-parabola flow } (+)}.$$

- For $a_{j_1 j_1 0}\alpha_{j_1 j_2}\prod_{s_2=1, s_2\neq s_1}^{3}(a_{j_1 j_1 s_1} - a_{j_1 j_1 s_2}) > 0$ and $a_{j_2 j_2 0}(a_{j_1 j_1 s_1} - a_{j_2 j_2 1})^3 < 0$,

$$(a_{j_1 j_1 s_1}, \dot{x}_{j_2}^*) = \underbrace{(\text{DP}, \text{nF})}_{\text{Down-parabola flow } (-)}.$$

- For $a_{j_1 j_1 0}\alpha_{j_1 j_2}\prod_{s_2=1, s_2\neq s_1}^{3}(a_{j_1 j_1 s_1} - a_{j_1 j_1 s_2}) < 0$ and $a_{j_2 j_2 0}(a_{j_1 j_1 s_1} - a_{j_2 j_2 1})^3 < 0$,

$$(a_{j_1 j_1 s_1}, \dot{x}_{j_2}^*) = \underbrace{(\text{UP}, \text{nF})}_{\text{up-parabola flow } (-)}.$$

($\mathrm{vi_{2b}}$) In phase space, $y_{j_1}^* = a_{j_2j_2 1}$,

$$\left.\frac{dx_{j_2}}{dx_{j_1}}\right|_{y_1^*=a_{j_2j_21}} = \left.\frac{a_{j_2j_20}(y_{j_1}-a_{j_2j_21})^3}{a_{j_1j_10}\prod_{s_1=1}^{3}(y_{j_1}-a_{j_1j_1s_1})}\right|_{y_1^*=a_{j_2j_21}} = 0,$$

$$\left.\frac{d^2x_{j_2}}{dx_{j_1}^2}\right|_{y_1^*=a_{j_2j_21}} = \left.\frac{d^3x_{j_2}}{dx_{j_1}^3}\right|_{y_1^*=a_{j_2j_21}} = 0.$$

If

$$\left.\frac{d^4x_{j_2}}{dx_{j_1}^4}\right|_{y_1^*=a_{j_2j_21}} = \frac{6a_{j_2j_20}\alpha_{j_1j_1}^3}{a_{j_1j_10}\prod_{s_1=1}^{3}(a_{j_2j_21}-a_{j_1j_1s_1})} > 0,$$

there is a third-order up-parabola flow at $y_1^* = a_{j_2j_21}$. If

$$\left.\frac{d^4x_{j_2}}{dx_{j_1}^4}\right|_{y_1^*=a_{j_2j_21}} = \frac{6a_{j_2j_20}\alpha_{j_1j_1}^3}{a_{j_1j_10}\prod_{s_1=1}^{3}(a_{j_2j_21}-a_{j_1j_1s_1})} < 0,$$

there is a third-order down-parabola flow at $y_1^* = a_{j_2j_21}$. With

$$\dot{x}_{j_1}^* = a_{j_1j_10}\prod_{s_1=1}^{3}(a_{j_2j_21}-a_{j_1j_1s_1}),$$

the third-order parabola flows are positive and negative for $\dot{x}_{j_1}^* > 0$ and $\dot{x}_{j_1}^* < 0$, respectively.

Therefore, the equilibrium flow of $y_{j_1}^* = a_{j_2j_21}$ for the x_{j_2}-direction has four cases with $\dot{x}_{j_1}^* = a_{j_1j_10}\prod_{s_1=1}^{3}(a_{j_2j_21}-a_{j_1j_1s_1})$ as in Eqs. (4.307)–(4.310):

- For $a_{j_1j_10}\prod_{s_1=1}^{3}(a_{j_2j_21}-a_{j_1j_1s_1}) > 0$ and $a_{j_2j_20}\alpha_{j_1j_1}^3 > 0$,

$$(\dot{x}_{j_1}^*, a_{j_2j_21}) = \underbrace{(\text{pF}, 3^{\text{rd}}\text{UP})}_{\text{third-order up-parabola flow }(+)}.$$

- For $a_{j_1j_10}\prod_{s_1=1}^{3}(a_{j_2j_21}-a_{j_1j_1s_1}) < 0$ and $a_{j_2j_20}\alpha_{j_1j_1}^3 > 0$,

$$(\dot{x}_{j_1}^*, a_{j_2j_21}) = \underbrace{(\text{nF}, 3^{\text{rd}}\text{DP})}_{\text{third-order down-parabola flow }(-)}.$$

- For $a_{j_1j_10}\prod_{s_1=1}^{3}(a_{j_2j_21}-a_{j_1j_1s_1}) > 0$ and $a_{j_2j_20}\alpha_{j_1j_1}^3 < 0$,

$$(\dot{x}_{j_1}^*, a_{j_2j_21}) = \underbrace{(\text{pF}, 3^{\text{rd}}\text{DP})}_{\text{third-order down-parabola flow }(+)}.$$

- For $a_{j_1j_10}\prod_{s_1=1}^{3}(a_{j_2j_21}-a_{j_1j_1s_1})<0$ and $a_{j_2j_20}\alpha_{j_1j_1}^{3}<0$,

$$(\dot{x}_{j_1}^{*},a_{j_2j_21})=\underbrace{(\mathrm{nF},3^{\mathrm{rd}}\mathrm{UP})}_{\text{third-order up-parabola flow }(-)}.$$

(vi$_3$) For $a_{j_2j_2l_1}=a_{j_1j_1s_1}$ $(s_1,s_2\in\{1,2,3\})$, the dynamical system becomes

$$\begin{aligned}
&\dot{x}_{j_1}=a_{j_1j_10}(y_{j_1}-a_{j_1j_1s_1})\prod_{s_2=1,s_2\neq s_1}^{3}(y_{j_1}-a_{j_1j_1s_2}),\\
&\dot{x}_{j_2}=a_{j_2j_20}(y_{j_1}-a_{j_2j_2l_1})(y_{j_1}-a_{j_2j_2l_2})^2,\\
&j_1,j_2\in\{1,2\};\ j_1\neq j_2;\ s_1,s_2\in\{1,2,3\},\ s_1\neq s_2.
\end{aligned}$$

For $a_{j_2j_2l_1}=a_{j_1j_1s_1}$, a dynamical system in the y_{j_1}-direction is constructed as

$$\begin{aligned}
\dot{y}_{j_1}=(y_{j_1}-a_{j_1j_1s_1})\Big[&\alpha_{j_1j_1}a_{j_1j_10}\prod_{s_2=1,s_2\neq s_1}^{3}(y_{j_1}-a_{j_1j_1s_2})\\
&+\alpha_{j_1j_2}a_{j_2j_20}(y_{j_1}-a_{j_2j_2l_2})^2\Big],
\end{aligned}$$

and a dynamical system in the $y_{j_1}^{\perp}$-direction is constructed as

$$\begin{aligned}
\dot{y}_{j_1}^{\perp}=(y_{j_1}-a_{j_1j_1s_1})\Big[&\beta_{j_1j_1}a_{j_1j_10}\prod_{s_2=1,s_2\neq s_1}^{3}(y_{j_1}-a_{j_1j_1s_2})\\
&+\beta_{j_1j_2}a_{j_2j_20}(y_{j_1}-a_{j_2j_2l_2})^2\Big],
\end{aligned}$$

and a G-function is defined as

$$\begin{aligned}
G_{j_1j_10}&=\frac{\partial}{\partial y_{j_1}}[\mathbf{n}^{\mathrm{T}}\cdot\mathbf{F}(y_{j_1},\mathbf{p})]\Big|_{y_{j_1}^{*}=a_{j_1j_1s_1}=a_{j_2j_2l_1}}\\
&=\alpha_{j_1j_1}a_{j_1j_10}\prod_{s_2=1,s_2\neq s_1}^{3}(a_{j_1j_1s_1}-a_{j_1j_1s_2})\\
&\quad+\alpha_{j_1j_2}a_{j_2j_20}(a_{j_2j_2l_1}-a_{j_2j_2l_2})^2.
\end{aligned}$$

For $y_{j_1}^{*}=a_{j_1j_1s_1}=a_{j_2j_2l_1}$, in phase space, if

$$\frac{dx_{j_1}}{dx_{j_2}}\bigg|_{y_1^{*}=a_{j_1j_1s_1}}=\frac{a_{j_1j_10}\prod_{s_2=1,s_2\neq s_1}^{3}(a_{j_1j_1s_1}-a_{j_1j_1s_2})}{a_{j_2j_20}(a_{j_2j_2l_1}-a_{j_2j_2l_2})^2}>0,$$

there is an increasing-inflection flow at $y_{j_1}^* = a_{j_1j_1s_1} = a_{j_2j_2l_1}$, and if

$$\left.\frac{dx_{j_1}}{dx_{j_2}}\right|_{y_1^*=a_{j_1j_1s_1}} = \frac{a_{j_1j_10}\prod_{s_2=1,s_2\neq s_1}^{3}(a_{j_1j_1s_1}-a_{j_1j_1s_2})}{a_{j_2j_20}(a_{j_2j_2l_1}-a_{j_2j_2l_2})^2} < 0,$$

there is a decreasing-inflection flow at $y_{j_1}^* = a_{j_1j_1s_1} = a_{j_2j_2l_1}$. Since the corresponding variational equation is

$$\begin{aligned}\Delta\dot{y}_{j_1} &= \Bigg[\alpha_{j_1j_1}a_{j_1j_10}\prod_{s_2=1,s_2\neq s_1}^{3}(a_{j_1j_1s_1}-a_{j_1j_1s_2})\\ &\quad + \alpha_{j_1j_2}a_{j_2j_20}(a_{j_2j_2l_1}-a_{j_2j_2l_2})^2\Bigg]\Delta y_{j_1}\\ &= G_{j_1j_10}\Delta y_{j_1},\end{aligned}$$

the inflection flows are source and sink for $G_{j_1j_10} > 0$ and $G_{j_1j_10} < 0$, respectively.

Therefore, the infinite-equilibrium flows of $y_{j_1}^* = a_{j_1j_1s_1} = a_{j_2j_2l_1}$ are the following properties as in Eqs. (4.315)–(4.318):

- For $a_{j_1j_10}\prod_{s_2=1,s_2\neq s_1}^{3}(a_{j_1j_1s_1}-a_{j_1j_1s_2})/a_{j_2j_20}(a_{j_2j_2l_1}-a_{j_2j_2l_2}) > 0$ and $G_{j_1j_10} > 0$,

$$(a_{j_1j_1s_1}, a_{j_2j_2l_1}) = \underbrace{(\text{II}, \text{SO})}_{\text{increasing-inflection source}}.$$

- For $a_{j_1j_10}\prod_{s_2=1,s_2\neq s_1}^{3}(a_{j_1j_1s_1}-a_{j_1j_1s_2})/a_{j_2j_20}(a_{j_2j_2l_1}-a_{j_2j_2l_2}) > 0$ and $G_{j_1j_10} < 0$,

$$(a_{j_1j_1s_1}, a_{j_2j_2l_1}) = \underbrace{(\text{II}, \text{SI})}_{\text{increasing-inflection sink}}.$$

- For $a_{j_1j_10}\prod_{s_2=1,s_2\neq s_1}^{3}(a_{j_1j_1s_1}-a_{j_1j_1s_2})/a_{j_2j_20}(a_{j_2j_2l_1}-a_{j_2j_2l_2}) < 0$ and $G_{j_1j_10} > 0$,

$$(a_{j_1j_1s_1}, a_{j_2j_2l_1}) = \underbrace{(\text{DI}, \text{SO})}_{\text{decreasing-inflection source}}.$$

- For $a_{j_1j_10}\prod_{s_2=1,s_2\neq s_1}^{3}(a_{j_1j_1s_1}-a_{j_1j_1s_2})/a_{j_2j_20}(a_{j_2j_2l_1}-a_{j_2j_2l_2}) < 0$ and $G_{j_1j_10} < 0$,

$$(a_{j_1j_1s_1}, a_{j_2j_2l_1}) = \underbrace{(\text{DI}, \text{SI})}_{\text{decreasing-inflection sink}}.$$

(vi$_4$) For $a_{j_2j_2l_2}=a_{j_1j_1s_1}$ $(s_1,s_2\in\{1,2,3\})$, the dynamical system becomes

$$
\begin{aligned}
\dot{x}_{j_1} &= a_{j_1j_10}(y_{j_1}-a_{j_1j_1s_1})\prod_{s_2=1,s_2\neq s_1}^{3}(y_{j_1}-a_{j_1j_1s_2}),\\
\dot{x}_{j_2} &= a_{j_2j_20}(y_{j_1}-a_{j_2j_2l_1})(y_{j_1}-a_{j_2j_2l_2})^2,\\
& j_1,j_2\in\{1,2\};\ j_1\neq j_2;\ s_1,s_2\in\{1,2,3\},\ s_1\neq s_2.
\end{aligned}
$$

For $a_{j_2j_2l_2}=a_{j_1j_1s_1}$, a dynamical system in the y_{j_1}-direction is constructed as

$$
\begin{aligned}
\dot{y}_{j_1} = (y_{j_1}-a_{j_1j_1s_1})\Big[&\alpha_{j_1j_1}a_{j_1j_10}\prod_{s_2=1,s_2\neq s_1}^{3}(y_{j_1}-a_{j_1j_1s_2})\\
&+\alpha_{j_1j_1}a_{j_2j_20}(y_{j_1}-a_{j_2j_2l_1})(y_{j_1}-a_{j_2j_2l_2})\Big],
\end{aligned}
$$

and a dynamical system in the $y_{j_1}^{\perp}$-direction is constructed as

$$
\begin{aligned}
\dot{y}_{j_1}^{\perp} = (y_{j_1}-a_{j_1j_1s_1})\Big[&\beta_{j_1j_1}a_{j_1j_10}\prod_{s_2=1,s_2\neq s_1}^{3}(y_{j_1}-a_{j_1j_1s_2})\\
&+\beta_{j_1j_1}a_{j_2j_20}(y_{j_1}-a_{j_2j_2l_1})(y_{j_1}-a_{j_2j_2l_2})\Big]
\end{aligned}
$$

and a G-function is defined as

$$
\begin{aligned}
G_{j_1j_10} &= \frac{\partial}{\partial y_{j_1}}[\mathbf{n}^{\mathrm{T}}\cdot\mathbf{F}(y_{j_1},\mathbf{p})]\Big|_{y_{j_1}^*=a_{j_1j_1s_1}=a_{j_2j_2l_2}}\\
&= \alpha_{j_1j_1}a_{j_1j_10}\prod_{s_2=1,s_2\neq s_1}^{3}(a_{j_1j_1s_1}-a_{j_1j_1s_2}).
\end{aligned}
$$

For $y_{j_1}^*=a_{j_2j_2l_2}=a_{j_1j_1s_1}$, in phase space,

$$
\frac{dx_{j_2}}{dx_{j_1}}\Big|_{y_1^*=a_{j_2j_2l_2}} = \frac{a_{j_2j_20}(y_{j_1}-a_{j_2j_2l_1})(y_{j_1}-a_{j_2j_2l_2})}{a_{j_1j_10}\prod_{s_2=1,s_2\neq s_1}^{3}(y_{j_1}-a_{j_1j_1s_2})}\Big|_{y_1^*=a_{j_2j_2l_2}} = 0.
$$

If

$$
\frac{d^2x_{j_2}}{dx_{j_1}^2}\Big|_{y_1^*=a_{j_2j_2l_2}} = \frac{a_{j_2j_20}\alpha_{j_1j_1}^2(a_{j_2j_2l_2}-a_{j_2j_2l_1})}{a_{j_1j_10}\alpha_{j_1j_1}\prod_{s_1=1}^{3}(a_{j_2j_2l_2}-a_{j_1j_1s_1})} > 0,
$$

there is an up-parabola flow at $y_1^* = a_{j_2j_2l_2} = a_{j_1j_1s_1}$. If

$$\left.\frac{d^2x_{j_2}}{dx_{j_1}^2}\right|_{y_1^*=a_{j_2j_2l_2}} = \frac{a_{j_2j_20}\alpha_{j_1j_1}^2(a_{j_2j_2l_2} - a_{j_2j_2l_1})}{a_{j_1j_10}\alpha_{j_1j_1}\prod_{s_1=1}^{3}(a_{j_2j_2l_2} - a_{j_1j_1s_1})} < 0,$$

there is a down-parabola flow at $y_1^* = a_{j_2j_2l_2} = a_{j_1j_1s_1}$. Since the corresponding variational equation is

$$\begin{aligned}\Delta\dot{y}_{j_1} &= \alpha_{j_1j_1}a_{j_1j_10}\prod_{s_2=1,s_2\neq s_1}^{3}(a_{j_1j_1s_1} - a_{j_1j_1s_2})\Delta y_{j_1}\\ &= G_{j_1j_10}\Delta y_{j_1},\end{aligned}$$

the parabola flows are source and sink for $G_{j_1j_10} > 0$ and $G_{j_1j_10} < 0$, respectively.

Therefore, the infinite-equilibrium flow of $y_{j_1}^* = a_{j_2j_2l_2} = a_{j_1j_1s_1}$ has the following properties as in Eqs. (4.323)–(4.326):

- For $\alpha_{j_1j_1}a_{j_1j_10}\prod_{s_2=1,s_2\neq s_1}^{3}(a_{j_1j_1s_1} - a_{j_1j_1s_2}) > 0$ and $a_{j_2j_20}\alpha_{j_1j_1}^2(a_{j_2j_2l_2} - a_{j_2j_2l_1}) > 0$,

$$(a_{j_1j_1s_1}, a_{j_2j_2l_2}) = \underbrace{(\text{SO}, \text{UP})}_{\text{up-parabola source}}.$$

- For $\alpha_{j_1j_1}a_{j_1j_10}\prod_{s_2=1,s_2\neq s_1}^{3}(a_{j_1j_1s_1} - a_{j_1j_1s_2}) < 0$ and $a_{j_2j_20}\alpha_{j_1j_1}^2(a_{j_2j_2l_2} - a_{j_2j_2l_1}) > 0$,

$$(a_{j_1j_1s_1}, a_{j_2j_2l_2}) = \underbrace{(\text{SI}, \text{DP})}_{\text{down-parabola sink}}.$$

- For $\alpha_{j_1j_1}a_{j_1j_10}\prod_{s_2=1,s_2\neq s_1}^{3}(a_{j_1j_1s_1} - a_{j_1j_1s_2}) > 0$ and $a_{j_2j_20}\alpha_{j_1j_1}^2(a_{j_2j_2l_2} - a_{j_2j_2l_1}) < 0$,

$$(a_{j_1j_1s_1}, a_{j_2j_2l_2}) = \underbrace{(\text{SO}, \text{DP})}_{\text{down-parabola source}}.$$

- For $\alpha_{j_1j_1}a_{j_1j_10}\prod_{s_2=1,s_2\neq s_1}^{3}(a_{j_1j_1s_1} - a_{j_1j_1s_2}) < 0$ and $a_{j_2j_20}\alpha_{j_1j_1}^2(a_{j_2j_2l_2} - a_{j_2j_2l_1}) < 0$,

$$(a_{j_1j_1s_1}, a_{j_2j_2l_2}) = \underbrace{(\text{SI}, \text{UP})}_{\text{up-parabola sink}}.$$

(vi$_5$) For $a_{j_2j_21} = a_{j_1j_1s_1}$ $(s_1, s_2 \in \{1,2,3\})$,

$$\begin{aligned}
\dot{x}_{j_1} &= a_{j_1j_10}(y_{j_1} - a_{j_1j_1s_1}) \prod_{s_2=1, s_2\neq s_1}^{3} (y_{j_1} - a_{j_1j_1s_2}),\\
\dot{x}_{j_2} &= a_{j_2j_20}(y_{j_1} - a_{j_2j_21})^3,\\
&j_1, j_2 \in \{1,2\};\ j_1 \neq j_2;\ s_1, s_2 \in \{1,2,3\},\ s_1 \neq s_2.
\end{aligned}$$

For $a_{j_2j_21} = a_{j_1j_1s_1}$, a dynamical system in the y_{j_1}-direction is constructed as

$$\begin{aligned}
\dot{y}_{j_1} = (y_{j_1} - a_{j_1j_1s_1})\Big[&\alpha_{j_1j_1}a_{j_1j_10} \prod_{s_2=1, s_2\neq s_1}^{3} (y_{j_1} - a_{j_1j_1s_2})\\
&+ \alpha_{j_1j_1}a_{j_2j_20}(y_{j_1} - a_{j_2j_21})^2\Big],
\end{aligned}$$

and a dynamical system in the $y_{j_1}^{\perp}$-direction is constructed as

$$\begin{aligned}
\dot{y}_{j_1}^{\perp} = (y_{j_1} - a_{j_1j_1s_1})\Big[&\beta_{j_1j_1}a_{j_1j_10} \prod_{s_2=1, s_2\neq s_1}^{3} (y_{j_1} - a_{j_1j_1s_2})\\
&+ \beta_{j_1j_1}a_{j_2j_20}(y_{j_1} - a_{j_2j_21})^2\Big],
\end{aligned}$$

and a G-function is defined as

$$\begin{aligned}
G_{j_1j_10} &= \frac{\partial}{\partial y_{j_1}}[\mathbf{n}^{\mathrm{T}} \cdot \mathbf{F}(y_{j_1}, \mathbf{p})]\Big|_{y_{j_1}^* = a_{j_1j_1s_1} = a_{j_2j_21}}\\
&= \alpha_{j_1j_1}a_{j_1j_10} \prod_{s_2=1, s_2\neq s_1}^{3} (a_{j_1j_1s_1} - a_{j_1j_1s_2}).
\end{aligned}$$

In phase space, at $y_1^* = a_{j_2j_21} = a_{j_1j_1s_1}$,

$$\begin{aligned}
\frac{dx_{j_2}}{dx_{j_1}}\Big|_{y_1^* = a_{j_2j_21}} &= \frac{a_{j_2j_20}(y_{j_1} - a_{j_2j_21})^2}{a_{j_1j_10}\prod_{s_1=1}^{3}(y_{j_1} - a_{j_1j_1s_1})}\Big|_{y_1^* = a_{j_2j_2l_2}} = 0,\\
\frac{d^2x_{j_2}}{dx_{j_1}^2}\Big|_{y_1^* = a_{j_2j_21}} &= 0.
\end{aligned}$$

If

$$\frac{d^3x_{j_2}}{dx_{j_1}^3}\Big|_{y_1^* = a_{j_2j_21}} = \frac{2a_{j_2j_20}\alpha_{j_1j_1}^3}{a_{j_1j_10}\alpha_{j_1j_1}\prod_{s_1=1}^{3}(a_{j_2j_21} - a_{j_1j_1s_1})} > 0,$$

there is an second-order increasing-inflection flow at $y_1^* = a_{j_2j_21} = a_{j_1j_1s_1}$. If

$$\left.\frac{d^3x_{j_2}}{dx_{j_1}^3}\right|_{y_1^*=a_{j_2j_21}} = \frac{2a_{j_2j_20}\alpha_{j_1j_1}^3}{a_{j_1j_10}\alpha_{j_1j_1}\prod_{s_1=1}^{3}(a_{j_2j_21}-a_{j_1j_1s_1})} < 0,$$

there is a second-order decreasing-inflection flow at $y_1^* = a_{j_2j_21} = a_{j_1j_1s_1}$. Since the corresponding variational equation is

$$\begin{aligned}\Delta\dot{y}_{j_1} &= \alpha_{j_1j_1}a_{j_1j_10}\prod_{s_2=1,s_2\neq s_1}^{3}(a_{j_1j_1s_1}-a_{j_1j_1s_2})\Delta y_{j_1}\\ &= G_{j_1j_10}\Delta y_{j_1},\end{aligned}$$

the inflection flows are positive and negative for $G_{j_1j_10} > 0$ and $G_{j_1j_10} < 0$, respectively.

Therefore, the infinite-equilibrium of $y_{j_1}^* = a_{j_2j_21} = a_{j_1j_1s_1}$ has the following properties as in Eqs. (4.331)–(4.334):

- For $\alpha_{j_1j_1}a_{j_1j_10}\prod_{s_2=1,s_2\neq s_1}^{3}(a_{j_1j_1s_1}-a_{j_1j_1s_2}) > 0$ and $a_{j_2j_20}\alpha_{j_1j_1}^3 > 0$,

$$(a_{j_1j_1s_1}, a_{j_2j_21}) = \underbrace{(\text{SO}, 2^{\text{nd}}\text{II})}_{\text{second-order increasing-inflection source}}.$$

- For $\alpha_{j_1j_1}a_{j_1j_10}\prod_{s_2=1,s_2\neq s_1}^{3}(a_{j_1j_1s_1}-a_{j_1j_1s_2}) < 0$ and $a_{j_2j_20}\alpha_{j_1j_1}^3 > 0$,

$$(a_{j_1j_1s_1}, a_{j_2j_21}) = \underbrace{(\text{SI}, 2^{\text{nd}}\text{DI})}_{\text{second-order decreasing-inflection sink}}.$$

- For $\alpha_{j_1j_1}a_{j_1j_10}\prod_{s_2=1,s_2\neq s_1}^{3}(a_{j_1j_1s_1}-a_{j_1j_1s_2}) > 0$ and $a_{j_2j_20}\alpha_{j_1j_1}^3 < 0$,

$$(a_{j_1j_1s_1}, a_{j_2j_21}) = \underbrace{(\text{SO}, 2^{\text{nd}}\text{DI})}_{\text{second-order decreasing-inflection source}}.$$

- For $\alpha_{j_1j_1}a_{j_1j_10}\prod_{s_2=1,s_2\neq s_1}^{3}(a_{j_1j_1s_1}-a_{j_1j_1s_2}) < 0$ and $a_{j_2j_20}\alpha_{j_1j_1}^3 < 0$,

$$(a_{j_1j_1s_1}, a_{j_2j_21}) = \underbrace{(\text{SI}, 2^{\text{nd}}\text{II})}_{\text{second-order increasing-inflection sink}}.$$

(vii) For

$$\Delta_{j_1j_1} < 0 \quad \text{and} \quad \Delta_{j_2j_2} > 0,$$

the dynamical system becomes

$$\begin{aligned}\dot{x}_{j_1} &= a_{j_1j_10}(y_{j_1} - a_{j_1j_11})[(y_{j_1} - a_{j_1j_1})^2 + b_{j_1j_1}],\\ \dot{x}_{j_2} &= a_{j_2j_20}(y_{j_1} - a_{j_2j_21})(y_{j_1} - a_{j_2j_22})(y_{j_1} - a_{j_2j_22}),\end{aligned}$$

where

$$\{a_{j_2j_21}, a_{j_2j_22}, a_{j_2j_23}\} = \text{sort}\{b_{j_2j_21}, b_{j_2j_22}, b_{j_2j_22}\},\ a_{j_2j_2l_1} < a_{j_2j_2l_2},\ l_1 < l_2;$$

$$b_{j_2j_22}, b_{j_2j_23} = -\frac{1}{2}(B_{j_2j_2} \pm \sqrt{\Delta_{j_2j_2}}), \quad a_{j_2j_20} = A_{j_2j_20};$$

$$a_{j_1j_11} = b_{j_1j_11}, a_{j_1j_1} = -\frac{1}{2}B_{j_1j_11}, b_{j_1j_1} = -\frac{1}{4}\Delta_{j_1j_1}, a_{j_1j_10} = A_{j_1j_10}.$$

(vii$_1$) For $a_{j_1j_11} \neq a_{j_2j_2l_1}(l_1, l_2 \in \{1, 2\})$, a dynamical system in the y_{j_1}-direction is constructed as

$$\begin{aligned}\dot{y}_{j_1} &= \alpha_{j_1j_1}a_{j_1j_10}(y_{j_1} - a_{j_1j_11})[(y_{j_1} - a_{j_1j_1})^2 + b_{j_1j_1}]\\ &\quad + \alpha_{j_1j_2}a_{j_2j_20}\prod_{l_1=1}^{3}(y_{j_1} - a_{j_2j_2l_1}),\end{aligned}$$

and a dynamical system in the $y_{j_1}^{\perp}$-direction is constructed as

$$\begin{aligned}\dot{y}_{j_1}^{\perp} &= \beta_{j_1j_1}a_{j_1j_10}(y_{j_1} - a_{j_1j_11})[(y_{j_1} - a_{j_1j_1})^2 + b_{j_1j_1}]\\ &\quad + \beta_{j_1j_2}a_{j_2j_20}\prod_{l_1=1}^{3}(y_{j_1} - a_{j_2j_2l_1}).\end{aligned}$$

(vii$_{1a}$) In phase space, at $y_1^* = a_{j_1j_11}$,

$$\left.\frac{dx_{j_1}}{dx_{j_2}}\right|_{y_1^*=a_{j_1j_11}} = \left.\frac{a_{j_1j_10}(y_{j_1} - a_{j_1j_11})[(y_{j_1} - a_{j_1j_1})^2 + b_{j_1j_1}]}{a_{j_2j_20}\prod_{l_1=1}^{3}(y_{j_1} - a_{j_2j_2l_1})}\right|_{y_1^*=a_{j_1j_11}} = 0.$$

If

$$\left.\frac{d^2x_{j_1}}{dx_{j_2}^2}\right|_{y_1^*=a_{j_1j_11}} = \frac{a_{j_1j_10}\alpha_{j_1j_2}[(a_{j_1j_11} - a_{j_1j_1})^2 + b_{j_1j_1}]}{a_{j_2j_20}\prod_{l_1=1}^{3}(a_{j_1j_11} - a_{j_2j_2l_1})} > 0,$$

there is an up-parabola flow at $y_1^* = a_{j_1j_11}$. If

$$\left.\frac{d^2x_{j_1}}{dx_{j_2}^2}\right|_{y_1^*=a_{j_1j_11}} = \frac{a_{j_1j_10}\alpha_{j_1j_2}[(a_{j_1j_11} - a_{j_1j_1})^2 + b_{j_1j_1}]}{a_{j_2j_20}\prod_{l_1=1}^{3}(a_{j_1j_11} - a_{j_2j_2l_1})} < 0,$$

there is a down-parabola flow at $y_1^* = a_{j_1j_11}$. With

$$\dot{x}_{j_2}^* = a_{j_2j_20}\prod_{l_1=1}^{3}(a_{j_1j_11} - a_{j_2j_2l_1}),$$

the parabola flows are positive and negative for $\dot{x}_{j_2}^* > 0$ and $\dot{x}_{j_2}^* < 0$, respectively.

Therefore, the flow of $y_1^* = a_{j_1j_11}$ with $\dot{x}_{j_2}^* = a_{j_2j_20}\prod_{l_1=1}^{3}(a_{j_1j_11}-a_{j_2j_2l_1})$ possesses the following properties as in Eqs. (4.340)–(4.343):

- For $\alpha_{j_1j_1}a_{j_1j_10} > 0$ and $a_{j_2j_20}\prod_{l_1=1}^{3}(a_{j_1j_11} - a_{j_2j_2l_1}) > 0$,
$$(a_{j_1j_11}, \dot{x}_{j_2}^*) = \underbrace{(\text{UP}, \text{pF})}_{\text{up-parabola flow }(+)}.$$
- For $\alpha_{j_1j_1}a_{j_1j_10} < 0$ and $a_{j_2j_20}\prod_{l_1=1}^{3}(a_{j_1j_11} - a_{j_2j_2l_1}) > 0$,
$$(a_{j_1j_11}, \dot{x}_{j_2}^*) = \underbrace{(\text{DP}, \text{pF})}_{\text{down-parabola flow }(+)}.$$
- For $\alpha_{j_1j_1}a_{j_1j_10} > 0$ and $a_{j_2j_20}\prod_{l_1=1}^{3}(a_{j_1j_11} - a_{j_2j_2l_1}) < 0$,
$$(a_{j_1j_11}, \dot{x}_{j_2}^*) = \underbrace{(\text{DP}, \text{nF})}_{\text{down-parabola flow }(-)}.$$
- For $\alpha_{j_1j_1}a_{j_1j_10} < 0$ and $a_{j_2j_20}\prod_{l_1=1}^{3}(a_{j_1j_11} - a_{j_2j_2l_1}) < 0$,
$$(a_{j_1j_11}, \dot{x}_{j_2}^*) = \underbrace{(\text{UP}, \text{nF})}_{\text{up-parabola flow }(-)}.$$

($\text{vii}_{1\text{b}}$) In phase space, at $y_1^* = a_{j_2j_2l_1}$,

$$\left.\frac{dx_{j_2}}{dx_{j_1}}\right|_{y_1^*=a_{j_2j_2l_1}} = \left.\frac{a_{j_2j_20}(y_{j_1} - a_{j_2j_2l_1})\prod_{l_2=1,l_2\neq l_1}^{3}(y_{j_1} - a_{j_2j_2l_2})}{a_{j_1j_10}(y_{j_1} - a_{j_1j_11})[(y_{j_1} - a_{j_1j_1})^2 + b_{j_1j_1}]}\right|_{y_1^*=a_{j_2j_2l_1}} = 0.$$

If

$$\left.\frac{d^2x_{j_2}}{dx_{j_1}^2}\right|_{y_1^*=a_{j_2j_2l_1}} = \frac{a_{j_2j_20}\alpha_{j_1j_1}\prod_{l_2=1,l_2\neq l_1}^{3}(a_{j_2j_2l_1} - a_{j_2j_2l_2})}{a_{j_1j_10}(a_{j_2j_2l_1} - a_{j_1j_11})[(a_{j_2j_2l_1} - a_{j_1j_1})^2 + b_{j_1j_1}]} > 0,$$

there is an up-parabola flow at $y_1^* = a_{j_2j_2l_1}$. If

$$\left.\frac{d^2x_{j_2}}{dx_{j_1}^2}\right|_{y_1^*=a_{j_2j_2l_1}} = \frac{a_{j_2j_20}\alpha_{j_1j_1}\prod_{l_2=1,l_2\neq l_1}^{3}(a_{j_2j_2l_1}-a_{j_2j_2l_2})}{a_{j_1j_10}(a_{j_2j_2l_1}-a_{j_1j_11})[(a_{j_2j_2l_1}-a_{j_1j_1})^2+b_{j_1j_1}]} < 0,$$

there is a down-parabola flow at $y_1^* = a_{j_2j_2l_1}$. With

$$\dot{x}_{j_1}^* = a_{j_1j_10}(a_{j_2j_2l_1}-a_{j_1j_11})[(a_{j_2j_2l_1}-a_{j_1j_1})^2+b_{j_1j_1}],$$

the parabola flows are positive and negative for $\dot{x}_{j_1}^* > 0$ and $\dot{x}_{j_1}^* < 0$, respectively.

Due to

$$(a_{j_2j_2l_1}-a_{j_1j_1})^2+b_{j_1j_1} > 0,$$

the flow of $y_1^* = a_{j_2j_2l_1}$ possesses the following properties as in Eqs. (4.344)–(4.347):

- For $a_{j_1j_10}(a_{j_2j_2l_1}-a_{j_1j_11}) > 0$ and $a_{j_2j_20}\alpha_{j_1j_1}\prod_{l_2=1,l_2\neq l_1}^{3}(a_{j_2j_2l_1}-a_{j_2j_2l_2}) > 0$,

$$(\dot{x}_{j_1}^*, a_{j_2j_2l_1}) = \underbrace{(\text{pF}, \text{UP})}_{\text{up-parabola flow }(+)}.$$

- For $a_{j_1j_10}(a_{j_2j_2l_1}-a_{j_1j_11}) < 0$ and $a_{j_2j_20}\alpha_{j_1j_1}\prod_{l_2=1,l_2\neq l_1}^{3}(a_{j_2j_2l_1}-a_{j_2j_2l_2}) > 0$,

$$(\dot{x}_{j_1}^*, a_{j_2j_2l_1}) = \underbrace{(\text{nF}, \text{DP})}_{\text{down-parabola flow }(-)}.$$

- For $a_{j_1j_10}(a_{j_2j_2l_1}-a_{j_1j_11}) > 0$ and $a_{j_2j_20}\alpha_{j_1j_1}\prod_{l_2=1,l_2\neq l_1}^{3}(a_{j_2j_2l_1}-a_{j_2j_2l_2}) < 0$,

$$(\dot{x}_{j_1}^*, a_{j_2j_2l_1}) = \underbrace{(\text{pF}, \text{DP})}_{\text{down-parabola flow }(+)}.$$

- For $a_{j_1j_10}(a_{j_2j_2l_1}-a_{j_1j_11}) < 0$ and $a_{j_2j_20}\alpha_{j_1j_1}\prod_{l_2=1,l_2\neq l_1}^{3}(a_{j_2j_2l_1}-a_{j_2j_2l_2}) < 0$,

$$(\dot{x}_{j_1}^*, a_{j_2j_2l_1}) = \underbrace{(\text{nF}, \text{UP})}_{\text{up-parabola flow }(-)}.$$

(vii$_2$) For $a_{j_1j_1 1} = a_{j_2j_2l_1}$, consider a dynamical system as

$$\dot{x}_{j_1} = a_{j_1j_1 0}(y_{j_1} - a_{j_1j_1 1})[(y_{j_1} - a_{j_1j_1})^2 + b_{j_1j_1}],$$

$$\dot{x}_{j_2} = a_{j_2j_2 0}(y_{j_1} - a_{j_2j_2l_1}) \prod_{l_2=1, l_2 \neq l_1}^{3} (y_{j_1} - a_{j_2j_2l_2}).$$

A dynamical system in the y_{j_1}-direction is constructed as

$$\dot{y}_{j_1} = (y_{j_1} - a_{j_1j_1 1})\Big\{\alpha_{j_1j_1} a_{j_1j_1 0}[(y_{j_1} - a_{j_1j_1})^2 + b_{j_1j_1}]$$

$$+ \alpha_{j_1j_2} a_{j_2j_2 0} \prod_{l_2=1, l_2 \neq l_1}^{3} (y_{j_1} - a_{j_2j_2l_2})\Big\},$$

and a dynamical system in the $y_{j_1}^{\perp}$-direction is constructed as

$$\dot{y}_{j_1}^{\perp} = (y_{j_1} - a_{j_1j_1 1})\Big\{\beta_{j_1j_1} a_{j_1j_1 0}[(y_{j_1} - a_{j_1j_1})^2 + b_{j_1j_1}]$$

$$+ \beta_{j_1j_2} a_{j_2j_2 0} \prod_{l_2=1, l_2 \neq l_1}^{3} (y_{j_1} - a_{j_2j_2l_2})\Big\}.$$

Define a G-function as

$$G_{j_1j_1 0} = \frac{\partial}{\partial y_{j_1}}[\mathbf{n}^{\mathrm{T}} \cdot \mathbf{F}(y_{j_1}, \mathbf{p})]\Big|_{y_{j_1}^{*} = a_{j_1j_1 1} = a_{j_2j_2l_1}}$$

$$= \alpha_{j_1j_1} a_{j_1j_1 0}[(a_{j_1j_1 1} - a_{j_1j_1})^2 + b_{j_1j_1}]$$

$$+ \alpha_{j_1j_2} a_{j_2j_2 0} \prod_{l_2=1, l_2 \neq l_1}^{3} (a_{j_2j_2l_1} - a_{j_2j_2l_2}).$$

In phase space, at $y_1^{*} = a_{j_1j_1 1} = a_{22l_1}$, if

$$\frac{dx_{j_2}}{dx_{j_1}}\Big|_{y_1^{*} = a_{j_2j_2l_1}} = \frac{a_{j_2j_2 0} \prod_{l_2=1, l_2 \neq l_1}^{3} (a_{j_2j_2l_1} - a_{j_2j_2l_2})}{a_{j_1j_1 0}[(a_{j_2j_2l_1} - a_{j_1j_1})^2 + b_{j_1j_1}]} > 0,$$

there is an increasing-inflection flow at $y_1^{*} = a_{j_2j_2l_1} = a_{j_1j_1 1}$, and if

$$\frac{dx_{j_2}}{dx_{j_1}}\Big|_{y_1^{*} = a_{j_2j_2l_1}} = \frac{a_{j_2j_2 0} \prod_{l_2=1, l_2 \neq l_1}^{3} (a_{j_2j_2l_1} - a_{j_2j_2l_2})}{a_{j_1j_1 0}[(a_{j_2j_2l_1} - a_{j_1j_1})^2 + b_{j_1j_1}]} < 0,$$

there is a decreasing-inflection flow at $y_1^* = a_{j_2j_2l_1} = a_{j_1j_11}$. The variational equation at $y_1^* = a_{j_2j_2l_1} = a_{j_1j_11}$ is

$$\begin{aligned}\Delta\dot{y}_{j_1} &= \Big\{\alpha_{j_1j_1}a_{j_1j_10}[(a_{j_1j_11} - a_{j_1j_1})^2 + b_{j_1j_1}] \\ &\quad + \alpha_{j_1j_2}a_{j_2j_20}\prod_{l_2=1,l_2\neq l_1}^{3}(a_{j_2j_2l_1} - a_{j_2j_2l_2})\Big\}\Delta y_{j_1} \\ &= G_{j_1j_10}\Delta y_{j_1}.\end{aligned}$$

The inflection flows are source and sink for $G_{j_1j_10} > 0$ and $G_{j_1j_10} < 0$, respectively.

Therefore, the infinite-equilibrium flow of $y_1^* = a_{j_2j_2l_1} = a_{j_1j_11}$ possesses the following properties as in Eqs. (4.352)–(4.355):

- For $a_{j_1j_10}/a_{j_2j_20}\prod_{l_2=1,l_2\neq l_1}^{3}(a_{j_2j_2l_1} - a_{j_2j_2l_2}) > 0$ and $G_{j_1j_10} > 0$,
$$(a_{j_1j_11}, a_{j_2j_2l_1}) = \underbrace{(\mathrm{II}, \mathrm{SO})}_{\text{increaing-inflection source}}.$$
- For $a_{j_1j_10}/a_{j_2j_20}\prod_{l_2=1,l_2\neq l_1}^{3}(a_{j_2j_2l_1} - a_{j_2j_2l_2}) > 0$ and $G_{j_1j_10} < 0$,
$$(a_{j_1j_11}, a_{j_2j_2l_1}) = \underbrace{(\mathrm{II}, \mathrm{SI})}_{\text{increaing-inflection sink}}.$$
- For $a_{j_1j_10}/a_{j_2j_20}\prod_{l_2=1,l_2\neq l_1}^{3}(a_{j_2j_2l_1} - a_{j_2j_2l_2}) < 0$ and $G_{j_1j_10} > 0$,
$$(a_{j_1j_11}, a_{j_2j_2l_1}) = \underbrace{(\mathrm{DI}, \mathrm{SO})}_{\text{decreaing-inflection source}}.$$
- For $a_{j_1j_10}/a_{j_2j_20}\prod_{l_2=1,l_2\neq l_1}^{3}(a_{j_2j_2l_1} - a_{j_2j_2l_2}) < 0$ and $G_{j_1j_10} < 0$,
$$(a_{j_1j_11}, a_{j_2j_2l_1}) = \underbrace{(\mathrm{DI}, \mathrm{SI})}_{\text{decreaing-inflection sink}}.$$

(viii) For

$$\Delta_{j_1j_1} = 0 \quad \text{and} \quad \Delta_{j_2j_2} > 0,$$

the dynamical system becomes

$$\begin{aligned}\dot{x}_{j_1} &= a_{j_1j_10}(y_{j_1} - a_{j_1j_1s_1})(y_{j_1} - a_{j_1j_1s_2})^2, \\ \dot{x}_{j_2} &= a_{j_2j_20}(y_{j_1} - a_{j_2j_21})(y_{j_1} - a_{j_2j_22})(y_{j_1} - a_{j_2j_23}), \\ &j_1, j_2 \in \{1, 2\};\ j_1 \neq j_2;\end{aligned}$$

where

$$\{a_{j_2j_21}, a_{j_2j_22}, a_{j_2j_23}\} = \text{sort}\{b_{j_2j_21}, b_{j_2j_22}, b_{j_2j_23}\},$$

$$a_{j_2j_2l_1} < a_{j_2j_2l_2}, l_1 < l_2;$$

$$b_{j_2j_22}, b_{j_2j_23} = -\frac{1}{2}(B_{j_2j_2} \pm \sqrt{\Delta_{j_2j_2}}), a_{j_2j_20} = A_{j_2j_20};$$

$$a_{j_1j_1s_1} = b_{j_1j_11},\ a_{j_1j_1s_2} = -\frac{1}{2}B_{j_1j_1},\ b_{j_1j_1} = 0,\ a_{j_1j_10} = A_{j_1j_10}.$$

(viii$_1$) For $a_{j_2j_2l_1} \neq a_{j_1j_1s_1}, a_{j_1j_1s_2}$, a dynamical system in the y_{j_1}-direction is constructed as

$$\dot{y}_{j_1} = \alpha_{j_1j_1}a_{j_1j_10}(y_{j_1} - a_{j_1j_1s_1})(y_{j_1} - a_{j_1j_1s_2})^2 + \alpha_{j_1j_2}a_{j_2j_20}\prod_{l_1=1}^{3}(y_{j_1} - a_{j_2j_2l_1}),$$

and a dynamical system in the $y_{j_1}^{\perp}$-direction is constructed as

$$\dot{y}_{j_1}^{\perp} = \beta_{j_1j_1}a_{j_1j_10}(y_{j_1} - a_{j_1j_1s_1})(y_{j_1} - a_{j_1j_1s_2})^2 + \beta_{j_1j_2}a_{j_2j_20}\prod_{l_1=1}^{3}(y_{j_1} - a_{j_2j_2l_1}).$$

(viii$_{1a}$) In phase space, at $y_1^* = a_{j_1j_1s_1}$,

$$\left.\frac{dx_{j_1}}{dx_{j_2}}\right|_{y_1^*=a_{j_1j_1s_1}} = \left.\frac{a_{j_1j_10}(y_{j_1} - a_{j_1j_1s_1})(y_{j_1} - a_{j_1j_1s_2})^2}{a_{j_2j_20}\prod_{l_1=1}^{3}(y_{j_1} - a_{j_2j_2l_1})}\right|_{y_1^*=a_{j_1j_1s_1}} = 0.$$

If

$$\left.\frac{d^2x_{j_1}}{dx_{j_2}^2}\right|_{y_1^*=a_{j_1j_1s_1}} = \frac{a_{j_1j_10}\alpha_{j_1j_2}(a_{j_1j_1s_1} - a_{j_1j_1s_2})^2}{a_{j_2j_20}\prod_{l_1=1}^{3}(a_{j_1j_1s_1} - a_{j_2j_2l_1})} > 0,$$

there is an up-parabola flow at $y_1^* = a_{j_1j_1s_1}$. If

$$\left.\frac{d^2x_{j_1}}{dx_{j_2}^2}\right|_{y_1^*=a_{j_1j_1s_1}} = \frac{a_{j_1j_10}\alpha_{j_1j_2}(a_{j_1j_1s_1} - a_{j_1j_1s_2})^2}{a_{j_2j_20}\prod_{l_1=1}^{3}(a_{j_1j_1s_1} - a_{j_2j_2l_1})} < 0,$$

there is a down-parabola flow at $y_1^* = a_{j_1j_1s_1}$. With

$$\dot{x}_{j_2}^* = a_{j_2j_20}\prod_{l_1=1}^{3}(a_{j_1j_1s_1} - a_{j_2j_2l_1}),$$

the parabola flows are positive and negative for $\dot{x}_{j_2}^* > 0$ and $\dot{x}_{j_2}^* < 0$, respectively.

Therefore, the flows of $y^*_{j_1} = a_{j_1j_1s_1} \neq a_{j_2j_21}$ for the x_{j_1}-direction have the following properties as in Eqs. (4.361)–(4.364):

- For $a_{j_1j_10}\alpha_{j_1j_2}(a_{j_1j_1s_1} - a_{j_1j_1s_2})^2 > 0$ and $a_{j_2j_20}\prod_{l_1=1}^{3}(a_{j_1j_1s_1} - a_{j_2j_2l_1}) > 0$,

$$(a_{j_1j_1s_1}, \dot{x}^*_{j_2}) = \underbrace{(\mathrm{UP}, \mathrm{pF})}_{\text{up-parabola flow }(+)}.$$

- For $a_{j_1j_10}\alpha_{j_1j_2}(a_{j_1j_1s_1} - a_{j_1j_1s_2})^2 < 0$ and $a_{j_2j_20}\prod_{l_1=1}^{3}(a_{j_1j_1s_1} - a_{j_2j_2l_1}) > 0$,

$$(a_{j_1j_1s_1}, \dot{x}^*_{j_2}) = \underbrace{(\mathrm{DP}, \mathrm{pF})}_{\text{down-parabola flow }(+)}.$$

- For $a_{j_1j_10}\alpha_{j_1j_2}(a_{j_1j_1s_1} - a_{j_1j_1s_2})^2 > 0$ and $a_{j_2j_20}\prod_{l_1=1}^{3}(a_{j_1j_1s_1} - a_{j_2j_2l_1}) < 0$,

$$(a_{j_1j_1s_1}, \dot{x}^*_{j_2}) = \underbrace{(\mathrm{DP}, \mathrm{nF})}_{\text{down-parabola flow }(-)}.$$

- For $a_{j_1j_10}\alpha_{j_1j_2}(a_{j_1j_1s_1} - a_{j_1j_1s_2})^2 < 0$ and $a_{j_2j_20}\prod_{l_1=1}^{3}(a_{j_1j_1s_1} - a_{j_2j_2l_1}) < 0$,

$$(a_{j_1j_1s_1}, \dot{x}^*_{j_2}) = \underbrace{(\mathrm{UP}, \mathrm{nF})}_{\text{up-parabola flow }(-)}.$$

(viii$_{1b}$) In phase space, at $y^*_1 = a_{j_1j_1s_2}$,

$$\left.\frac{dx_{j_1}}{dx_{j_2}}\right|_{y^*_1=a_{j_1j_1s_2}} = \left.\frac{a_{j_1j_10}(y_{j_1} - a_{j_1j_1s_1})(y_{j_1} - a_{j_1j_1s_2})^2}{a_{j_2j_20}\prod_{l_1=1}^{3}(y_{j_1} - a_{j_2j_2l_1})}\right|_{y^*_1=a_{j_1j_1s_2}} = 0,$$

$$\left.\frac{d^2x_{j_1}}{dx^2_{j_2}}\right|_{y^*_1=a_{j_1j_1s_2}} = 0.$$

If

$$\left.\frac{d^3x_{j_1}}{dx^3_{j_2}}\right|_{y^*_1=a_{j_1j_1s_2}} = \frac{2a_{j_1j_10}\alpha^2_{j_1j_2}(a_{j_1j_1s_2} - a_{j_1j_1s_1})}{a_{j_2j_20}\prod_{l_1=1}^{3}(a_{j_1j_1s_1} - a_{j_2j_2l_1})} > 0,$$

there is an increasing-inflection flow at $y_1^* = a_{j_1j_1s_2}$. If

$$\left.\frac{d^3x_{j_1}}{dx_{j_2}^3}\right|_{y_1^*=a_{j_1j_1s_2}} = \frac{2a_{j_1j_10}\alpha_{j_1j_2}^2(a_{j_1j_1s_2} - a_{j_1j_1s_1})}{a_{j_2j_20}\prod_{l_1=1}^3(a_{j_1j_1s_1} - a_{j_2j_2l_1})} < 0,$$

there is a decreasing-inflection flow at $y_1^* = a_{j_1j_1s_2}$. With

$$\dot{x}_{j_2}^* = a_{j_2j_20}\prod_{l_1=1}^3(a_{j_1j_1s_2} - a_{j_2j_2l_1}),$$

the inflection flows are positive and negative for $\dot{x}_{j_2}^* > 0$ and $\dot{x}_{j_2}^* < 0$, respectively.

Therefore, the flows of $y_{j_1}^* = a_{j_1j_1s_2} \neq a_{j_2j_21}$ for the x_{j_1}-direction with $a_{j_1j_11} \neq a_{j_2j_21}$ have the following properties as in Eqs. (4.365)–(4.368):

- For $a_{j_1j_10}\alpha_{j_1j_2}^2(a_{j_1j_1s_2} - a_{j_1j_1s_1}) > 0$ and $a_{j_2j_20}\prod_{l_1=1}^3(a_{j_1j_1s_2} - a_{j_2j_2l_1}) > 0$,
$$(a_{j_1j_1s_2}, \dot{x}_{j_2}^*) = \underbrace{(\text{II}, \text{pF})}_{\text{increasing-inflection flow }(-)}.$$
- For $a_{j_1j_10}\alpha_{j_1j_2}^2(a_{j_1j_1s_2} - a_{j_1j_1s_1}) > 0$ and $a_{j_2j_20}\prod_{l_1=1}^3(a_{j_1j_1s_2} - a_{j_2j_2l_1}) > 0$,
$$(a_{j_1j_1s_2}, \dot{x}_{j_2}^*) = \underbrace{(\text{DI}, \text{pF})}_{\text{decreasing-inflection flow }(+)}.$$
- For $a_{j_1j_10}\alpha_{j_1j_2}^2(a_{j_1j_1s_2} - a_{j_1j_1s_1}) > 0$ and $a_{j_2j_20}\prod_{l_1=1}^3(a_{j_1j_1s_2} - a_{j_2j_2l_1}) < 0$,
$$(a_{j_1j_1s_2}, \dot{x}_{j_2}^*) = \underbrace{(\text{DI}, \text{nF})}_{\text{decreasing-inflection flow }(-)}.$$
- For $a_{j_1j_10}\alpha_{j_1j_2}^2(a_{j_1j_1s_2} - a_{j_1j_1s_1}) < 0$ and $a_{j_2j_20}\prod_{l_1=1}^3(a_{j_1j_1s_2} - a_{j_2j_2l_1}) < 0$,
$$(a_{j_1j_1s_2}, \dot{x}_{j_2}^*) = \underbrace{(\text{II}, \text{nF})}_{\text{increasing-inflection flow }(-)}.$$

– The increasing and decreasing inflection flows are the appearing bifurcations for the up-parabola and down-parabola flows.

(viii$_{1c}$) In phase space, at $y_1^* = a_{j_2j_2l_1}$,

$$\left.\frac{dx_{j_2}}{dx_{j_1}}\right|_{y_1^*=a_{j_2j_2l_1}} = \left.\frac{a_{j_2j_20}(y_{j_1}-a_{j_2j_2l_1})\prod_{l_2=1,l_2\neq l_1}^{3}(y_{j_1}-a_{j_2j_2l_2})}{a_{j_1j_10}(y_{j_1}-a_{j_1j_1s_1})(y_{j_1}-a_{j_1j_1s_2})^2}\right|_{y_1^*=a_{j_2j_21}} = 0.$$

If

$$\left.\frac{d^2x_{j_2}}{dx_{j_1}^2}\right|_{y_1^*=a_{j_2j_2l_1}} = \frac{a_{j_2j_20}\alpha_{j_1j_1}\prod_{l_2=1,l_2\neq l_1}^{3}(a_{j_2j_2l_1}-a_{j_2j_2l_2})}{a_{j_1j_10}(a_{j_2j_2l_1}-a_{j_1j_1s_1})(a_{j_2j_2l_1}-a_{j_1j_1s_2})^2} > 0,$$

there is an up-parabola flow at $y_1^* = a_{j_2j_21}$. If

$$\left.\frac{d^2x_{j_2}}{dx_{j_1}^2}\right|_{y_1^*=a_{j_2j_2l_1}} = \frac{a_{j_2j_20}\alpha_{j_1j_1}\prod_{l_2=1,l_2\neq l_1}^{3}(a_{j_2j_2l_1}-a_{j_2j_2l_2})}{a_{j_1j_10}(a_{j_2j_2l_1}-a_{j_1j_1s_1})(a_{j_2j_2l_1}-a_{j_1j_1s_2})^2} < 0,$$

there is a down-parabola flow at $y_1^* = a_{j_2j_21}$. With

$$\dot{x}_{j_1}^* = a_{j_1j_10}(a_{j_2j_2l_1}-a_{j_1j_1s_1})(a_{j_2j_2l_1}-a_{j_1j_1s_2})^2,$$

the parabola flows are positive and negative for $\dot{x}_{j_1}^* > 0$ and $\dot{x}_{j_1}^* < 0$, respectively.

Therefore, the flow of $y_{j_1}^* = a_{j_2j_2l_1} \neq a_{j_1j_1s_1}, a_{j_1j_1s_2}$ for the x_{j_2}-direction has the following properties as in Eqs. (4.369)–(4.372):

- For $\dot{x}_{j_1}^* > 0$ and $a_{j_2j_20}\alpha_{j_1j_1}\prod_{l_2=1,l_2\neq l_1}^{3}(a_{j_2j_2l_1}-a_{j_2j_2l_2}) > 0$,
$$(\dot{x}_{j_1}^*, a_{j_2j_2l_1}) = \underbrace{(\text{pF},\text{UP})}_{\text{up-parabola flow }(+)}.$$
- For $\dot{x}_{j_1}^* < 0$ and $a_{j_2j_20}\alpha_{j_1j_1}\prod_{l_2=1,l_2\neq l_1}^{3}(a_{j_2j_2l_1}-a_{j_2j_2l_2}) > 0$,
$$(\dot{x}_{j_1}^*, a_{j_2j_2l_1}) = \underbrace{(\text{nF},\text{DP})}_{\text{down-parabola flow }(-)}.$$
- For $\dot{x}_{j_1}^* > 0$ and $a_{j_2j_20}\alpha_{j_1j_1}\prod_{l_2=1,l_2\neq l_1}^{3}(a_{j_2j_2l_1}-a_{j_2j_2l_2}) < 0$,
$$(\dot{x}_{j_1}^*, a_{j_2j_2l_1}) = \underbrace{(\text{pF},\text{DP})}_{\text{down-parabola flow }(+)}.$$
- For $\dot{x}_{j_1}^* < 0$ and $a_{j_2j_20}\alpha_{j_1j_1}\prod_{l_2=1,l_2\neq l_1}^{3}(a_{j_2j_2l_1}-a_{j_2j_2l_2}) < 0$,
$$(\dot{x}_{j_1}^*, a_{j_2j_2l_1}) = \underbrace{(\text{nF},\text{UP})}_{\text{up-parabola flow }(-)}.$$

(viii$_2$) For $a_{j_1j_1s_1}=a_{j_1j_1s_2}\equiv a_{j_1j_11}$, consider a system as

$$\dot{x}_{j_1}=a_{j_1j_10}(y_{j_1}-a_{j_1j_11})^3,$$
$$\dot{x}_{j_2}=a_{j_2j_20}(y_{j_1}-a_{j_2j_21})(y_{j_1}-a_{j_2j_22})(y_{j_1}-a_{j_2j_23}),$$
$$j_1,j_2\in\{1,2\};\ j_1\neq j_2.$$

For $a_{j_2j_2l_1}\neq a_{j_1j_11}$, a dynamical system in the y_{j_1}-direction is constructed as

$$\dot{y}_{j_1}=\alpha_{j_1j_1}a_{j_1j_10}(y_{j_1}-a_{j_1j_1s_1})(y_{j_1}-a_{j_1j_1s_2})^2 + \alpha_{j_1j_2}a_{j_2j_20}\prod_{l_1=1}^{3}(y_{j_1}-a_{j_2j_2l_1}),$$

and a dynamical system in the $y_{j_1}^{\perp}$-direction is constructed as

$$\dot{y}_{j_1}^{\perp}=\beta_{j_1j_1}a_{j_1j_10}(y_{j_1}-a_{j_1j_1s_1})(y_{j_1}-a_{j_1j_1s_2})^2 + \beta_{j_1j_2}a_{j_2j_20}\prod_{l_1=1}^{3}(y_{j_1}-a_{j_2j_2l_1}).$$

(viii$_{2a}$) In phase space, at $y_1^*=a_{j_1j_11}$,

$$\left.\frac{dx_{j_1}}{dx_{j_2}}\right|_{y_1^*=a_{j_1j_11}}=\left.\frac{a_{j_1j_10}(y_{j_1}-a_{j_1j_11})^3}{a_{j_2j_20}\prod_{l_1=1}^{3}(y_{j_1}-a_{j_2j_2l_1})}\right|_{y_1^*=a_{j_1j_11}}=0,$$

$$\left.\frac{d^2x_{j_1}}{dx_{j_2}^2}\right|_{y_1^*=a_{j_1j_11}}=\left.\frac{d^3x_{j_1}}{dx_{j_2}^3}\right|_{y_1^*=a_{j_1j_11}}=0.$$

If

$$\left.\frac{d^4x_{j_1}}{dx_{j_2}^4}\right|_{y_1^*=a_{j_1j_11}}=\frac{6a_{j_1j_10}\alpha_{j_1j_2}^3}{a_{j_2j_20}\prod_{l_1=1}^{3}(a_{j_1j_11}-a_{j_2j_2l_1})}>0,$$

there is a third-order up-parabola flow at $y_1^*=a_{j_1j_11}$. If

$$\left.\frac{d^4x_{j_1}}{dx_{j_2}^4}\right|_{y_1^*=a_{j_1j_11}}=\frac{6a_{j_1j_10}\alpha_{j_1j_2}^3}{a_{j_2j_20}\prod_{l_1=1}^{3}(a_{j_1j_11}-a_{j_2j_2l_1})}<0,$$

there is a third-order down-parabola flow at $y_1^*=a_{j_1j_11}$. With

$$\dot{x}_{j_2}^*=a_{j_2j_20}\prod_{l_1=1}^{3}(a_{j_1j_11}-a_{j_2j_2l_1}),$$

the third-order parabola flows are positive and negative for $\dot{x}_{j_2}^*>0$ and $\dot{x}_{j_2}^*<0$, respectively.

Therefore, the flow of $y_{j_1}^* = a_{j_1j_11} \neq a_{j_2j_21}$ with $\dot{x}_{j_2}^* = a_{j_2j_20}\prod_{l_1=1}^{3}(a_{j_1j_11} - a_{j_2j_2l_1})$ for the x_{j_1}-direction has the following properties as in Eqs. (4.376)–(4.379):

- For $a_{j_1j_10}\alpha_{j_1j_2}^3 > 0$ and $a_{j_2j_20}\prod_{l_1=1}^{3}(a_{j_1j_11} - a_{j_2j_2l_1}) > 0$,
$$(a_{j_1j_1s_1}, \dot{x}_{j_2}^*) = \underbrace{(3^{\text{rd}}\text{UP}, \text{pF})}_{\text{third-order up-parabola flow } (+)}.$$
- For $a_{j_1j_10}\alpha_{j_1j_2}^3 < 0$ and $a_{j_2j_20}\prod_{l_1=1}^{3}(a_{j_1j_11} - a_{j_2j_2l_1}) > 0$,
$$(a_{j_1j_1s_1}, \dot{x}_{j_2}^*) = \underbrace{(3^{\text{rd}}\text{DP}, \text{pF})}_{\text{third-order down-parabola flow } (+)}.$$
- For $a_{j_1j_10}\alpha_{j_1j_2}^3 > 0$ and $a_{j_2j_20}\prod_{l_1=1}^{3}(a_{j_1j_11} - a_{j_2j_2l_1}) < 0$,
$$(a_{j_1j_1s_1}, \dot{x}_{j_2}^*) = \underbrace{(3^{\text{rd}}\text{DP}, \text{nF})}_{\text{third-order down-parabola flow } (-)}.$$
- For $a_{j_1j_10}\alpha_{j_1j_2}^3 < 0$ and $a_{j_2j_20}\prod_{l_1=1}^{3}(a_{j_1j_11} - a_{j_2j_2l_1}) < 0$,
$$(a_{j_1j_1s_1}, \dot{x}_{j_2}^*) = \underbrace{(3^{\text{rd}}\text{UP}, \text{nF})}_{\text{third-order up-parabola flow } (-)}.$$

– The third-order up-parabola flows are the appearing bifurcations for up-parabola, down-parabola, and up-parabola flows.
– The third-order down-parabola flows are the appearing bifurcations for down-parabola, up-parabola, and down-parabola flows.

(viii$_{2b}$) In phase space, at $y_1^* = a_{j_2j_2l_1}$,
$$\left.\frac{dx_{j_2}}{dx_{j_1}}\right|_{y_1^*=a_{j_2j_2l_1}} = \left.\frac{a_{j_2j_20}(y_{j_1} - a_{j_2j_2l_1})\prod_{l_2=1,l_2\neq l_1}^{3}(y_{j_1} - a_{j_2j_2l_2})}{a_{j_1j_10}(y_{j_1} - a_{j_1j_11})^3}\right|_{y_1^*=a_{j_2j_2l_1}} = 0.$$
If
$$\left.\frac{d^2x_{j_2}}{dx_{j_1}^2}\right|_{y_1^*=a_{j_2j_2l_1}} = \frac{a_{j_2j_20}\alpha_{j_1j_1}\prod_{l_2=1,l_2\neq l_1}^{3}(a_{j_2j_2l_1} - a_{j_2j_2l_2})}{a_{j_1j_10}(a_{j_2j_2l_1} - a_{j_1j_11})^3} > 0,$$
there is an up-parabola flow at $y_1^* = a_{j_2j_2l_1}$. If
$$\left.\frac{d^2x_{j_2}}{dx_{j_1}^2}\right|_{y_1^*=a_{j_2j_2l_1}} = \frac{a_{j_2j_20}\alpha_{j_1j_1}\prod_{l_2=1,l_2\neq l_1}^{3}(a_{j_2j_2l_1} - a_{j_2j_2l_2})}{a_{j_1j_10}(a_{j_2j_2l_1} - a_{j_1j_11})^3} < 0,$$

there is a down-parabola flow at $y_1^* = a_{j_2j_2l_1}$. With

$$\dot{x}_{j_1}^* = a_{j_1j_10}(a_{j_2j_2l_1} - a_{j_1j_11})^3,$$

the inflection flows are positive and negative for $\dot{x}_{j_1}^* > 0$ and $\dot{x}_{j_1}^* < 0$, respectively.

Therefore, the flow of $y_{j_1}^* = a_{j_2j_2l_1} \neq a_{j_1j_11}$ for the x_{j_2}-direction has the following properties as in Eqs. (4.380)–(4.383):

- For $a_{j_1j_10}(a_{j_2j_2l_1} - a_{j_1j_11})^3 > 0$ and $a_{j_2j_20}\alpha_{j_1j_1}\prod_{l_2=1,l_2\neq l_1}^{3}(a_{j_2j_2l_1} - a_{j_2j_2l_2}) > 0$,
$$(\dot{x}_{j_1}^*, a_{j_2j_2l_1}) = \underbrace{(\mathrm{pF}, \mathrm{UP})}_{\text{up-parabola flow }(+)}.$$
- For $a_{j_1j_10}(a_{j_2j_2l_1} - a_{j_1j_11})^3 < 0$ and $a_{j_2j_20}\alpha_{j_1j_1}\prod_{l_2=1,l_2\neq l_1}^{3}(a_{j_2j_2l_1} - a_{j_2j_2l_2}) > 0$,
$$(\dot{x}_{j_1}^*, a_{j_2j_2l_1}) = \underbrace{(\mathrm{nF}, \mathrm{DP})}_{\text{down-parabola flow }(-)}.$$
- For $a_{j_1j_10}(a_{j_2j_2l_1} - a_{j_1j_11})^3 > 0$ and $a_{j_2j_20}\alpha_{j_1j_1}\prod_{l_2=1,l_2\neq l_1}^{3}(a_{j_2j_2l_1} - a_{j_2j_2l_2}) < 0$,
$$(\dot{x}_{j_1}^*, a_{j_2j_2l_1}) = \underbrace{(\mathrm{pF}, \mathrm{DP})}_{\text{down-parabola flow }(+)}.$$
- For $a_{j_1j_10}(a_{j_2j_2l_1} - a_{j_1j_11})^3 < 0$ and $a_{j_2j_20}\alpha_{j_1j_1}\prod_{l_2=1,l_2\neq l_1}^{3}(a_{j_2j_2l_1} - a_{j_2j_2l_2}) < 0$,
$$(\dot{x}_{j_1}^*, a_{j_2j_2l_1}) = \underbrace{(\mathrm{nF}, \mathrm{UP})}_{\text{up-parabola flow }(-)}.$$

(viii$_3$) For $a_{j_1j_1s_1} = a_{j_2j_2l_1}$, consider a dynamical system as

$$\dot{x}_{j_1} = a_{j_1j_10}(y_{j_1} - a_{j_1j_1s_1})(y_{j_1} - a_{j_1j_1s_2})^2,$$

$$\dot{x}_{j_2} = a_{j_2j_20}(y_{j_1} - a_{j_2j_2l_1})\prod_{l_2=1,l_2\neq l_1}^{3}(y_{j_1} - a_{j_2j_2l_2}).$$

The corresponding dynamical system in the y_{j_1}-direction is constructed as

$$\dot{y}_{j_1} = (y_{j_1} - a_{j_1j_1s_1})\Big[\alpha_{j_1j_1}a_{j_1j_10}(y_{j_1} - a_{j_1j_1s_2})^2 + \alpha_{j_1j_2}a_{j_2j_20}\prod_{l_2=1,l_2\neq l_1}^{3}(y_{j_1} - a_{j_2j_2l_2})\Big],$$

and the corresponding dynamical system in the $y_{j_1}^{\perp}$-direction is constructed as

$$\dot{y}_{j_1}^{\perp} = (y_{j_1} - a_{j_1j_1s_1})\Big[\beta_{j_1j_1}a_{j_1j_10}(y_{j_1} - a_{j_1j_1s_2})^2$$

$$+ \beta_{j_1j_2}a_{j_2j_20}\prod_{l_2=1,l_2\neq l_1}^{3}(y_{j_1} - a_{j_2j_2l_2})\Big].$$

Define a G-function as

$$\begin{aligned} G_{j_1j_10} &= \frac{\partial}{\partial y_{j_1}}[\mathbf{n}^{\mathrm{T}}\cdot\mathbf{F}(y_{j_1},\mathbf{p})]\Big|_{y_{j_1}^*=a_{j_1j_1s_1}=a_{j_2j_2l_1}} \\ &= \alpha_{j_1j_1}a_{j_1j_10}(a_{j_1j_1s_1} - a_{j_1j_1s_2})^2 \\ &\quad + \alpha_{j_1j_2}a_{j_2j_20}\prod_{l_2=1,l_2\neq l_1}^{3}(a_{j_2j_2l_1} - a_{j_2j_2l_2}). \end{aligned}$$

In phase space, at $y_1^* = a_{j_1j_1s_1} = a_{j_2j_2l_1}$, if

$$\frac{dx_{j_2}}{dx_{j_1}}\Big|_{y_1^*=a_{j_2j_2l_1}} = \frac{a_{j_2j_20}\prod_{l_2=1,l_2\neq l_1}^{3}(a_{j_2j_2l_1} - a_{j_2j_2l_2})}{a_{j_1j_10}(a_{j_1j_1s_1} - a_{j_1j_1s_2})^2} > 0,$$

there is an increasing-inflection flow at $y_1^* = a_{j_1j_1s_1} = a_{j_2j_2l_1}$, and if

$$\frac{dx_{j_2}}{dx_{j_1}}\Big|_{y_1^*=a_{j_2j_2l_1}} = \frac{a_{j_2j_20}\prod_{l_2=1,l_2\neq l_1}^{3}(a_{j_2j_2l_1} - a_{j_2j_2l_2})}{a_{j_1j_10}(a_{j_1j_1s_1} - a_{j_1j_1s_2})^2} < 0,$$

there is a decreasing-inflection flow at $y_1^* = a_{j_1j_1s_1} = a_{j_2j_2l_1}$. The variational equation in the y_{j_1}-direction is

$$\begin{aligned} \Delta\dot{y}_{j_1} &= \Big[\alpha_{j_1j_1}a_{j_1j_10}(a_{j_1j_1s_1} - a_{j_1j_1s_2})^2 \\ &\quad + \alpha_{j_1j_2}a_{j_2j_20}\prod_{l_2=1,l_2\neq l_1}^{3}(a_{j_2j_2l_1} - a_{j_2j_2l_2})\Big]\Delta y_{j_1} \\ &= G_{j_1j_10}\Delta y_{j_1}. \end{aligned}$$

The parabola flows in the y_{j_1}-direction are source and sink for $G_{j_1j_10} > 0$ and $G_{j_1j_10} < 0$, respectively.

Therefore, the infinite-equilibrium flows of $y_1^* = a_{j_1j_1s_1} = a_{j_2j_2l_1}$ possess the flowing properties as in Eqs. (4.388)–(4.391):

- For $a_{j_1j_10}(a_{j_1j_1s_1} - a_{j_1j_1s_2})^2/a_{j_2j_20}\prod_{l_2=1,l_2\neq l_1}^{3}(a_{j_2j_2l_1} - a_{j_2j_2l_2}) > 0$ and $G_{j_1j_10} > 0$,
$$(a_{j_1j_1s_1}, a_{j_2j_2l_1}) = \underbrace{(\text{II}, \text{SO})}_{\text{increasing-inflection source}}.$$
- For $a_{j_1j_10}(a_{j_1j_1s_1} - a_{j_1j_1s_2})^2/a_{j_2j_20}\prod_{l_2=1,l_2\neq l_1}^{3}(a_{j_2j_2l_1} - a_{j_2j_2l_2}) > 0$ and $G_{j_1j_10} < 0$,
$$(a_{j_1j_1s_1}, a_{j_2j_2l_1}) = \underbrace{(\text{II}, \text{SI})}_{\text{increasing-inflection sink}}.$$
- For $a_{j_1j_10}(a_{j_1j_1s_1} - a_{j_1j_1s_2})^2/a_{j_2j_20}\prod_{l_2=1,l_2\neq l_1}^{3}(a_{j_2j_2l_1} - a_{j_2j_2l_2}) < 0$ and $G_{j_1j_10} > 0$,
$$(a_{j_1j_1s_1}, a_{j_2j_2l_1}) = \underbrace{(\text{DI}, \text{SO})}_{\text{decreasing-inflection source}}.$$
- For $a_{j_1j_10}(a_{j_1j_1s_1} - a_{j_1j_1s_2})^2/a_{j_2j_20}\prod_{l_2=1,l_2\neq l_1}^{3}(a_{j_2j_2l_1} - a_{j_2j_2l_2}) < 0$ and $G_{j_1j_10} < 0$,
$$(a_{j_1j_1s_1}, a_{j_2j_2l_1}) = \underbrace{(\text{DI}, \text{SI})}_{\text{increasing-inflection sink}}.$$

– The inflection-source and sink infinite-equilibriums are the switching bifurcations for parabola flows in the x_{j_1}- and x_{j_2}-directions.

(viii$_4$) For $a_{j_1j_1s_2} = a_{j_2j_2l_1}$, consider a dynamical system as

$$\begin{aligned}
\dot{x}_{j_1} &= a_{j_1j_10}(y_{j_1} - a_{j_1j_1s_1})(y_{j_1} - a_{j_1j_1s_2})^2,\\
\dot{x}_{j_2} &= a_{j_2j_20}(y_{j_1} - a_{j_2j_2l_1})\prod_{l_2=1,l_2\neq l_1}^{3}(y_{j_1} - a_{j_2j_2l_2}),\\
& j_1, j_2 \in \{1,2\};\ j_1 \neq j_2.
\end{aligned}$$

The corresponding dynamical system in the y_{j_1}-direction is constructed as

$$\begin{aligned}
\dot{y}_{j_1} = (y_{j_1} - a_{j_1j_1s_2})\Big[&\alpha_{j_1j_1}a_{j_1j_10}(y_{j_1} - a_{j_1j_1s_1})(y_{j_1} - a_{j_1j_1s_2})\\
&+ \alpha_{j_1j_2}a_{j_2j_20}\prod_{l_2=1,l_2\neq l_1}^{3}(y_{j_1} - a_{j_2j_2l_2})\Big],
\end{aligned}$$

and the corresponding dynamical system in the $y_{j_1}^{\perp}$-direction is constructed as

$$\dot{y}_{j_1}^{\perp} = (y_{j_1} - a_{j_1j_1s_2})\Big[\beta_{j_1j_1}a_{j_1j_10}(y_{j_1} - a_{j_1j_1s_1})(y_{j_1} - a_{j_1j_1s_2})$$

$$+ \beta_{j_1j_2}a_{j_2j_20}\prod_{l_2=1,l_2\neq l_1}^{3}(y_{j_1} - a_{j_2j_2l_2})\Big].$$

Define a G-function as

$$G_{j_1j_10} = \frac{\partial}{\partial y_{j_1}}[\mathbf{n}^{\mathrm{T}}\cdot\mathbf{F}(y_{j_1},\mathbf{p})]\Big|_{y_{j_1}^{*}=a_{j_1j_1s_2}=a_{j_2j_2l_1}}$$

$$= \alpha_{j_1j_2}a_{j_2j_20}\prod_{l_2=1,l_2\neq l_1}^{3}(a_{j_2j_2l_1} - a_{j_2j_2l_2}).$$

In phase space, at $y_1^{*} = a_{j_1j_1s_2} = a_{j_2j_2l_1}$,

$$\frac{dx_{j_1}}{dx_{j_2}}\Big|_{y_1^{*}=a_{j_1j_1s_2}} = \frac{a_{j_1j_10}(y_1 - a_{j_1j_1s_1})(y_1 - a_{j_1j_1s_2})}{a_{j_2j_20}\prod_{l_2=1,l_2\neq l_1}^{3}(y_{j_1} - a_{j_2j_2l_2})}\Big|_{y_1^{*}=a_{j_1j_1s_2}} = 0.$$

If

$$\frac{d^2x_{j_1}}{dx_{j_2}^2}\Big|_{y_1^{*}=a_{j_1j_1s_2}} = \frac{a_{j_1j_10}a_{j_1j_2}(a_{j_1j_1s_2} - a_{j_1j_1s_1})}{a_{j_2j_20}\prod_{l_2=1,l_2\neq l_1}^{3}(a_{j_2j_2l_1} - a_{j_2j_2l_2})} > 0,$$

there is an up-parabola flow at $y_1^{*} = a_{j_1j_1s_2} = a_{j_2j_2l_1}$, and if

$$\frac{d^2x_{j_1}}{dx_{j_2}^2}\Big|_{y_1^{*}=a_{j_1j_1s_2}} = \frac{a_{j_1j_10}a_{j_1j_2}(a_{j_1j_1s_2} - a_{j_1j_1s_1})}{a_{j_2j_20}\prod_{l_2=1,l_2\neq l_1}^{3}(a_{j_2j_2l_1} - a_{j_2j_2l_2})} > 0,$$

there is a down-parabola flow at $y_1^{*} = a_{j_1j_1s_2} = a_{j_2j_2l_1}$. The variational equation in the y_{j_1}-direction is

$$\Delta\dot{y}_{j_1} = \alpha_{j_1j_2}a_{j_2j_20}\prod_{l_2=1,l_2\neq l_1}^{3}(a_{j_2j_2l_1} - a_{j_2j_2l_2})\Delta y_{j_1} = G_{j_1j_10}\Delta y_{j_1}.$$

The parabola flows in the y_{j_1}-direction are source and sink for $G_{j_1j_10} > 0$ and $G_{j_1j_10} < 0$, respectively.

Therefore, the infinite-equilibrium of $y_{j_1}^* = a_{j_1j_1s_2} = a_{j_2j_2l_1}$ has the following properties as in Eqs. (3.396)–(3.399):

- For $a_{j_1j_10}\alpha_{j_1j_2}^2(a_{j_1j_1s_2} - a_{j_1j_1s_1}) > 0$ and $\alpha_{j_1j_2}a_{j_2j_20}\prod_{l_2=1,l_2\neq l_1}^3(a_{j_2j_2l_1} - a_{j_2j_2l_2}) > 0$,
$$(a_{j_1j_1s_2}, a_{j_2j_2l_1}) = \underbrace{(\mathrm{UP}, \mathrm{SO})}_{\text{up-parabola source}}.$$
- For $a_{j_1j_10}\alpha_{j_1j_2}^2(a_{j_1j_1s_2} - a_{j_1j_1s_1}) < 0$ and $\alpha_{j_1j_2}a_{j_2j_20}\prod_{l_2=1,l_2\neq l_1}^3(a_{j_2j_2l_1} - a_{j_2j_2l_2}) > 0$,
$$(a_{j_1j_1s_2}, a_{j_2j_2l_1}) = \underbrace{(\mathrm{DP}, \mathrm{SO})}_{\text{down-parabola source}}.$$
- For $a_{j_1j_10}\alpha_{j_1j_2}^2(a_{j_1j_1s_2} - a_{j_1j_1s_1}) > 0$ and $\alpha_{j_1j_2}a_{j_2j_20}\prod_{l_2=1,l_2\neq l_1}^3(a_{j_2j_2l_1} - a_{j_2j_2l_2}) < 0$,
$$(a_{j_1j_1s_2}, a_{j_2j_2l_1}) = \underbrace{(\mathrm{DP}, \mathrm{SI})}_{\text{down-parabola sink}}.$$
- For $a_{j_1j_10}\alpha_{j_1j_2}^2(a_{j_1j_1s_2} - a_{j_1j_1s_1}) < 0$ and $\alpha_{j_1j_2}a_{j_2j_20}\prod_{l_2=1,l_2\neq l_1}^3(a_{j_2j_2l_1} - a_{j_2j_2l_2}) < 0$,
$$(a_{j_1j_1s_2}, a_{j_2j_2l_1}) = \underbrace{(\mathrm{UP}, \mathrm{SI})}_{\text{up-parabola source}}.$$

- The parabola source and sink infinite-equilibriums are the switching bifurcations for the increasing-inflection and parabola flows in the x_{j_1}- and x_{j_2}-directions.
- The inflection-saddle infinite-equilibriums are the appearing bifurcations from a parabola flow and x_{j_2}-direction to the up-parabola and down-parabola flows in the x_{j_1}-direction and a parabola flow in the x_{j_2}-direction.

(viii$_5$) For $a_{j_1j_11} = a_{j_2j_2l_2}$, consider a dynamical system as

$$\dot{x}_{j_1} = a_{j_1j_10}(y_{j_1} - a_{j_1j_11})^3,$$
$$\dot{x}_{j_2} = a_{j_2j_20}(y_{j_1} - a_{j_2j_2l_1})\prod_{l_2=1,l_2\neq l_1}^{3}(y_{j_1} - a_{j_2j_2l_2}),$$
$$j_1, j_2 \in \{1,2\};\ j_1 \neq j_2.$$

The corresponding dynamical system in the y_{j_1}-direction is constructed as

$$\dot{y}_{j_1} = (y_{j_1} - a_{j_1j_11})\Big[\alpha_{j_1j_1}a_{j_1j_10}(y_{j_1} - a_{j_1j_11})^2 + \alpha_{j_1j_2}a_{j_2j_20}\prod_{l_2=1,l_2\neq l_1}^{3}(y_{j_1} - a_{j_2j_2l_2})\Big],$$

and a dynamical system in the $y_{j_1}^{\perp}$-direction is constructed as

$$\dot{y}_{j_1}^{\perp} = (y_{j_1} - a_{j_1j_11})\Big[\beta_{j_1j_1}a_{j_1j_10}(y_{j_1} - a_{j_1j_11})^2 + \beta_{j_1j_2}a_{j_2j_20}\prod_{l_2=1,l_2\neq l_1}^{3}(y_{j_1} - a_{j_2j_2l_2})\Big].$$

Define a G-function as

$$G_{j_1j_10} = \frac{\partial}{\partial y_{j_1}}[\mathbf{n}^{\mathrm{T}} \cdot \mathbf{F}(y_{j_1}, \mathbf{p})]\Big|_{y_{j_1}^*=a_{j_1j_11}=a_{j_2j_2l_1}} = \alpha_{j_1j_2}a_{j_2j_20}\prod_{l_2=1,l_2\neq l_1}^{3}(a_{j_2j_2l_1} - a_{j_2j_2l_2}).$$

In phase space, at $y_{j_1}^* = a_{j_1j_11} = a_{j_2j_2l_1}$,

$$\frac{dx_{j_1}}{dx_{j_2}}\Big|_{y_1^*=a_{j_1j_11}} = \frac{a_{j_1j_10}(y_{j_1} - a_{j_1j_11})^2}{a_{j_2j_20}\prod_{l_2=1,l_2\neq l_1}^{3}(y_{j_1} - a_{j_2j_2l_2})}\Big|_{y_1^*=a_{j_1j_11}} = 0,$$

$$\frac{d^2x_{j_1}}{dx_{j_2}^2}\Big|_{y_1^*=a_{j_1j_11}} = 0.$$

If

$$\frac{d^3x_{j_1}}{dx_{j_2}^3}\Big|_{y_1^*=a_{j_1j_11}} = \frac{2a_{j_1j_10}\alpha_{j_1j_2}^3}{a_{j_2j_20}\alpha_{j_1j_2}\prod_{l_2=1,l_2\neq l_1}^{3}(a_{j_2j_2l_1} - a_{j_2j_2l_2})} > 0,$$

there is a second-order increasing-inflection flow at $y_{j_1}^* = a_{j_1j_11} = a_{j_2j_2l_1}$. If

$$\frac{d^3x_{j_1}}{dx_{j_2}^3}\Big|_{y_1^*=a_{j_1j_11}} = \frac{2a_{j_1j_10}\alpha_{j_1j_2}^3}{a_{j_2j_20}\alpha_{j_1j_2}\prod_{l_2=1,l_2\neq l_1}^{3}(a_{j_2j_2l_1} - a_{j_2j_2l_2})} < 0,$$

there is a second-order decreasing-inflection flow at $y^*_{j_1} = a_{j_1j_11} = a_{j_2j_2l_1}$. The variational equation in the y_{j_1}-direction is

$$\Delta\dot{y}_{j_1} = \alpha_{j_1j_2}a_{j_2j_20}\prod_{l_2=1,l_2\neq l_1}^{3}(a_{j_2j_2l_1} - a_{j_2j_2l_2})\Delta y_{j_1} = G_{j_1j_10}\Delta y_{j_1}.$$

The second-order inflection flows in the y_{j_1}-direction are source and sink for $G_{j_1j_10} > 0$ and $G_{j_1j_10} < 0$, respectively.

Therefore, the infinite-equilibrium of $y^*_{j_1} = a_{j_1j_11} = a_{j_2j_2l_1}$ has the following properties as in Eqs. (3.404)–(3.407):

- For $\alpha^3_{j_1j_2}a_{j_1j_10} > 0$ and $\alpha_{j_1j_2}a_{j_2j_20}\prod_{l_2=1,l_2\neq l_1}^{3}(a_{j_2j_2l_1} - a_{j_2j_2l_2}) > 0$,

$$(a_{j_1j_11}, a_{j_2j_2l_1}) = \underbrace{(2^{\text{nd}}\text{II}, \text{SO})}_{\text{second-order increasing-inflection source}}.$$

- For $\alpha^3_{j_1j_2}a_{j_1j_10} < 0$ and $\alpha_{j_1j_2}a_{j_2j_20}\prod_{l_2=1,l_2\neq l_1}^{3}(a_{j_2j_2l_1} - a_{j_2j_2l_2}) > 0$,

$$(a_{j_1j_11}, a_{j_2j_2l_1}) = \underbrace{(2^{\text{nd}}\text{DI}, \text{SO})}_{\text{second-order decreasing-inflection source}}.$$

- For $\alpha^3_{j_1j_2}a_{j_1j_10} > 0$ and $\alpha_{j_1j_2}a_{j_2j_20}\prod_{l_2=1,l_2\neq l_1}^{3}(a_{j_2j_2l_1} - a_{j_2j_2l_2}) < 0$,

$$(a_{j_1j_11}, a_{j_2j_2l_1}) = \underbrace{(2^{\text{nd}}\text{DI}, \text{SI})}_{\text{second-order decreasing-inflection sink}}.$$

- For $\alpha^3_{j_1j_2}a_{j_1j_10} < 0$ and $\alpha_{j_1j_2}a_{j_2j_20}\prod_{l_2=1,l_2\neq l_1}^{3}(a_{j_2j_2l_1} - a_{j_2j_2l_2}) < 0$,

$$(a_{j_1j_11}, a_{j_2j_2l_1}) = \underbrace{(2^{\text{nd}}\text{II}, \text{SI})}_{\text{second-order increasing-inflection sink}}.$$

– The second-order inflection-source and sink infinite-equilibriums are the switching bifurcations for third-order parabola and parabola flows in the x_{j_1}- and x_{j_2}-directions.
– The second-order inflection-source and sink infinite-equilibriums are the appearing bifurcations from a parabola flow and a parabola flow in the x_{j_1}- and x_{j_2}-directions to three parabola flows and a parabola flow in the x_{j_1}- and x_{j_2}-directions.

(ix) For

$$\Delta_{j_1j_1} > 0 \quad \text{and} \quad \Delta_{j_2j_2} > 0,$$

the dynamical system becomes

$$\begin{aligned}\dot{x}_{j_1} &= a_{j_1j_10}(y_{j_1} - a_{j_1j_11})(y_{j_1} - a_{j_1j_12})(y_{j_1} - a_{j_1j_13}),\\ \dot{x}_{j_2} &= a_{j_2j_20}(y_{j_1} - a_{j_2j_21})(y_{j_1} - a_{j_2j_22})(y_{j_1} - a_{j_2j_23}),\end{aligned}$$

where

$$\{a_{j_1j_11}, a_{j_1j_12}, a_{j_1j_13}\} = \text{sort}\{b_{j_1j_11}, b_{j_1j_12}, b_{j_1j_13}\}, \quad a_{j_1j_1s_1} < a_{j_1j_1s_2}, s_1 < s_2;$$

$$b_{j_1j_12}, b_{j_1j_13} = -\frac{1}{2}(B_{j_1j_11} \pm \sqrt{\Delta_{j_1j_1}}), \quad a_{j_1j_10} = A_{j_1j_10}.$$

$$\{a_{j_2j_21}, a_{j_2j_22}, a_{j_2j_23}\} = \text{sort}\{b_{j_2j_21}, b_{j_2j_22}, b_{j_2j_23}\}, \quad a_{j_2j_2l_1} < a_{j_2j_2l_2}, l_1 < l_2;$$

$$b_{j_2j_22}, b_{j_2j_23} = -\frac{1}{2}(B_{j_2j_21} \pm \sqrt{\Delta_{j_2j_2}}), a_{j_2j_20} = A_{j_2j_20}.$$

The corresponding dynamical system in the y_{j_1}-direction is constructed as

$$\dot{y}_{j_1} = \alpha_{j_1j_1}a_{j_1j_10}\prod_{s_1=1}^{3}(y_{j_1} - a_{j_1j_1s_1}) + \alpha_{j_1j_2}a_{j_2j_20}\prod_{l_1=1}^{3}(y_{j_1} - a_{j_2j_2l_1}),$$

and the corresponding dynamical system in the $y_{j_1}^{\perp}$-direction is constructed as

$$\dot{y}_{j_1}^{\perp} = \beta_{j_1j_1}a_{j_1j_10}\prod_{s_1=1}^{3}(y_{j_1} - a_{j_1j_1s_1}) + \beta_{j_1j_2}a_{j_2j_20}\prod_{l_1=1}^{3}(y_{j_1} - a_{j_2j_2l_1}).$$

(ix$_{1a}$) In phase space, at $y_1^* = a_{j_1j_1s_1}$,

$$\left.\frac{dx_{j_1}}{dx_{j_2}}\right|_{y_1^*=a_{j_1j_1s_1}} = \left.\frac{a_{j_1j_10}(y_{j_1} - a_{j_1j_1s_1})\prod_{s_2=1,s_2\neq s_1}^{3}(y_{j_1} - a_{j_1j_1s_2})}{a_{j_2j_20}\prod_{l_1=1}^{3}(y_{j_1} - a_{j_2j_2l_1})}\right|_{y_1^*=a_{j_1j_1s_1}} = 0.$$

If

$$\left.\frac{d^2x_{j_1}}{dx_{j_2}^2}\right|_{y_1^*=a_{j_1j_1s_1}} = \frac{a_{j_1j_10}\alpha_{j_1j_2}\prod_{s_2=1,s_2\neq s_1}^{3}(a_{j_1j_1s_1} - a_{j_1j_1s_2})}{a_{j_2j_20}\prod_{l_1=1}^{3}(a_{j_1j_1s_1} - a_{j_2j_2l_1})} > 0,$$

there is an up-parabola flow at $y_1^* = a_{j_1j_1s_1}$. If

$$\left.\frac{d^2x_{j_1}}{dx_{j_2}^2}\right|_{y_1^*=a_{j_1j_1s_1}} = \frac{a_{j_1j_10}\alpha_{j_1j_2}\prod_{s_2=1,s_2\neq s_1}^{3}(a_{j_1j_1s_1} - a_{j_1j_1s_2})}{a_{j_2j_20}\prod_{l_1=1}^{3}(a_{j_1j_1s_1} - a_{j_2j_2l_1})} < 0,$$

there is a down-parabola flow at $y_1^* = a_{j_1j_1s_1}$. With

$$\dot{x}_{j_2}^* = a_{j_2j_20} \prod_{l_1=1}^{3} (a_{j_1j_1s_1} - a_{j_2j_2l_1}),$$

the parabola flows are positive and negative for $\dot{x}_{j_2}^* > 0$ and $\dot{x}_{j_2}^* < 0$, respectively.

Therefore, the equilibrium flows of $y_{j_1}^* = a_{j_1j_1s_1}(s_1, s_2 \in \{1,2\}, s_1 \neq s_2)$ for the x_{j_1}-direction have the following properties as in Eqs. (4.413)–(4.416):

- For $a_{j_1j_10}\alpha_{j_1j_2}\prod_{s_2=1,s_2\neq s_1}^{3}(a_{j_1j_1s_1} - a_{j_1j_1s_2}) > 0$ and $a_{j_2j_20}\prod_{l_1=1}^{3}(a_{j_1j_1s_1} - a_{j_2j_2j_1}) > 0$,
$$(a_{j_1j_1s_1}, \dot{x}_{j_2}^*) = \underbrace{(\text{UP}, \text{pF})}_{\text{up-parabola flow } (+)}.$$
- For $a_{j_1j_10}\alpha_{j_1j_2}\prod_{s_2=1,s_2\neq s_1}^{3}(a_{j_1j_1s_1} - a_{j_1j_1s_2}) < 0$ and $a_{j_2j_20}\prod_{l_1=1}^{3}(a_{j_1j_1s_1} - a_{j_2j_2l_1}) > 0$,
$$(a_{j_1j_1s_1}, \dot{x}_{j_2}^*) = \underbrace{(\text{DP}, \text{pF})}_{\text{Down-parabola flow } (+)}.$$
- For $a_{j_1j_10}\alpha_{j_1j_2}\prod_{s_2=1,s_2\neq s_1}^{3}(a_{j_1j_1s_1} - a_{j_1j_1s_2}) > 0$ and $a_{j_2j_20}\prod_{l_1=1}^{3}(a_{j_1j_1s_1} - a_{j_2j_2l_1}) < 0$,
$$(a_{j_1j_1s_1}, \dot{x}_{j_2}^*) = \underbrace{(\text{DP}, \text{nF})}_{\text{Down-parabola flow } (-)}.$$
- For $a_{j_1j_10}\alpha_{j_1j_2}\prod_{s_2=1,s_2\neq s_1}^{3}(a_{j_1j_1s_1} - a_{j_1j_1s_2}) < 0$ and $a_{j_2j_20}\prod_{l_1=1}^{3}(a_{j_1j_1s_1} - a_{j_2j_2l_1}) < 0$,
$$(a_{j_1j_1s_1}, \dot{x}_{j_2}^*) = \underbrace{(\text{UP}, \text{nF})}_{\text{up-parabola flow } (-)}.$$

($\text{viii}_{1\text{b}}$) In phase space, at $y_1^* = a_{j_2j_2l_1}$,

$$\left.\frac{dx_{j_2}}{dx_{j_1}}\right|_{y_1^*=a_{j_2j_2l_1}} = \left.\frac{a_{j_2j_20}(y_{j_1} - a_{j_2j_2l_1})\prod_{l_2=1,l_2\neq l_1}^{3}(y_{j_1} - a_{j_2j_2l_2})}{a_{j_1j_10}\prod_{s_1=1}^{3}(y_{j_1} - a_{j_1j_1s_1})}\right|_{y_1^*=a_{j_2j_2l_1}} = 0.$$

If

$$\left.\frac{d^2x_{j_2}}{dx_{j_1}^2}\right|_{y_1^*=a_{j_2j_2l_1}} = \frac{a_{j_2j_20}\alpha_{j_1j_1}\prod_{l_2=1,l_2\neq l_1}^{3}(a_{j_2j_2l_1}-a_{j_2j_2l_2})}{a_{j_1j_10}\prod_{s_1=1}^{3}(a_{j_2j_2l_1}-a_{j_1j_1s_1})} > 0,$$

there is an up-parabola flow at $y_1^* = a_{j_2j_2l_1}$. If

$$\left.\frac{d^2x_{j_2}}{dx_{j_1}^2}\right|_{y_1^*=a_{j_2j_2l_1}} = \frac{a_{j_2j_20}\alpha_{j_1j_1}\prod_{l_2=1,l_2\neq l_1}^{3}(a_{j_2j_2l_1}-a_{j_2j_2l_2})}{a_{j_1j_10}\prod_{s_1=1}^{3}(a_{j_2j_2l_1}-a_{j_1j_1s_1})} < 0,$$

there is a down-parabola flow at $y_1^* = a_{j_2j_2l_1}$. With

$$\dot{x}_{j_1}^* = a_{j_1j_10}\prod_{s_1=1}^{3}(a_{j_2j_2l_1}-a_{j_1j_1s_1}),$$

the inflection flows are positive and negative for $\dot{x}_{j_1}^* > 0$ and $\dot{x}_{j_1}^* < 0$, respectively.

Therefore, the equilibrium flow of $y_{j_1}^* = a_{j_2j_2l_1}$ for the x_{j_2}-direction has the following properties as in Eqs. (4.417)–(4.420):

- For $a_{j_1j_10}\prod_{s_1=1}^{3}(a_{j_2j_2l_1}-a_{j_1j_1s_1}) > 0$ and $a_{j_2j_20}\alpha_{j_1j_1}\prod_{l_2=1,l_2\neq l_1}^{3}(a_{j_2j_2l_1}-a_{j_2j_2l_2}) > 0$,
$$(\dot{x}_{j_1}^*, a_{j_2j_2l_1}) = \underbrace{(\mathrm{pF}, \mathrm{UP})}_{\text{up-parabola flow }(+)}.$$

- For $a_{j_1j_10}\prod_{s_1=1}^{3}(a_{j_2j_2l_1}-a_{j_1j_1s_1}) < 0$ and $a_{j_2j_20}\alpha_{j_1j_1}\prod_{l_2=1,l_2\neq l_1}^{3}(a_{j_2j_2l_1}-a_{j_2j_2l_2}) > 0$,
$$(\dot{x}_{j_1}^*, a_{j_2j_2l_1}) = \underbrace{(\mathrm{nF}, \mathrm{DP})}_{\text{down-parabola flow }(-)}.$$

- For $a_{j_1j_10}\prod_{s_1=1}^{3}(a_{j_2j_2l_1}-a_{j_1j_1s_1}) > 0$ and $a_{j_2j_20}\alpha_{j_1j_1}\prod_{l_2=1,l_2\neq l_1}^{3}(a_{j_2j_2l_1}-a_{j_2j_2l_2}) < 0$,
$$(\dot{x}_{j_1}^*, a_{j_2j_2l_1}) = \underbrace{(\mathrm{pF}, \mathrm{DP})}_{\text{down-parabola flow }(+)}.$$

- For $a_{j_1j_10}\prod_{s_1=1}^{3}(a_{j_2j_2l_1}-a_{j_1j_1s_1}) < 0$ and $a_{j_2j_20}\alpha_{j_1j_1}\prod_{l_2=1,l_2\neq l_1}^{3}(a_{j_2j_2l_1}-a_{j_2j_2l_2}) < 0$,
$$(\dot{x}_{j_1}^*, a_{j_2j_2l_1}) = \underbrace{(\mathrm{nF}, \mathrm{UP})}_{\text{up-parabola flow }(-)}.$$

(ix$_2$) For $a_{j_2j_2l_1} = a_{j_1j_1s_1}$ $(s_1, s_2 \in \{1,2,3\}, l_1, l_2 \in \{1,2\})$, consider a dynamical system as

$$\dot{x}_{j_1} = a_{j_1j_10}(y_{j_1} - a_{j_1j_1s_1}) \prod_{s_2=1, s_2\neq s_1}^{3} (y_{j_1} - a_{j_1j_1s_2}),$$

$$\dot{x}_{j_2} = a_{j_2j_20}(y_{j_1} - a_{j_2j_2l_1}) \prod_{l_2=1, l_2\neq l_1}^{3} (y_{j_1} - a_{j_2j_2l_2}),$$

$$j_1, j_2 \in \{1,2\};\ j_1 \neq j_2;\ s_1, s_2 \in \{1,2,3\}, s_1 \neq s_2.$$

The corresponding dynamical system in the y_{j_1}-direction is constructed as

$$\dot{y}_{j_1} = (y_{j_1} - a_{j_1j_1s_1}) \Big[\alpha_{j_1j_1} a_{j_1j_10} \prod_{s_2=1, s_2\neq s_1}^{3} (y_{j_1} - a_{j_1j_1s_2})$$
$$+ \alpha_{j_1j_2} a_{j_2j_20} \prod_{l_2=1, l_2\neq l_1}^{3} (y_{j_1} - a_{j_2j_2l_2})\Big],$$

and the corresponding dynamical system in the $y_{j_1}^{\perp}$-direction is constructed as

$$\dot{y}_{j_1}^{\perp} = (y_{j_1} - a_{j_1j_1s_1}) \Big[\beta_{j_1j_1} a_{j_1j_10} \prod_{s_2=1, s_2\neq s_1}^{3} (y_{j_1} - a_{j_1j_1s_2})$$
$$+ \beta_{j_1j_2} a_{j_2j_20} \prod_{l_2=1, l_2\neq l_1}^{3} (y_{j_1} - a_{j_2j_2l_2})\Big].$$

Define a G-function as

$$G_{j_1j_10} = \frac{\partial}{\partial y_{j_1}} [\mathbf{n}^{\mathrm{T}} \cdot \mathbf{F}(y_{j_1}, \mathbf{p})]\Big|_{y_{j_1}^* = a_{j_1j_1s_1} = a_{j_2j_2l_1}}$$
$$= \alpha_{j_1j_1} a_{j_1j_10} \prod_{s_2=1, s_2\neq s_1}^{3} (a_{j_1j_1s_1} - a_{j_1j_1s_2})$$
$$+ \alpha_{j_1j_2} a_{j_2j_20} \prod_{l_2=1, l_2\neq l_1}^{3} (a_{j_2j_2l_1} - a_{j_2j_2l_2}).$$

In phase space, at $y_1^* = a_{j_1j_1s_1} = a_{j_2j_2l_1}$, if

$$\frac{dx_{j_2}}{dx_{j_1}}\Big|_{y_1^* = a_{j_2j_2l_1}} = \frac{a_{j_2j_20} \prod_{l_2=1, l_2\neq l_1}^{3} (a_{j_2j_2l_1} - a_{j_2j_2l_2})}{a_{j_1j_10} \prod_{s_2=1, s_2\neq s_1}^{3} (a_{j_1j_1s_1} - a_{j_1j_1s_2})} > 0,$$

there is an increasing-inflection flow at $y_1^* = a_{j_1j_1s_1} = a_{j_2j_2l_1}$, and if

$$\left.\frac{dx_{j_2}}{dx_{j_1}}\right|_{y_1^*=a_{j_2j_2l_1}} = \frac{a_{j_2j_20}\prod_{l_2=1,l_2\neq l_1}^{3}(a_{j_2j_2l_1}-a_{j_2j_2l_2})}{a_{j_1j_10}\prod_{s_2=1,s_2\neq s_1}^{3}(a_{j_1j_1s_1}-a_{j_1j_1s_2})} < 0,$$

there is a down-parabola flow at $y_1^* = a_{j_1j_1s_1} = a_{j_2j_2l_1}$. The variational equation in the y_{j_1}-direction is

$$\begin{aligned}\Delta\dot{y}_{j_1} &= \Big[\alpha_{j_1j_1}a_{j_1j_10}\prod_{s_2=1,s_2\neq s_1}^{3}(a_{j_1j_1s_1}-a_{j_1j_1s_2})\\ &\quad + \alpha_{j_1j_2}a_{j_2j_20}\prod_{l_2=1,l_2\neq l_1}^{3}(a_{j_2j_2l_1}-a_{j_2j_2l_2})\Big]\Delta y_{j_1}\\ &= G_{j_1j_10}\Delta y_{j_1}.\end{aligned}$$

The parabola flows in the y_{j_1}-direction are source and sink for in $G_{j_1j_10} > 0$ and $G_{j_1j_10} < 0$, respectively.

Therefore, the infinite-equilibriums of $y_1^* = a_{j_1j_1s_1} = a_{j_2j_2l_1}$ have the following properties as in Eqs. (4.425)–(4.428):

- For $a_{j_1j_10}\prod_{s_2=1,s_2\neq s_1}^{3}(a_{j_1j_1s_1} - a_{j_1j_1s_2})/a_{j_2j_20}\prod_{l_2=1,l_2\neq l_1}^{3}(a_{j_2j_2l_1} - a_{j_2j_2l_2}) > 0$ and $G_{j_1j_10} > 0$,

$$(a_{j_1j_1s_1}, a_{j_2j_2l_1}) = \underbrace{(\text{II}, \text{SO})}_{\text{increasing-inflection source}}.$$

- For $a_{j_1j_10}\prod_{s_2=1,s_2\neq s_1}^{3}(a_{j_1j_1s_1} - a_{j_1j_1s_2})/a_{j_2j_20}\prod_{l_2=1,l_2\neq l_1}^{3}(a_{j_2j_2l_1} - a_{j_2j_2l_2}) > 0$ and $G_{j_1j_10} < 0$,

$$(a_{j_1j_1s_1}, a_{j_2j_2l_1}) = \underbrace{(\text{II}, \text{SI})}_{\text{increasing-inflection sink}}.$$

- For $a_{j_1j_10}\prod_{s_2=1,s_2\neq s_1}^{3}(a_{j_1j_1s_1} - a_{j_1j_1s_2})/a_{j_2j_20}\prod_{l_2=1,l_2\neq l_1}^{3}(a_{j_2j_2l_1} - a_{j_2j_2l_2}) < 0$ and $G_{j_1j_10} > 0$,

$$(a_{j_1j_1s_1}, a_{j_2j_2l_1}) = \underbrace{(\text{DI}, \text{SO})}_{\text{decreasing-inflection source}}.$$

- For $a_{j_1j_10}\prod_{s_2=1,s_2\neq s_1}^{3}(a_{j_1j_1s_1} - a_{j_1j_1s_2})/a_{j_2j_20}\prod_{l_2=1,l_2\neq l_1}^{3}(a_{j_2j_2l_1} - a_{j_2j_2l_2}) < 0$ and $G_{j_1j_10} > 0$,

$$(a_{j_1j_1s_1}, a_{j_2j_2l_1}) = \underbrace{(\text{DI}, \text{SI})}_{\text{decreasing-inflection sink}}.$$

From the above local analysis of such a dynamical system, this theorem is proved. □

4.3 Singular Single-Linear-Bivariate Cubic Systems

In this section, dynamical systems with two cubic vector fields with a single bivariate function are discussed.

4.3.1 *Singular parabola and inflection flows*

Consider

$$\begin{aligned}\dot{x}_1 &= a_{110}(y_1 - a_{111})(y_1 - a_{112})^2,\\ \dot{x}_2 &= a_{220}(y_1 - a_{221})^3,\end{aligned} \tag{4.429}$$

where $a_{111} < a_{112} < a_{221}$ with $a_{211} > a_{112}$,

$$y_1 = \alpha_{11}x_1 + \alpha_{12}x_2. \tag{4.430}$$

There is a dynamical system in the y_1-direction as

$$\dot{y}_1 = \alpha_{11}a_{110}(y_1 - a_{111})(y_1 - a_{112})^2 + \alpha_{12}a_{220}(y_1 - a_{221})^3, \tag{4.431}$$

with a dynamical system in the $y_{j1}^{\perp}$-direction as

$$\dot{y}_1^{\perp} = \beta_{11}a_{110}(y_1 - a_{111})(y_1 - a_{112})^2 + \beta_{12}a_{220}(y_1 - a_{221})^3, \tag{4.432}$$

where

$$\begin{aligned}y_1^{\perp} &= \beta_{11}x_1 + \beta_{12}x_2,\\ \beta_{11}\alpha_{11} &+ \beta_{12}\alpha_{12} = 0.\end{aligned} \tag{4.433}$$

(i) For $a_{111} > a_{221}$ ($\alpha_{11} > 0, \alpha_{12} > 0$), presented in Fig. 4.1 are the one-dimensional flows of $(y_1^*, \dot{x}_2) = (a_{111}, \dot{x}_2), (a_{112}, \dot{x}_2)$ and $(\dot{x}_1, y_1^*) = (\dot{x}_1, a_{221})$. The corresponding properties for $\alpha_{11} > 0$ and $\alpha_{12} > 0$ are presented as follows:

$$\left\{\begin{matrix}(\dot{x}_1, a_{221})\\ (a_{111}, \dot{x}_2)\\ (a_{112}, \dot{x}_2)\end{matrix}\right\} = \left\{\begin{matrix}\underbrace{(\text{nF}, 3^{\text{rd}}\text{DP})}_{\text{third-order down-parabola flow }(-)}\\ \underbrace{(\text{UP}, \text{pF})}_{\text{up-parabola flow }(+)}\\ \underbrace{(\text{II}, \text{pF})}_{\text{increasing-inflection flow }(+)}\end{matrix}\right\}$$

$$\text{for} \quad a_{110} > 0 \quad \text{and} \quad a_{220} > 0, \tag{4.434}$$

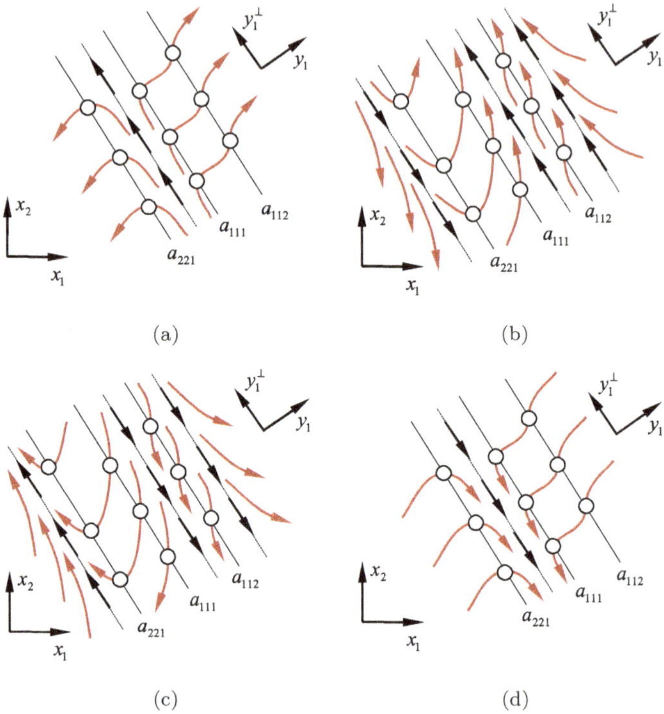

Fig. 4.1. Phase portraits ($a_{111} > a_{221}$ and $\alpha_{11} > 0, \alpha_{12} > 0$) for two-dimensional systems on the x_1-direction with $y_1^* = a_{111}, a_{112}$ and on the x_2-direction with $y_1^* = a_{221}$: (a) ($a_{110} > 0, a_{220} > 0$), (b) ($a_{110} < 0, a_{220} > 0$), (c) ($a_{110} > 0, a_{220} < 0$), (d) ($a_{110} < 0, a_{220} < 0$).

$$\left\{\begin{array}{c}(\dot{x}_1, a_{221})\\(a_{111}, \dot{x}_2)\\(a_{112}, \dot{x}_2)\end{array}\right\} = \left\{\begin{array}{c}\underbrace{(\mathrm{pF}, 3^{\mathrm{rd}}\mathrm{UP})}_{\text{third-order up-parabola flow }(+)}\\\underbrace{(\mathrm{DP}, \mathrm{pF})}_{\text{down-parabola flow }(+)}\\\underbrace{(\mathrm{DI}, \mathrm{pF})}_{\text{increasing-inflection flow }(+)}\end{array}\right\}$$

$$\text{for} \quad a_{110} < 0 \quad \text{and} \quad a_{220} > 0, \tag{4.435}$$

Table 4.1. Equilibrium flows of $\dot{y}_1 = 0$ $(\alpha_{11} > 0, \alpha_{12} > 0)$.

	$(\dot{x}_1, \dot{x}_2)$	y_1			
Figure 4.1	(a_{110}, a_{220})	$(-\infty, a_{221})$	(a_{221}, a_{111})	(a_{111}, a_{112})	(a_{112}, ∞)
(a)	$(+,+)$	$\dot{y}_1 < 0$ $(-,-)$	$\dot{y}_1^* = 0$ $(-,+)$	$\dot{y}_1 > 0$ $(+,+)$	$\dot{y}_1 > 0$ $(+,+)$
(b)	$(-,+)$	$\dot{y}_1^* = 0$ $(+,-)$	$\dot{y}_1 > 0$ $(+,+)$	$\dot{y}_1^* = 0$ $(-,+)$	$\dot{y}_1^* = 0$ $(-,+)$
(c)	$(+,-)$	$\dot{y}_1^* = 0$ $(-,+)$	$\dot{y}_1 < 0$ $(-,-)$	$\dot{y}_1^* = 0$ $(+,-)$	$\dot{y}_1^* = 0$ $(+,-)$
(d)	$(-,-)$	$\dot{y}_1 > 0$ $(+,+)$	$\dot{y}_1^* = 0$ $(-,+)$	$\dot{y}_1 < 0$ $(-,-)$	$\dot{y}_1 < 0$ $(-,-)$

$$\left\{\begin{array}{c}(\dot{x}_1, a_{221})\\(a_{111}, \dot{x}_2)\\(a_{112}, \dot{x}_2)\end{array}\right\} = \left\{\begin{array}{c}\underbrace{(\mathrm{nF}, 3^{\mathrm{rd}}\mathrm{UP})}_{\text{third-order up-parabola flow }(-)}\\\underbrace{(\mathrm{DP}, \mathrm{nF})}_{\text{down-parabola flow }(-)}\\\underbrace{(\mathrm{DI}, \mathrm{nF})}_{\text{increasing-inflection flow }(-)}\end{array}\right\}$$

$$\text{for} \quad a_{110} > 0 \quad \text{and} \quad a_{220} < 0, \tag{4.436}$$

$$\left\{\begin{array}{c}(\dot{x}_1, a_{221})\\(a_{111}, \dot{x}_2)\\(a_{112}, \dot{x}_2)\end{array}\right\} = \left\{\begin{array}{c}\underbrace{(\mathrm{pF}, 3^{\mathrm{rd}}\mathrm{DP})}_{\text{third-order down-parabola flow }(+)}\\\underbrace{(\mathrm{UP}, \mathrm{nF})}_{\text{up-parabola flow }(-)}\\\underbrace{(\mathrm{II}, \mathrm{nF})}_{\text{increasing-inflection flow }(-)}\end{array}\right\}$$

$$\text{for} \quad a_{110} < 0 \quad \text{and} \quad a_{220} < 0. \tag{4.437}$$

In addition, the equilibrium dynamics on the y_1-direction are listed in Table 4.1.

(ii) For $a_{111} > a_{221}$ $(\alpha_{11} > 0, \alpha_{12} > 0)$, presented in Fig. 4.2 are the one-dimensional flows of $(y_1^*, \dot{x}_2) = (a_{111}, \dot{x}_2), (a_{112}, \dot{x}_2)$ and $(\dot{x}_1, y_1^*) = (\dot{x}_1, a_{221})$. The corresponding properties for $\alpha_{11} > 0$ and $\alpha_{12} < 0$ are

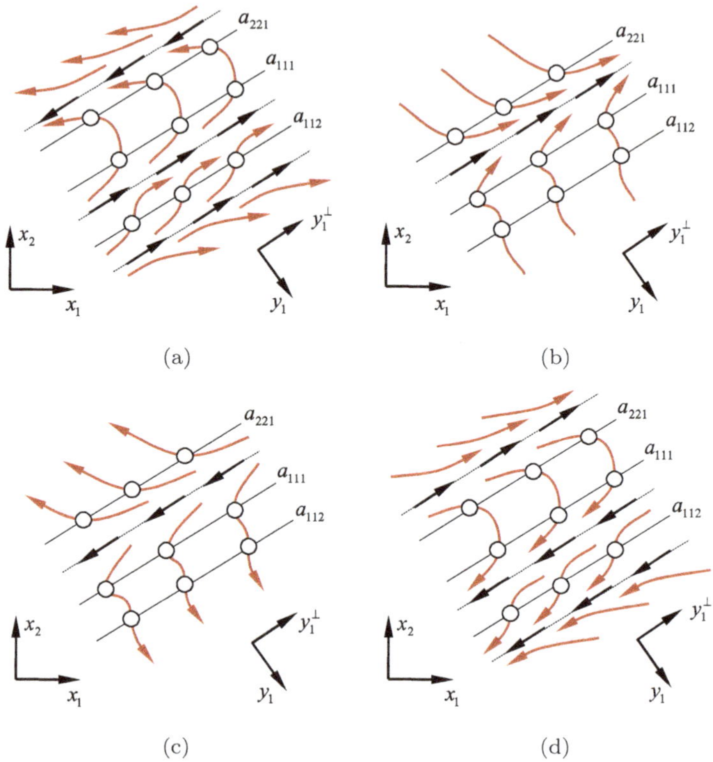

Fig. 4.2. Phase portraits ($a_{111} > a_{221}$ and $\alpha_{11} > 0, \alpha_{12} < 0$) for two-dimensional systems on the x_1-direction with $y_1^* = a_{111}, a_{112}$ and on the x_2-direction with $y_1^* = a_{221}$: (a) $(a_{110} > 0, a_{220} > 0)$, (b) $(a_{110} < 0, a_{220} > 0)$, (c) $(a_{110} > 0, a_{220} < 0)$, (d) $(a_{110} < 0, a_{220} < 0)$.

presented as follows:

$$\left\{ \begin{matrix} (\dot{x}_1, a_{221}) \\ (a_{111}, \dot{x}_2) \\ (a_{112}, \dot{x}_2) \end{matrix} \right\} = \left\{ \begin{matrix} \underbrace{(\text{nF}, 3^{\text{rd}}\text{DP})}_{\text{third-order down-parabola flow } (-)} \\ \underbrace{(\text{DP}, \text{pF})}_{\text{down-parabola flow } (+)} \\ \underbrace{(\text{II}, \text{pF})}_{\text{increasing-inflection flow } (+)} \end{matrix} \right\}$$

$$\text{for} \quad a_{110} > 0 \quad \text{and} \quad a_{220} > 0, \tag{4.438}$$

Table 4.2. Equilibrium flows of $\dot{y}_1 = 0$ ($\alpha_{11} > 0, \alpha_{12} < 0$).

	$(\dot{x}_1, \dot{x}_2)$	y_1			
Figure 4.2	(a_{110}, a_{220})	$(-\infty, a_{221})$	(a_{221}, a_{111})	(a_{111}, a_{112})	(a_{112}, ∞)
(a)	$(+,+)$	$\dot{y}_1^* = 0$ $(-,-)$	$\dot{y}_1 < 0$ $(-,+)$	$\dot{y}_1^* = 0$ $(+,+)$	$\dot{y}_1^* = 0$ $(+,+)$
(b)	$(-,+)$	$\dot{y}_1 > 0$ $(+,-)$	$\dot{y}_1^* = 0$ $(+,+)$	$\dot{y}_1 < 0$ $(-,+)$	$\dot{y}_1 < 0$ $(-,+)$
(c)	$(+,-)$	$\dot{y}_1 < 0$ $(-,+)$	$\dot{y}_1^* = 0$ $(-,-)$	$\dot{y}_1 > 0$ $(+,-)$	$\dot{y}_1 > 0$ $(+,-)$
(d)	$(-,-)$	$\dot{y}_1^* = 0$ $(+,+)$	$\dot{y}_1 > 0$ $(+,-)$	$\dot{y}_1^* = 0$ $(-,-)$	$\dot{y}_1^* = 0$ $(-,-)$

$$\left\{\begin{array}{c} (\dot{x}_1, a_{221}) \\ (a_{111}, \dot{x}_2) \\ (a_{112}, \dot{x}_2) \end{array}\right\} = \left\{\begin{array}{c} \underbrace{(\mathrm{pF}, 3^{\mathrm{rd}}\mathrm{UP})}_{\text{third-order up-parabola flow } (+)} \\ \underbrace{(\mathrm{UP}, \mathrm{pF})}_{\text{up-parabola flow } (+)} \\ \underbrace{(\mathrm{DI}, \mathrm{pF})}_{\text{increasing-inflection flow } (+)} \end{array}\right\}$$

$$\text{for} \quad a_{110} < 0 \quad \text{and} \quad a_{220} > 0, \tag{4.439}$$

$$\left\{\begin{array}{c} (\dot{x}_1, a_{221}) \\ (a_{111}, \dot{x}_2) \\ (a_{112}, \dot{x}_2) \end{array}\right\} = \left\{\begin{array}{c} \underbrace{(\mathrm{nF}, 3^{\mathrm{rd}}\mathrm{UP})}_{\text{third-order up-parabola flow } (-)} \\ \underbrace{(\mathrm{UP}, \mathrm{nF})}_{\text{up-parabola flow } (-)} \\ \underbrace{(\mathrm{DI}, \mathrm{nF})}_{\text{increasing-inflection flow } (-)} \end{array}\right\}$$

$$\text{for} \quad a_{110} > 0 \quad \text{and} \quad a_{220} < 0, \tag{4.440}$$

$$\left\{\begin{array}{c} (\dot{x}_1, a_{221}) \\ (a_{111}, \dot{x}_2) \\ (a_{112}, \dot{x}_2) \end{array}\right\} = \left\{\begin{array}{c} \underbrace{(\mathrm{pF}, 3^{\mathrm{rd}}\mathrm{DP})}_{\text{third-order down-parabola flow } (+)} \\ \underbrace{(\mathrm{DP}, \mathrm{nF})}_{\text{down-parabola flow } (-)} \\ \underbrace{(\mathrm{II}, \mathrm{nF})}_{\text{increasing-inflection flow } (-)} \end{array}\right\}$$

$$\text{for} \quad a_{110} < 0 \quad \text{and} \quad a_{220} < 0. \tag{4.441}$$

In addition, the equilibrium dynamics on the y_1-direction are listed in Table 4.2.

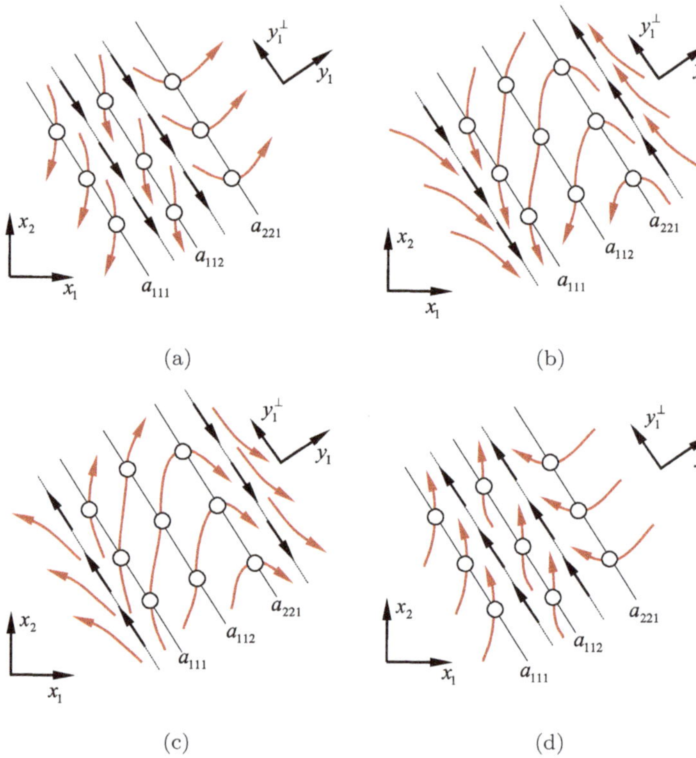

Fig. 4.3. Phase portraits ($a_{112} < a_{221}$ and $\alpha_{11} > 0, \alpha_{12} > 0$) for two-dimensional systems on the x_1-direction with $y_1^* = a_{111}, a_{112}$ and on the x_2-direction with $y_1^* = a_{221}$: (a) ($a_{110} > 0$, $a_{220} > 0$), (b) ($a_{110} < 0, a_{220} > 0$), (c) ($a_{110} > 0, a_{220} < 0$), (d) ($a_{110} < 0, a_{220} < 0$).

(iii) For $a_{112} < a_{221}$ ($\alpha_{11} > 0, \alpha_{12} > 0$), presented in Fig. 4.3 are the one-dimensional flows of $(y_1^*, \dot{x}_2) = (a_{111}, \dot{x}_2), (a_{112}, \dot{x}_2)$ and $(\dot{x}_1, y_1^*) = (\dot{x}_1, a_{221})$. The corresponding properties for $\alpha_{11} > 0$ and $\alpha_{12} > 0$ are presented as follows:

$$\left\{\begin{array}{c}(a_{111}, \dot{x}_2)\\(a_{112}, \dot{x}_2)\\(\dot{x}_1, a_{221})\end{array}\right\} = \left\{\begin{array}{c}\underbrace{(\mathrm{UP}, \mathrm{pF})}_{\text{up-parabola flow }(+)}\\ \underbrace{(\mathrm{II}, \mathrm{pF})}_{\text{increasing-inflection flow }(+)}\\ \underbrace{(\mathrm{pF}, 3^{\mathrm{rd}}\mathrm{UP})}_{\text{third-order up-parabola flow }(+)}\end{array}\right\}$$

$$\text{for } a_{110} > 0 \quad \text{and} \quad a_{220} > 0, \tag{4.442}$$

Table 4.3. Equilibrium flows of $\dot{y}_1 = 0$ $(\alpha_{11} > 0, \alpha_{12} > 0)$.

Figure 4.3	$(\dot{x}_1, \dot{x}_2)$ (a_{110}, a_{220})	y_1 $(-\infty, a_{111})$	(a_{111}, a_{112})	(a_{112}, a_{221})	(a_{221}, ∞)
(a)	$(+,+)$	$\dot{y}_1 < 0$ $(-,-)$	$\dot{y}_1^* = 0$ $(-,+)$	$\dot{y}_1^* = 0$ $(-,+)$	$\dot{y}_1 > 0$ $(+,+)$
(b)	$(-,+)$	$\dot{y}_1^* = 0$ $(+,-)$	$\dot{y}_1 < 0$ $(-,-)$	$\dot{y}_1 < 0$ $(-,-)$	$\dot{y}_1^* = 0$ $(-,+)$
(c)	$(+,-)$	$\dot{y}_1^* = 0$ $(-,+)$	$\dot{y}_1 > 0$ $(+,+)$	$\dot{y}_1 > 0$ $(+,+)$	$\dot{y}_1^* = 0$ $(+,-)$
(d)	$(-,-)$	$\dot{y}_1 > 0$ $(+,+)$	$\dot{y}_1^* = 0$ $(-,+)$	$\dot{y}_1^* = 0$ $(-,+)$	$\dot{y}_1 < 0$ $(-,-)$

$$\left\{\begin{array}{c}(a_{111}, \dot{x}_2)\\(a_{112}, \dot{x}_2)\\(\dot{x}_1, a_{221})\end{array}\right\} = \left\{\begin{array}{c}\underbrace{(\text{DP},\text{pF})}_{\text{down-parabola flow }(+)}\\\underbrace{(\text{DI},\text{pF})}_{\text{decreasing-inflection flow }(+)}\\\underbrace{(\text{nF}, 3^{\text{rd}}\text{DP})}_{\text{third-order down-parabola flow }(-)}\end{array}\right\}$$

for $a_{110} < 0$ and $a_{220} > 0$, (4.443)

$$\left\{\begin{array}{c}(a_{111}, \dot{x}_2)\\(a_{112}, \dot{x}_2)\\(\dot{x}_1, a_{221})\end{array}\right\} = \left\{\begin{array}{c}\underbrace{(\text{DP},\text{nF})}_{\text{down-parabola flow }(-)}\\\underbrace{(\text{DI},\text{nF})}_{\text{decreasing-inflection flow }(-)}\\\underbrace{(\text{pF}, 3^{\text{rd}}\text{DP})}_{\text{third-order down-parabola flow }(+)}\end{array}\right\}$$

for $a_{110} > 0$ and $a_{220} < 0$, (4.444)

$$\left\{\begin{array}{c}(a_{111}, \dot{x}_2)\\(a_{112}, \dot{x}_2)\\(\dot{x}_1, a_{221})\end{array}\right\} = \left\{\begin{array}{c}\underbrace{(\text{UP},\text{nF})}_{\text{up-parabola flow }(-)}\\\underbrace{(\text{II},\text{nF})}_{\text{increasing-inflection flow }(-)}\\\underbrace{(\text{pF}, 3^{\text{rd}}\text{UP})}_{\text{third-order up-parabola flow }(+)}\end{array}\right\}$$

for $a_{110} < 0$ and $a_{220} < 0$. (4.445)

In addition, the equilibrium dynamics on the y_1-direction are listed in Table 4.3.

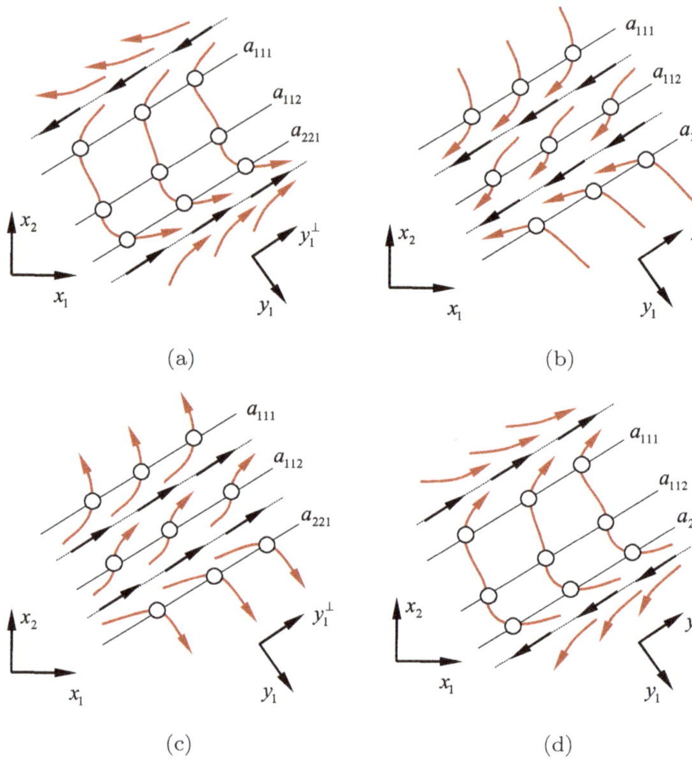

Fig. 4.4. Phase portraits ($a_{111} > a_{221}$ and $\alpha_{11} > 0, \alpha_{12} < 0$) for two-dimensional systems on the x_1-direction with $y_1^* = a_{111}, a_{112}$ and on the x_2-direction with $y_1^* = a_{221}$: (a) ($a_{110} > 0, a_{220} > 0$), (b) ($a_{110} < 0, a_{220} > 0$), (c) ($a_{110} > 0, a_{220} < 0$), (d) ($a_{110} < 0, a_{220} < 0$).

(iv) For $a_{112} < a_{221}$ ($\alpha_{11} > 0, \alpha_{12} < 0$), presented in Fig. 4.4 are the one-dimensional flows of $(y_1^*, \dot{x}_2) = (a_{111}, \dot{x}_2), (a_{112}, \dot{x}_2)$ and $(\dot{x}_1, y_1^*) = (\dot{x}_1, a_{221})$. The corresponding properties for $\alpha_{11} > 0$ and $\alpha_{12} < 0$ are presented as follows:

$$\left\{\begin{array}{c}(a_{111}, \dot{x}_2)\\(a_{112}, \dot{x}_2)\\(\dot{x}_1, a_{221})\end{array}\right\} = \left\{\begin{array}{c}\underbrace{(\text{DP,pF})}_{\text{down-parabola flow }(+)}\\\underbrace{(\text{II,pF})}_{\text{increasing-inflection flow }(+)}\\\underbrace{(\text{pF}, 3^{\text{rd}}\text{UP})}_{\text{third-order up-parabola flow }(+)}\end{array}\right\}$$

$$\text{for } a_{110} > 0 \quad \text{and} \quad a_{220} > 0, \tag{4.446}$$

Table 4.4. Equilibrium flows of $\dot{y}_1 = 0$ ($\alpha_{11} > 0, \alpha_{12} < 0$).

Figure 4.4	$(\dot{x}_1, \dot{x}_2)$ (a_{110}, a_{220})	y_1 $(-\infty, a_{111})$	(a_{111}, a_{112})	(a_{112}, a_{221})	(a_{221}, ∞)
(a)	$(+,+)$	$\dot{y}_1^* = 0$ $(-,-)$	$\dot{y}_1 < 0$ $(+,-)$	$\dot{y}_1 < 0$ $(+,-)$	$\dot{y}_1^* = 0$ $(+,+)$
(b)	$(-,+)$	$\dot{y}_1 > 0$ $(+,-)$	$\dot{y}_1^* = 0$ $(-,-)$	$\dot{y}_1^* = 0$ $(-,-)$	$\dot{y}_1 < 0$ $(-,+)$
(c)	$(+,-)$	$\dot{y}_1 < 0$ $(-,+)$	$\dot{y}_1^* = 0$ $(+,+)$	$\dot{y}_1^* = 0$ $(+,+)$	$\dot{y}_1 > 0$ $(+,-)$
(d)	$(-,-)$	$\dot{y}_1^* = 0$ $(+,+)$	$\dot{y}_1 > 0$ $(-,+)$	$\dot{y}_1 > 0$ $(-,+)$	$\dot{y}_1^* = 0$ $(-,-)$

$$\left\{\begin{array}{c}(a_{111}, \dot{x}_2)\\(a_{112}, \dot{x}_2)\\(\dot{x}_1, a_{221})\end{array}\right\} = \left\{\begin{array}{c}\underbrace{(\text{UP,pF})}_{\text{up-parabola flow }(+)}\\ \underbrace{(\text{DI,pF})}_{\text{decreasing-inflection flow }(+)}\\ \underbrace{(\text{nF}, 3^{\text{rd}}\text{DP})}_{\text{third-order down-parabola flow }(-)}\end{array}\right\}$$

$$\text{for } a_{110} < 0 \quad \text{and} \quad a_{220} > 0, \tag{4.447}$$

$$\left\{\begin{array}{c}(a_{111}, \dot{x}_2)\\(a_{112}, \dot{x}_2)\\(\dot{x}_1, a_{221})\end{array}\right\} = \left\{\begin{array}{c}\underbrace{(\text{UP,nF})}_{\text{up-parabola flow }(-)}\\ \underbrace{(\text{DI, nF})}_{\text{decreasing-inflection flow }(-)}\\ \underbrace{(\text{pF}, 3^{\text{rd}}\text{DP})}_{\text{third-order down-parabola flow }(+)}\end{array}\right\}$$

$$\text{for } a_{110} > 0 \quad \text{and} \quad a_{220} < 0, \tag{4.448}$$

$$\left\{\begin{array}{c}(a_{111}, \dot{x}_2)\\(a_{112}, \dot{x}_2)\\(\dot{x}_1, a_{221})\end{array}\right\} = \left\{\begin{array}{c}\underbrace{(\text{DP,nF})}_{\text{down-parabola flow }(-)}\\ \underbrace{(\text{II, nF})}_{\text{increasing-inflection flow }(-)}\\ \underbrace{(\text{nF}, 3^{\text{rd}}\text{UP})}_{\text{third-order up-parabola flow }(-)}\end{array}\right\}$$

$$\text{for } a_{110} < 0 \quad \text{and} \quad a_{220} < 0. \tag{4.449}$$

In addition, the equilibrium dynamics on the y_1-direction are listed in Table 4.4.

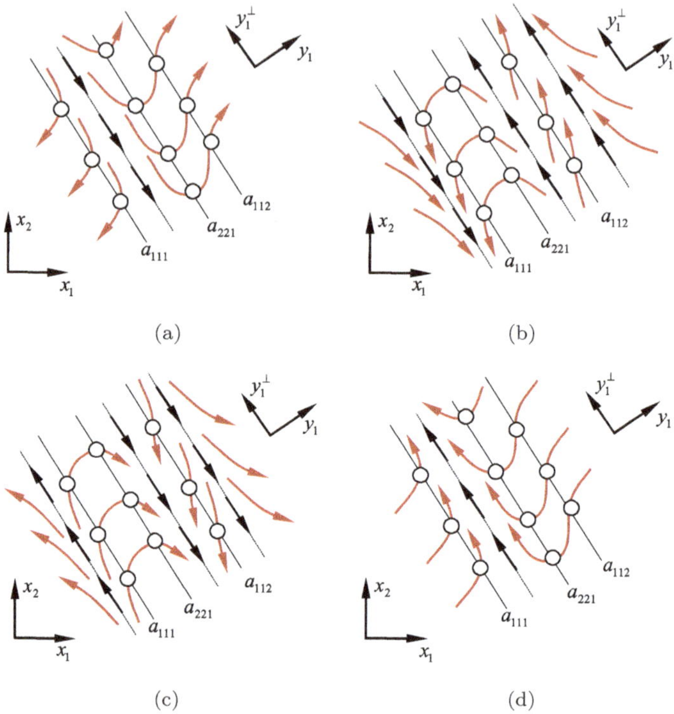

Fig. 4.5. Phase portraits ($a_{111} < a_{221} < a_{112}$ and $\alpha_{11} > 0, \alpha_{12} > 0$) for two-dimensional systems on the x_1-direction with $y_1^* = a_{111}, a_{112}$ and on the x_2-direction with $y_1^* = a_{221}$: (a) ($a_{110} > 0, a_{220} > 0$), (b) ($a_{110} < 0, a_{220} < 0$), (c) ($a_{110} > 0, a_{220} < 0$), (d) ($a_{110} < 0, a_{220} < 0$).

(v) For $a_{221} \in (a_{111}, a_{112})$ ($\alpha_{11} > 0, \alpha_{12} > 0$), presented in Fig. 4.5 are the one-dimensional flows of $(y_1^*, \dot{x}_2) = (a_{111}, \dot{x}_2), (a_{112}, \dot{x}_2)$ and $(\dot{x}_1, y_1^*) = (\dot{x}_1, a_{221})$. The corresponding properties for $\alpha_{11} > 0$ and $\alpha_{12} > 0$ are presented as follows:

$$\left\{ \begin{array}{c} (a_{111}, \dot{x}_2) \\ (\dot{x}_1, a_{221}) \\ (a_{112}, \dot{x}_2) \end{array} \right\} = \left\{ \begin{array}{c} \underbrace{(\mathrm{DP}, \mathrm{nF})}_{\text{down-parabola flow } (-)} \\ \underbrace{(\mathrm{pF}, 3^{\mathrm{rd}}\mathrm{UP})}_{\text{third-order up-parabola flow } (+)} \\ \underbrace{(\mathrm{II}, \mathrm{pF})}_{\text{increasing-inflection flow } (+)} \end{array} \right\}$$
$$\text{for } a_{110} > 0 \quad \text{and} \quad a_{220} > 0, \tag{4.450}$$

Table 4.5. Equilibrium flows of $\dot{y}_1 = 0$ $(\alpha_{11} > 0, \alpha_{12} > 0)$.

	$(\dot{x}_1, \dot{x}_2)$	y_1			
Figure 4.5	(a_{110}, a_{220})	$(-\infty, a_{111})$	(a_{111}, a_{221})	(a_{221}, a_{112})	(a_{112}, ∞)
(a)	$(+,+)$	$\dot{y}_1 < 0$ $(-,-)$	$\dot{y}_1^* = 0$ $(+,-)$	$\dot{y}_1 > 0$ $(+,+)$	$\dot{y}_1 > 0$ $(+,+)$
(b)	$(-,+)$	$\dot{y}_1^* = 0$ $(+,-)$	$\dot{y}_1 < 0$ $(-,-)$	$\dot{y}_1^* = 0$ $(-,+)$	$\dot{y}_1^* = 0$ $(-,+)$
(c)	$(+,-)$	$\dot{y}_1^* = 0$ $(-,+)$	$\dot{y}_1 > 0$ $(+,+)$	$\dot{y}_1^* = 0$ $(+,-)$	$\dot{y}_1^* = 0$ $(+,-)$
(d)	$(-,-)$	$\dot{y}_1 > 0$ $(+,+)$	$\dot{y}_1^* = 0$ $(-,+)$	$\dot{y}_1 < 0$ $(-,-)$	$\dot{y}_1 < 0$ $(-,-)$

$$\left\{\begin{matrix}(a_{111}, \dot{x}_2)\\(\dot{x}_1, a_{221})\\(a_{112}, \dot{x}_2)\end{matrix}\right\} = \left\{\begin{matrix}\underbrace{(\text{UP,nF})}_{\text{up-parabola flow }(-)}\\\underbrace{(\text{nF}, 3^{\text{rd}}\text{DP})}_{\text{third-order down-parabola flow }(-)}\\\underbrace{(\text{DI,pF})}_{\text{increasing-inflection flow }(+)}\end{matrix}\right\}$$

$$\text{for } a_{110} < 0 \quad \text{and} \quad a_{220} > 0, \tag{4.451}$$

$$\left\{\begin{matrix}(a_{111}, \dot{x}_2)\\(\dot{x}_1, a_{221})\\(a_{112}, \dot{x}_2)\end{matrix}\right\} = \left\{\begin{matrix}\underbrace{(\text{UP, pF})}_{\text{up-parabola flow }(+)}\\\underbrace{(\text{pF}, 3^{\text{rd}}\text{DP})}_{\text{third-order down-parabola flow }(+)}\\\underbrace{(\text{DI, nF})}_{\text{increasing-inflection flow }(-)}\end{matrix}\right\}$$

$$\text{for } a_{110} > 0 \quad \text{and} \quad a_{220} < 0, \tag{4.452}$$

$$\left\{\begin{matrix}(a_{111}, \dot{x}_2)\\(\dot{x}_1, a_{221})\\(a_{112}, \dot{x}_2)\end{matrix}\right\} = \left\{\begin{matrix}\underbrace{(\text{DP, pF})}_{\text{down-parabola flow }(+)}\\\underbrace{(\text{nF}, 3^{\text{rd}}\text{UP})}_{\text{third-order up-parabola flow }(-)}\\\underbrace{(\text{II, nF})}_{\text{increasing-inflection flow }(-)}\end{matrix}\right\}$$

$$\text{for } a_{110} < 0 \quad \text{and} \quad a_{220} < 0. \tag{4.453}$$

In addition, the equilibrium dynamics on the y_1-direction are listed in Table 4.5.

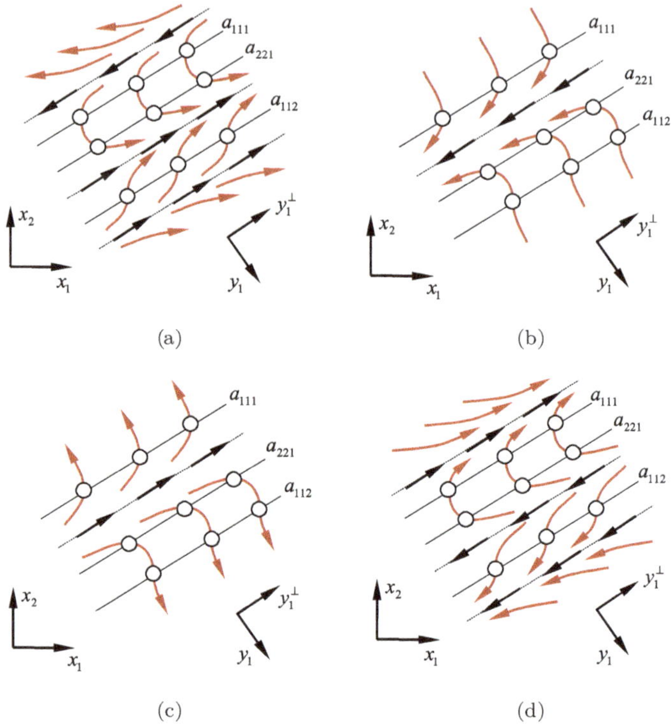

Fig. 4.6. Phase portraits ($a_{111} < a_{221} < a_{112}$ and $\alpha_{11} > 0, \alpha_{12} < 0$) for two-dimensional systems on the x_1-direction with $y_1^* = a_{111}, a_{112}$ and on the x_2-direction with $y_1^* = a_{221}$: (a) ($a_{110} > 0, a_{220} > 0$), (b) ($a_{110} < 0, a_{220} > 0$), (c) ($a_{110} > 0, a_{220} < 0$), (d) ($a_{110} < 0, a_{220} < 0$).

(vi) For $a_{221} \in (a_{111}, a_{112})$ ($\alpha_{11} > 0, \alpha_{12} < 0$), presented in Fig. 4.6 are the one-dimensional flows of $(y_1^*, \dot{x}_2) = (a_{111}, \dot{x}_2), (a_{112}, \dot{x}_2)$ and $(\dot{x}_1, y_1^*) = (\dot{x}_1, a_{221})$. The corresponding properties for $\alpha_{11} > 0$ and $\alpha_{12} < 0$ are presented as follows:

$$\left\{ \begin{array}{c} (a_{111}, \dot{x}_2) \\ (\dot{x}_1, a_{221}) \\ (a_{112}, \dot{x}_2) \end{array} \right\} = \left\{ \begin{array}{c} \underbrace{(\text{UP}, \text{nF})}_{\text{up-parabola flow }(-)} \\ \underbrace{(\text{pF}, 3^{\text{rd}}\text{UP})}_{\text{third-order up-parabola flow }(+)} \\ \underbrace{(\text{II}, \text{pF})}_{\text{increasing-inflection flow }(+)} \end{array} \right\}$$
$$\text{for } a_{110} > 0 \quad \text{and} \quad a_{220} > 0. \tag{4.454}$$

Table 4.6. Equilibrium flows of $\dot{y}_1 = 0$ ($\alpha_{11} > 0, \alpha_{12} < 0$).

Figure 4.6	$(\dot{x}_1, \dot{x}_2)$ (a_{110}, a_{220})	y_1 $(-\infty, a_{111})$	(a_{111}, a_{221})	(a_{221}, a_{112})	(a_{112}, ∞)
(a)	$(+,+)$	$\dot{y}_1^* = 0$ $(-,-)$	$\dot{y}_1 > 0$ $(+,-)$	$\dot{y}_1^* = 0$ $(+,+)$	$\dot{y}_1^* = 0$ $(+,+)$
(b)	$(-,+)$	$\dot{y}_1 > 0$ $(+,-)$	$\dot{y}_1^* = 0$ $(-,-)$	$\dot{y}_1 < 0$ $(-,+)$	$\dot{y}_1 < 0$ $(-,+)$
(c)	$(+,-)$	$\dot{y}_1 < 0$ $(-,+)$	$\dot{y}_1^* = 0$ $(+,+)$	$\dot{y}_1 > 0$ $(+,-)$	$\dot{y}_1 > 0$ $(+,-)$
(d)	$(-,-)$	$\dot{y}_1^* = 0$ $(+,+)$	$\dot{y}_1 < 0$ $(-,+)$	$\dot{y}_1^* = 0$ $(-,-)$	$\dot{y}_1^* = 0$ $(-,-)$

$$\left\{\begin{matrix}(a_{111}, \dot{x}_2)\\(\dot{x}_1, a_{221})\\(a_{112}, \dot{x}_2)\end{matrix}\right\} = \left\{\begin{matrix}\underbrace{(\text{DP}, \text{nF})}_{\text{down-parabola flow }(-)}\\\underbrace{(\text{nF}, 3^{\text{rd}}\text{DP})}_{\text{third-order down-parabola flow }(-)}\\\underbrace{(\text{DI}, \text{pF})}_{\text{increasing-inflection flow }(+)}\end{matrix}\right\}$$

$$\text{for } a_{110} < 0 \quad \text{and} \quad a_{220} > 0, \tag{4.455}$$

$$\left\{\begin{matrix}(a_{111}, \dot{x}_2)\\(\dot{x}_1, a_{221})\\(a_{112}, \dot{x}_2)\end{matrix}\right\} = \left\{\begin{matrix}\underbrace{(\text{DP}, \text{pF})}_{\text{down-parabola flow }(+)}\\\underbrace{(\text{pF}, 3^{\text{rd}}\text{DP})}_{\text{third-order down-parabola flow }(+)}\\\underbrace{(\text{DI}, \text{nF})}_{\text{increasing-inflection flow }(-)}\end{matrix}\right\}$$

$$\text{for } a_{110} < 0 \quad \text{and} \quad a_{220} > 0, \tag{4.456}$$

$$\left\{\begin{matrix}(a_{111}, \dot{x}_2)\\(\dot{x}_1, a_{221})\\(a_{112}, \dot{x}_2)\end{matrix}\right\} = \left\{\begin{matrix}\underbrace{(\text{UP}, \text{pF})}_{\text{up-parabola flow }(+)}\\\underbrace{(\text{nF}, 3^{\text{rd}}\text{UP})}_{\text{third-order up-parabola flow }(-)}\\\underbrace{(\text{II}, \text{nF})}_{\text{increasing-inflection flow }(-)}\end{matrix}\right\}$$

$$\text{for } a_{110} < 0 \quad \text{and} \quad a_{220} < 0, \tag{4.457}$$

In addition, the equilibrium dynamics on the y_1-direction are listed in Table 4.6.

4.3.2 *Inflection-saddle and parabola-sources and sinks*

(A) For $a_{111} = a_{221} < a_{112}$, consider

$$\begin{aligned} \dot{x}_1 &= a_{110}(y_1 - a_{111})(y_1 - a_{112})^2, \\ \dot{x}_2 &= a_{220}(y_1 - a_{221})^3, \end{aligned} \tag{4.458}$$

where

$$y_1 = \alpha_{11}x_1 + \alpha_{12}x_2. \tag{4.459}$$

There is a dynamical system in the y_1-direction as

$$\dot{y}_1 = \alpha_{11}a_{110}(y_1 - a_{111})(y_1 - a_{112})^2 + \alpha_{12}a_{220}(y_1 - a_{221})^3, \tag{4.460}$$

with a dynamical system in the $y_{j_1}^{\perp}$-direction as

$$\dot{y}_1 = \beta_{11}a_{110}(y_1 - a_{111})(y_1 - a_{112})^2 + \beta_{12}a_{220}(y_1 - a_{221})^3, \tag{4.461}$$

and a G-function is

$$G_{110} = \alpha_{11}a_{110}(a_{111} - a_{112})^2. \tag{4.462}$$

(i) For $a_{111} = a_{221}(\alpha_{11} > 0, \alpha_{12} > 0)$, the phase portraits for the two-dimensional system with the equilibrium of $y_1^* = a_{111}, a_{112}$ on the x_1-direction and with the equilibrium of $y_1^* = a_{221}$ on the x_2-direction are presented in Fig. 4.7. The corresponding properties are given as follows:

- For $a_{110} > 0$ and $a_{220} > 0$,

$$\left\{\begin{matrix}(a_{111}, a_{221}) \\ (a_{112}, \dot{x}_2^*)\end{matrix}\right\} = \left\{\begin{matrix}\underbrace{(2^{\text{nd}}\text{II}, \text{SO})}_{\text{second-order increasing-inflection sourxe}} \\ \underbrace{(\text{II}, \text{pF})}_{\text{increasing-inflection flow }(+)}\end{matrix}\right\} \text{ for } G_{110} > 0. \tag{4.463}$$

- For $a_{110} < 0$ and $a_{220} > 0$,

$$\left\{\begin{matrix}(a_{111}, a_{221}) \\ (a_{112}, \dot{x}_2^*)\end{matrix}\right\} = \left\{\begin{matrix}\underbrace{(2^{\text{nd}}\text{DI}, \text{SI})}_{\text{second-order decreasing-inflection sink}} \\ \underbrace{(\text{DI}, \text{pF})}_{\text{decreasing-inflection flow }(+)}\end{matrix}\right\} \text{ for } G_{110} < 0;$$

$$\left\{\begin{matrix}(a_{111}, a_{221}) \\ (a_{112}, \dot{x}_2^*)\end{matrix}\right\} = \left\{\begin{matrix}\underbrace{(2^{\text{nd}}\text{DI}, \text{SO})}_{\text{second-order decreasing-inflection source}} \\ \underbrace{(\text{DI}, \text{nF})}_{\text{decreasing-inflection flow }(-)}\end{matrix}\right\} \text{ for } G_{110} > 0. \tag{4.464}$$

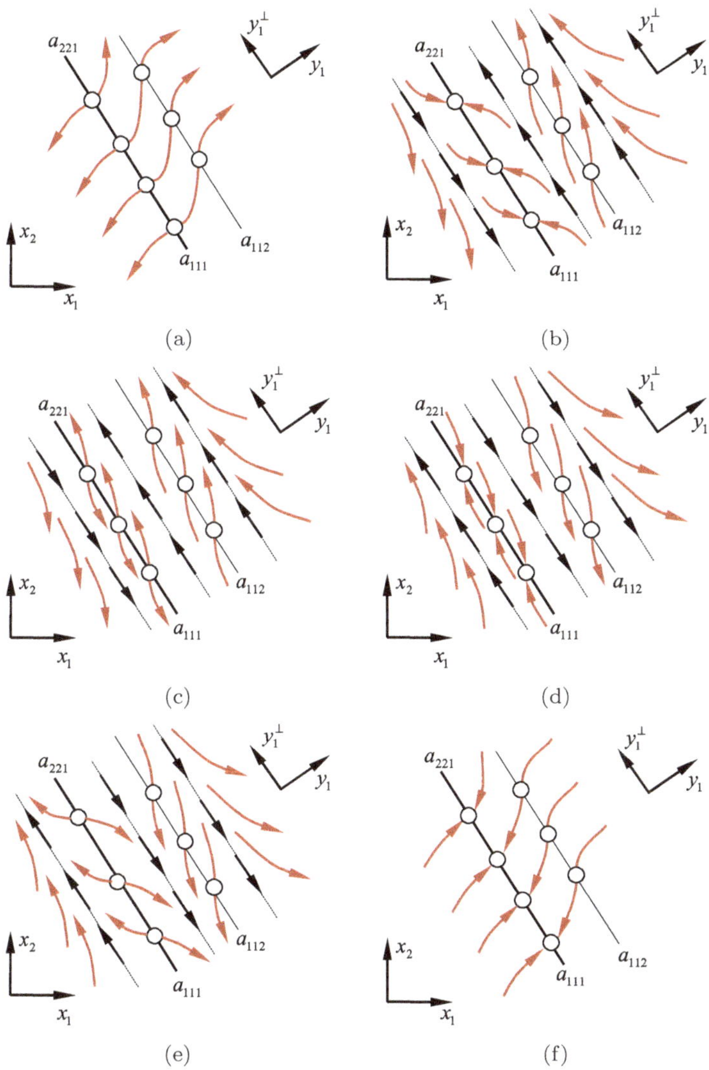

Fig. 4.7. Phase portraits ($a_{111} = a_{221}$ and $\alpha_{11} > 0, \alpha_{12} > 0$) for two-dimensional systems on the x_1-direction with $y_1^* = a_{111}, a_{112}$ and on the x_2-direction with $y_1^* = a_{221}$: (a) ($a_{110} > 0, a_{220} > 0$), (b) ($a_{110} < 0, a_{220} > 0$), (c) ($a_{110} > 0, a_{220} < 0$), (d) ($a_{110} < 0, a_{220} < 0$).

Table 4.7. Equilibrium flows of $\dot{y}_1 = 0$ for $a_{111} = a_{221}(\alpha_{11} > 0, \alpha_{12} > 0)$.

Figure 4.7	$(\dot{x}_1, \dot{x}_2)$	y_1		
	(a_{110}, a_{220})	$(-\infty, a_{111})$	(a_{111}, a_{112})	(a_{112}, ∞)
(a)	$(+,+)$	$\dot{y}_1 < 0$ $(-,-)$	$\dot{y}_1 > 0$ $(+,+)$	$\dot{y}_1 > 0$ $(+,+)$
(b,c)	$(-,+)$	$\dot{y}_1^* = 0$ $(+,-)$	$\dot{y}_1^* = 0$ $(-,+)$	$\dot{y}_1^* = 0$ $(-,+)$
(d,e)	$(+,-)$	$\dot{y}_1^* = 0$ $(-,+)$	$\dot{y}_1^* = 0$ $(+,-)$	$\dot{y}_1^* = 0$ $(+,-)$
(f)	$(-,-)$	$\dot{y}_1 > 0$ $(+,+)$	$\dot{y}_1 < 0$ $(-,-)$	$\dot{y}_1 < 0$ $(-,-)$

- For $a_{110} > 0$ and $a_{220} < 0$,

$$\left\{\begin{array}{c}(a_{111}, a_{221})\\(a_{112}, \dot{x}_2^*)\end{array}\right\} = \left\{\begin{array}{c}\underbrace{(2^{\mathrm{nd}}\mathrm{DI}, \mathrm{SO})}_{\text{second-order decreasing-inflection source}}\\\underbrace{(\mathrm{DI}, \mathrm{pF})}_{\text{decreasing-inflection flow } (+)}\end{array}\right\}$$

$$\text{for } G_{110} > 0;$$

$$\left\{\begin{array}{c}(a_{111}, a_{221})\\(a_{112}, \dot{x}_2^*)\end{array}\right\} = \left\{\begin{array}{c}\underbrace{(2^{\mathrm{nd}}\mathrm{DI}, \mathrm{SI})}_{\text{second-order decreasing-inflection sink}}\\\underbrace{(\mathrm{DI}, \mathrm{nF})}_{\text{decreasing-inflection flow } (-)}\end{array}\right\} \text{ for } G_{110} < 0. \tag{4.465}$$

- For $a_{110} < 0$ and $a_{220} < 0$,

$$\left\{\begin{array}{c}(a_{111}, a_{221})\\(a_{112}, \dot{x}_2^*)\end{array}\right\} = \left\{\begin{array}{c}\underbrace{(2^{\mathrm{nd}}\mathrm{II}, \mathrm{SI})}_{\text{second-order increasing-inflection sink}}\\\underbrace{(\mathrm{II}, \mathrm{nF})}_{\text{increasing-inflection flow } (-)}\end{array}\right\} \text{ for } G_{110} < 0. \tag{4.466}$$

The corresponding equilibrium flows of $\dot{y}_1 = 0$in phase space are listed in Table 4.7.

(ii) For $a_{111} = a_{221}(\alpha_{11} > 0, \alpha_{12} < 0)$, the phase portraits for the two-dimensional system with the equilibrium of $y_1^* = a_{111}, a_{112}$ on the x_1-direction and with the equilibrium of $y_1^* = a_{221}$ on the x_2-direction are presented in Fig. 4.8. The corresponding properties are given as follows:

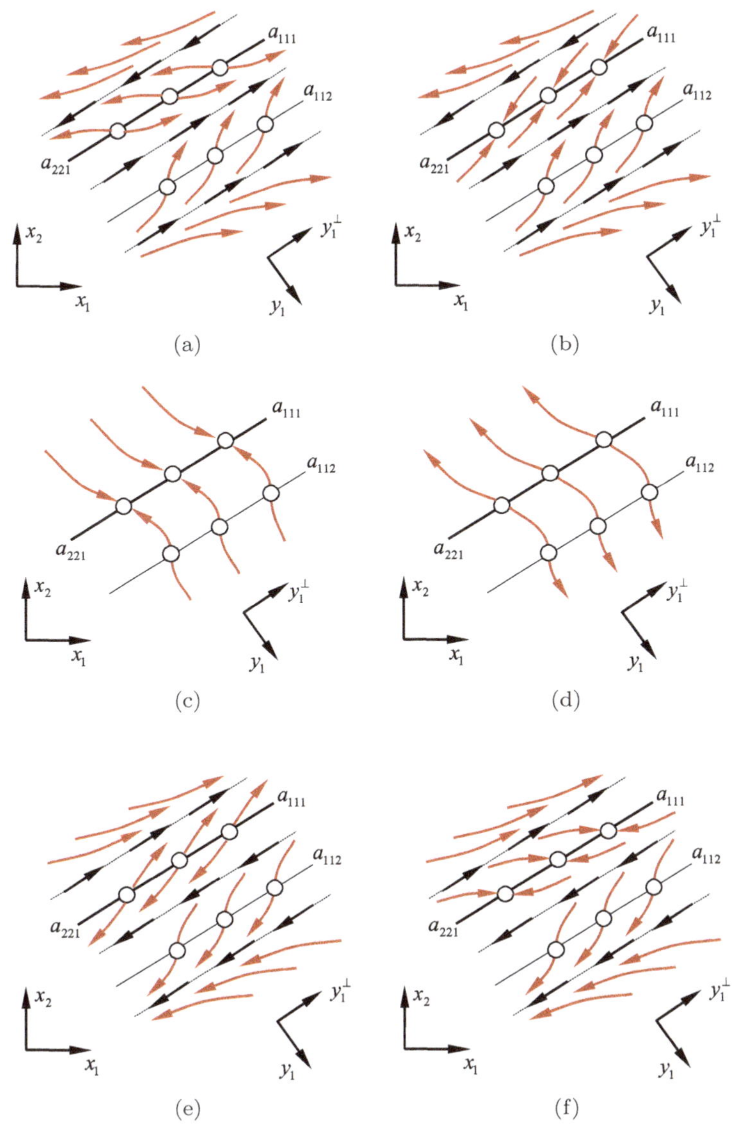

Fig. 4.8. Phase portraits ($a_{111} = a_{221}$ and $\alpha_{11} > 0, \alpha_{12} < 0$) for two-dimensional systems on the x_1-direction with $y_1^* = a_{111}, a_{112}$ and on the x_2-direction with $y_1^* = a_{221}$: (a) ($a_{110} > 0, a_{220} > 0$), (b) ($a_{110} < 0, a_{220} > 0$), (c) ($a_{110} > 0, a_{220} < 0$), (d) ($a_{110} < 0, a_{220} < 0$).

- For $a_{110} > 0$ and $a_{220} > 0$,

$$
\left\{\begin{array}{c}(a_{111}, a_{221})\\(a_{112}, \dot{x}_2^*)\end{array}\right\} = \left\{\begin{array}{c}\underbrace{(2^{\text{nd}}\text{II}, \text{SO})}_{\text{second-order increasing-inflection source}}\\ \underbrace{(\text{II}, \text{pF})}_{\text{increasing-inflection flow }(+)}\end{array}\right\} \quad \text{for } G_{110} > 0;
$$

$$
\left\{\begin{array}{c}(a_{111}, a_{221})\\(a_{112}, \dot{x}_2^*)\end{array}\right\} = \left\{\begin{array}{c}\underbrace{(2^{\text{nd}}\text{II}, \text{SI})}_{\text{second-order increasing-inflection sink}}\\ \underbrace{(\text{II}, \text{pF})}_{\text{increasing-inflection flow }(+)}\end{array}\right\} \quad \text{for } G_{110} < 0. \tag{4.467}
$$

- For $a_{110} < 0$ and $a_{220} > 0$,

$$
\left\{\begin{array}{c}(a_{111}, a_{221})\\(a_{112}, \dot{x}_2^*)\end{array}\right\} = \left\{\begin{array}{c}\underbrace{(2^{\text{nd}}\text{DI}, \text{SI})}_{\text{second-order decreasing-inflection sink}}\\ \underbrace{(\text{DI}, \text{pF})}_{\text{decreasing-inflection flow }(+)}\end{array}\right\} \quad \text{for } G_{110} < 0. \tag{4.468}
$$

- For $a_{110} > 0$ and $a_{220} < 0$,

$$
\left\{\begin{array}{c}(a_{111}, a_{221})\\(a_{112}, \dot{x}_2^*)\end{array}\right\} = \left\{\begin{array}{c}\underbrace{(2^{\text{nd}}\text{DI}, \text{SO})}_{\text{second-order decreasing-inflection source}}\\ \underbrace{(\text{DI}, \text{nF})}_{\text{decreasing-inflection flow }(-)}\end{array}\right\} \quad \text{for } G_{110} > 0. \tag{4.469}
$$

- For $a_{110} < 0$ and $a_{220} < 0$,

$$
\left\{\begin{array}{c}(a_{111}, a_{221})\\(a_{112}, \dot{x}_2^*)\end{array}\right\} = \left\{\begin{array}{c}\underbrace{(2^{\text{nd}}\text{II}, \text{SO})}_{\text{second-order increasing-inflection source}}\\ \underbrace{(\text{II}, \text{nF})}_{\text{increasing-inflection flow }(-)}\end{array}\right\} \quad \text{for } G_{110} > 0;
$$

$$
\left\{\begin{array}{c}(a_{111}, a_{221})\\(a_{112}, \dot{x}_2^*)\end{array}\right\} = \left\{\begin{array}{c}\underbrace{(2^{\text{nd}}\text{II}, \text{SI})}_{\text{second-order increasing-inflection sink}}\\ \underbrace{(\text{II}, \text{nF})}_{\text{increasing-inflection flow }(-)}\end{array}\right\} \quad \text{for } G_{110} < 0. \tag{4.470}
$$

The corresponding equilibrium flows of $\dot{y}_1 = 0$ in phase space are listed in Table 4.8.

Table 4.8. Equilibrium flows of $\dot{y}_1 = 0$ for $a_{111} = a_{221}$ $(\alpha_{11} > 0, \alpha_{12} < 0)$.

Figure 4.8	$(\dot{x}_1, \dot{x}_2)$	y_1		
	(a_{110}, a_{220})	$(-\infty, a_{111})$	(a_{111}, a_{112})	(a_{112}, ∞)
(a,b)	$(+,+)$	$\dot{y}_1^* = 0$ $(-,-)$	$\dot{y}_1^* = 0$ $(+,+)$	$\dot{y}_1^* = 0$ $(+,+)$
(c)	$(-,+)$	$\dot{y}_1 > 0$ $(+,-)$	$\dot{y}_1 < 0$ $(-,+)$	$\dot{y}_1 < 0$ $(-,+)$
(d)	$(+,-)$	$\dot{y}_1 > 0$ $(-,+)$	$\dot{y}_1 > 0$ $(+,-)$	$\dot{y}_1 > 0$ $(+,-)$
(e,f)	$(-,-)$	$\dot{y}_1^* = 0$ $(+,+)$	$\dot{y}_1^* = 0$ $(-,-)$	$\dot{y}_1^* = 0$ $(-,-)$

(B) For $a_{111} < a_{221} = a_{112}$, consider

$$\begin{aligned}\dot{x}_1 &= a_{110}(y_1 - a_{111})(y_1 - a_{112})^2,\\ \dot{x}_2 &= a_{220}(y_1 - a_{221})^3,\end{aligned} \tag{4.471}$$

where

$$y_1 = \alpha_{11}x_1 + \alpha_{12}x_2. \tag{4.472}$$

There is a dynamical system in the y_1-direction as

$$\dot{y}_1 = \alpha_{11}a_{110}(y_1 - a_{111})(y_1 - a_{112})^2 + \alpha_{12}a_{220}(y_1 - a_{221})^3, \tag{4.473}$$

with a dynamical system in the $y_{j_1}^{\perp}$-direction as

$$\dot{y}_1^{\perp} = \beta_{11}a_{110}(y_1 - a_{111})(y_1 - a_{112})^2 + \beta_{12}a_{220}(y_1 - a_{221})^3, \tag{4.474}$$

and a G-function is

$$G_{110}^{(1)} = \alpha_{11}a_{110}(a_{112} - a_{111}) + \alpha_{12}a_{220}. \tag{4.475}$$

(i) For $a_{111} = a_{221}(\alpha_{11} > 0, \alpha_{12} > 0)$, the phase portraits for the two-dimensional system with the equilibrium of $y_1^* = a_{111}, a_{112}$ on the x_1-direction and with the equilibrium of $y_1^* = a_{221}$ on the x_2-direction are presented in Fig. 4.9. The corresponding properties are given as follows:

- For $a_{110} > 0$ and $a_{220} > 0$,

$$\left\{\begin{matrix}(a_{111}, \dot{x}_2^*)\\ (a_{112}, a_{221})\end{matrix}\right\} = \left\{\begin{matrix}\underbrace{(\text{UP}, \text{pF})}_{\text{up-parabola flow }(+)}\\ \underbrace{(\text{UP}, \text{US})}_{\text{up-parabola upper-saddle}}\end{matrix}\right\}. \tag{4.476}$$

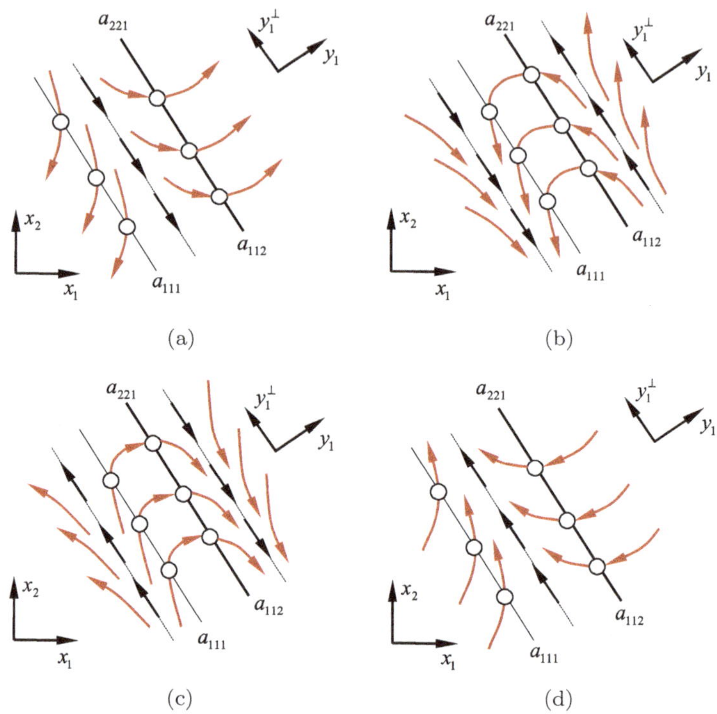

Fig. 4.9. Phase portraits ($a_{111} = a_{112}$ and $\alpha_{11} > 0, \alpha_{12} > 0$) for the two-dimensional system on the x_1-direction with $y_1^* = a_{111}, a_{112}$ and on the x_2-direction with $y_1^* = a_{221}$: (a) ($a_{110} > 0, a_{220} > 0$), (b,c) ($a_{110} < 0, a_{220} > 0$), (d,e) ($a_{110} > 0, a_{220} < 0$), (f) ($a_{110} < 0, a_{220} < 0$).

- For $a_{110} < 0$ and $a_{220} > 0$,

$$\left\{ \begin{array}{c} (a_{111}, \dot{x}_2^*) \\ (a_{112}, a_{221}) \end{array} \right\} = \left\{ \begin{array}{c} \underbrace{(\mathrm{DP}, \mathrm{nF})}_{\text{down-parabola flow } (-)} \\ \underbrace{(\mathrm{DP}, \mathrm{LS})}_{\text{down-parabola lower-saddle}} \end{array} \right\}. \tag{4.477}$$

- For $a_{110} > 0$ and $a_{220} < 0$,

$$\left\{ \begin{array}{c} (a_{111}, \dot{x}_2^*) \\ (a_{112}, a_{221}) \end{array} \right\} = \left\{ \begin{array}{c} \underbrace{(\mathrm{DP}, \mathrm{pF})}_{\text{down-parabola flow } (+)} \\ \underbrace{(\mathrm{DP}, \mathrm{US})}_{\text{down-parabola upper-saddle}} \end{array} \right\}. \tag{4.478}$$

Table 4.9. Equilibrium flows of $\dot{y}_1 = 0$ for $a_{111} = a_{221}(\alpha_{11} > 0, \alpha_{12} > 0)$.

Figure 4.9	$(\dot{x}_1, \dot{x}_2)$ (a_{110}, a_{220})	y_1 $(-\infty, a_{111})$	(a_{111}, a_{112})	(a_{112}, ∞)
(a)	$(+,+)$	$\dot{y}_1 < 0$ $(-,-)$	$\dot{y}_1^* = 0$ $(+,-)$	$\dot{y}_1 > 0$ $(+,+)$
(b)	$(-,+)$	$\dot{y}_1^* = 0$ $(+,-)$	$\dot{y}_1 < 0$ $(-,-)$	$\dot{y}_1^* = 0$ $(-,+)$
(c)	$(+,-)$	$\dot{y}_1^* = 0$ $(-,+)$	$\dot{y}_1 > 0$ $(+,+)$	$\dot{y}_1^* = 0$ $(+,-)$
(d)	$(-,-)$	$\dot{y}_1 > 0$ $(+,+)$	$\dot{y}_1^* = 0$ $(+,-)$	$\dot{y}_1 < 0$ $(-,-)$

- For $a_{110} < 0$ and $a_{220} < 0$,

$$\left\{ \begin{array}{c} (a_{111}, \dot{x}_2^*) \\ (a_{112}, a_{221}) \end{array} \right\} = \left\{ \begin{array}{c} \underbrace{(\mathrm{UP}, \mathrm{pF})}_{\text{up-parabola flow }(+)} \\ \underbrace{(\mathrm{UP}, \mathrm{LS})}_{\text{up-parabola lower-saddle}} \end{array} \right\}. \tag{4.479}$$

The corresponding equilibrium flows of $\dot{y}_1 = 0$ in phase space are in Table 4.9.

(ii) For $a_{111} = a_{221}(\alpha_{11} > 0, \alpha_{12} > 0)$, the phase portraits for the two-dimensional system with the equilibrium of $y_1^* = a_{111}, a_{112}$ on the x_1-direction and with the equilibrium of $y_1^* = a_{221}$ on the x_2-direction are presented in Fig. 4.10. The corresponding properties are given as follows:

- For $a_{110} > 0$ and $a_{220} > 0$,

$$\left\{ \begin{array}{c} (a_{111}, \dot{x}_2^*) \\ (a_{112}, a_{221}) \end{array} \right\} = \left\{ \begin{array}{c} \underbrace{(\mathrm{DP}, \mathrm{nF})}_{\text{down-parabola flow }(-)} \\ \underbrace{(\mathrm{UP}, \mathrm{US})}_{\text{up-parabola upper-saddle}} \end{array} \right\}. \tag{4.480}$$

- For $a_{110} < 0$ and $a_{220} > 0$,

$$\left\{ \begin{array}{c} (a_{111}, \dot{x}_2^*) \\ (a_{112}, a_{221}) \end{array} \right\} = \left\{ \begin{array}{c} \underbrace{(\mathrm{DP}, \mathrm{nF})}_{\text{down-parabola flow }(-)} \\ \underbrace{(\mathrm{DP}, \mathrm{LS})}_{\text{down-parabola lower-saddle}} \end{array} \right\}. \tag{4.481}$$

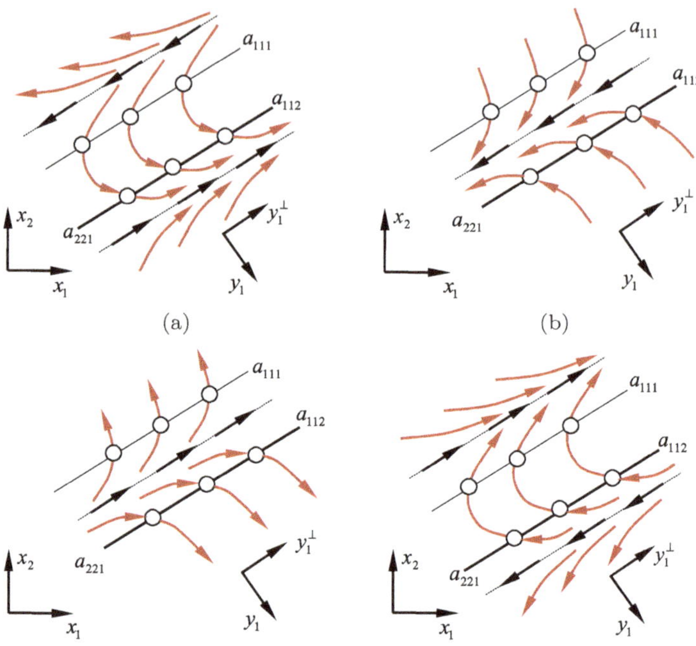

Fig. 4.10. Phase portraits ($a_{112} = a_{221}$ and $\alpha_{11} > 0, \alpha_{12} < 0$) for two-dimensional systems on the x_1-direction with $y_1^* = a_{111}, a_{112}$ and on the x_2-direction with $y_1^* = a_{221}$: (a,b) ($a_{110} > 0, a_{220} > 0$), (c) ($a_{110} < 0, a_{220} > 0$), (d) ($a_{110} > 0, a_{220} < 0$), (e,f) ($a_{110} < 0, a_{220} < 0$).

- For $a_{110} > 0$ and $a_{220} < 0$,

$$\left\{ \begin{matrix} (a_{111}, \dot{x}_2^*) \\ (a_{112}, a_{221}) \end{matrix} \right\} = \left\{ \begin{matrix} \underbrace{(\text{DP}, \text{pF})}_{\text{down-parabola flow } (+)} \\ \underbrace{(\text{DP}, \text{US})}_{\text{down-parabola upper-saddle}} \end{matrix} \right\}. \tag{4.482}$$

- For $a_{110} < 0$ and $a_{220} < 0$,

$$\left\{ \begin{matrix} (a_{111}, \dot{x}_2^*) \\ (a_{112}, a_{221}) \end{matrix} \right\} = \left\{ \begin{matrix} \underbrace{(\text{DP}, \text{nF})}_{\text{down-parabola flow } (-)} \\ \underbrace{(\text{UP}, \text{SI})}_{\text{up-parabola sink}} \end{matrix} \right\} \tag{4.483}$$

The corresponding equilibrium flows of $\dot{y}_1 = 0$ in phase space are in Table 4.10.

Table 4.10. Equilibrium flows of $\dot{y}_1 = 0$ for $a_{111} = a_{221}(\alpha_{11} > 0, \alpha_{12} < 0)$.

	$(\dot{x}_1, \dot{x}_2)$	y_1		
Figure 4.10	(a_{110}, a_{220})	$(-\infty, a_{111})$	(a_{111}, a_{112})	(a_{112}, ∞)
(a)	$(+,+)$	$\dot{y}_1^* = 0$ $(-,-)$	$\dot{y}_1 > 0$ $(+,-)$	$\dot{y}_1^* = 0$ $(+,+)$
(b)	$(-,+)$	$\dot{y}_1 < 0$ $(+,-)$	$\dot{y}_1^* = 0$ $(-,-)$	$\dot{y}_1 < 0$ $(-,+)$
(c)	$(+,-)$	$\dot{y}_1 > 0$ $(-,+)$	$\dot{y}_1^* = 0$ $(+,+)$	$\dot{y}_1 > 0$ $(+,-)$
(f)	$(-,-)$	$\dot{y}_1^* = 0$ $(+,+)$	$\dot{y}_1 < 0$ $(-,+)$	$\dot{y}_1^* = 0$ $(-,-)$

4.4 Singular Parabola Flows and Infinite-Equilibriums

In this section, the inflection and third-order parabola flows are presented, and the corresponding appearing bifurcations are discussed. The inflection and third-order flow switching dynamics are presented.

4.4.1 *Third-order parabola flows*

Consider

$$\begin{aligned} \dot{x}_1 &= a_{110}(y_1 - a_{111})^3, \\ \dot{x}_2 &= a_{220}(y_1 - a_{221})^3, \end{aligned} \tag{4.484}$$

with

$$y_1 = \alpha_{11}x_1 + \alpha_{12}x_2. \tag{4.485}$$

There is a dynamical system in the y_1-direction as

$$\dot{y}_1 = \alpha_{11}a_{110}(y_1 - a_{111})^3 + \alpha_{12}a_{220}(y_1 - a_{221})^3, \tag{4.486}$$

with a dynamical system in the $y_{j_1}^{\perp}$-direction as

$$\dot{y}_1^{\perp} = \beta_{11}a_{110}(y_1 - a_{111})^3 + \beta_{12}a_{220}(y_1 - a_{221})^3. \tag{4.487}$$

For $a_{111} < a_{221}(\alpha_{11} > 0, \alpha_{12} > 0)$, the phase portraits for two-dimensional systems with the equilibrium of $y_1^* = a_{111}$ on the x_1-direction and the equilibrium of $y_1^* = a_{221}$ on the x_2-direction are presented in Fig. 4.11. The corresponding properties are given as follows:

- For $a_{110} > 0$ and $a_{220} > 0$,

$$\left\{ \begin{matrix} (a_{111}, \dot{x}_2^*) \\ (\dot{x}_1^*, a_{221}) \end{matrix} \right\} = \left\{ \begin{matrix} \underbrace{(3^{\text{rd}}\text{DP}, \text{nF})}_{\text{third-order down-parabola flow } (-)} \\ \underbrace{(\text{pF}, 3^{\text{rd}}\text{UP})}_{\text{third-order up-parabola flow } (+)} \end{matrix} \right\}. \tag{4.488}$$

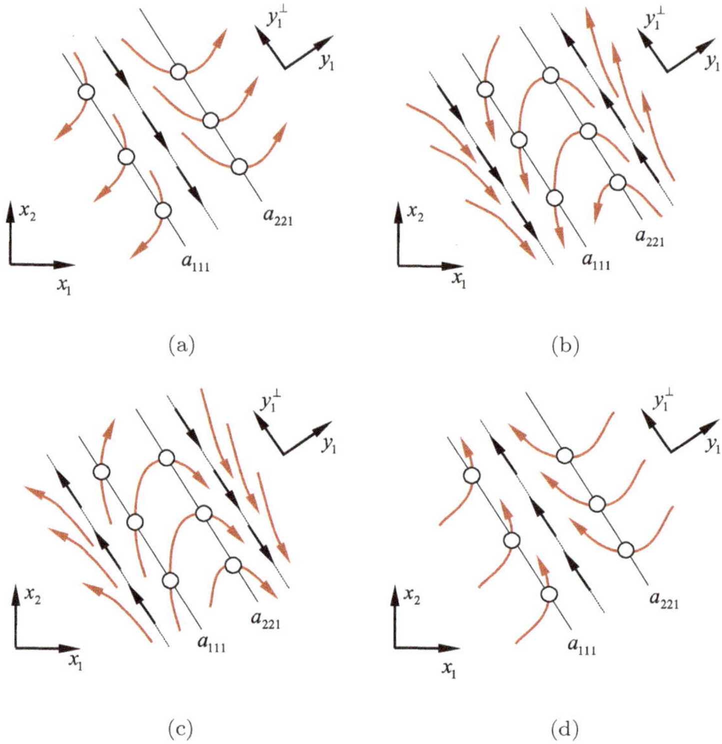

Fig. 4.11. Phase portraits ($a_{111} < a_{221}$ and $\alpha_{11} > 0, \alpha_{12} > 0$) for two-dimensional systems on the x_1-direction with $y_1^* = a_{111}$ and on the x_2-direction with $y_1^* = a_{221}$: (a) ($a_{110} > 0, a_{220} > 0$), (b) ($a_{110} < 0, a_{220} > 0$), (c) ($a_{110} > 0, a_{220} < 0$), (d) ($a_{110} < 0, a_{220} < 0$).

- For $a_{110} < 0$ and $a_{220} > 0$,

$$\left\{ \begin{matrix} (a_{111}, \dot{x}_2^*) \\ (\dot{x}_1^*, a_{221}) \end{matrix} \right\} = \left\{ \begin{matrix} \underbrace{(3^{\mathrm{rd}}\mathrm{UP}, \mathrm{nF})}_{\text{third-order up-parabola flow } (-)} \\ \underbrace{(\mathrm{nF}, 3^{\mathrm{rd}}\mathrm{DP})}_{\text{third-order down-parabola flow } (-)} \end{matrix} \right\}. \tag{4.489}$$

- For $a_{110} > 0$ and $a_{220} < 0$,

$$\left\{ \begin{matrix} (a_{111}, \dot{x}_2^*) \\ (\dot{x}_1^*, a_{221}) \end{matrix} \right\} = \left\{ \begin{matrix} \underbrace{(3^{\mathrm{rd}}\mathrm{UP}, \mathrm{pF})}_{\text{third-order up-parabola flow } (+)} \\ \underbrace{(\mathrm{pF}, 3^{\mathrm{rd}}\mathrm{DP})}_{\text{third-order down-parabola flow } (+)} \end{matrix} \right\}. \tag{4.490}$$

Table 4.11. Equilibrium flows of $\dot{y}_1 = 0$ for $a_{111} = a_{221}$ $(\alpha_{11} > 0, \alpha_{12} > 0)$.

	$(\dot{x}_1, \dot{x}_2)$	y_1		
Figure 4.11	(a_{110}, a_{220})	$(-\infty, a_{111})$	(a_{111}, a_{221})	(a_{221}, ∞)
(a)	$(+,+)$	$\dot{y}_1 < 0$ $(-,-)$	$\dot{y}_1^* = 0$ $(+,-)$	$\dot{y}_1 > 0$ $(+,+)$
(b)	$(-,+)$	$\dot{y}_1^* = 0$ $(+,-)$	$\dot{y}_1 < 0$ $(-,-)$	$\dot{y}_1^* = 0$ $(-,+)$
(c)	$(+,-)$	$\dot{y}_1^* = 0$ $(-,+)$	$\dot{y}_1 > 0$ $(+,+)$	$\dot{y}_1^* = 0$ $(+,-)$
(d)	$(-,-)$	$\dot{y}_1 < 0$ $(-,-)$	$\dot{y}_1^* = 0$ $(-,+)$	$\dot{y}_1 < 0$ $(-,-)$

- For $a_{110} < 0$ and $a_{220} < 0$,

$$\left\{ \begin{array}{c} (a_{111}, \dot{x}_2^*) \\ (\dot{x}_1^*, a_{221}) \end{array} \right\} = \left\{ \begin{array}{c} \underbrace{(3^{\text{rd}}\text{DP}, \text{pF})}_{\text{third-order down-parabola flow } (+)} \\ \underbrace{(\text{nF}, 3^{\text{rd}}\text{UP})}_{\text{third-order up-parabola flow } (-)} \end{array} \right\}. \tag{4.491}$$

The corresponding equilibrium flows of $\dot{y}_1 = 0$ in phase space are in Table 4.11.

(ii) For $a_{111} < a_{221}(\alpha_{11} > 0, \alpha_{12} < 0)$, the phase portraits for two-dimensional systems with the equilibrium of $y_1^* = a_{111}$ on the x_1-direction and the equilibrium of $y_1^* = a_{221}$ on the x_2-direction are presented in Fig. 4.12. The corresponding properties are given as follows:

- For $a_{110} > 0$ and $a_{220} > 0$,

$$\left\{ \begin{array}{c} (a_{111}, \dot{x}_2^*) \\ (\dot{x}_1^*, a_{221}) \end{array} \right\} = \left\{ \begin{array}{c} \underbrace{(3^{\text{rd}}\text{UP}, \text{nF})}_{\text{third-order up-parabola flow } (-)} \\ \underbrace{(\text{pF}, 3^{\text{rd}}\text{UP})}_{\text{third-order up-parabola flow } (+)} \end{array} \right\}. \tag{4.492}$$

- For $a_{110} < 0$ and $a_{220} > 0$,

$$\left\{ \begin{array}{c} (a_{111}, \dot{x}_2^*) \\ (\dot{x}_1^*, a_{221}) \end{array} \right\} = \left\{ \begin{array}{c} \underbrace{(3^{\text{rd}}\text{DP}, \text{nF})}_{\text{third-order down-parabola flow } (-)} \\ \underbrace{(\text{nF}, 3^{\text{rd}}\text{DP})}_{\text{third-order down-parabola flow } (-)} \end{array} \right\}. \tag{4.493}$$

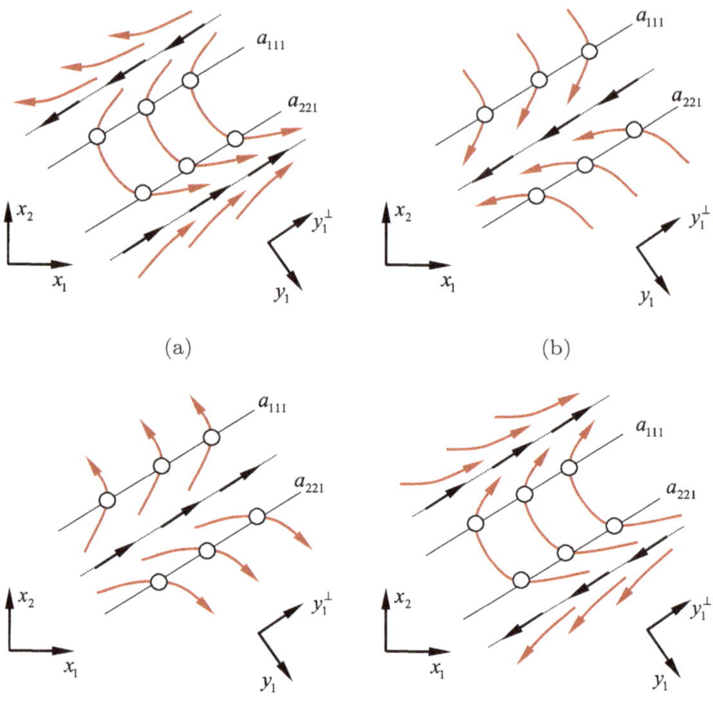

(a) (b) (c) (d)

Fig. 4.12. Phase portraits ($a_{111} < a_{221}$ and $\alpha_{11} > 0, \alpha_{12} < 0$) for two-dimensional systems on the x_1-direction with $y_1^* = a_{111}$ and on the x_2-direction with $y_1^* = a_{221}$: (a) ($a_{110} > 0, a_{220} > 0$), (b) ($a_{110} < 0, a_{220} > 0$), (c) ($a_{110} > 0, a_{220} < 0$), (d) ($a_{110} < 0, a_{220} < 0$).

- For $a_{110} > 0$ and $a_{220} < 0$,

$$\left\{ \begin{array}{c} (a_{111}, \dot{x}_2^*) \\ (\dot{x}_1^*, a_{221}) \end{array} \right\} = \left\{ \begin{array}{c} \underbrace{(3^{\mathrm{rd}}\mathrm{DP}, \mathrm{pF})}_{\text{third-order down-parabola flow } (+)} \\ \underbrace{(\mathrm{pF}, 3^{\mathrm{rd}}\mathrm{DP})}_{\text{third-order down-parabola flow } (+)} \end{array} \right\}. \tag{4.494}$$

- For $a_{110} < 0$ and $a_{220} < 0$,

$$\left\{ \begin{array}{c} (a_{111}, \dot{x}_2^*) \\ (\dot{x}_1^*, a_{221}) \end{array} \right\} = \left\{ \begin{array}{c} \underbrace{(3^{\mathrm{rd}}\mathrm{UP}, \mathrm{pF})}_{\text{third-order up-parabola flow } (-)} \\ \underbrace{(\mathrm{nF}, 3^{\mathrm{rd}}\mathrm{UP})}_{\text{third-order up-parabola flow } (-)} \end{array} \right\}. \tag{4.495}$$

Table 4.12. Equilibrium flows of $\dot{y}_1 = 0$ for $a_{111} < a_{221}$ $(\alpha_{11} > 0, \alpha_{12} < 0)$.

Figure 4.12	$(\dot{x}_1, \dot{x}_2)$ (a_{110}, a_{220})	y_1 $(-\infty, a_{111})$	(a_{111}, a_{221})	(a_{221}, ∞)
(a)	$(+,+)$	$\dot{y}_1^* = 0$ $(-,-)$	$\dot{y}_1 > 0$ $(+,-)$	$\dot{y}_1^* = 0$ $(+,+)$
(b)	$(-,+)$	$\dot{y}_1 > 0$ $(+,+)$	$\dot{y}_1^* = 0$ $(-,-)$	$\dot{y}_1 < 0$ $(-,+)$
(c)	$(+,-)$	$\dot{y}_1 < 0$ $(-,+)$	$\dot{y}_1^* = 0$ $(+,+)$	$\dot{y}_1 > 0$ $(+,-)$
(d)	$(-,-)$	$\dot{y}_1^* = 0$ $(+,+)$	$\dot{y}_1 < 0$ $(-,+)$	$\dot{y}_1^* = 0$ $(-,-)$

The corresponding equilibrium flows of $\dot{y}_1 = 0$ in phase space are in Table 4.12

The readers can complete a similar discussion for $a_{111} > a_{221}$ with $(\alpha_{11} > 0, \alpha_{12} > 0)$ and $(\alpha_{11} > 0, \alpha_{12} < 0)$.

4.4.2 *Inflection-third-order sources and sinks*

Consider for $a_{111} = a_{221}$,

$$\begin{aligned} \dot{x}_1 &= a_{110}(y_1 - a_{111})^3, \\ \dot{x}_2 &= a_{210}(y_1 - a_{221})^3. \end{aligned} \tag{4.496}$$

There is a dynamical system in the y_1-direction as

$$\dot{y}_1 = \alpha_{11}a_{110}(y_1 - a_{111})^3 + \alpha_{12}a_{220}(y_1 - a_{221})^3, \tag{4.497}$$

with a dynamical system in the $y_{j1}^{\perp}$-direction as

$$\dot{y}_1^{\perp} = \beta_{11}a_{110}(y_1 - a_{111})^3 + \beta_{12}a_{220}(y_1 - a_{221})^3, \tag{4.498}$$

and

$$G_{110}^{(2)} = \alpha_{11}a_{110}(y_1 - a_{111})^3 + \alpha_{12}a_{220}(y_1 - a_{221})^3. \tag{4.499}$$

For $a_{111} = a_{221}(\alpha_{11} > 0, \alpha_{12} > 0)$, the phase portraits for a two-dimensional system with the second-order parabola infinite-equilibrium of $y_1^* = a_{111} = a_{221}$ on the x_1- and x_2-directions are presented in Figs. 4.13(a)–(f) for the (II,3$^{\rm rd}$SO)-infinite-equilibrium $(a_{110} > 0, a_{220} > 0)$, the (DI,3$^{\rm rd}$SO) and (DI,3$^{\rm rd}$SI)-infinite-equilibriums $(a_{110} < 0, a_{220} > 0)$, the (DI,3$^{\rm rd}$SO) and (DI,3$^{\rm rd}$SI)-infinite-equilibriums $(a_{110} > 0, a_{220} < 0)$,

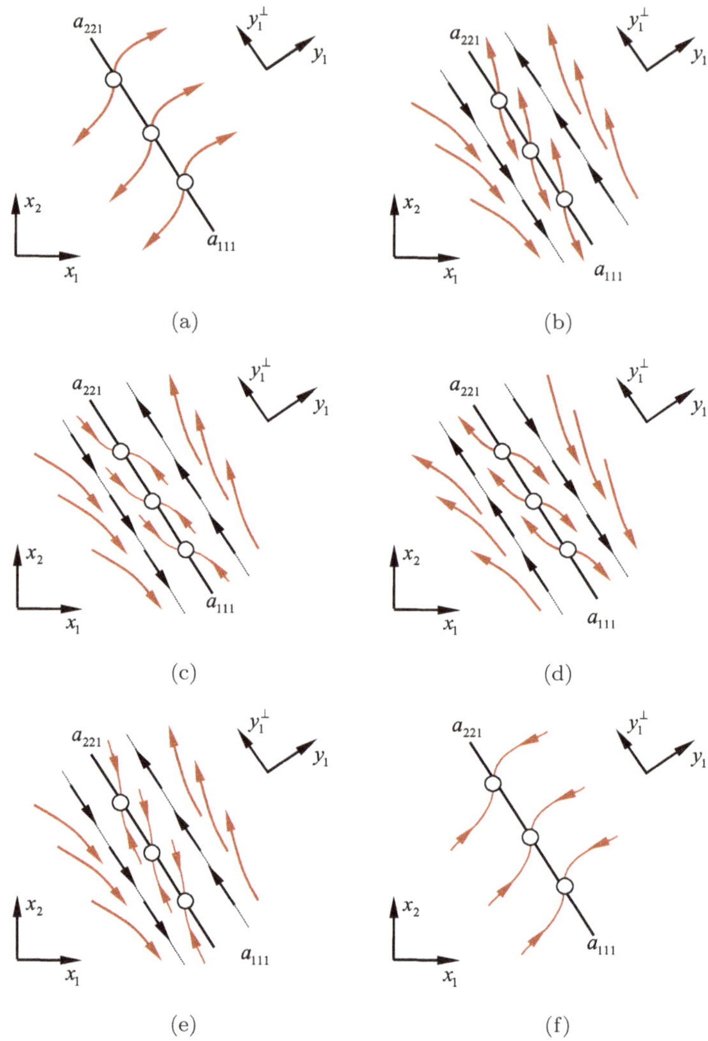

Fig. 4.13. Phase portraits ($a_{111} = a_{221}$ and $\alpha_{11} > 0, \alpha_{12} > 0$) for two-dimensional systems on the x_1-direction with $y_1^* = a_{111}$ and on the x_2-direction with $y_1^* = a_{221}$: (a) ($a_{110} > 0, a_{220} > 0$), (b) ($a_{110} < 0, a_{220} > 0$), (c) ($a_{110} > 0, a_{220} < 0$), (d) ($a_{110} < 0, a_{220} < 0$).

Table 4.13. Equilibrium flows of $\dot{y}_1 = 0$ for $a_{111} < a_{221}$ $(\alpha_{11} > 0, \alpha_{12} > 0)$.

Figure 4.13	$(\dot{x}_1, \dot{x}_2)$ (a_{110}, a_{220})	y_1 $(-\infty, a_{111})$	(a_{221}, ∞)
(a)	$(+,+)$	$\dot{y}_1 < 0$ $(-,-)$	$\dot{y}_1 > 0$ $(+,+)$
(b,c)	$(-,+)$	$\dot{y}_1^* = 0$ $(+,-)$	$\dot{y}_1^* = 0$ $(-,+)$
(d,e)	$(+,-)$	$\dot{y}_1^* = 0$ $(-,+)$	$\dot{y}_1^* = 0$ $(+,-)$
(f)	$(-,-)$	$\dot{y}_1 > 0$ $(+,+)$	$\dot{y}_1 < 0$ $(-,-)$

and the (II,3$^{\rm rd}$SI)-infinite-equilibrium $(a_{110} < 0, a_{220} < 0)$, respectively:

$$
(a_{111}, a_{221}) = \begin{cases}
\underbrace{(\mathrm{II}, 3^{\rm rd}\mathrm{SO})}_{\text{increasing-inflection third-order source}} \\
\quad \text{for } a_{j_1j_10} > 0,\ a_{j_1j_20} > 0,\ G^{(2)}_{j_1j_10} > 0, \\
\underbrace{(\mathrm{DI}, 3^{\rm rd}\mathrm{SO})}_{\text{decreasing-inflection third-order source}} \\
\quad \text{for } a_{j_1j_10} < 0,\ a_{j_1j_20} > 0,\ G^{(2)}_{j_1j_10} > 0, \\
\underbrace{(\mathrm{DI}, 3^{\rm rd}\mathrm{SI})}_{\text{decreasing-inflection third-order sink}} \\
\quad \text{for } a_{j_1j_10} < 0,\ a_{j_1j_20} > 0,\ G^{(2)}_{j_1j_10} < 0, \\
\underbrace{(\mathrm{DI}, 3^{\rm rd}\mathrm{SO})}_{\text{decreasing-inflection third-order source}} \\
\quad \text{for } a_{j_1j_10} > 0,\ a_{j_1j_20} < 0,\ G^{(2)}_{j_1j_10} > 0, \\
\underbrace{(\mathrm{DI}, 3^{\rm rd}\mathrm{SI})}_{\text{decreasing-inflection third-order sink}} \\
\quad \text{for } a_{j_1j_10} > 0,\ a_{j_1j_20} < 0,\ G^{(2)}_{j_1j_10} < 0, \\
\underbrace{(\mathrm{II}, 3^{\rm rd}\mathrm{SI})}_{\text{increasing-inflection third-order sink}} \\
\quad \text{for } a_{j_1j_10} > 0,\ a_{j_1j_20} < 0,\ G^{(2)}_{j_1j_10} > 0.
\end{cases} \tag{4.500}
$$

The corresponding equilibrium flows of $\dot{y}_1 = 0$ in phase space are in Table 4.13.

(ii) For $a_{111} = a_{221}(\alpha_{11} > 0, \alpha_{12} < 0)$, the phase portraits for a two-dimensional system with the third-order infinite-equilibrium of $y_1^* = a_{111} = a_{221}$ on the x_1-direction and on the x_2-direction are

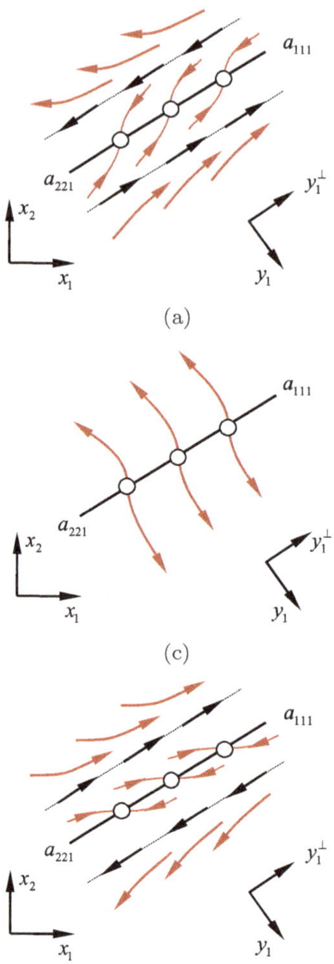

Fig. 4.14. Phase portraits ($a_{111} = a_{221}$ and $\alpha_{11} > 0, \alpha_{12} < 0$) for two-dimensional systems on the x_1-direction with $y_1^* = a_{111}$ and on the x_2-direction with $y_1^* = a_{221}$: (a,b) ($a_{110} > 0, a_{220} > 0$), (c) ($a_{110} < 0, a_{220} > 0$), (d) ($a_{110} > 0, a_{220} < 0$), (e,f) ($a_{110} < 0, a_{220} < 0$).

presented in Figs. 4.14(a)–(f) for the (II, 3rd SI) and (II, 3rd SO)-infinite-equilibriums ($a_{110} > 0, a_{220} < 0$), the (DI, 3rd SO)-infinite-equilibrium ($a_{110} < 0, a_{220} > 0$), the (DI, 3rdSI)-infinite-equilibrium ($a_{110} > 0$, $a_{220} <$ 0), and the (II, 3rd SI) and (II, 3rd SO)-infinite-equilibrium ($a_{110} <$

Table 4.14. Equilibrium flows of $\dot{y}_1 = 0$ for $a_{111} < a_{221}(\alpha_{11} > 0, \alpha_{12} < 0)$.

Figure 4.14	$(\dot{x}_1, \dot{x}_2)$ (a_{110}, a_{220})	y_1 $(-\infty, a_{111})$	(a_{221}, ∞)
(a,b)	$(+,+)$	$\dot{y}_1^* = 0$ $(-,-)$	$\dot{y}_1^* = 0$ $(+,+)$
(c)	$(-,+)$	$\dot{y}_1 > 0$ $(+,-)$	$\dot{y}_1 < 0$ $(-,+)$
(d)	$(+,-)$	$\dot{y}_1 < 0$ $(-,+)$	$\dot{y}_1 > 0$ $(+,-)$
(e,f)	$(-,-)$	$\dot{y}_1^* = 0$ $(+,+)$	$\dot{y}_1^* = 0$ $(-,-)$

$0, a_{220} < 0)$, respectively:

$$(a_{111}, a_{221}) = \begin{cases} \underbrace{(\mathrm{II}, 3^{\mathrm{rd}}\mathrm{SI})}_{\text{increasing-inflection third-order sink}} \\ \quad \text{for } a_{j_1j_10} > 0,\ a_{j_1j_20} > 0,\ G^{(2)}_{j_1j_10} < 0; \\ \underbrace{(\mathrm{II}, 3^{\mathrm{rd}}\mathrm{SO})}_{\text{increasing-inflection third-order source}} \\ \quad \text{for } a_{j_1j_10} > 0,\ a_{j_1j_20} > 0,\ G^{(2)}_{j_1j_10} > 0; \\ \underbrace{(\mathrm{DI}, 3^{\mathrm{rd}}\mathrm{SO})}_{\text{decreasing-inflection third-order source}} \\ \quad \text{for } a_{j_1j_10} < 0,\ a_{j_1j_20} > 0,\ G^{(2)}_{j_1j_10} > 0; \\ \underbrace{(\mathrm{DI}, 3^{\mathrm{rd}}\mathrm{SI})}_{\text{decreasing-inflection third-order sink}} \\ \quad \text{for } a_{j_1j_10} < 0,\ a_{j_1j_20} > 0,\ G^{(2)}_{j_1j_10} < 0; \\ \underbrace{(\mathrm{II}, 3^{\mathrm{rd}}\mathrm{SI})}_{\text{increasing-inflection third-order sink}} \\ \quad \text{for } a_{j_1j_10} < 0,\ a_{j_1j_20} > 0,\ G^{(2)}_{j_1j_10} < 0; \\ \underbrace{(\mathrm{II}, 3^{\mathrm{rd}}\mathrm{SO})}_{\text{increasing-inflection third-order source}} \\ \quad \text{for } a_{j_1j_10} < 0,\ a_{j_1j_20} < 0,\ G^{(2)}_{j_1j_10} > 0. \end{cases} \quad (4.501)$$

The corresponding equilibrium flows of $\dot{y}_1 = 0$ in phase space are in Table 4.14.

www.ingramcontent.com/pod-product-compliance
Ingram Content Group UK Ltd.
Pitfield, Milton Keynes, MK11 3LW, UK
UKHW061933100926
8556IPUK00007B/895

9 789819 818396